Advances in Experimental Medicine and Biology

This series of volumes focuses on concepts, techniques and recent advances in the field of proteomics, interactomics, metabolomics and systems biology. Recent advances in various 'omics' technologies enable quantitative monitoring of myriad various biological molecules in a high-throughput manner, and allow determination of their variation between different biological states on a genomic scale. Now that the sequencing of various genomes, from prokaryotes to humans, has provided the list and linear sequence of proteins and RNA that build living organisms, defining the complete set of interactions that sustain life constitutes one of the key challenges of the postgenomic era. This series is intended to cover experimental approaches for defining protein-protein, protein-RNA, protein-DNA and protein-lipid interactions; as well as theoretical approaches dealing with data analysis, integration and modeling and ethical issues.

More information about this series at http://www.springer.com/series/5584

Raul Delgado-Morales
Editor

Neuroepigenomics in Aging and Disease

Springer

Editor
Raul Delgado-Morales
Cancer Epigenetics and Biology
Program (PEBC)
Bellvitge Biomedical Research
Institute (IDIBELL)
CATALONIA
Spain

ISSN 0065-2598 ISSN 2214-8019 (electronic)
Advances in Experimental Medicine and Biology
ISBN 978-3-319-53888-4 ISBN 978-3-319-53889-1 (eBook)
DOI 10.1007/978-3-319-53889-1

Library of Congress Control Number: 2017939894

Printed on acid-free paper

This Springer imprint is published by Springer Nature
The registered company is Springer International Publishing AG
The registered company address is: Gewerbestrasse 11, 6330 Cham, Switzerland

Contents

Contributors

Vladimir Espinosa Angarica, Ph.D. Computational Biology Group, Luxembourg Centre for Systems Biomedicine, University of Luxembourg, Belvaux, Luxembourg

Juan Ausió Department of Biochemistry and Microbiology, University of Victoria, Victoria, BC, Canada

Andrew A. Bartlett, B.S. Department of Psychology, University of Massachusetts, Boston, MA, USA

Pavel Bashtrykov Institute of Biochemistry, Faculty of Chemistry, University of Stuttgart, Stuttgart, Germany

Silvia Bassi, Ph.D. The NeuroEpigenetics Laboratory, Centre for Integrative Biology, University of Trento, Povo (TN), Italy

Marta Biagioli, Ph.D. The NeuroEpigenetics Laboratory, Centre for Integrative Biology, University of Trento, Povo (TN), Italy

Amber N. Brown Department of Biological Science, Florida State University, Tallahassee, FL, USA

Diana E. Caballero-Hernandez CABIMER, Andalusian Center for Molecular Biology and Regenerative Medicine, Seville, Spain

School of Biological Sciences, Autonomous University of Nuevo Leon (UANL), Nuevo Leon, Mexico

Tanessa Call, B.A. Department of Basic Medical Sciences, University of Arizona College of Medicine-Phoenix, Phoenix, AZ, USA

Marta Cejudo-Guillén CABIMER, Andalusian Center for Molecular Biology and Regenerative Medicine, Seville, Spain

Department of Medical Biochemistry, Molecular Biology and Immunology, University of Seville Medical School, Seville, Spain

Kevin Czaplinski Center for Nervous System Disorders, Centers for Molecular Medicine, Stony Brook University, Stony Brook, NY, USA

Department of Anesthesiology, Stony Brook University, Stony Brook, NY, USA

Raul Delgado-Morales, Ph.D. Cancer Epigenetics and Biology Program (PEBC), Bellvitge Biomedical Research Institute (IDIBELL), Duran i Reynals Hospital, Barcelona, Catalonia, Spain

Department of Psychiatry and Neuropsychology, School for Mental Health and Neuroscience (MHeNs), Maastricht University, Maastricht, The Netherlands

Jan Deussing, Ph.D. Department of Stress Neurobiology and Neurogenetics, Max Planck Institute of Psychiatry, Munich, Bavaria, Germany

Mike Dragunow, Ph.D. Department of Pharmacology, Centre for Brain Research, University of Auckland, Auckland, New Zealand

Manel Esteller, M.D., Ph.D. Cancer Epigenetics and Biology Program (PEBC), Bellvitge Biomedical Research Institute (IDIBELL), Duran i Reynals Hospital, Barcelona, Catalonia, Spain

Department of Physiological Sciences II, School of Medicine, University of Barcelona, Barcelona, Catalonia, Spain

Institució Catalana de Recerca i Estudis Avançats (ICREA), Barcelona, Catalonia, Spain

Jian Feng Department of Biological Science, Florida State University, Tallahassee, FL, USA

Neuroscience Program, Florida State University, Tallahassee, FL, USA

Deveroux Ferguson, Ph.D. Department of Basic Medical Sciences, University of Arizona College of Medicine-Phoenix, Phoenix, AZ, USA

Jaime M. Franco CABIMER, Andalusian Center for Molecular Biology and Regenerative Medicine, Seville, Spain

Benjamin P. Garfinkel, Ph.D. The Edmond and Lily Safra Center of Brain Science, The Hebrew University of Jerusalem, Jerusalem, Israel

Juan Miguel Gomez-Zumaquero Genomic Unit, Malaga Institute of Biomedical Research (IBIMA), Malaga, Spain

Sònia Guil, Ph.D. Cancer Epigenetics and Biology Program (PEBC), Bellvitge Biomedical Research Institute (IDIBELL), Barcelona, Catalonia, Spain

Amy Guillaumet-Adkins, Ph.D. CNAG-CRG, Centre for Genomic Regulation (CRG), Barcelona Institute of Science and Technology (BIST), Universitat Pompeu Fabra (UPF), Barcelona, Catalonia, Spain

Sinan Gülöksüz, M.D., Ph.D. Department of Psychiatry and Neuropsychology, School for Mental Health and Neuroscience, Maastricht University, Maastricht, The Netherlands

Department of Psychiatry, Yale School of Medicine, Yale School of Medicine, New Haven, CT, USA

Holger Heyn, Ph.D. CNAG-CRG, Centre for Genomic Regulation (CRG), Barcelona Institute of Science and Technology (BIST), National Centre for Genomic Analysis (CNAG), Barcelona, Catalonia, Spain

Daniel van den Hove, Ph.D. Department of Psychiatry and Neuropsychology, School for Mental Health and Neuroscience (MHeNS), Maastricht University, Maastricht, The Netherlands

Laboratory of Translational Neuroscience, Department of Psychiatry, Psychosomatics and Psychotherapy, University of Wuerzburg, Wurzburg, Germany

Richard. G Hunter, B.S., Ph.D. Department of Psychology, University of Massachusetts, Boston, MA, USA

Mira Jakovcevski, Ph.D. Department of Stress Neurobiology and Neurogenetics, Max Planck Institute of Psychiatry, Munich, Bavaria, Germany

Albert Jeltsch Institute of Biochemistry, Faculty of Chemistry, University of Stuttgart, Stuttgart, Germany

Alba Jimenez-Pacheco CABIMER, Andalusian Center for Molecular Biology and Regenerative Medicine, Seville, Spain

Gunter Kenis, Ph.D. Department of Psychiatry and Neuropsychology, School for Mental Health and Neuroscience, Maastricht University, Maastricht, The Netherlands

Hee-Dae Kim, Ph.D. Department of Basic Medical Sciences, University of Arizona College of Medicine-Phoenix, Phoenix, AZ, USA

Edward Korzus, Ph.D. Department of Psychology and Neuroscience Program, University of California Riverside, Riverside, CA, USA

M. Magdalena Leal CABIMER, Andalusian Center for Molecular Biology and Regenerative Medicine, Seville, Spain

Francesca Di Leva, Ph.D. The NeuroEpigenetics Laboratory, Centre for Integrative Biology, University of Trento, Povo (TN), Italy

Rixing Lin, B.A. McGill Group for Suicide Studies, Douglas Mental Health University Institute, McGill University, Montreal, QC, Canada

Soledad Lopez CABIMER, Andalusian Center for Molecular Biology and Regenerative Medicine, Seville, Spain

Department of Medical Biochemistry, Molecular Biology and Immunology, University of Seville Medical School, Seville, Spain

Katie Lunnon, Ph.D., B.Sc. University of Exeter Medical School, RILD, Exeter, Devon, UK

Samantha Magazu, M.S. Department of Basic Medical Sciences, University of Arizona College of Medicine-Phoenix, Phoenix, AZ, USA

Rotem Maoz, B.Sc. The Edmond and Lily Safra Center of Brain Science, The Hebrew University of Jerusalem, Jerusalem, Israel

Alan Monziani The NeuroEpigenetics Laboratory, Centre for Integrative Biology, University of Trento, Povo (TN), Italy

Sebastian Moran, M.Sc. Cancer Epigenetics and Biology Program (PEBC), Bellvitge Biomedical Research Institute (IDIBELL), Duran i Reynalds Hospital, Barcelona, Catalonia, Spain

Pritika Narayan, B.M.Sc., P.G.Dip.Sci., Ph.D. School of Biological Sciences and Centre for Brain Research, University of Auckland, Auckland, New Zealand

Aida Obiols-Guardia, M.Sc. Cancer Epigenetics and Biology Program (PEBC), Bellvitge Biomedical Research Institute (IDIBELL), Barcelona, Catalonia, Spain

Tiago Fleming Outeiro, B.Sc., M.Sc., Ph.D. Department of NeuroDegeneration and Restorative Research, Center for Nanoscale Microscopy and Molecular Physiology of the Brain, University Medical Center Göttingen, Göttingen, Lower Saxony, Germany

Department of NeuroDegeneration and Restorative Research, University Medical Center Göttingen, Göttingen, Lower Saxony, Germany

Max Plank Institute for Experimental Medicine, Goettingen, Germany

Maria Angeliki S. Pavlou, B.Sc., M.Sc., Ph.D. Department of NeuroDegeneration and Restorative Research, Center for Nanoscale Microscopy and Molecular Physiology of the Brain, University Medical Center Göttingen, Göttingen, Lower Saxony, Germany

Department of NeuroDegeneration and Restorative Research, University Medical Center Göttingen, Göttingen, Lower Saxony, Germany

Alexia Martínez de Paz Department of Biochemistry and Microbiology, University of Victoria, Victoria, BC, Canada

Ehsan Pishva, M.D., Ph.D. Complex Disease Epigenetic Group, University of Exeter Medical School, Exeter, UK

Department of Psychiatry and Neuropsychology, School for Mental Health and Neuroscience (MHeNS), Maastricht University, Maastricht, The Netherlands

David Pozo CABIMER, Andalusian Center for Molecular Biology and Regenerative Medicine, Seville, Spain

Department of Medical Biochemistry, Molecular Biology and Immunology, University of Seville Medical School, Seville, Spain

Lotta-Katrin Pries, M.Sc., Ph.D. Department of Psychiatry and Neuropsychology, School for Mental Health and Neuroscience, Maastricht University, Maastricht, The Netherlands

Renzo J.M. Riemens, M.Sc. Cancer Epigenetics and Biology Program (PEBC), Bellvitge Biomedical Research Institute (IDIBELL), Duran i Reynals Hospital, Barcelona, Catalonia, Spain

Department of Psychiatry and Neuropsychology, School for Mental Health and Neuroscience (MHeNs), Maastricht University, Maastricht, The Netherlands

Institute of Human Genetics, Julius Maximilians University, Biozentrum, Wurzburg, Germany

Bart P.F. Rutten, M.D., Ph.D. Department of Psychiatry and Neuropsychology, School for Mental Health and Neuroscience (MHeNS), Maastricht University, Maastricht, The Netherlands

Rumani Singh, B.S., Ph.D. Department of Psychology, University of Massachusetts, Boston, MA, USA

Michelle T. Siu, Ph.D. Program in Genetics and Genome Biology, The Hospital for Sick Children, Toronto, ON, Canada

Rebecca G. Smith, Ph.D., B.Sc. University of Exeter Medical School, RILD, Exeter, Devon, UK

Edilene S. Soares Cancer Epigenetics and Biology Program (PEBC), Bellvitge Biomedical Research Institute (IDIBELL), Duran i Reynals Hospital, Barcelona, Catalonia, Spain

Antonio del Sol, Ph.D. Computational Biology Group, Luxembourg Centre for Systems Biomedicine, University of Luxembourg, Belvaux, Luxembourg

Hermona Soreq, Ph.D. The Edmond and Lily Safra Center of Brain Science, The Hebrew University of Jerusalem, Jerusalem, Israel

Howard Steiger, Ph.D. Eating Disorders Continuum, Douglas Institute, Montreal West Island Integrated University Health and Social Services Centre, Verdun, QC, Canada

Psychiatry Department, McGill University, Montréal, QC, Canada

Lea Thaler, Ph.D. Eating Disorders Continuum, Douglas Institute, Montreal West Island Integrated University Health and Social Services Centre, Verdun, QC, Canada

Psychiatry Department, McGill University, Montréal, QC, Canada

Elizabeth A. Thomas, Ph.D. Department of Molecular and Cellular Neuroscience, The Scripps Research Institute, La Jolla, CA, USA

Takshashila Tripathi, Ph.D. The NeuroEpigenetics Laboratory, Centre for Integrative Biology, University of Trento, Povo (TN), Italy

Gustavo Turecki, M.D., Ph.D. McGill Group for Suicide Studies, Douglas Mental Health University Institute, McGill University, Montreal, QC, Canada

Rosanna Weksberg, M.D., Ph.D. Program in Genetics and Genome Biology, The Hospital for Sick Children, Toronto, ON, Canada

Division of Clinical and Metabolic Genetics, The Hospital for Sick Children, Toronto, ON, Canada

Department of Paediatrics, University of Toronto, Toronto, ON, Canada

Institute of Medical Science, University of Toronto, Toronto, ON, Canada

Part I

Neurodevelopmental and Childhood Disorders

MeCP2, A Modulator of Neuronal Chromatin Organization Involved in Rett Syndrome

1

Alexia Martínez de Paz and Juan Ausió

Abstract

From an epigenetic perspective, the genomic chromatin organization of neurons exhibits unique features when compared to somatic cells. Methyl CpG binding protein 2 (MeCP2), through its ability to bind to methylated DNA, seems to be a major player in regulating such unusual organization. An important contribution to this uniqueness stems from the intrinsically disordered nature of this highly abundant chromosomal protein in neurons. Upon its binding to methylated/hydroxymethylated DNA, MeCP2 is able to recruit a plethora of interacting protein and RNA partners. The final outcome is a highly specialized chromatin organization wherein linker histones (histones of the H1 family) and MeCP2 share an organizational role that dynamically changes during neuronal development and that it is still poorly understood. MeCP2 mutations alter its chromatin-binding dynamics and/or impair the ability of the protein to interact with some of its partners, resulting in Rett syndrome (RTT). Therefore, deciphering the molecular details involved in the MeCP2 neuronal chromatin arrangement is critical for our understanding of the proper and altered functionality of these cells.

Keywords

MeCP2 • Chromatin • DNA methylation • Postmitotic neurons • Rett syndrome

A. Martínez de Paz • J. Ausió (✉)
Department of Biochemistry and Microbiology, University of Victoria,
Petch Building 258-260, Victoria, BC V8W 3P6, Canada
e-mail: jausio@uvic.ca

R. Delgado-Morales (ed.), *Neuroepigenomics in Aging and Disease*, Advances in Experimental Medicine and Biology, DOI 10.1007/978-3-319-53889-1_1

Abbreviations

AFM	Atomic force microscopy
ATRX	α-Thalassemia mental retardation X linked
ChIP-seq	Chromatin immunoprecipitation and sequencing
CREB	cAMP response element-binding protein
CTCF	CCCTC-binding factor
CTD	C-terminal domain
Dlk1	Delta-like 1 homolog (Drosophila)
Dlx	Distal-less homeobox
DNMT	DNA (cytosine-5)-methyltransferase 1
Gtl2	Gene trap locus 2
HDAC	Histone deacetylases
ICR	Imprinting control region
ID	Intervening domain
IDP	Intrinsically disordered protein
Igf2	Insulin-like growth factor 2
MBD	Methyl-binding domain
MeCp2	Methyl CpG binding protein
MoRF	Molecular recognition features
N-CoR	Nuclear receptor corepressor 1
NCP	Nucleosome core particle
NLS	nuclear localization signal
NRL	Nucleosome repeat length
NTD	N-terminal domain
PTM	Posttranslational modification
RTT	Rett syndrome
SIN3A	Switch-independent 3a
TET	Ten-eleven translocation
TRD	Transcriptional repressor domain
WDR	WW domain-binding region

1.1 Introduction

MeCP2 is a chromatin architectural protein that is ubiquitously expressed, but particularly abundant in postmitotic neurons [1]. MeCP2 was first discovered in 1992 by Adrian Bird and colleagues [2], and it gained further attention once it was found that mutations in the *Mecp2* gene give rise to Rett syndrome (RTT OMIM #312750) [3], an X-linked autism spectrum disorder considered to be one of the main causes of intellectual disability in girls [4]. Different observations highlight the important and distinctive role of MeCP2 in neurons: mice harboring an *Mecp2* gene deletion exclusively in neurons show a similar phenotype to that observed in complete knock out mice which mimics the RTT phenotype [5]. In addition, when

Mecp2 is re-expressed under the control of the neuron-specific Tau promoter in *Mecp2*-mutant mice, it can revert the pathologic symptoms [6]. Early characterizations using mouse cells ectopically expressing the protein showed its localization to the visible nuclear heterochromatic *foci*, comprising about 50% of the total CpGs and with high levels of methylation [7]. Moreover, it was shown that MeCP2 can block gene transcription through its binding to hypermethylated promoters [8]. These initial observations gave rise to the general view of the protein as a classic repressor which was able to block transcription in a methylation-dependent way. Nevertheless, numerous studies carried out during the past decade have revealed a much more complex involvement of MeCP2, far from the original simplistic idea, suggesting a broader role of the protein in global chromatin regulation [9–12].

The high levels of MeCP2 found in postmitotic neurons alongside its ability to bind to the different forms of methylated DNA (hydroxymethylation and other methylated sites in addition to CpGs) that are enriched in these cells [11, 13, 14] place the protein in a central position in the regulation of brain function. This chapter will discuss the distinctive features of neuronal chromatin and the role played by MeCP2 in its organization.

1.2 MeCP2: The Protein

MeCP2 is encoded by a gene located within Xq28 and is thus subject to X-chromosome inactivation [3]. The gene comprises four exons and during transcription can be alternatively spliced to include or skip exon 2, hence giving rise to two transcript variants (Fig. 1.1a). This results in two protein isoforms, MeCP2 E1 (aka MeCP2α or MeCP2 B) and MeCP2 E2 (aka MeCP2β or MeCP2 A). The former is translated from exon one and is the most abundant form of MeCP2 in brain. The translation initiation of MeCP2 E2 isoform lies on the second exon (Fig. 1.1b) [15].

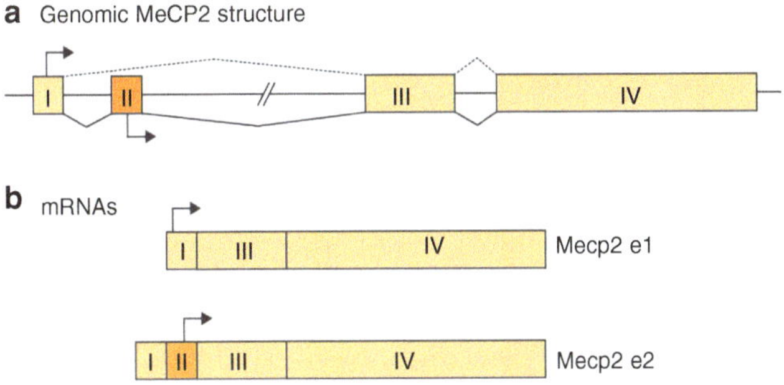

Fig. 1.1 Gene organization of MeCP2. (**a**) The MeCP2 gene consists of four exons (I, II, III and IV) and three introns. Exon II is an alternative exon that can be either skipped (*dash line*) or included (*solid line*) during transcription, leading to the synthesis of two final transcripts, Mecp2 e1 and e2, respectively. (**b**) MeCP2 transcripts e1 and e2 are translated from exon I and exon II, respectively (*arrows*)

These two isoforms are very similar in sequence, differing only in the N-terminal region, where MeCP2 E1 contains 24 unique amino acids encoded by exon 1, while MeCP2 E2 have 12 different amino acids encoded by exon 2 [16].

To understand the role of MeCP2 in the regulation of biological processes, it is important to consider that it belongs to the family of intrinsically disordered proteins (IDPs), whose members are characterized by having little or no structured elements such as alpha helices or beta sheets in solution [17]. These low levels of organization bestow such proteins with a high flexibility and extended interaction surface, within which different small linear motifs, called molecular recognition features (MoRFs), remain exposed for recognition purposes [18]. These features allow IDPs to bind to a plethora of molecular partners, and therefore they are thought to act as interaction "hubs" that play an important role in the regulation and coordination of signaling pathways [17, 19]. Remarkably, the expression level of these proteins is tightly regulated, and their functional importance is highlighted by the fact that their deregulation is commonly associated to physiological disorders such as neurological diseases or cancer [20]. In this regard, the maintenance of a very precise amount of MeCP2 seems to be critical for proper function of healthy neurons [21–24]. Indeed, RTT is mainly the result of MeCP2 loss of function mutations, and patients harboring an Xq28 duplication – covering the *Mecp2* locus – also exhibit related intellectual disability and neurological symptoms [25]. The larger the deviation from the physiological stoichiometric MeCP2 levels, the greater the severity of the pathological phenotype [26].

Despite the highly unstructured nature of MeCP2 (approximately 60–65% of the protein lacks secondary structure), trypsin digestion allows distinction of six well-defined structural domains (Fig. 1.2a) [27]: N-terminal domain (NTD), methyl-binding domain (MBD), intervening domain (ID), transcriptional repressor domain (TRD), and the C-terminal domains α and β (CTDα and CTDβ). These different domains encompass regions that mediate MeCP2 binding to DNA/chromatin, such as the MBD domain [28], adenine-thymine (AT) hooks [29], a DNA-binding domain [30] and binary chromatin-binding sites (Fig. 1.2b) [31]. In addition, two distinct regions allowing protein-protein interactions have been defined: the dimerizing domain [32] and the WW domain-binding region (WDR) (Fig. 1.2c) [33]. A great deal of attention has been given toward the MBD and TRD domains, since they are responsible and sufficient for mediating MeCP2 binding to methylated DNA and transcriptional repression, respectively [8, 34]. However, it is worth noting that the C-terminal domain is necessary for the proper maintenance of chromatin structure [30], and the full-length intact protein is crucial for brain function, since a great number of mutations associated to Rett syndrome can be found along the entire sequence [35].

The abovementioned conformational flexibility of MeCP2, together with the different domains and motifs present in its sequence, allows the protein to bind a wide variety of molecular partners. For instance, MeCP2 can interact with different proteins such as the transcriptional coactivator cAMP response element-binding protein (CREB), nuclear receptor corepressor 1 (N-CoR), and DNA (cytosine-5)-methyltransferase 1 (DNMT1) [9, 36, 37]. MeCP2 has also been

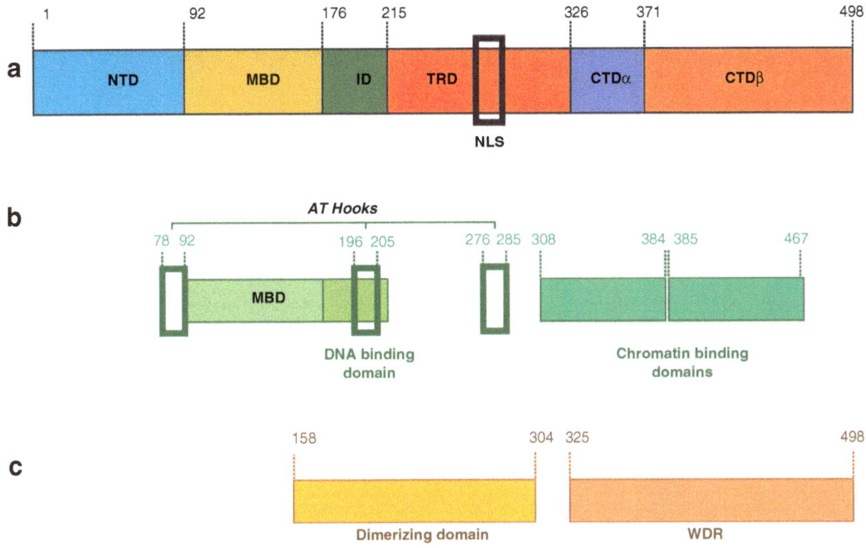

Fig. 1.2 MeCP2 protein structure. (**a**) MeCP2 domains: N-terminal domain (NTD), methyl-binding domain (MBD), intervening domain (ID), transcriptional repressor domain (TRD), and C-terminal domains α and β (CTDα and CTDβ). NLS: nuclear localization signal. (**b**) MeCP2 domains involved in DNA and chromatin interaction: *Black squares* represent the location of AT-hook domains, MBD, DNA-binding domain, and chromatin-binding domains. (**c**) Regions of MeCP2 involved in protein-protein interactions: the dimerizing domain and the WW domain-binding region (WDR)

shown to bind unmethylated DNA [38, 39] and to RNA [40], and although it was first discovered by its ability to bind to DNA methylated at cytosine followed by guanine (mCpG) dinucleotides [2], recent evidence indicates that the protein is also able to recognize other modified DNA forms. For instance, MeCP2 can interact with 5-hydroxymethylcytosine (5hmC) [13] and methylated DNA outside the context of CpG dinucleotides [mCH; H = adenine (A), thymine (T), or cytosine (C)] [14].

Such binding promiscuity has an important relevance within the brain context, where, in postmitotic neurons, MeCP2 is expressed close to stoichiometric levels with histone octamers and DNA methylation patterns are particularly complex and very different from those found in other somatic tissues [11, 41].

1.3 DNA Methylation and MeCP2 Binding in the Brain

DNA methylation is an epigenetic mark that results from the covalent binding of a methyl group to the 5′ carbon of the cytosine pyrimidine ring (adenines are also methylated in prokaryotes and plants) [42]. DNA methylation in animals was traditionally thought to be confined to CpG dinucleotides, and thus it has been the best characterized DNA methylation site. CpG methylation is important for different biological processes such as X-chromosome inactivation, genomic imprinting and silencing of transposable elements; therefore, it is generally considered to be a transcriptional repressive mark [43].

Among other epigenetic mechanisms, DNA methylation is thought to play a critical role in neurogenesis and neuronal plasticity, which is the fundamental constituent of learning and memory. Importantly, its dysregulation has been associated with different neurological disorders [44, 45]. In postmitotic neurons, DNA methylation within the context of CpGs is similar to that found in other tissues; between 60 and 90% of the CpG sites are methylated and are enriched in repetitive DNA sequences and intergenic regions, while regulatory elements like promoters or enhancers are depleted of methylated CpGs [41, 42]. However, mature neurons exhibit a high content of additional DNA modifications [14, 41]. In recent years, with the use of genome-wide single-base resolution sequencing methods, the existence of biologically relevant methylation in the CH context has also been uncovered [14]. Throughout synaptogenesis, which coincides with childhood and adolescence, mCH (mainly mCA) accumulate to become, together with 5hmCs, a major fraction of the bulk of methylated cytosines in postmitotic neurons [41]. Interestingly, during neuronal maturation, MeCP2 levels increase in parallel with the increase of these methylated species, reaching levels 30–60-fold higher than those of other cell types [11, 46]. These observations suggest that MeCP2 could bind to these DNA modifications and hence regulate the levels of general transcription in neurons. During the last few years, different groups have started to address this possibility [13, 14, 47, 48]. Using electrophoretic mobility shift assays (EMSAs), Gabel and colleagues performed a very insightful screen that showed that MeCP2 binds with a similarly high affinity to mCG and mCA, while its avidity for mCT or mCC was much weaker [47]. The observed preferential sites of MeCP2 binding were mCAs located along gene bodies (transcribed regions) of particularly long genes, which is generally associated with transcriptional repression [47]. Therefore, the high frequency of mCAs found in neurons, together with the affinity of MeCP2 for them, suggest that this modification could be a critical ligand of the protein.

The hydroxymethylated form of cytosine (5hmC) is an oxidized intermediate state originating during the process of active DNA demethylation and is carried out by the ten-eleven translocation (TET) proteins [13]. This modification is quite abundant in neurons, is recognized by MeCP2, and is associated with transcriptional activation. In agreement with previous studies [48, 49], Gabel et al. also described a lower affinity of MeCP2 for 5hmCG than for mCG or 5hmCA, although the latter is present in negligible amounts in the neuronal genome [41]. Overall, these observations suggest a model in which MeCP2 binds primarily to mCGs, and upon hydroxymethylation, MeCP2 binding affinity diminishes, causing the release of the protein and its relocation or degradation.

The ability of MeCP2 to bind to methylated cytosines is completely dependent on its MBD domain [14, 28]. Both in vitro and in vivo analyses have shown that, when this domain is mutated or lacking, the specificity of MeCP2 for methylated DNA and its overall DNA binding are impaired [50, 51]. Remarkably, crystallographic studies have shown that the protein does not directly recognize the modified cytosine; instead, it establishes hydrophilic interactions with water molecules located in the mayor groove of the methylated DNA around the cytosine moiety [28]. Of note, the binding capacity of MeCP2 to hydroxymethylated cytosines also seems to rely on the MBD domain, as the mutation R133C, present in patients with Rett syndrome, does not affect the binding of MeCP2 to mCpG, but impairs its interaction with 5hmC [13]. The MBD flanking regions, NTD and ID, are also very important for its binding to DNA. The NTD appears to play a critical role despite its inability to bind DNA itself, acting synergistically with the MBD to increase the binding affinity of this domain by a significant ten-fold [30].

1.4 MeCP2-Chromatin Binding

Most eukaryotic DNA is located inside the nucleus where it is associated with chromosomal proteins, resulting in a nucleoprotein complex called chromatin. The most basic structural unit of chromatin is the nucleosome core particle (NCP), which consists of 146 base pairs of DNA wrapped around a histone octamer (a tetramer $[H3\text{-}H4]_2$ and two H2A-H2B dimers) [52]. Adjacent nucleosomes are connected by a shorter free DNA region of variable length called linker DNA. The structure is stabilized by the presence of linker histones (of the histone H1 family) that bind to the linker DNA at the entry and exit sites of the NCP. Chromatin is highly dynamic and can undergo different degrees of compaction, from an extended conformation in euchromatin, characterized by spaced nucleosomes and readily accessible DNA, to the highly condensed metaphasic chromatin structure necessary for the appropriate chromosomal organization before cell division. Multiple in vitro studies have demonstrated that MeCP2 is able to bind to nucleosomes and fold chromatin by itself [31, 38, 53], and it has been proposed by Hansen et al. that this is most likely the result of a combinatorial interaction between the different MeCP2 binding domains with DNA/chromatin [18].

Early studies have demonstrated that transcriptional repression via DNA methylation was mediated by its association with chromatin and resulting compaction [54]. Soon after, MeCP2 was identified in Adrian Bird's lab, and the ability of its MBD domain to bind naked DNA harboring a single methylated CpG was described [2, 34]. These and other pioneering findings prompted Chandler et al. to investigate for the first time MeCP2-DNA binding within a nucleosome context [53]. Using mono-nucleosomes, they observed a binding of MeCP2 to the solution-exposed methyl CpGs located preferentially in the region of DNA at the nucleosome dyad axis and at its boundary regions. Furthermore, they found that such interactions were favored by the presence of linker DNA and, when bound, MeCP2 would protect nucleosomal DNA from nuclease digestion.

Subsequent in vitro studies aimed to determine the ability of MeCP2 to produce alterations in the chromatin structure using well-defined nucleosome arrays (a DNA template consisting of 12 tandem 208-bp repeats of the *Lytechinus* 5 S rDNA—a strong nucleosome positioning sequence—and chicken erythrocyte histone octamers) [55]. Using this system Georgel et al. described the ability of MeCP2 to induce chromatin condensation [38]. Combining electron microscopy and analytical ultracentrifuge analyses, this study showed that the majority of nucleosome arrays underwent a highly compacted organization when MeCP2 was added at nucleosome equimolar ratio. Moreover, these condensed particles could assemble into chromatin-like suprastructures at higher MeCP2/nucleosome molar ratios. Further studies provided new detail to the nature of MeCP2 interaction with chromatin [31, 56]. By digesting the nucleosome arrays with *micrococcal* nuclease (MNase), it was shown that a short stretch of 11 bp in the 5′ end of the DNA linker was preferentially protected by the protein; moreover, the presence of the DNA linker was essential for MeCP2 binding.

Recombinant versions of MeCP2 consisting of deletion mutants or harboring different Rett syndrome-causing mutations have been extensively used to ascertain the specific roles played by each domain in the interaction of the protein with DNA/chromatin. Using these approaches, it was shown early on by Chandler et al. that the MeCP2 C-terminal region is necessary for its interaction with nucleosomes and might be very important for the regulation of chromatin structure [53]. Several follow-up studies were in agreement with this initial observation [30, 31, 38]. Using EMSA it was determined that the mutant R294X, lacking both CTDα and CTDβ, is able to bind to naked DNA but exhibits an impaired interaction with nucleosomes [31]. More specifically, CTDβ is likely to be the one carrying nucleosome interacting regions, since it has been reported to bind to nucleosomes but not to dsDNA [30]. Moreover, and in line with the previous observations, the TRD-CTD domains were capable of condensing nucleosome arrays to similar levels of those achieved by the full-length MeCP2 [30]. Of note, the resulting structures obtained in this instance were different to those observed in the presence of an intact MeCP2 [31]. Electron microscopy revealed that the wild-type protein exhibited a tendency to link nucleosomes together. By contrast, the R249X mutant mediated compaction exclusively through DNA–DNA interactions, supporting again the notion of a C-terminal region controlling the chromatin binding.

1.5 Every Single Amino Acid Matters

As mentioned before, different mutations along the entire MeCP2 sequence have been found to result in a RTT phenotype, underscoring the functional importance of the entire protein. Some of these mutations affect regions that allow the protein to bind unmethylated DNA, as is the case of the R270X mutation which causes neonatal encephalopathy and early death [29]. In this study Baker and colleagues described how the said mutation and G273X are responsible for two extremely different phenotypes, despite their mere three-amino acid distance and their similar ability to disrupt MeCP2 transcriptional repression. The reason for the milder phenotype of G273X compared to R270X is that the latter but not the former disrupts a highly conserved AT-hook DNA-binding domain within the TRD region. This renders a protein with an impaired ability to bind to certain heterochromatic sequences and to compact chromatin [29]. AT-hooks are DNA-binding motifs with high affinity for adenine/thymine DNA sequences. These motifs are frequently present in other non-histone chromatin-associated proteins such as those of the high-mobility group (HMG). The authors suggested a model wherein MeCP2 could be specifically binding methylated DNA and subsequently making extensive contacts with the adjacent DNA and nucleosomes to alter the local structure. In agreement with this model, MeCP2 has been described to preferentially bind to CpGs containing a minimum of four adenine/thymidine nucleotides (AT $\geq$ 4), underscoring the importance of the AT-hook motifs present in the protein [57].

1.6 Teaming Up

The different domains of MeCP2 seem to work in a cooperative way. The protein appears to be divided into two fundamentally different functional regions: one encompassing the N-terminal domains, NTD-MBD-ID, and the second corresponding to the C-terminal regions TRD-CTDα-CTDβ [30]. In the first one, the MBD domain is responsible for the specific recognition and binding to hydroxymethylated/methylated DNA, and the surrounding NTD and ID domains might function as interaction stabilizers. In the second one, the C-terminal region is particularly involved in chromatin binding and folding. TRD-CTDα is responsible for binding to DNA, while CTDβ is required for chromatin compaction.

In addition to the cooperative activity of the different domains of MeCP2, the protein performs its functions by interacting with other partners, and different structural features of MeCP2 can modulate these interactions. A potential modulator is the increased structural complexity acquired upon binding of MeCP2 to specific targets, a characteristic process shown by IDPs [19]. Indeed, despite MeCP2 IDP nature, using hydrogen bond stabilizers such as trifluoroethanol (TFE), it was shown that 65% of the protein is able to acquire secondary structure [58]. It is tempting to speculate that, upon binding to hydroxymethylated/methylated DNA, MeCP2 possibly undergoes structural changes that could result in the exposure of different MoRFs to be recognized by other proteins, thus modulating its partner recruitment ability.

Another feature of IDPs is their great propensity to undergo posttranslational modifications (PTMs) that have potential to regulate their binding to different partners [19, 59]. For MeCP2, modifications such as phosphorylation, acetylation, ubiquitination, SUMOylation, and ribosylation have been described [60]. The specific function of most of these PTMs remains unknown, but some of them have been reported to regulate the interaction of MeCP2 with chromatin and determine the binding to different cofactors. In terms of its association with chromatin, MeCP2 poly(ADP-ribosyl)ation has recently been described to diminish its binding affinity for the chromocenters, and a role has been proposed for this PTM in the regulation of MeCP2-dependent heterochromatin aggregation [61]. Another recent example of a PTM that affects the MeCP2-chromatin interaction involves the developmentally regulated phosphorylation of S164 that has been shown to weaken MeCP2 binding

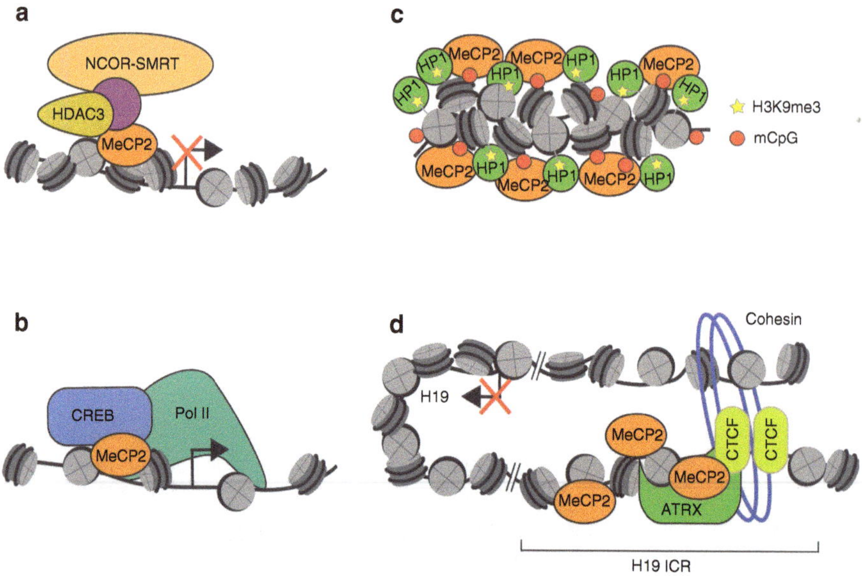

Fig. 1.3 MeCP2 regulatory function is mediated by its interaction with different partners. (**a**) MeCP2 interacts with corepressor complexes, such as NCOR-SMRT and HDACs, to promote chromatin compaction and repress transcription. (**b**) MeCP2 interacts with CREB to activate transcription. (**c**) MeCP2 interacts with HP1 to maintain the condensed and silenced state of pericentric heterochromatin, highly enriched in DNA methylation (mCpG) and tri-methylation of lysine 9 of histone H3 (H3K9me3). (**d**) MeCP2 partners with ATRX, cohesin, and CTCF at the *H19/Igr2* imprinted control region (ICR) to promote imprinting-related chromatin looping leading to transcriptional repression of *H19*

to chromatin [62]. With regard to PTMs affecting the association of MeCP2 with different cofactors, phosphorylation of S80 has been reported to mediate RNA-dependent interaction of MeCP2 with the Y-box binding protein 1 (YB-1), and phosphorylation of S229 has been shown to influence its association with heterochromatin protein 1 (HP1) [63].

The extreme flexibility of MeCP2 allows it to bind to DNA/chromatin in different ways and interact with a plethora of proteins to exert very diverse functions. This can be exemplified by the transcriptional repressor activity of MeCP2, through its interaction with the corepressor complexes N-CoR-SMRT (silencing mediator of retinoic acid and thyroid hormone receptor) and SIN3A (switch-independent 3a). These complexes bind to histone deacetylases (HDACs) to locally compact chromatin, hence blocking transcription (Fig. 1.3a) [64, 65]. Interestingly, the interaction with N-CoR is affected by phosphorylation of threonine 308 in the repressor domain of MeCP2 [36]. MeCP2 has also been shown to activate transcription by recruiting CREB1 to specific gene promoters (Fig. 1.3b) [9] and to participate in the regulation of the silenced chromatin state in pericentric regions by interacting with HP1 and enhancing its accumulation in heterochromatin (Fig. 1.3c) [66].

1.7 Higher-Order Chromatin Structures

The three-dimensional organization of chromatin in nuclei is not random; it is intimately linked to different functional processes and is highly conserved across cell types and species [67, 68]. In the past decade, the fast development of techniques, such as chromatin conformation capture in combination with high throughput sequencing, has improved our understanding of such complex organization. Chromatin is hierarchically organized, starting from a higher level where the interphasic chromosomes are distributed in specific territories or large megabase-sized topological domains, to lower hierarchical levels such as the loop domains, that are necessary to bring together regulatory regions that are located far apart in the linear genome [67–69].

Therefore, the concept of gene regulation must take into consideration a three-dimensional network of dynamic interactions that can also involve the participation of MeCP2, as there is some evidence of its role in the formation of higher-order chromatin structures. In vitro studies using nucleosome arrays, EM, and atomic force microscopy (AFM) techniques have shown that one of the most prominent motifs generated by the binding of MeCP2 to chromatin is the chromatin loop [10, 31, 38]. Moreover, an in vivo study carried out in mouse brains described the existence of an imprinted locus, the distal-less homeobox 5–6 (Dlx5–Dlx6), in which its transcriptional repressive state was mediated by the MeCP2-dependent establishment of an 11-kilobase chromatin loop [70]. This loop was absent in Mecp2-null mice and resulted in an increased expression of both genes. In addition, Dlx5 was described to exhibit loss of imprinting in lymphoblastoid cells derived from patients with Rett syndrome. Nevertheless, these results must be taken cautiously as they were subsequently challenged by another group [71], which observed biallelic expression of Dlx5 and Dlx6 in mice and claimed that MeCP2 was not involved in the maintenance of the imprinted locus. More recently, Kernohan and colleagues described the involvement of MeCP2, in association with cohesin and α-thalassemia mental retardation X linked (ATRX), in the control of imprinted genes in the mouse brain [72, 73]. Interestingly, ATRX and cohesin are also important regulators of chromatin organization and function, and their dysregulation results in the severe pathologies α-thalassemia mental retardation X linked and Cornelia de Lange, respectively. This underscores the critical relevance of chromatin organization in the nucleus. ATRX colocalizes with MeCP2, cohesin, and CCCTC-binding factor (CTCF) at the H19 imprinting control region (ICR), resulting in the regulation of the H19/Igf2 imprinted domain and a subset of other genes. The authors speculated that these proteins might control genomic imprinting by establishing higher-order chromatin structures, which have been described to be common regulatory mechanisms of genomic imprinting [74, 75]. Their results demonstrate that MeCP2 recruits ATRX to the H19 ICR, and, subsequently, ATRX modulates nucleosome positioning to extend the linker DNA and allow the binding of CTCF to the exposed binding sites (Fig. 1.3d). They observed that loss of MeCP2 decreases the number of chromatin interactions across the *H19/Igf2* and *Gtl2/Dlk1* imprinted domains. Such findings are in agreement with the well-documented role of cohesin in chromatin looping [76, 77] and are consistent with the ability of MeCP2 to assist in the formation of such structures in vitro [31].

1.8 Neuronal Chromatin: Histone H1 and MeCP2

During brain development, mitotically active cellular precursors evolve into specialized neuronal and glial cells. A complex neurodevelopmental program encompassing antiproliferative and neurogenic signals controls the differentiation process, which is overall strongly driven by epigenetic components such as chromatin remodelers, histone variants, histone PTMs, and DNA methylation, among other factors.

An interesting observation on neuronal chromatin structure was made back in the 1980s at Kuenzle's lab. They described how, during differentiation, mouse cortical neurons undergo a general decrease in the spacing between nucleosomes, the so-called nucleosome repeat length (NRL) [78]. The authors reported a shortening of the NRL from 200 bp in the prenatal stage to 170 bp after birth (Fig. 1.4a). By contrast, another type of highly abundant neurons in cerebellum, the granule cells, exhibited an increase from 165 bp before birth to 218 bp 30 days after birth, a final NRL similar to that of other adult cell types like glial or hepatic cells (Fig. 1.4a). Different labs followed up with these observations and characterized the distinct organization of neuronal chromatin. Research in the Jean Thomas lab began the characterization of neuronal and glial nuclear composition, and they described important differences between both cell types. They observed that neuronal nuclei are especially large in size, their chromatin is more dispersed and displays a higher transcriptional rate, and interestingly, the levels of histone H1 are 50% the amount which is present in glial cells [79].

The existence of MeCP2 was not known at that time, but the protein and its high abundance in brain were later discovered [2, 80]. As previously noted, many studies supported the view of MeCP2 as a classical transcriptional regulator. However, such notion started to be questioned. In different studies using mouse models of Rett syndrome and MeCP2 duplication syndrome, it was found that the gene expression variations were generally subtle and led to either the upregulation or the downregulation of a great number of genes [9, 81, 82]. Such observations are in agreement with the idea of MeCP2 being a chromatin structure regulator. Going back to the early studies on neuronal NRL, some tempting speculations can be made. Shahbazian and colleagues did an extensive study of MeCP2 levels in different brain regions and cell types [80], and a correlation between these and the Kuenzle's observations can now be stablished. Interestingly, in adult cortical neurons containing high levels of MeCP2, the NRL is shorter than in other cell types. In contrast, in other neurons, such as adult cerebellar granule cells that express extremely low amounts of MeCP2, or in different cell types like glial or hepatic cells also exhibiting negligible levels of MeCP2, the NRL is around 200 bp (Fig. 1.4a). During cortex development, the first neuronal precursors that stop dividing to start differentiating migrate to populate the deepest layer of the cortex, and once there, they become fully differentiated mature neurons (Fig. 1.4b). During this process, MeCP2 expression rises in these cells, initially displaying a diffuse nuclear distribution, until its later relocation to a typical punctate distribution that has been observed in rodent cells. Subsequent neuronal precursors that lose their dividing capacity will migrate to occupy more superficial layers, and MeCP2 will also increase its expression in these cells (Fig. 1.4b). Interestingly, immediately after birth, all cortical neurons start shortening their NRL

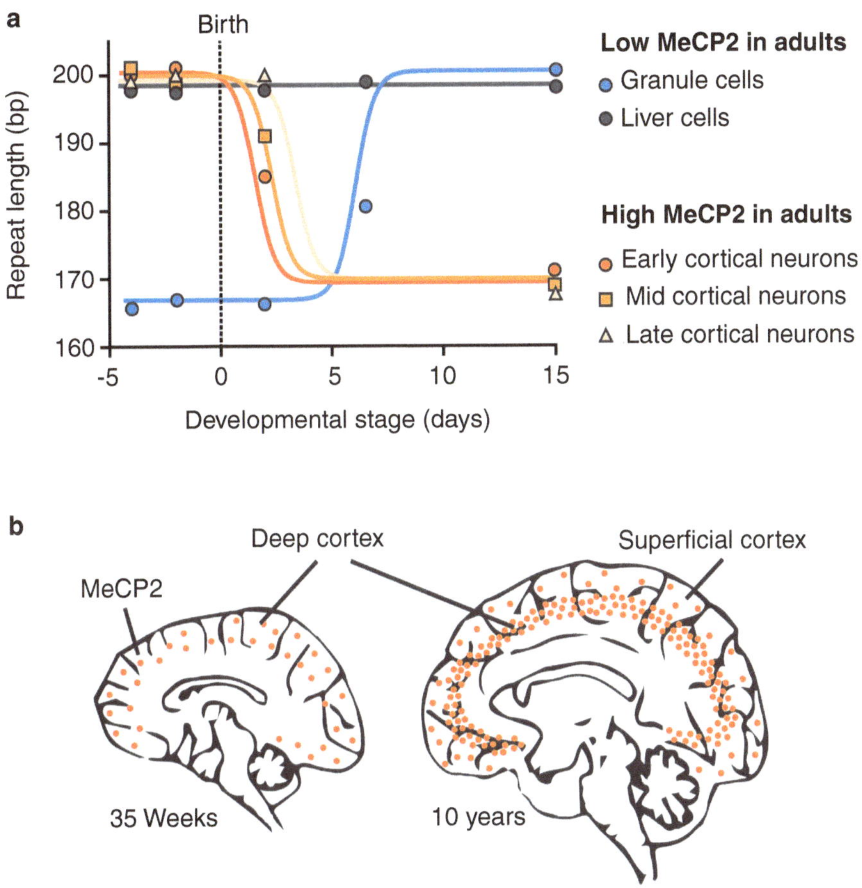

Fig. 1.4 Developmental neuronal chromatin reorganization and MeCP2 distribution during brain development. (**a**) Nucleosome repeat length (NRL) decreases during development in cortical neurons (*orange lines*). NRL decrease occurs postnatally following the order of neuronal differentiation (early differentiated neurons exhibit an early shortened NRL). Cerebellar granule cells show an opposite trend, with NRL increasing during development (*blue line*). Hepatic cells maintain a constant NRL during development (*gray line*). In the adult, neuronal cells containing high levels of MeCP2 (*orange lines*) show a reduced NRL compared to that of other brain cells with lower levels of the protein (*blue and gray lines*). (**b**) MeCP2 expression in 35-week- and 10-year-old human brain (*orange dots*). In the adult brain, early differentiated cortical neurons remain in the inner cortical layer where high levels of MeCP2 are detected. Newly differentiated neurons are subsequently located in more external layers where the high MeCP2 levels exhibit a later appearance (modified from [79])

in the same order in which they started their differentiation (e.g., neurons located in the inner cortical layer decrease their NRL first) (Fig. 1.4a). These observations suggest that MeCP2 could be involved in the developmental shaping of chromatin structure in neurons where the protein is present in higher levels.

More recently, a very interesting study by Skene et al. showed that MeCP2 in neurons is expressed at levels close to those of core histones; there is approximately one MeCP2 molecule every two nucleosomes, which would be consistent with the lower levels of H1 observed in neurons [11, 79]. Chromatin immunoprecipitation and sequencing (ChIP-seq) data from this and other studies [11, 12] show that MeCP2 is broadly distributed throughout the genome, exhibiting a binding profile similar to that of any other histone protein, therefore supporting its function as a core neuronal chromosomal protein. The global binding of MeCP2 alters chromatin structure by reducing histone acetylation and prevents spurious transcription of repetitive elements [11]. Moreover, in *Mecp2*-null neurons, there was a two-fold increase of the levels of H1 compared to wild-type neurons. This interrelationship between both proteins suggests that in neurons, where MeCP2 is highly expressed, it most likely replaces histone H1 and therefore exerts its role as an alternative highly specialized linker histone. This notion is in perfect agreement with increasing evidence describing diverse functional resemblances between both proteins, despite of their completely different primary structure. The association of MeCP2 with chromatin is known to exhibit a preference for linker DNA domains, especially when H1 is not present [83]. Both in vitro and in vivo studies have demonstrated the ability of MeCP2 to compete with histone H1 for the binding sites, both in the presence and absence of DNA methylation [8, 10, 83]. Indeed, it is able to displace up to a 40% of histone H1 when the underlying DNA is methylated. Furthermore, MeCP2 interacts with DNA cruciform structures with an affinity as high as that for mCpGs [84]. Interestingly, four-way junctions of cruciform DNA mimic the preferential binding site of the histone H1 at the entry and exit sites of DNA in the nucleosome. In vitro experiments show that, when bound to such sites, both H1 and MeCP2 lead to the formation of so-called "stem-like" structures and condense the chromatin fibers in a similar way [10, 85]. They facilitate the "zigzag" nucleosome organization necessary to achieve the higher levels of chromatin compaction that are involved in the formation of 30 nm DNA fibers. In addition, the dynamic behavior of both proteins is very similar, as fluorescence recovery after photobleaching (FRAP) experiments have demonstrated that the exchange rate is very fast and similar in both instances, suggesting that both of their bindings to chromatin are highly dynamic [86, 87].

All of this is heavily indicative of a very important role of MeCP2 in general organization of chromatin structure, especially in neurons, where this protein is so highly abundant.

1.9 Concluding Remarks

The neuronal functional regulation of chromatin is a complex process that involves a highly specialized and distinctive pattern of DNA methylation. MeCP2 seems to participate in this duality, being able to recognize and bind hydroxymethylated/methylated DNA and translate the message encrypted by these chemical modifications into different chromatin structural conformations. Despite the intensive research carried out on MeCP2 since its discovery, the detailed role of the protein in the interconnection between these structural features is still poorly understood.

It most likely will be key to our understanding of the mechanisms leading to different neurological diseases including and beyond Rett Syndrome.

Acknowledgments This work was supported by a Canadian Institutes of Health Research (CIHR) MOP-97878 grant to JA.

References

1. Guy J, Cheval H, Selfridge J, Bird A. The role of MeCP2 in the brain. Annu Rev Cell Dev Biol. 2011;27:631–52.
2. Lewis JD, Meehan RR, Henzel WJ, Maurer-Fogy I, Jeppesen P, Klein F, et al. Purification, sequence, and cellular localization of a novel chromosomal protein that binds to methylated DNA. Cell. 1992;69(6):905–14.
3. Amir RE, Van den Veyver IB, Wan M, Tran CQ, Francke U, Zoghbi HY. Rett syndrome is caused by mutations in X-linked MECP2, encoding methyl-CpG-binding protein 2. Nat Genet. 1999;23(2):185–8.
4. Percy AK, Lane JB. Rett syndrome: model of neurodevelopmental disorders. J Child Neurol. 2005;20(9):718–21.
5. Chen RZ, Akbarian S, Tudor M, Jaenisch R. Deficiency of methyl-CpG binding protein-2 in CNS neurons results in a Rett-like phenotype in mice. Nat Genet. 2001;27(3):327–31.
6. Luikenhuis S, Giacometti E, Beard CF, Jaenisch R. Expression of MeCP2 in postmitotic neurons rescues Rett syndrome in mice. Proc Natl Acad Sci U S A. 2004;101(16):6033–8.
7. Nan X, Tate P, Li E, Bird A. DNA methylation specifies chromosomal localization of MeCP2. Mol Cell Biol. 1996;16(1):414–21.
8. Nan X, Campoy FJ, Bird A. MeCP2 is a transcriptional repressor with abundant binding sites in genomic chromatin. Cell. 1997;88(4):471–81.
9. Chahrour M, Jung SY, Shaw C, Zhou X, Wong STC, Qin J, et al. MeCP2, a key contributor to neurological disease, activates and represses transcription. Science. 2008;320(5880):1224–9.
10. Ghosh RP, Horowitz-Scherer RA, Nikitina T, Shlyakhtenko LS, Woodcock CL. MeCP2 binds cooperatively to its substrate and competes with histone H1 for chromatin binding sites. Mol Cell Biol. 2010;30(19):4656–70.
11. Skene PJ, Illingworth RS, Webb S, Kerr ARW, James KD, Turner DJ, et al. Neuronal MeCP2 is expressed at near histone-octamer levels and globally alters the chromatin state. Mol Cell. 2010;37(4):457–68.
12. Cohen S, Gabel HW, Hemberg M, Hutchinson AN, Sadacca LA, Ebert DH, et al. Genome-wide activity-dependent MeCP2 phosphorylation regulates nervous system development and function. Neuron. 2011;72(1):72–85.
13. Mellén M, Ayata P, Dewell S, Kriaucionis S, Heintz N. MeCP2 binds to 5hmC enriched within active genes and accessible chromatin in the nervous system. Cell. 2012;151(7):1417–30.
14. Guo JU, Su Y, Shin JH, Shin J, Li H, Xie B, et al. Distribution, recognition and regulation of non-CpG methylation in the adult mammalian brain. Nat Neurosci. 2014;17(2):215–22.
15. Mnatzakanian GN, Lohi H, Munteanu I, Alfred SE, Yamada T, MacLeod PJM, et al. A previously unidentified MECP2 open reading frame defines a new protein isoform relevant to Rett syndrome. Nat Genet. 2004;36(4):339–41.
16. Ausió J, Martínez de Paz A, Esteller M. MeCP2: the long trip from a chromatin protein to neurological disorders. Trends Mol Med. 2014;20(9):487–98.
17. Das S, Mukhopadhyay D. Intrinsically unstructured proteins and neurodegenerative diseases: conformational promiscuity at its best. IUBMB Life. 2011;63(7):478–88.
18. Hansen JC, Ghosh RP, Woodcock CL. Binding of the Rett syndrome protein, MeCP2, to methylated and unmethylated DNA and chromatin. IUBMB Life. 2010;62(10):732–8.
19. Gsponer J, Futschik ME, Teichmann SA, Babu MM. Tight regulation of unstructured proteins: from transcript synthesis to protein degradation. Science. 2008;322(5906):1365–8.

20. Dyson HJ, Wright PE. Intrinsically unstructured proteins and their functions. Nat Rev Mol Cell Biol. 2005;6(3):197–208.
21. Van Esch H, Bauters M, Ignatius J, Jansen M, Raynaud M, Hollanders K, et al. Duplication of the MECP2 region is a frequent cause of severe mental retardation and progressive neurological symptoms in males. Am J Hum Genet. 2005;77(3):442–53.
22. del Gaudio D, Fang P, Scaglia F, Ward PA, Craigen WJ, Glaze DG, et al. Increased MECP2 gene copy number as the result of genomic duplication in neurodevelopmentally delayed males. Genet Med. 2006;8(12):784–92.
23. Collins AL, Levenson JM, Vilaythong AP, Richman R, Armstrong DL, Noebels JL, et al. Mild overexpression of MeCP2 causes a progressive neurological disorder in mice. Hum Mol Genet. 2004;13(21):2679–89.
24. Samaco RC, Nagarajan RP, Braunschweig D, LaSalle JM. Multiple pathways regulate MeCP2 expression in normal brain development and exhibit defects in autism-spectrum disorders. Hum Mol Genet. 2004;13(6):629–39.
25. Pohodich AE, Zoghbi HY. Rett syndrome: disruption of epigenetic control of postnatal neurological functions. Hum Mol Genet. 2015;24(R1):R10–6.
26. Chao H-T, Zoghbi HY. MeCP2: only 100% will do. Nat Neurosci. 2012;15(2):176–7.
27. Adams VH, McBryant SJ, Wade PA, Woodcock CL, Hansen JC. Intrinsic disorder and autonomous domain function in the multifunctional nuclear protein, MeCP2. J Biol Chem. 2007;282(20):15057–64.
28. Ho KL, McNae IW, Schmiedeberg L, Klose RJ, Bird AP, Walkinshaw MD. MeCP2 binding to DNA depends upon hydration at methyl-CpG. Mol Cell. 2008;29(4):525–31.
29. Baker SA, Chen L, Wilkins AD, Yu P, Lichtarge O, Zoghbi HY. An AT-hook domain in MeCP2 determines the clinical course of Rett syndrome and related disorders. Cell. 2013;152(5):984–96.
30. Ghosh RP, Nikitina T, Horowitz-Scherer RA, Gierasch LM, Uversky VN, Hite K, et al. Unique physical properties and interactions of the domains of methylated DNA binding protein 2. Biochemistry. 2010;49(20):4395–410.
31. Nikitina T, Shi X, Ghosh RP, Horowitz-Scherer RA, Hansen JC, Woodcock CL. Multiple modes of interaction between the methylated DNA binding protein MeCP2 and chromatin. Mol Cell Biol. 2007;27(3):864–77.
32. Becker A, Allmann L, Hofstätter M, Casà V, Weber P, Lehmkuhl A, et al. Direct homo- and hetero-interactions of MeCP2 and MBD2. PLoS One. 2013;8(1):e53730.
33. Buschdorf JP, Strätling WH. A WW domain binding region in methyl-CpG-binding protein MeCP2: impact on Rett syndrome. J Mol Med (Berl). 2004;82(2):135–43.
34. Nan X, Meehan RR, Bird A. Dissection of the methyl-CpG binding domain from the chromosomal protein MeCP2. Nucleic Acids Res. 1993;21(21):4886–92.
35. Bellini E, Pavesi G, Barbiero I, Bergo A, Chandola C, Nawaz MS, et al. MeCP2 post-translational modifications: a mechanism to control its involvement in synaptic plasticity and homeostasis? Front Cell Neurosci. 2014;8(August):236.
36. Ebert DH, Gabel HW, Robinson ND, Kastan NR, Hu LS, Cohen S, et al. Activity-dependent phosphorylation of MeCP2 threonine 308 regulates interaction with NCoR. Nature. 2013;499(7458):341–5.
37. Kimura H, Shiota K. Methyl-CpG-binding protein, MeCP2, is a target molecule for maintenance DNA methyltransferase, Dnmt1. J Biol Chem. 2003;278(7):4806–12.
38. Georgel PT, Horowitz-Scherer RA, Adkins N, Woodcock CL, Wade PA, Hansen JC. Chromatin compaction by human MeCP2. Assembly of novel secondary chromatin structures in the absence of DNA methylation. J Biol Chem. 2003;278(34):32181–8.
39. Fraga MF, Ballestar E, Montoya G, Taysavang P, Wade PA, Esteller M. The affinity of different MBD proteins for a specific methylated locus depends on their intrinsic binding properties. Nucleic Acids Res. 2003;31(6):1765–74.
40. Young JI, Hong EP, Castle JC, Crespo-Barreto J, Bowman AB, Rose MF, et al. Regulation of RNA splicing by the methylation-dependent transcriptional repressor methyl-CpG binding protein 2. Proc Natl Acad Sci U S A. 2005;102(49):17551–8.

41. Lister R, Mukamel EA, Nery JR, Urich M, Puddifoot CA, Johnson ND, et al. Global epigenomic reconfiguration during mammalian brain development. Science. 2013;341(6146):1237905.
42. Suzuki MM, Bird A. DNA methylation landscapes: provocative insights from epigenomics. Nat Rev Genet. 2008;9(6):465–76.
43. Cohen S, Greenberg ME. A bird's-eye view of MeCP2 binding. Mol Cell. 2010;37(4):451–2.
44. Jakovcevski M, Akbarian S. Epigenetic mechanisms in neurological disease. Nat Med. 2012;18(8):1194–204.
45. Zovkic IB, Guzman-Karlsson MC, Sweatt JD. Epigenetic regulation of memory formation and maintenance. Learn Mem. 2013;20(2):61–74.
46. Thambirajah AA, Ng MK, Frehlick LJ, Li A, Serpa JJ, Petrotchenko EV, et al. MeCP2 binds to nucleosome free (linker DNA) regions and to H3K9/H3K27 methylated nucleosomes in the brain. Nucleic Acids Res. 2012;40(7):2884–97.
47. Gabel HW, Kinde B, Stroud H, Gilbert CS, Harmin DA, Kastan NR, et al. Disruption of DNA-methylation-dependent long gene repression in Rett syndrome. Nature. 2015;522(7554): 89–93.
48. Spruijt CG, Gnerlich F, Smits AH, Pfaffeneder T, Jansen PWTC, Bauer C, et al. Dynamic readers for 5-(Hydroxy)methylcytosine and its oxidized derivatives. Cell. 2013;152(5):1146–59.
49. Khrapunov S, Warren C, Cheng H, Berko ER, Greally JM, Brenowitz M. Unusual characteristics of the DNA binding domain of epigenetic regulatory protein MeCP2 determine its binding specificity. Biochemistry. 2014;53(21):3379–91.
50. Ballestar E, Yusufzai TM, Wolffe AP. Effects of rett syndrome mutations of the Methyl-CpG binding domain of the transcriptional repressor MeCP2 on selectivity for association with methylated DNA. Biochemistry. 2000;39(24):7100–6.
51. Stuss DP, Cheema M, Ng MK, Martinezde Paz A, Williamson B, Missiaen K, et al. Impaired in vivo binding of MeCP2 to chromatin in the absence of its DNA methyl-binding domain. Nucleic Acids Res. 2013;41(9):4888–900.
52. Van Holde K. Chromatin. New York: Springer; 1988.
53. Chandler SP, Guschin D, Landsberger N, Wolffe AP. The methyl-CpG binding transcriptional repressor MeCP2 stably associates with nucleosomal DNA. Biochemistry. 1999;38(22):7008–18.
54. Buschhausen G, Wittig B, Graessmann M, Graessmann A. Chromatin structure is required to block transcription of the methylated herpes simplex virus thymidine kinase gene. Proc Natl Acad Sci U S A. 1987;84(5):1177–81.
55. Simpson RT, Thoma F, Brubaker JM. Chromatin reconstituted from tandemly repeated cloned DNA fragments and core histones: a model system for study of higher order structure. Cell. 1985;42(3):799–808.
56. Nikitina T, Ghosh RP, Horowitz-Scherer RA, Hansen JC, Grigoryev SA, Woodcock CL. MeCP2-chromatin interactions include the formation of chromatosome-like structures and are altered in mutations causing Rett syndrome. J Biol Chem. 2007;282(38):28237–45.
57. Klose RJ, Sarraf SA, Schmiedeberg L, SM MD, Stancheva I, Bird AP. DNA binding selectivity of MeCP2 due to a requirement for A/T sequences adjacent to methyl-CpG. Mol Cell. 2005;19(5):667–78.
58. Hite KC, Kalashnikova AA, Hansen JC. Coil-to-helix transitions in intrinsically disordered methyl CpG binding protein 2 and its isolated domains. Protein Sci. 2012;21(4):531–8.
59. Bah A, Forman-Kay JD. Modulation of intrinsically disordered protein function by post-translational modifications. J Biol Chem. 2016;291(13):6696–705.
60. Ausió J. MeCP2 and the enigmatic organization of brain chromatin. Implications for depression and cocaine addiction. Clin Epigenetics. 2016;8(1):58.
61. Becker A, Zhang P, Allmann L, Meilinger D, Bertulat B, Eck D, et al. Poly(ADP-ribosyl) ation of methyl CpG binding domain protein 2 regulates chromatin structure. J Biol Chem. 2016;291(10):M115.698357.
62. Stefanelli G, Gandaglia A, Costa M, Cheema MS, Di Marino D, Barbiero I, et al. Brain phosphorylation of MeCP2 at serine 164 is developmentally regulated and globally alters its chromatin association. Sci Rep. 2016;6(March):28295.

63. Gonzales ML, Adams S, Dunaway KW, LaSalle JM. Phosphorylation of distinct sites in MeCP2 modifies cofactor associations and the dynamics of transcriptional regulation. Mol Cell Biol. 2012;32(14):2894–903.
64. Lyst MJ, Ekiert R, Ebert DH, Merusi C, Nowak J, Selfridge J, et al. Rett syndrome mutations abolish the interaction of MeCP2 with the NCoR/SMRT co-repressor. Nat Neurosci. 2013;16(7):898–902.
65. Nan X, Ng HH, Johnson CA, Laherty CD, Turner BM, Eisenman RN, et al. Transcriptional repression by the methyl-CpG-binding protein MeCP2 involves a histone deacetylase complex. Nature. 1998;393(6683):386–9.
66. Agarwal N, Hardt T, Brero A, Nowak D, Rothbauer U, Becker A, et al. MeCP2 interacts with HP1 and modulates its heterochromatin association during myogenic differentiation. Nucleic Acids Res. 2007;35(16):5402–8.
67. Dixon JR, Selvaraj S, Yue F, Kim A, Li Y, Shen Y, et al. Topological domains in mammalian genomes identified by analysis of chromatin interactions. Nature. 2012;485(7398):376–80.
68. Rao SSP, Huntley MH, Durand NC, Stamenova EK, Bochkov ID, Robinson JT, et al. A 3D map of the human genome at kilobase resolution reveals principles of chromatin looping. Cell. 2014;159(7):1665–80.
69. Rowley MJ, Corces VG. The three-dimensional genome: principles and roles of long-distance interactions. Curr Opin Cell Biol. 2016;40:8–14.
70. Horike S, Cai S, Miyano M, Cheng J-F, Kohwi-Shigematsu T. Loss of silent-chromatin looping and impaired imprinting of DLX5 in Rett syndrome. Nat Genet. 2005;37(1):31–40.
71. Schüle B, Li HH, Fisch-Kohl C, Purmann C, Francke U. DLX5 and DLX6 expression is biallelic and not modulated by MeCP2 deficiency. Am J Hum Genet. 2007;81(3):492–506.
72. Kernohan KD, Jiang Y, Tremblay DC, Bonvissuto AC, Eubanks JH, Mann MRW, et al. ATRX partners with cohesin and MeCP2 and contributes to developmental silencing of imprinted genes in the brain. Dev Cell. 2010;18(2):191–202.
73. Kernohan KD, Vernimmen D, Gloor GB, Bérubé NG. Analysis of neonatal brain lacking ATRX or MeCP2 reveals changes in nucleosome density, CTCF binding and chromatin looping. Nucleic Acids Res. 2014;42(13):8356–68.
74. Murrell A, Heeson S, Reik W. Interaction between differentially methylated regions partitions the imprinted genes Igf2 and H19 into parent-specific chromatin loops. Nat Genet. 2004;36(8):889–93.
75. Court F, Camprubi C, Garcia C, Guillaumet-Adkins A, Sparago A, Seruggia D, et al. The PEG13-DMR and brain-specific enhancers dictate imprinted expression within the 8q24 intellectual disability risk locus. Epigenetics Chromatin. 2014;7(1):5.
76. Hadjur S, Williams LM, Ryan NK, Cobb BS, Sexton T, Fraser P, et al. Cohesins form chromosomal cis-interactions at the developmentally regulated IFNG locus. Nature. 2009;460(7253):410–3.
77. Mishiro T, Ishihara K, Hino S, Tsutsumi S, Aburatani H, Shirahige K, et al. Architectural roles of multiple chromatin insulators at the human apolipoprotein gene cluster. EMBO J. 2009;28(9):1234–45.
78. Jaeger AW, Kuenzle CC. The chromatin repeat length of brain cortex and cerebellar neurons changes concomitant with terminal differentiation. EMBO J. 1982;1(7):811–6.
79. Pearson EC, Bates DL, Prospero TD, Thomas JO. Neuronal nuclei and glial nuclei from mammalian cerebral cortex. Nucleosome repeat lengths, DNA contents and H1 contents. Eur J Biochem. 1984;144(2):353–60.
80. Shahbazian MD, Antalffy B, Armstrong DL, Zoghbi HY. Insight into Rett syndrome: MeCP2 levels display tissue- and cell-specific differences and correlate with neuronal maturation. Hum Mol Genet. 2002;11(2):115–24.
81. Tudor M, Akbarian S, Chen RZ, Jaenisch R. Transcriptional profiling of a mouse model for Rett syndrome reveals subtle transcriptional changes in the brain. Proc Natl Acad Sci U S A. 2002;99(24):15536–41.
82. Jordan C, Li HH, Kwan HC, Francke U. Cerebellar gene expression profiles of mouse models for Rett syndrome reveal novel MeCP2 targets. BMC Med Genet. 2007;8:36.

83. Ishibashi T, Thambirajah A, Ausió J. MeCP2 preferentially binds to methylated linker DNA in the absence of the terminal tail of histone H3 and independently of histone acetylation. FEBS Lett. 2008;582(7):1157–62.
84. Galvão TC, Thomas JO. Structure-specific binding of MeCP2 to four-way junction DNA through its methyl CpG-binding domain. Nucleic Acids Res. 2005;33(20):6603–9.
85. Bednar J, Horowitz RA, Grigoryev SA, Carruthers LM, Hansen JC, Koster AJ, et al. Nucleosomes, linker DNA, and linker histone form a unique structural motif that directs the higher-order folding and compaction of chromatin. Proc Natl Acad Sci U S A. 1998;95(24): 14173–8.
86. Misteli T, Gunjan A, Hock R, Bustin M, Brown DT. Dynamic binding of histone H1 to chromatin in living cells. Nature. 2000;408(6814):877–81.
87. Kumar A, Kamboj S, Malone BM, Kudo S, Twiss JL, Czymmek KJ, et al. Analysis of protein domains and Rett syndrome mutations indicate that multiple regions influence chromatin-binding dynamics of the chromatin-associated protein MECP2 in vivo. J Cell Sci. 2008;121 (Pt 7):1128–37.

2 The Role of Noncoding RNAs in Neurodevelopmental Disorders: The Case of Rett Syndrome

Aida Obiols-Guardia and Sònia Guil

Abstract

Current technologies have demonstrated that only a small fraction of our genes encode for protein products. The vast majority of the human transcriptome corresponds to noncoding RNA (ncRNA) of different size, localization, and expression profile. Despite the fact that a biological function remains yet to be determined for most ncRNAs, growing evidence points to their crucial regulatory roles at all stages in gene expression regulation, including transcriptional and posttranscriptional control, so that proper cell homeostasis seems to depend largely on a variety of ncRNA-mediated regulatory networks. This is particularly relevant in the human brain, which displays the richest repertoire of ncRNA species, and where several different ncRNA molecules are known to be involved in crucial steps for brain development and maturation. Rett syndrome is a neurodevelopmental disorder characterized by loss of function mutations in the X-linked gene encoding for methyl-CpG-binding protein 2 (*MeCP2*). MECP2 deficiency impacts globally on gene expression programs, mainly through its role as a transcriptional repressor, and growing data also points to an important dysregulation of the noncoding transcriptome in the disease. Here, we review the current knowledge on ncRNA alterations in Rett and explore links with other pathologies that might indicate the potential use of particular noncoding transcripts as therapeutical targets, tools, or disease biomarkers.

Keywords

ncRNA • miRNA • circRNA • Rett syndrome • Central nervous system • MeCP2

A. Obiols-Guardia, M.Sc. • S. Guil, Ph.D. (✉)
Cancer Epigenetics and Biology Program (PEBC), Bellvitge Biomedical Research Institute (IDIBELL), Barcelona, Catalonia, Spain
e-mail: aobiols@idibell.cat; sguil@idibell.cat

R. Delgado-Morales (ed.), *Neuroepigenomics in Aging and Disease*, Advances in Experimental Medicine and Biology, DOI 10.1007/978-3-319-53889-1_2

2.1 Introduction

Most of the human genome is transcribed as nonprotein-coding RNAs [1]. The development of massive parallel sequencing techniques has allowed a thorough survey of the noncoding transcriptome in a growing number of tissue types and developmental points, indicating that there exist a large number of noncoding transcripts (ncRNAs) whose roles are essentially unknown. However, the increasing number of studies about this "junk" DNA shows its crucial importance in the control of gene regulation, especially in the brain and neurological diseases [2]. Noncoding RNAs (ncRNAs) are conventionally classified as small (sncRNA, <200 nucleotides in length) or long (lncRNAs, >200 nucleotides in length) species, with the former group (including miRNAs, piRNAs, snoRNAs, snRNAs, among others) being the better characterized subtype. Most sncRNAs are involved in the biogenesis and modifications of other RNAs or are related to posttranscriptional regulation [3, 4]. By comparison, lncRNAs are thought to constitute the larger fraction of the noncoding transcriptome, but their precise biological roles are still largely unknown. Only a handful of examples are starting to be thoroughly characterized (including Xist, H19, HOTAIR), so that for the vast majority of them, only scarce information is available as to their functional relevance. Nonetheless, a growing body of evidence points to their involvement at all stages in fundamental cellular processes and specifically in gene expression regulation, from chromatin structural determinants to fine-tuning cytoplasmic controllers of mRNAs turnover, localization, and translation. In addition, different types of ncRNAs may interact and regulate each other, as well as mRNAs, to maintain a dynamic network of interactions essential for normal cell homeostasis (reviewed in [5]).

Comparatively, the brain displays the higher diversity of expressed ncRNAs, and it is also the organ with a higher number of tissue-specific lncRNAs, with regulated temporal and spatial expression patterns [6, 7]. Of note, the generally low evolutionarily conservation of lncRNAs is increased for the brain-specific lncRNAs, which show a higher degree of similarity in the patterns of expression during brain development [8, 9]. For these reasons, lncRNAs are thought to play crucial roles in brain maturation and function, and as a consequence, their dysregulation might be important during the onset or progression of a number of neurological diseases [10, 11]. In particular, a number of neurodevelopmental disorders are directly linked to the aberrant expression of some lncRNAs, including *SNORD115* and *SNORD116* in Prader-Willi syndrome [12] or the imprinted *KCNQ1OT1* and *H19* in Beckwith-Wiedemann syndrome [13]. However, the specific mechanisms involved are far from being completely understood, and further intense investigations will be required to give the complete picture of ncRNA-based regulatory functions in the central nervous system (CNS).

It is worth noting that, although most ncRNAs are linear molecules, a subgroup of them form a circle through covalent binding; circular RNAs (circRNAs) are expressed in a tissue- and organ-specific manner and derive mainly from noncanonical splicing of protein-coding pre-mRNAs [14]. Their functions remain largely unexplored, although one proposed mode of action is their ability to titrate out the pool of functional miRNAs (and also RNA binding proteins or other small RNAs) by acting as molecular "sponges" [15–17]. In the brain, circRNAs are highly abundant in synapses and more conserved than other noncoding RNAs [18], and their expression seems to increase during development of the CNS [19]. In addition, their presence in the bloodstream and their abundance in exosomes as circulating molecules have given them value as biomarkers for noninvasive diagnosis in a number of human diseases, including disorders of the CNS and degenerative diseases [20].

2.2 Rett Syndrome

Rett syndrome (RTT, OMIM 312750) is a postnatal progressive neurodevelopmental disorder that represents the second cause of mental retardation in females [21]. RTT is principally caused by de novo mutations in the X-linked gene encoding methyl-CpG-binding protein 2 *(MECP2)* [22]. The disease occurs almost exclusively in females, affecting one in every 10,000 live female births. The fact that *MECP2* is located in the X chromosome typically causes the premature death of males because of complete absence of functional protein [23]. However, in order to compensate the different gene dosage between males and females, cells are able to randomly inactivate one of their X chromosomes in girls, resulting in a mosaic of cells. For this reason, Rett girls dispose of functional *MECP2* in approximately 50% of their cells and can survive. Rett patients generally don't exhibit symptoms until 6–18 months of age [24]. From then on, the disorder may become evident when patients lose their verbal ability [25] and show severe cognitive impairment, stereotypic behaviors [26] as well as cardiac and respiratory abnormalities [27]. The life expectancy for patients is extremely reduced and sudden unexpected deaths are not rare, often due to cardiovascular and respiratory dysfunction [28]. The lack of an effective cure and even of treatment options to resolve the most disabling symptoms of the disorder urges a better understanding of its physiopathology [29, 30]. Current efforts to tackle RTT include both pharmacological and molecular genetics-based strategies, but their therapeutical success has been to date limited.

2.2.1 MeCP2

The genetic defect at the root of Rett syndrome is the occurrence of loss-of-function mutations in methyl-CpG-binding protein 2 gene (*MECP2*). MeCP2 is a basic nuclear protein classically considered as a transcriptional repressor [31] that is able to bind to DNA sequences at methylated cytosines in 5′CpG dinucleotides [32]. Nevertheless, several studies have shown that MeCP2 can play a variety of roles, as it might also act as a chromatin activator, RNA-splicing modulator [30], and may even take part in the nuclear microRNA cleavage step [33]. MeCP2 is thus a global transcription and translation regulator with profound effects on normal cell physiology. There are two isoforms of *MECP2*, long e1 and short e2, but it seems that the indispensable one for a normal neurodevelopment is e1 [34]. It is known that the expression of this gene changes with age, increasing its levels during postnatal development [35]. For that reason, it seems that MeCP2 is not essential for a normal brain development, but rather for the correct maintenance of the mature neurons [32].

Although MeCP2 is found in several tissues, its highest expression levels are in the brain [36], where the correct dose and function of the protein are crucial for the proper establishment and maturation of neuronal networks [32]. Thus, this syndrome is an example of how different mutations in one single gene can affect several pathways essential for the neural structure [37]. Although there is not sufficient data available to confirm if RTT is caused by MECP2 deficiency in specific cell types or by a global dysfunction in the entire brain [32], different analyses suggest the importance of MECP2 role in GABAergic neurons, whose deficiency can cause a shift in the excitatory/inhibitory balance, giving rise to Rett symptoms [38]. Also, recent studies have shown that loss of MeCP2 in astrocytes can contribute to the pathophysiology of the disease due to a negative impact of the glia over neurons [39], adding thus another area in which it is necessary to increase our knowledge of the disease. It is important to keep in mind that some symptoms can result from MECP2 dysfunction in other organs than CNS, too [40]. Also, there are differences in the severity of the syndrome depending on the particular mutation in the protein, being R106T, R168X, R270X, and R255X the most critical for the patients' phenotype [41].

2.2.2 Rett Syndrome Models

Despite the efforts to fully understand the molecular mechanisms underlying complex neurological diseases, we are still far from achieving this ambitious challenge. However, the roles of some important genetic and epigenetic networks in the brain are beginning to be uncovered, and most advances in this field have been possible, thanks to the generation of animal and human cell line models.

The first results derived from total *Mecp2* deletion using mice stem cells showed that the protein is indispensable for embryonic development [42]. It was in 2001 when the first *Mecp2* knockout (KO) mouse was obtained [43] using the Cre-loxP technology (bird strain). Also in 2006, Pelka et al. generated *Mecp2*-null mice erasing the methyl-CpG-binding domain and disrupting the transcriptional repression domain after a splice site modification (*Mecp2*^{tm1Tam}) [44]. Other animals where a truncated form of MeCP2 is expressed have also been generated [36, 45]. Besides, conditional knockout (CKO) mice models for Rett syndrome using the Cre-loxP approach have been designed too, restricting the protein deletion to specific types of neurons [38, 46–50]. These last animals are very useful for studying particular molecular pathways, but they do not recapitulate all features of RTT syndrome [51]. Recently, *Mecp2*-deficient mice using the CRISPR/Cas9 gene-editing technique [52] have been generated. This approach has several advantages over conventional genetic screens where, for example, RNA interference is used and off-target effects can disrupt the results [53, 54].

Certainly, the findings in animal models have proved essential and very helpful to elucidate possibly altered pathways in a wide range of diseases as well as to develop therapeutic and diagnostic strategies. However, the information that is obtained from these experiments cannot be directly translated to patients before verifying it in a human model. In this regard, the generation of induced pluripotent stem cells from human somatic cells (hiPSCs) in 2007 [55] was an important step for a more refined study of several diseases in a proper system, as well as to perform more personalized drug screenings [56]. hiPSCs have allowed us to increase our knowledge about molecular mechanisms in tissues that would not be accessible otherwise, such as the brain [57], and Rett syndrome is one of the disorders where the generation of such cells has been essential [58, 59]. Technology has thus enabled researchers to obtain in vitro derived patient-specific neurons that exhibit the genetic and phenotypic features of RTT [60], including lower synaptic activity or reduced spine density. Importantly, female hiPSCs also maintain an epigenetic memory of their tissue of origin such as a nonrandom X chromosome inactivation pattern [61], allowing the specific reproduction of the disease in each patient.

2.3 Rett Syndrome and ncRNAs

With this variety of models at hand, researchers have now a number of complementary options to study in depth the molecular and cellular mechanisms underlying the onset and progression of Rett syndrome and, most importantly, to envisage and develop efficient treatment strategies. One area of research insufficiently addressed to date is the detailed characterization of the changes in the noncoding transcriptome as a consequence of MECP2 dysfunction. Equally important is the full understanding of how RNA-based tools can help design and implement strategies that aim to recover normal MECP2 protein levels. For example, as a X chromosome-linked gene, one major area of therapeutic interest involves reactivation of the X chromosome across relevant cell types [62]. X-inactivation is established following a network of lncRNA-mediated regulatory events, with the X-inactive specific transcript (Xist) playing an indispensable role [63–65]. Intensive research aiming at developing tools for controlling the whole process is currently ongoing, which might help in RTT and other X-linked diseases.

The class of ncRNAs most commonly studied in normal and pathological states is miRNAs. Both in the normal brain and also in neurodevelopmental diseases, miRNAs' involvement has been frequently explored showing that their correct process and function are essential for the proper neural development [66–69]. In the context of Rett syndrome, changes in miRNA expression may reflect a direct (mostly through transcriptional derepression) or indirect deregulation following MECP2 loss of function, and there are several publications that indicate the relevance of the changes in miRNA expression profiles in the disease progress [70, 71]. For instance, upregulation of the miRNA cluster within the *Dlk1-Gtl2* imprinting domain in the cerebellum of KO mice might alter dendritogenesis and synapse formation [68, 72]. Another example of this regulation is miR-184, an imprinted miRNA whose transcription is repressed by MeCP2 under certain neuronal activity conditions [73]. It is also worth noting that, given the direct link between miRNAs and posttranscriptional regulation of protein-coding genes, many researchers evaluate the possibility that changes in some miRNAs levels could be causal in the deregulation of many mRNA transcripts in Mecp2 KO mice [74, 75]. In fact, miRNAs 30a, 381, and 495 are aberrantly upregulated in knockout mice cerebella at 6 weeks after birth and are known to target the 3′UTR of *Bdnf* [72], a neurotrophin that represents one of the most potent modulators of neuronal development [76]. BDNF is known to be aberrantly diminished in RTT individuals, and its restoration is the object of currently undergoing clinical trials [29].

Beyond its function as a transcriptional regulator, it has been shown that MeCP2 can also interact with and influence the activity of the pri-miRNA processing machinery under resting status in neurons [33]. Specifically, the phosphorylation of serine 80 in the C-terminal domain allows MeCP2 to act as a competitor with DROSHA for the binding to DGCR8 protein (see Fig. 2.1). Remarkably, neuronal depolarization triggers dephosphorylation at Ser80, indicating a neuronal activity-dependent function for MeCP2 as a Microprocessor modulator. When MeCP2 is bound to DGCR8, pri-miRNAs cannot be processed by DROSHA-DGCR8 complex and the pre-miRNAs are not generated. As a consequence, targets of these miRNAs and related molecular pathways are deregulated in a *mecp2*-null condition. Of note, and as an example, the authors demonstrated that targeted proteins of miR-134, an important miRNA for neural development and influenced by MeCP2 regulation, have an altered expression levels when MeCP2

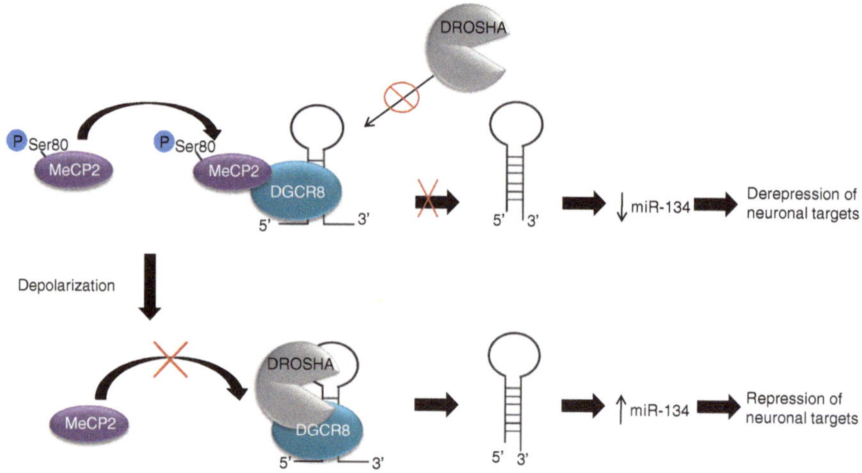

Fig. 2.1 MeCP2 controls Microprocessor activity through direct binding to DGCR8. In resting conditions, phosphorylated MeCP2 is able to bind to DGCR8 and prevent DROSHA cleavage, resulting in diminished levels of certain miRNAs. Upon neuronal depolarization, dephosphorylation of MeCP2 releases the protein and the pri-miRNAs are more efficiently processed by DROSHA/DGCR8

is suppressed [68]. These experiments highlighted the relevance of the C-terminus for the interaction of MeCP2 with DGCR8, uncovering a critical pathway in RTT patients with C-terminal deletions in MeCP2. Such deletions comprise a 15% of all genetic mutations identified among affected girls [77].

Another miRNA whose processing has been shown to be directly regulated by MeCP2 through its ability to interact with the Microprocessor complex is miR-199a. In this case, MeCP2 enhances the pri-miRNA to pre-miRNA cleavage by DROSHA [78]. MiR-199a targets inhibitors of the mechanistic target of rapamycin (mTOR) protein kinase pathway, which controls basic cellular processes involved in neuronal activity and synaptic connectivity (reviewed in [79]). mTOR signaling deficiency as a consequence of MeCP2 mutations has been linked to RTT pathogenesis [80, 81], whereas reactivation of the mTOR pathway has been shown to ameliorate the phenotype [60, 82]. This places miR-199a at the center of the MeCP2/mTOR axis and highlights the role of noncoding RNA in RTT pathophysiology.

MeCP2 mRNA itself is regulated at the posttranscriptional level by miRNAs, among others, miR-130a [83], miR-132 [71], miR-200a, and miR-302c [84]. Importantly, it is known that MeCP2 levels during development are also regulated by some miRNAs. The low expression of the protein during fetal stages suggests a specific posttranscriptional regulation of the messenger RNA (mRNA) through the binding of miRNAs such as miR-483-5p in the long 3′UTR [85]. Furthermore, several publications have shown the relevance of a correct fine-tuning between *MECP2* and miR-132 in postnatal periods [86, 87]. Also, there exists a reciprocal interaction between miR-7b and MeCP2 that could be crucial for the correct neural maturation and whose disruption can play an important role in the development of diseases as RTT [88]. Altogether, previous and future studies that provide a detailed knowledge of the link between *MECP2* and the miRNA population will increase our understanding of important networks affected in Rett syndrome and will help design better informed therapeutic assays.

In contrast to the growing information about changes in miRNA expression and RTT, the role of long noncoding RNAs in the pathophysiology of the disease is widely unknown, even though several publications have demonstrated the importance of lncRNAs in neuronal development [89–92]. Among the few studies on RTT, a lncRNA profiling performed in whole brain identified significant differences between wild-type and Mecp2-null mice. Despite the experimental caveat of not distinguishing between discrete brain regions, the results emphasized the upregulation of a specific lncRNA and the consequent downregulation of its host coding gene Gabrr2 in RTT mouse models [93]. Gabrr2 encodes a subunit of GABA-C receptor and contributes to the correct inhibitory GABAergic signaling [94]. Another study focused on the co-immunoprecipitation of Mecp2 with large protein complexes involved in different aspects of RNA processing and which show themselves an RNA dependency to form [95]. Among the different lncRNAs directly bound by Mecp2 in the mouse cerebellum, the study underscores the *retinal noncoding RNA 3* (*RNCR3*) transcript. Further research will be needed to clarify the relevance of this interaction, but one hypothesis is that the lncRNA *RNCR3* functions by targeting MeCP2 to specific chromatin sites (similar to the role proposed for *Evf2* lncRNA in the recruitment of Mecp2 to the *Dlx5/6* locus to act as a site-specific transcriptional repressor [96, 97]). Intriguingly, like in the previous case, RNCR3 is also required for normal GABAergic neuronal signaling [98]. Also, its primary transcript hosts the miRNA-124a, one of the most abundant miRNA in the CNS, which is required for hippocampal axogenesis [99].

Although not yet studied in RTT, one important lncRNA of potential therapeutical relevance is the brain-derived neurotrophic factor antisense transcript (*BDNF-AS*), which is in inverse correlation with *BDNF* mRNA [100]. As mentioned before, BDNF plays crucial roles in the development and plasticity of the CNS, and its levels have been known for many years to be severely decreased in RTT, with an amelioration of symptoms occurring upon restoration of BDNF expression [101, 102]. Regulation by the antisense lncRNA involves deposition of histone repressive marks that downregulate BDNF levels at the transcriptional level, so that strategies aimed at targeting the antisense transcript result in an increase in BDNF [100]. Recently, analysis of circulating lncRNAs has identified BDNF-AS as differentially expressed in patients of autism spectrum disorder [103], highlighting the interest of further studying the precise role of antisense-mediated BDNF expression dysregulation in RTT. This example illustrates the therapeutic potential of targeting individual ncRNAs that play pivotal roles in disease.

2.3.1 ncRNAs and Therapies

Since the discovery of MeCP2 as the principal affected gene in Rett syndrome [22], several research groups have focused their efforts on the development of therapies for this disorder. The demonstration of the phenotype rectification in mecp2-null mice where the protein was restored [104] uncovered the plasticity of neurons and the potential of Rett syndrome reversibility. Therefore, although there is not a treatment for RTT, several strategies have been designed in order to ameliorate the most of the symptoms.

One strategy to restore MeCP2 function includes the specific activation of the wild-type copy of *MECP2* on the inactive X chromosome. Indeed, the success of this technique was demonstrated in the case of Angelman syndrome [105, 106], caused by deficiency of the imprinted E3 ubiquitin ligase UBE3A. UBE3A is silenced by the antisense ncRNA UBE3A-ATS, and targeting of this lncRNA with antisense oligonucleotides (ASOs) reactivates Ube3a expression in neurons both in vivo and in vitro, concomitant with the amelioration of the disease phenotype. Furthermore, gene therapy also has been used in order to increase the expression of *MECP2* in affected cells. In this case, the aim is to enhance the protein levels through the transfection of a vector that expresses MeCP2, such as adeno-associated virus vectors, which are capable of crossing the blood-brain barrier and can transfer genes across the CNS [107, 108]. However, the principal goal in these therapies is to avoid the overexpression and toxicity caused by an excessive increase in the gene dosage [108].

Besides MeCP2, manipulation of its downstream targets has resulted in the improvement of several symptoms in mice models [109–112], being the amelioration of breathing dysfunction an example of a hopeful finding in order to increase the RTT patients' quality of life [113]. The principal pathways that the studies focused on are noradrenergic, serotonergic, glutamatergic, GABAergic, and cholinergic signaling as well as brain-derived neurotrophic factor (BDNF) pathway [29], whose relevance in the disease development has been widely demonstrated after the observation of reduced levels of *BDNF* mRNA and protein in MeCP2-null mice [101–103]. As mentioned above, one interesting avenue to be investigated is whether manipulations of *BDNF* antisense lncRNA have therapeutical value in RTT. Finally, it is worth mentioning that in animal models of Dravet syndrome (a neurodevelopmental disorder characterized by devastating epilepsies and with an overlapping phenotype with RTT), targeting of an antisense noncoding RNA of the sodium channel gene *SCN1A* with oligonucleotide-based compounds restores channel protein levels and ameliorates the phenotype [114].

2.4 Conclusions and Future Perspectives

Although of great potential therapeutic value, the field of ncRNA has been insufficiently addressed in Rett syndrome. Current knowledge is scarce and focused mainly on miRNAs, although even for this type of ncRNA the data available comes from a limited number of experimental systems and almost totally based in the mouse models. Even in mouse, we lack information about the role of ncRNAs in distinct brain circuits, neuronal subtypes, their localization in subcellular compartments, and their presence in the glia. The recent development of RTT human cell models, thanks to the advances in stem cell biology, will allow a necessary surge in the field of ncRNAs to better understand their roles and contribution to the physiopathology of the disorder and, most importantly, their use as potential therapeutical targets, tools, or biomarkers.

Acknowledgments This work was supported by the Ministerio de Economía y Competitividad (MINECO, grant number SAF2014-56894-R). A. O-G. is a recipient of a Predoctoral Research Grant from MINECO (file BES-2015-071452). We apologize to authors whose work could not be cited due to space constraints.

References

1. Flicek P, Ahmed I, Amode MR, Barrell D, Beal K, Brent S, et al. Ensembl 2013. Nucleic Acids Res. 2013;41:D48–55.
2. Qureshi IA, Mehler MF. Emerging roles of non-coding RNAs in brain evolution, development, plasticity and disease. Nat Rev Neurosci. 2012;13:528–41.
3. Mills JD, Chen BJ, Ueberham U, Arendt T, Janitz M. The antisense transcriptome and the human brain. J Mol Neurosci. 2016;58:1–15.

4. Moazed D. Small RNAs in transcriptional gene silencing and genome defence. Nature. 2009;457:413–20.
5. Guil S, Esteller M. RNA–RNA interactions in gene regulation: the coding and noncoding players. Trends Biochem Sci. 2015;40:248–56.
6. Derrien T, Johnson R, Bussotti G, Tanzer A, Djebali S, Tilgner H, et al. The GENCODE v7 catalog of human long noncoding RNAs: analysis of their gene structure, evolution, and expression. Genome Res. 2012;22:1775–89.
7. Washietl S, Kellis M, Garber M. Evolutionary dynamics and tissue specificity of human long noncoding RNAs in six mammals. Genome Res. 2014;24:616–28.
8. Chodroff RA, Goodstadt L, Sirey TM, Oliver PL, Davies KE, Green ED, et al. Long noncoding RNA genes: conservation of sequence and brain expression among diverse amniotes. Genome Biol. 2010;11:R72.
9. He Z, Bammann H, Han D, Xie G, Khaitovich P. Conserved expression of lincRNA during human and macaque prefrontal cortex development and maturation. RNA. 2014;20: 1103–11.
10. Briggs JA, Wolvetang EJ, Mattick JS, Rinn JL, Barry G. Mechanisms of long non-coding RNAs in mammalian nervous system development, plasticity, disease, and evolution. Neuron. 2015;88:861–77.
11. Aprea J, Calegari F. Long non-coding RNAs in corticogenesis: deciphering the non-coding code of the brain. EMBO J. 2015;34:2865–84.
12. Bortolin-Cavaille ML, Cavaille J. The SNORD115 (H/MBII-52) and SNORD116 (H/MBII-85) gene clusters at the imprinted Prader-Willi locus generate canonical box C/D snoRNAs. Nucleic Acids Res. 2012;40:6800–7.
13. Alders M, Blie J, vd Lip K, vd Bogaard R, Mannens M. Determination of KCNQ1OT1 and H19 methylation levels in BWS and SRS patients using methylation-sensitive high-resolution melting analysis. Eur J Hum Genet. 2009;17:467–73.
14. Chen I, Chen CY, Chuang TJ. Biogenesis, identification, and function of exonic circular RNAs. Wiley Interdiscip Rev RNA. 2015;6:563–79.
15. Memczak S, Jens M, Elefsinioti A, Torti F, Krueger J, Rybak A, et al. Circular RNAs are a large class of animal RNAs with regulatory potency. Nature. 2013;495:333–8.
16. Ashwal-Fluss R, Meyer M, Pamudurti NR, Ivanov A, Bartok O, Hanan M, et al. circRNA biogenesis competes with pre-mRNA splicing. Mol Cell. 2014;56:55–66.
17. Guo JU, Agarwal V, Guo H, Bartel DP. Expanded identification and characterization of mammalian circular RNAs. Genome Biol. 2014;15:409.
18. Rybak-Wolf A, Stottmeister C, Glažar P, Jens M, Pino N, Giusti S, et al. Circular RNAs in the mammalian brain are highly abundant, conserved, and dynamically expressed. Mol Cell. 2015;58:870–85.
19. Venø MT, Hansen TB, Venø ST, Clausen BH, Grebing M, Finsen B, et al. Spatio-temporal regulation of circular RNA expression during porcine embryonic brain development. Genome Biol. 2015;16:245.
20. Lukiw WJ. Circular RNA (circRNA) in Alzheimer's disease (AD). Front Genet. 2015; 4:307.
21. Rett A. On a unusual brain atrophy syndrome in hyperammonemia in childhood. Wien Med Wochenschr. 1966;116:723–6.
22. Amir RE, Van den Veyver IB, Wan M, Tran CQ, Francke U, Zoghbi HY, et al. Rett syndrome is caused by mutations in X-linked MECP2, encoding methyl-CpG-binding protein 2. Nat Genet. 1999;23:185–8.
23. Weng SM, Bailey ME, Cobb SR. Rett syndrome: from bed to bench. Pediatr Neonatol. 2011;52:309–16.
24. Chahrour M, Zoghbi HY. The story of Rett syndrome: from clinic to neurobiology. Neuron. 2007;56:422–37.
25. Balachandar V, Dhivya V, Gomathi M, Mohanadevi S, Venkatesh B, Geetha B. A review of Rett syndrome (RTT) with induced pluripotent stem cells. Stem Cell Invest. 2016;3:52.
26. Castro J, Mellios N, Sur M. Mechanisms and therapeutic challenges in autism spectrum disorders: insights from Rett syndrome. Curr Opin Neurol. 2013;26:154–9.

27. Weese-Mayer DE, Lieske SP, Boothby CM, Kenny AS, Bennett HL, Silvestri JM, et al. Autonomic nervous system dysregulation: breathing and heart rate perturbation during wakefulness in young girls with Rett syndrome. Pediatr Res. 2006;60:443–9.
28. Kerr AM, Armstrong DD, Prescott RJ, Doyle D, Kearney DL. Rett syndrome: analysis of deaths in the British survey. Eur Child Adolesc Psychiatry. 1997;6(Suppl 1):71–4.
29. Katz DM, Bird A, Coenraads M, Gray SJ, Menon DU, Philpot BD, et al. Rett syndrome: crossing the threshold to clinical translation. Trends Neurosci. 2016;39:100–13.
30. Lyst MJ, Bird A. Rett syndrome: a complex disorder with simple roots. Nat Rev Genet. 2015;16:261–75.
31. Nan X, Ng HH, Johnson CA, Laherty CD, Turner BM, Eisenman RN, et al. Transcriptional repression by the methyl-CpG-binding protein MeCP2 involves a histone deacetylase complex. Nature. 1998;393:386–9.
32. Guy J, Cheval H, Selfridge J, Bird A. The role of MeCP2 in the brain. Annu Rev Cell Dev Biol. 2011;27:631–52.
33. Cheng TL, Wang Z, Liao Q, Zhu Y, Zhou WH, Xu W, et al. MeCP2 suppresses nuclear microRNA processing and dendritic growth by regulating the DGCR8/Drosha complex. Dev Cell. 2014;28:547–60.
34. Itoh M, Tahimic CG, Ide S, Otsuki A, Sasaoka T, Noguchi S, et al. Methyl CpG-binding protein isoform MeCP2_e2 is dispensable for Rett syndrome phenotypes but essential for embryo viability and placenta development. J Biol Chem. 2012;287:13859–67.
35. Jain D, Singh K, Chirumamilla S, Bibat GM, Blue ME, Naidu SR, et al. Ocular MECP2 protein expression in patients with and without Rett syndrome. Pediatr Neurol. 2010;43:35–40.
36. Shahbazian MD, Antalffy B, Armstrong DL, Zoghbi HY. Insight into Rett syndrome: MeCP2 levels display tissue- and cell-specific differences and correlate with neuronal maturation. Hum Mol Genet. 2002;11:115–24.
37. Banerjee A, Castro J, Sur M. Rett syndrome: genes, synapses, circuits, and therapeutics. Front Psychiatry. 2012;3:34.
38. Chao HT, Chen H, Samaco RC, Xue M, Chahrour M, Yoo J, et al. Dysfunction in GABA signalling mediates autism-like stereotypies and Rett syndrome phenotypes. Nature. 2010;468: 263–9.
39. Lioy DT, Garg SK, Monaghan CE, Raber J, Foust KD, Kaspar BK, et al. A role for glia in the progression of Rett's syndrome. Nature. 2011;475:497–500.
40. Ehrhart F, Coort SL, Cirillo E, Smeets E, Evelo CT, Curfs LM. Rett syndrome—biological pathways leading from MECP2 to disorder phenotypes. Orphanet J Rare Dis. 2016;11:158.
41. Cuddapah VA, Pillai RB, Shekar KV, Lane JB, Motil KJ, Skinner SA, et al. Methyl-CpG-binding protein 2 (MECP2) mutation type is associated with disease severity in Rett syndrome. J Med Genet. 2014;51:152–8.
42. Tate P, Skarnes W, Bird A. The methyl-CpG binding protein MeCP2 is essential for embryonic development in the mouse. Nat Genet. 1996;12:205–8.
43. Guy J, Hendrich B, Holmes M, Martin JE, Bird A. A mouse Mecp2-null mutation causes neurological symptoms that mimic Rett syndrome. Nat Genet. 2001;27:322–6.
44. Pelka GJ, Watson CM, Radziewic T, Hayward M, Lahooti H, Christodoulou J, et al. Mecp2 deficiency is associated with learning and cognitive deficits and altered gene activity in the hippocampal region of mice. Brain. 2006;129:887–98.
45. Chen RZ, Akbarian S, Tudor M, Jaenisch R. Deficiency of methyl-CpG binding protein-2 in CNS neurons results in a Rett-like phenotype in mice. Nat Genet. 2001;27:327–31.
46. Fyffe SL, Neul JL, Samaco RC, Chao HT, Ben-Shachar S, Moretti P, et al. Deletion of Mecp2 in Sim1-expressing neurons reveals a critical role for MeCP2 in feeding behavior, aggression, and the response to stress. Neuron. 2008;59:947–58.
47. Samaco RC, Mandel-Brehm C, Chao HT, Ward CS, Fyffe-Maricich SL, Ren J, et al. Loss of MeCP2 in aminergic neurons causes cell-autonomous defects in neurotransmitter synthesis and specific behavioral abnormalities. Proc Natl Acad Sci U S A. 2009;106:21966–71.
48. Huang TW, Kochukov MY, Ward CS, Merritt J, Thomas K, Nguyen T, et al. Progressive changes in a distributed neural circuit underlie breathing abnormalities in mice lacking MeCP2. J Neurosci. 2016;36:5572–86.

49. Chao HT, Zoghbi HY, Rosenmund C. MeCP2 controls excitatory synaptic strength by regulating glutamatergic synapse number. Neuron. 2007;56:58–65.
50. Wan M, Lee SS, Zhang X, Houwink-Manville I, Song HR, Amir RE, et al. Rett syndrome and beyond: recurrent spontaneous and familial MECP2 mutations at CpG hotspots. Am J Hum Genet. 1999;65:1520–9.
51. Katz DM, Berger-Sweeney JE, Eubanks JH, Justice MJ, Neul JL, Pozzo-Miller L, et al. Preclinical research in Rett syndrome: setting the foundation for translational success. Dis Model Mech. 2012;5:733–45.
52. Tsuchiya Y, Minami Y, Umemura Y, Watanabe H, Ono D, Nakamura W, et al. Disruption of MeCP2 attenuates circadian rhythm in CRISPR/Cas9-based Rett syndrome model mouse. Genes Cells. 2015;20:992–1005.
53. Paquet D, Kwart D, Chen A, Sproul A, Jacob S, Teo S, et al. Efficient introduction of specific homozygous and heterozygous mutations using CRISPR/Cas9. Nature. 2016;533:125–9.
54. Xue HY, Ji LJ, Gao AM, Liu P, He JD, Lu XJ. CRISPR-Cas9 for medical genetic screens: applications and future perspectives. J Med Genet. 2016;53:91–7.
55. Takahashi K, Tanabe K, Ohnuki M, Narita M, Ichisaka T, Tomoda K, et al. Induction of pluripotent stem cells from adult human fibroblasts by defined factors. Cell. 2007;131:861–72.
56. Santostefano KE, Hamazaki T, Biel NM, Jin S, Umezawa A, Terada N. A practical guide to induced pluripotent stem cell research using patient samples. Lab Invest. 2015;95:4–13.
57. Russo FB, Cugola FR, Fernandes IR, Pignatari GC, Beltrao-Braga PC. Induced pluripotent stem cells for modeling neurological disorders. World J Transplant. 2015;5:209–21.
58. Cheung AY, Horvath LM, Grafodatskaya D, Pasceri P, Weksberg R, Hotta A, et al. Isolation of MECP2-null Rett syndrome patient hiPS cells and isogenic controls through X-chromosome inactivation. Hum Mol Genet. 2011;20:2103–15.
59. Hotta A, Cheung AY, Farra N, Garcha K, Chang WY, Pasceri P, et al. EOS lentiviral vector selection system for human induced pluripotent stem cells. Nat Protoc. 2009;4:1828–44.
60. Marchetto MC, Carromeu C, Acab A, Yu D, Yeo GW, Mu Y, et al. A model for neural development and treatment of Rett syndrome using human induced pluripotent stem cells. Cell. 2010;143:527–39.
61. Tchieu J, Kuoy E, Chin MH, Trinh H, Patterson M, Sherman SP, et al. Female human iPSCs retain an inactive X chromosome. Cell Stem Cell. 2010;7:329–42.
62. Bhatnagar S, Zhu X, Ou J, Lin L, Chamberlain L, Zhu LJ, et al. Genetic and pharmacological reactivation of the mammalian inactive X chromosome. Proc Natl Acad Sci U S A. 2014;111:12591–8.
63. Lee JT, Bartolomei MS. X-inactivation, imprinting, and long noncoding RNAs in health and disease. Cell. 2013;152:1308–23.
64. Richard JL, Ogawa Y. Understanding the complex circuitry of lncRNAs at the X-inactivation center and its implications in disease conditions. Curr Top Microbiol Immunol. 2016;394:1–27.
65. Vacca M, Della Ragione F, Scalabri F, D'Esposito M. X inactivation and reactivation in X-linked diseases. Semin Cell Dev Biol. 2016;56:78–87.
66. Fineberg SK, Kosik KS, Davidson BL. MicroRNAs potentiate neural development. Neuron. 2009;64:303–9.
67. Davis TH, Cuellar TL, Koch SM, Barker AJ, Harfe BD, McManus MT, et al. Conditional loss of Dicer disrupts cellular and tissue morphogenesis in the cortex and hippocampus. J Neurosci. 2008;28:4322–30.
68. Schratt GM, Tuebing F, Nigh EA, Kane CG, Sabatini ME, Kiebler M, et al. A brain-specific microRNA regulates dendritic spine development. Nature. 2006;439:283–9.
69. Kawase-Koga Y, Otaegi G, Sun T. Different timings of Dicer deletion affect neurogenesis and gliogenesis in the developing mouse central nervous system. Dev Dyn. 2009;238:2800–12.
70. Urdinguio RG, Fernandez AF, Lopez-Nieva P, Rossi S, Huertas D, Kulis M, et al. Disrupted microRNA expression caused by Mecp2 loss in a mouse model of Rett syndrome. Epigenetics. 2010;5:656–63.
71. Lyu JW, Yuan B, Cheng TL, Qiu ZL, Zhou WH. Reciprocal regulation of autism-related genes MeCP2 and PTEN via microRNAs. Sci Rep. 2016;6:20392.

72. Wu H, Tao J, Chen PJ, Shahab A, Ge W, Hart RP, et al. Genome-wide analysis reveals methyl-CpG-binding protein 2-dependent regulation of microRNAs in a mouse model of Rett syndrome. Proc Natl Acad Sci U S A. 2010;107:18161–6.
73. Nomura T, Kimura M, Horii T, Morita S, Soejima H, Kudo S, et al. MeCP2-dependent repression of an imprinted miR-184 released by depolarization. Hum Mol Genet. 2008;17:1192–9.
74. Ben-Shachar S, Chahrour M, Thaller C, Shaw CA, Zoghbi HY. Mouse models of MeCP2 disorders share gene expression changes in the cerebellum and hypothalamus. Hum Mol Genet. 2009;18:2431–42.
75. Chahrour M, Jung SY, Shaw C, Zhou X, Wong ST, Qin J, et al. MeCP2, a key contributor to neurological disease, activates and represses transcription. Science. 2008;320:1224–9.
76. Li W, Pozzo-Miller L. BDNF deregulation in Rett syndrome. Neuropharmacology. 2014;76 Pt C:737–46.
77. Bebbington A, Percy A, Christodoulou J, Ravine D, Ho G, Jacoby P, et al. Updating the profile of C-terminal MECP2 deletions in Rett syndrome. J Med Genet. 2010;47:242–8.
78. Tsujimura K, Irie K, Nakashima H, Egashira Y, Fukao Y, Fujiwara M, et al. miR-199a links MeCP2 with mTOR signaling and its dysregulation leads to Rett syndrome phenotypes. Cell Rep. 2015;12:1887–901.
79. Buffington SA, Huang W, Costa-Mattioli M. Translational control in synaptic plasticity and cognitive dysfunction. Annu Rev Neurosci. 2014;37:17–38.
80. Ricciardi S, Boggio EM, Grosso S, Lonetti G, Forlani G, Stefanelli G, et al. Reduced AKT/mTOR signaling and protein synthesis dysregulation in a Rett syndrome animal model. Hum Mol Genet. 2011;20:1182–96.
81. Li Y, Wang H, Muffat J, Cheng AW, Orlando DA, Lovén J, et al. Global transcriptional and translational repression in human-embryonic-stem-cell-derived Rett syndrome neurons. Cell Stem Cell. 2013;13:446–58.
82. Tropea D, Giacometti E, Wilson NR, Beard C, McCurry C, Fu DD, et al. Partial reversal of Rett Syndrome-like symptoms in MeCP2 mutant mice. Proc Natl Acad Sci U S A. 2009;106:2029–34.
83. Zhang Y, Chen M, Qiu Z, Hu K, McGee W, Chen X, et al. MiR-130a regulates neurite outgrowth and dendritic spine density by targeting MeCP2. Protein Cell. 2016;7:489–500.
84. Rodrigues DC, Kim DS, Yang G, Zaslavsky K, Ha KC, Mok RS, et al. MECP2 is post-transcriptionally regulated during human neurodevelopment by combinatorial action of RNA-binding proteins and miRNAs. Cell Rep. 2016;17:720–34.
85. Han K, Gennarino VA, Lee Y, Pang K, Hashimoto-Torii K, Choufani S, et al. Human-specific regulation of MeCP2 levels in fetal brains by microRNA miR-483-5p. Genes Dev. 2013;27:485–90.
86. Klein ME, Lioy DT, Ma L, Impey S, Mandel G, Goodman RH. Homeostatic regulation of MeCP2 expression by a CREB-induced microRNA. Nat Neurosci. 2007;10:1513–4.
87. Alvarez-Saavedra M, Antoun G, Yanagiya A, Oliva-Hernandez R, Cornejo-Palma D, Perez-Iratxeta C, et al. miRNA-132 orchestrates chromatin remodeling and translational control of the circadian clock. Hum Mol Genet. 2011;20:731–51.
88. Chen Y, Shin BC, Thamotharan S, Devaskar SU. Differential methylation of the micro-RNA 7b gene targets postnatal maturation of murine neuronal Mecp2 gene expression. Dev Neurobiol. 2014;74:407–25.
89. Lin N, Chang KY, Li Z, Gates K, Rana ZA, Dang J, et al. An evolutionarily conserved long noncoding RNA TUNA controls pluripotency and neural lineage commitment. Mol Cell. 2014;53:1005–19.
90. Ng SY, Bogu GK, Soh BS, Stanton LW. The long noncoding RNA RMST interacts with SOX2 to regulate neurogenesis. Mol Cell. 2013;51:349–59.
91. Mo CF, Wu FC, Tai KY, Chang WC, Chang KW, Kuo HC, et al. Loss of non-coding RNA expression from the DLK1-DIO3 imprinted locus correlates with reduced neural differentiation potential in human embryonic stem cell lines. Stem Cell Res Ther. 2015;6:1.

92. D'haene E, Jacobs EZ, Volders PJ, De Meyer T, Menten B, Vergult S. Identification of long non-coding RNAs involved in neuronal development and intellectual disability. Sci Rep. 2016;6:28396.
93. Petazzi P, Sandoval J, Szczesna K, Jorge OC, Roa L, Sayols S, et al. Dysregulation of the long non-coding RNA transcriptome in a Rett syndrome mouse model. RNA Biol. 2013;10: 1197–203.
94. Harvey VL, Duguid IC, Krasel C, Stephens GJ. Evidence that GABA rho subunits contribute to functional ionotropic GABA receptors in mouse cerebellar Purkinje cells. J Physiol. 2006;577:127–39.
95. Maxwell SS, Pelka GJ, Tam PP, El-Osta A. Chromatin context and ncRNA highlight targets of MeCP2 in brain. RNA Biol. 2013;10:1741–57.
96. Bond AM, Vangompel MJ, Sametsky EA, Clark MF, Savage JC, Disterhoft JF, et al. Balanced gene regulation by an embryonic brain ncRNA is critical for adult hippocampal GABA circuitry. Nat Neurosci. 2009;12:1020–7.
97. Berghoff EG, Clark MF, Chen S, Cajigas I, Leib DE, Kohtz JD. Evf2 (Dlx6as) lncRNA regulates ultraconserved enhancer methylation and the differential transcriptional control of adjacent genes. Development. 2013;140:4407–16.
98. Mercer TR, Qureshi IA, Gokhan S, Dinger ME, Li G, Mattick JS, et al. Long noncoding RNAs in neuronal-glial fate specification and oligodendrocyte lineage maturation. BMC Neurosci. 2010;11:14.
99. Sanuki R, Onishi A, Koike C, Muramatsu R, Watanabe S, Muranishi Y, et al. miR-124a is required for hippocampal axogenesis and retinal cone survival through Lhx2 suppression. Nat Neurosci. 2011;14:1125–34.
100. Modarresi F, Faghihi MA, Lopez-Toledano MA, Fatemi RP, Magistri M, Brothers SP, et al. Inhibition of natural antisense transcripts in vivo results in gene-specific transcriptional upregulation. Nat Biotechnol. 2012;30:453–9.
101. Chang Q, Khare G, Dani V, Nelson S, Jaenisch R. The disease progression of Mecp2 mutant mice is affected by the level of BDNF expression. Neuron. 2006;49:341–8.
102. Kline DD, Ogier M, Kunze DL, Katz DM. Exogenous brain-derived neurotrophic factor rescues synaptic dysfunction in Mecp2-null mice. J Neurosci. 2010;30:5303–10.
103. Wang Y, Zhao X, Ju W, Flory M, Zhong J, Jiang S, et al. Genome-wide differential expression of synaptic long noncoding RNAs in autism spectrum disorder. Transl Psychiatry. 2015;5:e660.
104. Guy J, Gan J, Selfridge J, Cobb S, Bird A. Reversal of neurological defects in a mouse model of Rett syndrome. Science. 2007;315:1143–7.
105. Huang HS, Allen JA, Mabb AM, King IF, Miriyala J, Taylor-Blake B, et al. Topoisomerase inhibitors unsilence the dormant allele of Ube3a in neurons. Nature. 2011;481:185–9.
106. Meng L, Ward AJ, Chun S, Bennett CF, Beaudet AL, Rigo F. Towards a therapy for Angelman syndrome by targeting a long non-coding RNA. Nature. 2015;518:409–12.
107. Gray SJ, Matagne V, Bachaboina L, Yadav S, Ojeda SR, Samulski RJ. Preclinical differences of intravascular AAV9 delivery to neurons and glia: a comparative study of adult mice and nonhuman primates. Mol Ther. 2011;19:1058–69.
108. Gadalla KK, Bailey ME, Spike RC, Ross PD, Woodard KT, Kalburgi SN, et al. Improved survival and reduced phenotypic severity following AAV9/MECP2 gene transfer to neonatal and juvenile male Mecp2 knockout mice. Mol Ther. 2013;21:18–30.
109. Deogracias R, Yazdani M, Dekkers MP, Guy J, Ionescu MC, Vogt KE, et al. Fingolimod, a sphingosine-1 phosphate receptor modulator, increases BDNF levels and improves symptoms of a mouse model of Rett syndrome. Proc Natl Acad Sci U S A. 2012;109:14230–5.
110. Johnson RA, Lam M, Punzo AM, Li H, Lin BR, Ye K, et al. 7,8-dihydroxyflavone exhibits therapeutic efficacy in a mouse model of Rett syndrome. J Appl Physiol. 1985;112: 704–10.
111. Kondo M, Gray LJ, Pelka GJ, Christodoulou J, Tam PP, Hannan AJ. Environmental enrichment ameliorates a motor coordination deficit in a mouse model of Rett syndrome—Mecp2 gene dosage effects and BDNF expression. Eur J Neurosci. 2008;27:3342–50.

112. Larimore JL, Chapleau CA, Kudo S, Theibert A, Percy AK, Pozzo-Miller L. Bdnf overexpression in hippocampal neurons prevents dendritic atrophy caused by Rett-associated MECP2 mutations. Neurobiol Dis. 2009;34:199–211.
113. Ramirez JM, Ward CS, Neul JL. Breathing challenges in Rett syndrome: lessons learned from humans and animal models. Respir Physiol Neurobiol. 2013;189:280–7.
114. Hsiao J, Yuan TY, Tsai MS, Lu CY, Lin YC, Lee ML, et al. Upregulation of haploinsufficient gene expression in the brain by targeting a long non-coding RNA improves seizure phenotype in a model of Dravet syndrome. EBioMedicine. 2016;9:257–77.

Rubinstein-Taybi Syndrome and Epigenetic Alterations

3

Edward Korzus

Abstract

Rubinstein-Taybi syndrome (RSTS) is a rare genetic disorder in humans characterized by growth and psychomotor delay, abnormal gross anatomy, and mild to severe mental retardation (Rubinstein and Taybi, Am J Dis Child 105:588–608, 1963, Hennekam et al., Am J Med Genet Suppl 6:56–64, 1990). RSTS is caused by de novo mutations in epigenetics-associated genes, including the cAMP response element-binding protein (CREBBP), the gene-encoding protein referred to as CBP, and the EP300 gene, which encodes the p300 protein, a CBP homologue. Recent studies of the epigenetic mechanisms underlying cognitive functions in mice provide direct evidence for the involvement of nuclear factors (e.g., CBP) in the control of higher cognitive functions. In fact, a role for CBP in higher cognitive function is suggested by the finding that RSTS is caused by heterozygous mutations at the CBP locus (Petrij et al., Nature 376:348–351, 1995). CBP was demonstrated to possess an intrinsic histone acetyltransferase activity (Ogryzko et al., Cell 87:953–959, 1996) that is required for CREB-mediated gene expression (Korzus et al., Science 279:703–707, 1998). The intrinsic protein acetyltransferase activity in CBP might directly destabilize promoter-bound nucleosomes, facilitating the activation of transcription. Due to the complexity of developmental abnormalities and the possible genetic compensation associated with this congenital disorder, however, it is difficult to establish a direct role for CBP in cognitive function in the adult brain. Although aspects of the clinical presentation in RSTS cases have been extensively studied, a spectrum of symptoms found in RSTS patients can be accessed only after birth, and, thus, prenatal

E. Korzus, Ph.D.
Department of Psychology and Neuroscience Program, University Of California Riverside, 900 University Ave, Riverside, CA 92521, USA
e-mail: edkorzus@ucr.edu

R. Delgado-Morales (ed.), *Neuroepigenomics in Aging and Disease*, Advances in Experimental Medicine and Biology, DOI 10.1007/978-3-319-53889-1_3

genetic tests for this extremely rare genetic disorder are seldom considered. Even though there has been intensive research on the genetic and epigenetic function of the CREBBP gene in rodents, the etiology of this devastating congenital human disorder is largely unknown.

Keywords
Rubinstein-Taybi syndrome • RSTS • CREBBP • CBP • EP300 • p300 • Memory • Epigenetic • Histone acetylation • HDAC

3.1 Introduction

The regulation of gene expression requires not only activation of transcription factors but also the recruitment of multifunctional coactivators that are independently regulated and utilized in a cell- and promoter-specific fashion to stimulate or repress transcription [1]. Dynamic changes in the organization of chromatin control gene expression and histone acetylation are one mechanism for the local and global control of chromatin structure [2, 3]. Studies have shown that chromatin acetylation at a region of ongoing transcription is essential for high-level gene expression [2, 3]. The cAMP response element-binding protein (CBP, which is encoded by the CREBBP gene) and its homologue p300 protein (encoded by the EP300 gene) are transcriptional coactivators [4, 5] that interact with multiple transcriptional regulators and facilitate the assembly of the basic transcriptional machinery [6] (Fig. 3.1). In addition to serving as molecular scaffolds, CBP and p300 each possess intrinsic histone acetyltransferase (HAT) activities [10] that can be specifically and directly inhibited by phosphorylation or by association with viral proteins [6, 11]. Owing to the ability of CBP and p300 to control the function of chromatin via mechanisms that involve the covalent modification of chromatin, resulting in long-lasting marks on histones, these proteins are defined as epigenetic "writers" (Fig. 3.2). More recently, the HAT activity of both CBP and p300 has been referred to as lysine acetyltransferase (KAT11) activity because it targets lysine not only on histones but also on nonhistone proteins [7]. CBP and p300 proteins share more than 70% sequence homology, but they retain some distinct cellular functions and behavioral functions and cannot always replace each other [8, 15, 16].

A role for the CBP gene in higher cognitive function is suggested by the finding that Rubinstein-Taybi syndrome (RSTS), a disorder in humans characterized

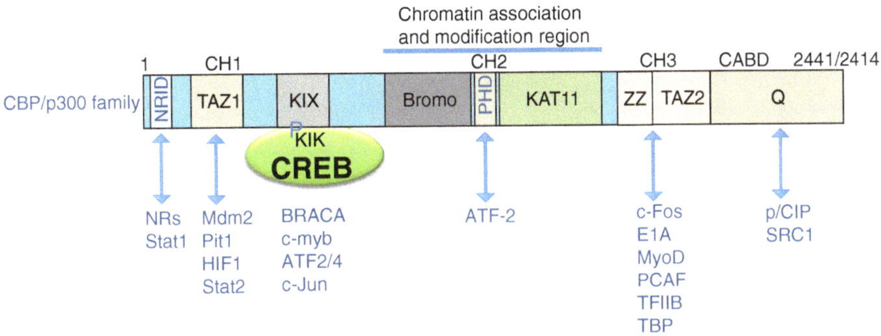

Fig. 3.1 Structure of CBP/p300 protein family. CBP and p300 proteins belong to the same family of coactivator of transcription and share 63% identity and 73% similarity with the highest homology as mapped to functional domains. A central part of CBP/p300 protein encompasses chromatin association and the modification region. The N-terminal and C-terminal regions, which include a variety of motives that provide a platform for specific protein-protein interactions, enable the formation of multiprotein complexes critical for a cellular signal- and promoter-specific gene expression regulation. CBP was discovered as a CREB-binding protein and a phosphorylation-dependent interaction between CREB's KID domain and CBP's KIX domain. Further research demonstrated that more than 400 proteins could interact with CBP/p300 proteins. The schematic shows examples of these interaction partners' interacting predominantly with regions containing zinc finger motives. The central region of CBP/p300 contains domains supporting chromatin-modifying functions. While bromodomain (Bromo) provides ability for chromatin recognition, the KIT11 domain has lysine acetyltransferase enzymatic activity targeting primary histone N-terminals and nonhistone nuclear proteins [7, 8]. For more detailed information, see the text. The diagram does not show proper proportions and is based on data from UniGene (*CREBBP*, NP_004371.2; *EP300*, NP_001420.2; and [8, 9]). CBP's known domains from the N-terminal are *NRID* nuclear receptor interaction domain, *TAZ1* transcriptional adapter zinc-binding domain 1, *KIX* CREB binding, *Bromo* bromodomain, *PHD* plant homeodomain, *KIT11* lysine acetyltransferase, *ZZ* zinc finger domain, *TAZ2*, *Q* polyglutamine stretch, *NRs* nuclear hormone receptors, *SRC* steroid receptor coactivator

by growth and psychomotor delay, abnormal gross anatomy, and mild to severe mental retardation [17, 18], is caused by heterozygous mutations at the CBP locus [19]. CBP is one of the major regulatory nuclear proteins that control gene expression associated with multiple critical cell functions during development and in the adult. Due to the complexity of developmental abnormalities and possible genetic compensation associated with this congenital disorder, however, it is difficult to establish a direct role for CBP in cognitive function in the adult brain.

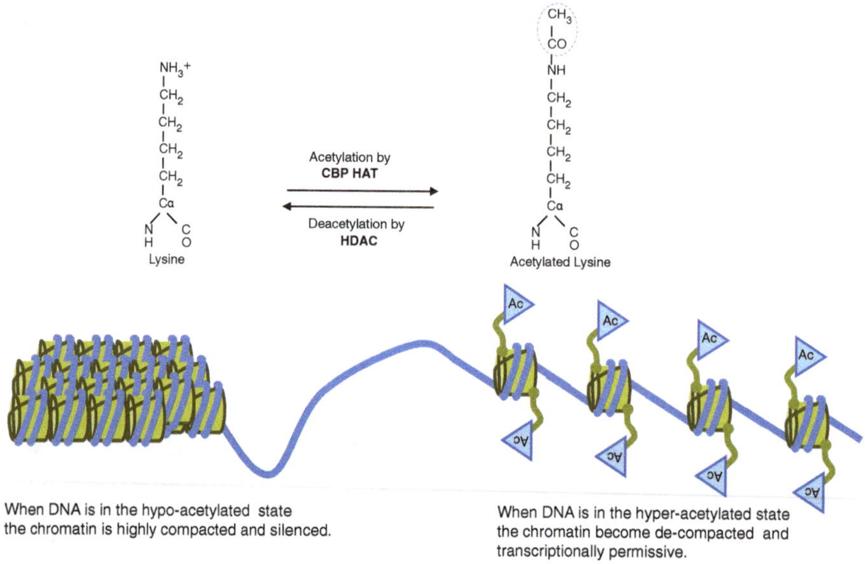

Fig. 3.2 Histone acetylation controls chromatin structure and function. Although CBP's function as a platform to recruit other required coactivators appears to be indispensable, the requirement for HAT activity is transcription unit specific and may depend on the structure of chromatin at a specific locus [12, 13]. Histone deacetylase (HDAC)-mediated hypo-acetylation of histones promotes a compact chromatin structure state, subsequently silencing transcription. Promoter-specific recruitment of chromatin remodeling factors, such as CBP HAT, facilitates de-compaction of the chromatin structure, where genes are accessible for large multiprotein complexes that mediate gene expression (i.e., RNA polymerase II holoenzyme). CBP HAT has been implicated in epigenetic mechanisms that control higher cognitive functions [14]

3.2 Genotype

Mutations in the human CREBBP gene were reported to be associated with RSTS (OMIM #180849, #613684), a haploinsufficiency disorder characterized by multiple developmental defects and severe mental retardation (Rubinstein and Taybi, 1963). RSTS is a rare genetic syndrome caused by de novo heterozygous mutations in epigenetic genes and was first described in 1963 [17]. RSTS is found in one case per 100,000 to 125,000 live births. In 1991, Petrij et al. demonstrated for the first time the genetic origin of RSTS by reporting de novo reciprocal (balanced) translocations with breakpoints in chromosomal regions 16p13.3 in RSTS patients [19]. The CREBBP gene (CREBBP; OMIM #600140) is located on chromosome 16p13.3. A mutation in the gene that encodes the CREBBP gene was reported in approximately 55% of RSTS cases (Fig. 3.3), defining RSTS type 1 (RSTS1; OMIM #180849) [19, 22, 23]. The CREBBP gene spans ~155 kb, and there are 31 exons in coding region. Transcription of the CREBBP gene proceeds from centromere to telomere with start codon located in exon 1 and stop codon in 31. ~10 kb CREBBP mRNA contains 7.3 kb of coding sequence. The CREBBP gene encodes a large protein (CBP) with a molecular weight of 26,531 Da that consists of 2441 amino acids.

In addition, 10% of cases of RSTS have been associated with mutations present in a CBP homologue, the E1A binding protein p300 [24–26] (Fig. 3.4). Protein p300 is encoded by the human EP300 gene (OMIM #602700) located at 22q13.2. Cases of mutated EP300 are described as RSTS type 2 (RSTS2; OMIM #613684). Mutations found in CREBBP and EP300 (i.e., frameshift, nonsense, splice site, and missense mutations) are heterozygous, rare, and de novo [27]. In addition, less frequently occurring large deletions (exonic or whole gene) or balanced inversions and translocations also have been characterized.

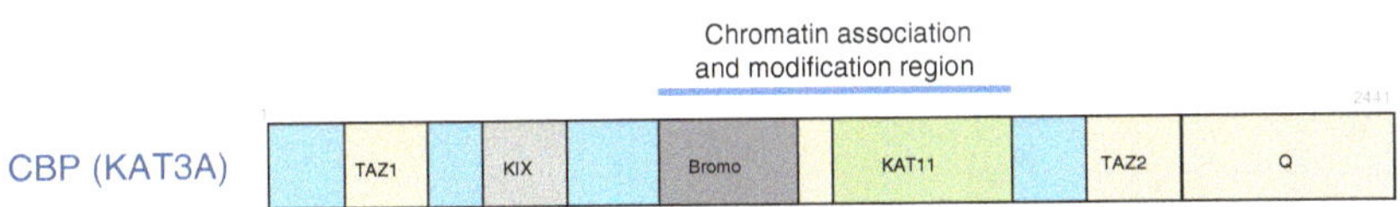

Exons:	1	2	3-4	5-9	10-16	17	18-30		31		
Amino acids:	1-362	363-432		587-667		1191-1194		1342-1649		1770-1843	1844-2441
# variants (point mutations) per domain	2	7	6	10	8	2		49		4	18

In addition to point mutations, 21 (differently sized) deletion mutation spanning CREBBP gene regions and adjacent sequences.

Fig. 3.3 CREBBP germline mutations in RSTS1 patients. CREBBP germline mutations account for 50–60% of RSTS cases. About 50% of mutations associated with RSTS1 have been mapped to chromatin association and the modification region in the CREBBP gene (Fig. 3.1). CREBBP germline mutations in RSTS patients include 106 point mutations and 21 deletions, such as exonic and whole-gene deletions, with some encompassing flanking genes. Only a few mosaic mutations in the CREBBP gene have been reported (not shown). Data used for analysis were reported in Leiden Open Variation Database [20], Gervasini et al. (2010) [21], and UniGene database (CREBBP: NP_004371.2)

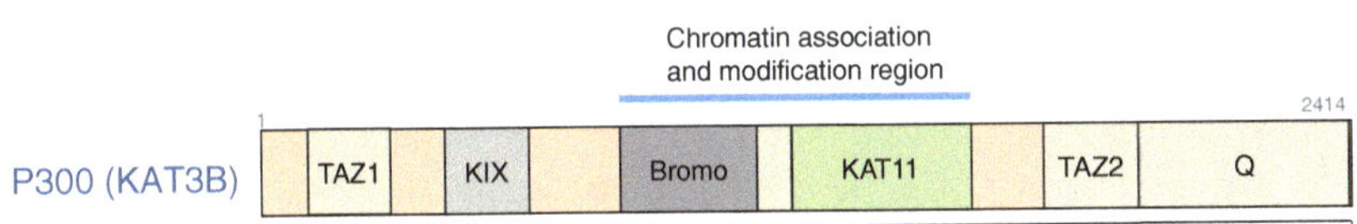

Exons:	1	2	3-4	5-9	10-16	17	18-30		31		
Amino acids:		347-416		566-646		1051-1158		1306-1649		1733-1806	1807-2414
# variants (point mutations) per domain	1	4		5	6			10		1	

In addition to point mutations, 1 whole gene deletion and 5 (differently sized) exonic deletion mutations spanning EP300 gene regions were found

Fig. 3.4 EP300 germline mutations in RSTS patients. To date, 34 EP300 mutations in known RSTS2 cases have been reported worldwide [25]. Mutations in the EP300 gene account for about 10% of all RSTS cases. EP300 germline mutations in RSTS patients include 27 point mutations, six exonic deletions, and one whole-gene deletion. About 40% of mutations associated with RSTS2 have been mapped to chromatin association and the modification region in the EP300 gene (Fig. 3.1). Data used for analysis were reported in the Leiden Open Variation Database [20], Negri et al. (2016) [25], and the UniGene database (EP300: NP_001420.2)

3.3 Clinical Phenotypes

There are no precise diagnostic criteria for RSTS, but there is a spectrum of clinical presentations with unique clinical hallmarks (Table 3.1). A number of studies have found characteristic developmental anomalies in patients with mutations in the CREBBP and EP300 genes, including growth and psychomotor delay, abnormal gross anatomy, and intellectual disabilities [17, 18, 28]. In general, phenotypes associated with mutations in EP300 are less marked than those found with CREBBP mutations. A high number of individuals with mutations in CREBBP and EP300 show broad thumbs and halluces (96% and 69%, respectively); facial abnormalities, such as a grimacing smile (94% and 47%, respectively); and intellectual disabilities (99% and 94%, respectively) (Table 3.1). Retarded motor and mental development is visible during the first year of life. The average IQ of individuals with RSTS varies between 35 and 50, but some patients show better performance [28].

Table 3.1 Comparison of symptoms found in RSTS individuals who carry mutations in CREBBP and EP300

Symptom	CREBBP[a] (*n* = 308)		EP300[b] (*n* = 52)		*p*-value[c]
Intellectual disability	99%	250/253	94%	48/51	NS
Severe	36%	33/92	7%	2/29	<0.005
Moderate	48%	44/92	31%	9/29	NS
Mild	14%	13/92	62%	18/29	<0.00005
Broad thumbs	96%	277/290	69%	36/52	<0.00005
Broad halluces	95%	221/233	81%	42/52	<0.005
Grimacing smile	94%	99/105	47%	21/45	<0.00005
Long eyelashes	89%	75/84	90%	44/49	NS
Columella below alae nasi	88%	195/222	92%	48/52	NS
Arched eyebrows	85%	71/84	65%	34/52	<0.05
Convex nasal ridge	81%	225/278	44%	23/52	<0.00005
Downslanted palpebral fissures	79%	208/263	56%	29/52	<0.001
Highly arched palate	77%	160/208	67%	30/45	NS
Hypertrichosis	76%	93/122	51%	23/45	<0.005
Postnatal growth retardation	75%	160/214	66%	33/50	NS
Micrognathia	61%	131/214	42%	22/52	<0.05
Microcephaly (OFC < 3rd centile)	54%	77/143	87%	45/52	<0.00005
Autism/autistiform behavior	49%	51/105	25%	12/49	<0.005
Angulated thumbs	49%	135/273	2%	1/51	<0.00005

[a]Data derived from [26, 135–137]; [b]Data derived from [28]; [c]The Fisher's exact test with $p < 0.05$ considered significant. From Fergelot et al. (2016) [28], modified with permission

3.3.1 Epigenetic Mechanisms Underlying RSTS

Extensive studies of CBP protein structure and functions, reported in more than 2000 manuscripts published since its discovery by Dr. Goodman's laboratory [4], have revealed a high level of complexity of structural features and that not all aspects of CBP molecular functions are understood (Fig. 3.1). CBP was discovered as the protein binding to CREB transcriptional factor, and the interaction between these two molecules depends on CREB phosphorylation at serine 133 (Ser 133) [4] located within the kinase-inducible domain (KIK). It turned out that CBP is a required coactivator for CREB-dependent transcriptional activation [29]. It is well documented that phosphorylation of CREB on serine 133 is the sufficient requirement for CREB-dependent transcriptional activation and involves direct interaction between the kinase-inducible domain interacting (KIX) domain (CBP) and phosphorylated KIK domain (CREB, Ser 133) [30].

CBP is a large, multifunctional transcriptional coactivator protein with molecular weight 26,531 Da that consists of 2441 amino acids. The CBP has been recognized as interacting with more than 400 nuclear proteins to mediate transcriptional activation from multiple promoters [31]. Through direct interactions with DNA-bound transcriptional factors and components of basal transcriptional machinery, including TBP, TFIIB, TFIIE, and TFIIF, CBP provides a platform for the generation of multiprotein complexes and serves as a bridge between transcriptional factors and the RNA polymerase II holoenzyme during transcriptional activation [32, 33]. CBP is recruited to the transcription sites via mechanisms that involve direct protein-protein interactions between the activation domain of transcription factors and CBP's multiple protein-protein interaction domains, including the transcriptional adapter zinc-binding (TAZ) domains (reviewed in [34]). The CBP structure includes four zinc-binding domains that are localized in the three cysteine-/histidine-rich regions (i.e., CH1, CH2, and CH3). The TAZ1 domain is mapped to CH1. CH2 is the plant homeodomain (PHD)-type zinc finger motive. The CH3 region comprises two independent zinc finger motives, ZZ and TAZ2. Both TAZ1 and TAZ2 possess four zinc finger motives. Three of these motives are identical, but the fourth zinc finger motive is substantially different between the two domains and is believed to provide recognition specificity, as both TAZ1 and TAZ2 bind to two different groups of nuclear factors (or different regions on the same molecule, e.g., p53). In fact, the CH1 and CH3 regions mediate the majority of CBP's protein-protein interactions. Although interaction with CBP's nuclear receptor interaction domain (NRID) provides a mechanism for nuclear receptor (i.e., steroid hormones) transcriptional activation, the KIX domain binds to phosphorylated CREB. In addition, the glutamine-rich (Q) coactivator binding domain (CABD) is another domain that is critical for the assembly of multiprotein complexes [35]. In addition to its function as a platform, CBP belongs to a family chromatin-modifying enzymes and enhances transcription by altering chromatin structure via histone acetylation (Fig. 3.2) [7, 10, 12, 36]. While bromodomain (Bromo) may be involved in chromatin recognition, the KIT11 domain is a lysine acetyltransferase able to transfer acetyl groups primary on histone N-terminals and also nonhistone nuclear proteins [7, 8].

In the eukaryotic cell nucleus, DNA is packaged by histones into nucleosomes, which are repeated subunits of chromatin. One of the central questions in the regulation of gene expression is how the transcriptional machinery gains access to DNA that is tightly packed in chromatin. Over the last two decades, our understanding of the causal relationship between histone acetylation and gene expression has been enhanced dramatically by the identification of intrinsic HAT domains in several newly discovered coactivators of transcription, including CBP, p300, GCN5, and PCAF [10, 37] (Fig. 3.2). The finding that transcriptional coactivators are HATs recruited to specific gene promoters by activated transcription factors are consistent with an idea of the targeted chromatin acetylation as a critical step in transcriptional activation [37]. Moreover, there is selectivity in the specific HAT activity required for the function of distinct classes of transcription factors [12]. The HAT activity of CBP, recruited directly by phosphorylated CREB, is selectively required for the transcriptional function of CREB, whereas the HAT activity of PCAF is indispensable only for nuclear receptor activity [12]. It is now known that CBP interacts with a variety of nuclear factors (Fig. 3.1), highlighting CBP's critical role in the regulation of a variety of cell programs during development and in the adult.

Histone acetylation is one of the major epigenetic mechanisms that controls chromatin structure and function in postmitotic mammalian neurons. As noted, CBP and p300 are critical for human and rodent development due to their ubiquitous expression patterns and their ability to control the functions of chromatin as epigenetic writers. In addition, they were found to bind to specific loci in daughter cells immediately following cell division and act as epigenetic chromatin "bookmarks" [38]. Owing to the high complexity of RSTS's etiology, genetically engineered mice have been employed to gain insight into the mechanisms underlying this devastating disorder.

Both acetyltransferases, CBP and p300, can enhance transcription by relaxing the structure of chromatin nucleosomes via a mechanism that involves the direct acetylation of lysines (K) located at the N-terminals of histones H2A, H2B, H3, and H4, including the acetylation of H4-K5, H3-K14, H3-K18, H3-K27, and H3-K56 [39, 40]. CBP and p300 can be involved in two major epigenetic mechanisms, including the control of global acetylation of chromatin and mediating local chromatin modification in a promoter-specific manner via specific interactions with DNA regulatory element-bound and activated transcriptional factors [41].

The following describes a study of two CBP mutant mice. In a CBP-deficient mutant mouse model, amino acids 29–265 of CBP were replaced with a targeting vector [42]. In second study, a CBP truncated mutant mouse model was generated by insertional mutagenesis, resulting in the expression of a truncated version of CBP (amino acids 1–1084) that contained the CREB-binding domain amino acids (462–661) [43, 44]. In both cases, the homozygous mutants died in utero between 8 and 10.5 days *post coitum* (d.p.c.) [44, 45]. The CBP+/−-deficient mice exhibited various developmental abnormalities, partially resembling RSTS in humans [42]. The CBP+/−-truncated mutant leads to classical RSTS anomalies, showing a much more severe phenotype, including deficiencies in learning and memory [43].

CBP and EP300 show very similar patterns of molecular functions, but their expression during development and in the adult does not fully overlap. Although it is believed that there are subtle differences between CBP and EP300, patients with mutations in CBP or EP300 do not present any marked phenotypic differences (Table 3.1). In addition to certain skeletal abnormalities found specifically in heterozygous CBP null mice, there are no marked phenotypic distinctions between CBP heterozygous null mice and EP300 heterozygous null mice [42, 43, 46]. Each of three mutants, homozygous CBP null mice, homozygous EP300 null mice, and double heterozygous mice for CBP and p300, presents strikingly similar phenotypes and dies in utero [46]. These findings lead to hypotheses that combined levels of CBP and EP300 direct developmental processes rather than that CBP and p300 each have very distinct developmental roles. Although this might be true in general, there is also evidence that CBP and p300 functions do not always compensate for each other (e.g., [16]).

3.3.2 Testing the Biological Function of CBP in Rodents: Epigenetics and Memory

Life depends on the fidelity of DNA replication and the decoding of DNA into RNA and protein. Although our ability to accumulate knowledge involves genetic mechanisms, the nature of cognitive processes, how information is encoded, and what controls functional brain anatomy are not obvious. We thus ask: To what extent does the blueprint of life affect who we are? What is the relationship between neural information processing and chromatin, the functional form of DNA?

The stabilization of learned information into long-term memories requires both new gene expression and alterations in synaptic strength. Whether nondeclarative or declarative memory systems [47] are examined in invertebrates or vertebrates, information is stored first in a transient short-term memory that can eventually be stabilized into long-term memory [48, 49]. A variety of inhibitors of protein and RNA synthesis have been shown to effectively block long-term memory without altering short-term memory [50, 51]. Environmental stimuli or high levels of neuronal activity are known to induce a variety of immediate-early genes, such as Fos, Jun, and Zif268, in many brain areas [52, 53]. In addition, genetic studies in mice suggest that Zif268, CREB, and c-Fos may be involved in memory formation and consolidation [54–59]. Thus, regulatory mechanisms that direct transcription subsequent to the molecular changes in neurons during transient memory formation play a pivotal role in the conversion of short- to long-term memory.

CRE-binding factors, such as CREB, seem to be conserved from mollusks to mammals, and their activity is regulated by both cAMP and calcium influx (reviewed in [60]). These CREB/ATF or CREM families of activators and repressors belong to the bZip transcription factor class. A proposed mechanism by which CREB activates its target promoters is based on the observation that PKA phosphorylates CREB at Ser-133 in response to elevated levels of cAMP. CREB mediates transcriptional induction upon its phosphorylation by PKA [61, 62] or a calcium-dependent nuclear kinase [63], followed by direct interaction with a coactivator of transcription, CBP [4], which facilitates the assembly of the basic transcriptional machinery. CREB has a bipartite transactivation domain that consists of a constitutive domain, Q2, and an inducible domain, KID. It has been shown that the Q2 domain can potentially interact with TFIID via an hTAF135 bridging protein. In contrast, phosphorylation of the KID domain (Ser-133) induces interaction with the KIX domain (CREB interaction domain in CBP). Although it is unclear how the transactivation occurs, it is believed that CBP/CREB-P complex formation enables direct association with the RNA polymerase II complex.

Studies in *Aplysia* demonstrate that the cAMP-signaling pathway appears to play a central role in memory encoding [64–66]. A single electric shock to the tail of the mollusk produces a transient enhancement of the gill-withdrawal reflex. This short-term memory could be transformed into long-term memory by applying multiple stimuli. Reconstitution of the neurons that mediate the gill-withdrawal reflex by co-culturing a single *Aplysia* sensory neuron with the motor neuron that mediates the reflex allows the study of the molecular and cellular mechanisms involved in this simple form of learning. In response to one pulse of serotonin, this synapse underwent short-term facilitation, while five repeated pulses of serotonin resulted in long-term facilitation. The physiological changes that accompanied presynaptic facilitation were observed after intracellular injection of the catalytic subunit of cAMP-dependent protein kinase A (PKA) into *Aplysia* sensory neurons [65], while inhibitors of PKA blocked both forms of facilitation. Taken together, the conversion of short- to long-term memory requires the removal of certain inhibitory constraints on the storage of long-term memory followed by an activation of CREB-controlled gene expression; both mechanisms are required for the stabilization of transient memory [64–70].

These observations in *Aplysia* are in agreement with reported genetic studies in fruit flies (*Drosophila*) on memory, which demonstrate that fruit flies that overexpress a CREB repressor transgene under the control of the inducible heat shock promoter and were tested for memory retention after Pavlovian olfactory learning showed drastically impaired long-term memory formation but without an effect on transient short-term memory [71]. Moreover, overexpression of a CREB activator decreased the number of training trials needed to establish long-term memory but did not affect short-term memory formation [71]. Thus, the level of active CREB was associated with the number of training trials required for long-term memory in *Drosophila* olfactory associative learning.

In 1973, Bliss and Lomo discovered that the synaptic connections within the rabbit hippocampus undergo long-term potentiation (LTP) [72], which encodes memories [73–75]. There are two classes of mechanisms for long-lasting modification of synaptic strength that follow different patterns of tetanic electrical stimulation: LTP (for mechanisms that increase synaptic strength) and long-term depression (LTD) for mechanisms that decrease synaptic strength [75]. Pharmacological studies have demonstrated that blockers of N-methyl-D-aspartate receptor (NMDAR) eliminated both hippocampal LTP and LTD [76, 77] and hippocampus-dependent spatial memory formation in rodents treated with an NMDAR antagonist [78].

Advances in the study of learning and memory, using the genetic approach, provide an opportunity to study the correlation between LTP and memory storage in mutant mice. Mutant mice that carry a targeted deletion of the gene encoding CaMKII, the fyn tyrosine kinase, PKCg, NMDAR, or CREB exhibited deficits in spatial memory when tested in the Morris water maze paradigm (which is known to depend on the integrity of the hippocampus) [54, 76–84]. These mutants also showed modified LTP between CA3 and CA1 neurons (Schaffer collateral LTP) tested in hippocampal slices [54, 76–84].

A number of studies in mammalian systems have shown that synaptic activity in neurons controls CBP's ability to function as a transcriptional coactivator [85–88]. Voltage-gated calcium channels, such as the N-methyl-D-aspartate receptor (NMDAR), or local calcium transients can induce CBP-dependent transactivation in cultured hippocampal and cortical neurons [85–87]. In addition, CBP itself is a target for calcium influx-induced CaM kinase IV via a mechanism that involves CBP phosphorylation at Ser301, a requirement for NMDA-induced transcription [87]. Although the question of whether NMDA-dependent transcription requires CBP HAT activity has not been addressed, these findings suggest that, in addition to the well-characterized function of transcription factors, such as CREB in NMDA-induced gene expression, coactivators, such as CBP, are regulated by a separate pathway that plays a critical role in the activation of gene expression. To investigate these functions of CBP, Korzus et al. generated transgenic mice that express a reversible CBP-lacking HAT activity in a subset of excitatory neurons in the adult [14]. These mutant mice exhibited long-term memory deficits, while the encoding of new information and short-term memory were both normal. The behavioral phenotype was due to an acute requirement for CBP HAT activity in the adult, as it is rescued by both suppression of the transgene expression and by administration of a histone deacetylase (HDAC) inhibitor. This mouse model eliminates data interpretation problems that result from developmental changes and mimics a heterozygous CBP mutation that causes severe mental retardation [18]. These data indicate that, independent of its developmental role, CBP HAT activity is essential for higher cognitive functions in adults. Furthermore, this research provides the first demonstration that CBP's HAT activity is required for brain information processing and demonstrates that histone modulation by deacetylase/acetylase activities affects specific processes of cognitive function. Thus, CBP-dependent transcriptional activation in response to synaptic signals appears to be regulated by multiple pathways through mechanisms that may underlie experience-induced neuronal gene expression during information encoding and memory formation (Fig. 3.5).

In independent studies, Kandel and collaborators [92] tested a developmental model for RSTS that carried only a single *crebbp* allele, referred to as the CBP+/−-deficient mice [42]. These CBP mutant mice exhibited certain typical RSTS developmental abnormalities [42], but brain gross anatomy, emotional responses, and working memory were spared [92]. These studies revealed that chromatin acetylation, memory, and synaptic plasticity were abnormal in CBP+/−-deficient mutant mice, providing further evidence that implicates CBP as an important component of the molecular mechanism underlying learning and memory. Remarkably, inhibiting

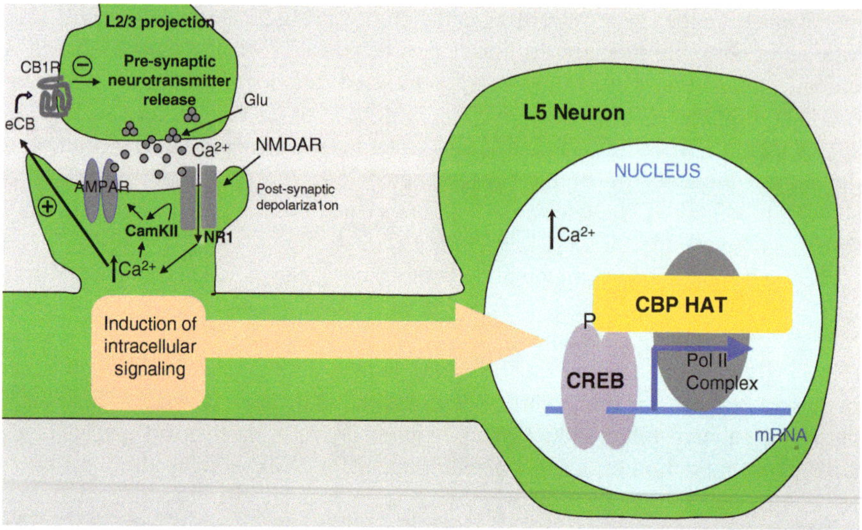

Fig. 3.5 Putative molecular mechanism underlying alterations of synaptic strength potentially associated with cognitive performance, including memory formation. Studies have revealed that a number of synaptic (e.g., NMDAR, CaMKII) and nuclear (e.g., CREB and CBP) molecules could be critical for long-term memory consolidation. Long-term potentiation, or LTP, is an induced increase in synaptic efficacy. Many believe that LTP is a laboratory model for learning and memory [89]. Involvement of glutamate receptors, such as NMDAR, in LTP was demonstrated by Susumu Tonagawa at a molecular level [90] and by Richard Morris in behavioral studies [91]. Neuronal activity induces glutamate release into the synaptic cleft. Glutamate acts on the α-amino-3-hydroxy-5-methyl-4-isoxazolepropionic acid receptors (AMPARs) and NMDARs. However, initially, Na^+ flows only through AMPAR because NMDAR is blocked by Mg. Postsynaptic depolarization removes the Mg^{2+} block, and then Ca^{2+} (and Na^+) can flow through NMDAR. The resultant rise in Ca^{2+} levels within the dendritic spine is the critical trigger for LTP. Ca^{2+} influx activates CaMKII through autophosphorylation, and activated CaMKII induces molecular changes in postsynaptic neurons, yielding a change in synaptic strength called LTP. Ca^{2+} influx in dendritic spines activates intracellular signaling pathways, directing CREB phosphorylation, which is required but not sufficient for NMDA-induced gene expression. Studies of *Drosophila*, *Aplysia*, and mice clearly demonstrated the requirement for CREB in long-term memory. Further studies implicated CREB's partner, CBP, as an obligatory component of the molecular mechanism underlying learning and memory [14, 92, 93]. AMPAR is α-amino-3-hydroxy-5-methyl-4-isoxazolepropionic acid receptor, NR1 is an obligatory subunit of postsynaptic NMDAR, CaMKII is Ca^{2+}/Calmodulin-dependent protein kinase, CREB is a transcription factor, and CBP is coactivator of transcription and histone acetyltransferase (HAT)

histone deacetyltransferase activity, the molecular counterpart of the histone acetylation function of CBP, ameliorated the memory impairment and the deficit in synaptic plasticity found in CBP+/−-deficient mutant mice.

Further studies of CBP function that involve different mouse models of RSTS that employs various CBP mutations and covalent histone modifications add substantially to the understanding of CBP function in memory consolidation [93–97]. Table 3.2 provides examples of genetic studies in mouse models. As seen in the table, disruption of the interaction between the transcriptional coactivator CBP and a gene promoter binding transcription factor, such as CREB, led to impaired synaptic plasticity, fear memory, and poor performance on tasks that tested the ability to

Table 3.2 Examples of mutations targeted to the CREBBP gene in a mouse model that revealed high complexity of this gene function in cell regulation, development, adult neurogenesis, and cognition

Mouse model	Genotype	Phenotype
Mutations that specifically target CBP HAT activity in adults:		
CBP{HAT-}$^{Exc. Neur}$ [93]	Transgenic/hemizygote (germline mutation)	Deficit in LTM
	Tetracycline-inducible dominant negative CBP transgene that lacks HAT activity expressed in CaMKIIα positive excitatory neurons (exc. Neur.) in the forebrain	Normal STM
		No effects on prenatal development
		No effects on postnatal development
CBP {HAT-}^{mPFC} [98, 99]	Viral-mediated gene transfer (somatic mutation) dominant negative CBP transgene that lacks HAT activity expressed in the medial prefrontal cortex (mPFC)	Deficit in LTM (limited to mPFC-dependent behavior)
		Normal STM
		Deficit in discrimination between safety and danger
		No effects on prenatal development
		No effects on postnatal development
Mutations that target CBP function as a "platform":		
CBP −/− [45]	Null mutant (germline mutation)	Death in utero
CBP **Δ/Δ** [44]	Null mutant (germline mutation)	Death in utero
CBP +/**Δ** [43]	Heterozygote (germline mutation)	Deficit in LTM
	Truncated allele that expresses dominant negative CBP truncated protein (aa1–1084)	Normal STM
		Skeletal abnormalities/ growth retardation
CBP +/− [42, 92, 100]	Heterozygote (germline mutation)	Deficit in LTM
	Normal STM	Deficit in neurogenesis
		Skeletal abnormalities/ growth retardation
CBP **Δ1** [93]	Transgenic/hemizygote (germline mutation)	Normal STM
	Transgenic mice that express dominant negative truncated CBP protein (aa1–1083) under control of CaMKIIα promoter transgene activated during postnatal brain development	Deficit in LTM
		No effects on prenatal development
CBP$^{KIX/KIX}$ [95, 101]	Homozygote (germline mutation) triple point mutation in the KIX domain	Deficit in LTM
	Focal homozygous k.o. No effects on prenatal development in the dorsal hippocampus	Normal STM
		No effects on postnatal development

remember familiar objects [94]. The ability of HDAC inhibitors to rescue some these phenotypes demonstrated that recruitment of CBP's HAT to specific gene promoters is critical for long-term memory consolidation [102].

The system-level effects of CBP hypofunction depend not only on whether the mutation was present during development or only in the adult but also on whether the locus of the mutation is critical. A number of studies have addressed this specific problem with great success. Studies have shown that CBP hypofunction targeted specifically to adult excitatory neurons [14], adult hippocampal neurons [95], adult excitatory neurons in the medial prefrontal cortex (mPFC) [98, 99], or intra-lateral amygdala infusion of c646 (a selective pharmacological inhibitor of p300/CBP HAT activity) in adult mice shortly following fear conditioning [103] resulted in selective impairment of synaptic plasticity in the targeted brain region [95, 103] and deficits in long-term memory consolidation [14, 95, 98, 99, 103]. Further, changes in histone acetylation in the mPFC were connected to the extinction of conditioned fear [104, 105], whereas intrahippocampal delivery of histone deacetylase inhibitors facilitated fear extinction [106].

CBP also has been implicated in adult neurogenesis, which has been linked to higher cognitive function and depression. The reduction of CBP targeted to the adult brain leads to abnormalities in environmental enrichment-induced neurogenesis [100], which provides strong evidence for the role of CBP in adult neurogenesis-dependent enhancement of adaptability toward novel experiences [107, 108].

A CBP deficiency that results from elimination of its multiple functional domains, including HAT, is associated with abnormal synaptic plasticity [92, 93], and altered histone acetylation in the hippocampus correlates with new learning experiences [92, 109]. Thus, chromatin dynamics appear to control memory consolidation via CBP-mediated acetylation during gene activation, altering chromatin structure and mediating gene-specific removal of epigenetically controlled repression.

Synaptic stimulation of excitatory neurons leads to increases in intracellular Ca^{2+} and modulation of gene expression [110, 111]. Ca^{2+} signaling to the nucleus has been studied extensively in neurons in culture, where it is clear that a primary Ca^{2+} signal can act both directly through CaM in the nucleus [112–114] and indirectly through the activation of other second-messenger pathways, such as cAMP [115] and MAP kinase [114, 116]. Mayford laboratory discovered that nuclear blocking of Ca^{2+}/calmodulin-mediated signaling in the nuclei of hippocampal neurons blocks stabilization of new memories but spares learning and short-term memory [117]. Thus, both impaired nuclear Ca^{2+} signaling in excitatory neurons [117] and a deficiency in CBP acetyltransferase activity in excitatory neurons [14] produce exactly the same biological effects, preventing memory consolidation processes (Fig. 3.5).

Understanding how chromatin is involved in neural information processing also is of interest. It has been established that CREB phosphorylation (a critical step for CBP recruitment) is insufficient to induce neuronal gene expression [87, 118]. A separate, still unknown, CBP-signaling pathway was shown to be required for synaptic plasticity-dependent neuronal gene expression. Conversely, Frey and Morris proposed that heterosynaptic coactivation of both glutamate and modulatory receptors may be necessary to trigger the upregulation of the relevant protein synthesis [119]. These findings suggest that multiple signaling pathways must be activated to allow neuronal activity-dependent gene expression and that such pathways are critical for memory consolidation by recruiting functional CBP HAT activity and possibly by altering chromatin structure (Fig. 3.6). Chromatin-based information processing provides an advanced threshold control, whereby the nuclear signal generated as a sum of individual synaptic signals determines the outcome. If transcription is specifically induced by combining synaptic signals in the presence (or absence) of other irrelevant signals, chromatin may function to filter neural information. Thus, the integration of brief synaptic signals by CBP in the nucleus leads to (a) reversible changes in chromatin structure that result in removal of gene-specific epigenetically controlled repression, (b) transient alterations in the gene expression that last a few hours after the initial learning experience, and, subsequently, (c) permanent learning-dependent changes in neuronal networks (Fig. 3.6).

The model presented in Fig. 3.6 has several implications. The gating and filtering of neural signals by chromatin suggests an additional layer of complexity for computational aspects of neural networks. The chromatin gating/filtering hypothesis provides a novel way to discriminate stabilized changes in neural networks. Another implication of this model has more practical use in biomedical research. This implication is that epigenetic mechanisms permit biological systems to respond and adapt to environmental stimuli by altering gene expression. Alternatively, environmental influences can provoke undesired and persistent epigenetic effects that may increase susceptibility to mental retardation, long-latency ailments of the nervous system, and unsuccessful aging. Recently, strong evidence has implicated epigenetic mechanisms based on chromatin remodeling as the cause of neuronal dysfunction [120, 121]. Future research should broaden the study of molecular mechanisms that are involved in the stabilization of learning-initiated changes in neuronal networks underlying memory consolidation and provide new avenues to investigate mental illnesses, including memory disorders and psychostimulant-elicited plasticity in the brain-reward system that underlies drug addiction.

The model presented in Fig. 3.6 is consistent with the prevailing hypothesis of learning and memory in the mammalian brain, which postulates that the ability to learn involves a selective and bidirectional modification of individual synapses, such that the same stimulus elicits different responses before and after learning [73]. Each learning experience initiates changes in specific synapses that, subsequently, translate into alterations of neuronal network properties and alter behavioral responses. During initial learning, protein synthesis inhibitors allow animals to learn and remember a given task for a few hours but severely impair memory after 24 h [51]. The link between gene expression and synapse alteration suggests

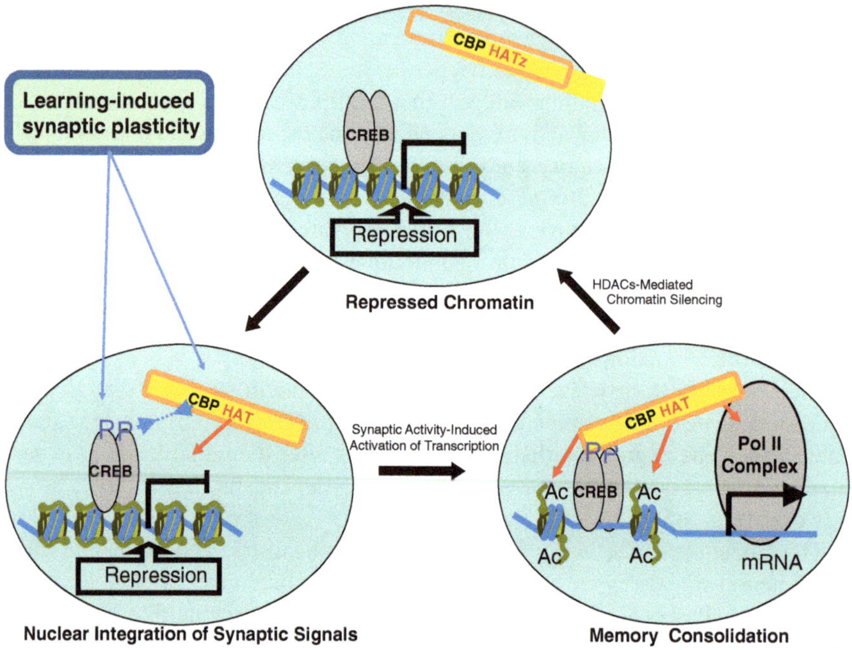

Fig. 3.6 Proposed model for CBP HAT involvement in long-term memory consolidation. Before activation, target genes are repressed by chromatin structure. Signal-inducing CREB phosphorylation must occur but is not sufficient. It is well known that CBP is independently regulated in response to NMDA. A second signal is required to remove chromatin repression. This can be accomplished by CBP-dependent histone acetylation. The covalent modification by HAT activity leaves long-lasting marks on chromatin at the target genes. This represents a very attractive mechanism for the regulation of long-lasting transcriptional changes associated with long-term synaptic and behavioral plasticity. The proposed model postulates that two signals—the first signal induces CREB and the second signal removes repression of target genes by chromatin acetylation—are required to occur during initial learning. This sort of the acetylation-mediated covalent modification by CBP could change requirements for subsequent transcriptional activation of genes in response to future signals. This would open a temporal window in which cellular signals, which would not recruit acetyltransferase, would nevertheless stimulate transcription required for memory consolidation. Chromatin opening at the target genes by acetylation would allow for prolonged transcription, even in the absence of an initial stimulus. After execution of "memory program," these transiently activated gene expression would shut down via default mechanisms controlled by ubiquitous HDACs, allowing for a homeostatic update of integrated circuits. Model initially presented in Korzus et al. (2004) [14], modified by author

that protein synthesis is required to permanently alter synapses modified by a learning experience [122]. Thus, the molecular neuronal changes during information acquisition have pivotal effects on gene expression, facilitating the conversion of short- to long-term memory. Recent research has revealed that the rate of transcription is a function of the physical state of chromatin and that chromatin plasticity mediated by chemical modification allows dynamic changes in gene expression without changing the DNA sequence [123, 124].

Chromatin functions are not limited to controlling the transmission of genetic and epigenetic information. Based on current data, it has been proposed that chromatin dynamics in individual neurons may shape neural information processing for what is or is not permanently stored. This hypothesis is consistent with the view that knowledge is not encoded in a simple molecular form but reflects the dynamics of neural interconnectivity throughout the brain. This hypothesis postulates that epigenetic regulation controls cognitive performance and may explain some of the cognitive phenotypes associated with epigenetic disorders, such as RSTS.

3.3.3 Potential Therapeutic Applications of Histone Deacetylase Inhibitors for RSTS

It is feasible to infer that epigenetic aberrations can be "healed" because the major feature of epigenetic changes is their reversibility. Whereas the specificity of these changes creates a major challenge, the global patterning of epigenetic marks can be controlled by generating an appropriate balance between histone acetylation and deacetylation (e.g., [14]). In addition, there is growing evidence that shows that a variety of available drugs are capable of reactivating epigenetically silenced genes, including a variety of HDAC inhibitors, including suberoylanilide hydroxamic acid (SAHA), valproic acid (VPA), and trichostatin A (TSA). It also has been demonstrated that DNA demethylation drugs, such as zebularine and 5-aza-20-deoxycytidine (5-ADC), are capable of unlocking previously inactive genes. Although some of these drugs are being used in cancer treatment therapies, epigenetic drugs have recently been considered for neurological and mental disorders [125]. Preclinical studies of mouse models for RSTS with reduced HAT activity demonstrated that some of the cognitive impairments could be ameliorated with HDAC inhibitor treatment [14, 92]. Recently, Lopez-Atalaya et al. reported that deficits in the levels of histone H2A and H2B acetylation in lymphoblastoid cell lines derived from nine patients with RSTS2 were rescued by treatment with the HDAC inhibitor TSA [126]. Although some HDAC inhibitors show positive effects on memory and synaptic plasticity [102, 106, 109, 127, 128], the HDAC3 gene was found to negatively regulate long-term memory [129]. In fact, HDAC inhibitors have been used in the past by psychiatrists as mood stabilizers and antiepileptics (e.g., VPA). Epigenetic regulators are being intensively investigated as a possible treatment for cancers [130], parasitics [131], inflammatory diseases [124] and, more recently, mental and neurological disorders, including RSTS [132–134].

References

1. Rosenfeld MG, Glass CK. Coregulator codes of transcriptional regulation by nuclear receptors. J Biol Chem. 2001;276(40):36865–8.
2. Neely KE, Workman JL. Histone acetylation and chromatin remodeling: which comes first? Mol Genet Metab. 2002;76(1):1–5.
3. Fischle W, Wang Y, Allis CD. Histone and chromatin cross-talk. Curr Opin Cell Biol. 2003;15(2):172–83.
4. Chrivia JC, Kwok RP, Lamb N, Hagiwara M, Montminy MR, Goodman RH. Phosphorylated CREB binds specifically to the nuclear protein CBP. Nature. 1993;365(6449):855–9.
5. Eckner R, Ewen ME, Newsome D, Gerdes M, DeCaprio JA, Lawrence JB, Livingston DM. Molecular cloning and functional analysis of the adenovirus E1A-associated 300-kD protein (p300) reveals a protein with properties of a transcriptional adaptor. Genes Dev. 1994;8(8):869–84.
6. Janknecht R. The versatile functions of the transcriptional coactivators p300 and CBP and their roles in disease. Histol Histopathol. 2002;17(2):657–68.
7. Allis CD, Berger SL, Cote J, Dent S, Jenuwien T, Kouzarides T, Pillus L, Reinberg D, Shi Y, Shiekhattar R, Shilatifard A, Workman J, Zhang Y. New nomenclature for chromatin-modifying enzymes. Cell. 2007;131(4):633–6.
8. Chan HM, La Thangue NB. p300/CBP proteins: HATs for transcriptional bridges and scaffolds. J Cell Sci. 2001;114(Pt 13):2363–73.
9. Wang F, Marshall CB, Ikura M. Transcriptional/epigenetic regulator CBP/p300 in tumorigenesis: structural and functional versatility in target recognition. Cell Mol Life Sci. 2013;70(21):3989–4008.
10. Ogryzko VV, Schiltz RL, Russanova V, Howard BH, Nakatani Y. The transcriptional coactivators p300 and CBP are histone acetyltransferases. Cell. 1996;87(5):953–9.
11. Yuan LW, Soh JW, Weinstein IB. Inhibition of histone acetyltransferase function of p300 by PKCdelta. Biochim Biophys Acta. 2002;1592(2):205–11.
12. Korzus E, Torchia J, Rose DW, Xu L, Kurokawa R, McInerney EM, Mullen TM, Glass CK, Rosenfeld MG. Transcription factor-specific requirements for coactivators and their acetyltransferase functions. Science. 1998;279(5351):703–7.
13. Puri PL, Avantaggiati ML, Balsano C, Sang N, Graessmann A, Giordano A, Levrero M. p300 is required for MyoD-dependent cell cycle arrest and muscle-specific gene transcription. EMBO J. 1997;16(2):369–83.
14. Korzus E, Rosenfeld MG, Mayford M. CBP histone acetyltransferase activity is a critical component of memory consolidation. Neuron. 2004;42(6):961–72.
15. Lopez-Atalaya JP, Valor LM, Barco A. Epigenetic factors in intellectual disability: the Rubinstein-Taybi syndrome as a paradigm of neurodevelopmental disorder with epigenetic origin. Prog Mol Biol Transl Sci. 2014;128:139–76.
16. Oliveira AM, Abel T, Brindle PK, Wood MA. Differential role for CBP and p300 CREB-binding domain in motor skill learning. Behav Neurosci. 2006;120(3):724–9.
17. Rubinstein JH, Taybi H. Broad thumbs and toes and facial abnormalities. Am J Dis Child. 1963;105:588–608.
18. Hennekam RC, Stevens CA, Van de Kamp JJ. Etiology and recurrence risk in Rubinstein-Taybi syndrome. Am J Med Genet Suppl. 1990;6:56–64.
19. Petrij F, Giles RH, Dauwerse HG, Saris JJ, Hennekam RC, Masuno M, Tommerup N, van Ommen GJ, Goodman RH, Peters DJ, et al. Rubinstein-Taybi syndrome caused by mutations in the transcriptional co-activator CBP. Nature. 1995;376(6538):348–51.
20. LOVD v.2.0: the next generation in gene variant databases (Updated February 12, 2016) [Internet]. Wiley-Liss, Inc. 2011. http://www.LOVD.nl.
21. CREBBP [Internet]. Jean-Loup Huret (Editor-in-Chief); INIST-CNRS (Publisher). 2010. http://atlasgeneticsoncology.org//Genes/CBPID42.html.
22. Coupry I, Roudaut C, Stef M, Delrue MA, Marche M, Burgelin I, Taine L, Cruaud C, Lacombe D, Arveiler B. Molecular analysis of the CBP gene in 60 patients with Rubinstein-Taybi syndrome. J Med Genet. 2002;39(6):415–21.

23. Bentivegna A, Milani D, Gervasini C, Castronovo P, Mottadelli F, Manzini S, Colapietro P, Giordano L, Atzeri F, Divizia MT, Uzielli ML, Neri G, Bedeschi MF, Faravelli F, Selicorni A, Larizza L. Rubinstein-Taybi syndrome: spectrum of CREBBP mutations in Italian patients. BMC Med Genet. 2006;7:77.
24. Roelfsema JH, White SJ, Ariyurek Y, Bartholdi D, Niedrist D, Papadia F, Bacino CA, den Dunnen JT, van Ommen GJ, Breuning MH, Hennekam RC, Peters DJ. Genetic heterogeneity in Rubinstein-Taybi syndrome: mutations in both the CBP and EP300 genes cause disease. Am J Hum Genet. 2005;76(4):572–80.
25. Negri G, Magini P, Milani D, Colapietro P, Rusconi D, Scarano E, Bonati MT, Priolo M, Crippa M, Mazzanti L, Wischmeijer A, Tamburrino F, Pippucci T, Finelli P, Larizza L, Gervasini C. From whole gene deletion to point mutations of EP300-positive Rubinstein-Taybi patients: new insights into the mutational spectrum and peculiar clinical hallmarks. Hum Mutat. 2016;37(2):175–83.
26. Spena S, Milani D, Rusconi D, Negri G, Colapietro P, Elcioglu N, Bedeschi F, Pilotta A, Spaccini L, Ficcadenti A, Magnani C, Scarano G, Selicorni A, Larizza L, Gervasini C. Insights into genotype-phenotype correlations from CREBBP point mutation screening in a cohort of 46 Rubinstein-Taybi syndrome patients. Clin Genet. 2015;88(5):431–40.
27. van Belzen M, Bartsch O, Lacombe D, Peters DJ, Hennekam RC. Rubinstein-Taybi syndrome (CREBBP, EP300). Eur J Hum Genet. 2011;19(1):121.
28. Fergelot P, Van Belzen M, Van Gils J, Afenjar A, Armour CM, Arveiler B, Beets L, Burglen L, Busa T, Collet M, Deforges J, de Vries BB, Dominguez Garrido E, Dorison N, Dupont J, Francannet C, Garcia-Minaur S, Gabau Vila E, Gebre-Medhin S, Gener Querol B, Genevieve D, Gerard M, Gervasini CG, Goldenberg A, Josifova D, Lachlan K, Maas S, Maranda B, Moilanen JS, Nordgren A, Parent P, Rankin J, Reardon W, Rio M, Roume J, Shaw A, Smigiel R, Sojo A, Solomon B, Stembalska A, Stumpel C, Suarez F, Terhal P, Thomas S, Touraine R, Verloes A, Vincent-Delorme C, Wincent J, Peters DJ, Bartsch O, Larizza L, Lacombe D, Hennekam RC. Phenotype and genotype in 52 patients with Rubinstein-Taybi syndrome caused by EP300 mutations. Am J Med Genet A. 2016;170(12): 3069–82.
29. Kwok RP, Lundblad JR, Chrivia JC, Richards JP, Bachinger HP, Brennan RG, Roberts SG, Green MR, Goodman RH. Nuclear protein CBP is a coactivator for the transcription factor CREB. Nature. 1994;370(6486):223–6.
30. Cardinaux JR, Notis JC, Zhang Q, Vo N, Craig JC, Fass DM, Brennan RG, Goodman RH. Recruitment of CREB binding protein is sufficient for CREB-mediated gene activation. Mol Cell Biol. 2000;20(5):1546–52.
31. Bedford DC, Kasper LH, Fukuyama T, Brindle PK. Target gene context influences the transcriptional requirement for the KAT3 family of CBP and p300 histone acetyltransferases. Epigenetics. 2010;5(1):9–15.
32. Shiama N. The p300/CBP family: integrating signals with transcription factors and chromatin. Trends Cell Biol. 1997;7(6):230–6.
33. Goodman RH, Smolik S. CBP/p300 in cell growth, transformation, and development. Genes Dev. 2000;14(13):1553–77.
34. De Guzman RN, Martinez-Yamout MA, Dyson HJ, Wright PE. Structure and function of the CBP/p300 TAZ domains. In: Iuchi S, Kuldell N, editors. Zinc finger proteins: from atomic contact to cellular function. Molecular biology intelligence unit. Georgetown, TX, New York: Landes Bioscience; Kluwer Academic/Plenum; 2005.
35. Demarest SJ, Martinez-Yamout M, Chung J, Chen H, Xu W, Dyson HJ, Evans RM, Wright PE. Mutual synergistic folding in recruitment of CBP/p300 by p160 nuclear receptor coactivators. Nature. 2002;415(6871):549–53.
36. Bannister AJ, Kouzarides T. The CBP co-activator is a histone acetyltransferase. Nature. 1996;384(6610):641–3.
37. Brownell JE, Allis CD. Special HATs for special occasions: linking histone acetylation to chromatin assembly and gene activation. Curr Opin Genet Dev. 1996;6(2):176–84.
38. Wong MM, Byun JS, Sacta M, Jin Q, Baek S, Gardner K. Promoter-bound p300 complexes facilitate post-mitotic transmission of transcriptional memory. PLoS One. 2014;9(6):e99989.

39. Das C, Lucia MS, Hansen KC, Tyler JK. CBP/p300-mediated acetylation of histone H3 on lysine 56. Nature. 2009;459(7243):113–7.
40. Jin Q, Yu LR, Wang L, Zhang Z, Kasper LH, Lee JE, Wang C, Brindle PK, Dent SY, Ge K. Distinct roles of GCN5/PCAF-mediated H3K9ac and CBP/p300-mediated H3K18/27ac in nuclear receptor transactivation. EMBO J. 2011;30(2):249–62.
41. McManus KJ, Hendzel MJ. Quantitative analysis of CBP- and P300-induced histone acetylations in vivo using native chromatin. Mol Cell Biol. 2003;23(21):7611–27.
42. Tanaka Y, Naruse I, Maekawa T, Masuya H, Shiroishi T, Ishii S. Abnormal skeletal patterning in embryos lacking a single Cbp allele: a partial similarity with Rubinstein-Taybi syndrome. Proc Natl Acad Sci U S A. 1997;94(19):10215–20.
43. Oike Y, Hata A, Mamiya T, Kaname T, Noda Y, Suzuki M, Yasue H, Nabeshima T, Araki K, Yamamura K. Truncated CBP protein leads to classical Rubinstein-Taybi syndrome phenotypes in mice: implications for a dominant-negative mechanism. Hum Mol Genet. 1999;8(3):387–96.
44. Oike Y, Takakura N, Hata A, Kaname T, Akizuki M, Yamaguchi Y, Yasue H, Araki K, Yamamura K, Suda T. Mice homozygous for a truncated form of CREB-binding protein exhibit defects in hematopoiesis and vasculo-angiogenesis. Blood. 1999;93(9):2771–9.
45. Tanaka Y, Naruse I, Hongo T, Xu M, Nakahata T, Maekawa T, Ishii S. Extensive brain hemorrhage and embryonic lethality in a mouse null mutant of CREB-binding protein. Mech Dev. 2000;95(1–2):133–45.
46. Yao TP, Oh SP, Fuchs M, Zhou ND, Ch'ng LE, Newsome D, Bronson RT, Li E, Livingston DM, Eckner R. Gene dosage-dependent embryonic development and proliferation defects in mice lacking the transcriptional integrator p300. Cell. 1998;93(3):361–72.
47. Squire LR. Memory and the hippocampus: a synthesis from findings with rats, monkeys, and humans [published erratum appears in Psychol Rev 1992 Jul;99(3):582]. Psychol Rev. 1992;99(2):195–231.
48. McGaugh JL, Hertz MJ. Memory consolidation. San Francisco: Albion; 1972.
49. Squire LR. Memory and brain. New York: Oxford; 1987.
50. Andrew RJ. The functional organization of phases of memory consolidation. In: Hinde RA, Beer C, Bunsel M, editors. Advances in the study of behaviour, vol. 11. New York: Academic; 1980. p. 337–67.
51. Davis HP, Squire LR. Protein synthesis and memory: a review. Psychol Bull. 1984;96(3): 518–59.
52. Worley PF, Cole AJ, Murphy TH, Christy BA, Nakabeppu Y, Baraban JM. Synaptic regulation of immediate-early genes in brain. Cold Spring Harb Symp Quant Biol. 1990;55:213–23.
53. Tischmeyer W, Grimm R. Activation of immediate early genes and memory formation. Cell Mol Life Sci. 1999;55(4):564–74.
54. Bourtchuladze R, Frenguelli B, Blendy J, Cioffi D, Schutz G, Silva AJ. Deficient long-term memory in mice with a targeted mutation of the cAMP- responsive element-binding protein. Cell. 1994;79(1):59–68.
55. Pittenger C, Huang YY, Paletzki RF, Bourtchouladze R, Scanlin H, Vronskaya S, Kandel ER. Reversible inhibition of CREB/ATF transcription factors in region CA1 of the dorsal hippocampus disrupts hippocampus-dependent spatial memory. Neuron. 2002;34(3):447–62.
56. Jones MW, Errington ML, French PJ, Fine A, Bliss TV, Garel S, Charnay P, Bozon B, Laroche S, Davis S. A requirement for the immediate early gene Zif268 in the expression of late LTP and long-term memories. Nat Neurosci. 2001;4(3):289–96.
57. Josselyn SA, Shi C, Carlezon Jr WA, Neve RL, Nestler EJ, Davis M. Long-term memory is facilitated by cAMP response element-binding protein overexpression in the amygdala. J Neurosci. 2001;21(7):2404–12.
58. Fleischmann A, Hvalby O, Jensen V, Strekalova T, Zacher C, Layer LE, Kvello A, Reschke M, Spanagel R, Sprengel R, Wagner EF, Gass P. Impaired long-term memory and NR2A-type NMDA receptor-dependent synaptic plasticity in mice lacking c-Fos in the CNS. J Neurosci. 2003;23(27):9116–22.
59. Kida S, Josselyn SA, de Ortiz SP, Kogan JH, Chevere I, Masushige S, Silva AJ. CREB required for the stability of new and reactivated fear memories. Nat Neurosci. 2002;5(4):348–55.

60. Brindle PK, Montminy MR. The CREB family of transcription activators. Curr Opin Genet Dev. 1992;2(2):199–204.
61. Gonzalez GA, Montminy MR. Cyclic AMP stimulates somatostatin gene transcription by phosphorylation of CREB at serine 133. Cell. 1989;59(4):675–80.
62. Gonzalez GA, Yamamoto KK, Fischer WH, Karr D, Menzel P, Biggs 3rd W, Vale WW, Montminy MR. A cluster of phosphorylation sites on the cyclic AMP-regulated nuclear factor CREB predicted by its sequence. Nature. 1989;337(6209):749–52.
63. Deisseroth K, Bito H, Tsien RW. Signaling from synapse to nucleus: postsynaptic CREB phosphorylation during multiple forms of hippocampal synaptic plasticity. Neuron. 1996;16(1): 89–101.
64. Byrne JH, Kandel ER. Presynaptic facilitation revisited: state and time dependence. J Neurosci. 1996;16(2):425–35.
65. Castellucci VF, Kandel ER, Schwartz JH, Wilson FD, Nairn AC, Greengard P. Intracellular injection of t he catalytic subunit of cyclic AMP- dependent protein kinase simulates facilitation of transmitter release underlying behavioral sensitization in Aplysia. Proc Natl Acad Sci U S A. 1980;77(12):7492–6.
66. Castellucci VF, Nairn A, Greengard P, Schwartz JH, Kandel ER. Inhibitor of adenosine 3':5'-monophosphate-dependent protein kinase blocks presynaptic facilitation in Aplysia. J Neurosci. 1982;2(12):1673–81.
67. Abel T, Martin KC, Bartsch D, Kandel ER. Memory suppressor genes: inhibitory constraints on the storage of long- term memory. Science. 1998;279(5349):338–41.
68. Alberini CM, Ghirardi M, Metz R, Kandel ER. C/EBP is an immediate-early gene required for the consolidation of long- term facilitation in Aplysia. Cell. 1994;76(6):1099–114.
69. Bartsch D, Casadio A, Karl KA, Serodio P, Kandel ER. CREB1 encodes a nuclear activator, a repressor, and a cytoplasmic modulator that form a regulatory unit critical for long-term facilitation. Cell. 1998;95(2):211–23.
70. Bartsch D, Ghirardi M, Skehel PA, Karl KA, Herder SP, Chen M, Bailey CH, Kandel ER. Aplysia CREB2 represses long-term facilitation: relief of repression converts transient facilitation into long-term functional and structural change. Cell. 1995;83(6):979–92.
71. Yin JC, Wallach JS, Del Vecchio M, Wilder EL, Zhou H, Quinn WG, Tully T. Induction of a dominant negative CREB transgene specifically blocks long-term memory in Drosophila. Cell. 1994;79(1):49–58.
72. Bliss TV, Gardner-Medwin AR. Long-lasting potentiation of synaptic transmission in the dentate area of the unanaestetized rabbit following stimulation of the perforant path. J Physiol (Lond). 1973;232(2):357–74.
73. Bliss TV, Collingridge GL. A synaptic model of memory: long-term potentiation in the hippocampus. Nature. 1993;361(6407):31–9.
74. Bliss TV, Richter-Levin G. Spatial learning and the saturation of long-term potentiation [comment]. Hippocampus. 1993;3(2):123–5.
75. Malenka RC. Synaptic plasticity in the hippocampus: LTP and LTD. Cell. 1994;78(4):535–8.
76. Collingridge GL, Kehl SJ, McLennan H. The antagonism of amino acid-induced excitations of rat hippocampal CA1 neurones in vitro. J Physiol (Lond). 1983;334:19–31.
77. Dudek SM, Bear MF. Homosynaptic long-term depression in area CA1 of hippocampus and effects of N-methyl-D-aspartate receptor blockade. Proc Natl Acad Sci U S A. 1992;89(10):4363–7.
78. Davis S, Butcher SP, Morris RG. The NMDA receptor antagonist D-2-amino-5-phosphonopentanoate (D-AP5) impairs spatial learning and LTP in vivo at intracerebral concentrations comparable to those that block LTP in vitro. J Neurosci. 1992;12(1):21–34.
79. Abeliovich A, Chen C, Goda Y, Silva AJ, Stevens CF, Tonegawa S. Modified hippocampal long-term potentiation in PKC gamma-mutant mice. Cell. 1993;75(7):1253–62.
80. Abeliovich A, Paylor R, Chen C, Kim JJ, Wehner JM, Tonegawa S. PKC gamma mutant mice exhibit mild deficits in spatial and contextual learning. Cell. 1993;75(7):1263–71.
81. Grant SG, O'Dell TJ, Karl KA, Stein PL, Soriano P, Kandel ER. Impaired long-term potentiation, spatial learning, and hippocampal development in fyn mutant mice [see comments]. Science. 1992;258(5090):1903–10.

82. Sakimura K, Kutsuwada T, Ito I, Manabe T, Takayama C, Kushiya E, Yagi T, Aizawa S, Inoue Y, Sugiyama H, et al. Reduced hippocampal LTP and spatial learning in mice lacking NMDA receptor epsilon 1 subunit. Nature. 1995;373(6510):151–5.
83. Silva AJ, Stevens CF, Tonegawa S, Wang Y. Deficient hippocampal long-term potentiation in alpha-calcium- calmodulin kinase II mutant mice. Science. 1992;257(5067):201–6.
84. Silva AJ, Wang Y, Paylor R, Wehner JM, Stevens CF, Tonegawa S. Alpha calcium/calmodulin kinase II mutant mice: deficient long-term potentiation and impaired spatial learning. Cold Spring Harb Symp Quant Biol. 1992;57:527–39.
85. Hardingham GE, Chawla S, Cruzalegui FH, Bading H. Control of recruitment and transcription-activating function of CBP determines gene regulation by NMDA receptors and L-type calcium channels. Neuron. 1999;22(4):789–98.
86. Hu SC, Chrivia J, Ghosh A. Regulation of CBP-mediated transcription by neuronal calcium signaling. Neuron. 1999;22(4):799–808.
87. Impey S, Fong AL, Wang Y, Cardinaux JR, Fass DM, Obrietan K, Wayman GA, Storm DR, Soderling TR, Goodman RH. Phosphorylation of CBP mediates transcriptional activation by neural activity and CaM kinase IV. Neuron. 2002;34(2):235–44.
88. Guan Z, Giustetto M, Lomvardas S, Kim JH, Miniaci MC, Schwartz JH, Thanos D, Kandel ER. Integration of long-term-memory-related synaptic plasticity involves bidirectional regulation of gene expression and chromatin structure. Cell. 2002;111(4):483–93.
89. Malenka RC, Nicoll RA. Long-term potentiation—a decade of progress? Science. 1999;285(5435):1870–4.
90. Tsien JZ, Huerta PT, Tonegawa S. The essential role of hippocampal CA1 NMDA receptor-dependent synaptic plasticity in spatial memory. Cell. 1996;87(7):1327–38.
91. Morris RG, Anderson E, Lynch GS, Baudry M. Selective impairment of learning and blockade of long-term potentiation by an N-methyl-D-aspartate receptor antagonist, AP5. Nature. 1986;319(6056):774–6.
92. Alarcon JM, Malleret G, Touzani K, Vronskaya S, Ishii S, Kandel ER, Barco A. Chromatin acetylation, memory, and LTP are impaired in CBP+/− mice: a model for the cognitive deficit in Rubinstein-Taybi syndrome and its amelioration. Neuron. 2004;42(6):947–59.
93. Wood MA, Kaplan MP, Park A, Blanchard EJ, Oliveira AM, Lombardi TL, Abel T. Transgenic mice expressing a truncated form of CREB-binding protein (CBP) exhibit deficits in hippocampal synaptic plasticity and memory storage. Learn Mem. 2005;12(2):111–9.
94. Wood MA, Attner MA, Oliveira AM, Brindle PK, Abel T. A transcription factor-binding domain of the coactivator CBP is essential for long-term memory and the expression of specific target genes. Learn Mem. 2006;13(5):609–17.
95. Barrett RM, Malvaez M, Kramar E, Matheos DP, Arrizon A, Cabrera SM, Lynch G, Greene RW, Wood MA. Hippocampal focal knockout of CBP affects specific histone modifications, long-term potentiation, and long-term memory. Neuropsychopharmacology. 2011;36(8):1545–56.
96. Valor LM, Pulopulos MM, Jimenez-Minchan M, Olivares R, Lutz B, Barco A. Ablation of CBP in forebrain principal neurons causes modest memory and transcriptional defects and a dramatic reduction of histone acetylation but does not affect cell viability. J Neurosci. 2011;31(5):1652–63.
97. Levenson JM, Sweatt JD. Epigenetic mechanisms: a common theme in vertebrate and invertebrate memory formation. Cell Mol Life Sci. 2006;63(9):1009–16.
98. Vieira PA, Korzus E. CBP-dependent memory consolidation in the prefrontal cortex supports object-location learning. Hippocampus. 2015;25(12):1532–40.
99. Vieira PA, Lovelace JW, Corches A, Rashid AJ, Josselyn SA, Korzus E. Prefrontal consolidation supports the attainment of fear memory accuracy. Learn Mem. 2014;21(8):394–405.
100. Lopez-Atalaya JP, Ciccarelli A, Viosca J, Valor LM, Jimenez-Minchan M, Canals S, Giustetto M, Barco A. CBP is required for environmental enrichment-induced neurogenesis and cognitive enhancement. EMBO J. 2011;30(20):4287–98.
101. Kasper LH, Boussouar F, Ney PA, Jackson CW, Rehg J, van Deursen JM, Brindle PK. A transcription-factor-binding surface of coactivator p300 is required for haematopoiesis. Nature. 2002;419(6908):738–43.

102. Vecsey CG, Hawk JD, Lattal KM, Stein JM, Fabian SA, Attner MA, Cabrera SM, McDonough CB, Brindle PK, Abel T, Wood MA. Histone deacetylase inhibitors enhance memory and synaptic plasticity via CREB:CBP-dependent transcriptional activation. J Neurosci. 2007;27(23):6128–40.
103. Maddox SA, Watts CS, Schafe GE. p300/CBP histone acetyltransferase activity is required for newly acquired and reactivated fear memories in the lateral amygdala. Learn Mem. 2013;20(2):109–19.
104. Stafford JM, Raybuck JD, Ryabinin AE, Lattal KM. Increasing histone acetylation in the hippocampus-infralimbic network enhances fear extinction. Biol Psychiatry. 2012;72(1):25–33.
105. Bredy TW, Wu H, Crego C, Zellhoefer J, Sun YE, Barad M. Histone modifications around individual BDNF gene promoters in prefrontal cortex are associated with extinction of conditioned fear. Learn Mem. 2007;14(4):268–76.
106. Lattal KM, Barrett RM, Wood MA. Systemic or intrahippocampal delivery of histone deacetylase inhibitors facilitates fear extinction. Behav Neurosci. 2007;121(5):1125–31.
107. Aimone JB, Deng W, Gage FH. Resolving new memories: a critical look at the dentate gyrus, adult neurogenesis, and pattern separation. Neuron. 2011;70(4):589–96.
108. Sahay A, Wilson DA, Hen R. Pattern separation: a common function for new neurons in hippocampus and olfactory bulb. Neuron. 2011;70(4):582–8.
109. Levenson JM, O'Riordan KJ, Brown KD, Trinh MA, Molfese DL, Sweatt JD. Regulation of histone acetylation during memory formation in the hippocampus. J Biol Chem. 2004;279(39):40545–59.
110. Bading H. Transcription-dependent neuronal plasticity the nuclear calcium hypothesis. Eur J Biochem. 2000;267(17):5280–3.
111. West AE, Chen WG, Dalva MB, Dolmetsch RE, Kornhauser JM, Shaywitz AJ, Takasu MA, Tao X, Greenberg ME. Calcium regulation of neuronal gene expression. Proc Natl Acad Sci U S A. 2001;98(20):11024–31.
112. Hardingham GE, Chawla S, Johnson CM, Bading H. Distinct functions of nuclear and cytoplasmic calcium in the control of gene expression. Nature. 1997;385(6613):260–5.
113. Deisseroth K, Heist EK, Tsien RW. Translocation of calmodulin to the nucleus supports CREB phosphorylation in hippocampal neurons. Nature. 1998;392(6672):198–202.
114. Hardingham GE, Arnold FJ, Bading H. Nuclear calcium signaling controls CREB-mediated gene expression triggered by synaptic activity. Nat Neurosci. 2001;4(3):261–7.
115. Chetkovich DM, Gray R, Johnston D, Sweatt JD. N-methyl-D-aspartate receptor activation increases cAMP levels and voltage-gated Ca2+ channel activity in area CA1 of hippocampus. Proc Natl Acad Sci U S A. 1991;88(15):6467–71.
116. Dolmetsch RE, Pajvani U, Fife K, Spotts JM, Greenberg ME. Signaling to the nucleus by an L-type calcium channel-calmodulin complex through the MAP kinase pathway. Science. 2001;294(5541):333–9.
117. Limback-Stokin K, Korzus E, Nagaoka-Yasuda R, Mayford M. Nuclear calcium/calmodulin regulates memory consolidation. J Neurosci. 2004;24(48):10858–67.
118. Chawla S, Hardingham GE, Quinn DR, Bading H. CBP: a signal-regulated transcriptional coactivator controlled by nuclear calcium and CaM kinase IV. Science. 1998;281(5382):1505–9.
119. O'Carroll CM, Morris RG. Heterosynaptic co-activation of glutamatergic and dopaminergic afferents is required to induce persistent long-term potentiation. Neuropharmacology. 2004;47(3):324–32.
120. Egger G, Liang G, Aparicio A, Jones PA. Epigenetics in human disease and prospects for epigenetic therapy. Nature. 2004;429(6990):457–63.
121. Tsankova NM, Berton O, Renthal W, Kumar A, Neve RL, Nestler EJ. Sustained hippocampal chromatin regulation in a mouse model of depression and antidepressant action. Nat Neurosci. 2006;9(4):519–25.
122. Frey U, Morris RG. Synaptic tagging and long-term potentiation [see comments]. Nature. 1997;385(6616):533–6.
123. Cosgrove MS, Wolberger C. How does the histone code work? Biochem Cell Biol. 2005;83(4):468–76.

124. Luger K. Dynamic nucleosomes. Chromosome Res. 2006;14(1):5–16.
125. Day JJ, Sweatt JD. Epigenetic treatments for cognitive impairments. Neuropsychopharmacology. 2012;37(1):247–60.
126. Lopez-Atalaya JP, Gervasini C, Mottadelli F, Spena S, Piccione M, Scarano G, Selicorni A, Barco A, Larizza L. Histone acetylation deficits in lymphoblastoid cell lines from patients with Rubinstein-Taybi syndrome. J Med Genet. 2012;49(1):66–74.
127. Bredy TW, Barad M. The histone deacetylase inhibitor valproic acid enhances acquisition, extinction, and reconsolidation of conditioned fear. Learn Mem. 2008;15(1):39–45.
128. Roozendaal B, Hernandez A, Cabrera SM, Hagewoud R, Malvaez M, Stefanko DP, Haettig J, Wood MA. Membrane-associated glucocorticoid activity is necessary for modulation of long-term memory via chromatin modification. J Neurosci. 2010;30(14):5037–46.
129. McQuown SC, Barrett RM, Matheos DP, Post RJ, Rogge GA, Alenghat T, Mullican SE, Jones S, Rusche JR, Lazar MA, Wood MA. HDAC3 is a critical negative regulator of long-term memory formation. J Neurosci. 2011;31(2):764–74.
130. Berdasco M, Esteller M. Genetic syndromes caused by mutations in epigenetic genes. Hum Genet. 2013;132(4):359–83.
131. Duraisingh MT, Horn D. Epigenetic regulation of virulence gene expression in parasitic protozoa. Cell Host Microbe. 2016;19(5):629–40.
132. Abel T, Zukin RS. Epigenetic targets of HDAC inhibition in neurodegenerative and psychiatric disorders. Curr Opin Pharmacol. 2008;8(1):57–64.
133. Graff J, Mansuy IM. Epigenetic dysregulation in cognitive disorders. Eur J Neurosci. 2009; 30(1):1–8.
134. Rudenko A, Tsai LH. Epigenetic modifications in the nervous system and their impact upon cognitive impairments. Neuropharmacology. 2014;80:70–82.
135. Schorry EK, Keddache M, Lanphear N, Rubinstein JH, Srodulski S, Fletcher D, Blough-Pfau RI, Grabowski GA. Genotype-phenotype correlations in Rubinstein-Taybi syndrome. Am J Med Genet A. 2008;146A(19):2512–9.
136. Rusconi D, Negri G, Colapietro P, Picinelli C, Milani D, Spena S, Magnani C, Silengo MC, Sorasio L, Curtisova V, Cavaliere ML, Prontera P, Stangoni G, Ferrero GB, Biamino E, Fischetto R, Piccione M, Gasparini P, Salviati L, Selicorni A, Finelli P, Larizza L, Gervasini C. Characterization of 14 novel deletions underlying Rubinstein-Taybi syndrome: an update of the CREBBP deletion repertoire. Hum Genet. 2015;134(6):613–26.
137. Wincent J, Luthman A, van Belzen M, van der Lans C, Albert J, Nordgren A, Anderlid BM. CREBBP and EP300 mutational spectrum and clinical presentations in a cohort of Swedish patients with Rubinstein-Taybi syndrome. Mol Genet Genomic Med. 2016;4(1): 39–45.
138. Shanmugam MK, Sethi G. Role of epigenetics in inflammation-associated diseases. Subcell Biochem. 2013;61:627–57.

Epigenetics of Autism Spectrum Disorder

4

Michelle T. Siu and Rosanna Weksberg

Abstract

Autism spectrum disorder (ASD), one of the most common childhood neurodevelopmental disorders (NDDs), is diagnosed in 1 of every 68 children. ASD is incredibly heterogeneous both clinically and aetiologically. The etiopathogenesis of ASD is known to be complex, including genetic, environmental and epigenetic factors. Normal epigenetic marks modifiable by both genetics and environmental exposures can result in epigenetic alterations that disrupt the regulation of gene expression, negatively impacting biological pathways important for brain development. In this chapter we aim to summarize some of the important literature that supports a role for epigenetics in the underlying molecular mechanism of ASD. We provide evidence from work in genetics, from environmental exposures and finally from more recent studies aimed at directly determining ASD-specific epigenetic patterns, focusing mainly on DNA methylation (DNAm). Finally, we briefly discuss some of the implications of current research on potential epigenetic targets for therapeutics and novel avenues for future work.

M.T. Siu, Ph.D.
Program in Genetics and Genome Biology, The Hospital for Sick Children,
555 University Ave, Toronto, ON M5G 1X8, Canada
e-mail: michelle.siu@sickkids.ca

R. Weksberg, M.D., Ph.D. (✉)
Program in Genetics and Genome Biology, The Hospital for Sick Children,
555 University Ave, Toronto, ON M5G 1X8, Canada

Division of Clinical and Metabolic Genetics, The Hospital for Sick Children,
555 University Ave, Toronto, ON M5G 1X8, Canada

Department of Paediatrics, University of Toronto, Toronto, ON M5S 1A1, Canada

Institute of Medical Science, University of Toronto, Toronto, ON M5S 1A8, Canada
e-mail: rweksb@sickkids.ca

R. Delgado-Morales (ed.), *Neuroepigenomics in Aging and Disease*, Advances in Experimental Medicine and Biology, DOI 10.1007/978-3-319-53889-1_4

Keywords
Autism spectrum disorder • Heterogeneity • Aetiology • Molecular mechanisms • Genetics • Epigenetics • DNAmethylation

Abbreviations

5-hmC	5-Hydroxymethylcytosine
5-mC	5-Methylcytosine
$\Delta\beta$	Difference in DNA methylation
ADHD	Attention deficit and hyperactivity disorder
ARID1B	AT-rich interaction domain 1B
ART	Assisted reproductive technologies
AS	Angelman syndrome
ASD	Autism spectrum disorder
ATP	Adenosine triphosphate
BA10	Brodmann area 10
BA19	Brodmann area 19
BA24	Brodmann area 24
BBB	Blood–brain barrier
BCL-2	BCL2, apoptosis regulator
BPA	Bisphenol A
BWS	Beckwith-Wiedemann syndrome
C11orf21	Chromosome 11 open reading frame 21
C1Q	Complement subcomponent C1q
CBL	Cerebellum
CDKL5	Cyclin-dependent kinase-like 5
CHARGE	Coloboma of the eye, heart defects, atresia of the nasal choanae, retardation of growth and/or development, genital and/or urinary abnormalities and ear abnormalities/deafness
CHD7	Chromodomain helicase DNA-binding protein 7
CHD8	Chromodomain helicase DNA-binding protein 8
CNV	Copy number variant
DGCR6	DiGeorge critical region 6
DGCR8	DiGeorge critical region 8
DLGAP1	DLG associated protein 1
DLGAP2	DLG associated protein 2
DNAm	DNA methylation
DNMT	DNA methyl transferases
FDR	False discovery rate
FMR1	Fragile X mental retardation 1
GAD65	Glutamic acid decarboxylase 65
GDM	Gestational diabetes mellitus
GOM	Gain of methylation

GRIN2B	Glutamate ionotropic receptor NMDA type subunit 2B
GxE	Gene by environment interactions
H3K27ac	Histone 3 lysine 27 acetylation
H3K4me3	Histone 3 lysine 4 trimethylation
HDAC	Histone deacetylase
HUWE1	HECT, UBA and WWE domain-containing 1, E3 ubiquitin protein
IC	Imprinting centre
ID	Intellectual disability
KDM6A	Lysine demethylase 6A
KMT2D	Lysine methyltransferase 2D
lncRNA	Long noncoding RNA
LOF	Loss of function
LOM	Loss of methylation
MBDs	Methyl-CpG-binding proteins
MECP2	Methyl-CpG-binding protein 2
miRNA	Micro-RNA
MZ	Monozygotic
NDD	Neurodevelopmental disorder
NRXN1	Neurexin 1
NSD1	Nuclear receptor SET (su(var)3–9, enhancer-of-zeste, trithorax) domain-containing protein-1 gene
OCM	One carbon metabolism
OR	Odds ratio
OXTR	Oxytocin receptor
PDD-NOS	Pervasive developmental disorder – not otherwise specified
PFC	Prefrontal cortex
PRRT1	Proline-rich transmembrane protein 1
PTEN	Phosphatase and tensin homolog
RELN	Reelin
RORA	Retinoic acid-related orphan receptor
RTT	Rett syndrome
SAM	*S*-Adenosyl methionine
SHANK3	SH3 and multiple ankyrin repeat domains 3
SNP	Single-nucleotide polymorphism
SNRPN	Small nuclear ribonucleoprotein polypeptide N
TC	Temporal cortex
TNF-α	Tumour necrosis factor alpha
TSPAN32	Tetraspanin 32
UBE3A	Ubiquitin protein ligase
UBE3A	Ubiquitin protein ligase E3A
UPD	Uniparental disomy
VPA	Valproic acid
VUS	Variant of unknown significance
WGS	Whole genome sequencing
ZFP57	ZFP57 zinc finger protein

4.1 Autism Spectrum Disorder (ASD) and Proposed Aetiologies

Autism spectrum disorder (ASD), one of the most common neurodevelopmental disorders (NDDs), is diagnosed in 1 of 68 children in the United States [1] with a 4:1 male-to-female sex ratio. ASD is comprised of a group of complex NDDs characterized by impaired social communication and repetitive behaviours (DSM-5). ASD also presents with a range of other features including morphological (e.g. macrocephaly), physiological (e.g. gastrointestinal, sleep problems) and psychiatric comorbidities (e.g. anxiety) [2]. Variable neuropathological features consistently described in some, but not all, cases of ASD include decreased size and number of Purkinje cells, abnormal neuronal migration, neurite outgrowth and branching and axonal guidance [3, 4]. These additional features may or may not be observed as part of the profile of syndromic cases (Sect. 4.4.1), where ASD is associated with either single-gene mutations or defined chromosomal/cytogenetic abnormalities. In May 2013, the clinical criteria for the diagnosis of ASD were redefined by the DSM-5 diagnostic manual. Several subtypes of ASD previously considered as distinct disorders (autistic disorder, childhood disintegrative disorder, pervasive developmental disorder—not otherwise specified (PDD-NOS) and Asperger syndrome) were then merged under a single umbrella diagnosis, ASD. The clinical criteria for the diagnosis of ASD are still evolving due in part to phenotypic variability and to clinical and aetiologic overlap with other NDDs (e.g. ~30% of ASD cases are comorbid with attention deficit and hyperactivity disorder [ADHD] symptoms) [5, 6]. Therefore, different NDDs are best represented not as distinct categories but as entities along a continuum with some convergence in the underlying genes and pathways. One of the greatest challenges in improving diagnosis and treatment for ASD is the degree of heterogeneity both clinically and aetiologically. It is therefore unsurprising that there are multiple proposed aetiologies and risk factors (Fig. 4.1) identified.

Commonly proposed physiological and metabolic causes of ASD consist of immune, oxidative stress and mitochondrial dysfunction. In some ASD cases, a pro-inflammatory state [8–12] is suggested by alterations in immune and inflammatory markers (e.g. cells of the innate and adaptive [cytokines, interleukins] immune system and abnormalities in microglia [immune cells of the brain]). Furthermore, it has been shown that immune changes in both peripheral tissues and brain can result in anxiety and impaired social behaviour [13–16]. Studies in humans and mouse models suggest that dysbiosis of the intestinal microbiome is a novel physiological contributor to ASD risk, impacting immune function and subsequently learning, memory and behaviour [17–20]. Further, there is emerging evidence, mostly in animal models, that this gut-brain axis is, in part, regulated by epigenetic mechanisms [21–23]. Enhanced oxidative stress, impaired antioxidative capacity and mitochondrial dysfunction have also been extensively reviewed [12, 24]. A number of independent studies support claims that the degree of immune disruption (e.g. cytokines) and mitochondrial (e.g. phosphocreatine) dysfunction are positively correlated with ASD symptom severity [12, 24, 25]. However, it is not known whether these abnormalities existed prior to ASD diagnosis or if they play a causal role.

Investigations of underlying molecular mechanisms in ASD aetiology include altered genetic and epigenetic regulation. Epigenetics has emerged as a vital

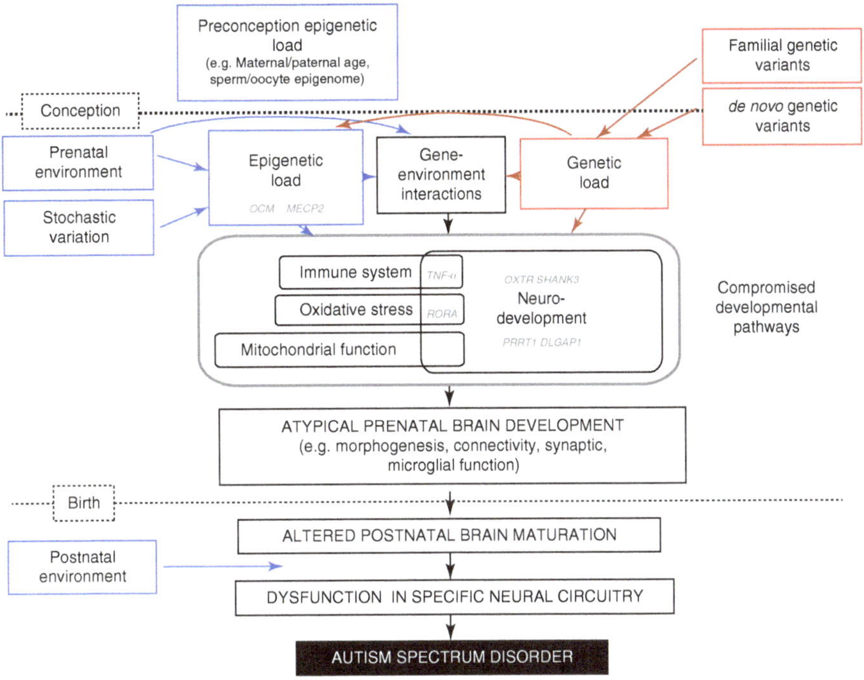

Fig. 4.1 Diagrammatic overview of how genetic, environmental and epigenetic factors interact in the aetiology of ASD. Epigenetic load (from preconception, prenatal environment and stochastic variation) and genetic load (from familial and *de novo* variation) interact to alter neurodevelopmental, immune, oxidative stress and mitochondrial pathways identified through studies of ASD genetics, physiology, expression and/or DNA methylation (targeted and genome-wide). Highlighted in *grey* is the putative involvement of specific genes or pathways mentioned in this review. Over a certain threshold of genetic and epigenetic dysfunction, development is compromised and can lead to adverse neurodevelopmental outcomes such as ASD. Further, postnatal environments may also contribute to severity of symptoms. Abbreviations: *DLGAP1* DLG-associated protein 1, *OCM* one carbon metabolism, *OXTR* oxytocin receptor, *MECP2* methyl-CpG-binding protein 2, *PRRT1* proline-rich transmembrane protein 1, *RORA* retinoic acid receptor-related orphan receptor alpha, *SHANK3* SH3 and multiple ankyrin repeat domains 3, *TNF-α* tumour necrosis factor alpha. Adapted from Fig. 1 in [7]

genome-wide regulatory layer that modulates the transcriptome, impacting transcription initiation, splicing processes and binding of transcription factors. Epigenetic regulation helps to determine the proper spatiotemporal expression of genes via a number of mechanisms including DNA methylation (DNAm), histone modifications and ATP-dependent chromatin remodelling. Epigenetics provides new avenues to investigate and refine risk estimates for NDDs beyond genetic risks alone. The failure to establish proper epigenetic marks may result in aberrant gene expression and, subsequently, various disease phenotypes. Genomic aberrations (e.g. mutations, insertions/deletions, copy number variants [CNVs]) of genes involved in epigenetic regulation ('epigenes') or dysregulation of epigenetically regulated genomic regions (e.g. imprinted genes/regions) can lead to epigenetic disruptions and, ultimately, NDDs. There are >600 confirmed and putative human epigenes [26], many of which are associated with NDDs such as intellectual disability (ID) and ASD (e.g. chromodomain helicase DNA-binding protein 8 [*CHD8*], DNA methyltransferase 3A [*DNMT3A*], HECT, UBA and WWE domain-containing 1, E3 ubiquitin protein [*HUWE1*]) [27, 28]. There is a growing body of evidence to show that there is substantial genetic overlap between risks for neuropsychiatric disorders and NDDs [29–32]. Many of these genes encode proteins involved in neuronal and synaptic pathways, while others are relevant to molecular pathways involved in epigenetic regulation [28, 32, 33] (Sects. 4.2, 4.3 and 4.4).

What is clear is that there is no single underlying cause of ASD. In order to understand the multifactorial aetiologies of ASD, we must better understand the natural history of molecular events and their regulation during critical periods of human development. ASD more likely arises as an interaction between genetic and environmental risk factors (GxE) mediated by epigenetic mechanisms. Ultimately, a better understanding of how genetics, epigenetics and environment collectively interact and contribute to ASD risk will allow us to better classify and diagnose the disorder and facilitate the application of precision-based medicine.

4.2 Genetics of ASD

The aetiology of ASD is known to have a strong genetic component. Early twin studies of heritability reported estimates of up to 90% heritability [34, 35]. In contrast, current estimates are closer to 10–30% according to data from more recent twin and family studies [36–38]. Next generation sequencing has significantly accelerated our understanding of genetic variability in individuals with ASD compared to the general population. ASD genomic risk variants are comprised of rare, *de novo* variants of large effect, independently of or in combination with more common and/or inherited variants of small effect [39]. Interestingly, it has recently been shown, using whole genome sequencing (WGS) of 85 ASD quartet families (parents and two affected siblings), that although ASD-relevant mutations were found in 42% of individuals with ASD, only 31% of sibling pairs carried the same variants, emphasizing the genetic heterogeneity of the disorder even within families [40]. The considerable variability of ASD necessitates genetic testing of large cohorts of patients on whole genome technologies (whole exome sequencing and WGS) which are becoming more affordable.

Despite the advantages of WGS, sequence variant classification (e.g. variants of unknown significance [VUS]) still poses a significant challenge. Such VUS are being reported at a faster rate than we are able to characterize them with respect to disease relevance. These studies underscore the aforementioned heterogeneity of the disorder; >200 ASD-risk genes have been identified [41–46]; SFARI gene: https://gene.sfari.org]. However, genomic aberrations are detected in only 25–40% of cases [40, 47, 48]. Further, no single mutation or CNV accounts for >1% of ASD cases and is variably penetrant with respect to the ASD phenotype. In addition to risk variants, six risk loci (1q21.1, 3q29, 7q11.23, 16p11.2, 15q11.2-13, 22q11.2) and several genetic syndromes (Sect. 4.4.1) are well known to be associated with ASD [47, 49, 50]. Unsurprisingly, few strong genotype-phenotype relationships have yet to be uncovered. Many genomic variants are recurrent, but rare (e.g. *CHD8* mutations; Sect. 4.4.1), and therefore more patients are required to better establish such relationships.

Several studies have shown that many ASD-risk genes are involved in converging pathways relevant to the biological bases of ASD. These include development and cell proliferation, neural development, synaptic function and, of particular interest to this chapter, chromatin modifiers and transcriptional regulators [48, 49, 51]. Importantly, many of these genes are expressed in the brain during embryonic development. These functional categories relate to the neurocognitive phenotype of ASD (and other NDDs) and can help us to understand the molecular mechanisms, such as epigenetics, that are perturbed in ASD. The study of ASD-associated genetic syndromes caused by mutations in epigenes will also aid in this endeavour (Sect. 4.4.1). It is becoming increasingly apparent that some ASD-risk genes and loci confer an increased risk for other neuropsychiatric and neurological disorders including ID, ADHD, schizophrenia, epilepsy, motor impairment and sleep disturbance [30, 47, 52].

The genomic architecture of ASD is further elucidated through examinations of large numbers of families and individuals with ASD [40, 43–46, 49]. Better genotype-phenotype relationships are being defined; for example, the presence of *de novo* loss-of-function (LOF) mutations or CNVs is associated with lower IQ [43, 49, 53, 54]. Higher mutational burden has also been correlated with certain ASD features such as seizures and head circumference, observed in subsets of individuals [49, 55]. Genomic studies of ASD have identified genomic features relevant to ASD aetiology [49]. For example, *de novo* variants are distributed in a non-random fashion, enriched in epigenetically relevant regions (e.g. simple repeats and DNase I hypersensitivity sites, marks of open chromatin) [56].

Epigenetic mechanisms can help fill the aetiologic knowledge gap where genetic information alone is insufficient to explain the aetiology of all ASD cases. Epigenetic outcomes, much like genetic outcomes, are also expected to be heterogeneous (Sect. 4.4.2). Combining genetic and epigenetic data is likely to provide a more comprehensive understanding of the molecular landscape of the aetiology of ASD. The discovery of consistent molecular (genotype, epigenotype) and biochemical associations with ASD or ASD subtype-specific phenotypes will allow clinicians to better classify individuals, facilitate earlier diagnosis, and improve prognosis. In parallel, gaining molecular insights into the disorder will also help us to identify more homogeneous subgroups of individuals, which will allow for better patient stratification for behavioural and pharmacological treatments.

4.3 Environmental Exposures and ASD Risk

Twin, adoption and sibling studies have defined the heterogeneous and complex etiopathogenic nature of ASD and have supported potential contributions of environmental factors to ASD risk. Results from such studies suggest that ASD aetiology can be attributed to ~50% genetic contribution and ~50% influenced by non-shared environmental factors [37, 57–60]. The epigenome acts as an interface between the genome and the environment, transforming the genome into a regulator of cell type and developmental time-specific transcription. The epigenome is programmed during embryonic/foetal development by multiple genetic factors including genes that encode DNA methyl transferases (DNMTs), histone deacetylases (HDACs) and chromatin remodelling factors. Epigenetic errors can arise as primary stochastic events or in response to genetic mutations and/or environmental exposures. At critical times during development, typical foetal programming can be dysregulated by gene mutations, environmental exposures or epigenetic errors potentially leading to adverse long-term health outcomes [61–63]. There have been extensive investigations during the critical period of maternal gestation to examine the effects of exposures to both exogenous and endogenous environmental factors, which will be summarized below.

4.3.1 Exogenous Environment

Smoking, alcohol, medications (e.g. valproic acid [VPA], selective serotonin reuptake inhibitors) and environmental chemicals (e.g. pesticides, metals, bisphenol A [BPA]) are the most commonly studied exogenous exposures in relation to adverse foetal neurodevelopmental outcomes. Gestational smoking and alcohol exposure studies are inconsistent with respect to ASD risk [64–67], likely due to differences in study cohorts and methodologies, but also because it is extremely difficult to accurately estimate levels and timing of exposure in the mothers and more critically in the foetus. These challenges are compounded with the fact that the mechanisms of action and effects of maternal vs. foetal metabolism are not well understood enough to directly infer causation. In general, the impact of complex GxE interactions requires further investigation. Integrating genetic, environmental and epigenetic datasets will enable us to better understand the synergistic effects of these interactions. Animal models are critical for such studies, allowing for more precise and quantitative manipulations of environmental exposures. Further, we need to be able to distinguish between the direct effects of the exposure itself (e.g. direct perturbation of neurodevelopmentally important genes by gestational exposure to maternal smoking) and the downstream effects that may result (e.g. the fact that maternal smoking has been linked to decreased birth weight and reduced in utero brain growth). Interestingly, several studies demonstrate that environmental exposures affect epigenetic marks. Altered DNAm or DNMT expression/activity has been reported in various tissues of both human and animal models following exposure to a variety of toxicants including alcohol, cigarette smoking, BPA and VPA [68–76]. Some of these DNAm alterations have been further associated with ASD-relevant endpoints (e.g. behavioural outcomes, neurite outgrowth, axon formation and in ASD-relevant brain regions) (Table 2 in [65]).

4.3.2 Endogenous Environment

Preconception environmental risks include maternal [37, 77] and paternal [77, 78] age, which have been positively associated with an increase in ASD risk (relative risk [RR] of 1.16 to >1.5), both independently and with a joint effect. Interestingly, genetics may partially explain this finding; a greater number of *de novo* mutations in ASD probands have been found as a function of paternal age [43, 46]. Age-related epigenetic changes such as altered DNAm observed in both sperm and oocytes [79–81] may also contribute to this association. The mechanisms by which these factors introduce enhanced ASD risk still need to be further explored.

Epidemiologic studies show that preterm birth, due to various causes, significantly increases (3–14-fold) the risk of developing ASD [82–84]. A recent study [85] tested for DNAm differences between preterm and term foetal placental tissues at preselected ASD candidate genes. These include *OXTR*, *SHANK3*, BCL2, apoptosis regulator (*BCL-2*) and *RORA* that are known to have altered DNAm in some ASD cases. A significant gain of methylation (GOM) was found only in *OXTR*. More studies are needed to understand whether this DNAm mark at *OXTR* has a functional impact on ASD risk.

There have been inconsistent reports regarding the risk of ASD following the use of assisted reproductive technologies (ART; e.g. *in vitro* fertilization, intracytoplasmic sperm injection) [86–88]. ART currently account for ~1.6% of live births in the United States and rates of use are increasing [89, 90]. Although results are controversial, epidemiological studies have shown a possible increase in the incidence of ASD in offspring conceived with ART [86, 87]. There are several reasons why ART should be carefully considered. First, there are inherent risks of ART including preterm labour, multiple births and low birth weight that, independently of ART, already confer an increased risk for ASD [91–93]. Second, ART are used during critical windows of gametogenesis and early embryogenesis, when epigenetic reprogramming is occurring [92, 94]. Previous studies in the aetiologically heterogeneous paediatric imprinting disorders (Beckwith-Wiedemann [BWS] and Angelman syndromes [AS]) demonstrate that ART have a significant impact on epigenetic outcomes. These disorders are caused by loss of methylation (LOM) at critical imprinted sites on chromosomes 11p15 and 15q11 in individuals with BWS or AS, respectively. Most molecular alterations identified in these subjects following ART arose from epigenetic rather than genetic alterations; LOM is increased in frequency following the use of ART [95–100]. Currently, it is unclear if a parallel effect of ART occurs in the context of the development of ASD.

The endogenous maternal gestational environment has been studied extensively with respect to ASD risk [101–103]. Modest increases in ASD risk have been found to be associated with certain perinatal complications. Two of the main hypotheses relate back to two proposed underlying aetiologies of ASD (Sect. 4.2)—oxidative stress and immune function. Epidemiological data have shown that hypoxia-related obstetric complications pose a significant increase in ASD risk (effect estimate >1.4) [91, 101]. The maternal immune system is also central to several associations, although they are yet to be well established or replicated. Infection during pregnancy [82, 104–106] (OR: 1.24–1.37) and autoimmune disorders [107] (OR: 1.34) are linked to an increased risk for ASD. Recent studies report the maternal production of antibodies against circulating foetal brain proteins, detected in ~20% of mothers of children with ASD, but only in 1% of mothers of neurotypical children [108, 109]. Replication and further studies are required to solidify the potential role of these antibodies as biomarkers and their functional effect on foetal outcome. Maternal metabolic factors such as obesity and gestational diabetes mellitus (GDM) have both been associated with an increased risk for ASD [110–113]. There are convincing data demonstrating a role for epigenetic regulation of these metabolic processes. One study of women of South Asian origin, who have a high risk of GDM, showed statistically significant DNAm differences in cord blood and placental tissue, identifying differentially methylated genes involved in embryonic development and intracellular metabolic processes [114]. Other studies have reported DNAm changes in general immune, metabolic and endocrine pathways in placenta and cord blood of infants exposed to GDM [115–117]. There are no existing data available yet to validate these individual studies in cohorts of ASD patients exposed to GDM or maternal obesity.

Prenatal maternal stress has been identified as a small but robust risk factor for ADHD and ASD [118]. Maternal stress has been correlated with offspring autistic traits [119]. General and social communication scores were associated with altered DNAm of *OXTR*, a recurring ASD-risk gene of interest (Sect. 4.4.2). Further, specific single-nucleotide polymorphism (SNP) genotypes of *OXTR* were found to be predictive of methylation outcomes. However, this study was unable to find a GxE interaction, where associations between maternal stress and autistic traits were not related to *OXTR* methylation or genotype. Maternal prenatal nutrition, specifically folate supplementation, has been shown to have a protective effect, showing a reduced risk of ASD (OR: 0.61) [120–122]. Folate is well known for its role in preventing neural tube defects. Its protective effect for ASD may derive from the role of folate in OCM. OCM recycles homocysteine to generate cysteine and methionine for the process of methylation and antioxidative capacity through the formation of *S*-adenosyl methionine (SAM), supporting epigenetic processes important for typical neurodevelopment.

In summary, it will be important to more thoroughly explore whether the associations between environment and ASD outcome may be reflected in stable epigenetic marks detectable in the foetus or neonate. This would require large sample sizes to achieve sufficient power and well-annotated exposure data. Ultimately, the goal is to discover replicable GxE effects associated with ASD, perhaps in parallel with prenatal genetic testing, that are not only statistically but also biologically significant.

4.4 The Direct Role of Epigenetics in ASD

4.4.1 Genetic Syndromes Involving Epigenes and ASD

Syndromic ASD accounts for ~10–15% of cases [123–125]. A significant number of such genetic syndromes involve mutations in epigenes and are associated with increased risk for ASD (Table 4.1). Some of these epigenes function as

Table 4.1 Genetic syndromes with known/putative epigenetic aetiology comorbid with ASD and/or ID

Syndrome	Aetiology	Epigenetic type	Epigenetic mechanism involved in disorder	Risk for ASD	Degree of ID
Rett	Mutations in *MECP2* and *CDKL5*	Reader	Improper reading and establishment of epigenetic marks by MECP2	>50% Rett females with ASD symptoms [141]. Mutations in *MECP2* found in ASD patients[127–129, 142, 143]	Severe to profound
Fragile X	CGG repeat expansion and subsequent DNA methylation of *FMR1* gene	Indirect; non-coding RNA	Reduced *FMR1* expression due to DNAm; *FMR1* involvement in RNA processing	60–67% in males, 23% in females[123, 144]	Severe to mild, majority in moderate range
22q11.2 deletion (DiGeorge)	1.5–3 Mb hemizygous deletion	Indirect, not well-defined; non-coding RNA	*DGCR8* found in 22q11.2del region involved in miRNA processing; *DGCR6*, also in 22q11.2del region, is imprinted	20–40% (based on DSM-IV criteria) [145, 146], <20% if using both clinical criteria and parental report [147]	Mild to moderate
Prader-Willi	Paternal deletions, maternal UPD at 15q11–13, deletions and mutations of IC, translocations disrupting *SNRPN*	Imprinted region	Lack of expression of paternally expressed genes from imprinted cluster at 15q11–13, due to GOM at paternal IC	19–36.5% [148–150]	Mild to moderate
Angelman	Maternal deletion, paternal UPD, deletions and epimutations at IC, mutations of *UBE3A*	Imprinted gene	Lack of expression of maternally expressed gene *UBE3A* in brain due to LOM at maternal IC	Not conclusive due to severe intellectual disability	Severe to profound
15q11–13 maternal duplication	Maternal duplications of 15q11–13 region	Unknown	Unknown	>85% [151]	Variable

(continued)

Table 4.1 (continued)

Syndrome	Aetiology	Epigenetic type	Epigenetic mechanism involved in disorder	Risk for ASD	Degree of ID
Sotos	Mutations in *NSD1*	Writer; specific histone modifications	*NSD1* encodes histone H3K36 methyltransferase, important for normal embryonic development	No clear estimate for risk, but >80% demonstrate some ASD clinical features [138]	Mild to severe
Kabuki	Mutations in *KMT2D*, *KDM6A*	Writer; specific histone modifications	*KMT2D* encodes histone H3K4 methyltransferase, *KDM6A* encodes a tri-/dimethylated histone H3 demethylase. Interaction with members of WAR complex (WDR5, RBBP5 and ASH2L), shown to be involved in histone methylation	Autism or autistic-like behaviour reported in several cases [152, 153], no risk estimate	Mild to severe
CHARGE	Mutations/ deletions in *CHD7*	Chromatin remodeller	Alters CHD7 binding of active chromatin, also interacts with members of WAR complex	15–50% [133–135]	Normal to severe
CHD8 mutations with ASD	Mutations/ deletions in *CHD8*	Chromatin remodeller	Alters CHD8 binding of active chromatin, regulates transcription through CTCF binding	>85% [131, 132]	Normal to profound
Turner syndrome	Monosomy for chromosome X	Potential imprinted gene(s)	Potential imprinted gene(s) on chromosome X	3% [154]	Usually no ID

Note: Epigenetic marks or mechanisms associated with each syndrome are described; there may be additional known/unknown mechanisms. Abbreviations: *ASH2L* ASH2 like histone lysine methyltransferase complex subunit, *CDKL5* cyclin-dependent kinase-like 5, *CHARGE* coloboma of the eye, heart defects, atresia of the nasal choanae, retardation of growth and/or development, genital and/or urinary abnormalities and ear abnormalities/deafness, *CHD7* chromodomain helicase DNA-binding protein 7, *CHD8* chromodomain helicase DNA-binding protein 8, *CTCF* CCCTC-binding factor, *DGCR6* DiGeorge critical region 6, *DGCR8* DiGeorge critical region 8, *FMR1* fragile X mental retardation 1, *GOM* gain of methylation, *IC* imprinting centre, *ID* intellectual disability, *KDM6A* lysine demethylase 6A, *KMT2D* lysine methyltransferase 2D, *LOM* loss of methylation, *MECP2* methyl-CpG-binding protein 2, *miRNA* micro-RNA, *NSD1* nuclear receptor SET (su(var)3–9, enhancer-of-zeste, trithorax) domain-containing protein-1, *RBBP5* RB binding protein 5, histone lysine methyltransferase complex subunit, *SNRPN* small nuclear ribonucleoprotein polypeptide N, *UBE3A* ubiquitin protein ligase E3A, *UPD* uniparental disomy, *WDR5* WD repeat domain 5

epigenetic writers (DNMTs, histone methyltransferases and acetyltransferases), erasers (HDACs, lysine demethylases), readers (proteins containing bromo-, chromo- or Tudor domains), chromatin remodelling factors (e.g. *CHD8*) and epigenetic regulators of imprinted regions (ZFP57 zinc finger protein). Others exert more indirect effects through OCM, noncoding RNA processing or recruitment of methyl-CpG-binding proteins (MBDs) to modify histones and regulate transcription. Genetic syndromes caused by mutations in epigenes are also highly comorbid with ID, sometimes making it difficult to estimate ASD risk. As data for WGS of larger numbers of well-phenotyped ASD cases become available, additional genetic syndromes involving epigenes may be identified.

Rett syndrome (RTT; OMIM 312750) has been described in detail in Chaps. 1 and 2 and therefore will not be discussed in detail here. ASD symptoms can appear in early infancy, but the clinical phenotype becomes more distinct as RTT features (e.g. loss of hand skills, deceleration of head growth) develop with age. Interestingly, a large proportion of patients (>70%) with milder RTT variants exhibit ASD-like features [126]. Rare *MECP2* mutations associated with ASD but not RTT have also been identified [127–130]. Typically, these mutations are found to be intronic and located in the 3′ untranslated region of the gene as opposed to LOF mutations which lead to RTT. The functional role that *MECP2* may play in ASD pathogenesis has yet to be identified.

Mutations (single-nucleotide variants and small indels) in the chromatin modifier gene *CHD8* have recently been described as a novel genetic syndrome with a strong association (>87%) with an ASD phenotype, amongst other common features such as macrocephaly (>80%), tall stature (86%) and gastrointestinal problems (80%) [131, 132]. Individuals with mutations in a related gene, chromodomain helicase DNA-binding protein 7 (*CHD7*; CHARGE syndrome, OMIM 214800), have a lesser but still significant risk (40%) for ASD [133–135]. The two genes have different interacting proteins and target binding sites [136], explaining at least in part the differences in phenotype.

Sotos syndrome (SS; OMIM 117550) is a congenital overgrowth disorder caused primarily (90%) by mutations in the nuclear receptor SET (su(var)3–9, enhancer-of-zeste, trithorax) domain-containing protein-1 gene (*NSD1*), a developmentally important histone methyltransferase. SS presents an example of a genetic syndrome with robust functionally relevant genome-wide epigenetic alterations [137]. Notably, reports show that >55% of SS patients display ASD symptomatology above clinical cutoffs [138–140]. It has recently been shown that individuals with SS have a specific blood DNAm signature that distinguishes individuals with pathogenic *NDS1* mutations from controls [137]. Further, this DNAm signature is able to classify *NSD1* VUS, which holds great potential for clinical application and molecular diagnostics. Examining a genetically homogeneous group of individuals as an approach for the study of ASD may eliminate some of the resultant epigenetic heterogeneity. The ability to refine and make more consistent molecular and/or phenotypic observations within subsets of individuals with ASD will help to establish causal roles for aetiologic factors.

4.4.2 Direct Assessment of Epigenetic Marks in ASD

Specific genomic alterations (e.g. mutations, CNVs) are known to confer increased risks for ASD, but these risks often are fairly broad ranging. Given the anticipated role of epigenetic dysregulation in ASD aetiology, multiple studies of different epigenetic marks in ASD cases have investigated stable epigenetic biomarkers either with or without an underlying genomic change. Stable biomarkers found in easily accessible, peripheral tissues such as blood would have a profound impact in the clinical diagnostic arena, especially if blood biomarkers were confirmed to reflect biomarkers in the brain. Examining cross-tissue markers, specifically brain vs. peripheral tissues, would help to further elucidate the underlying biological pathways involved in the aetiology of ASD. This section will focus mainly on assessments of the most stable and commonly studied epigenetic mark, DNAm (Table 4.2). We will also review data for epigenetic marks that are less frequently examined that will complement DNAm data in the future.

Table 4.2 Genome-wide DNA methylation studies in ASD

Reference	Sample population (*n*)	Tissue	Method	Findings
[155]	MZ twins discordant for ASD (3, male-male), unaffected siblings (2).	Lymphoblastoid cell lines	8.1 K CpG island microarray	**GOM** in *BCL-2* and *RORA*
[156]	MZ twins discordant and concordant for ASD/ASD severity (50 twin pairs)	Whole blood	Illumina 27 K	No global differences. Multiple DMVs (**GOM and LOM**) identified b/w discordant twin pairs, including *MBD4*
[157]	ASD cases (47), controls (48) born to mothers >35y.o.	Buccal epithelium	Illumina 450 K	Only 1 DMR passing FDR correction (*OR2L13*), **LOM** at promoter
[158]	Biological fathers of existing ASD child, collected in 1st or 2nd trimester of second pregnancy (44)	Sperm	CHARM 3.0 array, Illumina 450 K	**193** DMRs; overlapped sperm CHARM data with 450 K (75/193 probes) and post-mortem brain (18/75) data from [160]
[159]	ASD cases (9), unrelated controls (9)	Post-mortem brain (BA19)	Illumina 27 K	**No** significant DMVs; downregulation of expression of genes of mitochondrial phosphorylation, protein translation

Table 4.2 (continued)

Reference	Sample population (*n*)	Tissue	Method	Findings
[160]	ASD cases (19), unrelated controls (21)	Post-mortem brain (TC, PFC, CBL)	Illumina 450 K	4 significant DMRs (TC, CBL), Δβ range from 6.6% **LOM** to 15.8% **LOM**. 3/4 DMRs validated in multiple brain regions of independent ASD cases
[161]	ASD cases (23), unrelated controls (23)	Post-mortem brain (BA10, BA24)	Illumina 450 K	>5000 DMVs in BA10, >10,000 DMVs in BA24 ($q < 0.05$, $\Delta\beta > 5\%$): **LOM** in BA10 at sites related to immune function (e.g. *C1Q*, *TNF-α*), **GOM** at sites related to synaptic membrane (e.g. *DLGAP1*, *DLGAP2*)

Abbreviations: *Δβ* difference in DNAm, *BA10* Brodmann area 10, *BA19* Brodmann area 19, *BA24* Brodmann area 24, *BCL-2* BCL2, apoptosis regulator, *C1Q* complement C1q A chain, *CBL* cerebellum, *DLGAP1* and *DLGAP2* DLG-associated proteins 1 and 2, *DMR* differentially methylated region, *DMV* differentially methylated variant, *GOM* gain of methylation, *Illumina 27 K* Illumina Infinium HumanMethylation27 BeadChip array, *Illumina 450 K* Illumina Infinium HumanMethylation450 BeadChip array, *LOM* loss of methylation, *MBD4* methyl-binding domain 4, *MZ* monozygotic, *OR2L13* olfactory receptor family 2 subfamily L member 13, *PFC* prefrontal cortex, *RORA* retinoic acid-related orphan receptor, *TC* temporal cortex, *TNF-α* tumour necrosis factor alpha, *y.o.* years old

Although there is a solid rationale supporting a role for epigenetics in ASD molecular aetiology, there are relatively few studies that have directly measured epigenetic marks in ASD patients, especially when compared with the number of genetic studies available. Differentially methylated variants (DMVs) at specific CpG sites or differentially methylated regions (DMRs) spanning multiple CpGs have been measured in a variety of tissue types: lymphoblastoid cell lines [155], whole blood [156], buccal [157], sperm [158] and post-mortem brain [159–161]. The epigenome is characterized by cell-, tissue- and brain region-specific methylation patterns [161–167], making it impossible to directly compare data across these studies. However, as previously mentioned, from a biomarker and pathophysiological standpoint, it will be important to define intersecting ASD-specific DMVs/DMRs and pathways across cell types and tissues.

Genome-wide studies performed primarily using DNAm microarrays have yielded variable results for several reasons: differences in tissue type, ASD cohorts, methods (single site vs. region specific, i.e. DMVs vs. DMRs) and limited sample size (<50 cases) will affect the epigenetic output. These results emphasize the need for increased power in genome-wide DNAm studies focused on discovery of ASD-specific DNAm alterations across heterogeneous ASD groups. Most findings in ASD are reported with overwhelmingly modest effect sizes (<10% absolute difference), and some are statistically unreliable (e.g. without correction for multiple testing). Other possible confounding variables have yet to be addressed for their potential impact on DNAm outcome, including sex, age, post-mortem interval and cause of death (for post-mortem brain samples) and brain cell type-specific DNAm patterns, to name a few. Replication of these results in larger cohorts of ASD patients will strengthen the support for a role for dysregulation of DNAm in ASD neuropathology. Only two studies [160, 161] have shown replication of differentially methylated sites that were hypomethylated in the 3′ untranslated region of *PRRT1*, tetraspanin 32 (*TSPAN32*) and *C11orf21* in brains of ASD patients when compared with controls. For many of the other identified differentially methylated genes across all studies, there is no known function in the context of ASD. Others appear to be functionally relevant, with potential roles in brain electrophysiological function (e.g. *PRRT1*), immunity (e.g. *C1Q*, *TNF*-α) and/or involving known ASD-risk genes (e.g. AT-rich interaction domain 1B [*ARID1B*], glutamate ionotropic receptor NMDA type subunit 2B [*GRIN2B*], neurexin 1 [*NRXN1*], phosphatase and tensin homolog [*PTEN*]).

Several studies have focused on the targeted quantification of DNAm in promoters of ASD candidate genes (glutamic acid decarboxylase 65 [*GAD65*], *OXTR*, *SHANK3*, reelin (*RELN*), *UBE3A* and *MECP2*) [168–173] in different tissues (blood, specific regions in post-mortem brain). No differences between ASD and controls were found for DNAm of *GAD65* or *RELN* [173] nor for one of the *OXTR* studies [168]. The latter result did not agree with an earlier study covering an overlapping region of *OXTR* [169], where significant GOM was found at specific CpG sites overall and in a sex-specific manner. In summary, significant ASD-specific DMVs at *OXTR*, *SHANK3*, *UBE3A* and *MECP2* were identified at specific promoter CpG sites in each gene (i.e. not across all sites analysed) [168–172]. Absolute differences were found to be modest (~twofold) on average and did not affect all ASD cases equally. Overall, the variability in results observed in the genome-wide DNAm studies is also reflected in these targeted studies, for many of the same reasons (tissue type, unselected ASD cases examined, methods, sample size).

Only two studies to date have looked at differences in histone marks between individuals with ASD and neurotypical controls [174, 175]. The two studies are difficult to compare since each study differed in the histone marks examined (H3K4me3 vs. H3K27ac), brain regions (prefrontal cortex vs. prefrontal cortex, temporal cortex and cerebellum), patient cohorts and methodology. However, each study independently found ASD-specific patterns of histone mark methylation to varying degrees. Shulha *et al.* (2012) report that there were no global alterations of H3K4me3, but rather an expansion in the presence of H3K4me3 at specific genomic regions in ASD. Sun *et al.* (2016) describe brain region- and ASD-specific differentially acetylated regions. Differences were found to correspond to functionally relevant genes involved in synaptic transmission, neuronal connectivity, immunity and behaviour. Researchers are also just beginning to look at differential miRNA and long noncoding RNA (lncRNA) expression in ASD [176–179]. Thus far, there is a lack of consistent findings across these studies.

In spite of current limitations, the studies cited above will act as a catalyst for the study of ASD to identify epigenetic biomarkers for prediction or classification of individuals with ASD. Additionally, they have brought to light the many critical variables that need to be considered to improve study design and interpretation of data going forward.

4.5 Therapeutics

Identifying epigenetic targets with therapeutic potential exploits the dynamic and modifiable nature of epigenetic pathways, allowing for new approaches to ameliorate ASD symptoms. Pharmacologic agents could be used to target direct epigenetic regulators such as histone acetylation (HDACs) or to target indirect and/or downstream pathways (e.g. OCM). Some of these targets already have existing therapies/drugs for testing in clinical trials.

There are several significant challenges to this endeavour. Will general inhibition/activation of epigenetic processes be too disruptive of other mechanisms? Conversely, how can we develop more targeted (e.g. tissue and cell type, enzyme isoform-specific) epigenetic drugs? One of the most critical obstacles to overcome for the treatment of ASD neurobehavioural deficits is ensuring that a drug is able to pass the blood–brain barrier (BBB). Currently, HDAC and DNAm inhibitors have poor brain penetrance and potency, although some recent work is showing improvement in this area. Improved delivery systems are being tested with a novel HDAC inhibitor analogue [180] and image-guided (positron emission tomography) radiolabelled drug delivery [181]. However, the BBB issue may be bypassed by harnessing the therapeutic potential of the microbiome to effect downstream neurobehavioural outcomes.

Indirect targets and pathways that affect epigenetic mechanisms may also have pharmacologic potential. The importance of folate in maintaining proper SAM levels and therefore methyl donors (Sect. 4.4) is demonstrated in its apparent protective effect on ASD risk. The neuropeptide hormone oxytocin, which binds to *OXTR*, has been highlighted as a promising pharmacological agent in several clinical trials [182–185] for the treatment of certain neuropsychiatric disorders including ASD. Although results have been mixed, some positive results [186–188] support the use of oxytocin for improving specific deficits seen in ASD such as emotion recognition and eye gaze, as well as for its prosocial and anxiolytic properties.

The exploration of epigenetic therapeutic targets for ASD is in its infancy since researchers are still uncovering the molecular conundrum by which epigenetic mechanisms are perturbed in ASD. However, genomic and epigenomic insights are uncovering potential biological pathways that may be targeted for therapeutics. With more comprehensive classifications of ASD patients, we may identify subgroups of individuals that will be candidates for more precision-based therapies (e.g. pathways susceptible to environmental influences, immune, metabolic, chromatin modifiers, etc.).

4.6 Future Directions/Summary

There is still much to learn about ASD in the context of epigenetics. As technology advances, we may interrogate the genome and epigenome with higher resolution. This will allow researchers to refine DNAm studies with increased genomic coverage, to expand on our knowledge of ASD-specific histone marks and to explore the role of noncoding regions (e.g. enhancers, intergenic regions, noncoding RNAs).

Beyond CpG methylation, non-CpG methylation (CpH where H = A, C, or T) and 5-hydroxymethylcytosine (5-hmC) should not be overlooked. Both of these alternative types of methylation have been found to be important during neurogenesis [163, 189, 190] and thus could play a role in the pathogenesis of ASD. 5-hmC, an intermediate in the process of oxidative demethylation, is highly abundant in the brain relative to 5-methylcytosine (5mC) [191] and could reveal important regulatory brain region-specific epigenetic patterns.

One emerging method of better defining these mechanisms is to tackle the issue of heterogeneity in ASD by examining more homogeneous subsets of ASD patients based on various factors. Presented in this chapter were examples of environmental (preterm labour) and genetic (*NSD1* mutations in SS) stratification, which demonstrate the strength of this approach and are paving the road for the future of research into the aetiologies of ASD. It is clear from the current literature that many roads lead back to epigenetics; from genetics to structural, physiological and biochemical hallmarks of the disorder to environment, epigenetic mechanisms are intimately involved in interfacing with and regulating these aetiologic factors. The complexity of epigenetic mechanisms, its intermediary role bridging multifactorial risk factors through GxE interactions and its malleable nature underscore both the challenges of studying ASD in the context of epigenetics and the exciting potential for this area of research.

Although we were unable to touch upon these areas of research and knowledge in this chapter, there are several important additional topics to consider in the context of ASD aetiology and epigenetics.

Additional Reading
Sex bias: [192–194].
Animal models of ASD: [195–199].
Noncoding RNA and ASD: [200–203].

References

1. Christensen DL, Baio J, Braun KV, Bilder D, Charles J, Constantino JN, et al. Prevalence and Characteristics of Autism Spectrum Disorder Among Children Aged 8 Years — Autism and Developmental Disabilities Monitoring Network, 11 Sites, United States, 2012. MMWR Surveill Summ. 2016;65(No. SS-3)(No. SS-3):1–23.
2. Doshi-Velez F, Ge Y, Kohane I. Comorbidity clusters in autism spectrum disorders: an electronic health record time-series analysis. Pediatrics. 2014;133(1):e54–63.
3. Bailey A, Luthert P, Dean A, Harding B, Janota I, Montgomery M, et al. A clinicopathological study of autism. Brain. 1998;121(Pt 5):889–905.
4. Donovan AP, Basson MA. The neuroanatomy of autism - a developmental perspective. J Anat. 2016;230(1):4–15.
5. Rao PA, Landa RJ. Association between severity of behavioral phenotype and comorbid attention deficit hyperactivity disorder symptoms in children with autism spectrum disorders. Autism. 2014;18(3):272–80.
6. Ronald A, Simonoff E, Kuntsi J, Asherson P, Plomin R. Evidence for overlapping genetic influences on autistic and ADHD behaviours in a community twin sample. J Child Psychol Psychiatry. 2008;49(5):535–42.
7. Loke YJ, Hannan AJ, Craig JM. The Role of Epigenetic Change in Autism Spectrum Disorders. Front Neurol. 2015;6:107.

8. Gesundheit B, Rosenzweig JP, Naor D, Lerer B, Zachor DA, Prochazka V, et al. Immunological and autoimmune considerations of Autism Spectrum Disorders. J Autoimmun. 2013;44:1–7.
9. McDougle CJ, Landino SM, Vahabzadeh A, O'Rourke J, Zurcher NR, Finger BC, et al. Toward an immune-mediated subtype of autism spectrum disorder. Brain Res. 1617;2015:72–92.
10. Nardone S, Elliott E. The interaction between the immune system and epigenetics in the etiology of Autism Spectrum disorders. Front Neurosci. 2016;10:329.
11. Noriega DB, Savelkoul HF. Immune dysregulation in autism spectrum disorder. Eur J Pediatr. 2014;173(1):33–43.
12. Rossignol DA, Frye RE. A review of research trends in physiological abnormalities in autism spectrum disorders: immune dysregulation, inflammation, oxidative stress, mitochondrial dysfunction and environmental toxicant exposures. Mol Psychiatry. 2012;17(4):389–401.
13. Filiano AJ, Gadani SP, Kipnis J. Interactions of innate and adaptive immunity in brain development and function. Brain Res. 1617;2015:18–27.
14. Onore C, Careaga M, Ashwood P. The role of immune dysfunction in the pathophysiology of autism. Brain Behav Immun. 2012;26(3):383–92.
15. Rilett KC, Friedel M, Ellegood J, MacKenzie RN, Lerch JP, Foster JA. Loss of T cells influences sex differences in behavior and brain structure. Brain Behav Immun. 2015;46: 249–60.
16. Sidor MM, Halgren CR, Foster JA. The impact of early life immune challenge in behavior and microglia during postnatal development. Inflamm Cell Signal. 2014;1:51–60.
17. Bilbo SD, Nevison CD, Parker W. A model for the induction of autism in the ecosystem of the human body: the anatomy of a modern pandemic? Microb Ecol Health Dis. 2015;26: 26253.
18. Cao X, Lin P, Jiang P, Li C. Characteristics of the gastrointestinal microbiome in children with autism spectrum disorder: a systematic review. Shanghai Arch Psychiatry. 2013;25(6): 342–53.
19. Stilling RM, Dinan TG, Cryan JF. Microbial genes, brain & behaviour—epigenetic regulation of the gut-brain axis. Genes Brain Behav. 2014;13(1):69–86.
20. Zhang YJ, Li S, Gan RY, Zhou T, Xu DP, Li HB. Impacts of gut bacteria on human health and diseases. Int J Mol Sci. 2015;16(4):7493–519.
21. Alenghat T, Osborne LC, Saenz SA, Kobuley D, Ziegler CG, Mullican SE, et al. Histone deacetylase 3 coordinates commensal-bacteria-dependent intestinal homeostasis. Nature. 2013;504(7478):153–7.
22. Espallergues J, Teegarden SL, Veerakumar A, Boulden J, Challis C, Jochems J, et al. HDAC6 regulates glucocorticoid receptor signaling in serotonin pathways with critical impact on stress resilience. J Neurosci. 2012;32(13):4400–16.
23. Jochems J, Boulden J, Lee BG, Blendy JA, Jarpe M, Mazitschek R, et al. Antidepressant-like properties of novel HDAC6-selective inhibitors with improved brain bioavailability. Neuropsychopharmacology. 2014;39(2):389–400.
24. Rossignol DA, Frye RE. Evidence linking oxidative stress, mitochondrial dysfunction, and inflammation in the brain of individuals with autism. Front Physiol. 2014;5:150.
25. Rossignol DA, Frye RE. Mitochondrial dysfunction in autism spectrum disorders: a systematic review and meta-analysis. Mol Psychiatry. 2012;17(3):290–314.
26. Turinsky AL, Turner B, Borja RC, Gleeson JA, Heath M, Pu S, et al. DAnCER: disease-annotated chromatin epigenetics resource. Nucleic Acids Res. 2010;39(Database issue):D889–94.
27. van Bokhoven H. Genetic and epigenetic networks in intellectual disabilities. Annu Rev Genet. 2011;45:81–104.
28. McCarthy SE, Gillis J, Kramer M, Lihm J, Yoon S, Berstein Y, et al. De novo mutations in schizophrenia implicate chromatin remodeling and support a genetic overlap with autism and intellectual disability. Mol Psychiatry. 2014;19(6):652–8.
29. Carroll LS, Owen MJ. Genetic overlap between autism, schizophrenia and bipolar disorder. Genome Med. 2009;1(10):102.
30. Cukier HN, Dueker ND, Slifer SH, Lee JM, Whitehead PL, Lalanne E, et al. Exome sequencing of extended families with autism reveals genes shared across neurodevelopmental and neuropsychiatric disorders. Mol Autism. 2014;5(1):1.

31. Doherty JL, Owen MJ. Genomic insights into the overlap between psychiatric disorders: implications for research and clinical practice. Genome Med. 2014;6(4):29.
32. Network, Pathway Analysis Subgroup of Psychiatric Genomics C. Psychiatric genome-wide association study analyses implicate neuronal, immune and histone pathways. Nat Neurosci. 2015;18(2):199–209.
33. Kiser DP, Rivero O, Lesch KP. Annual research review: the (epi)genetics of neurodevelopmental disorders in the era of whole-genome sequencing—unveiling the dark matter. J Child Psychol Psychiatry. 2015;56(3):278–95.
34. Freitag CM. The genetics of autistic disorders and its clinical relevance: a review of the literature. Mol Psychiatry. 2007;12(1):2–22.
35. Smalley SL, Asarnow RF, Spence MA. Autism and genetics. A decade of research. Arch Gen Psychiatry. 1988;45(10):953–61.
36. Ozonoff S, Young GS, Carter A, Messinger D, Yirmiya N, Zwaigenbaum L, et al. Recurrence risk for autism spectrum disorders: a Baby Siblings Research Consortium study. Pediatrics. 2011;128(3):e488–95.
37. Sandin S, Lichtenstein P, Kuja-Halkola R, Larsson H, Hultman CM, Reichenberg A. The familial risk of autism. JAMA. 2014;311(17):1770–7.
38. Hallmayer J, Cleveland S, Torres A, Phillips J, Cohen B, Torigoe T, et al. Genetic heritability and shared environmental factors among twin pairs with autism. Arch Gen Psychiatry. 2011;68(11):1095–102.
39. Gratten J, Wray NR, Keller MC, Visscher PM. Large-scale genomics unveils the genetic architecture of psychiatric disorders. Nat Neurosci. 2014;17(6):782–90.
40. Yuen RK, Thiruvahindrapuram B, Merico D, Walker S, Tammimies K, Hoang N, et al. Whole-genome sequencing of quartet families with autism spectrum disorder. Nat Med. 2015;21(2):185–91.
41. Betancur C. Etiological heterogeneity in autism spectrum disorders: more than 100 genetic and genomic disorders and still counting. Brain Res. 2011;1380:42–77.
42. Devlin B, Scherer SW. Genetic architecture in autism spectrum disorder. Curr Opin Genet Dev. 2012;22(3):229–37.
43. Iossifov I, O'Roak BJ, Sanders SJ, Ronemus M, Krumm N, Levy D, et al. The contribution of de novo coding mutations to autism spectrum disorder. Nature. 2014;515(7526):216–21.
44. Iossifov I, Ronemus M, Levy D, Wang Z, Hakker I, Rosenbaum J, et al. De novo gene disruptions in children on the autistic spectrum. Neuron. 2012;74(2):285–99.
45. O'Roak BJ, Stessman HA, Boyle EA, Witherspoon KT, Martin B, Lee C, et al. Recurrent de novo mutations implicate novel genes underlying simplex autism risk. Nat Commun. 2014;5:5595.
46. Sanders SJ, Murtha MT, Gupta AR, Murdoch JD, Raubeson MJ, Willsey AJ, et al. De novo mutations revealed by whole-exome sequencing are strongly associated with autism. Nature. 2012;485(7397):237–41.
47. Jeste SS, Geschwind DH. Disentangling the heterogeneity of autism spectrum disorder through genetic findings. Nat Rev Neurol. 2014;10(2):74–81.
48. Pinto D, Delaby E, Merico D, Barbosa M, Merikangas A, Klei L, et al. Convergence of genes and cellular pathways dysregulated in autism spectrum disorders. Am J Hum Genet. 2014;94(5):677–94.
49. Sanders SJ, He X, Willsey AJ, Ercan-Sencicek AG, Samocha KE, Cicek AE, et al. Insights into Autism spectrum disorder genomic architecture and biology from 71 risk loci. Neuron. 2015;87(6):1215–33.
50. Marshall CR, Noor A, Vincent JB, Lionel AC, Feuk L, Skaug J, et al. Structural variation of chromosomes in autism spectrum disorder. Am J Hum Genet. 2008;82(2):477–88.
51. Lasalle JM. Autism genes keep turning up chromatin. OA Autism. 2013;1(2):14.
52. Li J, Cai T, Jiang Y, Chen H, He X, Chen C, et al. Genes with de novo mutations are shared by four neuropsychiatric disorders discovered from NPdenovo database. Mol Psychiatry. 2016;21(2):298.

53. Robinson EB, Samocha KE, Kosmicki JA, McGrath L, Neale BM, Perlis RH, et al. Autism spectrum disorder severity reflects the average contribution of de novo and familial influences. Proc Natl Acad Sci U S A. 2014;111(42):15161–5.
54. Samocha KE, Robinson EB, Sanders SJ, Stevens C, Sabo A, McGrath LM, et al. A framework for the interpretation of de novo mutation in human disease. Nat Genet. 2014;46(9):944–50.
55. Chaste P, Klei L, Sanders SJ, Murtha MT, Hus V, Lowe JK, et al. Adjusting head circumference for covariates in autism: clinical correlates of a highly heritable continuous trait. Biol Psychiatry. 2013;74(8):576–84.
56. Michaelson JJ, Shi Y, Gujral M, Zheng H, Malhotra D, Jin X, et al. Whole-genome sequencing in autism identifies hot spots for de novo germline mutation. Cell. 2012;151(7):1431–42.
57. Colvert E, Tick B, McEwen F, Stewart C, Curran SR, Woodhouse E, et al. Heritability of Autism spectrum disorder in a UK population-based twin sample. JAMA Psychiat. 2015;72(5):415–23.
58. Frazier TW, Thompson L, Youngstrom EA, Law P, Hardan AY, Eng C, et al. A twin study of heritable and shared environmental contributions to autism. J Autism Dev Disord. 2014;44(8):2013–25.
59. Gaugler T, Klei L, Sanders SJ, Bodea CA, Goldberg AP, Lee AB, et al. Most genetic risk for autism resides with common variation. Nat Genet. 2014;46(8):881–5.
60. Ronald A, Hoekstra RA. Autism spectrum disorders and autistic traits: a decade of new twin studies. Am J Med Genet B Neuropsychiatr Genet. 2011;156B(3):255–74.
61. El Hajj N, Schneider E, Lehnen H, Haaf T. Epigenetics and life-long consequences of an adverse nutritional and diabetic intrauterine environment. Reproduction. 2014;148(6):R111–20.
62. Roseboom TJ, Painter RC, van Abeelen AF, Veenendaal MV, de Rooij SR. Hungry in the womb: what are the consequences? Lessons from the Dutch famine. Maturitas. 2011;70(2):141–5.
63. Tobi EW, Goeman JJ, Monajemi R, Gu H, Putter H, Zhang Y, et al. DNA methylation signatures link prenatal famine exposure to growth and metabolism. Nat Commun. 2014;5:5592.
64. Dufour-Rainfray D, Vourc'h P, Tourlet S, Guilloteau D, Chalon S, Andres CR. Fetal exposure to teratogens: evidence of genes involved in autism. Neurosci Biobehav Rev. 2011;35(5): 1254–65.
65. Keil KP, Lein PJ. DNA methylation: a mechanism linking environmental chemical exposures to risk of autism spectrum disorders? Environ Epigenet. 2016;2(1).
66. Kim YS, Leventhal BL. Genetic epidemiology and insights into interactive genetic and environmental effects in autism spectrum disorders. Biol Psychiatry. 2015;77(1):66–74.
67. Lyall K, Schmidt RJ, Hertz-Picciotto I. Maternal lifestyle and environmental risk factors for autism spectrum disorders. Int J Epidemiol. 2014;43(2):443–64.
68. Laufer BI, Diehl EJ, Singh SM. Neurodevelopmental epigenetic etiologies: insights from studies on mouse models of fetal alcohol spectrum disorders. Epigenomics. 2013;5(5):465–8.
69. Laufer BI, Kapalanga J, Castellani CA, Diehl EJ, Yan L, Singh SM. Associative DNA methylation changes in children with prenatal alcohol exposure. Epigenomics. 2015;7(8):1259–74.
70. Laufer BI, Mantha K, Kleiber ML, Diehl EJ, Addison SM, Singh SM. Long-lasting alterations to DNA methylation and ncRNAs could underlie the effects of fetal alcohol exposure in mice. Dis Model Mech. 2013;6(4):977–92.
71. Lee KW, Richmond R, Hu P, French L, Shin J, Bourdon C, et al. Prenatal exposure to maternal cigarette smoking and DNA methylation: epigenome-wide association in a discovery sample of adolescents and replication in an independent cohort at birth through 17 years of age. Environ Health Perspect. 2015;123(2):193–9.
72. Portales-Casamar E, Lussier AA, Jones MJ, MacIsaac JL, Edgar RD, Mah SM, et al. DNA methylation signature of human fetal alcohol spectrum disorder. Epigenetics Chromatin. 2016;9:25.
73. Roullet FI, Lai JK, Foster JA. In utero exposure to valproic acid and autism—a current review of clinical and animal studies. Neurotoxicol Teratol. 2013;36:47–56.
74. Smith EG. Additional effect size measures helpful in understanding lithium and valproate trial results. Am J Psychiatry. 2012;169(1):97–8.

75. O'Brien E, Dolinoy DC, Mancuso P. Perinatal bisphenol A exposures increase production of pro-inflammatory mediators in bone marrow-derived mast cells of adult mice. J Immunotoxicol. 2014;11(3):205–12.
76. Yaoi T, Itoh K, Nakamura K, Ogi H, Fujiwara Y, Fushiki S. Genome-wide analysis of epigenomic alterations in fetal mouse forebrain after exposure to low doses of bisphenol A. Biochem Biophys Res Commun. 2008;376(3):563–7.
77. Sandin S, Schendel D, Magnusson P, Hultman C, Suren P, Susser E, et al. Autism risk associated with parental age and with increasing difference in age between the parents. Mol Psychiatry. 2016;21(5):693–700.
78. Frans EM, Lichtenstein P, Hultman CM, Kuja-Halkola R. Age at fatherhood: heritability and associations with psychiatric disorders. Psychol Med. 2016;46(14):2981–8.
79. Ge ZJ, Schatten H, Zhang CL, Sun QY. Oocyte ageing and epigenetics. Reproduction. 2015;149(3):R103–14.
80. Jenkins TG, Aston KI, Pflueger C, Cairns BR, Carrell DT. Age-associated sperm DNA methylation alterations: possible implications in offspring disease susceptibility. PLoS Genet. 2014;10(7):e1004458.
81. Milekic MH, Xin Y, O'Donnell A, Kumar KK, Bradley-Moore M, Malaspina D, et al. Age-related sperm DNA methylation changes are transmitted to offspring and associated with abnormal behavior and dysregulated gene expression. Mol Psychiatry. 2015;20(8):995–1001.
82. Atladottir HO, Schendel DE, Henriksen TB, Hjort L, Parner ET. Gestational age and autism spectrum disorder: trends in risk over time. Autism Res. 2016;9(2):224–31.
83. Kuzniewicz MW, Wi S, Qian Y, Walsh EM, Armstrong MA, Croen LA. Prevalence and neonatal factors associated with autism spectrum disorders in preterm infants. J Pediatr. 2014;164(1):20–5.
84. Leavey A, Zwaigenbaum L, Heavner K, Burstyn I. Gestational age at birth and risk of autism spectrum disorders in Alberta. Can J Pediatr. 2013;162(2):361–8.
85. Behnia F, Parets SE, Kechichian T, Yin H, Dutta EH, Saade GR, et al. Fetal DNA methylation of autism spectrum disorders candidate genes: association with spontaneous preterm birth. Am J Obstet Gynecol. 2015;212(4):533 e1–9.
86. Conti E, Mazzotti S, Calderoni S, Saviozzi I, Guzzetta A. Are children born after assisted reproductive technology at increased risk of autism spectrum disorders? A systematic review. Hum Reprod. 2013;28(12):3316–27.
87. Fountain C, Zhang Y, Kissin DM, Schieve LA, Jamieson DJ, Rice C, et al. Association between assisted reproductive technology conception and autism in California, 1997–2007. Am J Public Health. 2015;105(5):963–71.
88. Schieve LA, Fountain C, Boulet SL, Yeargin-Allsopp M, Kissin DM, Jamieson DJ, et al. Does Autism diagnosis age or symptom severity differ among children according to whether assisted reproductive technology was used to achieve pregnancy? J Autism Dev Disord. 2015;45(9):2991–3003.
89. Sunderam S, Kissin DM, Crawford SB, Folger SG, Jamieson DJ, Warner L, et al. Assisted Reproductive Technology Surveillance — United States, 2014. MMWR Surveill Summ. 2017;66(No. SS-6):1–24.
90. Schieve LA, Devine O, Boyle CA, Petrini JR, Warner L. Estimation of the contribution of non-assisted reproductive technology ovulation stimulation fertility treatments to US singleton and multiple births. Am J Epidemiol. 2009;170(11):1396–407.
91. Gardener H, Spiegelman D, Buka SL. Perinatal and neonatal risk factors for autism: a comprehensive meta-analysis. Pediatrics. 2011;128(2):344–55.
92. Grafodatskaya D, Cytrynbaum C, Weksberg R. The health risks of ART. EMBO Rep. 2013;14(2):129–35.
93. Savage T, Peek J, Hofman PL, Cutfield WS. Childhood outcomes of assisted reproductive technology. Hum Reprod. 2011;26(9):2392–400.
94. Iliadou AN, Janson PC, Cnattingius S. Epigenetics and assisted reproductive technology. J Intern Med. 2011;270(5):414–20.

95. DeBaun MR, Niemitz EL, Feinberg AP. Association of in vitro fertilization with Beckwith-Wiedemann syndrome and epigenetic alterations of LIT1 and H19. Am J Hum Genet. 2003;72(1):156–60.
96. Doornbos ME, Maas SM, McDonnell J, Vermeiden JP, Hennekam RC. Infertility, assisted reproduction technologies and imprinting disturbances: a Dutch study. Hum Reprod. 2007;22(9):2476–80.
97. Gicquel C, Gaston V, Mandelbaum J, Siffroi JP, Flahault A, Le Bouc Y. In vitro fertilization may increase the risk of Beckwith-Wiedemann syndrome related to the abnormal imprinting of the KCN1OT gene. Am J Hum Genet. 2003;72(5):1338–41.
98. Ludwig M, Katalinic A, Gross S, Sutcliffe A, Varon R, Horsthemke B. Increased prevalence of imprinting defects in patients with Angelman syndrome born to subfertile couples. J Med Genet. 2005;42(4):289–91.
99. Maher ER, Brueton LA, Bowdin SC, Luharia A, Cooper W, Cole TR, et al. Beckwith-Wiedemann syndrome and assisted reproduction technology (ART). J Med Genet. 2003;40(1):62–4.
100. Sutcliffe AG, Peters CJ, Bowdin S, Temple K, Reardon W, Wilson L, et al. Assisted reproductive therapies and imprinting disorders—a preliminary British survey. Hum Reprod. 2006;21(4):1009–11.
101. Froehlich-Santino W, Londono Tobon A, Cleveland S, Torres A, Phillips J, Cohen B, et al. Prenatal and perinatal risk factors in a twin study of autism spectrum disorders. J Psychiatr Res. 2014;54:100–8.
102. Guinchat V, Thorsen P, Laurent C, Cans C, Bodeau N, Cohen D. Pre-, peri- and neonatal risk factors for autism. Acta Obstet Gynecol Scand. 2012;91(3):287–300.
103. Schieve LA, Clayton HB, Durkin MS, Wingate MS, Drews-Botsch C. Comparison of perinatal risk factors associated with Autism Spectrum Disorder (ASD), Intellectual Disability (ID), and co-occurring ASD and ID. J Autism Dev Disord. 2015;45(8):2361–72.
104. Brown AS. Epidemiologic studies of exposure to prenatal infection and risk of schizophrenia and autism. Dev Neurobiol. 2012;72(10):1272–6.
105. Fang SY, Wang S, Huang N, Yeh HH, Chen CY. Prenatal infection and Autism Spectrum disorders in childhood: a population-based case-control study in Taiwan. Paediatr Perinat Epidemiol. 2015;29(4):307–16.
106. Lee BK, Magnusson C, Gardner RM, Blomstrom A, Newschaffer CJ, Burstyn I, et al. Maternal hospitalization with infection during pregnancy and risk of autism spectrum disorders. Brain Behav Immun. 2015;44:100–5.
107. Chen SW, Zhong XS, Jiang LN, Zheng XY, Xiong YQ, Ma SJ, et al. Maternal autoimmune diseases and the risk of autism spectrum disorders in offspring: a systematic review and meta-analysis. Behav Brain Res. 2016;296:61–9.
108. Braunschweig D, Krakowiak P, Duncanson P, Boyce R, Hansen RL, Ashwood P, et al. Autism-specific maternal autoantibodies recognize critical proteins in developing brain. Transl Psychiatry. 2013;3:e277.
109. Braunschweig D, Van de Water J. Maternal autoantibodies in autism. Arch Neurol. 2012;69(6):693–9.
110. Krakowiak P, Walker CK, Tancredi D, Hertz-Picciotto I, Van de Water J. Autism-specific maternal anti-fetal brain autoantibodies are associated with metabolic conditions. Autism Res. 2016;10(1):89–98.
111. Li M, Fallin MD, Riley A, Landa R, Walker SO, Silverstein M, et al. The association of maternal obesity and diabetes with autism and other developmental disabilities. Pediatrics. 2016;137(2):e20152206.
112. Li YM, Ou JJ, Liu L, Zhang D, Zhao JP, Tang SY. Association between maternal obesity and autism spectrum disorder in offspring: a meta-analysis. J Autism Dev Disord. 2016;46(1):95–102.
113. Xiang AH, Wang X, Martinez MP, Walthall JC, Curry ES, Page K, et al. Association of maternal diabetes with autism in offspring. JAMA. 2015;313(14):1425–34.

114. Finer S, Mathews C, Lowe R, Smart M, Hillman S, Foo L, et al. Maternal gestational diabetes is associated with genome-wide DNA methylation variation in placenta and cord blood of exposed offspring. Hum Mol Genet. 2015;24(11):3021–9.
115. Binder AM, LaRocca J, Lesseur C, Marsit CJ, Michels KB. Epigenome-wide and transcriptome-wide analyses reveal gestational diabetes is associated with alterations in the human leukocyte antigen complex. Clin Epigenetics. 2015;7:79.
116. Petropoulos S, Guillemin C, Ergaz Z, Dimov S, Suderman M, Weinstein-Fudim L, et al. Gestational diabetes alters offspring DNA methylation profiles in human and rat: identification of key pathways involved in endocrine system disorders, insulin signaling, diabetes signaling, and ILK signaling. Endocrinology. 2015;156(6):2222–38.
117. Ruchat SM, Hivert MF, Bouchard L. Epigenetic programming of obesity and diabetes by in utero exposure to gestational diabetes mellitus. Nutr Rev. 2013;71(Suppl 1):S88–94.
118. Ronald A, Pennell CE, Whitehouse AJ. Prenatal maternal stress associated with ADHD and autistic traits in early childhood. Front Psychol. 2010;1:223.
119. Rijlaarsdam J, Pappa I, Walton E, Bakermans-Kranenburg MJ, Mileva-Seitz VR, Rippe RC, et al. An epigenome-wide association meta-analysis of prenatal maternal stress in neonates: a model approach for replication. Epigenetics. 2016;11(2):140–9.
120. Schmidt RJ. Maternal folic acid supplements associated with reduced autism risk in the child. Evid Based Med. 2013;18(6):e53.
121. Schmidt RJ, Tancredi DJ, Ozonoff S, Hansen RL, Hartiala J, Allayee H, et al. Maternal periconceptional folic acid intake and risk of autism spectrum disorders and developmental delay in the CHARGE (CHildhood Autism Risks from Genetics and Environment) case-control study. Am J Clin Nutr. 2012;96(1):80–9.
122. Suren P, Roth C, Bresnahan M, Haugen M, Hornig M, Hirtz D, et al. Association between maternal use of folic acid supplements and risk of autism spectrum disorders in children. JAMA. 2013;309(6):570–7.
123. Harris SW, Hessl D, Goodlin-Jones B, Ferranti J, Bacalman S, Barbato I, et al. Autism profiles of males with fragile X syndrome. Am J Ment Retard. 2008;113(6):427–38.
124. Moretti P, Zoghbi HY. MeCP2 dysfunction in Rett syndrome and related disorders. Curr Opin Genet Dev. 2006;16(3):276–81.
125. Zafeiriou DI, Ververi A, Dafoulis V, Kalyva E, Vargiami E. Autism spectrum disorders: the quest for genetic syndromes. Am J Med Genet B Neuropsychiatr Genet. 2013;162B(4):327–66.
126. Renieri A, Mari F, Mencarelli MA, Scala E, Ariani F, Longo I, et al. Diagnostic criteria for the Zappella variant of Rett syndrome (the preserved speech variant). Brain Dev. 2009;31(3):208–16.
127. Beyer KS, Blasi F, Bacchelli E, Klauck SM, Maestrini E, Poustka A, et al. Mutation analysis of the coding sequence of the MECP2 gene in infantile autism. Hum Genet. 2002;111(4–5):305–9.
128. Carney RM, Wolpert CM, Ravan SA, Shahbazian M, Ashley-Koch A, Cuccaro ML, et al. Identification of MeCP2 mutations in a series of females with autistic disorder. Pediatr Neurol. 2003;28(3):205–11.
129. Coutinho AM, Oliveira G, Katz C, Feng J, Yan J, Yang C, et al. MECP2 coding sequence and 3'UTR variation in 172 unrelated autistic patients. Am J Med Genet B Neuropsychiatr Genet. 2007;144B(4):475–83.
130. Suter B, Treadwell-Deering D, Zoghbi HY, Glaze DG, Neul JL. Brief report: MECP2 mutations in people without Rett syndrome. J Autism Dev Disord. 2014;44(3):703–11.
131. Bernier R, Golzio C, Xiong B, Stessman HA, Coe BP, Penn O, et al. Disruptive CHD8 mutations define a subtype of autism early in development. Cell. 2014;158(2):263–76.
132. Merner N, Forgeot d'Arc B, Bell SC, Maussion G, Peng H, Gauthier J, et al. A de novo frameshift mutation in chromodomain helicase DNA-binding domain 8 (CHD8): a case report and literature review. Am J Med Genet A. 2016;170A(5):1225–35.
133. Hartshorne TS, Grialou TL, Parker KR. Autistic-like behavior in CHARGE syndrome. Am J Med Genet A. 2005;133A(3):257–61.

134. Johansson M, Rastam M, Billstedt E, Danielsson S, Stromland K, Miller M, et al. Autism spectrum disorders and underlying brain pathology in CHARGE association. Dev Med Child Neurol. 2006;48(1):40–50.
135. Smith IM, Nichols SL, Issekutz K, Blake K, Canadian Paediatric Surveillance P. Behavioral profiles and symptoms of autism in CHARGE syndrome: preliminary Canadian epidemiological data. Am J Med Genet A. 2005;133A(3):248–56.
136. Micucci JA, Sperry ED, Martin DM. Chromodomain helicase DNA-binding proteins in stem cells and human developmental diseases. Stem Cells Dev. 2015;24(8):917–26.
137. Choufani S, Cytrynbaum C, Chung BH, Turinsky AL, Grafodatskaya D, Chen YA, et al. NSD1 mutations generate a genome-wide DNA methylation signature. Nat Commun. 2015;6:10207.
138. Lane C, Milne E, Freeth M. Cognition and behaviour in Sotos syndrome: a systematic review. PLoS One. 2016;11(2):e0149189.
139. Lane C, Milne E, Freeth M. Characteristics of autism spectrum disorder in Sotos syndrome. J Autism Dev Disord. 2016;47(1):135–43.
140. Sheth K, Moss J, Hyland S, Stinton C, Cole T, Oliver C. The behavioral characteristics of Sotos syndrome. Am J Med Genet A. 2015;167A(12):2945–56.
141. Richards C, Jones C, Groves L, Moss J, Oliver C. Prevalence of autism spectrum disorder phenomenology in genetic disorders: a systematic review and meta-analysis. Lancet Psychiatry. 2015;2(10):909–16.
142. Lam CW, Yeung WL, Ko CH, Poon PM, Tong SF, Chan KY, et al. Spectrum of mutations in the MECP2 gene in patients with infantile autism and Rett syndrome. J Med Genet. 2000;37(12):E41.
143. Shibayama A, Cook Jr EH, Feng J, Glanzmann C, Yan J, Craddock N, et al. MECP2 structural and 3'-UTR variants in schizophrenia, autism and other psychiatric diseases: a possible association with autism. Am J Med Genet B Neuropsychiatr Genet. 2004;128B(1):50–3.
144. Crawford DC, Acuna JM, Sherman SL. FMR1 and the fragile X syndrome: human genome epidemiology review. Genet Med. 2001;3(5):359–71.
145. Niklasson L, Rasmussen P, Oskarsdottir S, Gillberg C. Neuropsychiatric disorders in the 22q11 deletion syndrome. Genet Med. 2001;3(1):79–84.
146. Niklasson L, Rasmussen P, Oskarsdottir S, Gillberg C. Autism, ADHD, mental retardation and behavior problems in 100 individuals with 22q11 deletion syndrome. Res Dev Disabil. 2009;30(4):763–73.
147. Angkustsiri K, Goodlin-Jones B, Deprey L, Brahmbhatt K, Harris S, Simon TJ. Social impairments in chromosome 22q11.2 deletion syndrome (22q11.2DS): autism spectrum disorder or a different endophenotype? J Autism Dev Disord. 2014;44(4):739–46.
148. Descheemaeker MJ, Govers V, Vermeulen P, Fryns JP. Pervasive developmental disorders in Prader-Willi syndrome: the Leuven experience in 59 subjects and controls. Am J Med Genet A. 2006;140(11):1136–42.
149. Veltman MW, Thompson RJ, Roberts SE, Thomas NS, Whittington J, Bolton PF. Prader-Willi syndrome—a study comparing deletion and uniparental disomy cases with reference to autism spectrum disorders. Eur Child Adolesc Psychiatry. 2004;13(1):42–50.
150. Bennett JA, Germani T, Haqq AM, Zwaigenbaum L. Autism spectrum disorder in Prader-Willi syndrome: a systematic review. Am J Med Genet A. 2015;167A(12):2936–44.
151. Hogart A, Wu D, Lasalle JM, Schanen NC. The comorbidity of autism with the genomic disorders of chromosome 15q11.2-q13. Neurobiol Dis. 2008;38(2):181–91.
152. Ho L, Crabtree GR. Chromatin remodelling during development. Nature. 2010; 463(7280):474–84.
153. Parisi L, Di Filippo T, Roccella M. Autism spectrum disorder in Kabuki syndrome: clinical, diagnostic and rehabilitative aspects assessed through the presentation of three cases. Minerva Pediatr. 2015;67(4):369–75.
154. Creswell C, Skuse DH. Autism in association with Turner syndrome: genetic implications for male vulnerability to pervasive developmental disorders. Neurocase. 1999;5(6):511–8.

155. Nguyen A, Rauch TA, Pfeifer GP, Hu VW. Global methylation profiling of lymphoblastoid cell lines reveals epigenetic contributions to autism spectrum disorders and a novel autism candidate gene, RORA, whose protein product is reduced in autistic brain. FASEB J. 2010;24(8):3036–51.
156. Wong CC, Meaburn EL, Ronald A, Price TS, Jeffries AR, Schalkwyk LC, et al. Methylomic analysis of monozygotic twins discordant for autism spectrum disorder and related behavioural traits. Mol Psychiatry. 2014;19(4):495–503.
157. Berko ER, Suzuki M, Beren F, Lemetre C, Alaimo CM, Calder RB, et al. Mosaic epigenetic dysregulation of ectodermal cells in autism spectrum disorder. PLoS Genet. 2014;10(5):e1004402.
158. Feinberg JI, Bakulski KM, Jaffe AE, Tryggvadottir R, Brown SC, Goldman LR, et al. Paternal sperm DNA methylation associated with early signs of autism risk in an autism-enriched cohort. Int J Epidemiol. 2015;44(4):1199–210.
159. Ginsberg MR, Rubin RA, Falcone T, Ting AH, Natowicz MR. Brain transcriptional and epigenetic associations with autism. PLoS One. 2012;7(9):e44736.
160. Ladd-Acosta C, Hansen KD, Briem E, Fallin MD, Kaufmann WE, Feinberg AP. Common DNA methylation alterations in multiple brain regions in autism. Mol Psychiatry. 2014;19(8):862–71.
161. Nardone S, Sams DS, Reuveni E, Getselter D, Oron O, Karpuj M, et al. DNA methylation analysis of the autistic brain reveals multiple dysregulated biological pathways. Transl Psychiatry. 2014;4:e433.
162. Hernandez DG, Nalls MA, Gibbs JR, Arepalli S, van der Brug M, Chong S, et al. Distinct DNA methylation changes highly correlated with chronological age in the human brain. Hum Mol Genet. 2011;20(6):1164–72.
163. Lister R, Mukamel EA, Nery JR, Urich M, Puddifoot CA, Johnson ND, et al. Global epigenomic reconfiguration during mammalian brain development. Science. 2013;341(6146):1237905.
164. Lokk K, Modhukur V, Rajashekar B, Martens K, Magi R, Kolde R, et al. DNA methylome profiling of human tissues identifies global and tissue-specific methylation patterns. Genome Biol. 2014;15(4):r54.
165. Roadmap Epigenomics C, Kundaje A, Meuleman W, Ernst J, Bilenky M, Yen A, et al. Integrative analysis of 111 reference human epigenomes. Nature. 2015;518(7539):317–30.
166. Slieker RC, Bos SD, Goeman JJ, Bovee JV, Talens RP, van der Breggen R, et al. Identification and systematic annotation of tissue-specific differentially methylated regions using the Illumina 450 k array. Epigenetics Chromatin. 2013;6(1):26.
167. Lowe R, Slodkowicz G, Goldman N, Rakyan VK. The human blood DNA methylome displays a highly distinctive profile compared with other somatic tissues. Epigenetics. 2015;10(4):274–81.
168. Elagoz Yuksel M, Yuceturk B, Faruk Karatas O, Ozen M, Dogangun B. The altered promoter methylation of oxytocin receptor gene in autism. J Neurogenet. 2016;30:280–4.
169. Gregory SG, Connelly JJ, Towers AJ, Johnson J, Biscocho D, Markunas CA, et al. Genomic and epigenetic evidence for oxytocin receptor deficiency in autism. BMC Med. 2009;7:62.
170. Jiang YH, Sahoo T, Michaelis RC, Bercovich D, Bressler J, Kashork CD, et al. A mixed epigenetic/genetic model for oligogenic inheritance of autism with a limited role for UBE3A. Am J Med Genet A. 2004;131(1):1–10.
171. Nagarajan RP, Hogart AR, Gwye Y, Martin MR, LaSalle JM. Reduced MeCP2 expression is frequent in autism frontal cortex and correlates with aberrant MECP2 promoter methylation. Epigenetics. 2006;1(4):e1–11.
172. Zhu L, Wang X, Li XL, Towers A, Cao X, Wang P, et al. Epigenetic dysregulation of SHANK3 in brain tissues from individuals with autism spectrum disorders. Hum Mol Genet. 2014;23(6):1563–78.
173. Zhubi A, Chen Y, Dong E, Cook EH, Guidotti A, Grayson DR. Increased binding of MeCP2 to the GAD1 and RELN promoters may be mediated by an enrichment of 5-hmC in autism spectrum disorder (ASD) cerebellum. Transl Psychiatry. 2014;4:e349.

174. Shulha HP, Cheung I, Whittle C, Wang J, Virgil D, Lin CL, et al. Epigenetic signatures of autism: trimethylated H3K4 landscapes in prefrontal neurons. Arch Gen Psychiatry. 2012;69(3):314–24.
175. Sun W, Poschmann J, Cruz-Herrera Del Rosario R, Parikshak NN, Hajan HS, Kumar V, et al. Histone acetylome-wide association study of autism spectrum disorder. Cell. 2016;167(5):1385–97. e11
176. Fregeac J, Colleaux L, Nguyen LS. The emerging roles of MicroRNAs in autism spectrum disorders. Neurosci Biobehav Rev. 2016;71:729–38.
177. Parikshak NN, Swarup V, Belgard TG, Irimia M, Ramaswami G, Gandal MJ, et al. Genome-wide changes in lncRNA, splicing, and regional gene expression patterns in autism. Nature. 2016;540:423–7.
178. Ziats MN, Rennert OM. Aberrant expression of long noncoding RNAs in autistic brain. J Mol Neurosci. 2013;49(3):589–93.
179. Wang Y, Zhao X, Ju W, Flory M, Zhong J, Jiang S, et al. Genome-wide differential expression of synaptic long noncoding RNAs in autism spectrum disorder. Transl Psychiatry. 2015;5:e660.
180. Grinshtein N, Rioseco CC, Marcellus R, Uehling D, Aman A, Lun X, et al. Small molecule epigenetic screen identifies novel EZH2 and HDAC inhibitors that target glioblastoma brain tumor-initiating cells. Oncotarget. 2016;7:59360–76.
181. Seo YJ, Kang Y, Muench L, Reid A, Caesar S, Jean L, et al. Image-guided synthesis reveals potent blood-brain barrier permeable histone deacetylase inhibitors. ACS Chem Nerosci. 2014;5(7):588–96.
182. Bakermans-Kranenburg MJ, van I Jzendoorn MH. Sniffing around oxytocin: review and meta-analyses of trials in healthy and clinical groups with implications for pharmacotherapy. Transl Psychiatry. 2013;3:e258.
183. Guastella AJ, Hickie IB. Oxytocin treatment, circuitry, and autism: a critical review of the literature placing oxytocin into the autism context. Biol Psychiatry. 2016;79(3):234–42.
184. Preti A, Melis M, Siddi S, Vellante M, Doneddu G, Fadda R. Oxytocin and autism: a systematic review of randomized controlled trials. J Child Adolesc Psychopharmacol. 2014;24(2):54–68.
185. Young LJ, Barrett CE. Neuroscience. Can oxytocin treat autism? Science. 2015; 347(6224):825–6.
186. Anagnostou E, Soorya L, Brian J, Dupuis A, Mankad D, Smile S, et al. Intranasal oxytocin in the treatment of autism spectrum disorders: a review of literature and early safety and efficacy data in youth. Brain Res. 2014;1580:188–98.
187. Anagnostou E, Soorya L, Chaplin W, Bartz J, Halpern D, Wasserman S, et al. Intranasal oxytocin versus placebo in the treatment of adults with autism spectrum disorders: a randomized controlled trial. Mol Autism. 2012;3(1):16.
188. Auyeung B, Lombardo MV, Heinrichs M, Chakrabarti B, Sule A, Deakin JB, et al. Oxytocin increases eye contact during a real-time, naturalistic social interaction in males with and without autism. Transl Psychiatry. 2015;5:e507.
189. Guo JU, Szulwach KE, Su Y, Li Y, Yao B, Xu Z, et al. Genome-wide antagonism between 5-hydroxymethylcytosine and DNA methylation in the adult mouse brain. Front Biol (Beijing). 2014;9(1):66–74.
190. Szulwach KE, Li X, Li Y, Song CX, Wu H, Dai Q, et al. 5-hmC-mediated epigenetic dynamics during postnatal neurodevelopment and aging. Nat Neurosci. 2011;14(12):1607–16.
191. Kriaucionis S, Heintz N. The nuclear DNA base 5-hydroxymethylcytosine is present in Purkinje neurons and the brain. Science. 2009;324(5929):929–30.
192. Lai MC, Baron-Cohen S, Buxbaum JD. Understanding autism in the light of sex/gender. Mol Autism. 2015;6:24.
193. Lai MC, Lombardo MV, Auyeung B, Chakrabarti B, Baron-Cohen S. Sex/gender differences and autism: setting the scene for future research. J Am Acad Child Adolesc Psychiatry. 2015;54(1):11–24.

194. Werling DM, Geschwind DH. Understanding sex bias in autism spectrum disorder. Proc Natl Acad Sci U S A. 2013;110(13):4868–9.
195. Ellegood J, Anagnostou E, Babineau BA, Crawley JN, Lin L, Genestine M, et al. Clustering autism: using neuroanatomical differences in 26 mouse models to gain insight into the heterogeneity. Mol Psychiatry. 2015;20(1):118–25.
196. Ey E, Leblond CS, Bourgeron T. Behavioral profiles of mouse models for autism spectrum disorders. Autism Res. 2011;4(1):5–16.
197. Moy SS, Nadler JJ. Advances in behavioral genetics: mouse models of autism. Mol Psychiatry. 2008;13(1):4–26.
198. Nakai N, Otsuka S, Myung J, Takumi T. Autism spectrum disorder model mice: focus on copy number variation and epigenetics. Sci China Life Sci. 2015;58(10):976–84.
199. Petrinovic MM, Kunnecke B. Neuroimaging endophenotypes in animal models of autism spectrum disorders: lost or found in translation? Psychopharmacology (Berl). 2014;231(6):1167–89.
200. Barry G. Integrating the roles of long and small non-coding RNA in brain function and disease. Mol Psychiatry. 2014;19(4):410–6.
201. Mellios N, Sur M. The emerging role of micrornas in schizophrenia and autism spectrum disorders. Front Psychiatry. 2012;3:39.
202. Roberts TC, Morris KV, Wood MJ. The role of long non-coding RNAs in neurodevelopment, brain function and neurological disease. Philos Trans R Soc Lond B Biol Sci. 2014;369(1652).
203. Wilkinson B, Campbell DB. Contribution of long noncoding RNAs to autism spectrum disorder risk. Int Rev Neurobiol. 2013;113:35–59.

Part II

Adolescence Brain Diseases

Eating Disorders and Epigenetics

5

Lea Thaler and Howard Steiger

Abstract

Eating disorders (EDs) are characterized by intense preoccupation with shape and weight and maladaptive eating practices. The complex of symptoms that characterize EDs often arise through the activation of latent genetic potentials by environmental exposures, and epigenetic mechanisms are believed to link environmental exposures to gene expression. This chapter provides an overview of genetic factors acting in the etiology of EDs. It then provides a background to the hypothesis that epigenetic mechanisms link stresses such as obstetric complications and childhood abuse as well as effects of malnutrition to eating disorders (EDs). The chapter then summarizes the emerging body of literature on epigenetics and EDs—mainly studies on DNA methylation in samples of anorexia and bulimia. The available evidence base suggests that an epigenetically informed perspective contributes in valuable ways to the understanding of why people develop EDs.

Keywords

Eating disorders • Anorexia • Bulimia • Binge eating • Epigenetics • Methylation

L. Thaler, Ph.D. (✉) • H. Steiger, Ph.D.
Eating Disorders Continuum, Douglas Institute, Montreal West Island Integrated University Health and Social Services Centre, 6875 LaSalle Blvd, Verdun, QC, Canada, H4H 1R3

Psychiatry Department, McGill University, Montréal, QC, Canada, H3A 1Y2
e-mail: lea.thaler@douglas.mcgill.ca; stehow@douglas.mcgill.ca

R. Delgado-Morales (ed.), *Neuroepigenomics in Aging and Disease*, Advances in Experimental Medicine and Biology, DOI 10.1007/978-3-319-53889-1_5

5.1 Introduction

Eating Disorders (EDs) are characterized by intense preoccupations with eating, shape, weight and body image, and maladaptive eating practices such as excessive caloric restraint, binge eating, and forms of purging including self-induced vomiting. The three most frequently seen types of EDs are anorexia nervosa (AN), bulimia nervosa (BN), and binge eating disorder (BED). AN is characterized by intense restriction of energy intake (leading to a markedly low body weight), as well as persistent behavior to avoid weight gain (even though already at low weight), or an intense fear of gaining weight. The syndrome includes two variants: AN, restricting type (AN-R), in which there is restriction of food intake but no binge eating or purging, and AN, binge eating/purging (AN-B/P) type, in which regular binge or purge episodes occur. BN is similarly characterized by binge eating followed by compensatory efforts (e.g., self-induced vomiting, laxative misuse, intensive exercise, or fasting), but in individuals with relatively normal (or above normal) weight. Individuals with both AN and BN are preoccupied with their weight or shape and place much importance on the influence of weight or shape on their self-evaluation. BED, like BN, is characterized by recurrent binge eating, but in the absence of compensatory behaviors (such as vomiting, exercise, or fasting), so that the syndrome is commonly associated with (or leads to) obesity.

In addition to behaviors related to eating, people affected by EDs show heterogeneous concurrent psychiatric symptoms that often include problems with mood (e.g., depression or lability) and anxiety (e.g., obsessive-compulsiveness, generalized anxiety, panic, social phobias, etc.). Some research has proposed a distinction between those individuals with EDs who restrict food intake without binge eating or purging (restricters) and those who do binge eat or purge (binger-purgers) [1]. Restricters (as exemplified by AN-R) display propensities toward perfectionism, preference for order, emotional constraint, and behavioral inhibition. In contrast, binger-purger ED variants (which include AN-B/P, BN, and BED) tend toward co-aggregation with traits of emotional instability, recklessness, and impulse dyscontrol and with propensities toward substance abuse and self-harm. In parallel, sufferers show variable degrees of exposure to such environmental adversities as perinatal/obstetric "insults," early childhood traumata, or later life stresses [2]. Despite such "prototypes," many people, sometimes with AN and sometimes BN, show negligible psychopathology—which raises the important point that it is not necessary to have marked psychopathology to develop an ED.

5.2 Genetics of EDs

EDs are multiply determined, with genetic factors playing an important role. Family studies demonstrate that EDs are strongly familial. Family members of those with AN are four times more likely to also have AN [3] and female relatives of AN patients are up to 11 times more likely to develop AN than individuals who do not have relatives with AN [4]. Individuals who have a relative with AN or BN are at elevated risk for developing either disorder [4, 5] which suggests some shared genetic diathesis for AN or BN. BED is also found to aggregate within families [6] and in a manner independent of obesity [7]. All EDs are highly heritable, with heritability estimates from twin studies of 0.48–0.74 for AN [8–11], 0.55–0.62 for BN [10, 12, 13], and 0.39–0.45 for BED [14, 15].

Current thinking has it that the complex of symptoms that characterizes EDs often arises through the activation of latent genetic potentials by environmental exposures. Molecular genetic studies, such as linkage and candidate gene studies, have shown that genes linked to such factors as (a) mood, anxiety, and impulse regulation [16–19]; (b) appetite, body weight, and related metabolic factors [20–22]; and (c) sex (e.g., genes influencing estrogen activity) [23–26] play roles in EDs. Genetic models are, however, insufficient on their own—as genetic liabilities, in all probability, need to be activated by environmental triggers which, in the case of EDs, are quite numerous. Evidence suggests that environmental risks for ED development include perinatal and obstetric insults [27, 28], early life stressors like childhood abuse [29], and later life stresses—among which are likely to figure stresses linked to too much caloric restraint [30]. For example, women with bulimia who are carriers of the 5HTTLPR S allele and survivors of severe childhood abuse display more pronounced dissocial behavior, characterized by novelty seeking, recklessness, or hostility [19]. Furthermore, carriers of the BCL1 C allele and who suffered childhood abuse are more likely to have BN than those with other allelic variations and who did not experience abuse [31]. Such findings could suggest environmental activation (via childhood abuse) of a genetically determined vulnerability.

5.3 Epigenetics and EDs

Epigenetics is a science that promises to help explain how environmental factors can "switch on" genetic susceptibilities. When speaking of epigenetic mechanisms, we refer to the process by which environmental exposures leave epigenetic "marks" on the genome that can influence later gene expression. It is in this manner that epigenetic mechanisms are thought to provide the physical substrates for gene-environment interactions that can shape the expression of latent genetic potentials.

The majority of the published studies that examine epigenetic alterations in the context of EDs focus on the role of DNA methylation, which involves the addition of methyl to genomic regions in which cytosine is followed by guanine—commonly called CpGs. Although variations occur, when gene promoters become methylated, this tends to reduce gene expression, with loss of function occurring directly (due to inhibition of the binding of transcription factors to recognition elements in the gene), or indirectly (via the recruitment of proteins that precipitate inactive chromatin). Evidence suggests that DNA methylation is influenced by diverse environmental exposures, including early life stressors, dietary factors, and even obstetric and perinatal insults [32–34].

5.4 Early Life Stresses

Stressors occurring during infancy and childhood can have profound influences on epigenetic programing. Meaney and colleagues have demonstrated that rat pups receiving low maternal care tended to display increased DNA methylation of the glucocorticoid receptor gene (*NR3C1*), decreased hippocampal glucocorticoid receptor mRNA expression, and increased stress reactivity as compared to pups who received high maternal care [35–37]. In humans, childhood abuse plays an important role in the epigenetic programming of genes related to stress reactivity and serotonin. For example, suicide victims who had experienced childhood abuse had higher overall methylation of an NR3C1 promoter and decreased levels of glucocorticoid receptor mRNA [38], and individuals who had experienced childhood sexual abuse show altered methylation of 5HTT [39] and SLC6A4 [40].

5.5 Obstetric and Perinatal Complications

Even before early life stressors can have an impact, epigenetic programming can occur in the womb. Mothers' and fathers' emotional distress [41] can affect the epigenetic status of their children and, as a result, affect the children's later physical development and emotional adjustment. Evidence gathered from naturalistic experiments occurring during natural disasters highlights how distress during gestation can cause epigenetic changes in offspring. During a somewhat recent natural disaster, children of mothers who were exposed to intense, third trimester gestational distress during very severe weather conditions—the 1998 Ice Storm in Quebec (regarded as one of Canada's worst natural disasters)—showed more ED symptoms at age 13 [28]. Indicating that the latter effect likely has epigenetic origins, in a separate study, Ice Storm exposed children showed altered DNA methylation patterns when compared to unexposed children [42]. Maternal depression during gestation can also cause epigenetic effects, such as increased methylation of offsprings' glucocorticoid receptor (NR3C1) gene at specific sites. Affected offspring show altered cortisol responses and increased stress reactivity [43, 44]. Fathers too can transmit epigenetic marks through the sperm line. For example, male mice raised in stressful environments produce more stressed offspring than do genetically identical mice raised under calmer conditions [45]. This holds true even when the father mice are physically separated from their children.

Maternal nutritional deficiency [46, 47] during pregnancy also plays an important role in epigenetic programming. For example, a 1944–1945 Nazi blockade of food supplies in Western Netherlands during the Second World War caused famine. The offspring of pregnant women who endured the famine showed adverse, epigenetically mediated effects upon their physical and mental health [48]. Children affected by this famine tended to have higher risk of obesity, smaller physical stature, and greater risk of various mental health problems.

5.6 Nutritional Factors and DNA Methylation

While maternal nutritional state can affect epigenetic programming, an individual's own nutritional state can also cause epigenetic changes, which in turn can be relevant to the development of various psychiatric disorders, including EDs [33, 34, 49]. Certain macronutrients are involved in the diverse methyl-transfer reactions upon which DNA methylation depends. Main micronutrients involved in such reactions include folate, vitamin B12, and choline [50], and an inadequate diet will produce deficiencies in these nutrients [51, 52]. The status of folate and B-vitamin is shown to affect DNA methylation and brain function at various life stages, spanning the prenatal period to old age [34, 53, 54].

In animals, folate/methyl-deficient diets (both of which are linked to depression) result in global hypermethylation of brain cells [55]. Other macronutrients such as choline and betaine impact DNA methylation [56] of B12. Anorexia has been associated with a condition called hyperhomocysteinemia (HHcy), caused by low dietary folate, choline, or vitamin B12 [34, 57]. HHcy is seen mainly in older AN patients and those with longer disease chronicity. The condition is reduced by daily folic acid supplementation [58].

5.7 Epigenetic Studies in ED Samples

Thus far, most epigenetic studies have focused on the promoter-specific methylation of the candidate genes that have been studied previously in EDs. The majority of the studies examined cases of AN, with fewer exploring epigenetics in BN or BN-spectrum cases. To date, no studies on the epigenetics of BED have been conducted.

Dopaminergic genes have been of interest due to their relationship with reward. A study of 22 patients with AN showed elevated mRNA expression of the dopamine transporter SLC6A3 gene, also known as DAT [59]. This elevated mRNA was due to hypermethylation of the gene's promoter region. The study also found hypermethylation of the DRD2 promoter in AN and hypermethylation of DAT in 24 patients with BN [59]. However, another study failed to find significant differences in the promoter-specific DNA methylation for DRD2, LEP, BDNF, and SLC6A4 genes in AN [60]. A study on women with BN and borderline personality disorder showed hypermethylation of the DRD2 promoter region [61].

Another candidate gene of interest is POMC due to its role in the regulation of appetite. A study comparing women with acute AN, those recovered from AN, and controls showed that expression of the functionally relevant long POMC mRNA transcript was significantly correlated with leptin levels and was higher in acute AN compared to recovered AN and controls [62]. Furthermore, methylation of single CpG residues in the E2F binding site was inversely related to POMC expression. In another study by the same group, no association between BMI or AN status (acute versus recovered) and *POMC* promoter methylation was seen, but hypomethylation was associated with cigarette smoking [63].

Other candidate genes have been studied in both AN and BN samples. Studies in AN reported hypermethylation of the alpha-synuclein gene (linked to sensitivity to dietary folate) and of the atrial natriuretic peptide gene, implicated in anxiety, depression and stress responses [64, 65]. A pilot study detected a number of CpG sites in the oxytocin receptor (OXTR) gene with higher than average methylation levels in 15 AN patients compared to controls. The methylation levels were negatively associated with BMI [66]. Another study showed higher levels of CNR1(CB1) receptor mRNA in the blood of 43 patients with either AN or BN compared to 26 healthy controls [67]. Another study on CB1 reported a downregulation of the gene's receptor mRNA in ED patients who also engage in self-injury [68]. In women with BN and specific comorbidities, our group has shown that women with BN and a history of suicidality show hypermethylation in the exon 1C region of the glucocorticoid receptor (GR) gene, and we found hypermethylation of specific CpG sites in the BDNF gene promoter region in women with BN with and without childhood abuse [69, 70]. In sum, evidence associates EDs with alterations in DNA methylation and especially in those ED sufferers displaying comorbid psychopathology or having experienced prior trauma.

Most recently, researchers have been examining genome-wide (GW) methylation in patients with EDs. In a study comparing 22 women with AN to 30 healthy controls, a pattern of significant global DNA hypomethylation was seen in the AN patients [65]. Another GW study of 32 adolescents with AN-R and 13 controls found whole-blood global DNA methylation to be reduced in the AN cases as compared to controls [71]. However, one small GW study observed no global differences when comparing methylation of LINE-1 repetitive elements and the H19 imprinting control region in DNA from peripheral blood obtained from ten women with AN and ten without [72]. The methods used in these preliminary studies investigate narrow genomic regions and do not support examinations into functional gene pathways. Recently, our group has applied microarray assays using the Infinium Illumina Human Methylation 450 K BeadChip, which provide the fullest coverage available of GW methylation. Our preliminary study used the 450 K BeadChip and DNA isolated from lymphocytes from 13 women with AN-R, 16 with AN-B/P and 15 normal weight controls [73]. False discovery rate (FDR) corrected comparisons identified 14 CpG probes on which significant between-group differences were present, including two CpGs associated with the NR1H3 gene, and three associated with the PXDNL gene. These genes have been implicated in histone acetylation and RNA modification (i.e., gene expression), cholesterol storage and lipid transport, and dopamine and glutamate. We also found associations between cumulative duration of illness and methylation levels at 142 probes, including genes related to liver function, immune function, metabolism, and behavior.

5.8 Conclusions and Future Directions

Although a number of preliminary studies on methylation in EDs have emerged, the precise contributions of epigenetic processes to ED development still need to be ascertained. Given evidence highlighting the importance of heredity, perinatal insults, developmental experiences, trauma, and social pressures for thinness in EDs, epigenetic mechanisms—by which such factors act to shape genetic expression—constitute a very plausible substrate for the convergence of such risks that ultimately contribute to ED development.

It has yet to be seen whether the findings in the above-mentioned studies in EDs will be replicated. Furthermore, a number of important limitations, such as small sample sizes, focus on candidate genes selected based on a priori hypotheses, use of blood and buccal cells as proxies to brain tissue, and methodological heterogeneity must be highlighted. In addition, future research should also explore epigenetic mechanisms other than DNA methylation, including histone modification, chromatin remodeling, and microRNA studies [74].

Epigenetic findings might contribute to the understanding of EDs in various ways. A thorough understanding of epigenetic principles could guide development of pharmacological or nutritional therapies that could be tailored to individual needs so as to increase response to standard treatment of EDs. Another way that epigenetic findings could contribute to the understanding of EDs would be that if a methylation "signature" associated with an active ED existed, this could inform clinical decision-making surrounding illness staging, entrenchment, and recovery.

In sum, epigenetically informed models help us realize that biological and environmental casual factors often act beyond the willful control of those affected by EDs. The consequence is that epigenetically informed perspectives may have a lot to offer in the sense of informing the understanding of why people develop EDs while contributing to genetically driven models that put less blame as compared to earlier etiological theories. Highlighting the potential benefits of communicating an epigenetically informed understanding of ED development to people with eating disorders, a recent study by Farrell and colleagues [75] reported that psychoeducation stressing epigenetic models of ED development (described as communicating a concept of "malleable biology") generated more prognostic optimism and self-efficacy for recovery in eating-disordered patients than did psychoeducational materials centered on singularly biological or psychological (cognitive-behavioral) concepts. We believe that an informed concept of the ways in which EDs represent the activation of real biological susceptibilities by real environmental triggers (as is inherent in an epigenetic perspective) promises to helps therapists and family members assume an optimally empathic and nonjudgmental stance with respect to eating-disordered people in their care and promotes maximal self-acceptance on the part of ED patients.

References

1. DaCosta M, Halmi KA. Classifications of anorexia nervosa: question of subtypes. Int J Eat Disord. 1992;11:305–13.
2. Steiger H, Coelho J, Thaler L, Van den Eynde F. Eating disorders. In: Blaney P, Krueger R, Millon T, editors. Oxford textbook of psychopathology, vol. 3. New York: Oxford University Press; 2015.
3. Steinhausen HC, Jakobsen H, Helenius D, Munk-Jørgensen P, Strober M. A nation-wide study of the family aggregation and risk factors in anorexia nervosa over three generations. Int J Eat Disord. 2015;48(1):1–8.

4. Strober M, Freeman R, Lampert C, Diamond J, Kaye W. Controlled family study of anorexia nervosa and bulimia nervosa: evidence of shared liability and transmission of partial syndromes. Am J Psychiatry. 2000;157(3):393–401.
5. Lilenfeld LR, Kaye WH, Greeno CG, Merikangas KR, Plotnicov K, Pollice C, et al. A controlled family study of anorexia nervosa and bulimia nervosa—psychiatric disorders in first-degree relatives and effects of proband comorbidity. Arch Gen Psychiatry. 1998;55(7):603–10.
6. Fowler SJ, Bulik CM. Family environment and psychiatric history in women with binge-eating disorder and obese controls. Behav Change. 1997;14(2):106–12.
7. Hudson JI, Lalonde JK, Berry JM, Pindyck LJ, Bulik CM, Crow SJ, et al. Binge-eating disorder as a distinct familial phenotype in obese individuals. Arch Gen Psychiatry. 2006;63(3):313–9.
8. Bulik CM, Sullivan PF, Tozzi F, Furberg H, Lichtenstein P, Pedersen NL. Prevalence, heritability, and prospective risk factors for anorexia nervosa. Arch Gen Psychiatry. 2006;63(3):305–12.
9. Klump KL, Miller K, Keel P, McGue M, Iacono W. Genetic and environmental influences on anorexia nervosa syndromes in a population–based twin sample. Psychol Med. 2001;31(4):737–40.
10. Bulik CM, Thornton LM, Root TL, Pisetsky EM, Lichtenstein P, Pedersen NL. Understanding the relation between anorexia nervosa and bulimia nervosa in a Swedish national twin sample. Biol Psychiatry. 2010;67(1):71–7.
11. Wade TD, Bulik CM, Neale M, Kendler KS. Anorexia nervosa and major depression: shared genetic and environmental risk factors. Am J Psychiatry. 2000;157(3):469–71.
12. Bulik CM, Sullivan PF, Kendler KS. Heritability of binge-eating and broadly defined bulimia nervosa. Biol Psychiatry. 1998;44(12):1210–8.
13. Trace SE, Thornton LM, Baker JH, Root TL, Janson LE, Lichtenstein P, et al. A behavioral-genetic investigation of bulimia nervosa and its relationship with alcohol use disorder. Psychiatry Res. 2013;208(3):232–7.
14. Javaras KN, Laird NM, Reichborn-Kjennerud T, Bulik CM, Pope HG, Hudson JI. Familiality and heritability of binge eating disorder: results of a case-control family study and a twin study. Int J Eat Disord. 2008;41(2):174–9.
15. Mitchell K, Neale M, Bulik C, Aggen S, Kendler K, Mazzeo S. Binge eating disorder: a symptom-level investigation of genetic and environmental influences on liability. Psychol Med. 2010;40(11):1899–906.
16. Di Bella D, Catalano M, Cavallini MC, Riboldi C, Bellodi L. Serotonin transporter linked polymorphic region in anorexia nervosa and bulimia nervosa. Mol Psychiatry. 2000;5(3):233.
17. Steiger H, Fichter M, Bruce KR, Joober R, Badawi G, Richardson J, et al. Molecular-genetic correlates of self-harming behaviors in eating-disordered women: findings from a combined Canadian–German sample. Prog Neuro-Psychopharmacol Biol Psychiatry. 2011;35(1):102–6.
18. Thaler L, Groleau P, Joober R, Bruce KR, Israel M, Badawi G, et al. Epistatic interaction between 5HTTLPR and TPH2 polymorphisms predicts novelty seeking in women with bulimia nervosa spectrum disorders. Psychiatry Res. 2013;208(1):101–3.
19. Steiger H, Richardson J, Joober R, Gauvin L, Israel M, Bruce KR, et al. The 5HTTLPR polymorphism, prior maltreatment and dramatic-erratic personality manifestations in women with bulimic syndromes. J Psychiatry Neurosci. 2007;32(5):354–62.
20. Moriya J, Takimoto Y, Yoshiuchi K, Shimosawa T, Akabayashi A. Plasma agouti-related protein levels in women with anorexia nervosa. Psychoneuroendocrinology. 2006;31(9):1057–61.
21. Ferron F, Considine RV, Peino R, Lado IG, Dieguez C, Casanueva FF. Serum leptin concentrations in patients with anorexia nervosa, bulimia nervosa and non-specific eating disorders correlate with the body mass index but are independent of the respective disease. Clin Endocrinol (Oxf). 1997;46(3):289–93.

22. Monteleone P, Tortorella A, Castaldo E, Di Filippo C, Maj M. The Leu72Met polymorphism of the ghrelin gene is significantly associated with binge eating disorder. Psychiatr Genet. 2007;17(1):13–6.
23. Versini A, Ramoz N, Le Strat Y, Scherag S, Ehrlich S, Boni C, et al. Estrogen receptor 1 gene (ESR1) is associated with restrictive anorexia nervosa. Neuropsychopharmacology. 2010;35(8):1818–25.
24. Eastwood H, Brown K, Markovic D, Pieri L. Variation in the ESR1 and ESR2 genes and genetic susceptibility to anorexia nervosa. Mol Psychiatry. 2002;7(1):86–9.
25. Rosenkranz K, Hinney A, Ziegler A, Hermann H, Fichter M, Mayer H, et al. Systematic mutation screening of the estrogen receptor beta gene in probands of different weight extremes: identification of several genetic variants. J Clin Endocrinol Metab. 1998;83(12):4524–7.
26. Nilsson M, Naessen S, Dahlman I, Hirschberg AL, Gustafsson J-Å, Dahlman-Wright K. Association of estrogen receptor β gene polymorphisms with bulimic disease in women. Mol Psychiatry. 2004;9(1):28–34.
27. Favaro A, Tenconi E, Santonastaso P. The interaction between perinatal factors and childhood abuse in the risk of developing anorexia nervosa. Psychol Med. 2010;40(4):657–65.
28. St-Hilaire A, Steiger H, Liu A, Laplante DP, Thaler L, Magill T, et al. A prospective study of effects of prenatal maternal stress on later eating-disorder manifestations in affected offspring: preliminary indications based on the project ice storm cohort. Int J Eat Disord. 2015;48(5):512–6.
29. Wonderlich SA, Brewerton TD, Jocic Z, Dansky BS, Abbott DW. Relationship of childhood sexual abuse and eating disorders. J Am Acad Child Adolesc Psychiatry. 1997;36(8): 1107–15.
30. Fairburn CG, Cooper Z, Shafran R. Cognitive behaviour therapy for eating disorders: a "transdiagnostic" theory and treatment. Behav Res Ther. 2003;41(5):509–28.
31. Steiger H, Gauvin L, Joober R, Israel M, Badawi G, Groleau P, et al. Interaction of the BclI glucocorticoid receptor polymorphism and childhood abuse in bulimia nervosa (BN): relationship to BN and to associated trait manifestations. J Psychiatr Res. 2012;46(2):152–8.
32. Dauncey MJ. Genomic and epigenomic insights into nutrition and brain disorders. Nutrients. 2013;5(3):887–914.
33. McGowan PO, Meaney MJ, Szyf M. Diet and the epigenetic (re)programming of phenotypic differences in behavior. Brain Res. 2008;1237:12–24.
34. Crider KS, Yang TP, Berry RJ, Bailey LB. Folate and DNA methylation: a review of molecular mechanisms and the evidence for folate's role. Adv Nutr Int Rev J. 2012;3(1):21–38.
35. Suderman M, McGowan PO, Sasaki A, Huang TC, Hallett MT, Meaney MJ, et al. Conserved epigenetic sensitivity to early life experience in the rat and human hippocampus. Proc Natl Acad Sci. 2012;109(Supplement 2):17266–72.
36. Liu D, Diorio J, Tannenbaum B, Caldji C, Francis D, Freedman A, et al. Maternal care, hippocampal glucocorticoid receptors, and hypothalamic–pituitary–adrenal responses to stress. Science. 1997;277(5332):1659–62.
37. Meaney MJ. Maternal care, gene expression, and the transmission of individual differences in stress reactivity across generations. Annu Rev Neurosci. 2001;24(1):1161–92.
38. McGowan PO, Sasaki A, D'Alessio AC, Dymov S, Labonte B, Szyf M, et al. Epigenetic regulation of the glucocorticoid receptor in human brain associates with childhood abuse. Nat Neurosci. 2009;12(3):342–8.
39. Beach SRH, Brody GH, Todorov AA, Gunter TD, Philibert RA. Methylation at 5HTT mediates the impact of child sex abuse on women's antisocial behavior: an examination of the Iowa Adoptee sample. Psychosom Med. 2011;73(1):83–7.
40. Beach SRH, Brody GH, Todorov AA, Gunter TD, Philibert RA. Methylation at SLC6A4 is linked to family history of child abuse: an examination of the Iowa Adoptee sample. Am J Med Genet B Neuropsychiatr Genet. 2010;153B(2):710–3.
41. O'Connor TG, Heron J, Golding J, Glover V. Maternal antenatal anxiety and behavioural/emotional problems in children: a test of a programming hypothesis. J Child Psychol Psychiatry. 2003;44(7):1025–36.

42. Cao-Lei L, Massart R, Suderman MJ, Machnes Z, Elgbeili G, Laplante DP, et al. DNA methylation signatures triggered by prenatal maternal stress exposure to a natural disaster: Project Ice Storm. PLoS One. 2014;9(9):e107653.
43. Oberlander TF, Weinberg J, Papsdorf M, Grunau R, Misri S, Devlin AM. Prenatal exposure to maternal depression, neonatal methylation of human glucocorticoid receptor gene (NR3C1) and infant cortisol stress responses. Epigenetics. 2008;3(2):97–106.
44. Weaver IC. Integrating early life experience, gene expression, brain development, and emergent phenotypes: unraveling the thread of nature via nurture. Adv Genet. 2014;86:277–307.
45. Alter MD, Gilani AI, Champagne FA, Curley JP, Turner JB, Hen R. Paternal transmission of complex phenotypes in inbred mice. Biol Psychiatry. 2009;66(11):1061–6.
46. MacLennan NK, James SJ, Melnyk S, Piroozi A, Jernigan S, Hsu JL, et al. Uteroplacental insufficiency alters DNA methylation, one-carbon metabolism, and histone acetylation in IUGR rats. Physiol Genomics. 2004;18(1):43–50.
47. Sinclair KD, Allegrucci C, Singh R, Gardner DS, Sebastian S, Bispham J, et al. DNA methylation, insulin resistance, and blood pressure in offspring determined by maternal periconceptional B vitamin and methionine status. Proc Natl Acad Sci. 2007;104(49):19351–6.
48. Franzek EJ, Sprangers N, Janssens A, Van Duijn CM, Van De Wetering BJ. Prenatal exposure to the 1944–45 Dutch 'hunger winter'and addiction later in life. Addiction. 2008;103(3): 433–8.
49. Strober M, Peris T, Steiger H. The plasticity of development: how knowledge of epigenetics may advance understanding of eating disorders. Int J Eat Disord. 2014;47(7):696–704.
50. Vanhees K, Vonhögen IG, van Schooten FJ, Godschalk RW. You are what you eat, and so are your children: the impact of micronutrients on the epigenetic programming of offspring. Cell Mol Life Sci. 2014;71(2):271–85.
51. Badun C, Evers S, Hooper M. Food security and nutritional concerns of parents in an economically disadvantaged community. J Can Diet Assoc. 1995;56:75–80.
52. Devaney BL, Assistance F. Assessing the nutrient intakes of vulnerable subgroups. Washington, DC: USDA Economic Research Service; 2005.
53. McGarel C, Pentieva K, Strain J, McNulty H. Emerging roles for folate and related B-vitamins in brain health across the lifecycle. Proc Nutr Soc. 2015;74(1):46–55.
54. Rampersaud GC, Kauwell GP, Hutson AD, Cerda JJ, Bailey LB. Genomic DNA methylation decreases in response to moderate folate depletion in elderly women. Am J Clin Nutr. 2000;72(4):998–1003.
55. Pogribny IP, Karpf AR, James SR, Melnyk S, Han T, Tryndyak VP. Epigenetic alterations in the brains of Fisher 344 rats induced by long-term administration of folate/methyl-deficient diet. Brain Res. 2008;1237:25–34.
56. Niculescu MD, Craciunescu CN, Zeisel SH. Dietary choline deficiency alters global and gene-specific DNA methylation in the developing hippocampus of mouse fetal brains. FASEB J. 2006;20(1):43–9.
57. Mandaviya PR, Stolk L, Heil SG. Homocysteine and DNA methylation: a review of animal and human literature. Mol Genet Metab. 2014;113(4):243–52.
58. Loria-Kohen V, Gómez-Candela C, Palma-Milla S, Amador-Sastre B, Hernanz A, Bermejo LM. A pilot study of folic acid supplementation for improving homocysteine levels, cognitive and depressive status in eating disorders. Nutr Hosp. 2013;28(3):807–15.
59. Frieling H, Römer KD, Scholz S, Mittelbach F, Wilhelm J, De Zwaan M, et al. Epigenetic dysregulation of dopaminergic genes in eating disorders. Int J Eat Disord. 2010;43(7):577–83.
60. Pjetri E, Dempster E, Collier DA, Treasure J, Kas MJ, Mill J, et al. Quantitative promoter DNA methylation analysis of four candidate genes in anorexia nervosa: a pilot study. J Psychiatr Res. 2013;47(2):280–2.
61. Groleau P, Joober R, Israel M, Zeramdini N, DeGuzman R, Steiger H. Methylation of the dopamine D2 receptor (DRD2) gene promoter in women with a bulimia-spectrum disorder: associations with borderline personality disorder and exposure to childhood abuse. J Psychiatr Res. 2014;48(1):121–7.

62. Ehrlich S, Weiss D, Burghardt R, Infante-Duarte C, Brockhaus S, Muschler MA, et al. Promoter specific DNA methylation and gene expression of POMC in acutely underweight and recovered patients with anorexia nervosa. J Psychiatr Res. 2010;44(13):827–33.
63. Ehrlich S, Roffman JL, Weiss D, Puls I, Burghardt R, Lehmkuhl U, et al. Smoking, but not malnutrition, influences promoter-specific DNA methylation of the proopiomelanocortin gene in patients with and without anorexia nervosa. Can J Psychiatry. 2012;57(3):168.
64. Frieling H, Bleich S, Otten J, Römer K, Kornhuber J, de Zwaan M, et al. Epigenetic downregulation of atrial natriuretic peptide but not vasopressin mRNA expression in females with eating disorders is related to impulsivity. Neuropsychopharmacology. 2008;33(11):2605–9.
65. Frieling H, Gozner A, Römer K, Lenz B, Bönsch D, Wilhelm J, et al. Global DNA hypomethylation and DNA hypermethylation of the alpha synuclein promoter in females with anorexia nervosa. Mol Psychiatry. 2007;12(3):229–30.
66. Kim Y-R, Kim J-H, Kim MJ, Treasure J. Differential methylation of the oxytocin receptor gene in patients with anorexia nervosa: a pilot study. PLoS One. 2014;9(2):e88673.
67. Frieling H, Albrecht H, Jedtberg S, Gozner A, Lenz B, Wilhelm J, et al. Elevated cannabinoid 1 receptor mRNA is linked to eating disorder related behavior and attitudes in females with eating disorders. Psychoneuroendocrinology. 2009;34(4):620–4.
68. Schroeder M, Eberlein C, de Zwaan M, Kornhuber J, Bleich S, Frieling H. Lower levels of cannabinoid 1 receptor mRNA in female eating disorder patients: association with wrist cutting as impulsive self-injurious behavior. Psychoneuroendocrinology. 2012;37(12):2032–6.
69. Steiger H, Labonté B, Groleau P, Turecki G, Israel M. Methylation of the glucocorticoid receptor gene promoter in bulimic women: associations with borderline personality disorder, suicidality, and exposure to childhood abuse. Int J Eat Disord. 2013;46(3):246–55.
70. Thaler L, Gauvin L, Joober R, Groleau P, de Guzman R, Ambalavanan A, et al. Methylation of BDNF in women with bulimic eating syndromes: associations with childhood abuse and borderline personality disorder. Prog Neuro-Psychopharmacol Biol Psychiatry. 2014;54:43–9.
71. Tremolizzo L, Conti E, Bomba M, Uccellini O, Rossi M, Marfone M, et al. Decreased whole-blood global DNA methylation is related to serum hormones in anorexia nervosa adolescents. World J Biol Psychiatry. 2014;15(4):327–33.
72. Saffrey R, Novakovic B, Wade TD. Assessing global and gene specific DNA methylation in anorexia nervosa: a pilot study. Int J Eat Disord. 2014;47(2):206–10.
73. Booij L, Casey KF, Antunes JM, Szyf M, Joober R, Israël M, et al. DNA methylation in individuals with anorexia nervosa and in matched normal-eater controls: a genome-wide study. Int J Eat Disord. 2015;48(7):874–82.
74. Yilmaz Z, Hardaway A, Bulik C. Genetics and epigenetics of eating disorders. Adv Genomics Genet. 2015;5:131–50.
75. Farrell NR, Lee AA, Deacon BJ. Biological or psychological? Effects of eating disorder psychoeducation on self-blame and recovery expectations among symptomatic individuals. Behav Res Ther. 2015;74:32–7.

6 Drug Addiction and DNA Modifications

Amber N. Brown and Jian Feng

Abstract

Drug addiction is a complex disorder which can be influenced by both genetic and environmental factors. Research has shown that epigenetic modifications can translate environmental signals into changes in gene expression, suggesting that epigenetic changes may underlie the causes and possibly treatment of substance use disorders. This chapter will focus on epigenetic modifications to DNA, which include DNA methylation and several recently defined additional DNA epigenetic changes. We will discuss the functions of DNA modifications and methods for detecting them, followed by a description of the research investigating the function and consequences of drug-induced changes in DNA methylation patterns. Understanding these epigenetic changes may provide us translational tools for the diagnosis and treatment of addiction in the future.

Keywords

drug addiction • DNA methylation • DNA modification • DNMT • TET • cocaine • alcohol • nicotine

A.N. Brown
Department of Biological Science, Florida State University, 319 Stadium Drive, Tallahassee, FL 32306, USA
e-mail: Brown@bio.fsu.edu

J. Feng (✉)
Department of Biological Science, Florida State University, 319 Stadium Drive, Tallahassee, FL 32306, USA

Neuroscience Program, Florida State University, 319 Stadium Drive, Tallahassee, FL 32306, USA
e-mail: feng@bio.fsu.edu

R. Delgado-Morales (ed.), *Neuroepigenomics in Aging and Disease*, Advances in Experimental Medicine and Biology, DOI 10.1007/978-3-319-53889-1_6

6.1 DNA Epigenetic Modifications

The nucleus of a mammalian cell houses approximately 2 m of negatively charged DNA. In order to package such a large amount of genetic material into a nucleus measuring ~10 μm across, multiple means of compaction are required. DNA is tightly wrapped around positively charged histone proteins to form the nucleosome, the founding unit of the DNA packaging material called chromatin [1]. The DNA and histone proteins can be chemically modified in numerous ways in order to change the binding relationship of the DNA-nucleosome complex. Tightly bound nucleosomal DNA is considered heterochromatin, which is generally transcriptionally inactive due to restricted access to DNA by transcriptional machinery. Loosely bound or nucleosome-free DNA is considered euchromatin, which is freely accessible to the transcriptional machinery and actively transcribed [2]. DNA itself can be covalently modified by a class of enzymes called DNA methyltransferases (DNMTs) which catalyze a reaction that adds a methyl group to the C5 position of a cytosine base (5mC) and is traditionally observed at cytosine-guanine dinucleotide (CpG) residues [3]. Replicative maintenance of DNA methylation, copying an existing 5mC onto the complementary DNA strand, following cell division, is accomplished by the maintenance DNA methyltransferase DNMT1 [4]. DNMT3a and DNMT3b are considered de novo methyltransferases, responsible for methylating previously unmethylated cytosines to establish a pattern of DNA methylation [5, 6].

Until recently, it wasn't known if or how DNA methylation was reversed. Several proteins (GADD45a, MBD2, DNMT3a, and DNMT3b) have been reported to catalyze DNA methylation, either by direct removal of methyl groups or via oxidation and repair by DNA repair processes. However, subsequent reports failed to substantiate these claims [7]. In 2009, it was demonstrated that the TET family of proteins (ten-eleven translocation proteins) oxidizes 5mC to 5-hydroxymethylcytosine (5hmC) [8, 9]. 5hmC can be further oxidized to 5-formylcytosine (5fC) and 5-carboxylcytosine, which can then be recognized by DNA damage pathways and repaired to unmethylated cytosine [10–15]. TET-mediated oxidation of mC to 5hmC can be an active process in the brain [16]. Interestingly, 5hmC is relatively stable [17] and present at higher levels in the brain than any other tissue [18], suggesting a specific function for 5hmC in the genetic regulation of neuronal function. Thus, we now recognize that DNA epigenetic modifications can be a labile mechanism for regulation of gene expression in non-mitotic neuronal populations (Fig. 6.1).

DNA methylation can have differential effects on transcriptional capacity, depending on the genomic context. 5mC within gene promoter regions is typically associated with a decrease in transcription, and these effects have been well studied. The presence of 5mC attracts methyl-binding domain proteins (MBD1-MBD4, MeCP2, and Kaiso) which, in turn, recruit repressor complexes [3] and histone deacetylases (HDACs) [19] to downregulate local transcriptional activity. Deacetylation of histone tails increases the affinity of the DNA-nucleosome interaction, thereby generating local regions of heterochromatin and decreased transcriptional capacity for the region. Extensive DNA methylation can result in complete

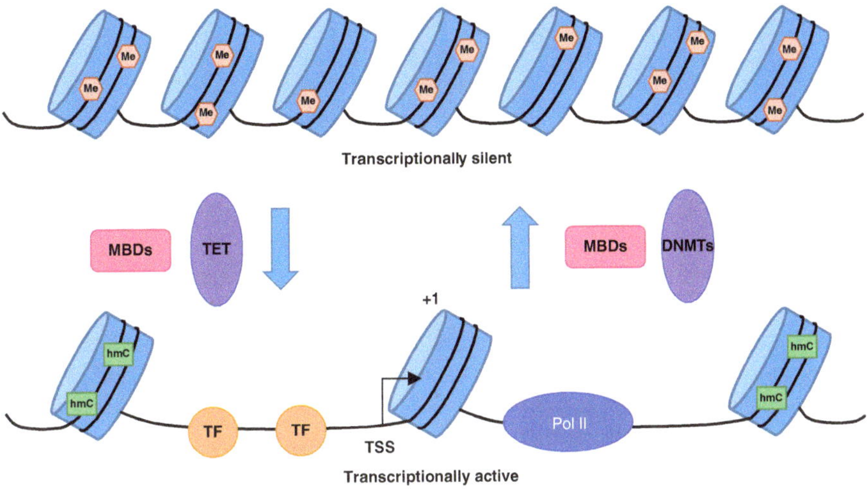

Fig. 6.1 DNA methylation and transcriptional activity (attached). When DNA is heavily methylated (*top*) by DNA methyl transferases (DNMTs), chromatin is tightly packaged and bound by methyl-CpG-binding domain (MBD) proteins, which recruit other heterochromatinizing proteins and render genic regions inaccessible and transcriptionally inactive. Genes lacking 5-methylcytosine (5mC) marks are more loosely packaged into chromatin, readily accessible by the transcriptional machinery and more likely to be transcriptionally active. TET proteins can oxidize 5-methylcytosine to 5-hydroxymethylcytosine (5hmC), which can also bind MBDs. However, in contrast to 5mC, 5hmC within enhancer and genic regions is associated with transcriptional activity

silencing of a gene or local cluster of genes, as has been shown during neuronal fate specification/development [20]. DNA methylation can also interfere with the specific binding of transcription factors, which can only bind to an unmethylated version of its binding site. 5mC within a gene body has been linked to active transcription [21], transcriptional elongation [22] and alternative splicing [23].

The functional consequences of 5hmC are just recently being recognized, but seem to be independent from those of 5mC. The mark is enriched within transcriptionally active genes, enhancers, and brain MBD proteins like MeCP2 which binds the mark with similar affinity as to 5mC [24, 25]. In fact, using mouse embryonic stem cells in a screen for CpG-binding proteins, researchers found few proteins bound preferentially to 5hmC. 5fC, however, was enriched for specific protein binding of several chromatin remodeling proteins and transcriptional regulators [26]. Whether 5hmC or even 5fC enrichment or depletion is correlated with transcription levels has yet to be determined, as results have varied depending on the model system and genomic context under investigation [14]. For example, the genomic distribution of 5hmC differs between neurons and embryonic stem cells. In neurons, 5hmC is enriched in gene bodies of expressed genes related to neuronal function [24], while in embryonic stem cells 5hmC is enriched at enhancers and depleted from transcription factor binding sites [27].

6.2 Addiction

Addiction is a relapsing neuropsychiatric disorder characterized by compulsive drug seeking with repeated and increased use, despite adverse consequences. Drugs of abuse include, but are not limited to, cocaine, nicotine, amphetamine, methamphetamine, heroin, morphine, and other opiates. Addictive drugs stimulate the brain's natural reward system through the release or synaptic accumulation of the neurotransmitter dopamine ([28, 29]). Stimulation of the reward system also engages learning responses in the brain. With repeated drug use, the dopamine-producing cells increasingly respond to drug-associated cues—environmental stimuli commonly experienced with drug use (people, places, smells, imagery)—such that the cues alone elicit a dopamine response and drive craving for the drug [30]. Whereas natural rewards would normally cause dopamine cells to stop firing once a reward is achieved, drugs of abuse override this process and continue to stimulate powerfully high amounts of dopamine release. The excessively rewarding effects of drugs often override the more balanced dopamine released by natural rewards. Eventually, natural rewards become less reinforcing, and motivation switches to achieving the elevated dopamine release generated by the drugs. As the brain adapts to elevated dopamine levels, tolerance to the drug begins to develop, wherein increasing amounts of the substance are required for the user to achieve the desired degree of euphoria. However, in the absence of the drug, the user experiences a hypodopaminergic, dysphoric state and may seek out the drug just to relieve the discomfort. Thus, addiction becomes a vicious cycle in which the user seeks to relieve the symptoms of the disease by engaging in the behaviors which initiated the disease to begin with [29, 31]. Addiction is a worldwide problem which significantly impacts the health, economic, and social fabric of billions of people. In order to relieve this burden, researchers have sought to understand the genetic and environmental causes of substance use disorders.

Addiction is a complex disease resulting from a combination of both genetic and environmental risk factors. It is estimated that only about 10% of people exposed to addictive drugs will experience a severe substance use disorder [32], while the remaining 90% have protective genetic and/or environmental factors. In order to better understand the genetic factors involved in addiction, human studies have been conducted in drug addicts, former drug addicts, and postmortem brains of addicts. Many such studies have identified associations between drug use and allelic variants which may predispose an individual to risk-taking or drug-seeking behaviors. These genes are often related to neurotransmitter function or synaptic plasticity and include serotonin transporter and receptors, dopamine transporter and receptors, opioid receptors, GABA receptors, and MAOA (reviewed by [33]).

6.3 Neuroepigenetics of Addiction

In neuroscience, epigenetic studies have begun to help explain how a genetically stable, nondividing population of neurons can make activity-dependent changes in gene expression of either transient or lasting duration. Changes in DNA methylation around synaptic plasticity genes of neurons and nonneuronal cell types of the brain accompany the acquisition and maintenance of memory [34, 35] and changes in hydroxymethylation levels correlated with transcriptional and behavioral outcomes have been identified following fear extinction [36] and stress [37]. The DNA methylation detection and quantitation methods commonly used in neuroepigenetic studies have recently been applied to the study of addiction. While there have been several candidate gene studies of DNA methylation changes following drugs of abuse (detailed below), few have explored genome-wide changes in DNA methylation (Table 6.1). High-throughput sequencing of DNA methylation analyses can provide a global view of such changes with a potential benefit at single-base pair resolution and, coupled with mRNA sequencing transcriptome profiling, can help researchers probe the associations between changes in DNA methylation and transcriptional outcomes observed in addiction models.

6.3.1 Human Studies

Several human epigenome-wide association studies (EWAS) have linked genome-wide DNA methylation changes in whole blood samples to cigarette smoking (reviewed by [69]). From these EWAS studies and locus-specific methylation studies, several candidate genes have been identified as harboring DNA methylation changes among cells isolated from smokers' blood samples: MAOA (monoamine oxidase A) [40], MAOB (monoamine oxidase B) [64], COMT (catechol o-methyltransferase) [65, 68], AHRR (aryl-hydrocarbon receptor repressor) [67], and POMC (proopiomelanocortin) [66]. Cigarette smoking has also been linked to changes in DNA methylation in several tissue and cell types; however, non-nicotinic chemicals present in cigarettes can also lead to DNA damage and changes in DNA methylation and gene expression related to inflammation or hypoxia [70–72], making analysis of the effects of cigarette smoking a complicated endeavor.

Alcohol dependence has also been associated with genome-wide changes in blood cell DNA methylation [73, 74], some of which have been shown to reverse with the progression of abstinence [48]. Gene-specific studies have also shown an association between alcohol dependence and hypermethylation of the DAT (dopamine transporter) promoter, HERP (homocysteine-induced endoplasmic reticulum protein) promoter, and α-synuclein promoter [38, 39, 42], while POMC promoter methylation has been linked to alcohol dependence [46] and craving in alcohol-dependent subjects [43]. In addition, the severity of alcoholics' drinking patterns was found to be negatively correlated to DNA methylation of a cluster of CpGs associated with the promoter region of the NR2B (NMDA receptor 2B) gene [41]. Using postmortem human brains, researchers found an association between alcohol dependence and differential DNA methylation within the 3′-UTR of the PDNY (prodynorphin) gene [44] as well as hypomethylation of endogenous retroviruses in the frontal cortex of alcoholics [45].

CpG sites within the BDNF (brain-derived neurotrophic factor) promoter of patient blood cells have been shown to be significantly associated with

Table 6.1 Alterations of DNA epigenetic modifications in addiction

Drug of abuse	Genomic region	Gene(s)	Direction of change	Associated mRNA/protein change	Species	Tissue/cell type	DNA methylation method	References
Alcohol	Promoter	α-Synuclein	Increased DNA methylation	Elevated homocysteine levels	Human	Whole blood	Restriction endonuclease/qPCR	[38]
Alcohol	Promoter	HERP	Increased DNA methylation	Increased HERP mRNA	Human	Whole blood	Restriction endonuclease/qPCR	[39]
Alcohol	Promoter	MAOA	Increased DNA methylation	N/A	Human women	Lymphoblast	Bisulfite/mass spectrometry	[40]
Alcohol	Promoter	NR2B	Decreased DNA methylation	Increased NR2B mRNA	Human	Peripheral blood cells	Bisulfite/sequencing	[41]
Alcohol	Promoter	DAT	Increased DNA methylation	N/A	Human	Leukocytes	Restriction endonuclease/qPCR	[42]
Alcohol	Promoter	POMC	Increased DNA methylation	N/A	Human	Whole blood	Bisulfite/sequencing	[43]
Alcohol	3′-UTR	PDNY	Increased DNA methylation	Increased PDNY mRNA	Human	Postmortem PFC	Bisulfite/sequencing	[44]
Alcohol	Genome-wide	Genome-wide	Decreased DNA methylation	Increased expression of ERVs and CG-rich genes	Human	Postmortem cortex	Microarray	[45]

Alcohol	Genome-wide	GABRB3, POMC, HTR3A, NCAM1, DRD4, MBD3, HTR2B, GRIN1	Increased DNA methylation	N/A	Human	Peripheral blood	Microarray	[46]
Alcohol	Genome-wide	Genome-wide	Decreased DNA methylation	N/A	Human	Lymphocytes	Microarray	[47]
Alcohol	Genome-wide	Genome-wide	Both—diminish with progression of abstinence	N/A	Human	Peripheral blood mononuclear cells	Microarray	[48]
Alcohol	Genome-wide	SYT2	Increased DNA methylation	Decreased mRNA for SYT2 and genes encoding proteins involved in neurotransmitter release	Rats	mPFC	Antibody affinity	[49]
Cocaine	Promoter	PP1c	Increased DNA methylation	Decreased PP1c mRNA	Mouse	NAc	Me-DIP/PCR	[50]
Cocaine	Promoter	CDKL5	Increased DNA methylation	Decreased Cdkl5 mRNA	Rats	Striatum	Bisulfite/qPCR	[51]

(continued)

Table 6.1 (continued)

Drug of abuse	Genomic region	Gene(s)	Direction of change	Associated mRNA/protein change	Species	Tissue/cell type	DNA methylation method	References
Cocaine	Genome-wide	Genome-wide	Increased DNA methylation	N/A	Mouse	NAc	Antibody affinity	[52]
Cocaine	Genome-wide	Genome-wide	Decreased DNA methylation	N/A	Mouse	PFC	LC-ESI-MS/MS	[53]
Cocaine	Promoters	Cocaine-responsive genes	Both	mRNA inversely correlated to DNA methylation changes in cocaine-responsive genes	Mouse	NAc	Me-DIP	[54]
Cocaine	Genome-wide	Genome-wide	Both	N/A	Rats	PFC	MBD Ultra-Seq	[55]
Cocaine	Genome-wide	Genome-wide	Both	Altered expression of alternate splicing isoforms	Mouse	NAc	Bisulfite/oxidative bisulfite/sequencing	[56]
Cocaine	Genome-wide	Genome-wide	Both	mRNA partly inversely correlated to DNA methylation changes	Rats	NAc	Microarray	[57]

Cocaine	Genome-wide	c-Fos	Decreased DNA methylation	Increased c-FOS mRNA	Rats	NAc	LC-ESI-MS/MS	[58]
Heroin/methadone	Promoter	OPRM1	Increased DNA methylation	N/A	Human	Lymphocytes	Bisulfite/sequencing	[59]
Heroin	Promoter	BDNF	Decreased DNA methylation	N/A	Human	Whole blood	Bisulfite/sequencing	[60]
Methamphetamine	Genome-wide	GLUA1, GLUA2	Decreased DNA methylation 5hmC	Decreased GLUA1, GLUA2 mRNA	Rats	NAc	5hmC microarray	[61]
Methamphetamine	Promoter and intronic	Immediate early genes	Both	Decreased mRNA	Mouse	PFC, hippocampus	Bisulfite/sequencing	[62]
Methamphetamine	Genome-wide	KCNMA1, KCNN1, KCNN2	Increased 5hmC	Increased mRNA for KCNMA1, KCNN1, KCNN2	Rats	NAc	hMe-DIP/sequencing	[63]
Methamphetamine	Promoter	BDNF	Decreased DNA methylation	N/A	Human	Whole blood	Bisulfite/sequencing	[60]
Nicotine	Promoter	MAOA	Increased DNA methylation	N/A	Human women	Lymphoblast	Bisulfite/mass spectrometry	[40]

(continued)

Table 6.1 (continued)

Drug of abuse	Genomic region	Gene(s)	Direction of change	Associated mRNA/protein change	Species	Tissue/cell type	DNA methylation method	References
Nicotine	Promoter	MAOB	Decreased DNA methylation	Decreased MAOA protein	Human	Platelet and plasma	Bisulfite/sequencing	[64]
Nicotine	promoter	COMT	Increased DNA methylation	N/A	Human	Whole blood	Bisulfite/sequencing	[65]
Nicotine	Promoter	POMC	Decreased DNA methylation	N/A	Human	Peripheral blood mononuclear cells	Bisulfite/sequencing	[66]
Nicotine	Genome-wide	AHRR	Decreased DNA methylation	N/A	Human	Whole blood	Microarray	[67]
Nicotine	Promoter	COMT	Increased DNA methylation	N/A	Human adolescents	Whole blood	Bisulfite/mass spectrometry	[68]

methamphetamine and heroin addiction [60], and methadone-maintained former heroin addicts have increased DNA methylation at the OPRM1 (opioid receptor mu 1) promoter, leading to a decrease in OPRM1 gene expression in lymphocytes [59]. Exposure to social stressors can even lead to addiction-related changes in DNA methylation patterns. One group showed that lower socioeconomic status during adolescence is associated with increased blood cell DNA methylation in the promoter of the serotonin transporter gene, predicting changes in risk-related brain functions and predisposing these individuals to an increased addiction susceptibility [75].

6.3.2 Animal Studies

While human studies can provide insight into some of the genes involved in the process of addiction, controlled animal studies are necessary to fully investigate and manipulate experimental conditions to display detailed underpinnings. To date, much of the research on addiction has utilized rodent models of exposure. One animal model to human addiction is the self-administration (SA) model, wherein a rodent is trained to press a lever or a button to receive an intravenous infusion of a drug. This model best recapitulates the addiction process, as the animals will seek out the drug more frequently and persistently. Given the cost, time, and technical challenges related to the SA model, many researchers apply intraperitoneal (i.p.) drug injections, and although this model may not engage the brain regions involved in the choices an addict makes, it can successfully elucidate the direct behavioral, chemical, and genetic effects of the drug. Investigators using i.p. drug administration also employ a behavioral conditioning paradigm called conditioned place preference (CPP) to assess an animal's preference for a drug based on their preference to be in the same context or environment as where the drug was administered. These models are considered the standards in addiction research today, and their utilization makes for a more translational approach to understand the disease.

In the 2000s, epigenetic studies of psychostimulant exposure provided a hint that changes in DNA methylation may be occurring. In 2006 it was reported that following 10 days of i.p. cocaine injections, methyl-binding proteins MeCP2 and MBD1 were significantly induced in the caudate-putamen, frontal cortex, and dentate gyrus of adult rats. These changes were accompanied by an increase in HDAC2 (histone deacetylase 2) and deacetylated histones, presumably leading to reduced transcription [76]. It was subsequently shown that cocaine-induced MeCP2 was accompanied by increased MeCP2 binding at the Cdkl5 promoter and repression of the Cdkl5 gene in the striatum of cocaine-treated rats. In order to examine DNA methylation changes, DNA was subjected to sodium bisulfite treatment. Using this method, only unmethylated cytosines are converted to uracil. Subsequent comparison of untreated and bisulfite treated DNA can reveal which cytosines are methylated or unmethylated at single-base pair resolution [77] either at the single-locus level or genome-wide. Using bisulfite-converted DNA and Cdkl5-specific primers, it was shown that DNA methylation at the Cdkl5 promoter was inversely correlated with transcription of Cdkl5 mRNA [51]. Cdkl5, like MeCP2, is mutated in some forms of the autism-like Rett Syndrome [78]. However, its role in the action of cocaine is still unknown. Similar results were observed studying rats self-administering cocaine; MeCP2 expression was increased in multiple brain reward regions, and knockdown of MeCP2 or pharmacologically inhibiting DNMTs with trichostatin A (a histone deacetylase inhibitor known to induce DNA demethylation [79]) attenuated cocaine self-administration [80–82] and amphetamine reward [83].

In 2010, two papers provided thorough investigations into the complex interactions of MeCP2, BDNF, and a specific microRNA, miR-212. Using a rat self-administration model of cocaine addiction, first, it was shown that expression of miR-212 is increased in the dorsal striatum of rats with extended access to cocaine self-administration and that miR-212 expression was inversely correlated with cocaine intake [84]. However, miR-212 is located in a genomic region dense in CpG islands and may be subject to regulation by MeCP2. Therefore, researchers investigated the interaction between MeCP2, miR-212, and cocaine intake in the same rat self-administration model. They found that miR-212 and MeCP2 expression are inversely correlated with one another; knockdown of MeCP2 increases miR-212 expression, and overexpression of miR-212 inhibits MeCP2 expression. MeCP2 is a known regulator of BDNF [85], which is known to promote sensitivity to cocaine [86]. It was also demonstrated that miR-212 also regulates BDNF expression *indirectly* through repression of MeCP2, such that a complicated feedback loop between BDNF, miR-212, and MeCP2 serves to regulate cocaine-taking behavior [81].

As it became recognized that DNA methylation plays a role in addiction, it was further demonstrated that repeated cocaine administration altered DNMT3a transcription (but not DNMT3b) in the mouse NAc [52]. Interestingly, the changes observed were time dependent; DNMT3a was upregulated 4 h after the last cocaine dose, but was subsequently downregulated 24 h later. Following a 28-day period of withdrawal from either i.p. cocaine or SA, DNMT3a was again found to be upregulated. When DNMT3a was overexpressed in the NAc, mice showed a decreased preference for cocaine in the CPP paradigm. These behavioral changes were accompanied by an increase in DNA methylation, as assayed by an ELISA-like colorimetric assay. In this assay, an antibody to 5mC recognizes methylated DNA, and a secondary antibody produces a color which is proportional to the amount of methylated DNA (Epigentek, Farmingdale, NY). Preference for cocaine could be attenuated by pharmacological inhibition using a DNMT inhibitor, RG108 [52]. The persistent induction of DNMT3a after a month of abstinence from cocaine may be of particular relevance to understanding the molecular susceptibility to relapse and warrants further investigation for potential therapeutic interventions.

In contrast to the previous study, another group reported that, when administered acutely, a single 15 mg/kg injection of cocaine was shown to upregulate both DNMT3a and DNMT3b in the mouse NAc [50]. This prompted an investigation of the DNA methylation status of NAc tissue using an immunoprecipitation-based method called Me-DIP (methylated DNA immunoprecipitation). This technique utilizes an antibody to 5mC to isolate methylated DNA from a pool of fragmented DNA [87]. Downstream analyses of Me-DIP fragments can be used for single-locus

PCR, microarray, or sequencing. The authors found that acute and repeated cocaine resulted in DNA hypermethylation and increased MeCP2 binding to the PP1c promoter, resulting in downregulation of the PP1c gene [50], as was seen with Cdkl5 [51]. Pharmacologically blocking DNMT activity decreased cocaine-induced PP1c hypermethylation and gene expression changes while delaying the development of cocaine-induced behavioral sensitization. However, the opposite effect was seen at the immediate early gene, FosB—DNA became hypomethylated and MeCP2 binding was decreased following a single cocaine injection [50]. Therefore, cocaine may not cause global changes in DNA methylation in a nonspecific manner. Rather, specific genes or networks of genes appear to be co-regulated at the level of chromatin following drug exposure. For example, in 2015, two groups found that chronic methamphetamine or alcohol consumption increased DNA methylation at CpG sites in synaptic plasticity-related genes, resulting in downregulation of associated mRNAs in rat frontal cortex [49, 62].

With increasing evidence that DNA methylation plays an important role in the progression of addiction, withdrawal, and relapse, the possibility of using the methyl donor methionine as a therapeutic gained interest. Pretreatment with methionine has been shown to reduce cocaine-conditioned place preference (CPP) in mice [52]. However, it is unknown if these effects were due to a genuine increase in DNA methylation or some other effects of methionine, as the DNA methylation status was not evaluated under these conditions [52].

Another group compared the rewarding effects of cocaine, morphine, and food using the CPP procedure and evaluated resulting changes in global DNA methylation by LC-ESI-MS/MS (liquid chromatography-electrospray ionization tandem mass spectrometry) [53]. In this method, LC is used to separate 5mC from the other nucleotides, and ESI-MS/MS can detect and quantify 5mC with high specificity and sensitivity [88]. This method can provide reliable quantitation of global DNA methylation levels with very low amounts of input DNA, but cannot be used to determine specific methylation patterns. Using this method, researchers found that cocaine, but not food or morphine, decreased DNA methylation and DNMT3b expression in the mouse prefrontal cortex. Treatment with methionine before and during the CPP procedure blocked the cocaine-induced decrease in DNMT3b expression and DNA methylation and attenuated cocaine preference, but had no effects on the establishment of food or morphine preference [53].

Conversely, it was shown that pretreatment of mice with methionine for 7 days significantly potentiated the development of cocaine-induced locomotor sensitization. NAc whole-genome gene expression profiling revealed that repeated SAM treatment affected cocaine-induced gene expression, nonspecifically dampening the cocaine response, in part due to decreased methyltransferase activity via downregulation of *Dnmt3a* mRNA. Using Me-DIP, they found specific hypo- and hypermethylation in the promoters of cocaine-responsive genes in the nucleus accumbens [54].

In 2015, another group similarly examined these changes in the nucleus accumbens of cocaine-sensitized and self-administering rats with or without methionine pretreatment. They showed that methionine pretreatment can upregulate DNMT3a and DNMT3b, and LC-ESI-MS/MS revealed global DNA hypomethylation in the NAc of cocaine-treated rats. The treatment blocked locomotor sensitization and reduced cocaine-primed reinstatement of self-administration. Conversely, the cocaine-induced upregulation and hypomethylation of c-Fos was reduced in rats receiving methionine, [58] again demonstrating that cocaine-induced changes in DNA methylation (as well as methionine-reversed changes) are likely gene-specific events. While the locomotor-sensitizing effects of methionine differ between the [54] study and the [58] study, this is possibly due to the differing routes of cocaine administration, as experimenter-administered injections do not engage the same circuits in the brain as does the self-administration model. Nevertheless, they show promise for nutritional supplementation with agents like methionine as a potential method of promoting or restoring a healthy methylome.

Not only does the experimental paradigm differentially affect DNA methylation, but abstinence and withdrawal also have characteristic changes in DNA methylation patterns. Using MBD Ultra-Seq, a method in which DNA fragments immunoprecipitated by MBD antibodies are sequenced [36], researchers found that 29 regions of the genome were differentially methylated in the medial prefrontal cortex of cocaine self-administering rats, but not in response to experimenter-administered cocaine. Furthermore, an additional 28 regions became differentially methylated during forced abstinence or withdrawal from cocaine [55]. In a similar study using Me-DIP coupled with a custom tiling microarray, it was found that, in addition to significant DNA methylation changes in the NAc during withdrawal from cocaine self-administration, cue-induced cocaine seeking (a model of a relapse paradigm) caused broad, time-dependent enhancement of DNA methylation alterations which were, in part, negatively correlated to gene expression. In addition, intra-NAc injections of DNMT inhibitor RG108, ESR1 agonist propyl pyrazole triol, and CDK5 inhibitor roscovitine each reduced or completely abolished cue-induced cocaine seeking [57]. These data show that DNA methylation and downstream targets of DNA methylation are viable targets for the treatment of drug craving and addiction.

With the advancement of molecular genetic techniques, researchers are now able to differentiate between different types of DNA methylation, namely, 5mC and 5hmC, which had previously been indistinguishable and lumped together using older methods. In the last few years, 5hmC has become recognized as a functional DNA modification that may lead to DNA demethylation. Using Me-DIP and hMe-DIP (hydroxymethylcytosine DNA immunoprecipitation), researchers showed that chronic methamphetamine treatment decreased enrichment of 5mC and 5hmC at the GluA1 and GluA2 genes while conversely increasing MeCP2 binding and decreasing GluA1 and GluA2 gene expression in rat striatum [61]. In addition, methamphetamine-addicted rats show differential 5hmC patterns in the nucleus accumbens, as determined using hMe-DIP sequencing. These changes were primarily concentrated in intergenic regions. However, differential 5hmC changes within gene bodies correlated with increased transcription of that gene product [63].

The TET1 enzyme, which is responsible for the oxidative conversion of methylated cytosine to hydroxymethylated cytosine, was shown to be downregulated in the nucleus accumbens of mice treated with cocaine as well [56]. This downregulation of TET1 was also found in the same brain region of cocaine addicts, when examined postmortem. Using bisulfite and oxidative bisulfite sequencing, 5hmC was elevated within enhancer and coding regions of the genome. When TET1 function was overexpressed or knocked down, it negatively regulates cocaine reward-type behaviors. Specifically, these intragenic changes in 5hmC increased expression of alternate splicing isoforms of many genes with important roles in addiction and could persist for at least one month following drug exposure [56].

6.4 Multigenerational Effects of Drug Exposure

Recent work has demonstrated that exposure to various chemical and environmental stressors can also cause changes in DNA methylation and transcriptional output, which can be transmitted to subsequent generations. Several groups have shown that parental exposure to drugs of abuse can have significant behavioral, biochemical, and neuroanatomical effects on the offspring (reviewed by [89]). Epigenetic mechanisms, such as DNA methylation and histone modifications, have been attributed to many such effects. For example, children exposed to cigarette smoke in utero also have altered patterns of DNA methylation within repetitive DNA elements LINE1 and AluYb8, which persisted through at least age 6 [90]. Rats exposed to cocaine during prenatal development have altered patterns of hippocampal DNA methylation with corresponding changes in transcriptional output [91].

Drug exposure during embryonic development not only exposes the developing fetus (F1) to the effects of the drug but also exposes the germ cells (F2) to these effects as well. Similarly, parental drug use exposes *their* germ cells, effectively exposing the F1 generation. Adolescent rat exposure to cannabinoid receptor agonist WIN 55,212–2 or THC caused genome-wide changes in male and female F1's DNA methylation status, associated changes in gene expression, and enhanced F1 offspring's sensitivity to morphine [92–95]. Research has revealed that altered patterns of DNA methylation can be transgenerationally inherited *beyond* the exposed generations (F3 for embryonic exposure and F2 for parental exposure) [96, 97]. This was shown for animal models in which the parents were exposed to chemical and environmental stressors such as stress [98], plastics and endocrine disruptors [99–101], pesticides, jet fuel, and dioxin [102, 103]. The epigenetic effects of prenatal exposure to the endocrine disruptor vinclozolin were shown to be transmitted through DNA methylation in the male germ cells [104]. Rodents self-administering cocaine show decreased DNMT1 in the seminiferous tubules [105] and males who consume heavy amount of alcohol have a reduction in hypermethylated, paternally imprinted regions of the sperm genome [106], indicating that cocaine and alcohol may also have DNA methylation effects on the male germ line which could be transmitted to subsequent generations.

Conclusion

The state of neuroepigenetic addiction research has progressed to a point where we can apply cell-specific, high-throughput technologies to determine drug-specific effects on DNA methylation and corresponding transcriptional and behavioral output. Thorough understanding of the mechanisms that drive the addiction process will enable researchers to develop diagnostic biomarkers and better therapeutic strategies for treatment and prevention of substance use disorders. As demonstrated with the transgenerational studies, efforts toward combating drug use and addiction will contribute to furthering the health and fitness of the worldwide population for generations to come.

References

1. Bednar J, Horowitz RA, Grigoryev SA, Carruthers LM, Hansen JC, Koster AJ, et al. Nucleosomes, linker DNA, and linker histone form a unique structural motif that directs the higher-order folding and compaction of chromatin. Proc Natl Acad Sci U S A. 1998;95(24):14173–8.
2. Jenuwein T, Allis CD. Translating the histone code. Science. 2001;293(5532):1074–80.
3. Jaenisch R, Bird A. Epigenetic regulation of gene expression: how the genome integrates intrinsic and environmental signals. Nat Genet. 2003;33(Suppl):245–54.
4. Pradhan S, Bacolla A, Wells RD, Roberts RJ. Recombinant human DNA (cytosine-5) methyltransferase. I. Expression, purification, and comparison of de novo and maintenance methylation. J Biol Chem. 1999;274(46):33002–10.
5. Chen T, Ueda Y, Dodge JE, Wang Z, Li E. Establishment and maintenance of genomic methylation patterns in mouse embryonic stem cells by Dnmt3a and Dnmt3b. Mol Cell Biol. 2003;23(16):5594–605.
6. Okano M, Bell DW, Haber DA, Li E. DNA methyltransferases Dnmt3a and Dnmt3b are essential for de novo methylation and mammalian development. Cell. 1999;99(3):247–57.
7. Ooi SK, Bestor TH. The colorful history of active DNA demethylation. Cell. 2008; 133(7):1145–8.
8. Ito S, D'Alessio AC, Taranova OV, Hong K, Sowers LC, Zhang Y. Role of Tet proteins in 5mC to 5hmC conversion, ES-cell self-renewal and inner cell mass specification. Nature. 2010;466(7310):1129–33.
9. Tahiliani M, Koh KP, Shen Y, Pastor WA, Bandukwala H, Brudno Y, et al. Conversion of 5-methylcytosine to 5-hydroxymethylcytosine in mammalian DNA by MLL partner TET1. Science. 2009;324(5929):930–5.
10. He YF, Li BZ, Li Z, Liu P, Wang Y, Tang Q, et al. Tet-mediated formation of 5-carboxylcytosine and its excision by TDG in mammalian DNA. Science. 2011;333(6047):1303–7.
11. Ito S, Shen L, Dai Q, Wu SC, Collins LB, Swenberg JA, et al. Tet proteins can convert 5-methylcytosine to 5-formylcytosine and 5-carboxylcytosine. Science. 2011;333(6047): 1300–3.
12. Wu H, Zhang Y. Tet1 and 5-hydroxymethylation: a genome-wide view in mouse embryonic stem cells. Cell Cycle. 2011a;10(15):2428–36.
13. Wu H, Zhang Y. Mechanisms and functions of Tet protein-mediated 5-methylcytosine oxidation. Genes Dev. 2011b;25(23):2436–52.
14. Branco MR, Ficz G, Reik W. Uncovering the role of 5-hydroxymethylcytosine in the epigenome. Nat Rev Genet. 2012;13(1):7–13.
15. Pastor WA, Aravind L, Rao A. TETonic shift: biological roles of TET proteins in DNA demethylation and transcription. Nat Rev Mol Cell Biol. 2013;14(6):341–56.

16. Guo JU, Su Y, Zhong C, Ming GL, Song H. Emerging roles of TET proteins and 5-hydroxymethylcytosines in active DNA demethylation and beyond. Cell Cycle. 2011;10(16):2662–8.
17. Bachman M, Uribe-Lewis S, Yang X, Williams M, Murrell A, Balasubramanian S. 5-Hydroxymethylcytosine is a predominantly stable DNA modification. Nat Chem. 2014;6(12):1049–55.
18. Globisch D, Munzel M, Muller M, Michalakis S, Wagner M, Koch S, et al. Tissue distribution of 5-hydroxymethylcytosine and search for active demethylation intermediates. PLoS One. 2010;5(12):e15367.
19. Jones PL, Veenstra GJ, Wade PA, Vermaak D, Kass SU, Landsberger N, et al. Methylated DNA and MeCP2 recruit histone deacetylase to repress transcription. Nat Genet. 1998;19(2):187–91.
20. Hsieh J, Gage FH. Epigenetic control of neural stem cell fate. Curr Opin Genet Dev. 2004;14(5):461–9.
21. Hellman A, Chess A. Gene body-specific methylation on the active X chromosome. Science. 2007;315(5815):1141–3.
22. Jones PA. Functions of DNA methylation: islands, start sites, gene bodies and beyond. Nat Rev Genet. 2012;13(7):484–92.
23. Maunakea AK, Chepelev I, Cui K, Zhao K. Intragenic DNA methylation modulates alternative splicing by recruiting MeCP2 to promote exon recognition. Cell Res. 2013;23(11):1256–69.
24. Mellen M, Ayata P, Dewell S, Kriaucionis S, Heintz N. MeCP2 binds to 5hmC enriched within active genes and accessible chromatin in the nervous system. Cell. 2012;151(7):1417–30.
25. Stroud H, Feng S, Morey Kinney S, Pradhan S, Jacobsen SE. 5-Hydroxymethylcytosine is associated with enhancers and gene bodies in human embryonic stem cells. Genome Biol. 2011;12(6):R54.
26. Iurlaro M, Ficz G, Oxley D, Raiber EA, Bachman M, Booth MJ, et al. A screen for hydroxymethylcytosine and formylcytosine binding proteins suggests functions in transcription and chromatin regulation. Genome Biol. 2013;14(10):R119.
27. Yu M, Hon GC, Szulwach KE, Song CX, Zhang L, Kim A, et al. Base-resolution analysis of 5-hydroxymethylcytosine in the mammalian genome. Cell. 2012;149(6):1368–80.
28. Nestler EJ. Is there a common molecular pathway for addiction? Nat Neurosci. 2005;8(11):1445–9.
29. Koob GF, Volkow ND. Neurocircuitry of addiction. Neuropsychopharmacology. 2010; 35(1):217–38.
30. Hyman SE. Addiction: a disease of learning and memory. Am J Psychiatry. 2005; 162(8):1414–22.
31. Volkow ND, Koob GF, McLellan AT. Neurobiologic advances from the brain disease model of addiction. N Engl J Med. 2016;374(4):363–71.
32. Warner LA, Kessler RC, Hughes M, Anthony JC, Nelson CB. Prevalence and correlates of drug use and dependence in the United States. Results from the National Comorbidity Survey. Arch Gen Psychiatry. 1995;52(3):219–29.
33. Blum K, Oscar-Berman M, Demetrovics Z, Barh D, Gold MS. Genetic Addiction Risk Score (GARS): molecular neurogenetic evidence for predisposition to Reward Deficiency Syndrome (RDS). Mol Neurobiol. 2014;50(3):765–96.
34. Day JJ, Sweatt JD. DNA methylation and memory formation. Nat Neurosci. 2010;13(11):1319–23.
35. Halder R, Hennion M, Vidal RO, Shomroni O, Rahman RU, Rajput A, et al. DNA methylation changes in plasticity genes accompany the formation and maintenance of memory. Nat Neurosci. 2016;19(1):102–10.
36. Li X, Baker-Andresen D, Zhao Q, Marshall V, Bredy TW. Methyl CpG binding domain ultra-sequencing: a novel method for identifying inter-individual and cell-type-specific variation in DNA methylation. Genes Brain Behav. 2014;13(7):721–31.

37. Li SS, Papale LA, Zhang Q, Madrid A, Chen L, Chopra P, et al. Genome-wide alterations in hippocampal 5-hydroxymethylcytosine links plasticity genes to acute stress. Neurobiol Dis. 2016;86:99–108.
38. Bonsch D, Lenz B, Kornhuber J, Bleich S. DNA hypermethylation of the alpha synuclein promoter in patients with alcoholism. Neuroreport. 2005;16(2):167–70.
39. Bleich S, Lenz B, Ziegenbein M, Beutler S, Frieling H, Kornhuber J, et al. Epigenetic DNA hypermethylation of the HERP gene promoter induces down-regulation of its mRNA expression in patients with alcohol dependence. Alcohol Clin Exp Res. 2006;30(4):587–91.
40. Philibert RA, Gunter TD, Beach SR, Brody GH, Madan A. MAOA methylation is associated with nicotine and alcohol dependence in women. Am J Med Genet B Neuropsychiatr Genet. 2008;147B(5):565–70.
41. Biermann T, Reulbach U, Lenz B, Frieling H, Muschler M, Hillemacher T, et al. N-methyl-D-aspartate 2b receptor subtype (NR2B) promoter methylation in patients during alcohol withdrawal. J Neural Transm (Vienna). 2009;116(5):615–22.
42. Hillemacher T, Frieling H, Hartl T, Wilhelm J, Kornhuber J, Bleich S. Promoter specific methylation of the dopamine transporter gene is altered in alcohol dependence and associated with craving. J Psychiatr Res. 2009;43(4):388–92.
43. Muschler MA, Hillemacher T, Kraus C, Kornhuber J, Bleich S, Frieling H. DNA methylation of the POMC gene promoter is associated with craving in alcohol dependence. J Neural Transm (Vienna). 2010;117(4):513–9.
44. Taqi MM, Bazov I, Watanabe H, Sheedy D, Harper C, Alkass K, et al. Prodynorphin CpG-SNPs associated with alcohol dependence: elevated methylation in the brain of human alcoholics. Addict Biol. 2011;16(3):499–509.
45. Ponomarev I, Wang S, Zhang L, Harris RA, Mayfield RD. Gene coexpression networks in human brain identify epigenetic modifications in alcohol dependence. J Neurosci. 2012;32(5):1884–97.
46. Zhang H, Herman AI, Kranzler HR, Anton RF, Zhao H, Zheng W, et al. Array-Based Profiling of DNA Methylation Changes Associated with Alcohol Dependence. Alcohol Clin Exp Res. 2013;37:E108–E115. doi:10.1111/j.1530-0277.2012.01928.x.
47. Zhang R, Miao Q, Wang C, Zhao R, Li W, Haile CN, et al. Genome-wide DNA methylation analysis in alcohol dependence. Addict Biol. 2013;18(2):392–403.
48. Philibert RA, Penaluna B, White T, Shires S, Gunter T, Liesveld J, et al. A pilot examination of the genome-wide DNA methylation signatures of subjects entering and exiting short-term alcohol dependence treatment programs. Epigenetics. 2014;9(9):1212–9.
49. Barbier E, Tapocik JD, Juergens N, Pitcairn C, Borich A, Schank JR, et al. DNA methylation in the medial prefrontal cortex regulates alcohol-induced behavior and plasticity. J Neurosci. 2015;35(15):6153–64.
50. Anier K, Malinovskaja K, Aonurm-Helm A, Zharkovsky A, Kalda A. DNA methylation regulates cocaine-induced behavioral sensitization in mice. Neuropsychopharmacology. 2010;35(12):2450–61.
51. Carouge D, Host L, Aunis D, Zwiller J, Anglard P. CDKL5 is a brain MeCP2 target gene regulated by DNA methylation. Neurobiol Dis. 2010;38(3):414–24.
52. LaPlant Q, Vialou V, Covington 3rd HE, Dumitriu D, Feng J, Warren BL, et al. Dnmt3a regulates emotional behavior and spine plasticity in the nucleus accumbens. Nat Neurosci. 2010;13(9):1137–43.
53. Tian W, Zhao M, Li M, Song T, Zhang M, Quan L, et al. Reversal of cocaine-conditioned place preference through methyl supplementation in mice: altering global DNA methylation in the prefrontal cortex. PLoS One. 2012;7(3):e33435.
54. Anier K, Zharkovsky A, Kalda A. S-adenosylmethionine modifies cocaine-induced DNA methylation and increases locomotor sensitization in mice. Int J Neuropsychopharmacol. 2013;16(9):2053–66.
55. Baker-Andresen D, Zhao Q, Li X, Jupp B, Chesworth R, Lawrence AJ, et al. Persistent variations in neuronal DNA methylation following cocaine self-administration and protracted abstinence in mice. Neuroepigenetics. 2015;4:1–11.

56. Feng J, Shao N, Szulwach KE, Vialou V, Huynh J, Zhong C, et al. Role of Tet1 and 5-hydroxymethylcytosine in cocaine action. Nat Neurosci. 2015;18(4):536–44.
57. Massart R, Barnea R, Dikshtein Y, Suderman M, Meir O, Hallett M, et al. Role of DNA methylation in the nucleus accumbens in incubation of cocaine craving. J Neurosci. 2015;35(21):8042–58.
58. Wright KN, Hollis F, Duclot F, Dossat AM, Strong CE, Francis TC, et al. Methyl supplementation attenuates cocaine-seeking behaviors and cocaine-induced c-Fos activation in a DNA methylation-dependent manner. J Neurosci. 2015;35(23):8948–58.
59. Nielsen DA, Yuferov V, Hamon S, Jackson C, Ho A, Ott J, et al. Increased OPRM1 DNA methylation in lymphocytes of methadone-maintained former heroin addicts. Neuropsychopharmacology. 2009;34(4):867–73.
60. Xu X, Ji H, Liu G, Wang Q, Liu H, Shen W, et al. A significant association between BDNF promoter methylation and the risk of drug addiction. Gene. 2016;584(1):54–9.
61. Jayanthi S, McCoy MT, Chen B, Britt JP, Kourrich S, Yau HJ, et al. Methamphetamine downregulates striatal glutamate receptors via diverse epigenetic mechanisms. Biol Psychiatry. 2014;76(1):47–56.
62. Cheng MC, Hsu SH, Chen CH. Chronic methamphetamine treatment reduces the expression of synaptic plasticity genes and changes their DNA methylation status in the mouse brain. Brain Res. 2015;1629:126–34.
63. Cadet JL, Brannock C, Krasnova IN, Jayanthi S, Ladenheim B, McCoy MT, et al. Genome-wide DNA hydroxymethylation identifies potassium channels in the nucleus accumbens as discriminators of methamphetamine addiction and abstinence. Mol Psychiatry. 2016. doi:0.1038/mp.2016.48.
64. Launay JM, Del Pino M, Chironi G, Callebert J, Peoc'h K, Megnien JL, et al. Smoking induces long-lasting effects through a monoamine-oxidase epigenetic regulation. PLoS One. 2009;4(11):e7959.
65. Xu Q, Ma JZ, Payne TJ, Li MD. Determination of Methylated CpG Sites in the Promoter Region of Catechol-O-Methyltransferase (COMT) and their Involvement in the Etiology of Tobacco Smoking. Front Psych. 2010;1:16.
66. Ehrlich S, Walton E, Roffman JL, Weiss D, Puls I, Doehler N, et al. Smoking, but not malnutrition, influences promoter-specific DNA methylation of the proopiomelanocortin gene in patients with and without anorexia nervosa. Can J Psychiatry. 2012;57(3):168–76.
67. Zeilinger S, Kuhnel B, Klopp N, Baurecht H, Kleinschmidt A, Gieger C, et al. Tobacco smoking leads to extensive genome-wide changes in DNA methylation. PLoS One. 2013;8(5):e63812.
68. van der Knaap LJ, Schaefer JM, Franken IH, Verhulst FC, van Oort FV, Riese H. Catechol-O-methyltransferase gene methylation and substance use in adolescents: the TRAILS study. Genes Brain Behav. 2014;13(7):618–25.
69. Gao X, Jia M, Zhang Y, Breitling LP, Brenner H. DNA methylation changes of whole blood cells in response to active smoking exposure in adults: a systematic review of DNA methylation studies. Clin Epigenetics. 2015;7:113.
70. Breitling LP, Yang R, Korn B, Burwinkel B, Brenner H. Tobacco-smoking-related differential DNA methylation: 27 K discovery and replication. Am J Hum Genet. 2011;88(4):450–7.
71. Sandoval J, Heyn H, Moran S, Serra-Musach J, Pujana MA, Bibikova M, et al. Validation of a DNA methylation microarray for 450,000 CpG sites in the human genome. Epigenetics. 2011;6(6):692–702.
72. Shenker N, Flanagan JM. Intragenic DNA methylation: implications of this epigenetic mechanism for cancer research. Br J Cancer. 2012;106(2):248–53.
73. Philibert RA, Plume JM, Gibbons FX, Brody GH, Beach SR. The impact of recent alcohol use on genome wide DNA methylation signatures. Front Genet. 2012;3:54.
74. Zhang H, Herman AI, Kranzler HR, Anton RF, Zhao H, Zheng W, et al. Array-based profiling of DNA methylation changes associated with alcohol dependence. Alcohol Clin Exp Res. 2013;37(Suppl 1):E108–15.

75. Swartz JR, Hariri AR, Williamson DE. An epigenetic mechanism links socioeconomic status to changes in depression-related brain function in high-risk adolescents. Mol Psychiatry. 2017;22(2):209–14.
76. Cassel S, Carouge D, Gensburger C, Anglard P, Burgun C, Dietrich JB, et al. Fluoxetine and cocaine induce the epigenetic factors MeCP2 and MBD1 in adult rat brain. Mol Pharmacol. 2006;70(2):487–92.
77. Frommer M, McDonald LE, Millar DS, Collis CM, Watt F, Grigg GW, et al. A genomic sequencing protocol that yields a positive display of 5-methylcytosine residues in individual DNA strands. Proc Natl Acad Sci U S A. 1992a;89(5):1827–31.
78. Mari F, Azimonti S, Bertani I, Bolognese F, Colombo E, Caselli R, et al. CDKL5 belongs to the same molecular pathway of MeCP2 and it is responsible for the early-onset seizure variant of Rett syndrome. Hum Mol Genet. 2005;14(14):1935–46.
79. Ou JN, Torrisani J, Unterberger A, Provencal N, Shikimi K, Karimi M, et al. Histone deacetylase inhibitor Trichostatin A induces global and gene-specific DNA demethylation in human cancer cell lines. Biochem Pharmacol. 2007;73(9):1297–307.
80. Host L, Dietrich JB, Carouge D, Aunis D, Zwiller J. Cocaine self-administration alters the expression of chromatin-remodelling proteins; modulation by histone deacetylase inhibition. J Psychopharmacol. 2011;25(2):222–9.
81. Im HI, Hollander JA, Bali P, Kenny PJ. MeCP2 controls BDNF expression and cocaine intake through homeostatic interactions with microRNA-212. Nat Neurosci. 2010;13(9):1120–7.
82. Romieu P, Host L, Gobaille S, Sandner G, Aunis D, Zwiller J. Histone deacetylase inhibitors decrease cocaine but not sucrose self-administration in rats. J Neurosci. 2008;28(38):9342–8.
83. Deng JV, Rodriguiz RM, Hutchinson AN, Kim IH, Wetsel WC, West AE. MeCP2 in the nucleus accumbens contributes to neural and behavioral responses to psychostimulants. Nat Neurosci. 2010;13(9):1128–36.
84. Hollander JA, Im HI, Amelio AL, Kocerha J, Bali P, Lu Q, et al. Striatal microRNA controls cocaine intake through CREB signalling. Nature. 2010;466(7303):197–202.
85. Chang Q, Khare G, Dani V, Nelson S, Jaenisch R. The disease progression of Mecp2 mutant mice is affected by the level of BDNF expression. Neuron. 2006;49(3):341–8.
86. Graham DL, Edwards S, Bachtell RK, DiLeone RJ, Rios M, Self DW. Dynamic BDNF activity in nucleus accumbens with cocaine use increases self-administration and relapse. Nat Neurosci. 2007;10(8):1029–37.
87. Weber M, Davies JJ, Wittig D, Oakeley EJ, Haase M, Lam WL, et al. Chromosome-wide and promoter-specific analyses identify sites of differential DNA methylation in normal and transformed human cells. Nat Genet. 2005;37(8):853–62.
88. Song L, James SR, Kazim L, Karpf AR. Specific method for the determination of genomic DNA methylation by liquid chromatography-electrospray ionization tandem mass spectrometry. Anal Chem. 2005;77(2):504–10.
89. Yohn NL, Bartolomei MS, Blendy JA. Multigenerational and transgenerational inheritance of drug exposure: the effects of alcohol, opiates, cocaine, marijuana, and nicotine. Prog Biophys Mol Biol. 2015;118(1–2):21–33.
90. Breton CV, Byun HM, Wenten M, Pan F, Yang A, Gilliland FD. Prenatal tobacco smoke exposure affects global and gene-specific DNA methylation. Am J Respir Crit Care Med. 2009;180(5):462–7.
91. Novikova SI, He F, Bai J, Cutrufello NJ, Lidow MS, Undieh AS. Maternal cocaine administration in mice alters DNA methylation and gene expression in hippocampal neurons of neonatal and prepubertal offspring. PLoS One. 2008;3(4):e1919.
92. Byrnes JJ, Johnson NL, Schenk ME, Byrnes EM. Cannabinoid exposure in adolescent female rats induces transgenerational effects on morphine conditioned place preference in male offspring. J Psychopharmacol. 2012;26(10):1348–54.
93. Szutorisz H, Egervari G, Sperry J, Carter JM, Hurd YL. Cross-generational THC exposure alters the developmental sensitivity of ventral and dorsal striatal gene expression in male and female offspring. Neurotoxicol Teratol. 2016;58:107–14.

94. Vassoler FM, Johnson NL, Byrnes EM. Female adolescent exposure to cannabinoids causes transgenerational effects on morphine sensitization in female offspring in the absence of in utero exposure. J Psychopharmacol. 2013;27(11):1015–22.
95. Watson CT, Szutorisz H, Garg P, Martin Q, Landry JA, Sharp AJ, et al. Genome-wide DNA methylation profiling reveals epigenetic changes in the rat nucleus accumbens associated with Cross-generational effects of adolescent THC exposure. Neuropsychopharmacology. 2015;40(13):2993–3005.
96. Anway MD, Skinner MK. Epigenetic transgenerational actions of endocrine disruptors. Endocrinology. 2006;147(6 Suppl):S43–9.
97. Skinner MK. What is an epigenetic transgenerational phenotype? F3 or F2. Reprod Toxicol. 2008;25(1):2–6.
98. Dietz DM, Laplant Q, Watts EL, Hodes GE, Russo SJ, Feng J, et al. Paternal transmission of stress-induced pathologies. Biol Psychiatry. 2011;70(5):408–14.
99. Anway MD, Cupp AS, Uzumcu M, Skinner MK. Epigenetic transgenerational actions of endocrine disruptors and male fertility. Science. 2005;308(5727):1466–9.
100. Guerrero-Bosagna C, Settles M, Lucker B, Skinner MK. Epigenetic transgenerational actions of vinclozolin on promoter regions of the sperm epigenome. PLoS One. 2010;5(9): e13100.
101. Manikkam M, Tracey R, Guerrero-Bosagna C, Skinner MK. Plastics derived endocrine disruptors (BPA, DEHP and DBP) induce epigenetic transgenerational inheritance of obesity, reproductive disease and sperm epimutations. PLoS One. 2013;8(1):e55387.
102. Manikkam M, Tracey R, Guerrero-Bosagna C, Skinner MK. Pesticide and insect repellent mixture (permethrin and DEET) induces epigenetic transgenerational inheritance of disease and sperm epimutations. Reprod Toxicol. 2012;34(4):708–19.
103. Tracey R, Manikkam M, Guerrero-Bosagna C, Skinner MK. Hydrocarbons (jet fuel JP-8) induce epigenetic transgenerational inheritance of obesity, reproductive disease and sperm epimutations. Reprod Toxicol. 2013;36:104–16.
104. Anway MD, Memon MA, Uzumcu M, Skinner MK. Transgenerational effect of the endocrine disruptor vinclozolin on male spermatogenesis. J Androl. 2006;27(6):868–79.
105. He F, Lidow IA, Lidow MS. Consequences of paternal cocaine exposure in mice. Neurotoxicol Teratol. 2006;28(2):198–209.
106. Ouko LA, Shantikumar K, Knezovich J, Haycock P, Schnugh DJ, Ramsay M. Effect of alcohol consumption on CpG methylation in the differentially methylated regions of H19 and IG-DMR in male gametes: implications for fetal alcohol spectrum disorders. Alcohol Clin Exp Res. 2009;33(9):1615–27.

Drug Addiction and Histone Code Alterations

7

Hee-Dae Kim, Tanessa Call, Samantha Magazu, and Deveroux Ferguson

Abstract

Acute and prolonged exposure to drugs of abuse induces changes in gene expression, synaptic function, and neural plasticity in brain regions involved in reward. Numerous genes are involved in this process, and persistent changes in gene expression coincide with epigenetic histone modifications and DNA methylation. Histone modifications are attractive regulatory mechanisms, which can encode complex environmental signals in the genome of postmitotic cells, like neurons. Recently, it has been demonstrated that specific histone modifications are involved in addiction-related gene regulatory mechanisms, by a diverse set of histone-modifying enzymes and readers. These histone modifiers and readers may prove to be valuable pharmacological targets for effective treatments for drug addiction.

Keywords

Addiction • Drug of abuse • Epigenetics • Histone modification • Reward circuitry

H.-D. Kim, Ph.D. • T. Call, B.A. • S. Magazu, M.S. • D. Ferguson, Ph.D. (✉)
Department of Basic Medical Sciences, University of Arizona College of Medicine-Phoenix, Phoenix, AZ 85004, USA
e-mail: heedaekim@email.arizona.edu; tcall1@asu.edu; smagazu@email.arizona.edu; dferguson@email.arizona.edu

R. Delgado-Morales (ed.), *Neuroepigenomics in Aging and Disease*, Advances in Experimental Medicine and Biology, DOI 10.1007/978-3-319-53889-1_7

7.1 Introduction

Addiction can be characterized as compulsive drug use despite severe negative consequence. Drugs of abuse exert a substantial public health and financial cost to society, and up to 10% of the US population currently suffer from addiction and are in need of treatment. According to the World Health Organization (WHO), drug-related deaths account for 12.4% of mortality worldwide. Estimates from the National Institute of Drug Abuse (NIDA) approximate a $500 billion dollar economic loss to the USA annually as a result of drug abuse. Currently, there are few treatments for addiction to psychostimulants; thus, there is a significant need to discover and develop innovative therapeutics for addiction.

Acute and chronic exposure to drugs of abuse induces changes in gene expression, synaptic function, and neural plasticity throughout the brain, particularly in regions implicated in reward. Many immediate early genes and neural plasticity-related genes are involved in this process, and persistent changes in gene expression coincide with epigenetic regulation including histone modifications, DNA methylation, and noncoding RNAs [1]. Among them, histone modifications are potent regulatory mechanisms which can encode complex environmental signals in the genome of postmitotic cells, like neurons. Histone modifications, including acetylation and methylation [2], are known to be involved in addiction-related gene regulatory mechanisms.

7.1.1 Epigenetics and Histone Code

DNA molecules are highly organized into chromatin structure in eukaryotic cells, and the nucleosome is the basic unit of chromatin. The nucleosome core is composed of a histone octamer with two histone H2A, H2B, H3, and H4 subunits (Fig. 7.1a), and 146-bp-long nucleotides are wrapped around the core. Nucleosome core subunits are well-conserved, highly basic proteins, possessing lysine (K) and arginine (R) residues, thus ensuring coordinated interactions between the core and DNA molecules in sequence-independent manner. Relatively limited information, especially in vivo and neuronal systems, are available on H1 linker histones because of their sequence heterogeneity and subunit variability, although they seem to have a unique role in the regulation of higher-order structures and chromatin remodeling [3].

Posttranslational modifications (acetylation, methylation, phosphorylation, ubiquitination, sumoylation, citrullination, and ADP-ribosylation) occur at the specific residues of each histone tail (Fig. 7.1b), and these dynamic covalent modifications regulate overall chromatin structure and facilitate the recruitment of effector proteins. Furthermore, combinations and cross talks between these modifications, referred as "histone code" [4], considerably increase their regulatory capacity and provide means to specifically modulate various physiological and pathological conditions.

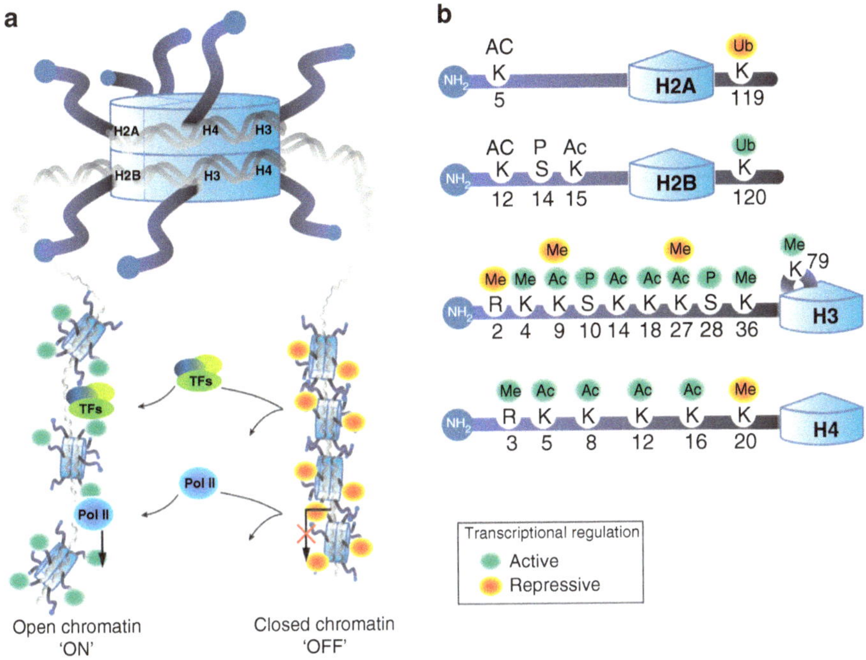

Fig. 7.1 Nucleosome structure and histone modifications. (**a**) A nucleosome, histone octamer, consists of H2A, H2B, H3, and H4 subunits, and specific amino acid residues undergo posttranslational modifications like acetylation (Ac), methylation (Me), phosphorylation (P), and ubiquitination (Ub). (**b**) Major lysine (K), serine (S), and arginine (R) residues for the histone modifications are indicated on each histone subunit tail (H3K79 located in core domain). The known functions of each modification in transcriptional regulation are indicated with color codes: *green* (transcriptionally active) or *orange* (repressive)

7.1.2 Histone Acetylation

Histone acetylation is a well-characterized modification and is correlated with transcriptional activation of target genes. It is generally accepted that acetylation of histone residues (usually lysine, K) neutralizes the positive charge and elicits loosening of nucleosome structure, which increases accessibility of the genome and subsequent transcriptional activation. A variety of histone acetyltransferases (HATs) and histone deacetylases (HDACs) regulate histone acetylation at specific residues of each histone subunit. HATs, like CREB-binding protein (CBP), p300, and GCN5, transfer an acetyl group from a cofactor, acetyl-CoA, to target residues. HDACs can be divided into four classes, and classes I, II, and IV are Zn^{2+}-dependent HDAC (class I, HDAC 1–3 and 8; class II, HDAC 4–7, 9, and 10; class IV, HDAC11), whereas class III HDAC sirtuins (Sirt1-7) use NAD^{+} as a cofactor. The acetylation of H3K9 and H3K14 residues is well-known markers of transcriptional activation of genes [5].

7.1.3 Histone Methylation

Lysine (K) and arginine (R) residues of H3 and H4 subunits are methylated by lysine methyltransferases (KMTs) and arginine methyltransferases (PRMTs), respectively. KMTs and PRMTs transfer up to 2~3 methyl groups (mono-, di-, or trimethylation) to a specific histone residue from S-adenosylmethionine (SAM), and several lysine demethylases (KDMs) can reverse lysine methylations [6]. PRMTs are divided into two functional types: type I PRMTs [1–4, 6, 8], which catalyze asymmetric dimethylarginines, and type II PRMTs [5, 7, 9], which catalyze symmetric dimethylarginines. Histone methylations can be either active or repressive transcriptional markers depending on the modification sites. Trimethylated H3K4 modification around transcription start site is a well-established transcriptional activation marker, and H3K36 and H3K79 (reside in core) are also related to gene activation. In contrast, methylation of H3K9, H3K27, and H4K20 residues is associated with transcriptional repression [6].

7.1.4 Histone Phosphorylation

Histone phosphorylation is involved in transcriptional regulation, DNA replication, and damage repair.

The phosphorylation of serine (S), threonine (T), and tyrosine (Y) residues of each histone tail is mediated by specific kinases and phosphatases. S10, T11, and S28 residues of H3 subunit are strongly associated with transcriptional activation, while phosphorylation of H2A and H2B rather seems to involve in other regulatory mechanisms, like DNA repair and mitosis/meiosis [7].

7.1.5 Histone Ubiquitination and Sumoylation

Lysine residues of H2A and H2B can be modified by mono- or poly-ubiquitination (preferentially mono-ubiquitination), which correlate with transcriptional regulation and DNA repair mechanisms. Specific E3 ligases and deubiquitination enzymes regulate each target residue. The mono-ubiquitination of H2A K119 elicits gene silencing [8], whereas the H2B K120 is involved in transcriptional initiation and elongation by inducing di- and trimethylation of H3K4 [9]. Sumoylation also occurs on lysine residues in a similar manner to ubiquitination and seems to counteract acetylation and ubiquitination by competing for the same lysine residues, inducing transcriptional repression [10].

7.1.6 Other Modifications and Combinatorial Effects

Additional modifications, like ADP-ribosylation, citrullination (deimination), and O-GlcNAcylation, are also involved in histone modification mechanisms.

ADP-ribosylation occurs on lysine, arginine, and glutamate residues, by poly-ADP-ribose polymerases (PARPs), and induces histone H4 acetylation [11]. Poly(ADP-ribose) polymerase-1 (PARP-1), the founding member of the PARP family, is a catalyst for NAD^+-dependent synthesis of a polymer called poly-ADP-ribose (PAR) on target proteins [12]. Citrullination is the conversion of arginine residues to citrullines, catalyzed by a peptidyl deiminase PAD14, and this process neutralizes the positive charge of arginine or removes methylated arginine [13, 14]. O-linked β-*N*-acetylglucosamine (O-GlcNAc) modifies serine and threonine residues of core histone subunits, and the histone O-GlcNAcylation seems to associate with gene activation in mammals [15].

Thus far, the combinatorial effects of histone modifications have been reported as examples of the "histone code." For example, phosphorylation of H3S10 residues tends to associate with acetylation of nearby H3K9 and H3K14 sites and subsequently induces gene activation [16]. In addition, there are also interactions with DNA methylation, transcription factors, and chromatin remodeling complexes. Histone methylation is closely linked with DNA methylation [17], and certain H3 acetylation seems to be modulated by DNA methylation in the hippocampus [18]. Competition between other modifications at the same sites is also common, such as H3K9 which is subjected to either acetylation or methylation. These complex interactions may mediate the addiction process; thus, there is a growing need to analyze the genome-wide changes in the brain using bioinformatics tools [19, 20].

7.2 Histone Modifications and Addiction

Several studies have revealed altered histone modifications in the reward circuitry (Fig. 7.2a), upon drug administrations. Various histone modifications at specific genes have been identified via pharmacological and genetic (virus, transgenic mice, etc.) approaches that alter activity or expression of histone-modifying enzymes [2]. Among the various modifications, histone acetylation and methylation have been studied extensively.

7.2.1 Histone Acetylation

As previously mentioned, it is widely accepted that histone acetylation induces transcriptional activation and studies have identified altered acetylation patterns upon exposure to drugs of abuse with increases of global histone acetylation (histone H3 and/or H4 acetylation) [2] (Fig. 7.2b). Additionally, treatment with HDAC inhibitors increases behavioral and neural responses of various drugs of abuse [21–24]. A genome-wide study showed that increased acetylation is observed at various gene loci, much more than genes with decreased acetylation, although the expressions of the target genes do not always necessarily correlate with transcriptional activation [25].

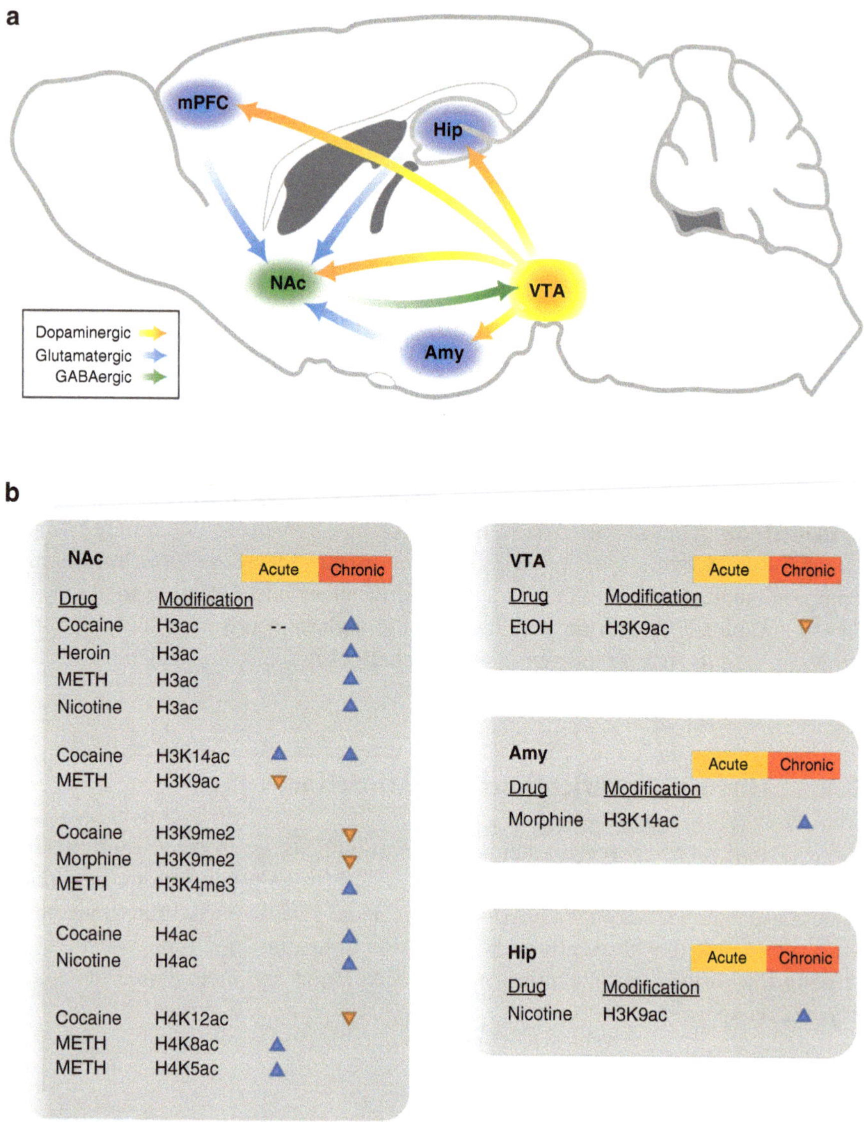

Fig. 7.2 Brain reward circuitry and histone modification changes upon drug administrations. (**a**) Major projections between brain regions in the reward circuitry: dopaminergic projections from ventral tegmental area (VTA) to nucleus accumbens (NAc), medial prefrontal cortex (mPFC), hippocampus (Hip), and amygdala (Amy); glutamatergic innervation to the NAc from mPFC, Hip, and Amy; GABAergic projections from NAc to VTA (D1-MSN-mediated direct pathway and D2-MSN-mediated indirect pathway; not indicated separately). (**b**) General changes of histone modifications in the reward circuitry upon acute or chronic drug treatment. *Arrows* indicate increase (*blue*), decrease (*red*), or no changes (*dashed line*) in each histone modification. *METH* methamphetamine, *EtOH* ethanol

Cocaine administration generally increases global acetylation levels and regulates the expression of immediate early genes and plasticity-related genes [23–25]. For example, acute and chronic (including the self-administration model) cocaine injections differentially modulate *c-Fos*, *FosB*, and *Bdnf* genes in the striatum [24]. Acute cocaine injection increases global histone acetylation at the *c-Fos* (H3 and H4 acetylation) and *FosB* (H4 acetylation only) promoters, but in the chronic administration scheme, acetylation was attenuated except for H3 hyperacetylation at the *FosB* promoter. Chronic exposure to cocaine induces H3 hyperacetylation at the *Bdnf* and *Cdk5* gene promoters. Indeed, cocaine-affected genes show differential pattern of H3 and H4 acetylation according to genome-wide analysis in the NAc [25], and the pattern seems to be variable between drugs of abuse [21]. But it is also noteworthy that the global H3 acetylation level in the shell region of NAc shows positive correlation with cocaine responses, but it's not the case in H4 acetylation [23].

As "writers" and "erasers" of acetylation, specific HATs and HDACs directly regulate histone acetylation at specific lysine residues. Immediate early genes, neurotrophic factors, and synaptic proteins are direct targets of HATs and HDACs [2]. Cocaine administration increases CBP and histone H4 acetylation on *FosB* promoter in the mouse striatum, followed by increased expression of *FosB*, and reduced CBP expression in CBP haploinsufficient (+/−) mice attenuates cocaine-induced locomotor activity [26]. CBP +/− mice also show reduced LTP induction in the hippocampus (dentate gyrus) upon nicotine-primed cocaine injection which is accompanied by increased H3K9 acetylation [27], with similar changes observed in the NAc and amygdala regions [22, 28]. In addition, focal CBP knockout in the NAc reduces histone acetylation and alters *c-Fos* expression (increase in chronic, but not in acute cocaine) and attenuates behavioral responses upon cocaine injection [29]. In the VTA region, CBP and histone H3 acetylation are increased at *Bdnf* promoter regions after cocaine self-administration followed by forced withdrawal [30]. Because other HATs, other than CBP, have been identified and are known to be involved in the acetylation of each histone residue [31], additional studies are required to reveal specific mechanisms within the context of addiction-related research.

Regulatory mechanisms by HDACs are rather complicated. Several studies have revealed that HDAC inhibition using pharmacological or genetic methods increases effects of drug of abuse and activation elicits reverse effects [23, 24, 26, 32, 33]. When specific HDACs were targeted, HDAC3 [34] or HDAC5 [33] deletion in the NAc enhances cocaine-induced behavioral changes (locomotor activity, conditioned place preference), whereas overexpression of HDAC4 reduces drug motivation [23, 24]. Modulation of HDAC2 or HDAC9 has no effect on cocaine responses [33, 35]. These results are consistent with the general notion that the increased acetylation enhances drug effects.

But prolonged inhibition of HDAC1 in the NAc increases not only global H3 acetylation but also H3 methylation (H3K9me2) and blunts cocaine-induced behavioral changes [35]. The proposed mechanism is that HDAC1 downregulation increases H3 acetylation at KMT promoters (*Ehmt2* and *Suv39h1*), and the increase

KMTs suppresses cocaine-induced gene expression via histone methylation. It is noteworthy that HDAC1 knockout affects expression of other HDACs (HDAC2, HDAC5, and Sirt1), although the relevance of this compensatory regulation is still elusive. Also, Malvaez et al. [36] showed that HDAC3 inhibition only during CPP extinction training enhances rapid and persistent extinction of cocaine-associated memory. These studies illustrate that results from the coadministration of addictive drugs and epigenetic modulators are inevitably time and context dependent and can be paradoxical between different experimental schemes. An additional, noncanonical mechanism was recently reported by Tsai et al. [37] where cocaine was reported to suppress monoamine oxidase B (*MAOB*) expression in the NAc and PFC via sigma-1 receptor-mediated recruitments of class I HDACs (HDAC1–HDAC3) at the *MAOB* promoter. Here, it is noteworthy that this epigenetic mechanism is a dopamine transporter-independent pathway, which cocaine acts as a sigma-1 receptor agonist.

Recently published data have established an essential role for class III HDACs, sirtuins, as epigenetic chromatin remodeler in cocaine and morphine reward [25, 38, 39]. Ferguson et al. [38] reported that chronic cocaine and morphine administration increases *Sirt1* expression, a well-characterized sirtuin, the NAc via drug-induced binding of ΔFosB at the promoter region, and increased SIRT1 enhances expression of plasticity-related genes, spine density, and drug-induced behavioral changes. ChIP-seq methods revealed that cocaine modifies genome-wide SIRT1 binding and H4K16 acetylation patterns in the NAc, in addition to the increase of SIRT1 expression followed by Foxo3a activation [39]. Both changes are closely associated with global changes of gene expression and CPP induction in chronic cocaine administration. Chronic methamphetamine treatment also induces *Sirt1* and *Sirt2* expressions in the striatum, and they can modulate downstream genes via modifying H4K16 acetylation levels of downstream genes like AMPAR subunits [40]. It is also of interest that sirtuin modulates ethanol-induced addiction-like behaviors via acting on H3K9 acetylation in *Drosophila*, implicating conserved functions of SIRT1 in drug addiction [41].

7.2.2 Histone Methylation

Histone methylation is important in regulating normal cognitive function and the development of several psychiatric disorders [42]. Methylation can be either permissive or repressive depending on the lysine or arginine residue targeted and be dynamically regulated by methyltransferases (KMTs) and demethylases (KDMs) upon drug administration.

Cocaine administration alters multiple histone modifications including methylation and demethylation of histone tails. G9a is the core subunit of multimeric repressive KMT complex, which primarily acts to catalyze the dimethylation of histone H3 at lysine 9 (H3K9me2) throughout the genome. G9a is responsible for many changes following repeated cocaine exposure and leads to a decrease in repressive methylation and, therefore, an enhanced euchromatin state [43]. Repeated cocaine exposure results in G9a catalyzing repressive histone modification; both G9a and its target H3K9me2 levels are repressed in NAc [44]. In response to repeated cocaine exposure, H3K9me2 is reduced globally in NAc [45]. This decreased methylation in the NAc is likely due to cocaine-induced downregulation of G9a and G9a-like protein (*GLP*) [1, 45]. Blockade, or knockdown, of G9a through genetic or pharmacological manipulations in the NAc potentiates behavioral responses to cocaine, whereas overexpressing G9a results in opposite behavioral effects as well as an increase in dendritic arborization of neurons within the NAc [44].

Another gene regulated by cocaine and histone methylation is cyclin-dependent kinase 5 (*Cdk5*), a gene highly expressed throughout the nervous system, and is increased following repeated cocaine treatment [46]. Cdk5 is associated with hypomethylation of histone H3 in the NAc [44, 46]. *Cdk5*-targeted H3K9me2 attenuates cocaine-induced locomotor behavior as well as conditioned place preference (CPP). Inhibition of Cdk5 through pharmacological methods potentiates cocaine-induced locomotor behavior however; epigenetic suppression of endogenous *Cdk5* through histone methylation reduces behavioral responses to cocaine [46]. Repeated cocaine treatment or transgenic overexpression of ΔFosB, the transcription factor that controls *Cdk5* expression, leads to a ~50% increase of *Cdk5* levels in the NAc [47].

In addition to lysine methylation, recent research has focused on methylation of histone arginine by protein arginine *N*-methyltransferase (PRMT) [48]. Repeated injections of cocaine result in an increase of PRMT1, and PRMT1 regulates histone H4 arginine 3 asymmetric demethylation (H4R3me2a) in the NAc [48]. The upregulation of H4R3me2a following repeated cocaine injections controls expression of *Cdk5* and *CaMKII*. The increase in PRMT1 from repeated cocaine exposure results in a strengthening of CPP, and this upregulation is long lasting, up to 7 days following cocaine withdrawal but returning to baseline after 14 days [48]. Conversely, expression of PRMT6 is downregulated following repeated cocaine injections [48, 49]. Damez-Werno et al. [49] show that asymmetric demethylation of R2 on histone H3 (H3R2me2a), the histone target of PRMT6, is decreased in the NAc of mice, rats, and human postmortem samples following repeated cocaine exposure. It is of interest that cocaine also modulates histone arginine methylation (H3R2me2a) in the NAc in a cell-type-specific manner [49]. D2-MSN-specific decrease of H3R2me2a levels upregulates *Srcin1* expression, and subsequent suppression of Src signaling is closely linked with reduced cocaine reward.

Amphetamine research, specifically methamphetamine (METH), has shown several changes in histone methylation associated with the memory of drug exposure. An enzyme involved in trimethylation of histone H3 at K4 (H3K4me3), KMT2A, was upregulated following repeated METH injections [50]. KMT2A was found to be necessary for repeated METH-associated memory formation and maintenance. In addition, the demethylation of H3K4 by KDM5C was shown to be necessary for maintenance of METH CPP [50, 51]. Methylation of transcriptionally repressive residues is unchanged or decreased after repeated METH exposure [50]. Following repeated METH treatments, trimethylation of H3K4me3, which is often associated with active transcription, was increased in the NAc in conjunction with CPP [50]. An increase of H3K4me2, a transcriptionally active chromatin modifier, was observed following a single METH treatment in mice [50]. Conversely, during METH self-administration, no increase of H3K4me3 binding was seen [51, 52]. These conflicting reports are not surprising as there are often differences in enzyme activation between researcher-administered and self-administered drugs.

Sun et al. [53] show that, following repeated morphine administration, but not acute, both G9a and H3K9me2 are downregulated approximately 25% in the NAc. Mice that received 5–7 days of once-daily i.p. injections (20 mg/kg) resulted in a downregulation of G9a and its associated mark H3K9me2. This was not seen following acute, 1–3 days, treatment. In addition, the effect was dose dependent, lower doses having little to no effect, and did not result from the number of injections as indicated by a lack of increase of G9a and H3K9me2 following six treatments over 2.5 days [53]. This morphine-induced downregulation of H3K9me2 is brain region specific; no changes in dorsal striatum were recorded [53]. Downregulation of G9a in NAc results in enhanced locomotor sensitization and delays development of analgesic tolerance to morphine [53]. Overexpression of G9a in NAc results in opposition of morphine reward and locomotor sensitization as well as promotes analgesic tolerance and naloxone-precipitated withdrawal [53]. Morphine also induces H3K27me3 at the *Bdnf* promoter, and subsequent blocking of CREB binding results in downregulation of BDNF in VTA region, enhancing morphine CPP [54]. Nurr1, an upstream regulator of *Bdnf*, also shows increased H3K27me3 levels at the promoter regions, and the decreased NURR1 contributes to the *Bdnf* downregulation and behavioral effects of morphine.

Recently, genome-wide histone methylation patterns were analyzed with ChIP-seq and RNA-seq methods [55]. Feng et al. [55] reported on the genome-wide analysis of several histone methylation marks in the NAc after chronic cocaine treatment. Using sophisticated bioinformatics tool, they predicted one hub gene for cocaine responses, splicing factor *A2bp1*, and validate its functional relevance with molecular and behavioral tests: *A2bp1* knockdown significantly decreases cocaine reward. Subsequent binding motif analysis revealed that A2BP1 regulates neurite and synapse formation-related genes through the binding at cocaine-regulated H3K4me3 sites.

7.2.3 Other Modifications

Histone phosphorylation. It has been known that many drugs of abuse modulate histone phosphorylation level. Among them, cocaine strongly increases global histone H3 phosphorylation levels in the striatum [21, 56]. MSK1 regulates histone H3S10, a typical transcriptional activation marker, and CREB phosphorylation levels in the striatum upon cocaine treatment, and induces transcription of immediate early genes, like c-Fos and pDyn [56]. Direct stimulation of D1R increases H3S10 phosphorylation in the striatum. Nuclear DARPP-32 regulates histone H3S10 phosphorylation via D1R signaling cascade and elicits long-term transcriptional effects [57]. The induction of H3S10 phosphorylation in D1-MSNs after cocaine administration was also demonstrated using cell-type-specific histone modification analysis by FACS [58]. Although extensive studies are needed on the regulation of histone phosphorylation, it seems that the histone H3 phosphorylation is involved in the triggering of addiction states through immediate early gene activation.

Histone poly-ADP-ribosylation. PARP-1 is a ubiquitous nuclear protein found abundantly in most cell types that binds to histones, DNA, and other proteins [59]. Recently PARP-1 has been shown to play two important roles in gene transcription regulation: a histone modifier and a component of enhancer/promoter regulatory complexes [60]. PARP-1 activity has been reported to play a crucial role in learning and memory as well as neuronal viability [61, 62]. In the context of drug addiction, upregulation of PARP-1 in the NAc was shown to enhance behavioral responses to cocaine in mice and rats given i.p. injections of cocaine or self-administration, respectively [63]. Conversely, downregulation of PARP-1 results in a decrease of behavioral response to cocaine [63]. According to genome-wide ChIP-seq analysis, this regulation by PARP-1 in the NAc following cocaine administration occurs at H1 and H3, especially H3K4me3-enriched transcription start sites, and suggests, in this instance, that chronic cocaine exposure leads to a permissive chromatin structure through selective histone PARylation [63].

7.3 Conclusions and Future Work

Many addictive drugs modulate target genes through histone modifications in brain region-specific and time-delimited manner. Interestingly, drug-induced pathways share common histone modifications and target factors including immediate early genes and neurotrophic factors, although the direction of modifications and effects can be opposite sometimes. But again, global changes of histone modification do not necessarily correlate with changes of gene expression and are rather complicated because of combinatorial effects and cross talk between various histone modifications and intertwined regulations between target genes, including histone modifiers themselves [25]. Thus, although the transcriptional effects of many histone modification sites are known, it is still hard to predict gene expressions and behavioral effects of each modulation. Because of the complex nature of addiction-related gene network, sophisticated genome-wide epigenomic studies on each drug addiction model are required to understand addiction mechanisms and find druggable target genes [19]. Recently, combined analysis using ChIP-seq and RNA-seq methods has defined novel regulatory pathways and hub genes as putative drug targets [55, 63]. This method also can identify additional regulatory mechanisms of histone modifications like alternative splicing at the hundreds of target gene loci in holistic manner [55]. The cross talk with other epigenetic mechanisms, like DNA methylation, noncoding RNA, and chromatin remodeling, also can be addressed with this method.

Addiction is considered as a learning and memory process, which shares various plasticity-related components, but acts in abnormal manners upon drug administration [64]. In this regard, metaplasticity is a very attractive hypothesis to explain addiction processes [65, 66]. Recently, Cahill et al. [67] presented strong evidence from signaling pathways that linked to behavioral changes. They reported that transient increase of Rab1b gene expression, which accompanies increased H3K4me3 at the *Rab1b* promoter, induces the metaplastic state in the NAc after chronic cocaine injection. In this respect, Malvaez et al.'s [36] report that HDAC3 inhibition during CPP extinction period enhances extinction of cocaine associated memory can be explained as an example of induced metaplasticity by modulating histone modifications directly, via acetylation. Interestingly, authors reported that the other memory task (object location memory test) is also enhanced, because there is no selectivity in systemic treatment of HDAC3 inhibitor, which induces metaplastic state.

One of the interesting features of drug injection studies is that histone modifications and gene expression patterns induced by acute and chronic drug exposure are variable throughout time domain; some genes show acute response and some do not, with or without connections between histone modification and gene expression. The "nonresponsive genes" with robust histone modifications can be under the process of priming or desensitization. *Bdnf* gene, which promotes synapse formation, shows relatively early, robust induction of H3 acetylation after cocaine treatment (priming), but the *Bdnf* expression increases gradually after withdrawal [24, 30, 68]. Many other genes like *Mef2d* and *Hdac4* need to be studied further in this respect [25]. We don't know how the delayed response to the histone modifications occurs, but it is possible that specific epigenetic "reader" proteins are involved in this process. Recently, Sartor et al. [69] found that a histone acetylation reader, BRD4, modulates *Bdnf* expression in cocaine addiction. Although the detail mechanisms are still illusive and need to be validated, one can postulate that these kinds of regulation also can be acting on induction or suppression of different phases of metaplasticity and so considered as druggable targets combined with extinction process to make "good" memory.

Although we usually analyze functionally distinct brain regions in experiments, a "single brain region" contains many types of neurons and nonneuronal cells, and their functions in addiction can be very different, sometimes opposite. For example, it has been well known that D1- and D2-MSNs in the NAc show opposite actions in cocaine reward [49, 70, 71]. Because of the cellular heterogeneity, we should consider the possibility that mixed population of cells in brain samples can blunt cell-type-specific changes during data analysis. Recently, researchers have addressed cell-type-specific actions of drugs using various experimental approaches: cell-type-specific analysis of gene expression and/or histone modifications using FACS [58, 72, 73] and transgenic mice models like TRAP [71] and RiboTag mice [49, 74].

One of the major pitfalls in histone modification research is that the ability to modify epigenetic chromatin states in a precise locus-specific manner has been limited. Currently, CRISPR and zinc-finger protein-based cell-type-specific epigenome modification strategies are available to modulate histone modifications in a gene locus-specific manner [46, 75–77]. The ability to directly regulate epigenomic states using site-specific genome editing approaches provides investigators with the necessary tools to explore the causal relationship between epigenetic mechanisms and the pathogenesis of drug addictions. Recently, Heller et al. [46, 77] showed that zinc-finger protein fused with G9a elicits target gene-directed epigenetic modulation and subsequent behavioral changes in addiction and depression models. This kind of precise editing of target gene epigenome will be valuable tools to validate causal relations between histone modification and gene regulation and dissect complex gene regulatory network in drug addiction. Combining with genome-wide analysis and cell-type-specific approaches, this will advance studies on the addiction process.

Many psychiatric disorders share common brain structures, circuits, and molecular mechanisms, including epigenetic histone modifications of many genes [42, 78]. Indeed, various immediate early genes, signaling molecules, neurotrophic factors, and histone modifiers are involved in both reward-related behaviors, addiction, and depression, although the relations between them are very complicated and unexpected: some factors modulate both natural reward and drug reward in the same or opposite direction [78]. Although much more studies are needed to reveal the relations, it would be valuable to study how the mechanisms are shared or differentiated for each reward process.

In conclusion, persistent alteration in gene expression and neuronal activity is closely related to changes in histone modification upon the administration of drugs of abuse. Modulation of these histone modifications can alter drug-induced abnormal behavioral changes. Recent methodological progress will provide valuable tools to reveal drug-specific histone modification pathways and hub genes, governing drug responses, which can be valuable therapeutic targets.

Acknowledgments This work was supported by grants from NARSAD and NIMH.

References

1. Feng J, Nestler EJ. Epigenetic mechanisms of drug addiction. Curr Opin Neurobiol. 2013;23(4):521–8.
2. Walker DM, Cates HM, Heller EA, Nestler EJ. Regulation of chromatin states by drugs of abuse. Curr Opin Neurobiol. 2015;30:112–21.
3. Izzo A, Schneider R. The role of linker histone H1 modifications in the regulation of gene expression and chromatin dynamics. Biochim Biophys Acta. 2016;1859(3):486–95.
4. Jenuwein T, Allis CD. Translating the histone code. Science. 2001;293(5532):1074–80.
5. Graff J, Tsai LH. Histone acetylation: molecular mnemonics on the chromatin. Nat Rev Neurosci. 2013;14(2):97–111.
6. Alam H, Gu B, Lee MG. Histone methylation modifiers in cellular signaling pathways. Cell Mol Life Sci. 2015;72(23):4577–92.
7. Rossetto D, Avvakumov N, Cote J. Histone phosphorylation: a chromatin modification involved in diverse nuclear events. Epigenetics. 2012;7(10):1098–108.
8. Wang H, Wang L, Erdjument-Bromage H, Vidal M, Tempst P, Jones RS, et al. Role of histone H2A ubiquitination in Polycomb silencing. Nature. 2004;431(7010):873–8.
9. Kim J, Guermah M, McGinty RK, Lee JS, Tang Z, Milne TA, et al. RAD6-mediated transcription-coupled H2B ubiquitylation directly stimulates H3K4 methylation in human cells. Cell. 2009;137(3):459–71.
10. Nathan D, Ingvarsdottir K, Sterner DE, Bylebyl GR, Dokmanovic M, Dorsey JA, et al. Histone sumoylation is a negative regulator in *Saccharomyces cerevisiae* and shows dynamic interplay with positive-acting histone modifications. Genes Dev. 2006;20(8):966–76.

11. Cohen-Armon M, Visochek L, Rozensal D, Kalal A, Geistrikh I, Klein R, et al. DNA-independent PARP-1 activation by phosphorylated ERK2 increases Elk1 activity: a link to histone acetylation. Mol Cell. 2007;25(2):297–308.
12. D'Amours D, Desnoyers S, D'Silva I, Poirier GG. Poly(ADP-ribosyl)ation reactions in the regulation of nuclear functions. Biochem J. 1999;342(Pt 2):249–68.
13. Cuthbert GL, Daujat S, Snowden AW, Erdjument-Bromage H, Hagiwara T, Yamada M, et al. Histone deimination antagonizes arginine methylation. Cell. 2004;118(5):545–53.
14. Wang Y, Wysocka J, Sayegh J, Lee YH, Perlin JR, Leonelli L, et al. Human PAD4 regulates histone arginine methylation levels via demethylimination. Science. 2004;306(5694):279–83.
15. Gambetta MC, Muller J. A critical perspective of the diverse roles of O-GlcNAc transferase in chromatin. Chromosoma. 2015;124(4):429–42.
16. Cheung P, Tanner KG, Cheung WL, Sassone-Corsi P, Denu JM, Allis CD. Synergistic coupling of histone H3 phosphorylation and acetylation in response to epidermal growth factor stimulation. Mol Cell. 2000;5(6):905–15.
17. Du J, Johnson LM, Jacobsen SE, Patel DJ. DNA methylation pathways and their crosstalk with histone methylation. Nat Rev Mol Cell Biol. 2015;16(9):519–32.
18. Miller CA, Campbell SL, Sweatt JD. DNA methylation and histone acetylation work in concert to regulate memory formation and synaptic plasticity. Neurobiol Learn Mem. 2008; 89(4):599–603.
19. Maze I, Shen L, Zhang B, Garcia BA, Shao N, Mitchell A, et al. Analytical tools and current challenges in the modern era of neuroepigenomics. Nat Neurosci. 2014;17(11):1476–90.
20. Tweedie-Cullen RY, Brunner AM, Grossmann J, Mohanna S, Sichau D, Nanni P, et al. Identification of combinatorial patterns of post-translational modifications on individual histones in the mouse brain. PLoS One. 2012;7(5):e36980.
21. Sanchis-Segura C, Lopez-Atalaya JP, Barco A. Selective boosting of transcriptional and behavioral responses to drugs of abuse by histone deacetylase inhibition. Neuropsychopharmacology. 2009;34(13):2642–54.
22. Levine A, Huang Y, Drisaldi B, Griffin Jr EA, Pollak DD, Xu S, et al. Molecular mechanism for a gateway drug: epigenetic changes initiated by nicotine prime gene expression by cocaine. Sci Transl Med. 2011;3(107):107ra9.
23. Wang L, Lv Z, Hu Z, Sheng J, Hui B, Sun J, et al. Chronic cocaine-induced H3 acetylation and transcriptional activation of CaMKIIalpha in the nucleus accumbens is critical for motivation for drug reinforcement. Neuropsychopharmacology. 2010;35(4):913–28.
24. Kumar A, Choi KH, Renthal W, Tsankova NM, Theobald DE, Truong HT, et al. Chromatin remodeling is a key mechanism underlying cocaine-induced plasticity in striatum. Neuron. 2005;48(2):303–14.
25. Renthal W, Kumar A, Xiao G, Wilkinson M, Covington 3rd HE, Maze I, et al. Genome-wide analysis of chromatin regulation by cocaine reveals a role for sirtuins. Neuron. 2009;62(3):335–48.
26. Levine AA, Guan Z, Barco A, Xu S, Kandel ER, Schwartz JH. CREB-binding protein controls response to cocaine by acetylating histones at the fosB promoter in the mouse striatum. Proc Natl Acad Sci U S A. 2005;102(52):19186–91.
27. Huang YY, Levine A, Kandel DB, Yin D, Colnaghi L, Drisaldi B, et al. D1/D5 receptors and histone deacetylation mediate the Gateway effect of LTP in hippocampal dentate gyrus. Learn Mem. 2014;21(3):153–60.
28. Huang YY, Kandel DB, Kandel ER, Levine A. Nicotine primes the effect of cocaine on the induction of LTP in the amygdala. Neuropharmacology. 2013;74:126–34.
29. Malvaez M, Mhillaj E, Matheos DP, Palmery M, Wood MA. CBP in the nucleus accumbens regulates cocaine-induced histone acetylation and is critical for cocaine-associated behaviors. J Neurosci. 2011;31(47):16941–8.
30. Schmidt HD, Sangrey GR, Darnell SB, Schassburger RL, Cha JH, Pierce RC, et al. Increased brain-derived neurotrophic factor (BDNF) expression in the ventral tegmental area during

cocaine abstinence is associated with increased histone acetylation at BDNF exon I-containing promoters. J Neurochem. 2012;120(2):202–9.
31. Drazic A, Myklebust LM, Ree R, Arnesen T. The world of protein acetylation. Biochim Biophys Acta. 2016;1864(10):1372–401.
32. Malvaez M, Sanchis-Segura C, Vo D, Lattal KM, Wood MA. Modulation of chromatin modification facilitates extinction of cocaine-induced conditioned place preference. Biol Psychiatry. 2010;67(1):36–43.
33. Renthal W, Maze I, Krishnan V, Covington 3rd HE, Xiao G, Kumar A, et al. Histone deacetylase 5 epigenetically controls behavioral adaptations to chronic emotional stimuli. Neuron. 2007;56(3):517–29.
34. Rogge GA, Singh H, Dang R, Wood MA. HDAC3 is a negative regulator of cocaine-context-associated memory formation. J Neurosci. 2013;33(15):6623–32.
35. Kennedy PJ, Feng J, Robison AJ, Maze I, Badimon A, Mouzon E, et al. Class I HDAC inhibition blocks cocaine-induced plasticity by targeted changes in histone methylation. Nat Neurosci. 2013;16(4):434–40.
36. Malvaez M, McQuown SC, Rogge GA, Astarabadi M, Jacques V, Carreiro S, et al. HDAC3-selective inhibitor enhances extinction of cocaine-seeking behavior in a persistent manner. Proc Natl Acad Sci U S A. 2013;110(7):2647–52.
37. Tsai SY, Chuang JY, Tsai MS, Wang XF, Xi ZX, Hung JJ, et al. Sigma-1 receptor mediates cocaine-induced transcriptional regulation by recruiting chromatin-remodeling factors at the nuclear envelope. Proc Natl Acad Sci U S A. 2015;112(47):E6562–70.
38. Ferguson D, Koo JW, Feng J, Heller E, Rabkin J, Heshmati M, et al. Essential role of SIRT1 signaling in the nucleus accumbens in cocaine and morphine action. J Neurosci. 2013;33(41):16088–98.
39. Ferguson D, Shao N, Heller E, Feng J, Neve R, Kim HD, et al. SIRT1-FOXO3a regulate cocaine actions in the nucleus accumbens. J Neurosci. 2015;35(7):3100–11.
40. Jayanthi S, McCoy MT, Chen B, Britt JP, Kourrich S, Yau HJ, et al. Methamphetamine downregulates striatal glutamate receptors via diverse epigenetic mechanisms. Biol Psychiatry. 2014;76(1):47–56.
41. Engel GL, Marella S, Kaun KR, Wu J, Adhikari P, Kong EC, et al. Sir2/Sirt1 links acute inebriation to presynaptic changes and the development of alcohol tolerance, preference, and reward. J Neurosci. 2016;36(19):5241–51.
42. Pena CJ, Bagot RC, Labonte B, Nestler EJ. Epigenetic signaling in psychiatric disorders. J Mol Biol. 2014;426(20):3389–412.
43. Maze I, Feng J, Wilkinson MB, Sun H, Shen L, Nestler EJ. Cocaine dynamically regulates heterochromatin and repetitive element unsilencing in nucleus accumbens. Proc Natl Acad Sci U S A. 2011;108(7):3035–40.
44. Maze I, Covington 3rd HE, Dietz DM, LaPlant Q, Renthal W, Russo SJ, et al. Essential role of the histone methyltransferase G9a in cocaine-induced plasticity. Science. 2010;327(5962):213–6.
45. Robison AJ, Nestler EJ. Transcriptional and epigenetic mechanisms of addiction. Nat Rev Neurosci. 2011;12(11):623–37.
46. Heller EA, Hamilton PJ, Burek DD, Lombroso SI, Pena CJ, Neve RL, et al. Targeted epigenetic remodeling of the Cdk5 gene in nucleus accumbens regulates cocaine- and stress-evoked behavior. J Neurosci. 2016;36(17):4690–7.
47. Bibb JA, Chen J, Taylor JR, Svenningsson P, Nishi A, Snyder GL, et al. Effects of chronic exposure to cocaine are regulated by the neuronal protein Cdk5. Nature. 2001;410(6826):376–80.
48. Li Y, Zhu R, Wang W, Fu D, Hou J, Ji S, et al. Arginine methyltransferase 1 in the nucleus accumbens regulates behavioral effects of cocaine. J Neurosci. 2015;35(37):12890–902.
49. Damez-Werno DM, Sun H, Scobie KN, Shao N, Rabkin J, Dias C, et al. Histone arginine methylation in cocaine action in the nucleus accumbens. Proc Natl Acad Sci U S A. 2016;113(34):9623–8.

50. Aguilar-Valles A, Vaissiere T, Griggs EM, Mikaelsson MA, Takacs IF, Young EJ, et al. Methamphetamine-associated memory is regulated by a writer and an eraser of permissive histone methylation. Biol Psychiatry. 2014;76(1):57–65.
51. Godino A, Jayanthi S, Cadet JL. Epigenetic landscape of amphetamine and methamphetamine addiction in rodents. Epigenetics. 2015;10(7):574–80.
52. Krasnova IN, Chiflikyan M, Justinova Z, McCoy MT, Ladenheim B, Jayanthi S, et al. CREB phosphorylation regulates striatal transcriptional responses in the self-administration model of methamphetamine addiction in the rat. Neurobiol Dis. 2013;58:132–43.
53. Sun H, Maze I, Dietz DM, Scobie KN, Kennedy PJ, Damez-Werno D, et al. Morphine epigenomically regulates behavior through alterations in histone H3 lysine 9 dimethylation in the nucleus accumbens. J Neurosci. 2012;32(48):17454–64.
54. Koo JW, Mazei-Robison MS, LaPlant Q, Egervari G, Braunscheidel KM, Adank DN, et al. Epigenetic basis of opiate suppression of Bdnf gene expression in the ventral tegmental area. Nat Neurosci. 2015;18(3):415–22.
55. Feng J, Wilkinson M, Liu X, Purushothaman I, Ferguson D, Vialou V, et al. Chronic cocaine-regulated epigenomic changes in mouse nucleus accumbens. Genome Biol. 2014;15(4):R65.
56. Brami-Cherrier K, Valjent E, Herve D, Darragh J, Corvol JC, Pages C, et al. Parsing molecular and behavioral effects of cocaine in mitogen- and stress-activated protein kinase-1-deficient mice. J Neurosci. 2005;25(49):11444–54.
57. Stipanovich A, Valjent E, Matamales M, Nishi A, Ahn JH, Maroteaux M, et al. A phosphatase cascade by which rewarding stimuli control nucleosomal response. Nature. 2008;453(7197):879–84.
58. Jordi E, Heiman M, Marion-Poll L, Guermonprez P, Cheng SK, Nairn AC, et al. Differential effects of cocaine on histone posttranslational modifications in identified populations of striatal neurons. Proc Natl Acad Sci U S A. 2013;110(23):9511–6.
59. Kraus WL. Transcriptional control by PARP-1: chromatin modulation, enhancer-binding, coregulation, and insulation. Curr Opin Cell Biol. 2008;20(3):294–302.
60. Kraus WL, Lis JT. PARP goes transcription. Cell. 2003;113(6):677–83.
61. Ha HC, Snyder SH. Poly(ADP-ribose) polymerase-1 in the nervous system. Neurobiol Dis. 2000;7(4):225–39.
62. Fontan-Lozano A, Suarez-Pereira I, Horrillo A, del-Pozo-Martin Y, Hmadcha A, Carrion AM. Histone H1 poly[ADP]-ribosylation regulates the chromatin alterations required for learning consolidation. J Neurosci. 2010;30(40):13305–13.
63. Scobie KN, Damez-Werno D, Sun H, Shao N, Gancarz A, Panganiban CH, et al. Essential role of poly(ADP-ribosyl)ation in cocaine action. Proc Natl Acad Sci U S A. 2014;111(5):2005–10.
64. Nestler EJ. Common molecular and cellular substrates of addiction and memory. Neurobiol Learn Mem. 2002;78(3):637–47.
65. Dong Y, Nestler EJ. The neural rejuvenation hypothesis of cocaine addiction. Trends Pharmacol Sci. 2014;35(8):374–83.
66. Creed MC, Luscher C. Drug-evoked synaptic plasticity: beyond metaplasticity. Curr Opin Neurobiol. 2013;23(4):553–8.
67. Cahill ME, Bagot RC, Gancarz AM, Walker DM, Sun H, Wang ZJ, et al. Bidirectional synaptic structural plasticity after chronic cocaine administration occurs through Rap1 small GTPase signaling. Neuron. 2016;89(3):566–82.
68. Grimm JW, Lu L, Hayashi T, Hope BT, Su TP, Shaham Y. Time-dependent increases in brain-derived neurotrophic factor protein levels within the mesolimbic dopamine system after withdrawal from cocaine: implications for incubation of cocaine craving. J Neurosci. 2003;23(3):742–7.
69. Sartor GC, Powell SK, Brothers SP, Wahlestedt C. Epigenetic readers of lysine acetylation regulate cocaine-induced plasticity. J Neurosci. 2015;35(45):15062–72.
70. Lobo MK, Covington 3rd HE, Chaudhury D, Friedman AK, Sun H, Damez-Werno D, et al. Cell type-specific loss of BDNF signaling mimics optogenetic control of cocaine reward. Science. 2010;330(6002):385–90.

71. Maze I, Chaudhury D, Dietz DM, Von Schimmelmann M, Kennedy PJ, Lobo MK, et al. G9a influences neuronal subtype specification in striatum. Nat Neurosci. 2014;17(4):533–9.
72. Koo JW, Lobo MK, Chaudhury D, Labonte B, Friedman A, Heller E, et al. Loss of BDNF signaling in D1R-expressing NAc neurons enhances morphine reward by reducing GABA inhibition. Neuropsychopharmacology. 2014;39(11):2646–53.
73. Guez-Barber D, Fanous S, Golden SA, Schrama R, Koya E, Stern AL, et al. FACS identifies unique cocaine-induced gene regulation in selectively activated adult striatal neurons. J Neurosci. 2011;31(11):4251–9.
74. Chandra R, Francis TC, Konkalmatt P, Amgalan A, Gancarz AM, Dietz DM, et al. Opposing role for Egr3 in nucleus accumbens cell subtypes in cocaine action. J Neurosci. 2015;35(20):7927–37.
75. Thakore PI, Black JB, Hilton IB, Gersbach CA. Editing the epigenome: technologies for programmable transcription and epigenetic modulation. Nat Methods. 2016;13(2):127–37.
76. Hilton IB, D'Ippolito AM, Vockley CM, Thakore PI, Crawford GE, Reddy TE, et al. Epigenome editing by a CRISPR-Cas9-based acetyltransferase activates genes from promoters and enhancers. Nat Biotechnol. 2015;33(5):510–7.
77. Heller EA, Cates HM, Pena CJ, Sun H, Shao N, Feng J, et al. Locus-specific epigenetic remodeling controls addiction- and depression-related behaviors. Nat Neurosci. 2014;17(12):1720–7.
78. Russo SJ, Nestler EJ. The brain reward circuitry in mood disorders. Nat Rev Neurosci. 2013;14(9):609–25.

Anxiety and Epigenetics

8

Andrew A. Bartlett, Rumani Singh, and Richard G. Hunter

Abstract

Anxiety disorders are highly prevalent psychiatric disorders often comorbid with depression and substance abuse. Twin studies have shown that anxiety disorders are moderately heritable. Yet, genome-wide association studies (GWASs) have failed to identify gene(s) significantly associated with diagnosis suggesting a strong role for environmental factors and the epigenome. A number of anxiety disorder subtypes are considered "stress related." A large focus of research has been on the epigenetic and anxiety-like behavioral consequences of stress. Animal models of anxiety-related disorders have provided strong evidence for the role of stress on the epigenetic control of the hypothalamic-pituitary-adrenal (HPA) axis and of stress-responsive brain regions. Neuroepigenetics may continue to explain individual variation in susceptibility to environmental perturbations and consequently anxious behavior. Behavioral and pharmacological interventions aimed at targeting epigenetic marks associated with anxiety may prove fruitful in developing treatments.

Keywords

Anxiety • Epigenetic • Stress • Glucocorticoid • Plasticity • Hippocampus • Amygdala • Prefrontal cortex • Histone • DNA methylation • Noncoding RNA

A.A. Bartlett • R. Singh, Ph.D. • R.G. Hunter, Ph.D. (✉)
Department of Psychology, University of Massachusetts,
100 Morrissey Blvd, Boston, MA 02125, USA
e-mail: Andrew.Bartlett001@umb.edu; Rumani.Singh@umb.edu; Richard.Hunter@umb.edu

R. Delgado-Morales (ed.), *Neuroepigenomics in Aging and Disease*, Advances in Experimental Medicine and Biology, DOI 10.1007/978-3-319-53889-1_8

8.1 Introduction

Anxiety disorders (ADs) are among the most common psychiatric disorders, occurring in roughly a third of the US population. They are also highly comorbid with depression and substance abuse disorders, and the pathogenesis of AD is likely highly interrelated [1]. While anxiety disorders are heritable and genetic factors play a role in anxiety disorders, most of the risk of these disorders is environmental in nature [2]. Stress, particularly in early life, substance abuse, circadian, and microbiota have all been shown to have an influence on risk of anxiety disorders [3–6]. Further, it is likely that anxious phenotypes are influenced by more subtle factors such as the interplay between an anxious parent and a child whose early life is defined in part by adapting to that parent's behavior [7, 8]. Indeed the latter case is emblematic of one of the important distinctions between heritability, which can include epigenetic mechanisms, both behavioral and molecular, and the strictly genetic inheritance with which heritability, in general, is often conflated. ADs are moderately heritable with most of the disorders in the classification showing heritability in the range of 30% [9]. GWASs have generally not met the criterion of genome-wide significance, and candidate gene approaches have also been relatively unsuccessful [10]. However some genetic polymorphisms do show replicable associations with AD, for example, the glucocorticoid receptor chaperone FKBP5 has been associated with risk of post-traumatic stress disorder (PTSD) in individuals with a history of child abuse in an African–American sample, and the same sample also demonstrated a female-specific association with PTSD and the ADCYPAP1R1 receptor for the neuropeptide PACAP [11, 12]. The catechol-O-methyltransferase (COMT) valine158methionine polymorphism has been repeatedly implicated in risk of panic disorder, though with different alleles imparting risk in European versus Asian populations [13]. Even these findings point to the role of other contextual factors like ancestry and sex as influences on the underlying genetics, developmental context also appears to influence the expression of genetic risk, as a study in a Swedish cohort has shown that different risk factors act at different times across adolescence and early adulthood [14].

While environmental factors like stress are clearly significant in many anxiety disorders, their effects can vary wildly across individuals. The role of the environment is most clear with PTSD, where most individuals are resilient and only a fraction go on to develop the disorder after a trauma exposure [15, 16]. The question of differential susceptibility in AD is another to which genetic explanations thus far fall short.

8.2 The Neuroanatomy of Anxiety Disorders

Human anxiety is defined by emotional symptoms as well as behavioral and physiological phenotypes. Much of the work in understanding the underlying neuroanatomy involved in anxiety-related pathology has been done using animal models. Specifically, the focus has been on conserved endocrine systems and brain regions that identify or respond to environmental threats. For example, noxious stimuli may result in freezing behavior, sympathetic nervous system activation, and subsequent endocrine response in both the rat and the human. Across species, the limbic system and the prefrontal cortex appear to be crucial for regulating threat recognition and response. The hormonal response to threat appears, likewise, remarkably similar and feature highly conserved signaling pathways.

Within the limbic system are a number of structures necessary for threat response and assessment. The amygdala, for instance, appears to be necessary for fear responses. Patients with Urbach-Wiethe disease have compromised amygdala function and report loss of feelings of fear [17]. In rodents, fear conditioning models pair a benign stimulus, the conditioned stimulus (CS), with a noxious stimulus, the unconditioned stimulus (US). A frequently used example of a US is a foot shock which elicits freezing behavior, an unconditioned response (UR). After pairing of the CS with the US, the CS alone can elicit this freezing behavior. This freezing is referred to as the conditioned response (CR). Lesioning the amygdala has been shown to obliterate freezing behavior, the CR, in conditioned rats [18, 19]. Stimulation of the amygdala during CS presentation produces subsequent freezing behavior to the CS without US pairing [20]. During encoding of fear memories, hippocampal inputs to the amygdala appear to be necessary for CS-US pairing for contextual clues [21]. Within the amygdala, various subnuclei have been shown to regulate different processes. The central amygdala (CeA) appears to regulate CR expression through projections to the periaqueductal gray (PAG) [22]. The lateral amygdala (LA) appears to receive CS and US inputs through cortical and thalamic innervations [23]. The stimulation of a subpopulation of neurons in the LA, when paired with presentation of the CS, appears to be sufficient to generate a CR. The basal amygdala (BA) appears to have a dual role in both CR expression and suppression [24]. Two distinct populations of neurons were identified in the BA, one innervated by the hippocampus and the other innervated by the PFC [24, 25]. During extinction of the CS-US pairing, the PFC appears to inhibit the BA and attenuate freezing CR [26–28]. The pairing of the CS-US improves prediction of the US allowing for rapid behavioral response. However, when the CS fails to correctly predict the US, the association must not continue to persist else anxiety or avoidance for the US has now generalized to benign stimuli. These regions are critical to avoidant and anxiety-like behavior. The dysregulation of these circuits may lead to recurrent avoidance or anxious behavior to inappropriate stimuli similar to the definition of human anxiety.

8.3 The Neuroendocrine Axis in Anxiety Disorders

The HPA axis is a critical component of the acute stress response. In response to a stressor, the body must divert resources appropriately in order to efficiently address the challenge at hand. In part, this tentative balance is achieved through the activation of the HPA axis. The HPA axis is a negative feedback loop that begins with the release of arginine-vasopressin (AVP) and corticotrophin-releasing factor (CRF) into the pituitary portal from the paraventricular nucleus (PVN) in the hypothalamus. This release promotes the production of proopiomelanocortin (POMC) in the pituitary. POMC is subsequently converted to adrenocorticotropic hormone (ACTH) and released into the bloodstream. The adrenal gland produces corticosteroids in the adrenal cortex as a consequence of ACTH. Corticosteroids are released into the blood and bind to the mineralocorticoid (MR) and glucocorticoid receptor (GR). In the PVN, pituitary and hippocampus GRs inhibit the production of CRF resulting in negative feedback loop.

The adrenal gland also produces two other hormones, epinephrine and norepinephrine, from the adrenal medulla in response to ACTH. These hormones do not engage in a self-regulating negative feedback loop but are indirectly regulated through the actions of GRs. The role of these hormones is to control the response of the body and the peripheral nervous system, for instance, reducing digestion and immune function while increasing heart rate and blood pressure acutely. Interestingly, pharmacological interventions targeting norepinephrine receptors have proved effective reducing phobias during fear memory reconsolidation [29]. These findings suggest that the autonomic nervous system may remain a potential area for research and intervention in stress-related anxiety disorders such as phobias and PTSD.

8.4 Epigenetic Factors

The relative prominence of the environment and the moderate contribution of genetic factors to the pathogenesis of anxiety disorders have made the study of these disorders through the lens of epigenetics a fruitful avenue of research in recent years. Many molecular epigenetic mechanisms have now been implicated in AD, including DNA and histone modification as well as noncoding RNA (ncRNA). Epigenetics, in the strict molecular sense, refers to regulation of DNA sequences that does not involve alteration of actual base composition. Transcription and other genomic functions are regulated directly through epigenetic modifications that typically annotate DNA and its associated histones via acetylation, methylation, and phosphorylation. These epigenetic marks are tightly linked to chromatin state as complex of DNA, RNA, and protein. Open chromatin is associated with active transcription, whereas closed chromatin is associated with transcriptional silencing. Epigenetic marks that define the epigenotype include DNA methylation and various modifications (e.g., methylation, acetylation) of histone proteins that are complexed with DNA. DNA methylation occurs at cytosines of CpG dinucleotides and is catalyzed by enzymes of the DNA methyltransferase family. DNA methylation may inhibit gene expression by direct interaction with factors that repress transcription or, indirectly, through recruitment of methyl-CpG binding proteins (MeCP2 and MBDs) complexed with enzymes that modify histone proteins. These modifications can transform chromatin from an active to a repressed state, or vice versa.

The role of the epigenome in etiology of anxiety disorders and variations in behavior and neurological status can now be investigated. Of particular importance in epigenetics research is the fact that epigenetic marks are modifiable both in the germ line and in somatic tissues by genetic, environmental, and stochastic factors. Each cell in the human body possesses not only a genotype, identical in all somatic cells of an organism, but also an epigenotype that is highly variable among the different tissues of an individual. Errors or alterations in epigenotype can occur as primary stochastic events or secondarily in response to either genetic mutations (e.g. transposition events) or environmental exposures. Therefore a discussion of potential epigenetic etiologies of anxiety disorders necessarily involves both genetic and environmental factors. Dysregulation of genes that control epigenetic mechanisms leads to a number of "epigenetic syndromes" falling into two groups. Those with changes in genes regulating epigenetic marks include enzymes such as DNA methyltransferases, methyl-binding proteins, and enzymes that affect histone modification. The second category involves genes that are regulated by epigenetic marks, for example, imprinted genes.

8.5 Epigenetics in Animal Models of Anxiety

Twin studies of generalized anxiety disorder have failed to identify either a genetic basis for or strongly heritable component of the disorder [30]. This class of mental health disorders is often comorbid with addiction [31]. Both involve pathological behaviors that have a neurobiological basis. Over the last decade, increasing focus has been placed on how gene-environment interactions mediated by epigenetic molecular mechanisms might improve our understanding of the disease. Though environmental influences including trauma and substance abuse are known contributors to anxiety, it is difficult or impossible at present to examine molecular epigenetic changes in the central nervous system of clinical populations, and given the tissue-specific nature of epigenetic mechanisms, accessible peripheral tissues such as blood or epithelial cells may not reflect the changes present in the brain. For these reasons, animal models have been employed to mimic the signs of anxiety. While symptoms, such as intrusive thoughts, are impossible to model in rodents or nonhuman primates, sophisticated paradigms have been used to model aspects of social anxiety, general anxiety, and more broadly anxious temperament. In rodent models, common behavioral paradigms to assess anxiety-like behaviors include the elevated plus maze (EPM), light/dark box (LD), open field test (OFT), social defeat (SD), and the social interaction test (SIT). The EPM consists of two arms of open platforms and two arms of closed platforms featuring three walls. The EPM is based on an innate fear of heights and open spaces such that rodents prefer the closed platforms to the open platforms. After quickly equilibrating to the testing arena, less anxious rodents will explore and spend increasing time on the open platforms. The LD box consists of two connected chambers, one illuminated while the other is not. The natural preference of the rodent is the dark chamber; however, given time less anxious rodents again will explore and spend increasing time in the light chamber. The OFT is a square testing arena with four walls. In novel settings, rodents prefer to remain unexposed to predators, in this case, close to the wall. After exposure, less anxious animals will cross the arena exploring and spend increasingly more time in the center. SD paradigms vary to some extent but primarily involve repeated exposure of a rodent to another dominating rodent. The exposed rodents display depressive-like symptoms but also social avoidance. Social avoidance is most commonly measured using SIT. SIT is conducted in a two-chamber arena separated by a wall to prevent contact but allow for other sensory exchanges, i.e., visual cues, odor, and ultrasonic vocalizations. More anxious, socially avoidant rodents will spend less time in the area closest to the neighboring chamber after habituation. These consist of the major testing paradigms used to proximate anxiety in the rodent and have allowed for a more comprehensive understanding of the neuroepigenetic regulation of anxiolytic behavior.

Natural variation in susceptibility to clinical anxiety has been subject to increased scrutiny in recent years. Early animal work suggested that gene-environment interactions likely mediated anxiety outcomes as SD paradigms among other stressors produced anxious phenotypes. Notably, an early study showed that susceptibility to SD, as measured by reduced interaction time in the SIT, was correlated with DNA methylation of CpG islands in the promoter of the *CRH* gene in paraventricular nucleus (PVN) [32]. Natural variation in maternal care during the first week of life was shown to differentially pattern the methylation of nr3c1 promoter of offspring, modification that persisted into adulthood and corresponded to reduced glucocorticoid receptor expression and enhanced HPA axis activation to an acute stressor [33]. These offspring were later characterized as displaying differential anxiety-like behaviors as a consequence of maternal care received as measured by the EPM and OFT [34, 35]. In adult mice, voluntary exercise has been demonstrated to increase nr3c1 expression while reducing miRNA-124, known to inhibit nr3c1, expression [36]. Though in contradiction to other findings, voluntary exercise decreased time in the open arms of the EPM suggesting an increasingly anxious phenotype. Recently, long noncoding RNA (lncRNA) expression of gomafu in the prefrontal cortex (PFC) has been shown to regulate time spent in the center of the OFT and grooming time to suggest that expression of this lncRNA is necessary for reducing anxiety-like behaviors [37]. Likewise, loss-of-function l3mbtl1, null mice show reduced latency to enter the light chamber in the LD box and increased time spent in the center of the OFT [38]. As l3mbtl1 codes for a methylated lysine domain histone-binding protein, a so-called chromatin reader, this suggests that histone lysine methylation is required for regulating anxiety-like behavior. In this vein, tlr4 null mice did not show increased synaptic enrichment of NR1 following in the short term following repeated ethanol exposure nor increased GluR1 enrichment in the long term in the mPFC compared to similarly treated wild-type controls. These tlr4 mice failed to show mPFC enrichment of acetylated-H4 at the promoter of fosB and BDNF in response to ethanol exposure. This observation suggests that tlr4 is necessary for histone H4 acetylation at fosB and BDNF following ethanol exposure and appears to be necessary for ethanol-induced increases in anxiety-like behavior as indicated by time spent in the open arms of the EPM [39]. In contrast, others have shown that acute ethanol exposure reduces amygdalar miRNA-494 subsequently increasing Cited2, CBP, and p300 expression. These changes were associated with increased H3 acetylation in the central amygdala and anxiolysis [40].

8.6 Transgenerational Epigenetics

Transgenerational epigenetic can be either direct inheritance of mRNAs, protein, or DNA modification via the germline or indirect "inheritance" such that the feed-forward phenotypic profile of the parent can lead to changes in either noncoding RNA expression, histone modification, or DNA methylation. Indirect inheritance was shown by Weaver et al. (2004) where cross-fostering experiments suggested that maternal care alone determined GR 1-7 promoter methylation in offspring hippocampi [33]. Morgan and Bale (2011), in a case of direct inheritance, showed that prenatal stress can lead to alterations in stress sensitivity and miRNA expression in the brains of male offspring [41]. These effects persist for several generations suggesting direct inheritance of paternal miRNAs or DNA methylation via sperm.

Transgenerational effects have been consistently observed in the offspring of Holocaust survivors [42–44]. Maternal PTSD of these survivors has been predictive of offspring PTSD risk and increased corticosteroid sensitivity. In specific importance to this chapter, offspring of Holocaust survivors were found to be at a far greater risk of developing an anxiety disorder compared to control, age-matched offspring born to Jewish parents [42]. At this date, the number of generations out to which this inheritance persists and affects offspring of survivors remains unknown. Transgenerational non-genomic transmission of both maternal behavior and HPA axis activation in rats was initially demonstrated by Meaney et al. [45]. The same group showed that glucocorticoid sensitivity and anxiety-like behavior are patterned by maternal care and can persist out for several generations [33, 35]. The level of maternal care during the first week of life patterned the methylation of the GR 1-7 promoter and subsequently GR expression in the hippocampus. These phenotypes can be reversed however by cross-fostering offspring of low-licking and grooming dams with high-licking and grooming dams. In anxious adults of low-licking and grooming dams, the phenotype can be reversed by supplication of an HDAC inhibitor to the hippocampus [34, 35]. Conversely, in low-anxiety adults of high-licking and grooming dams, the phenotype can be reversed by infusion of a methyl donor to the hippocampus [35]. Interestingly, maternal care has also been shown to affect peripheral oxytocin receptor (OXTR) methylation status in rats [46]. A recent clinical study also found that peripheral *OXTR* methylation was associated with increased frequency of anxiety and depression [47]. Genome-wide methylation analysis in infants of mothers with depression and/or anxiety revealed a number of CpG islands to be differentially methylated [48]. Similarly, increased methylation of the *BDNF* gene in blood of adults has been linked to lower maternal care and interpersonal violence-related PTSD [49, 50]. In addition, poor maternal care and anxiety has been linked to risk of diabetes and metabolic syndrome in bonnet macaque offspring [51, 52]. In high- and low-anxiety bred rats, increased H3K9me3 accumulation was found at both the GR and *FGF2* promoters in the hippocampus [53]. This group also found differences in DNA methylation of the *FGF2* promoter in the hippocampus between high- and low-anxiety rats. High-anxiety rats had reduced DNA methylation and methyl-binding protein association at the *FGF2* promoter, which presumably was permissive for increased FGF2 expression [53]. This group also showed that FGF2 increases H3K9me3 association with both the GR promoter and its own. This demonstrates a potential mechanism by which early-life perturbations independent of maternal care can contribute to anxiety-like behavior across generations.

8.7 Neuroepigenetic Effects of Early Stress on Anxious Behaviors

Early-life stress has been demonstrated repeatedly to pattern stress reactivity and anxious behavior. These changes persist beyond the time frame of the initial stressor and often long into adulthood. The prenatal effects of stress lead to dysregulation of the HPA axis associated mainly with changes GR expression [35]. Though these findings were first reported in animal studies. Recently, these findings have been recapitulated in longitudinal human studies. For instance, maternal prenatal anxiety has been shown to predict internalizing and anxiety scores on the child behavior checklist in the infant [49]. Further, differences in global DNA methylation were observed at a number of CpG sites in neonatal cord blood of mothers affected by anxiety during gestation [48]. Likewise, maternal PTSD has been shown to associate with both increased glucocorticoid sensitivity in the offspring of Holocaust survivors and increased offspring diagnosed with anxiety disorders [42]. Maternal PTSD has also been demonstrated to be predictive of offspring PTSD and presumably through inherited stress reactivity [43, 44]. These findings suggest that both the prenatal environment and stress/trauma history may recruit epigenetic processes in the intergenerational transmission of HPA axis dysregulation and anxiogenic consequences. However, consideration of allostatic load must be of concern as severe and mild stress have opposing roles on physiology and behavior. Allostatic load is the cumulative effect of multiple stressors taking into consideration severity, duration, and ability to cope with stressors [54, 55]. Consider the effects of a severe uncontrollable stressor, for example, maternal separation, on stress sensitivity in contrast to a mild controllable stressor such as voluntary exercise. While maternal separation sensitizes the HPA axis of the infant, voluntary exercise can promote resiliency to future stressors [56, 57].

8.7.1 Prenatal Stress

In utero exposure to maternal stress and corticosteroids patterns the HPA axis of infants ultimately altering synaptic connectivity, function, and behavioral responses specifically those involved in stress adaptation [58–60]. Prenatal restraint stress has been shown to impair offspring brain function and development reducing HPA axis feedback and altering neuroplasticity [61]. Prenatal stress and glucocorticoid treatment produce lasting behavioral changes such as spatial learning impairment and increased anxiety-like behavior [58, 59]. In addition, mild stressors, for instance, postnatal handling, have been shown to reduce these deficits as well as attenuate

HPA axis sensitivity [58, 59]. Prenatal stress does so by altering synaptic connectivity, neurogenesis, and chromatin structure in stress-sensitive regions of the brain, for example, in the PFC where offspring of maternally stressed dams show reduced dendritic spine complexity and density [60]. Similarly, in both rodent and nonhuman primate models, prenatal stress retards hippocampal neurogenesis in the dentate gyrus. Prenatal stress has been linked to increased methylation of the GR 1-7 promoter in the hippocampus as well as reduced methylation of the CRF promoter in the hypothalamus and amygdala of male but not female mice [62]. These sex-specific changes have been linked to differential expression of DNA methyltransferase 1 (DNMT1), though the changes responsible for this dichotomized expression remain unknown. Elliot et al. (2010) first ascribed natural variation in social interaction following social defeat in adults to be due in part to the methylation of the CRH promoter in the hypothalamus. Mice susceptible to social defeat show increased social anxiety and reduced CRH promoter methylation in the PVN [32]. The methylation status of the CRF promoter in PVN helps to explain natural variability in the susceptibility of mice to social defeat and consequently social anxiety. Prenatal stress had previously been shown to differentially affect CRF release in the PVN [63]. Interestingly, a subsequent study found that prenatal restraint stress increased both anxious behavior and corticosterone release in response to stress while reducing CRF promoter methylation at the same CpG islands noted by Elliot et al. (2010) [64]. Prenatal restraint stress has also been shown to increase methylation of the *REELIN* promoter in the PFC perhaps linking changes in synaptic connectivity observed there to underlying molecular influences [65]. *REELIN* is an important neuroplasticity gene, known to be epigenetically regulated by fear conditioning [66]. Similarly, prenatal exposure to maternal depression and anxiety has been linked to increased *NR3C1* 1F promoter methylation and increased salivary cortisol following exposure to a stressor in infants [67, 68]. Maternal anxiety has been linked to differential methylation of a number of other genes in cord blood including *IGF2* and *H19* [69]. In fact, distress during pregnancy has been linked to placental methylation of a number of stress-related genes including *HSD11B2*, *NR3C1*, and *FKBP5* [70]. Other perturbations, including maternal diet and paternal exposure to drugs of abuse such as cocaine and ethanol, have been shown to alter cortical gene expression through changes in the epigenetic machinery and affect anxiolytic behavior in the offspring [71–73]. Importantly, mild postnatal stressors have been shown to reverse the effects of prenatal stress as well as promote resiliency [58, 59, 74, 75]. Given the association between maternal stress and anxiety, these findings provide evidence for the efficacy of behavioral therapy and alike as an early-life intervention [7, 76].

8.7.2 Early-Life Stress

The vast majority of studies of early-life stress focus on the epigenetic consequences of the interactions within the mother-infant dyad. Both maternal care and separation have been demonstrated to both alter HPA axis stress reactivity and adult anxiety behaviors of the infant through lasting changes to the epigenomes [35, 77]. Specifically, maternal separation has been shown to sensitize offspring HPA axis activation early-life interventions including environmental enrichment attenuate this effect [78]. Poor rearing conditions have been shown to increase CRF release from the PVN and amygdala as well as hypermethylate the GR 1-7 promoter in the hippocampus [33, 79]. Conversely, good maternal care and rearing conditions have been demonstrated to hypomethylate the GR 1-7 promoter in the hippocampus, produce efficient stress responses, and reduce anxiety-like behaviors [35, 80–82]. The GR 1F promoter is the human ortholog of the rodent 1-7 promoter [83]. Hypermethylation of the 1F promoter in the brains of suicide victims was associated with childhood abuse [84]. The findings of McGowan et al. (2009) were later expanded to include the 1-B, 1-C, and 1-H promoters as well [85]. Other groups have failed to replicate some of these findings, however [86]. The McGowan group has also shown that hippocampal ribosomal RNA expression is reduced in suicide victims suggesting reduced hippocampal protein synthesis [87]. Childhood adversity has also been linked to increased 1F promoter methylation in peripheral cells as well [88, 89]. Methylation patterns as a consequence of childhood abuse overwhelmingly persist into adulthood [90]. Early postnatal stress followed by subsequent adult chronic stress has been linked to reduced hippocampal plasticity and increased anxiety-like behaviors [91]. Maternal separation has been shown to reduce amygdalar neurotensin receptor 1 (NTSR1) expression through increased methylation of the *NTSR1* promoter. Microinfusion of NTSR1 receptor agonist increased conditioned freezing responses, while an agonist reduced this behavior suggesting an epigenetic molecular mechanism sufficient for increasing anxiety-like behavior [92]. Similarly, maternal separation has been linked to increased HPA activation to environmental stressors in adult offspring [93]. More recently, however, this finding was both replicated and associated with hypomethylation of the POMC, thc gene encoding the precursor for ACTH, in the pituitary [94]. As HPA axis dysregulation has been associated with anxiety-like outcomes, again these findings suggest a critical role of these molecular influences as a consequence of stress in the context of anxiety outcomes. Clinical work has recently shown that early childhood trauma affects CpG methylation in both the promoter and gene proper of the 5-HT3Ar in blood [95]. Interestingly, this locus is downstream of GR response element which showed altered CpG methylation associated with emotional neglect and CpG methylation associated with anxiety-related behaviors.

Adolescence represents another postnatal life stage sensitive to the epigenetic effects of stress [96–100]. For instance, chronic variable stress during adolescence reduces hippocampal volume and spatial cognition, these effects persisting into adulthood [101]. Isolation rearing in adolescent mice reduces the expression of 5-α-reductase I, the rate-limiting enzyme for allopregnanolone, a hormone shown to reduce depressive- and anxiety-like symptoms in rodents [102, 103]. Isolated juveniles show increased CpG methylation upstream of the transcription start site of the *SRD5A1* gene, which codes for this enzyme; one of these islands was demonstrated to be sufficient to reduce expression in the PFC [102]. In adolescent rhesus monkeys, anxious temperament is associated with increased methylation and reduced expression of the *BCL11A* and *JAG1* genes, associated with neuroplasticity, in the amygdala [104]. Similarly, these findings have been supported by recent clinical work identifying a correlation between *NR3C1*, the gene coding for the glucocorticoid receptor, 1F promoter methylation in blood, and internalizing symptoms [105]. Moreover, these adolescents showing increased 1F promoter methylation and displaying internalizing behavior also had higher concentrations of cortisol upon waking. These findings, in tandem, indicate a significant role of neuroplasticity and HPA axis regulation in stress-sensitive regions of the brain, notably the hippocampus, amygdala, PVN, and PFC, during adolescence and may underscore potential individual variations that contribute to anxious susceptibility. These epigenetic predispositions may be compounded by other environmental perturbations such as exposure to drugs of abuse. Intermittent alcohol exposure, for instance, has been shown to increase HDAC activity in the rodent amygdala [106]. These changes were also associated with reduced time spent in the open arms of the EPM and in the light compartment of the dark/light box into adulthood. Further alcohol-exposed adults had reductions in the number of spines and increased alcohol intake. Conversely, acute alcohol exposure during adolescence produces similar changes in anxiety-like behaviors while decreasing HDAC activity in the rodent amygdala [107]. In summation both predisposition and environmental perturbation may work in synchrony during adolescence to dysregulate both transcription and synaptic integrity in the amygdala and ultimately help shape entrain anxious behavior.

8.8 Stress in Adulthood

Stress induces lasting changes in heterochromatin structure, ultimately changing neuronal plasticity and behavior. The hippocampus, PFC, and amygdala are targets of glucocorticoids. As these regions help regulate spatial memory, executive function, and fear responses, respectively, they are of the utmost importance in the context of anxiety. These regions are extremely sensitive to both acute and chronic stressors and express a large number of epigenetic enzymes and display profound structural changes at the synaptic level in response to environmental stressors. Stressors often produce some type of learning, the spatial and contextual components of which are presumed to be coded by the hippocampus and the cue-based components coded by the amygdalar [24, 108]. The reconsolidation and extinction of these associations are mediated by the PFC. Dysregulation of these memories may fail to attenuate improper responses to environmental stimuli, much like the symptoms of anxiety. Fear conditioning is widely used to study learning and neuroplastic consequences thereof as well as to model symptoms of a number of anxiety disorders as well as other stress-related disorders such as post-traumatic stress disorder [109, 110]. Epigenetics has been thought to be a potential basis of memory on the molecular level [111–113]. Initially, Sweatt et al. (2004) first demonstrated the role of hippocampal histone acetylation during fear memory formation [114]. Miller and Sweatt (2007) later showed that fear conditioning upregulated expression of hippocampal DNMT3A and 3B, and that DNMT activity there was required for fear memory consolidation [66]. Hippocampal methylation of reelin, PP1, and BDNF was also changed by fear conditioning [115]. Interestingly, reelin and BDNF have well-established roles in dendritic remodeling, and PP1 codes for a phosphatase that acts at histone H3S10 [116, 117]. Presumably these are the grounds for its role in memory as the dual acetylation-phosphorylation H3 mark was enriched at the *BDNF* locus in the hippocampus. Others have found similarly that both histone modification and DNA methylation play critical roles in the amygdala in memory reconsolidation and consolidation, respectively [118]. Tsai et al. (2007) established that both environmental enrichment and HDAC inhibition were sufficient for restoring deficits in memory and synaptic connectivity in a mouse model of neuronal cell loss [119]. A later study by the same group (2009) identified HDAC2 to be necessary for the negative impacts on memory [120]. Recall of recent memories results HDAC2 dissociation from the chromatin, which causes increases in H3 acetylation and increased expression of immediate early genes [121]. Recall of less recent memories do not produce such profound changes in HDAC activity. Yet, HDAC inhibition during reconsolidation of remote fear memories allows for H3 acetylation, increased immediate early gene expression, and neuroplastic changes [121]. This suggests that epigenetic control of chromatin structure regulates neuroplastic changes underpinning behavioral outputs related to fear memories.

Social defeat represents another type of stress-based learning producing an anxious phenotype in the defeated. Social defeat is a well-characterized animal model of a number of psychiatric disorders including modeling symptoms of depression and anxiety [122]. The Nestler group was early in demonstrating that social defeat affects hippocampal chromatin signatures [123]. They showed that chronic social defeat increased H3K27me3 repression of the *BDNF* promoter in the hippocampus. Also, the accumulation of this repressive mark was mitigated by antidepressant treatment, inhibiting HDAC2, resulting in increases in H3 acetylation and H3K4 methylation, both marks promoting transcription [123, 124]. The same group also showed that chronic stress or cocaine exposure altered HDAC activity in the nucleus accumbens [125]. DNMT3A expression increases as a consequence of chronic defeat and decreases as a consequence of chronic cocaine which were associated with synaptic changes as well in the same nuclei [126]. Interestingly, natural variation to susceptibility to social defeat has been associated with distinct methylation signatures of the CRF promoter in the PVN [32]. Resilient animals also show increased H3K9me3 and K3K27me3 in the nucleus accumbens [127, 128]. The levels of accumbal H3K9me3 also change in response to cocaine exposure as well as dendritic morphology [129]. Acute stress and chronic antidepressant treatment have also been shown to increase H3K9me3 levels in the hippocampus [130]. This repressive mark appears to accumulate selectively at repetitive elements, specifically retrotransposons (for review see Lapp & Hunter, 2016) in the genome [131, 132]. Interestingly, Alu and LINE1 retrotransposons appear upregulated in PTSD veterans compared to combat deployed controls [133]. Socially defeated animals also show increased basal corticosterone in circulation, reduced time spent in open arms of the EPM and in the light component of the light/dark box, as well as reduced hippocampal H3 acetylation and increased HDAC5 expression [134]. These deficits, however, were rescued by a moderate, involuntary exercise regiment, a mild stressor [134]. Voluntary exercise, a mild and controllable stressor, alone has been shown to have anxiolytic effects in addition to reducing hippocampal expression of the histone H2 variant H2A.z and increasing expression of mitochondrial-related genes *TFAM* and *NDUFA6* in the same region [135]. These recent findings hark back to the importance of allostasis and suggest an epigenetic underpinning of anxious behaviors. Further, it has been suggested that stress opens up "windows of epigenetic plasticity" that are unique to the stressor and elicit dynamic effects based on previous stress history [136, 137]. The recent work of the McEwen laboratory has provided strong evidence for this nuanced view of the epigenetic effects of stress. While chronic restraint stress resulted in reduced time spend in the light component of the light/dark box, only a novel acute stressor led to persistent reduction in time spend in the light compartment. These differences corresponded to changes in hippocampal long-term potentiation and NMDA receptor expression [136, 137]. Acute restraint stress exposure has also been shown to convert DNA methylation through the addition of a hydroxyl group of *NR3C1* promoter in the hippocampus [138]. More recently, hyper-hydroxymethylation has been observed in regions associated with neuronal plasticity following acute restraint stress in the hippocampus [139].

8.9 Prospects for an Epigenetic Pharmacology of Anxiety

Epigenetic interventions have proven effective in animal models of anxiety and stress, and some psychiatric drugs, such as the mood stabilizer valproate, have known epigenetic effects (valproate is an HDAC inhibitor). Thus, it would appear that the prospects for epigenetic therapies for anxiety disorders are fairly high.

Most pharmacologic studies of drugs with epigenetic activities have focused on histone acetylation, with the HDACs being the major targets. In fear extinction models, which have substantial relevance to human AD, a variety of HDAC inhibitors have been shown to be effective in enhancing extinction [140]. Similarly, HDAC inhibition reversed the group differences in maternal behavior and adult stress reactivity observed by Weaver in his landmark paper on epigenetic programming of maternal behavior [33]. Similarly, the same phenotype and associated anxious behavior could be reversed with central infusion of the methyl donor S-adenosyl-methionine (SAMe) in adult animals [34, 35]. A number of studies have found SAMe to be more effective than placebo in the treatment of depression, though other well-designed trials have had negative results [141, 142]. A recent Cochrane collaboration review concluded that there was not strong evidence for the efficacy of SAMe in depression but that further research was warranted [143]. Studies of the efficacy of SAMe in the treatment of anxiety symptoms, however, are very limited to date. DNA methyltransferase inhibitors such as zebularine, *N*-phthaloyl-L-tryptophan and 5-aza-deoxycytidine have been shown to interfere with fear memory formation in preclinical models [66, 118]. To date, little clinical work has been done with this class of drugs, likely due to concerns about side effects, which are significant for some of these agents.

The study of epigenetic drug targets for anxiety remains in its infancy, and many questions remain to be adequately researched. One such question is whether these agents actually offer superior outcomes to existing treatments. Another is whether they might be used in combination with both other drugs and behavioral interventions to additive or even synergistic effect. Nonetheless, molecular epigenetics offers a novel class of potential drug targets for disorders like AD which have historically had relatively few molecular mechanisms with which to work.

Conclusions

Epigenetic mechanisms play a clear mechanistic role in animal models of anxiety, and human epigenetic studies suggest that these observations are generalizable to clinical populations. Indeed, some effort has already been made to translate the preclinical findings in the field into the clinic. Nonetheless, significant questions, particularly those relating to the time course and nature of epigenetic changes in humans, remain to be answered. Beyond the borders of what might now be regarded as "classical" epigenetics, novel molecular mechanisms of epigenomic, genomic, and epitranscriptomic plasticity are being revealed in the brain in behavioral contexts relevant to anxiety disorders. Transposons, which are mobile elements of the genome, have been shown to be regulated by stress exposure in both humans and animal models [131, 144, 145]. The mitochondria, which contains its own genome, shows transcriptional regulation in response to stress, and its function in the nucleus accumbens has been linked to anxiety phenotypes and social subordination in mice [146–148]. Even more intriguingly, covalent modification of RNA in the prefrontal cortex, the methylation of adenosine, has been shown to associate with the development of fear memory in mice [149]. This epitranscriptomic effect points to yet another layer of molecular complexity that will need to be incorporated into our models of anxiety, both normal and pathologic in model systems and in the clinic.

While neuroepigenetics is a relatively young science, it is already clear that it has relevance to our understanding of AD. Indeed, it has begun to produce usable translational findings for the treatment of disorders, like depression, which are highly comorbid with numerous anxiety disorders. There is ample reason to believe that neuroepigenetic mechanisms will continue to be a fruitful area of research into the biology of anxiety and AD.

References

1. Kessler RC, Berglund P, Demler O, Jin R, Merikangas KR, Walters EE. Lifetime prevalence and age-of-onset distributions of DSM-IV disorders in the National Comorbidity Survey Replication. Arch Gen Psychiatry. 2005;62:593–602.
2. Smoller JW. The genetics of stress-related disorders: PTSD, depression, and anxiety disorders. Neuropsychopharmacology. 2016;41:297–319.

3. Hunter RG, McEwen BS. Stress and anxiety across the lifespan: structural plasticity and epigenetic regulation. Epigenomics. 2013;5:177–94.
4. Lai HMX, Cleary M, Sitharthan T, Hunt GE. Prevalence of comorbid substance use, anxiety and mood disorders in epidemiological surveys, 1990–2014: a systematic review and meta-analysis. Drug Alcohol Depend. 2015;154:1–13.
5. Coles ME, Schubert JR, Nota JA. Sleep, circadian rhythms, and anxious traits. Curr Psychiatry Rep. 2015;17:73.
6. Mayer EA, Knight R, Mazmanian SK, Cryan JF, Tillisch K. Gut microbes and the brain: paradigm shift in neuroscience. J Neurosci. 2014;34:15490–6.
7. DiCorcia JA, Tronick E. Quotidian resilience: exploring mechanisms that drive resilience from a perspective of everyday stress and coping. Neurosci Biobehav Rev. 2011;35:1593–602.
8. Tronick E, Hunter RG. Waddington, dynamic systems, and epigenetics. Front Behav Neurosci. 2016;10:107.
9. Hettema JM, Neale MC, Kendler KS. A review and meta-analysis of the genetic epidemiology of anxiety disorders. Am J Psychiatry. 2001;158:1568–78.
10. Perez JA, Otowa T, Roberson-Nay R, Hettema JM. In: Charney DS, Nestler EJ, Sklar P, Buxbaum JD, editors. Genetics of anxiety disorders in neurobiology of mental illness. New York: OUP; 2013.
11. Binder EB, Bradley RG, Liu W, Epstein MP, Deveau TC, Mercer KB, et al. Association of FKBP5 polymorphisms and childhood abuse with risk of posttraumatic stress disorder symptoms in adults. JAMA. 2008;299:1291–305.
12. Ressler KJ, Mercer KB, Bradley B, Jovanovic T, Mahan A, Kerley K, et al. Post-traumatic stress disorder is associated with PACAP and the PAC1 receptor. Nature. 2011;470:492–7.
13. Domschke K, Deckert J, O'donovan MC, Glatt SJ. Meta-analysis of COMT val158met in panic disorder: ethnic heterogeneity and gender specificity. Am J Med Genet B Neuropsychiatr Genet. 2007;144B:667–73.
14. Kendler KS, Gardner CO, Annas P, Neale MC, Eaves LJ, Lichtenstein P. A longitudinal twin study of fears from middle childhood to early adulthood: evidence for a developmentally dynamic genome. Arch Gen Psychiatry. 2008;65:421–9.
15. Feder A, Nestler EJ, Charney DS. Psychobiology and molecular genetics of resilience. Nat Rev Neurosci. 2009;10:446–57.
16. Griffiths BB, Hunter RG. Neuroepigenetics of stress. Neuroscience. 2014;275:420–35.
17. Feinstein JS, Adolphs R, Damasio A, Tranel D. The human amygdala and the induction and experience of fear. Curr Biol. 2011;21:34–8.
18. Blanchard DC, Blanchard RJ. Innate and conditioned reactions to threat in rats with amygdaloid lesions. J Comp Physiol Psychol. 1972;81:281–90.
19. Kim JJ, Rison RA, Fanselow MS. Effects of amygdala, hippocampus, and periaqueductal gray lesions on short- and long-term contextual fear. Behav Neurosci. 1993;107:1093–8.
20. Weingarten H, White N. Exploration evoked by electrical stimulation of the amygdala of rats. Physiol Psychol. 2013;6:229–35.
21. Maren S, Fanselow MS. Synaptic plasticity in the basolateral amygdala induced by hippocampal formation stimulation in vivo. J Neurosci. 1995;15:7548–64.
22. LeDoux JE, Iwata J, Cicchetti P, Reis DJ. Different projections of the central amygdaloid nucleus mediate autonomic and behavioral correlates of conditioned fear. J Neurosci. 1988;8:2517–29.
23. Campese VD, Kim J, Lázaro-Muñoz G, Pena L, LeDoux JE, Cain CK. Lesions of lateral or central amygdala abolish aversive Pavlovian-to-instrumental transfer in rats. Front Behav Neurosci. 2014;8:161.
24. Herry C, Ciocchi S, Senn V, Demmou L, Müller C, Lüthi A. Switching on and off fear by distinct neuronal circuits. Nature. 2008;454:600–6.
25. Ciocchi S, Herry C, Grenier F, Wolff SBE, Letzkus JJ, Vlachos I, et al. Encoding of conditioned fear in central amygdala inhibitory circuits. Nature. 2010;468:277–82.

26. Milad MR, Quirk GJ. Neurons in medial prefrontal cortex signal memory for fear extinction. Nature. 2002;420:70–4.
27. Milad MR, Vidal-Gonzalez I, Quirk GJ. Electrical stimulation of medial prefrontal cortex reduces conditioned fear in a temporally specific manner. Behav Neurosci. 2004;118:389–94.
28. Quirk GJ, Likhtik E, Pelletier JG, Paré D. Stimulation of medial prefrontal cortex decreases the responsiveness of central amygdala output neurons. J Neurosci. 2003;23:8800–7.
29. Soeter M, Kindt M. An abrupt transformation of phobic behavior after a post-retrieval amnesic agent. Biol Psychiatry. 2015;78:880–6.
30. Mackintosh M-A, Gatz M, Wetherell JL, Pedersen NL. A twin study of lifetime generalized anxiety disorder (GAD) in older adults: genetic and environmental influences shared by neuroticism and GAD. Twin Res Hum Genet. 2006;9:30–7.
31. Lüthi A, Lüscher C. Pathological circuit function underlying addiction and anxiety disorders. Nat Neurosci. 2014;17:1635–43.
32. Elliott E, Ezra-Nevo G, Regev L, Neufeld-Cohen A, Chen A. Resilience to social stress coincides with functional DNA methylation of the Crf gene in adult mice. Nat Neurosci. 2010;13:1351–3.
33. Weaver ICG, Cervoni N, Champagne FA, D'Alessio AC, Sharma S, Seckl JR, et al. Epigenetic programming by maternal behavior. Nat Neurosci. 2004;7:847–54.
34. Weaver ICG, Champagne FA, Brown SE, Dymov S, Sharma S, Meaney MJ, et al. Reversal of maternal programming of stress responses in adult offspring through methyl supplementation: altering epigenetic marking later in life. J Neurosci. 2005;25:11045–54.
35. Weaver ICG, Meaney MJ, Szyf M. Maternal care effects on the hippocampal transcriptome and anxiety-mediated behaviors in the offspring that are reversible in adulthood. Proc Natl Acad Sci U S A. 2006;103:3480–5.
36. Pan-Vazquez A, Rye N, Ameri M, McSparron B, Smallwood G, Bickerdyke J, et al. Impact of voluntary exercise and housing conditions on hippocampal glucocorticoid receptor, miR-124 and anxiety. Mol Brain. 2015;8:40.
37. Spadaro PA, Flavell CR, Widagdo J, Ratnu VS, Troup M, Ragan C, et al. Long noncoding RNA-directed epigenetic regulation of gene expression is associated with anxiety-like behavior in mice. Biol Psychiatry. 2015;78:848–59.
38. Shen EY, Jiang Y, Mao W, Futai K, Hock H, Akbarian S. Cognition and mood-related behaviors in L3mbtl1 null mutant mice. PLoS One. 2015;10:e0121252.
39. Montesinos J, Pascual M, Rodríguez-Arias M, Miñarro J, Guerri C. Involvement of TLR4 in the long-term epigenetic changes, rewarding and anxiety effects induced by intermittent ethanol treatment in adolescence. Brain Behav Immun. 2016;53:159–71.
40. Teppen TL, Krishnan HR, Zhang H, Sakharkar AJ, Pandey SC. The potential role of amygdaloid microRNA-494 in alcohol-induced anxiolysis. Biol Psychiatry. 2016;80:711–9.
41. Morgan CP, Bale TL. Early prenatal stress epigenetically programs dysmasculinization in second-generation offspring via the paternal lineage. J Neurosci. 2011;31:11748–55.
42. Lehrner A, Bierer LM, Passarelli V, Pratchett LC, Flory JD, Bader HN, et al. Maternal PTSD associates with greater glucocorticoid sensitivity in offspring of Holocaust survivors. Psychoneuroendocrinology. 2014;40:213–20.
43. Yehuda R, Bierer LM. Transgenerational transmission of cortisol and PTSD risk. Prog Brain Res. 2008;167:121–35.
44. Yehuda R, Bell A, Bierer LM, Schmeidler J. Maternal, not paternal, PTSD is related to increased risk for PTSD in offspring of Holocaust survivors. J Psychiatr Res. 2008;42:1104–11.
45. Francis D, Diorio J, Liu D, Meaney MJ. Nongenomic transmission across generations of maternal behavior and stress responses in the rat. Science. 1999;286:1155–8.
46. Beery AK, McEwen LM, MacIsaac JL, Francis DD, Kobor MS. Natural variation in maternal care and cross-tissue patterns of oxytocin receptor gene methylation in rats. Horm Behav. 2016;77:42–52.

47. Chagnon YC, Potvin O, Hudon C, Préville M. DNA methylation and single nucleotide variants in the brain-derived neurotrophic factor (BDNF) and oxytocin receptor (OXTR) genes are associated with anxiety/depression in older women. Front Genet. 2015;6:230.
48. Non AL, Binder AM, Kubzansky LD, Michels KB. Genome-wide DNA methylation in neonates exposed to maternal depression, anxiety, or SSRI medication during pregnancy. Epigenetics. 2014;9:964–72.
49. Sharp H, Hill J, Hellier J, Pickles A. Maternal antenatal anxiety, postnatal stroking and emotional problems in children: outcomes predicted from pre- and postnatal programming hypotheses. Psychol Med. 2015;45:269–83.
50. Moser DA, Paoloni-Giacobino A, Stenz L, Adouan W, Manini A, Suardi F, et al. BDNF methylation and maternal brain activity in a violence-related sample. PLoS One. 2015;10:e0143427.
51. Kaufman D, Smith ELP, Gohil BC, Banerji M, Coplan JD, Kral JG, et al. Early appearance of the metabolic syndrome in socially reared bonnet macaques. J Clin Endocrinol Metab. 2005;90:404–8.
52. Kaufman D, Banerji MA, Shorman I, Smith ELP, Coplan JD, Rosenblum LA, et al. Early-life stress and the development of obesity and insulin resistance in juvenile bonnet macaques. Diabetes. 2007;56:1382–6.
53. Chaudhury S, Aurbach EL, Sharma V, Blandino P, Turner CA, Watson SJ, et al. FGF2 is a target and a trigger of epigenetic mechanisms associated with differences in emotionality: partnership with H3K9me3. Proc Natl Acad Sci U S A. 2014;111:11834–9.
54. McEwen BS, Stellar E. Stress and the individual. Mechanisms leading to disease. Arch Intern Med. 1993;153:2093–101.
55. McEwen BS. Allostasis and allostatic load: implications for neuropsychopharmacology. Neuropsychopharmacology. 2000;22:108–24.
56. Francis DD, Meaney MJ. Maternal care and the development of stress responses. Curr Opin Neurobiol. 1999;9:128–34.
57. Wosiski-Kuhn M, Stranahan AM. Opposing effects of positive and negative stress on hippocampal plasticity over the lifespan. Ageing Res Rev. 2012;11:399–403.
58. Vallée M, MacCari S, Dellu F, Simon H, Le Moal M, Mayo W. Long-term effects of prenatal stress and postnatal handling on age-related glucocorticoid secretion and cognitive performance: a longitudinal study in the rat. Eur J Neurosci. 1999;11:2906–16.
59. Vallée M, Mayo W, Dellu F, Le Moal M, Simon H, Maccari S. Prenatal stress induces high anxiety and postnatal handling induces low anxiety in adult offspring: correlation with stress-induced corticosterone secretion. J Neurosci. 1997;17:2626–36.
60. Murmu MS, Salomon S, Biala Y, Weinstock M, Braun K, Bock J. Changes of spine density and dendritic complexity in the prefrontal cortex in offspring of mothers exposed to stress during pregnancy. Eur J Neurosci. 2006;24:1477–87.
61. Maccari S, Morley-Fletcher S. Effects of prenatal restraint stress on the hypothalamus-pituitary-adrenal axis and related behavioural and neurobiological alterations. Psychoneuroendocrinology. 2007;32(Suppl 1):S10–5.
62. Mueller BR, Bale TL. Sex-specific programming of offspring emotionality after stress early in pregnancy. J Neurosci. 2008;28:9055–65.
63. Fujioka T, Sakata Y, Yamaguchi K, Shibasaki T, Kato H, Nakamura S. The effects of prenatal stress on the development of hypothalamic paraventricular neurons in fetal rats. Neuroscience. 1999;92:1079–88.
64. Xu L, Sun Y, Gao L, Cai Y-Y, Shi S-X. Prenatal restraint stress is associated with demethylation of corticotrophin releasing hormone (CRH) promoter and enhances CRH transcriptional responses to stress in adolescent rats. Neurochem Res. 2014;39:1193–8.
65. Palacios-García I, Lara-Vásquez A, Montiel JF, Díaz-Véliz GF, Sepúlveda H, Utreras E, et al. Prenatal stress down-regulates Reelin expression by methylation of its promoter and induces adult behavioral impairments in rats. PLoS One. 2015;10:e0117680.
66. Miller CA, Sweatt JD. Covalent modification of DNA regulates memory formation. Neuron. 2007;53:857–69.

67. Oberlander TF, Weinberg J, Papsdorf M, Grunau R, Misri S, Devlin AM. Prenatal exposure to maternal depression, neonatal methylation of human glucocorticoid receptor gene (NR3C1) and infant cortisol stress responses. Epigenetics. 2008;3:97–106.
68. Hompes T, Izzi B, Gellens E, Morreels M, Fieuws S, Pexsters A, et al. Investigating the influence of maternal cortisol and emotional state during pregnancy on the DNA methylation status of the glucocorticoid receptor gene (NR3C1) promoter region in cord blood. J Psychiatr Res. 2013;47:880–91.
69. Mansell T, Novakovic B, Meyer B, Rzehak P, Vuillermin P, Ponsonby A-L, et al. The effects of maternal anxiety during pregnancy on IGF2/H19 methylation in cord blood. Transl Psychiatry. 2016;6:e765.
70. Monk C, Feng T, Lee S, Krupska I, Champagne FA, Tycko B. Distress during pregnancy: epigenetic regulation of placenta glucocorticoid-related genes and fetal neurobehavior. Am J Psychiatry. 2016;173:705–13.
71. Tyagi E, Zhuang Y, Agrawal R, Ying Z, Gomez-Pinilla F. Interactive actions of Bdnf methylation and cell metabolism for building neural resilience under the influence of diet. Neurobiol Dis. 2015;73:307–18.
72. Vassoler FM, White SL, Schmidt HD, Sadri-Vakili G, Pierce RC. Epigenetic inheritance of a cocaine-resistance phenotype. Nat Neurosci. 2013;16:42–7.
73. Liang F, Diao L, Liu J, Jiang N, Zhang J, Wang H, et al. Paternal ethanol exposure and behav ioral abnormities in offspring: associated alterations in imprinted gene methylation. Neuropharmacology. 2014;81:126–33.
74. Lemaire V, Lamarque S, Le Moal M, Piazza P-V, Abrous DN. Postnatal stimulation of the pups counteracts prenatal stress-induced deficits in hippocampal neurogenesis. Biol Psychiatry. 2006;59:786–92.
75. Fujioka T, Fujioka A, Tan N, Chowdhury GM, Mouri H, Sakata Y, et al. Mild prenatal stress enhances learning performance in the non-adopted rat offspring. Neuroscience. 2001;103:301–7.
76. Brand SR, Brennan PA. Impact of antenatal and postpartum maternal mental illness: how are the children? Clin Obstet Gynecol. 2009;52:441–55.
77. Kember RL, Dempster EL, Lee THA, Schalkwyk LC, Mill J, Fernandes C. Maternal separation is associated with strain-specific responses to stress and epigenetic alterations to Nr3c1, Avp, and Nr4a1 in mouse. Brain Behav. 2012;2:455–67.
78. Francis DD, Diorio J, Plotsky PM, Meaney MJ. Environmental enrichment reverses the effects of maternal separation on stress reactivity. J Neurosci. 2002;22:7840–3.
79. Plotsky PM, Thrivikraman KV, Nemeroff CB, Caldji C, Sharma S, Meaney MJ. Long-term consequences of neonatal rearing on central corticotropin-releasing factor systems in adult male rat offspring. Neuropsychopharmacology. 2005;30:2192–204.
80. Caldji C, Diorio J, Meaney MJ. Variations in maternal care in infancy regulate the development of stress reactivity. Biol Psychiatry. 2000;48:1164–74.
81. Tang AC, Reeb-Sutherland BC, Yang Z, Romeo RD, McEwen BS. Neonatal novelty-induced persistent enhancement in offspring spatial memory and the modulatory role of maternal self-stress regulation. J Neurosci. 2011;31:5348–52.
82. Akers KG, Yang Z, DelVecchio DP, Reeb BC, Romeo RD, McEwen BS, et al. Social competitiveness and plasticity of neuroendocrine function in old age: influence of neonatal novelty exposure and maternal care reliability. PLoS One. 2008;3:e2840.
83. Turner JD, Muller CP. Structure of the glucocorticoid receptor (NR3C1) gene 5′ untranslated region: identification, and tissue distribution of multiple new human exon 1. J Mol Endocrinol. 2005;35:283–92.
84. McGowan PO, Sasaki A, D'Alessio AC, Dymov S, Labonté B, Szyf M, et al. Epigenetic regulation of the glucocorticoid receptor in human brain associates with childhood abuse. Nat Neurosci. 2009;12:342–8.
85. Labonte B, Yerko V, Gross J, Mechawar N, Meaney MJ, Szyf M, et al. Differential glucocorticoid receptor exon 1(B), 1(C), and 1(H) expression and methylation in suicide completers with a history of childhood abuse. Biol Psychiatry. 2012;72:41–8.

86. Alt SR, Turner JD, Klok MD, Meijer OC, Lakke EAJF, Derijk RH, et al. Differential expression of glucocorticoid receptor transcripts in major depressive disorder is not epigenetically programmed. Psychoneuroendocrinology. 2010;35:544–56.
87. McGowan PO, Sasaki A, Huang TCT, Unterberger A, Suderman M, Ernst C, et al. Promoter-wide hypermethylation of the ribosomal RNA gene promoter in the suicide brain. PLoS One. 2008;3:e2085.
88. Tyrka AR, Price LH, Marsit C, Walters OC, Carpenter LL. Childhood adversity and epigenetic modulation of the leukocyte glucocorticoid receptor: preliminary findings in healthy adults. PLoS One. 2012;7:e30148.
89. Romens SE, McDonald J, Svaren J, Pollak SD. Associations between early life stress and gene methylation in children. Child Dev. 2015;86:303–9.
90. Suderman M, Borghol N, Pappas JJ, Pinto Pereira SM, Pembrey M, Hertzman C, et al. Childhood abuse is associated with methylation of multiple loci in adult DNA. BMC Med Genet. 2014;7:13.
91. Eiland L, Ramroop J, Hill MN, Manley J, McEwen BS. Chronic juvenile stress produces corticolimbic dendritic architectural remodeling and modulates emotional behavior in male and female rats. Psychoneuroendocrinology. 2012;37:39–47.
92. Toda H, Boku S, Nakagawa S, Inoue T, Kato A, Takamura N, et al. Maternal separation enhances conditioned fear and decreases the mRNA levels of the neurotensin receptor 1 gene with hypermethylation of this gene in the rat amygdala. PLoS One. 2014;9:e97421.
93. Pesonen A-K, Räikkönen K. The lifespan consequences of early life stress. Physiol Behav. 2012;106:722–7.
94. Wu Y, Patchev AV, Daniel G, Almeida OFX, Spengler D. Early-life stress reduces DNA methylation of the Pomc gene in male mice. Endocrinology. 2014;155:1751–62.
95. Perroud N, Zewdie S, Stenz L, Adouan W, Bavamian S, Prada P, et al. Methylation of serotonin receptor 3a in ADHD, borderline personality, and bipolar disorders: link with severity of the disorders and childhood maltreatment. Depress Anxiety. 2016;33:45–55.
96. McGorry PD, Purcell R, Goldstone S, Amminger GP. Age of onset and timing of treatment for mental and substance use disorders: implications for preventive intervention strategies and models of care. Curr Opin Psychiatry. 2011;24:301–6.
97. Blakemore S-J. Development of the social brain during adolescence. Q J Exp Psychol (Hove). 2008;61:40–9.
98. Blakemore S-J. Development of the social brain in adolescence. J R Soc Med. 2012;105:111–6.
99. Gunnar MR, Wewerka S, Frenn K, Long JD, Griggs C. Developmental changes in hypothalamus–pituitary–adrenal activity over the transition to adolescence: normative changes and associations with puberty. Dev Psychopathol. 2009;21:69–85.
100. Romeo RD, Bellani R, Karatsoreos IN, Chhua N, Vernov M, Conrad CD, et al. Stress history and pubertal development interact to shape hypothalamic-pituitary-adrenal axis plasticity. Endocrinology. 2006;147:1664–74.
101. Isgor C, Kabbaj M, Akil H, Watson SJ. Delayed effects of chronic variable stress during peripubertal-juvenile period on hippocampal morphology and on cognitive and stress axis functions in rats. Hippocampus. 2004;14:636–48.
102. Araki R, Nishida S, Hiraki Y, Matsumoto K, Yabe T. DNA methylation of the GC box in the promoter region mediates isolation rearing-induced suppression of srd5a1 transcription in the prefrontal cortex. Neurosci Lett. 2015;606:135–9.
103. Reddy DS, Kulkarni SK. Differential anxiolytic effects of neurosteroids in the mirrored chamber behavior test in mice. Brain Res. 1997;752:61–71.
104. Alisch RS, Chopra P, Fox AS, Chen K, White ATJ, Roseboom PH, et al. Differentially methylated plasticity genes in the amygdala of young primates are linked to anxious temperament, an at risk phenotype for anxiety and depressive disorders. J Neurosci. 2014;34:15548–56.
105. Dadds MR, Moul C, Hawes DJ, Mendoza Diaz A, Brennan J. Individual differences in childhood behavior disorders associated with epigenetic modulation of the cortisol receptor gene. Child Dev. 2015;86:1311–20.

106. Pandey SC, Sakharkar AJ, Tang L, Zhang H. Potential role of adolescent alcohol exposure-induced amygdaloid histone modifications in anxiety and alcohol intake during adulthood. Neurobiol Dis. 2015;82:607–19.
107. Sakharkar AJ, Zhang H, Tang L, Baxstrom K, Shi G, Moonat S, et al. Effects of histone deacetylase inhibitors on amygdaloid histone acetylation and neuropeptide Y expression: a role in anxiety-like and alcohol-drinking behaviours. Int J Neuropsychopharmacol. 2014;17:1207–20.
108. Marschner A, Kalisch R, Vervliet B, Vansteenwegen D, Büchel C. Dissociable roles for the hippocampus and the amygdala in human cued versus context fear conditioning. J Neurosci. 2008;28:9030–6.
109. Zovkic IB, Sweatt JD. Epigenetic mechanisms in learned fear: implications for PTSD. Neuropsychopharmacology. 2013;38:77–93.
110. Sultan FA, Day JJ. Epigenetic mechanisms in memory and synaptic function. Epigenomics. 2011;3:157–81.
111. Griffith JS, Mahler HR. DNA ticketing theory of memory. Nature. 1969;223:580–2.
112. Crick F. Memory and molecular turnover. Nature. 1984;312:101.
113. Holliday R. Is there an epigenetic component in long-term memory? J Theor Biol. 1999;200:339–41.
114. Levenson JM, O'Riordan KJ, Brown KD, Trinh MA, Molfese DL, Sweatt JD. Regulation of histone acetylation during memory formation in the hippocampus. J Biol Chem. 2004;279:40545–59.
115. Lubin FD, Roth TL, Sweatt JD. Epigenetic regulation of BDNF gene transcription in the consolidation of fear memory. J Neurosci. 2008;28:10576–86.
116. Koshibu K, Gräff J, Beullens M, Heitz FD, Berchtold D, Russig H, et al. Protein phosphatase 1 regulates the histone code for long-term memory. J Neurosci. 2009;29:13079–89.
117. Koshibu K, Gräff J, Mansuy IM. Nuclear protein phosphatase-1: an epigenetic regulator of fear memory and amygdala long-term potentiation. Neuroscience. 2011;173:30–6.
118. Maddox SA, Schafe GE. Epigenetic alterations in the lateral amygdala are required for reconsolidation of a Pavlovian fear memory. Learn Mem. 2011;18:579–93.
119. Fischer A, Sananbenesi F, Wang X, Dobbin M, Tsai L-H. Recovery of learning and memory is associated with chromatin remodelling. Nature. 2007;447:178–82.
120. Guan J-S, Haggarty SJ, Giacometti E, Dannenberg J-H, Joseph N, Gao J, et al. HDAC2 negatively regulates memory formation and synaptic plasticity. Nature. 2009;459:55–60.
121. Gräff J, Joseph NF, Horn ME, Samiei A, Meng J, Seo J, et al. Epigenetic priming of memory updating during reconsolidation to attenuate remote fear memories. Cell. 2014;156:261–76.
122. Nestler EJ, Hyman SE. Animal models of neuropsychiatric disorders. Nat Neurosci. 2010;13:1161–9.
123. Tsankova NM, Berton O, Renthal W, Kumar A, Neve RL, Nestler EJ. Sustained hippocampal chromatin regulation in a mouse model of depression and antidepressant action. Nat Neurosci. 2006;9:519–25.
124. Tsankova NM, Kumar A, Nestler EJ. Histone modifications at gene promoter regions in rat hippocampus after acute and chronic electroconvulsive seizures. J Neurosci. 2004;24:5603–10.
125. Renthal W, Maze I, Krishnan V, Covington HE, Xiao G, Kumar A, et al. Histone deacetylase 5 epigenetically controls behavioral adaptations to chronic emotional stimuli. Neuron. 2007;56:517–29.
126. LaPlant Q, Vialou V, Covington HE, Dumitriu D, Feng J, Warren BL, et al. Dnmt3a regulates emotional behavior and spine plasticity in the nucleus accumbens. Nat Neurosci. 2010;13:1137–43.
127. Wilkinson MB, Xiao G, Kumar A, LaPlant Q, Renthal W, Sikder D, et al. Imipramine treatment and resiliency exhibit similar chromatin regulation in the mouse nucleus accumbens in depression models. J Neurosci. 2009;29:7820–32.

128. Covington HE, Maze I, Sun H, Bomze HM, DeMaio KD, Wu EY, et al. A role for repressive histone methylation in cocaine-induced vulnerability to stress. Neuron. 2011;71:656–70.
129. Maze I, Covington HE, Dietz DM, LaPlant Q, Renthal W, Russo SJ, et al. Essential role of the histone methyltransferase G9a in cocaine-induced plasticity. Science. 2010; 327:213–6.
130. Hunter RG, McCarthy KJ, Milne TA, Pfaff DW, McEwen BS. Regulation of hippocampal H3 histone methylation by acute and chronic stress. Proc Natl Acad Sci U S A. 2009;106:20912–7.
131. Hunter RG, Murakami G, Dewell S, Seligsohn M, Baker ME, Datson NA, et al. Acute stress and hippocampal histone H3 lysine 9 trimethylation, a retrotransposon silencing response. Proc Natl Acad Sci U S A. 2012;109:17657–62.
132. Lapp HE, Hunter RG. The dynamic genome: transposons and environmental adaptation in the nervous system. Epigenomics. 2016;8:237–49.
133. Rusiecki JA, Chen L, Srikantan V, Zhang L, Yan L, Polin ML, et al. DNA methylation in repetitive elements and post-traumatic stress disorder: a case-control study of US military service members. Epigenomics. 2012;4:29–40.
134. Patki G, Solanki N, Atrooz F, Ansari A, Allam F, Jannise B, et al. Novel mechanistic insights into treadmill exercise based rescue of social defeat-induced anxiety-like behavior and memory impairment in rats. Physiol Behav. 2014;130:135–44.
135. Aguiar AS, Stragier E, da Luz SD, Remor AP, Oliveira PA, Prediger RD, et al. Effects of exercise on mitochondrial function, neuroplasticity and anxio-depressive behavior of mice. Neuroscience. 2014;271:56–63.
136. Gray JD, Rubin TG, Hunter RG, McEwen BS. Hippocampal gene expression changes underlying stress sensitization and recovery. Mol Psychiatry. 2014;19:1171–8.
137. Nasca C, Zelli D, Bigio B, Piccinin S, Scaccianoce S, Nisticò R, et al. Stress dynamically regulates behavior and glutamatergic gene expression in hippocampus by opening a window of epigenetic plasticity. Proc Natl Acad Sci U S A. 2015;112:14960–5.
138. Li S, Papale LA, Kintner DB, Sabat G, Barrett-Wilt GA, Cengiz P, et al. Hippocampal increase of 5-hmC in the glucocorticoid receptor gene following acute stress. Behav Brain Res. 2015;286:236–40.
139. Li S, Papale LA, Zhang Q, Madrid A, Chen L, Chopra P, et al. Genome-wide alterations in hippocampal 5-hydroxymethylcytosine links plasticity genes to acute stress. Neurobiol Dis. 2016;86:99–108.
140. Singewald N, Schmuckermair C, Whittle N, Holmes A, Ressler KJ. Pharmacology of cognitive enhancers for exposure-based therapy of fear, anxiety and trauma-related disorders. Pharmacol Ther. 2015;149:150–90.
141. De Berardis D, Orsolini L, Serroni N, Girinelli G, Iasevoli F, Tomasetti C, et al. A comprehensive review on the efficacy of S-Adenosyl-L-methionine in major depressive disorder. CNS Neurol Disord Drug Targets. 2016;15:35–44.
142. Mischoulon D, Price LH, Carpenter LL, Tyrka AR, Papakostas GI, Baer L, et al. A double-blind, randomized, placebo-controlled clinical trial of S-adenosyl-L-methionine (SAMe) versus escitalopram in major depressive disorder. J Clin Psychiatry. 2014;75:370–6.
143. Galizia I, Oldani L, Macritchie K, Amari E, Dougall D, Jones TN, et al. S-adenosyl methionine (SAMe) for depression in adults. Cochrane Database Syst Rev. 2016;10:CD011286.
144. Hunter RG, Gagnidze K, McEwen BS, Pfaff DW. Stress and the dynamic genome: steroids, epigenetics, and the transposome. Proc Natl Acad Sci U S A. 2015;112:6828–33.
145. Ponomarev I, Wang S, Zhang L, Harris RA, Mayfield RD. Gene coexpression networks in human brain identify epigenetic modifications in alcohol dependence. J Neurosci. 2012;32:1884–97.
146. Hunter RG, Seligsohn M, Rubin TG, Griffiths BB, Ozdemir Y, Pfaff DW, et al. Stress and corticosteroids regulate rat hippocampal mitochondrial DNA gene expression via the glucocorticoid receptor. Proc Natl Acad Sci U S A. 2016;113:9099–104.

147. Du J, Wang Y, Hunter R, Wei Y, Blumenthal R, Falke C, et al. Dynamic regulation of mitochondrial function by glucocorticoids. Proc Natl Acad Sci U S A. 2009;106:3543–8.
148. Hollis F, van der Kooij MA, Zanoletti O, Lozano L, Cantó C, Sandi C. Mitochondrial function in the brain links anxiety with social subordination. Proc Natl Acad Sci U S A. 2015;112:15486–91.
149. Widagdo J, Zhao Q-Y, Kempen M-J, Tan MC, Ratnu VS, Wei W, et al. Experience-dependent accumulation of N6-methyladenosine in the prefrontal cortex is associated with memory processes in mice. J Neurosci. 2016;36:6771–7.

Part III

Brain Disorders During Adulthood

Histone Modifications in Major Depressive Disorder and Related Rodent Models

9

Jan M. Deussing and Mira Jakovcevski

Abstract

Major depressive disorder (MDD) is a multifactorial disease, weakly linked to multiple genetic risk factors. In contrast to that, environmental factors and "gene × environment" interaction between specific risk genes and environmental factors, such as severe or early stress exposure, have been strongly linked to MDD vulnerability. Stressors can act on the interface between an organism and the environment, the epigenome. The molecular foundation for the impact of stressors on the risk to develop MDD is based on the hormonal stress response itself: the glucocorticoid receptor (GR, encoded by *NR3C1*). *NR3C1* can directly interact with the epigenome in the cell nucleus. Besides DNA methylation, histone modifications have been reported to be crucial targets for the interaction with the stress response system. Here, we review critical findings on the impact of the most relevant histone modifications, i.e. histone acetylation and methylation, in the context of MDD and related animal models. We discuss new treatment options which have been based on these findings, including histone deacetylase inhibitors (HDACis) and drugs targeting specific histone marks, closely linked to psychiatric disease. In this context we talk about contemporary and future approaches required to fully understand (1) the epigenetics of stress-related disease and (2) the mode of action of potential MDD drugs targeting histone modifications. This includes harnessing the unprecedented potentials of genome-wide analysis of the epigenome and transcriptome, in a cell type-specific manner, and the use of epigenome editing technologies to clearly link epigenetic marks on specific genomic loci to functional relevance.

J.M. Deussing, Ph.D. • M. Jakovcevski, Ph.D. (✉)
Department of Stress Neurobiology and Neurogenetics, Max Planck Institute of Psychiatry, Kraepelinstr. 2, 80804 Munich, Bavaria, Germany
e-mail: deussing@psych.mpg.de; mira_jakovcevski@psych.mpg.de

R. Delgado-Morales (ed.), *Neuroepigenomics in Aging and Disease*, Advances in Experimental Medicine and Biology, DOI 10.1007/978-3-319-53889-1_9

Keywords
Epigenome editing • Histone H3 lysine 4 methylation (H3K4me) • Histone H3 lysine 27 trimethylation (H3K27me3) • Histone H3 lysine 27 acetylation (H3K27ac) • Hypothalamus–pituitary–adrenal (HPA) axis • Fluorescence-activated cell sorting (FACS) • Human post-mortem brain • Massive parallel sequencing • Mouse brain

9.1 Introduction

Major depressive disorder (MDD) is a highly prevalent and debilitating disorder with a high risk of treatment resistance and reoccurrence of relapse episodes therefore not only causing human suffering and loss of individual life quality but also laying a major burden on affected families and society. All of which make the search for better treatment options inevitable [1–4].

Clinical and scientific classifications of MDD comprise a wide range of typical depressive symptoms such as overwhelming sadness and despair in conjunction with sleep disturbance and secondary cognitive difficulties. These symptoms present with different severity and may belong to diverse disease trajectories [5, 6], which makes the classification of this multifactorial disease very complex. Among the different factors implicated in the pathogenesis of MDD is an array of various risk genes for which point mutations have been linked to the disease in gene association studies [7–9]. For MDD these associations are rather weak to moderate which might be partially explained by the fact that MDD is in stark contrast to monogenetic disease likely to be caused by an entire set of genes. Yet, a confounding variable that could interfere with the low significance of previous data may be the lack of stratification of cohorts, e.g. by age of disease onset [9]. On the other hand, there is no doubt that the vulnerability to develop MDD is closely linked to environmental factors. Here, the experience of severe stress such as early childhood trauma or traumatic events later in life presents the most important environmental variable [10, 11]. Stressors may then interact with MDD risk genes through activation of the stress response system. Thus, the impact of "gene x environment" interactions is likely to account for the weak genetic associations for MDD and be causal for the poor to missing heritability of the disease [8, 12] as supported by results from epidemiological twin studies [13, 14].

9.2 Stress Response and Epigenetics

Typically, stressors activate the hypothalamus–pituitary–adrenal (HPA) axis to allow the organism to cope with the stressor by releasing cortisol which helps the body to mobilize energy. After the stressor has ceased, negative feedback mechanisms will bring back the HPA axis to baseline resting levels [15–17]. It may not come as a surprise that point deletions/mutations and/or related expression levels of genes that are involved in the regulation of the HPA axis have been stronger associated with depression-related disease. Among them are key components of the HPA axis such as the corticotropin-releasing factor (CRF, encoded by *CRH*) which is released upon stress exposure from hypothalamic structures and its type 1 receptor (*CRHR1*) [18–20]. Likewise, a strong association with MDD has been reported for the glucocorticoid receptor (GR, encoded by *NR3C1*, [21, 22]). Interestingly, *NR3C1* is not only the major receptor for the glucocorticoid cortisol, but can act as a transcription factor that shuttles upon activation, e.g. during stress response, from the cytoplasm to the nucleus [23]. Since transcription factors can mediate the expression of multiple genes, either directly or by changing the chromatin structure and histone landscape, the pivotal role of *NR3C1* in stress response might be causal for the diverse number of different MDD-associated genes. Furthermore, the weak associations of each of the risk genes with the disease might be explained by the fact that most studies did not take stress exposure as a potential mediator of gene expression and MDD development into account. Importantly, beyond its role in transcriptional regulation, based on its function as a transcription factor, *NR3C1* has been reported to interact with and shape the histone landscape [24, 25]. The close interaction of the stress response system with the histone landscape builds one of the molecular platforms for the interaction of the environment with the (epi)genome of an organism (Fig. 9.1). Therefore, stressors, mostly either severe, repeated ones or when the stress exposure is experienced in specific vulnerable developmental periods such as early childhood, leave "epigenetic scars". Such molecular scars may, either when they accumulate or when they occur in combination with mutations on MDD risk genes, set off the molecular and behavioural disease trajectory.

9.3 The Histone Landscape

Before reviewing studies supporting the idea that stress and the histone landscape do interact, we will summarize the key principles of histone modifications. Basically, modifications to the histone landscape may change gene expression by changing the conformation of chromatin in the cell's nucleus. Chromatin is composed of DNA and histone proteins, with the smallest unit being the nucleosome. The nucleosome consists of an octamer of the four different histone proteins (histones H2A, H2B, H3 and H4) which are represented in two copies each. An approximately 147-base pair long segment of DNA is wrapped around these histone proteins [26]. The histone proteins are equipped with histone tails that can be chemically modified which in turn changes the conformation of the chromatin, with the two extreme conformation states of open chromatin ("euchromatin") or condensed chromatin ("heterochromatin"), either allowing for or hampering transcription. Histone modifications such as histone methylation or acetylation on specific lysines are either associated with active/open or silenced/closed chromatin [27]. Histone acetylation is mostly linked to active transcription, while the vast majority of histone methylation marks are linked to repression of transcription. All of these histone lysine modifications are mediated enzymatically by histone lysine deacetylases (HDAC/KDAC and

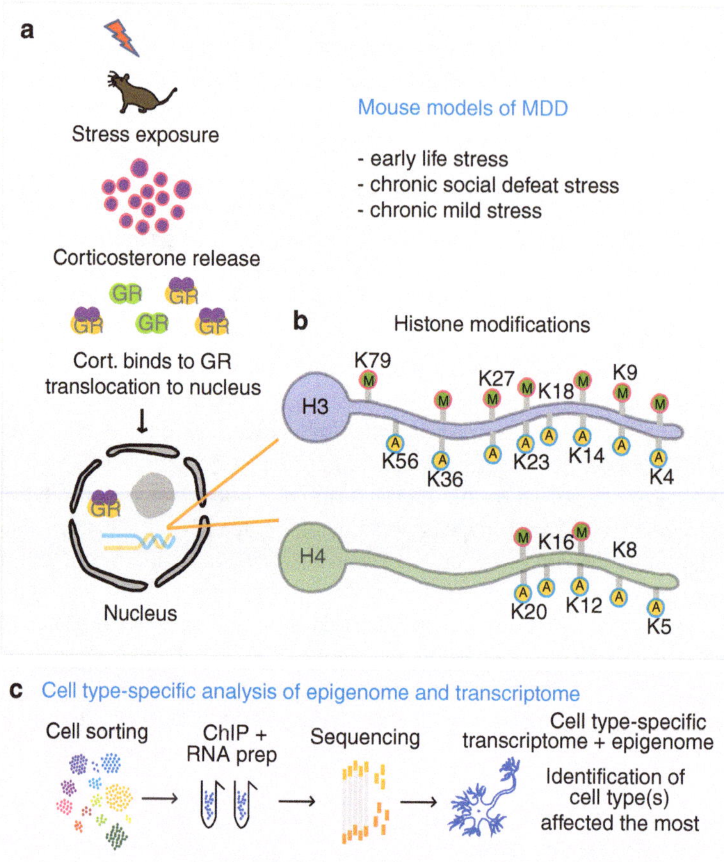

Fig. 9.1 (**a**) Molecular impact of stressors for the development of major depressive disorder (MDD). Exposure to stressors is considered to be the important switch which might determine if a person with a single-nucleotide polymorphism (SNP) in a risk gene will develop MDD. In rodent models of MDD, chronic stressors are used to induce depression-like phenotypes. These models allow us to study molecular aspects of the stress response and treatment strategies in a systematic manner. The cartoon on the left side of the panel illustrates the basics of the molecular stress response. Upon stress exposure, corticosterone is released from the adrenal glands into the bloodstream and gets transported throughout the body, including the brain. In the brain, corticosterone enters neural cells and binds to glucocorticoid receptors (GRs, encoded by *Nr3c1*) in their cytoplasm. Once corticosterone is bound to GR, the complex will be translocated into the cell nucleus, where GR/*Nr3c1* functions according to its role as a transcription factor and interacts with the epigenome. Depending on the duration and intensity of the stressor, this interaction will cause transient to long-lasting changes to the epigenome, including a set of different histone modifications. (**b**) Acetylation (*yellow lollipop symbols*) and methylation (*green lollipop symbols*) of lysines on histones H3 (upper cartoon, *blue*) and H4 (lower cartoon, *green*) are the best studied histone modifications with respect to MDD, its treatment and relevant mouse models. (**c**) Since histone modification profiles are typically cell type-specific, future studies will need to address this aspect by probing the histone landscape using chromatin immunoprecipitation (ChIP) in conjunction with next-generation sequencing in specific cell types. It will be of high interest to compare these data sets with the related transcriptome data in the same specific cell type to test if the genome-wide histone and transcriptional signatures might overlap better than in data sets derived from mixed tissues

sirtuins), histone lysine acetyltransferases (HAT/KAT) and histone lysine methyltransferases (HMT/KMT) and histone lysine demethylases (HDM/KDM). Obviously, depending on the combination of different histone marks ("histone code") on a defined genomic locus, the chromatin conformation will be in permissive, silenced or poised states [28].

9.4 Histone Modifications and MDD

When stressors interact now via the release of glucocorticoids and activation of glucocorticoid receptors with histones, this may indeed modify the histone landscape and thus the epigenome. To what extent that occurs in animal models of MDD or in MDD patient brains will be discussed in the following sections of this review, along with findings on how histone modifications relate to changes in gene expression and ultimately to adaption or maladaptation of the organism to the stress exposure.

9.4.1 Findings from Animal Models of MDD

The connection between stress, depression-related symptoms and histone modifications can be studied in a very controlled manner in animal models. Here, rodents are exposed to different forms of stressors which induce anxiety and depression-related behaviour, learned helplessness and anhedonia. Another common type of study in the field is the analysis of behavioural and molecular effects following antidepressant treatment in either naïve or previously stressed animals. Using similar strategies, earlier studies demonstrated a bidirectional relationship between histone modifications in rodent models of MDD and MDD treatment strategies. For example, in rats that had been treated with experimental electroconvulsive therapy, a treatment option effective in human patients with severe treatment-resistant MDD, upregulation of the antidepressant neurotrophin *Bdnf* (brain-derived neurotrophic factor) was accompanied by changes in global histone 3 (H3) and histone 4 (H4) acetylation at *Bdnf* promoters [29]. Comparable effects on mRNA levels of *Bdnf* were achieved by treatment of chronically stressed mice with the antidepressant drug imipramine [30]. Vice versa, reduced hippocampal mRNA levels of *Bdnf* and a reduction of global histone H3 and histone H4 acetylation at *Bdnf* promoters were reported after stress exposure [31]. Similar results have been demonstrated using different stress protocols and testing other brain regions. For instance, chronic social defeat stress increases H3 acetylation in the nucleus accumbens (NAc, [32]) and in the infralimbic medial prefrontal cortex [33]. Besides global levels of acetylation on histone H3, also changes on bulk acetylation levels for more specific histone modifications were demonstrated after stress exposure. Regulation of histone H3 lysine 14 acetylation (H3K14ac) levels [32] or histone H4 lysine 12 acetylation (H4K12ac, [34]) and acetylation of several other lysines (5, 8, 16) on histone H4 [35] have been reported after chronic social defeat stress. Interestingly, histone-modifying enzymes involved in the regulation of acetylation such as histone deacetylases [30, 32, 36], sirtuins [37] and histone acetyltransferases [38] accompanied the changes in global and bulk histone acetylation levels.

Likewise, in parallel to prominent findings for acetylation, an important role for histone methylation and related enzymes has been reported. Jiang et al. [39] demonstrated that overexpression of the histone methyltransferase *Setdb1* in the mouse forebrain led to an antidepressant phenotype, including decreased anhedonia, decreased learned helplessness and reduced levels of depression-like behaviour in the tail suspension and forced swim tests. Resonating well with these findings, expression levels of the histone methyltransferase G9/*Ehmt2*, associated like *Setdb1* with methylation of histone H3 lysine K9 (H3K9meX), were downregulated after chronic stress [40]. Furthermore, levels of G9/*Ehmt2* were demonstrated to determine the intensity of stress reactivity in a bidirectional manner, with low levels being linked to higher stress reactivity and high levels correlating with stress resilience [41]. Interestingly, Uchida et al. [36] reported that increased levels of pan histone acetylation on the *Gdnf* (glial cell-derived neurotrophic factor) promoter, induced by chronic mild stress, were accompanied by reduced levels of the repressive mark histone H3 lysine 27 trimethylation (H3K27me3) and the active methylation mark histone H3 lysine 4 trimethylation (H3K4me3). These results indicate that it may be important to consider the analysis of both, histone acetylation and methylation, in parallel.

9.4.2 Similar Findings in Human Post-mortem Brain of MDD Patients

A few studies have addressed histone modifications or the expression of histone-modifying enzymes in human post-mortem brains of MDD patients. Robison et al. [42] demonstrated reduced global/pan histone H3 acetylation at the calcium-/calmodulin-dependent protein kinase II alpha (*CaMK2A*) promoter in the nucleus accumbens (NAc) of MDD patients that had been on antidepressant medication with the selective serotonin reuptake inhibitor (SSRI) fluoxetine at the time of death. Consistent with a decrease in this active histone modification on the *CaMK2A*, important for synaptic plasticity, these findings were accompanied by reduced mRNA expression levels of *CaMK2A* and increased occupation levels of the repressive histone mark H3 lysine 9 dimethylation (H3K9me2). Since SSRI treatment led to the same paradoxical downregulation of *Camk2a* in the mouse model, the effects are likely to be attributed to the medication [42]. In a similar study design, *RAC1*, a small Rho GTPase crucial for synaptic structure and typically downregulated by chronic social defeat stress in the mouse, was found to be downregulated on the mRNA level in the NAc of nonmedicated MDD post-mortem brain. This reduction of *RAC1* mRNA correlated well with a reduction in histone H3 pan acetylation and an increase in the repressive mark histone H3 lysine K27 trimethylation (H3K27me3) [43]. Several studies report changes in histone modifications in the prefrontal cortex (PFC) in post-mortem tissue of depressed patients. The active mark H3K4me3 (histone H3 lysine 4 trimethylation) was increased at the promoters of several synapsin genes (*SYN1a,b, SYN2b*) in conjunction with upregulation of the related mRNAs in PFC of depressed patients which died by suicide [44]. In another cohort of suicide victims, diagnosed with major depression, which either had a history of antidepressant medication or did not undergo any MDD medication, *BDNF* mRNA was shown to be downregulated in non-treated subjects with MDD, whereas it was upregulated following antidepressant medication. Compared to the non-treated subjects, MDD patients with medications had a lower occupation of the respective *BDNF* promoter with the repressive histone mark H3K27me3 [45]. It should be noted that these results perfectly reflect aforementioned results on *Bdnf* expression levels after stress exposure and antidepressant treatment in the rodent model. Likewise, differential expression levels of histone deacetylases were found in post-mortem brain of patient with MDD [32, 46].

9.4.3 Therapeutic Implications

Studies either in mouse models or in human post-mortem brain have linked both histone acetylation and histone methylation to stress exposure or MDD. Quite a few researchers characterize their genomic loci of interest more vigorously by investigating several different histone modifications including histone methylation and acetylation. Therefore, treatment options targeting both histone methylation and acetylation are well conceivable. However, not too many studies have investigated the relevance of levels of histone methyltransferases and other enzymes important for histone methylation levels in stress response and MDD, in contrast to the number of studies on depression-related regulation of HDACs. Similarly, studies on transgenic mice testing the effects of expression levels for histone methyltransferases on stress vulnerability are rather sparse. Interestingly, Wei et al. [40] suggested that the regulation of the methyltransferase G9/E*hmt2*, at least in their study, was a secondary effect due to upregulation of the histone deacetylase *Hdac2*. We recently reported that neuronal levels of the histone methyltransferase *Mll1*, a methyltransferase relevant for maintenance of H3K4me levels [47], are critical for dopaminergic signalling in the mouse forebrain and nucleus accumbens formation and convey a phenotype consistent with increased anxiety, but not depression-like behaviour [48]. This study supports the view that histone methylation-based drugs, such as those targeting *Mll1*, might be promising at least for treatment of anxiety disorders [48]. One could speculate that other histone methylation targets may have some merits in treating depression-like phenotypes. Nevertheless, potential drugs targeting histone methylation are not yet as readily available as drugs that act on histone deacetylase enzymes.

Therefore, it might not come as a surprise that targeting histone acetylation has become the epigenetic treatment approach of choice. Different histone deacetylase inhibitors with variable pharmacokinetics and affinities for the 11 different histone deacetylases (HDAC1–11) are readily available and have been tested for their antidepressant properties in animal models of MDD. Treatment of rats that had been raised under the adverse conditions of limited maternal care, a model for early childhood trauma and neglect in humans, with the histone deacetylase inhibitor trichostatin A ameliorated the anxiety behaviour, typically seen in adult animals [49]. Similar antidepressant effects of HDACis have been observed after chronic restraint stress in adult mice [50]. Here, anhedonia as measured in the sucrose preference test, depression-like behaviour (tail suspension test) and anxiety (light/dark test) had been rescued after treatment with sodium butyrate. Likewise, a comparable rescue was observed after treatment with the histone deacetylase inhibitor MS-275 in mice after social defeat stress [32]. Moreover, in conjunction with antidepressant drugs, i.e. sodium butyrate together with the selective serotonin reuptake inhibitor (SSRI) fluoxetine [51] or trichostatin A in combination with fluoxetine [52], it decreased depression-like behaviours, suggesting the use of HDACis as a valuable add-on therapy.

Since similar changes to the histone landscape and to related histone modifying enzymes have been reported in human post-mortem brain and in mouse models, and the effectiveness of HDACis has been tested in animal models of depression, the use of HDACis in severe or treatment-resistant depression in patients has been proposed [53].

9.5 Where Do We Go from Here?

At present HDACis might be only a treatment option for severe MDD. In contrast, valproic acid, which largely functions as an HDACi, is commonly used in therapy of bipolar disorder [54, 55] and thus may indeed improve depression-related symptoms in patients. However, at present, available HDACis might still be too unspecific and have side effects, and many have unknown off-targets [55, 56]. Nevertheless, targeting histone acetylation in the treatment of MDD has not only opened new avenues to a better understanding of the pathophysiology of stress-related disorders but might as well be a promising treatment strategy. Now, it will be important to find the essential molecular mechanisms of how HDACis work to reduce off-targets through increased specificity. This task will add multiple levels of complexity to studies addressing histone modifications in the context of stress and MDD. Future studies need to:

1. Validate the importance of single or a combination of multiple histone deacetylases for the stress response to target the right HDACs with the specific inhibitor.
2. Identify the brain regions where HDACs are changed the most after stress exposure or in MDD.
3. Profile the epigenetic landscapes in MDD more vigorously.
4. Identify the major cell types undergoing expression changes of HDACs and transcriptomic and epigenomic remodelling.
5. Test functionality of the histone landscape for MDD phenotypes and probe whether HDACis can reinstate a proper histone landscape.

Extended validation of single or multiple HDACs in their function for shaping the stress response can be achieved by testing knockout mouse models for their susceptibility to react to the impact of stress exposure. Mapping of the distribution of HDACs in MDD and after stress response in the human or mouse brain can be orchestrated by standard gene expression profiling techniques in post-mortem brain or novel imaging approaches using radioactively-labelled HDAC inhibitors in the living brain [57]. In the following paragraphs, we will elaborate on how future genome-wide profiling of the epigenome in a preferentially cell type-specific manner will be useful in understanding MDD and stress-related pathologies. Ultimately, those findings might largely contribute to the development of more refined drugs targeting the relevant histone marks in the most defected cell types. Similarly, testing the actual functional relevance of specific histone marks at defined loci with epigenome editing-based strategies will contribute at large to this aim.

9.5.1 Genome-Wide Profiling of the Epigenetic Landscape in MDD and Related Models

With the availability of next-generation sequencing technologies for genome-wide analyses of the epigenome and the transcriptome, we got the right tools to study a multifactorial disease like MDD.

Here, a complete picture of which genes are deregulated after stress exposure in the animal model or in post-mortem brain of patients with MDD and which histone marks show differential occupation on which genes in MDD patients and in the related animal model may be revealed. Besides these obvious characterizations, it is important to connect the two layers of analysis to understand the mechanisms behind the transcriptional changes and to understand if all of the changes on the epigenome are directly related to gene expression changes. A few studies have used these techniques to either map the histone landscape for specific acetylation or methylation marks after chronic social defeat stress [58] or to profile the transcriptome after stress exposure [59]. However, it is more than obvious, even though both data sets are from the same lab, investigate changes in the same brain region and are even stratified by stress resilient and susceptible animals, that transcriptional changes only sparsely overlap with the epigenetic histone profile. This is surprising, since early studies (reviewed above in this chapter) were suggesting that acetylation changes and mRNA expression changes of the specific gene showcased in each study were perfectly matching. Likewise, it is interesting that the overlap between transcriptional and histone profile changes did not correlate better at any of the investigated time points after stress exposure. The authors screened gene expression changes after acute or chronic social defeat stress and after exposure to a combination of both. This finding is highly interesting, since one might hypothesize that either gene expression changes and changes to the histone landscape might correlate already at baseline or, in contrast to that, the epigenetic landscape might be primed by chronic stress for transcriptional changes in response to future acute stress events. Without doubt, the relationship between stress-induced transcriptional changes and the epigenome might be more complex, and we may lack a few layers in our current analysis. It has been demonstrated that specific combinations of histone modifications might be a good predictor of gene expression, e.g. the analysis of the active marks H3K4me1 (histone H3 lysine 4 mono-methylation) and H3K27 acetylation in combination with the repressive mark H3K27me3 (histone H3 lysine 4 trimethylation) has been shown to indicate if a gene is in active, silenced or poised state [60]. Such an approach will also be very promising for the investigation of the relationship between gene expression and the epigenome in stress response and MDD to identify the transcripts that really matter.

9.5.2 Cell Type-Specific Changes

One of the most striking characteristics of epigenetic mechanisms is their cell type specificity. In particular both the expression of histone-modifying enzymes and epigenetic signatures differ between cell types in the nervous system ([61–64]). This is of high relevance in the context of MDD, since other cell types than neurons, including microglia and astrocytes, contribute to the stress response and likely to MDD pathology [65–67]. Similarly, antidepressants target different cell types in the brain [65, 67, 68].

There is no doubt that research into epigenetics of depression will greatly profit from a cell type-specific approach, including the analysis of the epigenome and the respective transcriptome in selected cell types (Fig. 9.1). Specific methodology may comprise the analysis of FACSorted (fluorescence-activated cell sorting) nuclei for epigenome profiling and the use of antibody-based cellular panning [69, 70], genetic tagging of RNA in specific cell types or single-cell RNA sequencing to analyse cell type-specific transcriptomes [64, 71–73]. These findings will greatly promote the understanding of the connection between histone modifications and transcription. Given that epigenetic mechanisms and adaptive/maladaptive changes in stress-related disorders are cell type-specific, it is well conceivable that in the common bulk analysis of brain tissue, we may lose resolution in the analysis of the epigenome and the transcriptome and miss potential changes in either parameter and might overlook an overlap between both. Most importantly, fine mapping of molecular changes implicated in stress response and MDD will refine the development of drugs targeting epigenetic changes more specifically with less side effects.

9.5.3 Testing Functionality of Changes to the Epigenome

Finally, it will be important to test whether changes to the epigenome are indeed functional for depression-related phenotypes. With the development of novel mutagenesis tools such as CRISPR-Cas9 (clustered regularly interspaced short palindromic repeats) or TALENs (transcription activator-like effector nucleases), epigenome editing has evolved into a realistic option to test true causality in the mouse model. To date, a number of histone marks can be specifically edited, including the repressive mark H3K9me3 [74] and the active marks H3K27ac [75] and H3K4me2 [76] among others. Typically, most of these studies were initially established in defined cell culture systems, but there are good reasons to apply similar tools in vivo. Heller et al. [77] successfully targeted the active marks H3K9/K14ac and the repressive mark H3K9me2 at the *Cdk5* (cyclin-dependent kinase 5) locus in the mouse nucleus accumbens using a zinc finger protein (ZFP)-based engineering method employing transcription factors to edit the histone mark of interest. Authors delivered the construct into the mouse brain using a viral vector and tested the mice, among other behaviours, for their resilience towards social defeat stress which was found to be increased when *Cdk5* was equipped with higher levels of the H3K9/K14 acetylation mark. Similarly, same authors successfully edited in an earlier study the *Fosb* promoter in the same brain region using a comparable approach. Here, an H3K9me2 occupation increase at the selected *Fosb* locus was causal for a more pro-depressant phenotype [78]. Both studies strongly suggest functionality of the histone marks at the tested loci. The next step would be to test functional histone modifications for more loci and probe if those may get targeted by novel HDACis.

9.6 Outlook and Perspective

While basic mechanisms of a strong interconnection between stress responses, stress-related disease such as MDD and histone modifications have been demonstrated, safe treatment strategies based on epigenetic mechanisms are not established to date. Authors demonstrated in this review multiple avenues that will be crucial for translating epigenetic findings into actual drugs with significantly reduced side effects. In summary, future research will need to address true functional connections between gene expression, histone modifications and MDD and/or stress response.

Acknowledgement The work of Dr. Mira Jakovcevski (M. J.) is supported by a NARSAD Young Investigator Grant from the Brain and Behavior Research Foundation. M. J. is an "Attias Family Foundation Investigator". The work of Jan M. Deussing is supported by the German Federal Ministry of Education and Research, within the framework of the e:Med research and funding concept (IntegraMent: Integrated Understanding of Causes and Mechanisms in Mental Disorders; FKZ 01ZX1314H). Authors thank Dr. Dagmar Pommereit for help with the literature screening.

References

1. Kessler RC, Chiu WT, Demler O, Merikangas KR, Walters EE. Prevalence, severity, and comorbidity of 12-month DSM-IV disorders in the national comorbidity survey replication. Arch Gen Psychiatry. 2005;62:617–27.
2. Krishnan V, Nestler EJ. Linking molecules to mood: new insight into the biology of depression. Am J Psychiatry. 2010;167:1305–20. doi:10.1176/appi.ajp.2009.10030434.
3. Murray CJ, Lopez AD. Global mortality, disability, and the contribution of risk factors: global burden of disease study. Lancet. 1997a;349:1436–42.
4. Murray CJ, Lopez AD. Alternative projections of mortality and disability by cause 1990-2020: global burden of disease study. Lancet. 1997b;349:1498–504.

5. Bekhuis E, Schoevers RA, van Borkulo CD, Rosmalen JG, Boschloo L. The network structure of major depressive disorder, generalized anxiety disorder and somatic symptomatology. Psychol Med. 2016;46:2989–98.
6. Ready RE, Mather MA, Santorelli GD, Santospago BP. Apathy, alexithymia, and depressive symptoms: Points of convergence and divergence. Psychiatry Res. 2016;244:306–11. doi:10.1016/j.psychres.2016.07.046.
7. CONVERGE consortium. Sparse whole-genome sequencing identifies two loci for major depressive disorder. Nature. 2015;523:588–91. doi:10.1038/nature14659.
8. Klengel T, Binder EB. Epigenetics of stress-related psychiatric disorders and gene × environment interactions. Neuron. 2015;86:1343–57. doi:10.1016/j.neuron.2015.05.036.
9. Power RA, Tansey KE, Buttenschøn HN, Cohen-Woods S, Bigdeli T, Hall LS, et al. Genome-wide association for major depression through age at onset stratification: major depressive disorder working group of the psychiatric genomics consortium. Biol Psychiatry. 2016; doi:10.1016/j.biopsych.2016.05.010.
10. Stein MB, Jang KL, Taylor S, Vernon PA, Livesley WJ. Genetic and environmental influences on trauma exposure and posttraumatic stress disorder symptoms: a twin study. Am J Psychiatry. 2002;159:1675–81.
11. Sullivan PF, Neale MC, Kendler KS. Genetic epidemiology of major depression: review and meta-analysis. Am J Psychiatry. 2000;157:1552–62.
12. Maher B. Personal genomes: the case of the missing heritability. Nature. 2008;456:18–21. doi:10.1038/456018a.
13. Kendler KS. Twin studies of psychiatric illness: an update. Arch Gen Psychiatry. 2001;58:1005–14.
14. Kendler KS, Neale MC, Kessler RC, Heath AC, Eaves LJ. Familial influences on the clinical characteristics of major depression: a twin study. Acta Psychiatr Scand. 1992;86:371–8.
15. de Kloet ER, Vreugdenhil E, Oitzl MS, Joels M. Brain corticosteroid receptor balance in health and disease. Endocr Rev. 1998;19:269–301.
16. Korte SM. Corticosteroids in relation to fear, anxiety and psychopathology. Neurosci Biobehav Rev. 2001;25:117–42.
17. Sandi C. Stress, cognitive impairment and cell adhesion molecules. Nat Rev Neurosci. 2004;5:917–30.
18. Binder EB, Nemeroff CB. The CRF system, stress, depression and anxiety-insights from human genetic studies. Mol Psychiatry. 2010;15:574–88. doi:10.1038/mp.2009.141.
19. Holsboer F, Ising M. Central CRH system in depression and anxiety-evidence from clinical studies with CRH1 receptor antagonists. Eur J Pharmacol. 2008;583:350–7. doi:10.1016/j.ejphar.2007.12.032.
20. Hsu DT, Mickey BJ, Langenecker SA, Heitzeg MM, Love TM, Wang H, et al. Variation in the corticotropin-releasing hormone receptor 1 (CRHR1) gene influences fMRI signal responses during emotional stimulus processing. J Neurosci. 2012;32:3253–60. doi:10.1523/JNEUROSCI.5533-11.2012.
21. de Kloet ER. Hormones, brain and stress. Endocr Regul. 2003;37:51–68.
22. van Rossum EF, Binder EB, Majer M, Koper JW, Ising M, Modell S, et al. Polymorphisms of the glucocorticoid receptor gene and major depression. Biol Psychiatry. 2006;59:681–8.
23. Kino T. Glucocorticoid receptor. In: De Groot LJ, Beck-Peccoz P, Chrousos G, Dungan K, Grossman A, Hershman JM, Koch C, McLachlan R, New M, Rebar R, Singer F, Vinik A, Weickert MO, editors. South Dartmouth (MA): Endotext.MDText.com, Inc; 2013. P. 2000
24. Muratcioglu S, Presman DM, Pooley JR, Grøntved L, Hager GL, Nussinov R, et al. Structural modeling of GR interactions with the SWI/SNF chromatin remodeling complex and C/EBP. Biophys J. 2015;109:1227–39. doi:10.1016/j.bpj.2015.06.044.
25. Voss TC, Hager GL. Dynamic regulation of transcriptional states by chromatin and transcription factors. Nat Rev Genet. 2014;15:69–81. doi:10.1038/nrg3623.
26. Luger K, Mäder AW, Richmond RK, Sargent DF, Richmond TJ. Crystal structure of the nucleosome core particle at 2.8 A resolution. Nature. 1997;389:251–60.

27. Kouzarides T. Chromatin modifications and their function. Cell. 2007;128:693–705.
28. Strahl BD, Allis CD. The language of covalent histone modifications. Nature. 2000;403: 41–5.
29. Tsankova NM, Kumar A, Nestler EJ. Histone modifications at gene promoter regions in rat hippocampus after acute and chronic electroconvulsive seizures. J Neurosci. 2004;24: 5603–10.
30. Tsankova NM, Berton O, Renthal W, Kumar A, Neve RL, Nestler EJ. Sustained hippocampal chromatin regulation in a mouse model of depression and antidepressant action. Nat Neurosci. 2006;9:519–25.
31. Fuchikami M, Morinobu S, Kurata A, Yamamoto S, Yamawaki S. Single immobilization stress differentially alters the expression profile of transcripts of the brain-derived neurotrophic factor (BDNF) gene and histone acetylation at its promoters in the rat hippocampus. Int J Neuropsychopharmacol. 2009;12:73–82. doi:10.1017/S1461145708008997.
32. Covington 3rd HE, Maze I, LaPlant QC, Vialou VF, Ohnishi YN, Berton O, et al. Antidepressant actions of histone deacetylase inhibitors. J Neurosci. 2009;29:11451–60. doi:10.1523/JNEUROSCI.1758-09.2009.
33. Hinwood M, Tynan RJ, Day TA, Walker FR. Repeated social defeat selectively increases δFosB expression and histone H3 acetylation in the infralimbic medial prefrontal cortex. Cereb Cortex. 2011;21:262–71. doi:10.1093/cercor/bhq080.
34. Montagud-Romero S, Montesinos J, Pascual M, Aguilar MA, Roger-Sanchez C, Guerri C, et al. Up-regulation of histone acetylation induced by social defeat mediates the conditioned rewarding effects of cocaine. Prog Neuropsychopharmacol Biol Psychiatry. 2016;70:39–48. doi:10.1016/j.pnpbp.2016.04.016.
35. Kenworthy CA, Sengupta A, Luz SM, Ver Hoeve ES, Meda K, Bhatnagar S, et al. Social defeat induces changes in histone acetylation and expression of histone modifying enzymes in the ventral hippocampus, prefrontal cortex, and dorsal raphe nucleus. Neuroscience. 2014;264:88–98. doi:10.1016/j.neuroscience.2013.01.024.
36. Uchida S, Hara K, Kobayashi A, Otsuki K, Yamagata H, Hobara T, et al. Epigenetic status of Gdnf in the ventral striatum determines susceptibility and adaptation to daily stressful events. Neuron. 2011;69:359–72. doi:10.1016/j.neuron.2010.12.023.
37. Abe-Higuchi N, Uchida S, Yamagata H, Higuchi F, Hobara T, Hara K, et al. Hippocampal sirtuin 1 signaling mediates depression-like behavior. Biol Psychiatry. 2016; doi:10.1016/j.biopsych.2016.01.009.
38. Jiang WG, Li SX, Liu JF, Sun Y, Zhou SJ, Zhu WL, et al. Hippocampal CLOCK protein participates in the persistence of depressive-like behavior induced by chronic unpredictable stress. Psychopharmacology (Berl). 2013;227:79–92. doi:10.1007/s00213-012-2941-4.
39. Jiang Y, Jakovcevski M, Bharadwaj R, Connor C, Schroeder FA, Lin CL, et al. Setdb1 histone methyltransferase regulates mood-related behaviors and expression of the NMDA receptor subunit NR2B. J Neurosci. 2010;30:7152–67. doi:10.1523/JNEUROSCI.1314-10.2010.
40. Wei J, Xiong Z, Lee JB, Cheng J, Duffney LJ, Matas E, et al. Histone modification of Nedd4 ubiquitin ligase controls the loss of AMPA receptors and cognitive impairment induced by repeated stress. J Neurosci. 2016;36:2119–23. doi:10.1523/JNEUROSCI.3056-15.2016.
41. Covington 3rd HE, Maze I, Sun H, Bomze HM, DeMaio KD, Wu EY, et al. A role for repressive histone methylation in cocaine-induced vulnerability to stress. Neuron. 2011;71:656–70. doi:10.1016/j.neuron.2011.06.007.
42. Robison AJ, Vialou V, Sun HS, Labonte B, Golden SA, Dias C, et al. Fluoxetine epigenetically alters the CaMKIIα promoter in nucleus accumbens to regulate ΔFosB binding and antidepressant effects. Neuropsychopharmacology. 2014;39:1178–86. doi:10.1038/npp.2013.319.
43. Golden SA, Christoffel DJ, Heshmati M, Hodes GE, Magida J, Davis K, et al. Epigenetic regulation of RAC1 induces synaptic remodeling in stress disorders and depression. Nat Med. 2013;19:337–44. doi:10.1038/nm.3090.
44. Cruceanu C, Alda M, Nagy C, Freemantle E, Rouleau GA, Turecki G. H3K4 tri-methylation in synapsin genes leads to different expression patterns in bipolar disorder and major depression. Int J Neuropsychopharmacol. 2013;1:289–99. doi:10.1017/S1461145712000363.

45. Chen ES, Ernst C, Turecki G. The epigenetic effects of antidepressant treatment on human prefrontal cortex BDNF expression. Int J Neuropsychopharmacol. 2011;14:427–9. doi: 10.1017/S146114571000 1422.
46. Hobara T, Uchida S, Otsuki K, Matsubara T, Funato H, Matsuo K, et al. Altered gene expression of histone deacetylases in mood disorder patients. J Psychiatr Res. 2010;44:263–70. doi:10.1016/j.jpsychires.2009.08.015.
47. Del Rizzo PA, Trievel RC. Substrate and product specificities of SET domain methyltransferases. Epigenetics. 2011;6:1059–67. doi:10.4161/epi.6.9.16069.
48. Shen EY, Jiang Y, Javidfar B, Kassim B, Loh YH, Ma Q, et al. Neuronal deletion of Kmt2a/Mll1 histone methyltransferase in ventral striatum is associated with defective spike-timing dependent striatal synaptic plasticity, altered response to dopaminergic drugs and increased anxiety. Neuropsychopharmacology. 2016;41:3103–13. doi:10.1038/npp.2016.144.
49. Weaver IC, Meaney MJ, Szyf M. Maternal care effects on the hippocampal transcriptome and anxiety-mediated behaviors in the offspring that are reversible in adulthood. Proc Natl Acad Sci U S A. 2006;103:3480–5.
50. Han A, Sung YB, Chung SY, Kwon MS. Possible additional antidepressant-like mechanism of sodium butyrate: targeting the hippocampus. Neuropharmacology. 2014;81:292–302. doi:10.1016/j.neuropharm.2014.02.017.
51. Schroeder FA, Lin CL, Crusio WE, Akbarian S. Antidepressant-like effects of the histone deacetylase inhibitor, sodium butyrate, in the mouse. Biol Psychiatry. 2007;62:55–64.
52. Schmauss C. An HDAC-dependent epigenetic mechanism that enhances the efficacy of the anti-depressant drug fluoxetine. Sci Rep. 2015; doi:10.1038/srep08171.
53. Fuchikami M, Yamamoto S, Morinobu S, Okada S, Yamawaki Y, Yamawaki S. The potential use of histone deacetylase inhibitors in the treatment of depression. Prog Neuropsychopharmacol Biol Psychiatry. 2016;64:320–4. doi:10.1016/j.pnpbp.2015.03.010.
54. Houtepen LC, van Bergen AH, Vinkers CH, Boks MP. DNA methylation signatures of mood stabilizers and antipsychotics in bipolar disorder. Epigenomics. 2016;8:197–208. doi:10.2217/epi.15.98.
55. Phiel CJ, Zhang F, Huang EY, Guenther MG, Lazar MA, Klein PS. Histone deacetylase is a direct target of valproic acid, a potent anticonvulsant, mood stabilizer, and teratogen. J Biol Chem. 2001;276:36734–41.
56. Schroeder FA, Lewis MC, Fass DM, Wagner FF, Zhang YL, Hennig KM, Gale J, Zhao WN, Reis S, Barker DD, Berry-Scott E, Kim SW, Clore EL, Hooker JM, Holson EB, Haggarty SJ, Petryshen TL. A selective HDAC 1/2 inhibitor modulates chromatin and gene expression in brain and alters mouse behavior in two mood-related tests. PLoS One. 2013; doi:10.1126/scitranslmed.aaf7551.
57. Wey HY, Gilbert TM, Zürcher NR, She A, Bhanot A, Taillon BD, et al. Insights into neuroepigenetics through human histone deacetylase PET imaging. Sci Transl Med. 2016; doi:10.1126/scitranslmed. aaf7551.
58. Dias C, Feng J, Sun H, Shao NY, Mazei-Robison MS, Damez-Werno D, et al. β-catenin mediates stress resilience through Dicer1/microRNA regulation. Nature. 2014;516:51–5. doi:10.1038/ nature 13976.
59. Bagot RC, Cates HM, Purushothaman I, Lorsch ZS, Walker DM, Wang J, et al. Circuit-wide transcriptional profiling reveals brain region-specific gene networks regulating depression susceptibility. Neuron. 2016;90:969–83. doi:10.1016/j.neuron.2016.04.015.
60. Rada-Iglesias A, Bajpai R, Swigut T, Brugmann SA, Flynn RA, Wysocka J. A unique chromatin signature uncovers early developmental enhancers in humans. Nature. 2011;470:279–83. doi:10.1038/nature09692.
61. MacDonald JL, Roskams AJ. Histone deacetylases 1 and 2 are expressed at distinct stages of neuro-glial development. Dev Dyn. 2008;237:2256–67. doi:10.1002/dvdy.21626.
62. Shulha HP, Cheung I, Whittle C, Wang J, Virgil D, Lin CL, et al. Epigenetic signatures of autism: trimethylated H3K4 landscapes in prefrontal neurons. Arch Gen Psychiatry. 2012;69:314–24. doi:10.1001/archgenpsychiatry.2011.151.

63. Heinz S, Romanoski CE, Benner C, Glass CK. The selection and function of cell type-specific enhancers. Nat Rev Mol Cell Biol. 2015;16:144–54. doi: 10.1038/nrm3949.
64. Jakovcevski M, Akbarian S, Di Benedetto B. Pharmacological modulation of astrocytes and the role of cell type-specific histone modifications for the treatment of mood disorders. Curr Opin Pharmacol. 2016;26:61–6. doi:10.1016/j.coph.2015.10.002.
65. Lima A, Sardinha VM, Oliveira AF, Reis M, Mota C, Silva MA, et al. Astrocyte pathology in the prefrontal cortex impairs the cognitive function of rats. Mol Psychiatry. 2014;19:834–41. doi:10.1038/mp.2013.182.
66. Mayhew J, Beart PM, Walker FR. Astrocyte and microglial control of glutamatergic signalling: a primer on understanding the disruptive role of chronic stress. J Neuroendocrinol. 2015;27:498–506. doi:10.1111/jne.12273.
67. Sanacora G, Banasr M. From pathophysiology to novel antidepressant drugs: glial contributions to the pathology and treatment of mood disorders. Biol Psychiatry. 2013;73:1172–9. doi:10.1016/ j.biopsych.2013.03.032.
68. Bessa JM, Morais M, Marques F, Pinto L, Palha JA, Almeida OF, et al. Stress-induced anhedonia is associated with hypertrophy of medium spiny neurons of the nucleus accumbens. Transl Psychiatry. 2013; doi:10.1038/tp.2013.39.
69. Jakovcevski M, Ruan H, Shen EY, Dincer A, Javidfar B, Ma Q, et al. Neuronal Kmt2a/Mll1 histone methyltransferase is essential for prefrontal synaptic plasticity and working memory. J Neurosci. 2015;35:5097–108. doi:10.1523/JNEUROSCI.3004-14.2015.
70. Zhang Y, Chen K, Sloan SA, Bennett ML, Scholze AR, O'Keeffe S, et al. An RNA-sequencing transcriptome and splicing database of glia, neurons, and vascular cells of the cerebral cortex. J Neurosci. 2014;34:11929–47. doi:10.1523/JNEUROSCI.1860-14.2014.
71. Gerfen CR, Paletzki R, Heintz N. GENSAT BAC cre-recombinase driver lines to study the functional organization of cerebral cortical and basal ganglia circuits. Neuron. 2013;80:1368–83. doi:10.1016/j.neuron.2013.10.016.
72. Sanz E, Yang L, Su T, Morris DR, McKnight GS, Amieux PS. Cell-type-specific isolation of ribosome-associated mRNA from complex tissues. Proc Natl Acad Sci U S A. 2009;106:13939–44. doi:10.1073/pnas.0907143106.
73. Zeisel A, Muñoz-Manchado AB, Codeluppi S, Lönnerberg P, La Manno G, Juréus A, et al. Brain structure. Cell types in the mouse cortex and hippocampus revealed by single-cell RNA-seq. Science. 2015;347:1138–42. doi:10.1126/science.aaa1934.
74. Falahi F, Huisman C, Kazemier HG, van der Vlies P, Kok K, Hospers GA, et al. Towards sustained silencing of HER2/neu in cancer by epigenetic editing. Mol Cancer Res. 2013;11:1029–39. doi: 10.1158/1541-7786.MCR-12-0567.
75. Hilton IB, D'Ippolito AM, Vockley CM, Thakore PI, Crawford GE, Reddy TE, et al. Epigenome editing by a CRISPR-Cas9-based acetyltransferase activates genes from promoters and enhancers. Nat Biotechnol. 2015;33:510–7. doi:10.1038/nbt.3199.
76. Mendenhall EM, Williamson KE, Reyon D, Zou JY, Ram O, Joung JK, et al. Locus-specific editing of histone modifications at endogenous enhancers. Nat Biotechnol. 2013;31:1133–6. doi:10.1038/nbt.2701.
77. Heller EA, Hamilton PJ, Burek DD, Lombroso SI, Peña CJ, Neve RL, et al. Targeted epigenetic remodeling of the Cdk5 gene in nucleus accumbens regulates cocaine- and stress-evoked behavior. J Neurosci. 2016;36:4690–7. doi:10.1523/JNEUROSCI.0013-16.2016.
78. Heller EA, Cates HM, Peña CJ, Sun H, Shao N, Feng J, et al. Locus-specific epigenetic remodeling controls addiction- and depression-related behaviors. Nat Neurosci. 2014;17:1720–7. doi:10.1038/nn.3871.

DNA Methylation in Major Depressive Disorder

10

Ehsan Pishva, Bart P.F. Rutten, and Daniel van den Hove

Abstract

Epigenetic mechanisms regulate gene expression, influencing protein levels and ultimately shaping phenotypes during life. However, both stochastic epigenetic variations and environmental reprogramming of the epigenome might influence neurodevelopment and ageing, and this may contribute to the origins of mental ill-health. Studying the role of epigenetic mechanisms is challenging, as genotype-, tissue- and cell type-dependent epigenetic changes have to be taken into account, while the nature of mental disorders also poses significant challenges

E. Pishva, M.D., Ph.D. (✉)
Complex Disease Epigenetic Group, University of Exeter Medical School, RILD Building, RD&E Hospital Wonford, Barrack Road, Exeter EX2 5DW, UK

Department of Psychiatry and Neuropsychology, School for Mental Health and Neuroscience (MHeNS), Maastricht University, Universiteitssingel 50, Maastricht 6200 MD, The Netherlands
e-mail: e.pishva@exeter.ac.uk; e.pishva@maastrichtuniversity.nl

B.P.F. Rutten, M.D., Ph.D.
Department of Psychiatry and Neuropsychology, School for Mental Health and Neuroscience (MHeNS), Maastricht University, Universiteitssingel 50, Maastricht 6200 MD, The Netherlands
e-mail: b.rutten@maastrichtuniversity.nl

D. van den Hove, Ph.D.
Department of Psychiatry and Neuropsychology, School for Mental Health and Neuroscience (MHeNS), Maastricht University, Universiteitssingel 50, Maastricht 6200 MD, The Netherlands

Laboratory of Translational Neuroscience, Department of Psychiatry, Psychosomatics and Psychotherapy, University of Wuerzburg, Fuechsleinstrasse 15, Wurzburg 97080, Germany
e-mail: d.vandenhove@maastrichtuniversity.nl

R. Delgado-Morales (ed.), *Neuroepigenomics in Aging and Disease*, Advances in Experimental Medicine and Biology, DOI 10.1007/978-3-319-53889-1_10

for linking them with biological profiles. In this chapter, we summarise the current evidence suggesting the role of DNA methylation as a key epigenetic mechanism in major depressive disorder.

Keywords
Epigenetics • DNA methylation • Major depressive diorders • Antidepressants

Major depressive disorder (MDD) is a prevalent mood disorder characterised by persistent low mood accompanied by significant morbidity and mortality. The lifetime prevalence of MDD is 15–20% and women suffer from MDD about twice more often than men [1]. MDD is a complex multifactorial disorder, with both genetic and environmental factors playing an important role in its development. The heritability is estimated to be approximately 37%, and numerous links have been made between genetic variation and clinical depression [2]. However, DNA sequence variations cannot fully explain the susceptibility to MDD as detecting strong and replicable genetic associations with the development and course of clinical depression have proven difficult. Moreover, exposure to known environmental risk factors for MDD such as childhood adversities does not always generate the disorder. Therefore, the development and course of MDD are thought to be explained by gene-environment (GxE) interactions, where the effect of the environment depends on a person's genotype or, equivalently, the effect of a person's genotype depends on the environment [3]. Thus, it can be suggested that individuals genetically vulnerable to mental illness might undergo structural brain changes, imbalances in multiple neurotransmitter systems, alterations in neurotrophic signalling, and neuroendocrine abnormalities when exposed to harmful environmental factors at critical times during neurodevelopment.

A consistent line of evidence in human and rodent studies has shown that environmental factors regulate gene transcription and epigenetic mechanisms emerged as prime candidates for mediating GxE interactions in several brain regions [4]. In this chapter, we focus on DNA methylation, representing the most studied epigenetic mechanisms in psychiatric research. We will first summarise the current stage of knowledge within this field indicating the potential contribution of DNA methylation to MDD and discuss the results of the most replicated human studies on DNA methylation alterations in relation to environmental exposures associated with MDD. Finally, we will discuss the current challenges and perspectives in the field of epigenetic research on MDD.

10.1 DNA Methylation and MDD

Several methods have been applied to investigate MDD-associated and stress-induced alterations in DNA methylation. In the following sections, we will discuss the observed candidate and methylome-wide associations in MDD.

10.1.1 Candidate Gene Studies

Most DNA methylation studies thus far used candidate gene approaches. Impairments in stress response pathways, neurotrophic signalling and monoaminergic systems are well-known processes involved in the pathogenesis of MDD. Therefore, the focus of candidate gene methylation studies was predominantly on the promoter sequences of genes involved in these biological pathways. Here, we will address a selection of the main findings of the first studies on candidate genes implicated in MDD.

10.1.2 Stress Reactivity Genes

Given the role of the hypothalamic-pituitary-adrenal (HPA) axis and early life experiences and the aetiology of MDD, a series of studies investigated the DNA methylation patterns of the NR3C1 gene. NR3C1 encodes the glucocorticoid receptor (GR), which is known for its regulatory role in dampening the activity of the HPA axis. Environmental reprogramming of NR3C1 gene expression has been shown in both the brain and periphery. Both human and rodent studies suggested that early life trauma is associated with significant hypermethylation in the promoter region of the alternate exon 1_F (humans) or 1_7 (rodents) within the hippocampus and, subsequently, reduced GR expression [5, 6]. Increased methylation levels of the NR3C1 promoter have been reported in lymphocytes of newborn prenatally exposed to maternal depression [7, 8]. One study examined a functional association between methylation statuses of NR3C1 gene promoter and cortisol as the end product of the stress pathway. A decreased cortisol response to the dexamethasone/corticotropin-releasing hormone (CRH) test was associated with increased methylation levels exon 1_F NR3C1 gene promoter in leukocytes [9].

FK506-binding protein 5 (FKBP5), a member of the immunophilin protein family, functionally interacts with GRs and is linked to environmental stress exposure and MDD. FKBP5 protein reduces glucocorticoid-binding affinity [10], while cortisol and its binding to GR induce FKBP5 expression [11]. Several studies have shown the interactive effects of polymorphisms in FKBP5 and early life adversities predicting MDD [12]. Decreased allele-specific methylation of FKBP5 has been observed in peripheral blood cells of subjects who have experienced childhood abuse, whereas a similar effect has been observed in a neuronal progenitor cell line after exposure to GR agonists [13].

10.1.3 Neurotrophic Signalling Genes

Brain-derived neurotrophic factor (BDNF) is a member of the neurotrophin family and has been shown to regulate the development, plasticity and survival of dopaminergic, cholinergic and serotonergic neurons. The neurotrophic hypothesis of depression suggests that low levels of BDNF, concomitant with reduced neuronal and synaptic plasticity, are associated with clinical depression [14]. The BDNF protein is abundant in the brain and periphery and known to be able to cross the blood-brain barrier [15]. Chronic stress-induced decreases in BDNF and, in particular, antidepressant-induced increases in BDNF have been extensively studied in relation to the development and course of MDD [16]. Therefore, the correlation between BDNF methylation status and MDD has already been a focus of interest for years [17]. Differential regulation of BDNF exons I, IV and IX expression has been reported repeatedly, accompanied by changes in (allele-specific) DNA methylation of the corresponding promoters within various brain regions as well in the blood of patients with MDD [18]. The majority of studies have found increased methylation levels at different loci within the BDNF gene in MDD patients compared to controls in both brain and the periphery [19]. However, the direction of the effects is not always congruent [20].

To explore the functional relevance of DNA methylation variation and MDD, some studies examined the association between DNA methylation at the BDNF gene and structural changes in the brain of MDD patients. As such, in patients with MDD, the prefrontal and occipital cortices have been indicated as regions in which BDNF promoter hypermethylation is associated with reduced cortical thickness [21]. Another study showed a correlation between BDNF promoter hypermethylation and reduced white-matter integrity in individuals with MDD [22]. Significant associations have been found between BDNF promoter VI hypermethylation and a history of suicidal attempts and suicidal ideation [23]. In two subsequent studies, the same group found a significant association between BDNF promoter VI methylation levels and late-life depression [24], as well as depression related to breast cancer [25].

Methylation status of BDNF promoters has also been studied as a predictor for antidepressant treatment. Differential methylation of BDNF promoter I was observed when comparing responders and nonresponders to electroconvulsive therapy (ECT) [26]. Likewise, the methylation state of a CpG site within the exon IV promoter region of BDNF has been suggested to predict responses to antidepressant pharmacotherapy [27].

10.1.4 Monoaminergic Transporter Genes

The monoamine hypothesis of depression highlights the importance of serotonin (5-HT) turnover and transmission in synaptic cleft in the pathophysiology of major depression [28]. In particular, genetic variation in SLC6A4, the gene encoding the serotonin transporter (5-HTT or SERT), is a well-known candidate gene studied in MDD. 5-HTT is responsible for the reuptake of 5-HT into the presynaptic neuron and the primary target of antidepressants to normalise 5-HT levels and, consequently, the clinical symptoms of depression. SLC6A4 polymorphisms and their interaction with environmental stress have been studied extensively in association with MDD [29]. However, studies on the link between genetic variation of SLC6A4, life events and MDD are characterised by inconsistent findings, which may be partly explained by the additional and possibly interdependent role of epigenetic variation. Hypermethylation in the promoter region of SLC6A4 in lymphoblast cell lines was accompanied by lower gene expression and associated with vulnerability to MDD, concomitant with a complex interaction with genetic variation in SLC6A4 [30]. Clearly, these findings await replication.

Increased buccal cell methylation levels have been observed in short-allele carriers of the SLC6A4 gene in association with depressive symptoms [31]. In addition, increased SLC6A4 promoter methylation status was significantly associated with childhood adversities, a family history of depression, but not antidepressant treatment outcomes [32]. However, another study showed that lower average SLC6A4 CpG methylation was associated with an impaired antidepressant treatment response [33]. Thus far, all studies with regard to the association between DNA methylation within the SLC6A4 gene have been performed on peripheral blood tissues. A number of studies have examined the association between SLC6A4 promoter methylation and structural and functional changes in the brain. Using positron emission tomography (PET), decreased 5-HT synthesis in the orbitofrontal cortex was shown to be correlated to increased SLC6A4 promoter methylation [34]. In addition, increased methylation of the SLC6A4 promoter has been shown to be associated with increased hippocampal volume assessed by voxel-based morphometry [35]. Increased SLC6A4 methylation has furthermore been associated with childhood trauma and decreased hippocampal volume [36].

10.2 Methylome-Wide Association Studies

Methylome-wide association studies (MWAS) using different platforms have shown distinguished patterns of DNA methylation in MDD (see Table 10.1). A study comparing post-mortem frontal cortex tissue from MDD patients and healthy controls identified 224 differentially methylated regions (DMRs). These regions were highly enriched for genes involved in neuronal growth and development. However, the technique that has been used to measure the methylation levels for this study could not provide information at a single CpG resolution [37]. Another MWAS on the prefrontal cortex of suicidal completers with depression and sudden death controls showed 115 DMRs mainly related to astrocytic functioning [38]. Findings from MWAS performed on blood samples of individuals with a lifetime history of depression compared to nondepressed controls provided evidence for the involvement of inflammatory pathways previously implicated in depression [39]. Another genome-wide DNA methylation profiling of peripheral leukocytes of medication-free MDD patients identified 363 hypomethylated CpG sites [40]. In this study, three CpG sites residing in DGKH, GSK3B and SGK1 genes have been previously implicated in MDD, and they could show a significant inverse correlation between GSK3B promoter DNA methylation and expression. A methylome-wide study that incorporated an environmental risk factor reported an association

Table 10.1 Summary of MWAS and twin studies

References	Study characteristics	Tissue type	Platform	Main findings
[39]	33 lifetime history of MDD, 67 C	Peripheral blood	HM27	– Increased methylation in genomic regions related to brain development while decreased methylation in regions related to lipoprotein
[37]	39 MDD, 26 C	Post-mortem frontal cortex	CHARM (3.5 million CpGs)	– 224 candidate regions with methylation differences more than 10%. PRIMA1**:** the best replicated and validated
[38]	Discovery 76 MDD, 45C; replication cohort: 22 MDD,17 C	Prefrontal cortex	MBD2-Sequencing	– GRIK2 and BEGAIN as the two most differentially methylated regions (DMRs) between suicide completers and sudden death control
[40]	20 medication-free patients with MDD and 19 C Replication: 12 medication-free patients with MDD, 12 C	Peripheral leukocytes	HM450	– A significant inverse correlation between the GSK3B promoter DNA methylation and expression was observed
[41]	94 maltreated children (35% with MDD), 96 C	Saliva	HM450	– 3 genome-wide–significant predictors of depression: ID3, GRIN1, TPPP
[45]	50 MZ pairs discordant for MDD (27 UK, 23 Australia) Case-control replication: 118 MDD, 236 C	Whole blood	MeDIP-Sequencing	– Hypermethylation within the coding region of ZBTB20 – Replicated in an independent cohort of 356 case-control individuals
[43]	18 MZ pairs discordant for MDD Case-control replication: 14 MDD, 15 C	Buccal cells Post-mortem brain (cerebellum)	HM450	– STK32C hypomethylation associated with MDD – Validated in post-mortem brain case-control individuals
[44]	17 MZ twin pairs (four concordant, six discordant, seven healthy)	Peripheral blood	HM450	– Differentially DNA methylated in the WDR26 gene associated with MDD

Table 10.1 (continued)

References	Study characteristics	Tissue type	Platform	Main findings
[51]	171 MZ twin pairs	White blood cells (100) Prefrontal cortex (71)	8.1 K CpG island microarrays	– Differentially modified regions exhibited a highly significant number of overlaps across the tested tissues and previous studies

HM27 Illumina Infinium Human Methylation 27, *HM450* Illumina Infinium Human Methylation450 Beadchip, *CHARM* comprehensive high-throughput arrays for relative methylation, *C* control, *MDD* major depressive disorder, *MBD2* methylation-binding domain-2, *MeDIP* methylated DNA immunoprecipitation

between a history of childhood maltreatment and changes in saliva DNA methylation within three genes (ID3, GRIN1 and TPPP), of which the degree of methylation predicted depressive symptoms. These genes are known to be involved in stress-related neuroendocrine pathways and neural plasticity [41]. A recent genome-wide profiling of cortical brain regions (BA11 and BA25) from MDD suicide cases compared to controls identified a DMR, upstream of the PSORS1C3 noncoding gene, which is consistently hypomethylated across both cortical brain regions in MDD patients [42].

10.2.1 Twin Studies

Phenotypic discordance in monozygotic (MZ) twin pairs provides a valuable tool to differentiate genetic from non-genetic causes of diseases. Utilising MZ twin designs allows ruling out DNA sequence variations as a confounding source for epigenetic studies. While the genomic content in MZ twins is almost identical, the discordance in phenotypes can be attributed to non-shared environmental and stochastic factors. It has been suggested that different levels of DNA methylation at specific loci within MZ pairs measured as differentially methylated positions (DMPs) can be linked to the environmental causes of disease, while the changes in methylation variance, measured as variably methylated probes (VMPs), may be related to stochasticity. There is increasing evidence suggesting that epigenetic variation between MZ pairs plays a role in the aetiology of MDD. However, using buccal and blood cells, studying MZ pairs discordant for MDD has not identified any methylome-wide significant loci after correcting for multiple testing, which is most likely the result of a lack of statistical power. Interestingly, several of the most differentially methylated genes have previously been associated with the pathogenesis of MDD (see Table 10.1) [43, 44]. A meta-analysis of 50 MZ pairs, all female discordant for MDD, using 8.1 K human CpG island microarrays, identified 17 DMRs with genome-wide significance, and some of these genes have previously been associated with MDD (see Table 10.1) [45].

10.3 DNA Methylation and Antidepressant Treatment

There is increasing evidence suggesting that the effectiveness of antidepressants might be associated with DNA methylation status of certain genes. One example is the finding that decreased DNA methylation of SLC6A4 promotor region was associated with impaired antidepressant treatment response [33]. On the contrary, another study showed that less improvement in clinical symptoms of depression was correlated with higher methylation percentage at SLC6A4 gene promotor region [32].

Epigenetic mechanisms have also been proposed to mediate the mechanism of action of antidepressants. Alterations in DNA methylation and enzymes catalysing the methylation processes have been suggested to be involved in therapeutic effects of tricyclic antidepressants (TCA), selective serotonin reuptake inhibitor (SSRI) and valproate (VPA) [46, 47].

A number of MWA studies investigated the methylation changes in newborns exposed to maternal antidepressant use during pregnancy. Exposure to antidepressants, regardless of the medication family, was shown to be associated with differential methylation of CpG sites in the TNFRSF21 and CHRNA2 genes [48]. Another genome-wide methylation study identified increased methylation levels at CpG sites in CYP2E1, EVA1 and SLMAP in SSRIs exposed neonates [49]. Moreover, various studies compared the methylation changes associated with maternal use of antidepressants in candidate genes such as NR3C1, BDNF and SLC6A4 [27, 50]. However, results were not conclusive and none have been replicated in genome-wide array-based approaches.

10.4 Discussion and Future Perspectives

Over the past few years, a considerable number of epigenome-wide and candidate gene studies have been performed to identify a relation between DNA methylation and the pathogenesis of MDD. Current state of evidence on the link between DNA methylation and MDD indicates limited overlap between the identified genes in different studies. Moreover, the significant associations identified in candidate gene studies have not been observed in MWAS. These inconsistent findings can be explained by a broad range of methodological concerns in the field of epigenetic psychiatry. Evidently, limited access to human brain tissue is a major challenge in studying the specific epigenetic patterns in mental illnesses. Moreover, DNA methylation is characterised by the addition of a methyl or hydroxymethyl group to the C5 position of cytosine. DNA hydroxymethylation can both act as a stable functional epigenetic mark and may also serve as an intermediate mark in the active process of DNA demethylation. However, the large majority of current commercially available techniques that have been used to study epigenetic variation in MDD do not distinguish between DNA methylation and other related epigenetic marks, e.g. DNA hydroxymethylation, limiting the power of such approaches. Furthermore, true epigenetic variations can be influenced by state- or disease-specific differences in the cellular composition of both brain and peripheral blood tissues. In addition, significant findings remain vulnerable to confounding factors such as age, gender and smoking, which might cause false association discoveries between MDD and DNA methylation marks. Similarly, disease-related factors like medication, heterogeneity in diagnostic tools, the onset and course of MDD in different studies might explain the observed inconsistency in findings. Moreover, thus far, no MWAS of MDD has been integrated with genomic data. As such, genetic variations known to predict variations in the methylome should be taken into account in future approaches. Larger sample sizes and meta-analytical approaches can be utilised to boost statistical power in order to identify methylome-wide significant genes in association with MDD or the effects of antidepressants. Evidently, heterogeneity between studies, though strengthening the value of observed converging evidence, at the same time challenges the consolidation of less robust findings within individual datasets. Replication studies of epigenetic findings are therefore highly necessary. In addition, it will be crucial to move from associations and correlations towards examining causality. For instance, in vivo epigenetic editing of identified targets, as well as using longitudinal study designs in human cohorts would significantly improve our understanding about the causal relationship between changes in DNA methylation and the development and course of MDD.

References

1. Kessler RC, Bromet EJ. The epidemiology of depression across cultures. Annu Rev Public Health. 2013;34:119–38. doi:10.1146/annurev-publhealth-031912-114409.
2. Sullivan PF, Neale MC, Kendler KS. Genetic epidemiology of major depression: review and meta-analysis. Am J Psychiatry. 2000;157(10):1552–62. doi:10.1176/appi.ajp.157.10.1552.
3. Duncan LE, Keller MC. A critical review of the first 10 years of candidate gene-by-environment interaction research in psychiatry. Am J Psychiatry. 2011;168(10):1041–9. doi:10.1176/appi.ajp.2011.11020191.
4. Jaenisch R, Bird A. Epigenetic regulation of gene expression: how the genome integrates intrinsic and environmental signals. Nat Genet. 2003;33:245–54. doi:10.1038/ng1089.
5. McGowan PO, Sasaki A, D'Alessio AC, Dymov S, Labonte B, Szyf M, et al. Epigenetic regulation of the glucocorticoid receptor in human brain associates with childhood abuse. Nat Neurosci. 2009;12(3):342–8. doi:10.1038/nn.2270.
6. Weaver IC, Cervoni N, Champagne FA, D'Alessio AC, Sharma S, Seckl JR, et al. Epigenetic programming by maternal behavior. Nat Neurosci. 2004;7(8):847–54. doi:10.1038/nn1276.
7. Oberlander TF, Weinberg J, Papsdorf M, Grunau R, Misri S, Devlin AM. Prenatal exposure to maternal depression, neonatal methylation of human glucocorticoid receptor gene (NR3C1) and infant cortisol stress responses. Epigenetics. 2008;3(2):97–106.
8. Palma-Gudiel H, Cordova-Palomera A, Eixarch E, Deuschle M, Fananas L. Maternal psychosocial stress during pregnancy alters the epigenetic signature of the glucocorticoid receptor gene promoter in their offspring: a meta-analysis. Epigenetics. 2015;10(10):893–902. doi:10.1080/15592294.2015.1088630.
9. Tyrka AR, Parade SH, Welch ES, Ridout KK, Price LH, Marsit C, et al. Methylation of the leukocyte glucocorticoid receptor gene promoter in adults: associations with early adversity and depressive, anxiety and substance-use disorders. Transl Psychiatry. 2016;6(7):e848. doi:10.1038/tp.2016.112.
10. Denny WB, Valentine DL, Reynolds PD, Smith DF, Scammell JG. Squirrel monkey immunophilin FKBP51 is a potent inhibitor of glucocorticoid receptor binding. Endocrinology. 2000;141(11):4107–13. doi:10.1210/endo.141.11.7785.

11. Binder EB. The role of FKBP5, a co-chaperone of the glucocorticoid receptor in the pathogenesis and therapy of affective and anxiety disorders. Psychoneuroendocrinology. 2009;34(Suppl 1):S186–95. doi:10.1016/j.psyneuen.2009.05.021.
12. Zannas AS, Binder EB. Gene-environment interactions at the FKBP5 locus: sensitive periods, mechanisms and pleiotropism. Genes Brain Behav. 2014;13(1):25–37. doi:10.1111/gbb.12104.
13. Klengel T, Binder EB. Gene × environment interactions in the prediction of response to antidepressant treatment. Int J Neuropsychopharmacol. 2013;16(3):701–11. doi:10.1017/S1461145712001459.
14. Duman RS. Role of neurotrophic factors in the etiology and treatment of mood disorders. Neuromol Med. 2004;5(1):11–25. doi:10.1385/NMM:5:1:011.
15. Pan W, Banks WA, Fasold MB, Bluth J, Kastin AJ. Transport of brain-derived neurotrophic factor across the blood–brain barrier. Neuropharmacology. 1998;37(12):1553–61.
16. Lee BH, Kim YK. The roles of BDNF in the pathophysiology of major depression and in antidepressant treatment. Psychiatry Investig. 2010;7(4):231–5. doi:10.4306/pi.2010.7.4.231.
17. Boulle F, van den Hove DL, Jakob SB, Rutten BP, Hamon M, van Os J, et al. Epigenetic regulation of the BDNF gene: implications for psychiatric disorders. Mol Psychiatry. 2012;17(6):584–96. doi:10.1038/mp.2011.107.
18. Januar V, Ancelin ML, Ritchie K, Saffery R, Ryan J. BDNF promoter methylation and genetic variation in late-life depression. Transl Psychiatry. 2015;5:e619. doi:10.1038/tp.2015.114.
19. Keller S, Sarchiapone M, Zarrilli F, Videtic A, Ferraro A, Carli V, et al. Increased BDNF promoter methylation in the Wernicke area of suicide subjects. Arch Gen Psychiatry. 2010;67(3):258–67. doi:10.1001/archgenpsychiatry.2010.9.
20. Fuchikami M, Morinobu S, Segawa M, Okamoto Y, Yamawaki S, Ozaki N, et al. DNA methylation profiles of the brain-derived neurotrophic factor (BDNF) gene as a potent diagnostic biomarker in major depression. PLoS One. 2011;6(8):e23881. doi:10.1371/journal.pone.0023881.
21. Na KS, Won E, Kang J, Chang HS, Yoon HK, Tae WS, et al. Brain-derived neurotrophic factor promoter methylation and cortical thickness in recurrent major depressive disorder. Sci Rep. 2016;6:21089. doi:10.1038/srep21089.
22. Choi S, Han KM, Won E, Yoon BJ, Lee MS, Ham BJ. Association of brain-derived neurotrophic factor DNA methylation and reduced white matter integrity in the anterior corona radiata in major depression. J Affect Disord. 2015;172:74–80. doi:10.1016/j.jad.2014.09.042.
23. Kang HJ, Kim JM, Lee JY, Kim SY, Bae KY, Kim SW, et al. BDNF promoter methylation and suicidal behavior in depressive patients. J Affect Disord. 2013;151(2):679–85.
24. Kang HJ, Kim JM, Bae KY, Kim SW, Shin IS, Kim HR, et al. Longitudinal associations between BDNF promoter methylation and late-life depression. Neurobiol Aging. 2015;36(4):1764.e1–7. doi:10.1016/j.neurobiolaging.2014.12.035.
25. Kang HJ, Kim JM, Kim SY, Kim SW, Shin IS, Kim HR, et al. A longitudinal study of BDNF promoter methylation and depression in breast cancer. Psychiatry Investig. 2015;12(4):523–31. doi:10.4306/pi.2015.12.4.523.
26. Kleimann A, Kotsiari A, Sperling W, Groschl M, Heberlein A, Kahl KG, et al. BDNF serum levels and promoter methylation of BDNF exon I, IV and VI in depressed patients receiving electroconvulsive therapy. J Neural Transm (Vienna). 2015;122(6):925–8. doi:10.1007/s00702-014-1336-6.
27. Tadic A, Muller-Engling L, Schlicht KF, Kotsiari A, Dreimuller N, Kleimann A, et al. Methylation of the promoter of brain-derived neurotrophic factor exon IV and antidepressant response in major depression. Mol Psychiatry. 2014;19(3):281–3. doi:10.1038/mp.2013.58.
28. Owens MJ, Nemeroff CB. Role of serotonin in the pathophysiology of depression: focus on the serotonin transporter. Clin Chem. 1994;40(2):288–95.
29. Caspi A, Sugden K, Moffitt TE, Taylor A, Craig IW, Harrington H, et al. Influence of life stress on depression: moderation by a polymorphism in the 5-HTT gene. Science. 2003;301(5631):386–9. doi:10.1126/science.1083968.

30. Philibert RA, Sandhu H, Hollenbeck N, Gunter T, Adams W, Madan A. The relationship of 5HTT (SLC6A4) methylation and genotype on mRNA expression and liability to major depression and alcohol dependence in subjects from the Iowa Adoption Studies. Am J Med Genet B Neuropsychiatr Genet. 2008;147B(5):543–9. doi:10.1002/ajmg.b.30657.
31. Olsson CA, Foley DL, Parkinson-Bates M, Byrnes G, McKenzie M, Patton GC, et al. Prospects for epigenetic research within cohort studies of psychological disorder: a pilot investigation of a peripheral cell marker of epigenetic risk for depression. Biol Psychol. 2010;83(2):159–65. doi:10.1016/j.biopsycho.2009.12.003.
32. Kang HJ, Kim JM, Stewart R, Kim SY, Bae KY, Kim SW, et al. Association of SLC6A4 methylation with early adversity, characteristics and outcomes in depression. Prog Neuropsychopharmacol Biol Psychiatry. 2013;44:23–8. doi:10.1016/j.pnpbp.2013.01.006.
33. Domschke K, Tidow N, Schwarte K, Deckert J, Lesch KP, Arolt V, et al. Serotonin transporter gene hypomethylation predicts impaired antidepressant treatment response. Int J Neuropsychopharmacol. 2014;17(8):1167–76. doi:10.1017/S146114571400039x.
34. Wang D, Szyf M, Benkelfat C, Provencal N, Turecki G, Caramaschi D, et al. Peripheral SLC6A4 DNA methylation is associated with in vivo measures of human brain serotonin synthesis and childhood physical aggression. PLoS One. 2012;7(6):e39501. doi:10.1371/journal.pone.0039501.
35. Dannlowski U, Kugel H, Redlich R, Halik A, Schneider I, Opel N, et al. Serotonin transporter gene methylation is associated with hippocampal gray matter volume. Hum Brain Mapp. 2014;35(11):5356–67. doi:10.1002/hbm.22555.
36. Booij L, Szyf M, Carballedo A, Frey EM, Morris D, Dymov S, et al. DNA methylation of the serotonin transporter gene in peripheral cells and stress-related changes in hippocampal volume: a study in depressed patients and healthy controls. PLoS One. 2015; 10(3):e0119061.
37. Sabunciyan S, Aryee MJ, Irizarry RA, Rongione M, Webster MJ, Kaufman WE, et al. Genome-wide DNA methylation scan in major depressive disorder. PLoS One. 2012;7(4):e34451. doi:10.1371/journal.pone.0034451.
38. Nagy C, Suderman M, Yang J, Szyf M, Mechawar N, Ernst C, et al. Astrocytic abnormalities and global DNA methylation patterns in depression and suicide. Mol Psychiatry. 2015;20(3):320–8. doi:10.1038/mp.2014.21.
39. Uddin M, Koenen KC, Aiello AE, Wildman DE, de los Santos R, Galea S. Epigenetic and inflammatory marker profiles associated with depression in a community-based epidemiologic sample. Psychol Med. 2011;41(5):997–1007. doi:10.1017/S0033291710001674.
40. Numata S, Ishii K, Tajima A, Iga J, Kinoshita M, Watanabe S, et al. Blood diagnostic biomarkers for major depressive disorder using multiplex DNA methylation profiles: discovery and validation. Epigenetics. 2015;10(2):135–41. doi:10.1080/15592294.2014.1003743.
41. Weder N, Zhang H, Jensen K, Yang BZ, Simen A, Jackowski A, et al. Child abuse, depression, and methylation in genes involved with stress, neural plasticity, and brain circuitry. J Am Acad Child Adolesc Psychiatry. 2014;53(4):417–24.e5. doi:10.1016/j.jaac.2013.12.025.
42. Murphy TM, Crawford B, Dempster EL, Hannon E, Burrage J, Turecki G, et al. Methylomic profiling of cortex samples from completed suicide cases implicates a role for PSORS1C3 in major depression and suicide. Transl Psychiatry. 2017;7(1):e989. doi:10.1038/tp.2016.249.
43. Dempster EL, Wong CC, Lester KJ, Burrage J, Gregory AM, Mill J, et al. Genome-wide methylomic analysis of monozygotic twins discordant for adolescent depression. Biol Psychiatry. 2014;76(12):977–83. doi:10.1016/j.biopsych.2014.04.013.
44. Cordova-Palomera A, Fatjo-Vilas M, Gasto C, Navarro V, Krebs MO, Fananas L. Genome-wide methylation study on depression: differential methylation and variable methylation in monozygotic twins. Transl Psychiatry. 2015;5:e557. doi:10.1038/tp.2015.49.
45. Davies MN, Krause L, Bell JT, Gao F, Ward KJ, Wu HL, et al. Hypermethylation in the ZBTB20 gene is associated with major depressive disorder. Genome Biol. 2014;15(4):R56. doi:10.1186/gb-2014-15-4-r56.

46. Zimmermann N, Zschocke J, Perisic T, Yu S, Holsboer F, Rein T. Antidepressants inhibit DNA methyltransferase 1 through reducing G9a levels. Biochem J. 2012;448(1):93–102. doi:10.1042/BJ20120674.
47. Detich N, Bovenzi V, Szyf M. Valproate induces replication-independent active DNA demethylation. J Biol Chem. 2003;278(30):27586–92. doi:10.1074/jbc.M303740200.
48. Schroeder JW, Smith AK, Brennan PA, Conneely KN, Kilaru V, Knight BT, et al. DNA methylation in neonates born to women receiving psychiatric care. Epigenetics. 2012;7(4):409–14. doi:10.4161/epi.19551.
49. Gurnot C, Martin-Subero I, Mah SM, Weikum W, Goodman SJ, Brain U, et al. Prenatal antidepressant exposure associated with CYP2E1 DNA methylation change in neonates. Epigenetics. 2015;10(5):361–72. doi:10.1080/15592294.2015.1026031.
50. Zhao JY, Goldberg J, Bremner JD, Vaccarino V. Association between promoter methylation of serotonin transporter gene and depressive symptoms: a monozygotic twin study. Psychosom Med. 2013;75(6):523–9. doi:10.1097/PSY.0b013e3182924cf4.
51. Oh G, Wang SC, Pal M, Chen ZF, Khare T, Tochigi M, et al. DNA modification study of major depressive disorder: beyond locus-by-locus comparisons. Biol Psychiatry. 2015;77(3):246–55. doi:10.1016/j.biopsych.2014.06.016.

Noncoding RNAs in Depression

11

Rixing Lin and Gustavo Turecki

Abstract

Major depressive disorder (MDD) is a common psychiatric disorder affecting millions of people worldwide, yet its etiology remains elusive. The last decades have seen great advances in our understanding of the genome structure and functional organization. Noncoding RNAs (ncRNAs) are RNAs that do not code for proteins but have important regulatory roles. The investigation of ncRNAs as regulators of gene expression has been a topic of growing interest in health research, including in studies investigating etiological and therapeutic factors in major depression. Several different species of ncRNAs have been identified in association to and have shown to be dysregulated in depressed individuals or in animal models of depression. This review will detail the complex relation between ncRNAs and major depression and the studies that propose mechanisms and pathways that specific ncRNAs may be involved in major depression.

Keywords

Noncoding RNA • Major depressive disorder • Antidepressant response • Biomarkers

R. Lin, B.A. • G. Turecki, M.D., Ph.D. (✉)
McGill Group for Suicide Studies, Douglas Mental Health University Institute, McGill University, 6875 Boulevard LaSalle, Montreal, QC, Canada, H4H1R3
e-mail: rixing.lin@mail.mcgill.ca; gustavo.turecki@mcgill.ca

R. Delgado-Morales (ed.), *Neuroepigenomics in Aging and Disease*, Advances in Experimental Medicine and Biology, DOI 10.1007/978-3-319-53889-1_11

11.1 Introduction

Major depressive disorder (MDD) is a common mental illness with an estimated 350 million people affected around the world [1]. Symptoms of MDD include constant depressed mood, loss of interest in previously pleasurable activities, significant weight loss or gain, insomnia or hypersomnia, feelings of worthlessness, and, in the worst case, recurrent thoughts of death and suicide [2]. Despite the prevalence and burden of MDD, our current understanding of the pathophysiology remains incomplete.

The functional activity of a gene gives rise to most biological processes. Thus, investigating gene functional changes associated with MDD and the molecular factors that modulate this change should provide us insight on molecular mechanisms associated with MDD. The last decades have seen great advances on our understanding of the genome structure and functional organization [3–9]. A major finding showed that a vast majority of the mammalian genome produces RNA; however, only an estimated 1% of the mammalian genome is capable of encoding protein [10]. RNAs that show no evidence of coding for any protein are known as noncoding RNAs (ncRNAs) and have been shown to carry out biologically relevant functions including regulatory roles as well as housekeeping roles [11]. There has also been an increasing interest exploring ncRNAs that have been proposed to have regulatory functions, such as regulating gene expression.

ncRNAs are further classified into small or long ncRNAs based on a nucleotide length threshold of 200 nucleotides. Small ncRNAs are less than 200 nucleotides, with distinct subgroups including microRNA (miRNA), small interfering RNA (siRNA), small nuclear RNA (snRNA), small nucleolar RNA (snoRNAs), and piwi-interacting RNA (piRNA). Long ncRNAs are defined as those greater than 200 nucleotides with no evidence of protein-coding ability. Ribosomal RNA (rRNA) and transfer RNA (tRNA) are also well-known ncRNAs that serve housekeeping functions in RNA translation. ncRNAs have been proposed to have a variety of functions but in general are mainly involved in posttranscriptional gene regulation. Among ncRNAs, miRNAs and long ncRNAs are the only species that have been studied in MDD. Thus, the potential role of all other species of small ncRNAs has yet to be explored in psychiatry. However, a species of small ncRNAs called snoRNAs is promising because these molecules have a relatively well-understood mechanism of action. In this review, we will discuss previous work and potential avenues investigating ncRNAs in MDD.

11.2 miRNA

miRNAs are single-stranded ncRNAs that are 17–22 nucleotides long and usually transcribed by RNA polymerase II [12, 13]. A primary miRNA (pri-miRNA) transcript results from miRNA gene transcription. This pri-miRNA is typically 1 kb in length and exhibits a 33–35 base-pair stem structure with a terminal loop and single-stranded RNA strands on both the 5′- and 3′-end of the stem-loop structure.

The pri-miRNA is further processed in the nucleus by Drosha, a class 2 RNase III enzyme, which cleaves the pri-miRNA releasing a 60–65-nucleotide hairpin-structured RNA called the precursor miRNA (pre-miRNA) [14]. Pre-miRNAs are then transported out of the nucleus via exportin-5 for further processing into mature miRNAs by Dicer, an endonuclease cytoplasmic RNase III enzyme. After the 20–22-nucleotide mature miRNA is released from the pre-miRNA, it is loaded into an AGO family protein complex forming an effector complex called RNA-induced silencing complex (RISC). One of the miRNA strands, called the guide strand, remains bound to the AGO protein, while the other, called the passenger strand is degraded [12, 13].

miRNAs have two key roles in the regulation of mRNA. They act by either inducing degradation or translational silencing of its mRNA target. Target recognition is primarily determined by a stretch of six nucleotides near the 5′-end of the miRNA known as the seed region [13]. miRNA primarily bind via their seed region to mRNA at the 3′UTR region [13]. Moreover, because of the short length of complementarity, a single miRNA can be predicted to target multiple gene transcripts, and a single transcript can be targeted by several miRNAs [15]. Throughout the years, miRNA have demonstrated a clear role and importance as their own class of gene-regulatory molecules, and thus dysregulation of miRNAs is believed to play a major role in disease etiology. The field of miRNA in MDD has been growing rapidly, and several studies have been conducted showing associations between miRNA and MDD or have investigated miRNA on treatment response. These investigations have focused on expression profiles of miRNA in hopes of discovering biomarkers that could predict antidepressant outcomes or to gain insight into MDD at the molecular level [16–18].

miRNAs circulate in blood and other bodily fluids and are stably transported in exosomes, double-lipid layered vesicles, which suggests that circulating miRNAs may act as molecular signals between different cells and tissues [19]. Moreover, circulating miRNA levels could change with physiological changes or according to treatment [20–22]. These characteristics provide support for miRNAs as potential biomarkers of disease, and it is tempting to speculate, in the case of MDD, that circulating miRNAs may reflect biological changes occurring in the brain as a result of the disease or antidepressant treatment effects. Studies in MDD have investigated the expression of miRNAs in postmortem brain and peripheral blood tissues; however, whether miRNAs detected in the periphery reflect the brain miRNA changes remains to be determined.

11.3 Other ncRNAs

Whereas miRNAs have been highly investigated with a number of interesting findings already reported, there have been very few studies on other species of ncRNAs. Compared to miRNAs, which have a clearly defined direct impact on mRNA, our understanding of how other ncRNA species affect gene expression is not completely understood. Below is a brief introduction of other ncRNAs that have an emerging presence in the field of MDD research.

11.3.1 snoRNAs

snoRNAs are 60–100-nucleotide ncRNAs with two distinct classes: the box C/D and box H/ACA snoRNAs. The box C/D snoRNA forms a stem-bulge-stem structure and has a conserved box C motif (RUGAUGA) located on the 5′-terminal and a conserved box D motif (CUGA) near the 3′-terminal. Box C/D snoRNAs primarily act through methylation of its target rRNA [55, 56]. The box H/ACA snoRNA forms a hairpin-hinge-hairpin-tail structure and has a conserved box H motif (ANANNA; N is any nucleotide) located on the hinge region and a conserved box ACA trinucleotide sequence on the tail region. Box H/ACA snoRNAs act primarily through pseudouridylation of its target rRNA [55, 56]. Both box C/D and box H/ACA snoRNAs form ribonucleoprotein (snoRNP) complexes by associating with partner proteins. The box motifs themselves act by guiding the binding of partner proteins, while the protein partners catalyze the specific chemical modifications. More recently, there has been evidence gathering that snoRNAs may also be involved in pre-mRNA splicing where it competes with canonical splicing machinery, blocking their activity [57, 58]. More specifically, snoRNA HBII-52 has been shown to be a brain-specific snoRNA that is involved in the regulation of alternative splicing of the serotonin receptor 2C [58].

11.3.2 lncRNAs

Long ncRNAs (lncRNAs) are molecularly very similar to mRNAs, as most lncRNAs undergo polyadenylation, receive a 5′-CAP, and often undergo splicing events [59]. lncRNAs are located both within and between protein-coding genes and range in size from single to multiple exon transcripts [60]. lncRNAs have many functions such as regulating transcription by recruitment of transcription factors, modulating mRNA processing, providing DNA scaffold for higher order complexes, aiding chromatin structure via recruitment of histone and chromatin modifiers, and acting as a miRNA sponge, among others [61–63]. Multiple independent studies have shown that lncRNAs have very distinct expression patterns between different cell and tissue types and also throughout developmental stages [64]. The specific expression patterns suggest that lncRNAs have a strong regulatory role in determining cell fate and overall development [65]. Recently, lncRNAs have gained popularity in central nervous system research since an investigation into tissue-specific expression of lncRNAs revealed that they are most abundant in the brain [65]. Moreover, within the brain, lncRNAs have been shown to have specific expression levels, in both the developing brain and adult brain, over time and across different brain regions [64].

11.4 ncRNAs in MDD

To date, the majority of published studies exploring ncRNAs in MDD have been on miRNAs, with only one study reported on lncRNAs. In addition, many of these investigations were conducted using rodent models of depression, in which direct manipulation and access to brain tissues is possible, allowing for thorough characterization and follow-up on findings. In this chapter, we will review the findings reported in animal models of depression as well as those reported in human MDD, and in suicidal behavior as MDD is closely associated with suicide and the majority of suicides occur during an episode of major depression [23]. To aid the reader in understanding the vast amount of information, the findings are grouped and presented according to the genes that the miRNA purportedly targets, all of which have been previously identified as being associated with MDD. Before proceeding further, we will first provide some important background on methodological approaches to studying ncRNAs.

11.5 Methodological Approaches

Approaches to measuring ncRNA levels range from high-throughput techniques, including RNA sequencing and microarrays, to lower-throughput and more sensitive measures, such as quantitative reverse transcriptase PCR (qRT-PCR) [24]. High-throughput techniques are commonly used for initial discovery of differentially expressed ncRNAs, e.g., between MDD individuals compared to psychiatrically normal control group. ncRNAs that are found to be significantly differentially expressed require technical validation using more specific quantitative techniques (i.e., qRT-PCR or targeted sequencing) to increase confidence and replicate the differential expression of the candidate ncRNAs. Further follow-up experiments typically include the identification of molecular mechanisms, which is tailored to the specific ncRNA species that are being profiled since each has a specific function.

For miRNA characterization, in silico target analysis is usually the first step, where potential targets for the candidate miRNA are determined via complementation between the miRNA seed region and the 3′UTR region of an mRNA [25]. After in silico target prediction, follow-up in vitro experiments are required to validate miRNA-mRNA interactions, which are commonly demonstrated using luciferase assays. miRNA knockdown assays are also performed followed by gene expression profiling of predicted targets, which together provide further evidence for a relationship between the miRNA and its target gene, since miRNA reduce target mRNA levels [26].

11.6 Findings in miRNAs Targeting Monoamines

11.6.1 Serotonin

Serotonin (5HT) is produced by neurons in the midbrain raphe nuclei (RN), which projects extensively throughout the brain. Most currently available antidepressants present direct or indirect serotonergic effects. miR-16 has been predicted to target the 3′UTR of serotonin transporter (SERT) transcripts and has been investigated in the regulation of serotonin [27, 28]. Using 1C11 neuroectodermal cells, which have the potential to differentiate into serotonergic or noradrenergic neurons, Baudry et al. [27] showed that miR-16 levels were elevated in cells that followed the noradrenergic pathway, whereas miR-16 levels remained steady in cells that followed the serotonergic pathway. Furthermore, the expression profile of miR-16 was lower in mice RN when compared to the noradrenergic neuron-rich region, locus coeruleus [27]. This gives evidence that miR-16 may play a role in silencing SERT transcripts during noradrenergic differentiation. In addition, miR-16 knockdown restored SERT protein expression in 1C11 noradrenergic cells and had no effect on noradrenergic-associated function. miR-16 knockdown also induced noradrenergic cells to exhibit serotonergic metabolism such as synthesizing, storing, and degrading 5-HT [27]. Fluoxetine infusion into mice RN induced a 2.5-fold increase in miR-16 levels, suggesting that SSRIs may be regulating SERT expression through miR-16 in the RN. Direct infusion of fluoxetine into the locus coeruleus failed to induce any change in miR-16 expression; however, infusion of fluoxetine into the RN resulted in a 30% reduction of miR-16 in the locus coeruleus. This observed downregulation of miR-16 in the locus coeruleus was subsequently accompanied by increased SERT expression and tryptophan hydroxylase (rate-limiting enzyme in 5-HT synthesis) activity [27]. This effect may be a result of the bilateral connection between the RN and the locus coeruleus. This study shows that miR-16 can have distal effects on different brain regions where serotonergic neurons project, can be targeted by SSRIs to regulate SERT expression, and can mediate an adaptive response to serotonergic neuron responses.

miR-135 was studied in relation to serotonergic neuron activity and was also shown to regulate SERT and Htr1a (5HT 1A auto receptor) in the RN of mice as confirmed by a luciferase assay showing that miR-135 targeting SERT 3′UTR and Htr1a 3′UTR resulted in a 30–40% reduction of translation of these transcripts [29]. Further analysis revealed the existence of three highly conserved variants of miR-135: miR-135a-1, miR-135a-2, and miR-135b. Chronic social defeat stress in mice did not alter miR-135a levels in the RN; however, imipramine (non-specific 5HT reuptake inhibitor) administration chronically or acutely, both in stressed and non-stressed mice, significantly increased miR-135a expression [29]. Moreover, administration of fluoxetine (SSRI) resulted in elevated miR-135a levels. However, administration of reboxetine (norepinephrine reuptake inhibitor) had no effect on miR-135a expression. miR-135a was also shown to be significantly lower in human RN of suicide completers compared to controls, further giving support for a specific role of miR-135a in the serotonin system [29]. miR-135a overexpression specifically in 5HT neurons was found to reduce anxiety-like and MDD-like behaviors in mice, while miR-135a knockdown in the RN increased anxiety-like behaviors and decreased the response to antidepressant treatment [29]. In the same study, miR-135a was measured in humans and was found to be decreased in peripheral blood in depressed patients compared to controls. Interestingly, miR-135a blood levels showed a statistically significant increase after cognitive behavioral therapy but not after SSRI antidepressant treatment. These results demonstrate that miR-135a is responsive to antidepressants, has a regulatory role in serotonergic neurons, and could also act as a biomarker for a depressive state and antidepressant response.

11.6.2 Dopamine

The mesolimbic dopamine system plays an important role in regulation of motivation, reward, attention, and mood. Dopamine (DA) is the most abundant catecholaminergic neurotransmitter in the mammalian brain and is involved in incentive-motivational pathways as well as in response to aversive stimuli [30]. Lowered DA neurotransmission has been shown to play a role in MDD and may be a result of diminished DA release from presynaptic neurons, changes in receptor number/function, and intracellular signal processing dysfunction [31]. There has been some level of evidence suggesting that an upregulation of dopamine receptor D2 (DRD2) may be associated with MDD [31].

In a rat model of depression, induced either by maternal deprivation (MD) and/or chronic unpredictable stress (CUS), increased binding of DRD2 in the nucleus accumbens (NAc) and striatum was reported as well as increased striatal DRD2 mRNA levels in adult rats that were subjected to MD and increased accumbal DRD2 protein levels in adult rats subjected to CUS [32]. This increased expression of DRD2 may be the result of dysregulated miRNA expression in the NAc and striatum. Accordingly, miR-9 was found to be downregulated in NAc of CUS rats and in striatum of MD rats, which suggests that miR-9 in the NAc is disrupted by stress in adulthood (modeled by CUS rats), whereas miR-9 in the striatum is disrupted by early-life adversity (modeled by MD rats) [32]. Furthermore, when rats were exposed to MD as well as CUS, miR-9 was more significantly downregulated in both the NAc and striatum. Treatment with the antidepressant, escitalopram (SSRI), did not reverse the downregulation of miR-9 expression indicating that miR-9 may be specific to stress susceptibility and not escitalopram response [32]. To show that miR-9 directly targets and interacts with DRD2, miR-9 mimics were transfected into 5H-SY5Y cells, which resulted in significantly reduced DRD2 protein expression. Additionally, a luciferase assay with DRD2 3′UTR reporter and miR-9 mimics significantly reduced luciferase reporter (DRD2 3′UTR) activity [32]. miR-9 was also found to be negatively correlated with DRD2 protein levels [32]. This gives evidence that miR-9 downregulation plays a role in the disrupted expression of DRD2 in the NAc and striatum.

The same study also reported miR-326 as being upregulated in NAc and downregulated in the striatum of CUS rats and upregulated in the striatum of MD rats [32]. miR-326 expression was found to be positively correlated with DRD2 protein levels in the NAc and the striatum; however, it is not correlated with DRD2 mRNA levels. Interestingly, the disrupted expression profile of miR-326 was reversed after 4 weeks of escitalopram treatment, suggesting that miR-326 activity is not a direct regulation of DRD2 gene expression but may be involved in escitalopram response [32].

11.7 Findings in miRNAs Targeting the Polyaminergic Systems

The polyamine stress response (PSA) has been an area of interest in mood disorders, anxiety disorders, and suicide [33, 34]. Polyamines are ubiquitous aliphatic molecules with two or more amine groups and are primarily comprised of spermine, spermidine, putrescine, and agmatine [34, 35]. Polyamines have several functions including regulation of gene transcription, posttranscriptional modifications, and regulating the expression or release of neurotransmitters including catecholamines, glutamate, GABA, and nitric oxide [35]. Furthermore, polyamine levels are tightly regulated by control of their biosynthesis, transport, and catabolism [35, 36]. Polyamine biosynthesis has several rate-limiting enzymes including spermidine/spermine N1-acetyltransferase (SAT1), spermine oxidase (SMOX), ornithine decarboxylase (ODC), and S-adenosylmethionine (SAM), among others [36]. The PSA is activated after exposure to stressful stimuli resulting in elevated levels of putrescine and agmatine in the brain as well as in peripheral tissue [34, 35, 37]. Moreover, the elevated levels of putrescine and agmatine in the brain have been shown to be beneficial and play a role in the adaptive PSA [38]. Thus the reported significant downregulation of rate-limiting enzymes in polyamine catabolism, SAT1 and SMOX, in suicide completers may be involved in the dysregulation of the PSA, affecting its adaptability [36, 39–42]. Furthermore, this downregulation is thought to be due in part by miRNA-induced gene suppression. miR-124a, miR-34a, miR-34c-5p, miR-497, and miR-873 have been predicted to target both SAT1 and SMOX [43]. The co-targeting of SAT1 and SMOX suggests that these miRNAs interact in a network to regulate SAT1 and SMOX in the PSA. Furthermore, expression analysis of the five candidate miRNAs showed a statistically significant upregulation in BA44 of suicide completers. A correlational analysis between the expression profile of the candidate miRNAs with the expression profile of SAT1 and SMOX showed a statistically significant negative correlation between SAT1 and miR-34c-5p and miR-320c, and a statistically significant negative correlation was observed between SMOX and miR-139-5p and miR-320c [43]. These findings further support that miRNAs may be playing a role in the regulation of genes involved in the PSA, and dysregulation of miRNAs in the case of MDD may lead to downstream effects in gene expression and ultimately affect the adaptability of the PSA.

11.8 Findings in miRNAs Targeting the Glutamatergic System

miR-1202 is primate specific and enriched in brains and targets glutamate metabotropic receptor 4 (GRM4) [44]. GRM4 is a group III glutamate receptor and has been implicated in the regulation of anxiety-related behaviors [45, 46]. Moreover, in cellular models, miR-1202 mediate the antidepressant response, as chronic treatment of citalopram (SSRI) or imipramine (TCA) on neural progenitor cells (NPCs) results in an upregulation of miR-1202 and a decrease in expression of GRM4 mRNA [44]. This effect seems to be specific to antidepressant drugs, since the same effect could not be observed with drugs from other classes of medications commonly used in psychiatric treatment. In addition, in depressed patients, those who responded to antidepressant treatment showed a decrease in peripheral blood levels of miR-1202 at baseline compared to nonresponders and controls. Furthermore, after 8 weeks of antidepressant treatment, the responder group showed an upregulation of miR-1202 [44]. miR-335 is another miRNA that also targets GRM4, and similar results to those reported for miR-1202 were observed after treatment with the antidepressant citalopram; miR-335 expression was upregulated, and GRM4 expression was decreased in peripheral blood [47]. These studies give evidence that SSRIs may regulate the altered miRNA expression in depressed patients, which elicit downstream effects on gene expression providing an adaptive response to the effects of MDD.

11.9 Global miRNA Changes in MDD

A study by Smalheiser and colleagues, which was one of the first studies to profile miRNA in depression, examined genome-wide miRNA expression in the prefrontal cortex (PFC) of postmortem human brains and found a global downregulation of miRNAs of approximately 17% in depressed subjects compared to controls [48]. Twenty-one miRNAs showed significant downregulation, and 16 miRNAs did not reach statistical significance but were downregulated by approximately 30%. Moreover, the majority of these 16 miRNAs shared the same seed sequences and are encoded at a chromosomal locus near other miRNAs that reported statistical significances [48]. Target predictions performed for the 37 miRNAs revealed several mRNA targets. Furthermore, predicted mRNA targets showed overlap between miRNAs. Vascular endothelial growth factor (VEGF), anti-apoptotic protein B-cell/lymphoma 2 (BCL2), and DNA methyltransferase 3B (DNMT3B) are three validated genes that were previously implicated in MDD and were further explored in this study. miR-20b, miR-20a, miR-34a, and miR-34b are predicted to target VEGF, a growth factor associated with depression, and has been shown to be significantly elevated in peripheral blood of humans with MDD compared to controls [49]. miR-34a is predicted to target BCL2, whose protein levels were found to be downregulated in depressed subjects. miR-148b is predicted to target DNMT3B, whose protein levels were upregulated in depressed subjects. BCL2 and DNMT3B both showed positive correlations with their predicted miRNA counterparts, miR-34a and miR-148b, respectively. If miR-34a and miR-148b acted by suppressing mRNA transcripts, then a negative correlation would have been expected; however, the observed positive correlation suggests that miR-34a and miR-148b may be co-regulated with their mRNA targets [49, 50]. An additional miRNome study found miR-508-3p, a primate-specific miRNA, and miR-152-3p to be downregulated in postmortem PFC tissue from MDD subjects [51].

The miRNome has also been profiled in peripheral blood collected from subjects with MDD. Using a microarray assay, seven miRNAs were found to show differential expression in MDD patients when compared to controls: four miRNAs (let-7a-5p, let-7d-5p, let-7f-5p, and miR-1915-3p) were significantly downregulated in MDD subjects, and three miRNAs (miR-199a-5p, miR-24-3p, and miR-425-3p) were significantly upregulated in MDD subjects [52]. let-7a-5p, let-7d-5p, let-7f-5-p, miR-24-3p, and miR-425-3p were validated using qRT-PCR, which confirmed the results observed from the microarray. Interestingly, miR-425-3p was also found to be significantly upregulated in peripheral blood of MDD subjects in a previous external study [52, 53]. Additionally, in a previous study by the same group, let-7d-5p and let-7f-5p showed increased levels in peripheral blood after chronic treatment of escitalopram (SSRI) [52, 54]. This provides evidence that let-7d-5p and let-7f-5p not only plays a role in MDD pathogenesis but may also be involved in SSRI antidepressant effects. Additional studies have also profiled miRNA expression in peripheral blood using microarrays and found nine upregulated miRNAs and one downregulated miRNA [16]. Validation using qRT-PCR in a separate MDD group validated miR-26b, miR-4485, miR-1972, miR-4498, and miRNA-4743 as they were found to be significantly upregulated in MDD subjects compared to controls [16]. As the reported studies used microarrays with defined miRNA probes, it is quite possible that more recently described miRNAs may not have been examined. Thus, future studies exploring miRNomes should consider using RNA sequencing to increase coverage. With that said, exploring global miRNA expression profiles in the brain and peripheral blood would be a good option for hypothesis-free discovery studies investigating miRNAs associated with MDD.

11.10 Future Directions

ncRNAs serve essential and diverse roles in cell function and survival. Thus, in recent years, there has been much interest in the investigation of the role of ncRNAs in disease etiology, including MDD. Due to the relative limitations in both knowledge of molecular function of most types of ncRNAs, and availability of good assays, a large majority of ncRNA studies have focused on miRNA. More recently we are beginning to see studies that are profiling other ncRNAs as we gain more insight on the function of ncRNAs other than miRNAs and as technology progresses and new assays are created. lncRNAs have recently started to gain traction in the field of psychiatry [66], and there has been one notable study focusing on lncRNAs in MDD [67]. This study profiled the expression of 34,834 lncRNAs and 39,224 mRNAs in peripheral blood from depressed patients and healthy controls and found 2007 lncRNAs significantly differentially expressed. Of these, the majority were upregulated in MDD subjects. The study also looked at the expression profiles of mRNA and found 157 mRNAs that had a 1.5-fold change and were significantly upregulated in MDD subjects. A co-expression analysis of the differentially expressed lncRNAs and mRNAs was performed to obtain insight into their possible interactions. In order to identify lncRNA networks specific to MDD, a subset of differentially expressed mRNA that were previously identified to be associated with the etiology of MDD were used for co-expression analysis. The lncRNAs chr10:874695-874794, chr10:75873456-75873642, and chr3:47048304-47048512 were found to be associated with four MDD related genes, while no association was observed in control subjects. This is the first and currently the only study that has explored lncRNAs in MDD. Clearly more studies need to be conducted investigating lncRNAs in MDD before mechanism can be identified in association with MDD.

Another promising species of ncRNA are snoRNAs. As mentioned above, snoRNAs have been shown to have direct effects on posttranscriptional regulation. Profiling species of ncRNAs other than miRNAs has proven to be difficult since clear mechanistic effects are still elusive and methodological techniques are still being developed.

It is clear that ncRNAs play a major role in the regulation of gene expression. Although much advancement has been made in the field of MDD, the underlying molecular mechanism still remains unclear. Investigating ncRNAs has shown to be a promising avenue to pursue in MDD research and has led to further insight into the molecular mechanisms underlying MDD. However, the majority of studies on ncRNAs have focused solely on miRNAs, and it is likely that different species of ncRNAs may be interacting within a large network. Furthermore, it is unlikely that a single ncRNA can solely explain the dysregulated gene expression associated with MDD, and it is quite probable that several molecular mechanisms are at play. Continuing to further our understanding of the functions and mechanisms of different species of ncRNAs will help to develop future ncRNA investigations in MDD. As more findings on ncRNAs in MDD continued to be generated, we will be able to make better conclusions on how the vast majority of ncRNAs may be working together in a meta-network.

References

1. World Health Organization. Depression 2016. http://www.who.int/mediacentre/factsheets/fs369/en/.
2. American Psychiatric Association. Diagnostic and statistical manual of mental disorders. 5th ed. Washington, DC: APA; 2013.
3. Sullivan PF, de Geus EJ, Willemsen G, James MR, Smit JH, Zandbelt T, et al. Genome-wide association for major depressive disorder: a possible role for the presynaptic protein piccolo. Mol Psychiatry. 2009;14(4):359–75.
4. Sullivan PF, Neale MC, Kendler KS. Genetic epidemiology of major depression: review and meta-analysis. Am J Psychiatry. 2000;157(10):1552–62.
5. Lewis CM, Ng MY, Butler AW, Cohen-Woods S, Uher R, Pirlo K, et al. Genome-wide association study of major recurrent depression in the U.K. population. Am J Psychiatry. 2010;167(8):949–57.
6. Muglia P, Tozzi F, Galwey NW, Francks C, Upmanyu R, Kong XQ, et al. Genome-wide association study of recurrent major depressive disorder in two European case-control cohorts. Mol Psychiatry. 2010;15(6):589–601.
7. Rietschel M, Mattheisen M, Frank J, Treutlein J, Degenhardt F, Breuer R, et al. Genome-wide association-, replication-, and neuroimaging study implicates HOMER1 in the etiology of major depression. Biol Psychiatry. 2010;68(6):578–85.
8. Shi J, Potash JB, Knowles JA, Weissman MM, Coryell W, Scheftner WA, et al. Genome-wide association study of recurrent early-onset major depressive disorder. Mol Psychiatry. 2011;16(2):193–201.
9. Fernandez-Pujals AM, Adams MJ, Thomson P, McKechanie AG, Blackwood DH, Smith BH, et al. Epidemiology and heritability of major depressive disorder, stratified by age of onset, sex, and illness course in Generation Scotland: Scottish Family Health Study (GS:SFHS). PLoS One. 2015;10(11):e0142197.
10. Clark MB, Amaral PP, Schlesinger FJ, Dinger ME, Taft RJ, Rinn JL, et al. The reality of pervasive transcription. PLoS Biol. 2011;9(7):e1000625. discussion e1102

11. Pauli A, Rinn JL, Schier AF. Non-coding RNAs as regulators of embryogenesis. Nat Rev Genet. 2011;12(2):136–49.
12. Ha M, Kim VN. Regulation of microRNA biogenesis. Nat Rev Mol Cell Biol. 2014;15(8):509–24.
13. Wahid F, Shehzad A, Khan T, Kim YY. MicroRNAs: synthesis, mechanism, function, and recent clinical trials. Biochim Biophys Acta. 2010;1803(11):1231–43.
14. Winter J, Jung S, Keller S, Gregory RI, Diederichs S. Many roads to maturity: microRNA biogenesis pathways and their regulation. Nat Cell Biol. 2009;11(3):228–34.
15. Hausser J, Zavolan M. Identification and consequences of miRNA-target interactions—beyond repression of gene expression. Nat Rev Genet. 2014;15(9):599–612.
16. Fan HM, Sun XY, Guo W, Zhong AF, Niu W, Zhao L, et al. Differential expression of microRNA in peripheral blood mononuclear cells as specific biomarker for major depressive disorder patients. J Psychiatr Res. 2014;59:45–52.
17. Wan Y, Liu Y, Wang X, Wu J, Liu K, Zhou J, et al. Identification of differential microRNAs in cerebrospinal fluid and serum of patients with major depressive disorder. PLoS One. 2015;10(3):e0121975.
18. Serafini G, Pompili M, Hansen KF, Obrietan K, Dwivedi Y, Shomron N, et al. The involvement of microRNAs in major depression, suicidal behavior, and related disorders: a focus on miR-185 and miR-491-3p. Cell Mol Neurobiol. 2014;34(1):17–30.
19. Valadi H, Ekstrom K, Bossios A, Sjostrand M, Lee JJ, Lotvall JO. Exosome-mediated transfer of mRNAs and microRNAs is a novel mechanism of genetic exchange between cells. Nat Cell Biol. 2007;9(6):654–9.
20. Chen X, Ba Y, Ma L, Cai X, Yin Y, Wang K, et al. Characterization of microRNAs in serum: a novel class of biomarkers for diagnosis of cancer and other diseases. Cell Res. 2008;18(10):997–1006.
21. Gilad S, Meiri E, Yogev Y, Benjamin S, Lebanony D, Yerushalmi N, et al. Serum microRNAs are promising novel biomarkers. PLoS One. 2008;3(9):e3148.
22. Mitchell PS, Parkin RK, Kroh EM, Fritz BR, Wyman SK, Pogosova-Agadjanyan EL, et al. Circulating microRNAs as stable blood-based markers for cancer detection. Proc Natl Acad Sci U S A. 2008;105(30):10513–8.
23. Hardt J, Bernert S, Matschinger H, Angermeier MC, Vilagut G, Bruffaerts R, et al. Suicidality and its relationship with depression, alcohol disorders and childhood experiences of violence: results from the ESEMeD study. J Affect Disord. 2015;175:168–74.
24. Lopez JP, Diallo A, Cruceanu C, Fiori LM, Laboissiere S, Guillet I, et al. Biomarker discovery: quantification of microRNAs and other small non-coding RNAs using next generation sequencing. BMC Med Genomics. 2015;8:35.
25. Witkos TM, Koscianska E, Krzyzosiak WJ. Practical aspects of microRNA target prediction. Curr Mol Med. 2011;11(2):93–109.
26. Guo H, Ingolia NT, Weissman JS, Bartel DP. Mammalian microRNAs predominantly act to decrease target mRNA levels. Nature. 2010;466(7308):835–40.
27. Baudry A, Mouillet-Richard S, Schneider B, Launay JM, Kellermann O. miR-16 targets the serotonin transporter: a new facet for adaptive responses to antidepressants. Science. 2010;329(5998):1537–41.
28. Song MF, Dong JZ, Wang YW, He J, Ju X, Zhang L, et al. CSF miR-16 is decreased in major depression patients and its neutralization in rats induces depression-like behaviors via a serotonin transmitter system. J Affect Disord. 2015;178:25–31.
29. Issler O, Haramati S, Paul ED, Maeno H, Navon I, Zwang R, et al. MicroRNA 135 is essential for chronic stress resiliency, antidepressant efficacy, and intact serotonergic activity. Neuron. 2014;83(2):344–60.
30. He M, Yan H, Duan ZX, Qu W, Gong HY, Fan ZL, et al. Genetic distribution and association analysis of DRD2 gene polymorphisms with major depressive disorder in the Chinese Han population. Int J Clin Exp Pathol. 2013;6(6):1142–9.
31. Dunlop BW, Nemeroff CB. The role of dopamine in the pathophysiology of depression. Arch Gen Psychiatry. 2007;64(3):327–37.

32. Zhang Y, Wang Y, Wang L, Bai M, Zhang X, Zhu X. Dopamine receptor D2 and associated microRNAs are involved in stress susceptibility and resistance to escitalopram treatment. Int J Neuropsychopharmacol. 2015;(8):18.
33. Gross JA, Turecki G. Suicide and the polyamine system. CNS Neurol Disord Drug Targets. 2013;12(7):980–8.
34. Fiori LM, Turecki G. Implication of the polyamine system in mental disorders. J Psychiatry Neurosci. 2008;33(2):102–10.
35. Turecki G. The molecular bases of the suicidal brain. Nat Rev Neurosci. 2014;15(12):802–16.
36. Fiori LM, Wanner B, Jomphe V, Croteau J, Vitaro F, Tremblay RE, et al. Association of polyaminergic loci with anxiety, mood disorders, and attempted suicide. PLoS One. 2010;5(11):e15146.
37. Limon A, Mamdani F, Hjelm BE, Vawter MP, Sequeira A. Targets of polyamine dysregulation in major depression and suicide: Activity-dependent feedback, excitability, and neurotransmission. Neurosci Biobehav Rev. 2016;66:80–91.
38. Piletz JE, Aricioglu F, Cheng JT, Fairbanks CA, Gilad VH, Haenisch B, et al. Agmatine: clinical applications after 100 years in translation. Drug Discov Today. 2013;18(17-18):880–93.
39. Sequeira A, Gwadry FG, Ffrench-Mullen JM, Canetti L, Gingras Y, Casero Jr RA, et al. Implication of SSAT by gene expression and genetic variation in suicide and major depression. Arch Gen Psychiatry. 2006;63(1):35–48.
40. Fiori LM, Bureau A, Labbe A, Croteau J, Noel S, Merette C, et al. Global gene expression profiling of the polyamine system in suicide completers. Int J Neuropsychopharmacol. 2011;14(5):595–605.
41. Klempan TA, Rujescu D, Merette C, Himmelman C, Sequeira A, Canetti L, et al. Profiling brain expression of the spermidine/spermine N1-acetyltransferase 1 (SAT1) gene in suicide. Am J Med Genet B Neuropsychiatr Genet. 2009;150B(7):934–43.
42. Chen GG, Fiori LM, Moquin L, Gratton A, Mamer O, Mechawar N, et al. Evidence of altered polyamine concentrations in cerebral cortex of suicide completers. Neuropsychopharmacology. 2010;35(7):1477–84.
43. Lopez JP, Fiori LM, Gross JA, Labonte B, Yerko V, Mechawar N, et al. Regulatory role of miRNAs in polyamine gene expression in the prefrontal cortex of depressed suicide completers. Int J Neuropsychopharmacol. 2014;17(1):23–32.
44. Lopez JP, Lim R, Cruceanu C, Crapper L, Fasano C, Labonte B, et al. miR-1202 is a primate-specific and brain-enriched microRNA involved in major depression and antidepressant treatment. Nat Med. 2014;20(7):764–8.
45. Pilc A, Chaki S, Nowak G, Witkin JM. Mood disorders: regulation by metabotropic glutamate receptors. Biochem Pharmacol. 2008;75(5):997–1006.
46. Davis MJ, Iancu OD, Acher FC, Stewart BM, Eiwaz MA, Duvoisin RM, et al. Role of mGluR4 in acquisition of fear learning and memory. Neuropharmacology. 2013;66:365–72.
47. Li J, Meng H, Cao W, Qiu T. MiR-335 is involved in major depression disorder and antidepressant treatment through targeting GRM4. Neurosci Lett. 2015;606:167–72.
48. Smalheiser NR, Lugli G, Rizavi HS, Torvik VI, Turecki G, Dwivedi Y. MicroRNA expression is down-regulated and reorganized in prefrontal cortex of depressed suicide subjects. PLoS One. 2012;7(3):e33201.
49. Tseng PT, Cheng YS, Chen YW, Wu CK, Lin PY. Increased levels of vascular endothelial growth factor in patients with major depressive disorder: A meta-analysis. Eur Neuropsychopharmacol. 2015;25(10):1622–30.
50. Dwivedi Y. Pathogenetic and therapeutic applications of microRNAs in major depressive disorder. Prog Neuropsychopharmacol Biol Psychiatry. 2016;64:341–8.
51. Smalheiser NR, Lugli G, Zhang H, Rizavi H, Cook EH, Dwivedi Y. Expression of microRNAs and other small RNAs in prefrontal cortex in schizophrenia, bipolar disorder and depressed subjects. PLoS One. 2014;9(1):e86469.
52. Maffioletti E, Cattaneo A, Rosso G, Maina G, Maj C, Gennarelli M, et al. Peripheral whole blood microRNA alterations in major depression and bipolar disorder. J Affect Disord. 2016;200:250–8.

53. Belzeaux R, Bergon A, Jeanjean V, Loriod B, Formisano-Treziny C, Verrier L, et al. Responder and nonresponder patients exhibit different peripheral transcriptional signatures during major depressive episode. Transl Psychiatry. 2012;2:e185.
54. Bocchio-Chiavetto L, Maffioletti E, Bettinsoli P, Giovannini C, Bignotti S, Tardito D, et al. Blood microRNA changes in depressed patients during antidepressant treatment. Eur Neuropsychopharmacol. 2013;23(7):602–11.
55. Bratkovic T, Rogelj B. Biology and applications of small nucleolar RNAs. Cell Mol Life Sci. 2011;68(23):3843–51.
56. Dupuis-Sandoval F, Poirier M, Scott MS. The emerging landscape of small nucleolar RNAs in cell biology. Wiley Interdiscip Rev RNA. 2015;6(4):381–97.
57. Falaleeva M, Pages A, Matuszek Z, Hidmi S, Agranat-Tamir L, Korotkov K, et al. Dual function of C/D box small nucleolar RNAs in rRNA modification and alternative pre-mRNA splicing. Proc Natl Acad Sci U S A. 2016;113(12):E1625–34.
58. Kishore S, Stamm S. The snoRNA HBII-52 regulates alternative splicing of the serotonin receptor 2C. Science. 2006;311(5758):230–2.
59. Ulitsky I, Bartel DP. lincRNAs: genomics, evolution, and mechanisms. Cell. 2013;154(1):26–46.
60. Briggs JA, Wolvetang EJ, Mattick JS, Rinn JL, Barry G. Mechanisms of long non-coding RNAs in mammalian nervous system development, plasticity, disease, and evolution. Neuron. 2015;88(5):861–77.
61. Salmena L, Poliseno L, Tay Y, Kats L, Pandolfi PP. A ceRNA hypothesis: the Rosetta Stone of a hidden RNA language? Cell. 2011;146(3):353–8.
62. Wang KC, Chang HY. Molecular mechanisms of long noncoding RNAs. Mol Cell. 2011;43(6):904–14.
63. Geisler S, Coller J. RNA in unexpected places: long non-coding RNA functions in diverse cellular contexts. Nat Rev Mol Cell Biol. 2013;14(11):699–712.
64. Aprea J, Calegari F. Long non-coding RNAs in corticogenesis: deciphering the non-coding code of the brain. EMBO J. 2015;34(23):2865–84.
65. Washietl S, Kellis M, Garber M. Evolutionary dynamics and tissue specificity of human long noncoding RNAs in six mammals. Genome Res. 2014;24(4):616–28.
66. Huang X, Luo YL, Mao YS, Ji JL. The link between long noncoding RNAs and depression. Prog Neuropsychopharmacol Biol Psychiatry. 2016.
67. Liu Z, Li X, Sun N, Xu Y, Meng Y, Yang C, et al. Microarray profiling and co-expression network analysis of circulating lncRNAs and mRNAs associated with major depressive disorder. PLoS One. 2014;9(3):e93388.

DNA Methylation in Schizophrenia

12

Lotta-Katrin Pries, Sinan Gülöksüz, and Gunter Kenis

Abstract

Schizophrenia is a highly heritable psychiatric condition that displays a complex phenotype. A multitude of genetic susceptibility loci have now been identified, but these fail to explain the high heritability estimates of schizophrenia. In addition, epidemiologically relevant environmental risk factors for schizophrenia may lead to permanent changes in brain function. In conjunction with genetic liability, these environmental risk factors—likely through epigenetic mechanisms—may give rise to schizophrenia, a clinical syndrome characterized by florid psychotic symptoms and moderate to severe cognitive impairment. These pathophysiological features point to the involvement of epigenetic processes. Recently, a wave of studies examining aberrant DNA modifications in schizophrenia was published. This chapter aims to comprehensively review the current findings, from both candidate gene studies and genome-wide approaches, on DNA methylation changes in schizophrenia.

Keywords

DNA methylation • Epigenetics • Schizophrenia • Psychosis

L.-K. Pries, M.Sc. • G. Kenis, Ph.D (✉)
Department of Psychiatry & Neuropsychology, School for Mental Health and Neuroscience, Maastricht University, Maastricht, The Netherlands
e-mail: lotta.pries@maastrichtuniversity.nl; g.kenis@maastrichtuniversity.nl

S. Gülöksüz, M.D., Ph.D
Department of Psychiatry & Neuropsychology, School for Mental Health and Neuroscience, Maastricht University, Maastricht, The Netherlands

Department of Psychiatry, Yale School of Medicine, New Haven, CT, USA
e-mail: sinan.guloksuz@maastrichtuniversity.nl; sinan.guloksuz@yale.edu

R. Delgado-Morales (ed.), *Neuroepigenomics in Aging and Disease*, Advances in Experimental Medicine and Biology, DOI 10.1007/978-3-319-53889-1_12

12.1 Introduction

After decades of research, our knowledge on the pathoetiology of schizophrenia—a heterogeneous phenotype characterized by reality distortion (e.g., hallucinations and delusions), affective dysregulation (e.g., negative symptoms and depression), and cognitive impairment—remains limited [1]. The complexity of schizophrenia not only arises from the heterogeneity of the phenotype but also from its multifactorial nature: genetic, developmental, and environmental [2]. However, we might be on the verge of reaching a tipping point where the pieces finally fall into place. Based on early clinical observations and findings from classical twin studies indicating a high heritability in schizophrenia, the field of psychiatry has heavily invested in genetics research to identify the underlying genetic background. While candidate genetic association studies mostly yielded inconsistent results, the recent surge of consecutive genome-wide association (GWA) studies led to the replicable identification of 108 genomic loci [3]. Further, a recent study converging neurobiology and genomics uncovered the biological significance of several loci, which might pave the way for a better understanding of the neurobiological mechanisms in schizophrenia, such as synaptic pruning [4, 5]. Findings from GWA studies suggest that the genetic background of schizophrenia is composed of a multitude of common single nucleotide variants (SNVs), each contributing to the overall liability with small effect sizes. Nevertheless, heritability estimates from GWA studies fall short of those predicted by classical twin studies [6], suggesting that other forms of genetic variation might be implicated. Recent evidence indeed shows a high burden of both rare SNVs [7] and rare copy number variants in patients with schizophrenia [8]. However, even the combination of common and rare variants only explains a fraction of the total heritability estimate.

Although a full picture of schizophrenia's genetic architecture has yet to be established, a long-held view states that common genetic variants make individuals vulnerable to environmental risks [9]. Indeed, a wide range of environmental factors has been implicated in the development of schizophrenia, including maternal stress and infection during prenatal development, perinatal events (i.e., obstetric complications), childhood trauma, minority status, growing up in an urban environment, and cannabis use during early adolescence. Exposure to these environmental factors impacts on critical developmental periods of relevant cognitive, affective, and social capabilities of which the timing correlates with specific maturational processes of the brain [10]: insults during fetal development influence overall neuronal organization, while early life and adolescent adverse experiences interfere with axonal growth, synaptogenesis, and synaptic pruning [11].

As environmental insults can induce enduring changes in gene expression through epigenetic mechanisms, it is suggested that exposure to adversity alters the meticulous regulation of gene expression networks involved in neuronal development and synaptic configuration, particularly when these neurodevelopmental processes are most active [12, 75]. Interestingly, pathway and gene ontology analyses, of both GWA and rare variant studies, also signal the involvement of genes related to neurodevelopment and synaptic (re)modeling processes in schizophrenia

[3, 7, 13]. Therefore, both genetic and environmental factors interfere with the processes that shape intracortical and cortical-subcortical communication networks, creating an at-risk mental state through an aberrant development of neurocognitive functions required to appropriately interpret and respond to the social world [11, 14], which are impaired in schizophrenia [15]. In addition, genes and environment may work in an interactive manner where the sensitivity of brain maturation processes to adversity is determined by genetic factors.

Overall, evidence suggests that epigenetic factors may play a role in schizophrenia. However, identifying these epigenetic factors is challenging, given the temporal and tissue-specific nature of epigenetic processes. Nevertheless, recent studies report differences in epigenetic modifications between patients with schizophrenia and control subjects, in both peripheral blood and postmortem brain tissue. Here, we present an overview of studies examining DNA methylation status in candidate genes, as well as those employing a genome-wide approach.

12.2 Candidate Genes

A priori hypotheses and analyses of candidate gene methylation profiles related to schizophrenia have focused mainly on genes associated with specific neurotransmitter-related systems: gamma-aminobutyric acid (GABA), glutamate, serotonin, and dopamine [16, 17]. An overview of these studies is presented in Table 12.1, and the most important analyzed genes are summarized below.

12.2.1 GABAergic System

GABA, the mammalian inhibitory neurotransmitter, has been linked to the schizophrenia phenotype and related cognitive impairments [18, 19]. In several studies, schizophrenia was associated with decreased GABA levels and reduced glutamic acid decarboxylase 67 (GAD67, the enzyme that converts glutamate to GABA) expression in multiple brain regions (e.g., the thalamus, ventral striatum, or medial temporal lobe [20–22]. However, to understand the pathology-related malfunction of the GABAergic system, researchers recently started focusing on methylation phenotypes. In the first methylome-wide analysis of major psychosis, Mill et al. [23] found a link between schizophrenia and methylation signatures of GABA-related genes, such as MARLIN-1, KCNJ6, and HELT. Other studies predominantly focused on GAD1 and reelin (RELN) as the most important GABAergic candidate genes. Augmented methylation levels of the RELN gene are associated with hypo-expression of the RELN protein, which mediates neuronal migration during development [24, 25]. Increased methylation levels of RELN promoter regions in different brain areas of patients with schizophrenia were found in several studies [26, 27], but were not confirmed by other reports [28]. GAD1 was shown to be downregulated in different brain regions of patients with schizophrenia [29]. In prefrontal cortex samples, the GAD1 promoter was differentially methylated in

Table 12.1 Summary of candidate gene studies

Neurotransmitter-related systems	Candidate gene(s)	First author	Tissue	Method	Sample	Significant methylation	Expression—methylation relation	Methylation associated with covariates
Dopaminergic	DRD2	Zhang et al. (2007) [40]	Whole blood	BS	48 discordant MZ for SZ (drug naïve)	No difference	—	No association with sex, age on admission, or age of onset of illness
	DRD1, DRD2, DRD4, and DRD5 promoter	Kordi-Tamandani et al. (2013) [38]	Whole blood	MSP	80 SZ 71 C	↓ *DRD4*, *DRD5*, and *DRD2*	Inversely DRD4, DRD5, and DRD2	—
	DRD2 (upstream exon 1: CpG positions 1–7)	Yoshino et al. (2016) [39]	Whole blood	PS	50 SZ 18 drug-free SZ 50 C	↓ CpG2, CpG4, CpG7 (in SZ) ↓ CpG 1–3 and CpG 5–7 (in drug-free SZ)	—	Association with age (in SZ) No association with age of onset, duration of illness, chlorpromazine equivalent, or smoking
	DRD4 promoter COMT (rs4680–val108/158met)	Mill et al. (2008) [28]	Frontal cortex	BS and PS	35 SZ 35 BD 35 C	No difference	—	—

	MB-COMT	Abdolmaleky et al. (2006) [42]	Frontal lobe	MSP and BS	35 SZ 35 BD 35 C 5 SZ (male) 5 C (male)	↓ (Especially in the left frontal lobe of SZ and BD) Val158Met polymorphism was related to RELN promoter methylation	Inversely MB-COMT and DRD1	Association with alcohol abuse (in SZ) No association with drug use, smoking, brain pH, postmortem interval, or age of onset
	MB-COMT	Nohesara et al. (2011) [43]	Saliva	BS and quantitative MSP	20 SZ 20 BD 20 first-degree relatives 25 C	↓ SZ and BD No difference Val/Met polymorphism of MB-COMT	—	Association with age (in SZ) No association with sex, marital status, or education
	S-COMT (putative promoter region/ parts of the MB-COMT coding region)	Murphy et al. (2005) [44]	Whole blood Brain samples	BS	20 SZ 3 concordant MZ for SZ 31 C 1 C (31 brain regions)	No difference (the different tissues show similar methylation pattern)	—	—
	S-COMT (from +20,729 to +20,802)	Melas et al. (2012) [37]	Whole blood	BS and PS	177 SZ 171 C	↑ S-COMT	—	—

(continued)

Table 12.1 (continued)

Neurotransmitter-related systems	Candidate gene(s)	First author	Tissue	Method	Sample	Significant methylation	Expression—methylation relation	Methylation associated with covariates
GABAergic	GAD1 promoter	Huang et al. (2007) [30]	Prefrontal cortex	BS of DNA purified from H3K27me3 and H3K4me3 immunoprecipitation	14 SZ 14 C	↓ Repressive H3K27me3 chromatin fraction	Positively	—
	GAD1 intron 1 and intron 3 and promoter RELN promoter	Mill et al. (2008) [23]	Frontal cortex	BS and PS	35 SZ 35 BD 35 C	No difference	—	—
	GAD1 regulatory network (set of 27 genes that regulate GAD1) [77]	Ruzicka et al. (2015) [31]	Hippocampus	HM450	8 SZ 8 BD 8 C	↑ MSX1 ↓↑ CCND2 ↓↑ DAXX	Inversely Positively and inversely Positively and inversely	—
	RELN promoter	Abdolmaleky et al. (2005) [26]	Frontal lobe	BS-MSP	5 SZ (male) 5 C (male)	↑	Inversely	—
	RELN promoter	Grayson et al. (2005) [27]	Occipital cortex Prefrontal cortex (BA 9 or 10)	BS	10 SZ 10 C 5 SZ 5 C	↑ Sites −134 to 139	—	No association with age of onset, age, sex, postmortem interval, drug equivalents, method of death, or brain hemisphere

	RELN promoter (positions +131, +229 and +227)	Tamura et al. (2007) [78]	Forebrains	Methylation-sensitive restriction (BsshII) and PCR	35 SZ 35 BD 35 C	No difference	Inversely (when the sample was prepared ≤18 h after death)	Association with age in controls No association with age of onset, sex, lifetime alcohol and drug use, and suicide status
	RELN promoter	Tochigi et al. (2008) [28]	Prefrontal cortex	PS and BS	14 SZ 13 C	No differences	—	—
	RELN promoter	Bönsch et al. (2012) [67]	Whole blood	Methylation-sensitive restriction (HpaII) and PCR	20 twin pairs discordant for SZ 8 twin pairs concordant for SZ 42 C twin pairs	No difference	—	Association with sex and twin type
Glutamatergic	GMR2, GMR5, GMR8, and GRIA3	Kordi-Tamandani et al. (2013) [38]	Whole blood	MSP	81 SZ 71 C	↓ GRM2 and GRM5 No difference GRIA3 and GMR8	Inversely GRM2, GRM5, and GRIA3	—
	GRIN2B	Mill et al. (2008) [23]	Frontal cortex	BS and PS	35 SZ 35 BD 35 C	No difference	—	—

(continued)

Table 12.1 (continued)

Neurotransmitter-related systems	Candidate gene(s)	First author	Tissue	Method	Sample	Significant methylation	Expression—methylation relation	Methylation associated with covariates
Serotonergic	*5HTR1A* promoter	Carrard et al. (2011) [34]	Leukocytes	HRM	40 SZ 58 BD 67 C	↑ For SZ and BD	—	No association with age or sex alone but in combination
	HTR2A promoter	Abdolmaleky et al. (2011) [35]	Frontal lobe	BS and quantitative MSP	35 SZ 35 BD 35 C	↑ −1438A/G polymorphic site ↓ T102C polymorphic site	Inversely −1438A/G polymorphic site Positively T102C polymorphic site	Associated with age of onset and age (in controls)
	HTR2A promoter and exon-1	Ghadirivasfi et al. (2011) [79]	Saliva	BS and quantitative MSP	63 SZ 92 BD 27 first-degree relatives 76 C	↓ T102C polymorphic site (in SZ, BD, and relatives)	—	Association with age (in SZ) No association with sex, marital status, smoking, alcohol abuse, and education
	5-HTT promoter (from +425 to +490)	Melas et al. (2012) [37]	Whole blood	BS and PS	177 SZ 171 C	No difference	—	—

	5-HTT promoter	Abdolmaleky et al. (2014) [36]	Frontal lobe Saliva	HM27 and HM450, followed by BS and quantitative MSP	35 SZ 35 BD 35 C 30 SZ 20 BD 20 first-degree relatives 30 C	↑ Especially in drug-free SZ (in both tissues)	Inversely (In SZ)	Association with age (in brain sample of controls) and duration of antipsychotic drug use
Other	ARVCF Intron 3 BDNF (rs6265–val66met) DTNBP1 Intron 1 and promoter MTHFR promoter NRG1 intron 1 and promoter	Mill et al. (2008) [23]	Frontal cortex	BS and PS	35 SZ 35 BD 35 C	No difference	—	—
	BDNF promoter IV	Kordi-Tamandani et al. (2012) [41]	Whole blood	MSP	80 drug-free SZ 71 C	↓	Inversely	—
	BDNF promoter I BDNF promoter IV	Ikegame et al. (2013) [47]	Whole blood	PS	100 SZ 100 C	↑ Promoter I No difference Promoter IV	—	Association with sex (promoter IV in controls and CpG I-72 in SZ) No association with age (in controls)

(continued)

Table 12.1 (continued)

Neurotransmitter-related systems	Candidate gene(s)	First author	Tissue	Method	Sample	Significant methylation	Expression—methylation relation	Methylation associated with covariates
	BDNF promoter (−219 to +160) and intron region (+664 to +945)	Keller et al. (2014) [49]	Prefrontal cortex and striatum	MassARRAY and PS	15 SZ 15 C	↓ −93 CpGs (in the prefrontal cortex)	No difference	—
	BDNF (areas within the promoter and the border between intron 3 and exon IV)	Çöpoğlu et al. (2016) [48]	Whole blood	MSP	49 SZ 65 C	No difference	—	Association with duration of illness (CpG island-1) No association with sex or age
	DTNBP1 promoter (CpGs neighboring a SP1 binding site)	Abdolmaleky et al. (2015) [80]	Saliva	BS and quantitative MSP	20 SZ 10 drug-naïve SZ 15 first-degree relatives 30 C	↑ Especially in drug-naïve SZ	—	Association with age of onset No association with sex
	MAOA (between promoter and exon 1)	Chen et al. (2012)	Whole blood	BS	371 SZ (paranoid) 288 C	↑ In male SZ	—	—

	SOX10	Iwamoto et al. (2005) [82]	Prefrontal cortex (BA10)	BS	13 SZ 15 C	↑	Inversely SOX10	—
	SOX10 promoter	Bönsch et al. (2012) [67]	Whole blood	Methylation-sensitive restriction (HpaII) and PCR	20 twin pairs discordant for SZ 8 twin pairs concordant for SZ 42 C twin pairs	No difference	—	Association with sex No association with twin type

BA Brodmann area, *BS* bisulfite sequencing, *C* control, *HM27* Illumina Infinium HumanMethylation27 BeadChip, *HM450* Illumina Infinium HumanMethylation450 BeadChip, *HRM* high-resolution melt analysis, *MSP* methylation-specific polymerase chain reaction, *MZ* monozygotic twins, *PS* pyrosequencing, *SZ* schizophrenia

repressive chromatin fractions of the proximal GAD1 promoter [30], while others showed that several genes of a network that regulates GAD1 expression (especially MSX1, CCND2, and DAXX) were differentially methylated in the hippocampus of patients versus controls [31].

12.2.2 Serotonergic System

There is some evidence that the serotonergic system is involved in schizophrenia, especially as serotonergic mechanisms appear to play a role in the action of second-generation antipsychotic drugs [32, 33]. Furthermore, several studies show schizophrenia-specific methylation profiles for serotonergic receptor and transporter genes.

For instance, in leukocytes, Carrard et al. [34] examined the pathology-related methylation profile of the promoter region of the 5HT1A receptor and found an association between schizophrenia and hypermethylated CpGs. In another study, Abdolmaleky et al. [35] evaluated methylation profiles of the serotonin receptor type-2 (HTR2A) promoter in postmortem brain samples of patients with schizophrenia, bipolar disorder, and matched controls. Hyper- and hypomethylated CpG sites were found in regions flanking the −1438 A/G and the T102C polymorphic sites, respectively [35].

In postmortem brain and saliva samples, hypermethylation of the serotonin transporter (5-HTT) promoter loci were found in patients with schizophrenia compared to matched control cases [36]. This effect was more pronounced in drug-naïve individuals. However, Melas et al. [37] observed no difference in 5-HTT promoter methylation levels in leukocytes of patients with schizophrenia compared to controls.

12.2.3 Dopaminergic System

The dopamine hypothesis of schizophrenia has been extensively studied, and methylation differences were revealed for several dopaminergic genes, especially for the various dopamine receptors (DRDs). Differentially methylation levels and mRNA expression of DRD2, DRD4, and DRD5 were discovered when comparing blood DNA of patients with schizophrenia and control subjects [38]. Yoshino et al. [39] also found pathology-related differentially methylation levels in regions upstream of DRD2. More specifically, methylation levels in leukocyte samples of drug-naïve patients as well as medicated patients were lower than those of matched controls. In another study, however, a link between site-specific DRD2 methylation frequencies in peripheral leukocytes and schizophrenia could not be found [40]. As well, after evaluating blood samples in a case-control study, researchers observed no association between pathology and DRD1 methylation as well as DRD1 expression [38, 41].

Another important candidate gene within the dopaminergic framework is catechol-O-methyltransferase (COMT), an enzyme related to dopamine degradation. Methylation levels and polymorphism (Val/Met) of the COMT gene were found to be associated with DRD1 expression in patients with schizophrenia [42]. In a case-control study, Abdolmaleky et al. [42] investigated the association between DRD1 expression and membrane-bound (MB) COMT promoter methylation in frontal lobes. In patients, hypo-methylation of MB-COMT promoter was linked to hyper-expression of MB-COMT and eventually reduced DRD1 expression, especially in the left frontal lobe. Later on, the association between schizophrenia and decreased MB-COMT promoter methylation was replicated in saliva samples [43]. However, Mill et al. [23] did not find pathology-related methylation signatures in COMT promoter areas in frontal cortex tissue. The COMT Val/Val genotype was associated with schizophrenia [42], which was further related to hypo-methylation of the MB-COMT gene and hypermethylation of the RELN promoter [42]. Concerning the soluble (S) isoform of COMT, results about its relationship with schizophrenia are mixed. On the one hand, schizophrenia was associated with hypermethylation of the S-COMT gene in leukocytes [37]. On the other hand, Murphy et al. [44] did not find differentially methylated S-COMT signatures in blood samples of patients in comparison to control cases.

12.2.4 Glutamatergic System

The involvement of the glutamatergic system in the pathophysiology of schizophrenia, the main excitatory neurotransmitter system in the brain, received substantial support [45]. Furthermore, in brain and blood samples, several differentially methylated glutamatergic genes were associated with schizophrenia, such as the glutamate receptor genes GMR2, GMR5 [46], GRIA2, and NR3B [23] and the vesicular glutamate transporter genes VGLUT1 and VGLUT2 [23].

12.2.5 Brain-Derived Neurotrophic Factor

Attention is often drawn to psychosis-related methylation profiles of the brain-derived neurotrophic factor (BDNF) gene. However, only some studies show a link between schizophrenia and BDNF methylation levels. In case-control studies of peripheral blood cells, schizophrenia was linked to modest BDNF promoter I hypermethylation [47] and reduced BDNF promoter IV methylation frequency [41]. However, the study by Ikegame et al. [47] did not detect methylation differences in promoter IV. Furthermore, no link between schizophrenia and methylation marks of CpG-islands within the BDNF promoter and the border between intron 3 and exon IV was found [48]. In post-mortem brain tissue, Mill et al. [23] observed no methylation differences between patients with schizophrenia and control cases in BDNF regions, including areas in exons I, III, VI, and IX. Furthermore, in prefrontal cortex and striatum samples, the promoter region (spanning CpGs from −219 to +160) and the intron region (including CpGs within +664 and +945) were analyzed in a case-control study [49]. Only the −93 CpGs in the prefrontal cortex was significantly related to schizophrenia [49].

12.3 Methylome-Wide Analyses

Alongside candidate gene studies, methylome-wide association (MWA) studies are increasingly used as an unbiased explorative strategy to identify differentially methylated genomic areas in schizophrenia. The technology for genome-wide methylation analyses has dramatically changed over the last years, with concomitant improvements in sensitivity and the number of CpG sites examined. Following on the work by Mill et al. [23] who were the first to conduct a schizophrenia MWA study, subsequent studies have employed diverse techniques, platforms, and analysis strategies to examine methylation patterns in different brain areas and blood. Therefore, in view of the methodological heterogeneity of the studies, comparing results at the level of single CpG sites, genes or even differentially methylated regions (DMRs) are difficult and lead to, not surprisingly, a low number of direct replications. Nevertheless, some degree of overlap exists between studies that, when combined with pathway and gene ontology analyses, renders new insight into the neurobiological processes involved in schizophrenia. Here, we summarize the most important findings from methylome-wide analyses of schizophrenia.

A detailed overview of the evaluated studies is presented in Table 12.2, which includes information on sample size, methods and platforms, main findings, top ranked genes, and results from gene enrichment and/or gene network analyses.

Methylome-wide analyses typically yield a large number of differentially methylated CpG positions (DMPs) and corresponding annotated genes. As presented in Table 12.2, there is little or no overlap in the top gene hits (derived from the most significant DMPs) between studies on schizophrenia. Nevertheless, comparing the lists of significant genes of each study indicates that several genes are replicated in different studies. For example, differential methylation of the following genes was associated with schizophrenia in at least four independent studies: PRKD2 [50–53], NOTCH4 [50, 52, 54, 55], NCOR2 [50, 52, 54, 55], PTPRN2 [50, 52, 54–56], FOXN3 [50, 52–54], FOXP1 [50, 52, 53, 57], MYT1L [50, 51, 53, 54], PTGER4 [50, 52, 54, 58], RPP21 [23, 50, 52, 54], TRERF1 [23, 50, 53, 54], WDR20 [50, 52, 54, 55], ADCY6 [50, 52, 54, 58], AUTS2 [23, 50, 52, 54], and C7orf50 [50, 52, 54, 55]. Interestingly, these genes were found significant across diverse tissue types. In addition, some genes identified through methylome-wide analyses were also supported by candidate gene studies, such as COMT [50, 52, 59], GAD1, RELN [53], and BDNF [52].

Thus, while top significant findings differ between studies, there seems to be some degree of overlap when looking at the bulk of genes with a differential methylation signal. To examine the relationship between these genes and their role in biological processes, the pool of differentially methylated genes is often interrogated with pathway and gene ontology enrichment analyses. Table 12.2 includes a summary of such analyses from schizophrenia MWA studies. It is apparent that a recurrent finding, especially from blood-based MWA studies, is the involvement of immune system functioning [52, 53, 60]. This corroborates results from GWA studies that report associations between schizophrenia and genomic regions that contain immune-related genes, including, but not limited to, the MHC locus [3, 61]. An

Table 12.2 Summary of methylome-wide analyses

First author	Tissue	Method	Sample	Main findings
Dempster et al. (2011) [58]	Whole blood Brain	HM27 EpiTYPER® (Sequenom Inc., CA, USA)	22 discordant MZ : 11 SZ 11 BD Replication sample 15 SZ 15 BD 15 Control	Numerous CpGs were differentially methylated *Top ranked genes* PUS3, SYNGR2, KDELR1, PDK3, PPARGC1A (for SZ), and ST6GALNAC1 (for psychosis, replicated in brain samples) *Enrichment analyses* Nervous system development and function; psychological disorders; dopamine receptor signaling and glutamate receptor signaling; nervous system development and function
Kinoshita et al. (2013) [55]	Whole blood	HM450	24 SZ (drug-free) 23 C 3 discordant MZ (SZ)	Diagnostic differences in DNA methylation in 10,747 and 15,872 CpG sites, respectively, for samples one and two
Aberg et al. (2014) [53]	Whole blood	Methyl-CpG binding domain (MBD) protein-based enrichment and single-end sequencing Bisulfite pyrosequencing	Discovery sample 759 SZ 738 C Replication samples (top findings) 178 SZ 182 C 561 SZ 582 C	Differential methylation of CpG sites overlapping 112 genes *Top ranked genes* FAM63B, ARHGAP26, CTAGE11P, TBC1D22A, and intergenic (only FAM63B remained significant in replication samples after control for multiple testing) *Enrichment analyses* Hypoxia; immune system
Kinoshita et al. (2014) [54]	Whole blood	HM450	63 SZ 42 C	Diagnostic differences in DNA methylation in 2552 CpG sites *Enrichment analyses* Transcription-related functions (e.g., transcription regulator activity, regulation of transcription, and regulation of transcription from RNA polymerase II promoter)

(continued)

Table 12.2 (continued)

First author	Tissue	Method	Sample	Main findings
Liu et al. (2014) [60]	Whole blood	HM27	98 SZ 108 C Validation sample 325 SZ 394 C	Methylation levels in 20 CpG sites were significantly associated with SZ status (16 were supported after validation, and 11 of those were related to reality distortion symptoms) *Top ranked genes* CD244, LAX1, PRF1, FAM173A, CBFA2T3, CD7, TCN1 (in both samples and additional in expression analysis associated with SZ) *Enrichment analyses* Inflammatory response
Hannon et al. (2016) [65]	Whole blood	HM450	353 SZ 322 C Replication sample 414 SZ 433 C Replication sample 96 MZ	25 DMPs were significantly associated with SZ status (not supported by the second replication sample) *Top ranked genes* FAM126A, PPTC7, GYG1, SIK3, USP36, EHD1, ITGA11, FAR2, IL15, GNB5, TRAF3IP3, MED22, AIM2, FBXO10, DDO, TNFAIP8 *Enrichment analyses* Immune function, neuronal proliferation, brain development
van den Oord et al. (2016) [57]	Buffy coat of whole blood Brain (BA10)	Methyl-CpG binding domain protein-enriched genome sequencing (MBD-seq)	712 SZ 696 C Replication sample 26 SZ 13 BD 27 C Validation sample (blood samples) 370 SZ 377 C	Seven CpG SNPs reached methylome-wide significance *Top ranked genes* IL1RAP and ENC1

Chen et al. (2014) [59]	Cerebellum	HM27	39 SZ 36 BD 43 C	Differential methylation in 488 CpG sites were associated with SZ status SLC16A7, PIK3R1, BTN3A3, and NHLH1 were differentially expressed and methylated in patients (SZ and BD)
Pidsley et al. (2014) [51]	Prefrontal cortex Cerebellum	HM450	20 SZ 23 C 21 SZ 23 C Replication sample 18 SZ 15 C	DMPs in four genes were significantly associated with SZ status in the prefrontal cortex *Top ranked genes* GSDMD, RASA3, HTR5A, PPFIA1 *Enrichment analyses* Schizophrenia and other neuropsychiatric disorders; nervous system development and function: neuron development; synaptic transmission; calcium ion binding
Jaffe et al. (2016) [64]	Prefrontal cortex (dorsolateral prefrontal cortex and BA46 and BA9)	HM450	108 SZ 136 C	Diagnostic differences in DNA methylation in 2104 CpGs *Top ranked genes* CD164, COPZ2, SUGT1, HAT1, TYW1B *Enrichment analyses* Embryo development, cell fate commitment, nervous system differentiation
Numata et al. (2014) [83]	Dorsolateral prefrontal cortex	HM27	106 SZ 110 C	Diagnostic differences in DNA methylation in 107 CpG sites
Mill et al. (2008) [23]	Frontal cortex	Microarray-based DNA methylation profiling	35 SZ 35 BD 35 C	Methylation differences at different loci associated with SZ (related to 51 genes in men and 37 genes in women) *Top ranked genes* RPP21 and KEL (in men and women with SZ status and in women with BD status) *Enrichment analyses* Brain development, mitochondrial functions

(continued)

Table 12.2 (continued)

First author	Tissue	Method	Sample	Main findings
Wockner et al. (2014) [50]	Frontal cortex	HM450	24 SZ 24 C	Methylation differences at probes associated with 2929 different genes were related to SZ *Top ranked genes* TRIM26, EPN2, BAHCC1, IQCG, TRIM2
Xiao et al. (2014) [84]	Frontal cortex and anterior cingulate (BA9 and BA24)	MeDIP sequencing	5 SZ 7 BP 6 C	4985 and 3867 DMRs discovered in BA9 and BA24 of SZ, respectively One gene (hsa-mir-4266) was hypermethylated in both brain regions of SZ patients *Enrichment analyses* Guidance and signaling; multicellular organismal development, signaling, axon guidance, oligodendrocyte differentiation
Zhao et al. (2015) [66]	Brain (BA9)	MeDIP sequencing	5 SZ 7 BD 6 C	10,961 DMRs were associated with SZ status

BA Brodmann area, *BD* bipolar disorder, *C* control, *DMPs* differentially methylated positions, *DMRs* differentially methylated regions, *HM27* Illumina Infinium HumanMethylation27 BeadChip, *HM450* Illumina Infinium HumanMethylation450 BeadChip, *MeDIP* methylated DNA immunoprecipitation, *MZ* monozygotic twins, *SZ* schizophrenia

association between immune dysregulation and schizophrenia has long been put forward, although the causal direction of this relationship is still debated [62]. Nevertheless, observations that severe inflammation and autoimmune disease constitute critical risk factors to the development of psychosis support a causal link [63]. Another recurrent result, from blood- and brain-based MWA studies, is the association between schizophrenia and neurodevelopmental processes [23, 51, 52, 58]. In this regard, recent studies show that CpGs, of which methylation levels change during fetal development, are enriched in schizophrenia susceptibility loci derived from GWA analyses [64, 65].

Furthermore, as methylation marks are especially interesting in respect to their dynamic impact on gene expression and transcription, some researchers investigated the interplay between methylation status and gene expression in patients with schizophrenia. For example, Liu et al. [60] identified 16 CpG sites to be differentially methylated in their case-control analysis. Interrogating the expression of the 16 corresponding genes in an independent dataset showed that seven genes were differentially expressed in schizophrenia patients, with five showing an inverse correlation between methylation marks and gene expression. In another approach, Chen et al. [59] used different datasets to explore the association between psychosis-related methylation levels and gene expression in the cerebellum. They focused on *cis*-regions within a 50Kb radius of the transcription start site (TSS) of matching genes and found strong support for the association between methylation and gene expression in four genes, of which three (PIK3R1, BTN3A3, NHLH1) showed an inverse and one (SLC16A7) a positive association. Next to such *cis*-region-related effects on gene expression, likely through interference with transcription factor binding, methylation in other genomic regions may indirectly impact gene expression by influencing the expression of regulatory noncoding RNAs. In this context, Zhao et al. [66] reported that schizophrenia-related differentially methylated sites mapped predominantly to intronic regions containing microRNAs which target differentially expressed genes.

Besides the analysis of particular genes or differentially methylated positions or regions, methylation signatures can also be explored by analyzing global methylation differences. For instance, Bönsch et al. [67] investigated whole blood samples of monozygotic twin pairs and found that twins with schizophrenia showed significant lower global methylation levels compared to healthy control twins. In the same study, it was found that drug-free patients had lower methylation levels than patients taking typical antipsychotics [67]. Supporting those findings, Melas et al. [37] found increased global methylation levels in leukocytes of patients with schizophrenia, and their analyses showed that 11% of the variance could be explained by antipsychotic medication as well as disease onset. However, there are also researchers who did not find a relationship between global methylation levels and schizophrenia status. Specifically, Dempster et al. [58] did not find a difference in global methylation between affected and unaffected twins [58]. Also, Bromberg et al. [68] did not find a main effect of global methylation status and diagnosis outcome. However, they observed an interaction between diagnosis and smoking behavior on the difference in global methylation between cases and controls. These studies show that global methylation levels in schizophrenia are likely influenced by concomitant environmental factors such as age, gender, medication, and smoking behavior.

Finally, combining data collected through GWA and MWA analyses may provide more accurate and disorder-specific information on the involvement of genomic loci in schizophrenia. By their nature, GWA analyses yield many loci due to linkage disequilibrium, and integrating GWA data with those of MWA studies may narrow down the results to more specific findings and may also give better insight into the associated molecular mechanisms [69]. Supporting this approach, studies confirm that there is an overlap between differentially methylated regions and genetic susceptibility loci found through GWA analyses [52, 69, 70]. Kumar et al. [69] found significant overlap in MWA and GWA results, leading to the replicable identification of three genes (SDCCAG8, CREB1, ATXN7) that have previously been implicated in schizophrenia [69]. Similarly, Hannon et al. [65] examined whether the genomic regions identified by the latest GWA study of the Psychiatric GWAS Consortium (PGC) [3] overlapped with differentially methylated DMPs from their MWA study [65]. Out of the 105 identified GWA regions, 25 showed significant overlap with MWA data. Overlapping signals from genetic and methylation analyses may indicate that changes in methylation mediate the association between genetic variation and schizophrenia susceptibility, but does not prove a causal link. It is also possible that both methylation and genetic variation independently disturb transcription factor binding or interfere with other regulatory processes. A more direct relationship between individual genotypes and methylation status is assessed by examining whether the DNA methylation pattern at a particular locus is influenced by sequence variation in a proximal or distal genomic region. The latter represent so-called methylation quantitative trait loci (mQTL), which are now being explored in the context of schizophrenia [71]. For example, building on their observed overlap between GWAS and MWAS signals, Hannon et al. [65] established mQTL pairs (genetic variation—methylation) within schizophrenia-associated regions from the PGC GWAS [3] and identified 66 pairs to be associated with both schizophrenia and DNA methylation. Seven of those regions were also retrieved from a similar analysis in a postmortem brain sample and thus represent highly interesting candidate loci for schizophrenia. Interestingly, three regions are annotated to the AS3MT gene, of which an alternative transcript was recently described to be upregulated in brains of schizophrenic patients and to play a role in establishing a neuronal fate during stem cell differentiation [5]. Taking it even one step further, van Eijk et al. [72] combined mQTL analysis with gene expression data. The authors calculated mQTLs of CpGs that were differentially methylated and of which the corresponding genes were differentially expressed in schizophrenia patients versus controls. The identified loci were strongly enriched in the PGC GWA signals, indicating that mQTLs derived from combined differential methylation and gene expression data are more sensitive to reveal susceptibility loci for schizophrenia [72]. The discovered genes are CALHM1, HLA-C, PRRT1, and MRPL41, of which the first two have previously been associated with schizophrenia while the role of the latter two awaits further research.

12.4 Discussion

The multifactorial nature of schizophrenia, with developmental and environmental components superimposed on genetic liability, and in addition to the timing of disease onset, points to the involvement of epigenetic processes. In this chapter we reviewed the extensive recent efforts that compared DNA methylation signatures in patients versus control subjects. Differences in technical procedures and tissue type (including the diversity in brain areas examined) hamper direct comparison of results. Nevertheless, a modest degree of overlap in top ranked genes of methylome-wide analyses, their partial replication in candidate gene studies, and similarity in functional annotations of gene enrichment analyses indicate that the current data highlight interesting targets for further research on the neurobiology of schizophrenia. Of all the factors that contribute to variability in study outcomes (e.g., methodology, platforms, sample size), tissue type and its cellular composition deserve special attention. Obviously, only peripheral sources of DNA, such as blood or saliva, can be used to assess methylation patterns in large cohorts of living patients. Since DNA methylation status is cell-type specific, differences in leukocyte composition, either due to disease status or other factors, influence the results of blood-based methylation patterns. For example, as shown by Kinoshita et al. [54], adjusting for cell-type proportions dramatically impacts the number and identity of discovered DMPs. Recently, methods to estimate cell-type composition in datasets from MWA analyses (derived from both brain and blood samples) have been developed and allow adjusting for this kind of heterogeneity [73, 74]). Until now, only the most recent studies have incorporated this adjustment which may partly explain differential results. Further refinements in analyses strategies and algorithms will improve the validity of future findings.

Another factor contributing to outcome variability is medication status. Most studies included patients that were on antipsychotic medication. Additionally, if studies include information on medication status, sample sizes are generally small (e.g., see [23, 60]). Nonetheless, the confounding effect of antipsychotics should not be ignored given that several studies showed that antipsychotic use is associated with increased global methylation levels [37, 67] and with methylation level of specific gene promoters, such as the MEK1 gene [23].

Finally, while the focus of the current chapter was drawn toward case-control studies investigating DNA methylation in association with the diagnosis of schizophrenia, future studies should chart methylation signatures within the framework of transdiagnostic psychoses-related phenotypes, symptom dimensions, and intermediary phenotypes. In this regard, altered methylation patterns have been described to be associated with symptoms [60], cognitive functioning [75], and brain volume difference [76].

In conclusion, schizophrenia is associated with altered DNA methylation patterns in blood cells and brain tissue. Although specific findings are variable between studies, overlapping results from candidate gene analyses and genome-wide methylation screenings show differential methylation in genes related to neurotransmitter systems and point to dysregulation in immune system functioning and neurodevelopmental processes. Interestingly, the latter findings converge with data from genetic association studies in schizophrenia. Combining genetic and epigenetic approaches may therefore lead to new discoveries that broaden our understanding of the biological underpinnings of schizophrenia.

References

1. van Os J, Kapur S. Schizophrenia. Lancet. 2009;374(9690):635–45. doi:10.1016/s0140-6736(09)60995-8.
2. Owen MJ, Sawa A, Mortensen PB. Schizophrenia. Lancet. 2016;388(10039):86–97. doi:10.1016/s0140-6736(15)01121-6.
3. Ripke S, Neale BM, Corvin A, Walters JT, Farh K-H, Holmans PA, et al. Biological insights from 108 schizophrenia-associated genetic loci. Nature. 2014;511(7510):421. doi:10.1038/nature13595.
4. Sekar A, Bialas AR, de Rivera H, Davis A, Hammond TR, Kamitaki N, et al. Schizophrenia risk from complex variation of complement component 4. Nature. 2016;530(7589):177–83. doi:10.1038/nature16549.
5. Li M, Jaffe AE, Straub RE, Tao R, Shin JH, Wang Y, et al. A human-specific AS3MT isoform and BORCS7 are molecular risk factors in the 10q24. 32 schizophrenia-associated locus. Nat Med. 2016;22(6):649–56. doi:10.1038/nm.4096.
6. Lee SH, Ripke S, Neale BM, Faraone SV, Purcell SM, Perlis RH, et al. Genetic relationship between five psychiatric disorders estimated from genome-wide SNPs. Nat Genet. 2013;45(9):984–94. doi:10.1038/ng.2711.
7. Genovese G, Fromer M, Stahl EA, Ruderfer DM, Chambert K, Landén M, et al. Increased burden of ultra-rare protein-altering variants among 4,877 individuals with schizophrenia. Nat Neurosci. 2016;19(11):1433–41. doi:10.1038/nn.4402.
8. Marshall C, Howrigan D, Merico D, Thiruvahindrapuram B, Wu W, Greer D et al. A contribution of novel CNVs to schizophrenia from a genome-wide study of 41,321 subjects. bioRxiv. 2016:040493.
9. Van Os J, Rutten BP, Poulton R. Gene-environment interactions in schizophrenia: review of epidemiological findings and future directions. Schizophr Bull. 2008;34(6):1066–82. doi:10.1093/schbul/sbn117.
10. van Os J, Kenis G, Rutten BP. The environment and schizophrenia. Nature. 2010;468(7321):203–12. doi:10.1038/nature09563.
11. Insel TR. Rethinking schizophrenia. Nature. 2010;468(7321):187–93. doi:10.1038/nature09552.
12. Horváth S, Mirnics K. Schizophrenia as a disorder of molecular pathways. Biol Psychiatry. 2015;77(1):22–8.
13. Hall J, Trent S, Thomas KL, O'Donovan MC, Owen MJ. Genetic risk for schizophrenia: convergence on synaptic pathways involved in plasticity. Biol Psychiatry. 2015;77(1):52–8.
14. Cannon TD. How Schizophrenia Develops: Cognitive and Brain Mechanisms Underlying Onset of Psychosis. Trends Cogn Sci. 2015;19(12):744–56. doi:10.1016/j.tics.2015.09.009.
15. Green MF, Horan WP, Lee J. Social cognition in schizophrenia. Nat Rev Neurosci. 2015;16(10):620–31. doi:10.1038/nrn4005.
16. Pishva E, Kenis G, van den Hove D, Lesch KP, Boks MP, van Os J, et al. The epigenome and postnatal environmental influences in psychotic disorders. Soc Psychiatry Psychiatr Epidemiol. 2014;49(3):337–48. doi:10.1007/s00127-014-0831-2.
17. Teroganova N, Girshkin L, Suter CM, Green MJ. DNA methylation in peripheral tissue of schizophrenia and bipolar disorder: a systematic review. BMC Genet. 2016;17(1):1. doi:10.1186/s12863-016-0332-2.

18. Benes FM, Berretta S. GABAergic interneurons: implications for understanding schizophrenia and bipolar disorder. Neuropsychopharmacology. 2001;25(1):1–27.
19. Lewis DA, Hashimoto T, Volk DW. Cortical inhibitory neurons and schizophrenia. Nat Rev Neurosci. 2005;6(4):312–24.
20. Bird E, Barnes J, Iversen L, Spokes E, Mackay AP, Shepherd M. Increased brain dopamine and reduced glutamic acid decarboxylase and choline acetyl transferase activity in schizophrenia and related psychoses. The Lancet. 1977;310(8049):1157–9.
21. Perry T, Buchanan J, Kish S, Hansen S. γ-Aminobutyric-acid deficiency in brain of schizophrenic patients. Lancet. 1979;313(8110):237–9.
22. Spokes EG, Garrett NJ, Rossor MN, Iversen LL. Distribution of GABA in post-mortem brain tissue from control, psychotic and Huntington's chorea subjects. J Neurol Sci. 1980;48(3):303–13.
23. Mill J, Tang T, Kaminsky Z, Khare T, Yazdanpanah S, Bouchard L, et al. Epigenomic profiling reveals DNA-methylation changes associated with major psychosis. Am J Hum Genet. 2008;82(3):696–711. doi:10.1016/j.ajhg.2008.01.008.
24. Chen Y, Sharma RP, Costa RH, Costa E, Grayson DR. On the epigenetic regulation of the human reelin promoter. Nucleic Acids Res. 2002;30(13):2930–9.
25. Costa E, Grayson DR, Guidotti A. Epigenetic downregulation of GABAergic function in schizophrenia: potential for pharmacological intervention? Mol Interv. 2003;3(4):220.
26. Abdolmaleky HM, Cheng KH, Russo A, Smith CL, Faraone SV, Wilcox M, et al. Hypermethylation of the reelin (RELN) promoter in the brain of schizophrenic patients: a preliminary report. Am J Med Genet B Neuropsychiatr Genet. 2005;134(1):60–6. doi:10.1002/ajmg.b.30140.
27. Grayson DR, Jia X, Chen Y, Sharma RP, Mitchell CP, Guidotti A, et al. Reelin promoter hypermethylation in schizophrenia. Proc Natl Acad Sci U S A. 2005;102(26):9341–6. doi:10.1073/pnas.0503736102.
28. Tochigi M, Iwamoto K, Bundo M, Komori A, Sasaki T, Kato N, et al. Methylation status of the reelin promoter region in the brain of schizophrenic patients. Biol Psychiatry. 2008;63(5):530–3. doi:10.1016/j.biopsych.2007.07.003.
29. Mitchell AC, Jiang Y, Peter C, Akbarian S. Transcriptional regulation of GAD1 GABA synthesis gene in the prefrontal cortex of subjects with schizophrenia. Schizophr Res. 2015;167(1):28–34. doi:10.1016/j.schres.2014.10.020.
30. Huang HS, Akbarian S. GAD1 mRNA expression and DNA methylation in prefrontal cortex of subjects with schizophrenia. PLoS One. 2007;2(8):e809. doi:10.1371/journal.pone.0000809.
31. Ruzicka WB, Subburaju S, Benes FM. Circuit-and Diagnosis-Specific DNA Methylation Changes at γ-Aminobutyric Acid–Related Genes in Postmortem Human Hippocampus in Schizophrenia and Bipolar Disorder. JAMA Psychiat. 2015;72(6):541–51.
32. Selvaraj S, Arnone D, Cappai A, Howes O. Alterations in the serotonin system in schizophrenia: a systematic review and meta-analysis of postmortem and molecular imaging studies. Neurosci Biobehav Rev. 2014;45:233–45.
33. Meltzer HY, Li Z, Kaneda Y, Ichikawa J. Serotonin receptors: their key role in drugs to treat schizophrenia. Prog Neuropsychopharmacol Biol Psychiatry. 2003;27(7):1159–72.
34. Carrard A, Salzmann A, Malafosse A, Karege F. Increased DNA methylation status of the serotonin receptor 5HTR1A gene promoter in schizophrenia and bipolar disorder. J Affect Disord. 2011;132(3):450–3. doi:10.1016/j.jad.2011.03.018.
35. Abdolmaleky HM, Yaqubi S, Papageorgis P, Lambert AW, Ozturk S, Sivaraman V, et al. Epigenetic dysregulation of HTR2A in the brain of patients with schizophrenia and bipolar disorder. Schizophr Res. 2011;129(2):183–90. doi:10.1016/j.schres.2011.04.007.
36. Abdolmaleky HM, Nohesara S, Ghadirivasfi M, Lambert AW, Ahmadkhaniha H, Ozturk S, et al. DNA hypermethylation of serotonin transporter gene promoter in drug naive patients with schizophrenia. Schizophr Res. 2014;152(2):373–80. doi:10.1016/j.schres.2013.12.007.

37. Melas PA, Rogdaki M, Ösby U, Schalling M, Lavebratt C, Ekström TJ. Epigenetic aberrations in leukocytes of patients with schizophrenia: association of global DNA methylation with antipsychotic drug treatment and disease onset. FASEB J. 2012;26(6):2712–8.
38. Kordi-Tamandani DM, Sahranavard R, Torkamanzehi A. Analysis of association between dopamine receptor genes' methylation and their expression profile with the risk of schizophrenia. Psychiatr Genet. 2013;23(5):183–7. doi:10.1097/YPG.0b013e328363d6e1.
39. Yoshino Y, Kawabe K, Mori T, Mori Y, Yamazaki K, Numata S, et al. Low methylation rates of dopamine receptor D2 gene promoter sites in Japanese schizophrenia subjects. World J Biol Psychiatry. 2016;17:449–56. doi:10.1080/15622975.2016.1197424.
40. Zhang AP, Yu J, Liu JX, Zhang HY, Du YY, Zhu JD, He G, Li XW, Gu NF, Feng GY, He L. The DNA methylation profile within the 5′-regulatory region of DRD2 in discordant sib pairs with schizophrenia. Schizophr Res. 2007;90(1):97–103. doi:10.1016/j.schres.2006.11.007.
41. Kordi-Tamandani DM, Sahranavard R, Torkamanzehi A. DNA methylation and expression profiles of the brain-derived neurotrophic factor (BDNF) and dopamine transporter (DAT1) genes in patients with schizophrenia. Mol Biol Rep. 2012;39(12):10889–93. doi:10.1007/s11033-012-1986-0.
42. Abdolmaleky HM, Cheng KH, Faraone SV, Wilcox M, Glatt SJ, Gao F, et al. Hypomethylation of MB-COMT promoter is a major risk factor for schizophrenia and bipolar disorder. Hum Mol Genet. 2006;15(21):3132–45. doi:10.1093/hmg/ddl253.
43. Nohesara S, Ghadirivasfi M, Mostafavi S, Eskandari MR, Ahmadkhaniha H, Thiagalingam S, et al. DNA hypomethylation of MB-COMT promoter in the DNA derived from saliva in schizophrenia and bipolar disorder. J Psychiatr Res. 2011;45(11):1432–8. doi:10.1016/j.jpsychires.2011.06.013.
44. Murphy BC, O'Reilly RL, Singh SM. Site-specific cytosine methylation in S-COMT promoter in 31 brain regions with implications for studies involving schizophrenia. Am J Med Genet B Neuropsychiatr Genet. 2005;133(1):37–42. doi:10.1002/ajmg.b.30134.
45. Howes O, McCutcheon R, Stone J. Glutamate and dopamine in schizophrenia: an update for the 21st century. J Psychopharmacol. 2015;29(2):97–115.
46. Kordi-Tamandani DM, Dahmardeh N, Torkamanzehi A. Evaluation of hypermethylation and expression pattern of GMR2, GMR5, GMR8, and GRIA3 in patients with schizophrenia. Gene. 2013;515(1):163–6. doi:10.1016/j.gene.2012.10.075.
47. Ikegame T, Bundo M, Sunaga F, Asai T, Nishimura F, Yoshikawa A, et al. DNA methylation analysis of BDNF gene promoters in peripheral blood cells of schizophrenia patients. Neurosci Res. 2013;77(4):208–14.
48. Çöpoğlu ÜS, İğci M, Bozgeyik E, Kokaçya MH, İğci YZ, Dokuyucu R, et al. DNA Methylation of BDNF Gene in Schizophrenia. Med Sci Monit. 2016;22:397.
49. Keller S, Errico F, Zarrilli F, Florio E, Punzo D, Mansueto S, et al. DNA methylation state of BDNF gene is not altered in prefrontal cortex and striatum of schizophrenia subjects. Psychiatry Res. 2014;220(3):1147–50.
50. Wockner LF, Noble EP, Lawford BR, Young RM, Morris CP, Whitehall VL, et al. Genome-wide DNA methylation analysis of human brain tissue from schizophrenia patients. Transl Psychiatry. 2014;4:e339. doi:10.1038/tp.2013.111.
51. Pidsley R, Viana J, Hannon E, Spiers H, Troakes C, Al-Saraj S, et al. Methylomic profiling of human brain tissue supports a neurodevelopmental origin for schizophrenia. Genome Biol. 2014;15(10):483. doi:10.1186/s13059-014-0483-2.
52. Hannon E, Dempster E, Viana J, Burrage J, Smith AR, Macdonald R, et al. An integrated genetic-epigenetic analysis of schizophrenia: evidence for co-localization of genetic associations and differential DNA methylation. Genome Biol. 2016;17(1):176. doi:10.1186/s13059-016-1041-x.
53. Aberg KA, McClay JL, Nerella S, Clark S, Kumar G, Chen W, et al. Methylome-wide association study of schizophrenia: Identifying blood biomarker signatures of environmental insults. JAMA Psychiat. 2014;71(3):255–64. doi:10.1001/jamapsychiatry.2013.3730.

54. Kinoshita M, Numata S, Tajima A, Ohi K, Hashimoto R, Shimodera S, et al. Aberrant DNA methylation of blood in schizophrenia by adjusting for estimated cellular proportions. Neuromolecular Med. 2014;16(4):697–703. doi:10.1007/s12017-014-8319-5.
55. Kinoshita M, Numata S, Tajima A, Shimodera S, Ono S, Imamura A, et al. DNA methylation signatures of peripheral leukocytes in schizophrenia. Neuromolecular Med. 2013;15(1):95–101. doi:10.1007/s12017-012-8198-6.
56. Castellani CA, Laufer BI, Melka MG, Diehl EJ, O'Reilly RL, Singh SM. DNA methylation differences in monozygotic twin pairs discordant for schizophrenia identifies psychosis related genes and networks. BMC Med Genomics. 2015;8(1):17. doi:10.1186/s12920-015-0093-1.
57. van den Oord EJ, Clark SL, Xie LY, Shabalin AA, Dozmorov MG, Kumar G, et al. A Whole Methylome CpG-SNP Association Study of Psychosis in Blood and Brain Tissue. Schizophr Bull. 2016;42(4):1018–26. doi:10.1093/schbul/sbv182.
58. Dempster EL, Pidsley R, Schalkwyk LC, Owens S, Georgiades A, Kane F, et al. Disease-associated epigenetic changes in monozygotic twins discordant for schizophrenia and bipolar disorder. Hum Mol Genet. 2011;20(24):4786–96. doi:10.1093/hmg/ddr416.
59. Chen C, Zhang CL, Cheng LJ, Reilly JL, Bishop JR, Sweeney JA, et al. Correlation between DNA methylation and gene expression in the brains of patients with bipolar disorder and schizophrenia. Bipolar Disord. 2014;16(8):790–9. doi:10.1111/bdi.12255.
60. Liu J, Chen J, Ehrlich S, Walton E, White T, Perrone-Bizzozero N, et al. Methylation patterns in whole blood correlate with symptoms in schizophrenia patients. Schizophr Bull. 2014;40(4):769–76. doi:10.1093/schbul/sbt080.
61. Network and Pathway Analysis Subgroup of Psychiatric Genomics Consortium. Psychiatric genome-wide association study analyses implicate neuronal, immune and histone pathways. Nat Neurosci. 2015;18(2):199–209. doi:10.1038/nn.3922. http://www.nature.com/neuro/journal/v18/n2/abs/nn.3922.html—Supplementary-information
62. Khandaker GM, Cousins L, Deakin J, Lennox BR, Yolken R, Jones PB. Inflammation and immunity in schizophrenia: implications for pathophysiology and treatment. Lancet Psychiatry. 2015;2(3):258–70. doi:10.1016/S2215-0366(14)00122-9.
63. Benros ME, Nielsen PR, Nordentoft M, Eaton WW, Dalton SO, Mortensen PB. Autoimmune diseases and severe infections as risk factors for schizophrenia: a 30-year population-based register study. Am J Psychiatry. 2011;168(12):1303–10. doi:10.1176/appi.ajp.2011.11030516.
64. Jaffe AE, Gao Y, Deep-Soboslay A, Tao R, Hyde TM, Weinberger DR, et al. Mapping DNA methylation across development, genotype and schizophrenia in the human frontal cortex. Nat Neurosci. 2016;19(1):40–7. doi:10.1038/nn.4181.
65. Hannon E, Spiers H, Viana J, Pidsley R, Burrage J, Murphy TM, et al. Methylation QTLs in the developing brain and their enrichment in schizophrenia risk loci. Nat Neurosci. 2016;19(1):48–54. doi:10.1038/nn.4182.
66. Zhao H, Xu J, Pang L, Zhang Y, Fan H, Liu L, et al. Genome-wide DNA methylome reveals the dysfunction of intronic microRNAs in major psychosis. BMC Med Genomics. 2015;8(1):62. doi:10.1186/s12920-015-0139-4.
67. Bönsch D, Wunschel M, Lenz B, Janssen G, Weisbrod M, Sauer H. Methylation matters? Decreased methylation status of genomic DNA in the blood of schizophrenic twins. Psychiatry Res. 2012;198(3):533–7. doi:10.1016/j.psychres.2011.09.004.
68. Bromberg A, Levine J, Nemetz B, Belmaker RH, Agam G. No association between global leukocyte DNA methylation and homocysteine levels in schizophrenia patients. Schizophr Res. 2008;101(1–3):50–7. doi:10.1016/j.schres.2008.01.009.
69. Kumar G, Clark SL, McClay JL, Shabalin AA, Adkins DE, Xie L, et al. Refinement of schizophrenia GWAS loci using methylome-wide association data. Hum Genet. 2015;134(1):77–87. doi:10.1007/s00439-014-1494-5.
70. Castellani CA, Melka MG, Gui JL, O'Reilly RL, Singh SM. Integration of DNA sequence and DNA methylation changes in monozygotic twin pairs discordant for schizophrenia. Schizophr Res. 2015;169(1–3):433–40. doi:10.1016/j.schres.2015.09.021.

71. Hoffmann A, Ziller M, Spengler D. The future is the past: methylation QTLs in schizophrenia. Genes (Basel). 2016;7(12) doi:10.3390/genes7120104.
72. van Eijk KR, de Jong S, Strengman E, Buizer-Voskamp JE, Kahn RS, Boks MP, et al. Identification of schizophrenia-associated loci by combining DNA methylation and gene expression data from whole blood. Eur J Hum Genet. 2015;23(8):1106–10. doi:10.1038/ejhg.2014.245.
73. Houseman EA, Accomando WP, Koestler DC, Christensen BC, Marsit CJ, Nelson HH, et al. DNA methylation arrays as surrogate measures of cell mixture distribution. BMC Bioinformatics. 2012;13:86. doi:10.1186/1471-2105-13-86.
74. Guintivano J, Aryee MJ, Kaminsky ZA. A cell epigenotype specific model for the correction of brain cellular heterogeneity bias and its application to age, brain region and major depression. Epigenetics. 2013;8(3):290–302. doi:10.4161/epi.23924.
75. Grove TB, Burghardt KJ, Kraal A, Dougherty R, Taylor S, Ellingrod V. Oxytocin receptor (OXTR) methylation and cognition in psychotic disorders. Mol Neuropsychiatry. 2016;2(3):151–60. doi:10.1159/000448173.
76. Rubin LH, Connelly JJ, Reilly JL, Carter CS, Drogos LL, Pournajafi-Nazarloo H, et al. Sex and diagnosis-specific associations between dna methylation of the oxytocin receptor gene with emotion processing and temporal-limbic and prefrontal brain volumes in psychotic disorders. Biol Psychiatry Cogn Neurosci Neuroimaging. 2016;1(2):141–51.
77. Benes FM, Lim B, Matzilevich D, Walsh JP, Subburaju S, Minns M. Regulation of the GABA cell phenotype in hippocampus of schizophrenics and bipolars. Proc Natl Acad Sci. 2007;104(24):10164–9. doi:10.1073/pnas.0703806104.
78. Tamura Y, Kunugi H, Ohashi J, Hohjoh H. Epigenetic aberration of the human REELIN gene in psychiatric disorders. Mol Psychiatry. 2007;12(6):593–600. doi:10.1038/sj.mp.4001965.
79. Ghadirivasfi M, Nohesara S, Ahmadkhaniha HR, Eskandari MR, Mostafavi S, Thiagalingam S, et al. Hypomethylation of the serotonin receptor type-2A Gene (HTR2A) at T102C polymorphic site in DNA derived from the saliva of patients with schizophrenia and bipolar disorder. Am J Med Genet B Neuropsychiatr Genet. 2011;156(5):536–45.
80. Abdolmaleky HM, Pajouhanfar S, Faghankhani M, Joghataei MT, Mostafavi A, Thiagalingam S. Antipsychotic drugs attenuate aberrant DNA methylation of DTNBP1 (dysbindin) promoter in saliva and post-mortem brain of patients with schizophrenia and Psychotic bipolar disorder. Am J Med Genet B Neuropsychiatr Genet. 2015;168(8):687–96. doi:10.1002/ajmg.b.32361.
81. Chen Y, Zhang J, Zhang L, Shen Y, Xu Q. Effects of MAOA promoter methylation on susceptibility to paranoid schizophrenia. Hum Genet. 2012;131(7):1081–7. doi:10.1007/s00439-011-1131-5.
82. Iwamoto K, Bundo M, Yamada K, Takao H, Iwayama-Shigeno Y, Yoshikawa T, et al. DNA methylation status of SOX10 correlates with its downregulation and oligodendrocyte dysfunction in schizophrenia. J Neurosci. 2005;25(22):5376–81. doi:10.1523/JNEUROSCI.0766-05.2005.
83. Numata S, Ye T, Herman M, Lipska BK. DNA methylation changes in the postmortem dorsolateral prefrontal cortex of patients with schizophrenia. Front Genet. 2014;5:280. doi:10.3389/fgene.2014.00280.
84. Xiao Y, Camarillo C, Ping Y, Arana TB, Zhao H, Thompson PM, et al. The DNA methylome and transcriptome of different brain regions in schizophrenia and bipolar disorder. PLoS One. 2014;9(4):e95875. doi:10.1371/journal.pone.0095875.

Histone Posttranslational Modifications in Schizophrenia

13

Elizabeth A. Thomas

Abstract

Schizophrenia is a complex neuropsychiatric disorder with high heritability; however, family and twin studies have indicated that environmental factors also play important roles in the etiology of disease. Environmental triggers exert their influence on behavior via epigenetic mechanisms. Epigenetic modifications, such as histone acetylation and methylation, as well as DNA methylation, can induce lasting changes in gene expression and have therefore been implicated in promoting the behavioral and neuronal behaviors that characterize this disorder. Importantly, because epigenetic processes are potentially reversible, they might serve as targets in the design of novel therapies in psychiatry. This chapter will review the current information regarding histone modifications in schizophrenia and the potential therapeutic relevance of such marks.

Keywords

Epigenetic • Psychiatric • CNS • Therapeutic

E.A. Thomas, Ph.D.
Department of Molecular and Cellular Neuroscience, The Scripps Research Institute, 10550 N. Torrey Pines Rd, SP2030, La Jolla, CA 92037, USA
e-mail: bthomas@scripps.edu

R. Delgado-Morales (ed.), *Neuroepigenomics in Aging and Disease*, Advances in Experimental Medicine and Biology, DOI 10.1007/978-3-319-53889-1_13

Abbreviations

HAT	histone acetyltransferase
HDAC	histone deacetylase
KDMs	lysine demethylases
KMT	lysine methyltransferases
polyI:C	polyinosinic-polycytidylic acid
PRMT	protein arginine methyltransferases

13.1 Introduction

Schizophrenia, affecting ~1% of the population worldwide, is a devastating, heterogeneous psychiatric disorder that typically begins in late adolescence or early adulthood. The manifestations of schizophrenia include a diversity of both positive symptoms, such as delusions, hallucinations, disordered thinking and bizarre behavior, and negative symptoms, which include social withdrawal, lack of motivation, poverty of speech, and affective blunting [1]. In addition, many individuals with schizophrenia experience difficulties with depression and substance abuse, factors that contribute to the 10%–15% lifetime incidence of suicide in this disorder [2]. Studies have shown that schizophrenia is associated with high heritability [3]; however, family and twin studies have indicated that environmental factors also play critical roles in the etiology of disease [4, 5]. Monozygotic twins show a 45%–50% concordance rate for schizophrenia, compared to 10%–15% for dizygotic twins. The 50%–55% discordance rate for schizophrenia cases who share identical genes strongly suggests that environmental factors play a key role in disease etiology. Environmental triggers include stress, viral infections, nutritional deficits, hypoxia, neurotoxins, and complications during pregnancy and birth, among others [6]. Epigenetic mechanisms of gene regulation represent a means to explain how environmental triggers can give rise to changes in disease states. It is known that schizophrenia is associated with dysregulated gene expression, as evident from microarray and RNA-sequencing studies demonstrating thousands of genes altered in expression in postmortem brain tissue from patients with schizophrenia (reviewed in [7, 8]). Although the origins of these expression changes are not clear.

During the last several years, there has been increased interest in the epigenetic origins of psychiatric diseases, including schizophrenia [9–12]. DNA methylation and histone acetylation/methylation are two of the most widely studied epigenetic marks in human disorders. DNA methylation changes in schizophrenia have been reviewed previously [13, 14]. This chapter will focus on what is known about histone modifications in relation to schizophrenia. Further, the available symptomatic treatments in schizophrenia are only partially successful, and therefore the development of new therapeutics, based on an understanding of the epigenetic etiology and pathogenesis of schizophrenia, is essential and will be discussed.

13.2 Epigenetics and Histone Modifications

The term "epigenetics" encompasses any change in gene activity not associated with a change in DNA sequence [15]. Epigenetic gene regulation involves post-replication and posttranslational modifications of DNA and histones that lead to lasting changes in chromatin structure, which results in alterations of gene transcription [16]. There is a diversity of epigenetic marks that govern gene regulation, many of which operate in context-dependent manners to control gene expression. The basic unit of chromatin is the nucleosome, which consists of 147 base pairs of DNA wrapped 1.6 times around an octamer of core histone proteins, H2A, H2B, H3, and H4 [17]. The amino-terminal tails of these core histones contain amino acid residues that are sites for acetylation, methylation, phosphorylation, and ubiquitination (Fig. 13.1); these posttranslational modifications alter histone interactions with DNA to determine the transcriptionally active status of the chromatin as well as accessibility to transcription factors [16]. These specific patterns of modified histones are often referred to as the "histone code" and correspond to various states of chromatin and to the activation or repression of distinct sets of genes [18].

Posttranslational modifications of histone proteins can occur on many different residues (Fig. 13.1), with some residues being sites for more than one type of modification. For example, lysine (K) residues can be either acetylated or methylated. Different types of modifications can also occur on different amino acid residues. For example, methylation can occur on K or arginine (R) residues. Further, the target

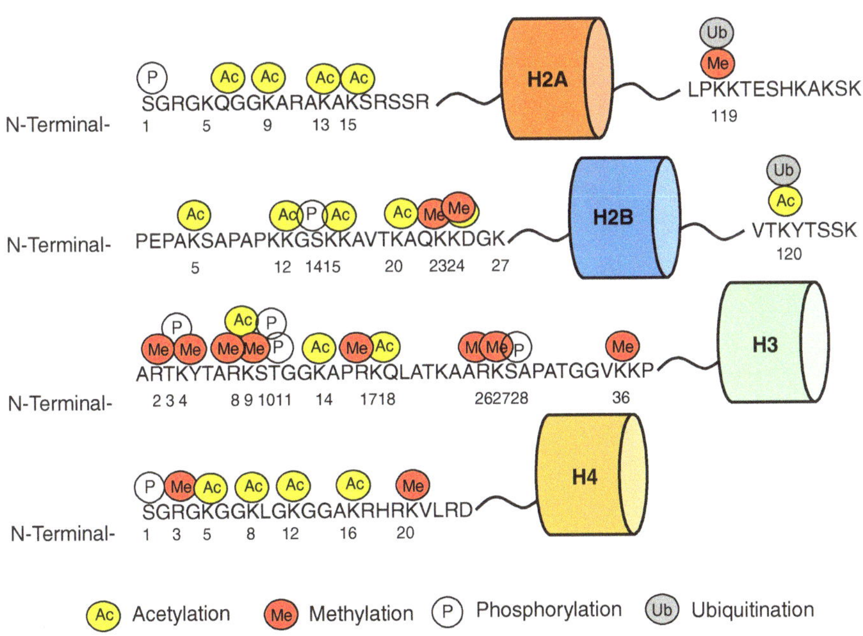

Fig. 13.1 Different posttranslational modifications present on histones

amino acids of histone tails can be mono-methylated, bi-methylated, or tri-methylated, making their functional status even more complex. In most species, histone H3 is primarily acetylated at K9, K14, K18, K23, and K56; methylated at R2 and K4, K9, K27, K36, and K79; and phosphorylated at serines (S) 10 and S28 and threonine (T) 3 and T11. Histone H4 is most commonly acetylated at K5, K8, K12, and K16, methylated at R3 and K20, and phosphorylated at S1. Improvements in the quantitative detection of distinct histone modifications would lead to a better understanding of the complex epigenetic regulation of gene expression.

13.2.1 Histone Acetylation

One of the best-studied histone posttranslational modifications is acetylation, the transfer of an acetyl group from acetyl coenzyme A to the lysine (K) side chain in the acceptor histone. Histone acetylation and deacetylation of histones are modulated by the actions of two opposing enzymes, histone acetyltransferases (HATs) and histone deacetylases (HDACs) [16, 18]. The correlation between histone acetylation and increased transcription has been known for many years, whereby increases in HAT activity promote acetylation of histone proteins leading to increased gene transcription by creating a more open conformation of chromatin (Fig. 13.2). In contrast, HDAC activity involves removing the acetyl group from histones, which

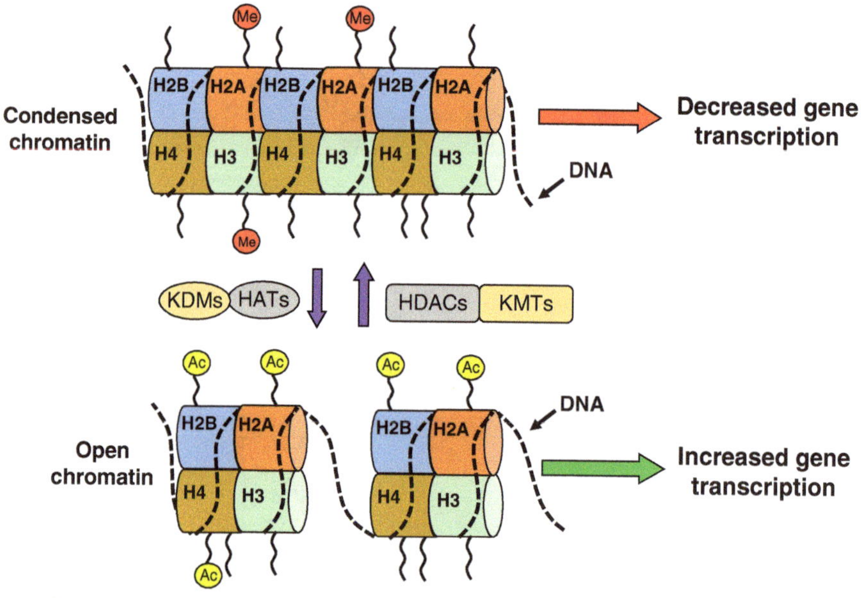

Fig. 13.2 Schematic depiction of histone factors that regulate condensed and open chromatin. The dynamic state of histone acetylation/deacetylation is regulated by histone acetyltransferase (HAT) and histone deacetylase (HDAC) enzymes, whereby histone methylation is governed broadly by lysine methyltransferases (KMTs) and lysine demethylases (KDMs)

results in a decrease in the space between the nucleosome and the DNA that is wrapped around it, resulting in condensation of chromatin structure and ensuing repression of gene expression (Fig. 13.2). However, the precise mechanisms of transcriptional regulation are likely to be more complex and involve many other chromatin-related proteins.

HATs makes up a large family of proteins, with diverse members comprising five families based on structural and functional similarities of their catalytic domains. These include the Gcn5-related N-acetyltransferase (GNAT) superfamily members, MYST proteins, p300/CREB-binding protein (CBP) HATs, general transcription factor HATs characterized by the TAF250 domain, and the steroid receptor coactivators/nuclear receptor coactivator family [19]. The specificity for acetylation of particular lysine residues for individual HAT enzymes is not clear.

The HDAC enzymes represent a related family of proteins with sequence and structural similarities [20]. In humans, the HDAC family of enzymes has 18 subtypes, which are divided into four main classes, classes I–IV [21]. HDACs exist in large multiprotein complexes, and evidence suggests that most, if not all, HDAC enzymes require interaction with other HDACs or proteins for optimal function [22, 23]. HDACs lack a DNA-binding motif, and one function of HDAC-interacting proteins is recruitment to their chromatin targets [24]. Acetylation of histones is a relatively transient mark, which is vital for precise temporal transcriptional control.

13.2.2 Histone Methylation

The enzymatic methylation of histones is performed by lysine methyltransferases (KMTs) and protein arginine (R) methyltransferases (PRMTs), with S-adenosyl-l-methionine as the methyl donor [25]. Histone methylation can involve the transfer of up to three methyl groups, thus resulting in mono-, di-, or tri-methylated lysine, respectively, and in mono- or di- methylated arginine. More than 50 different histone methyltransferases have been identified in humans, and unlike the HAT enzymes, they are specific for the K or R residues which they modify. There are two different families of lysine methyltransferases divided on the basis of their catalytic domain sequence: the DOT1-like proteins and the SET domain-containing proteins [26], which are further divided into four families, which include SET1, SET2, SUV39, and RIZ.

Another layer of specificity exists with histone methyltransferases, as many enzymes are specific only for particular residues. For example, SET1, MLL, and SMYD3 are histone methyltransferases that catalyze methylation of histone H3 at K4 in mammalian cells. Similarly, ESET, G9a, and SETDB1 catalyze methylation of histone H3 at K9. G9a and polycomb group enzymes such as EZH2 are histone methyltransferases that catalyze methylation of histone H3 at K27. Methylation at either H3K9 and H3K27 methylation mediates heterochromatin formation leading to gene silencing at euchromatic sites. Histone methyltransferases are also components of large, multiprotein nuclear complexes that contain other histone-modifying enzymes and other regulatory proteins including HATs, HDACs, and DNA methyltransferases.

Histone demethylation is the removal of methyl groups in modified histone proteins via histone lysine demethylases (KDMs). The discovery of histone demethylase enzymes demonstrates that histone methylation is not a permanent modification, like previously thought. To date, two classes of KDM have been described: the amine oxidase-type lysine-specific demethylases 1 and 2 (LSD1 and LSD2, also known as KDM1A and KDM1B, respectively) and the Jumonji C (JmjC) domain-containing histone demethylases. The latter consist of a group which contains over 30 members and can be divided, based on the JmjC-domain homology, into seven subfamilies (KDM2–KDM8) [26]. The specific amino acid residue and degree of methylation determines the demethylation enzyme. For example, on histone H3, mono- and di-methylated lysine 4 are demethylated by LSD1, whereby mono- and di-methylation of lysine 9 are demethylated by JMJD1 and tri-methylated lysine 9 is demethylated by JMJD2 [27]. Inhibition of histone demethylases may lead to histone re-methylation at specific residues important for chromatin dynamics and gene expression. Depending on the position and nature of the methylated residues, histone methylation can have positive as well as negative impacts on gene expression (Table 13.1).

Table 13.1 The most common mammalian histone modifications and their effects on chromatin

Modification	Histone	Residue	Effects of transcription
Acetylation	H2A	K5	Activation
	H2B	K5, K12, K15, K20	Activation
	H3	K4, K9, K14, K18,	Activation
	H3	K23, K36,	Activation
	H3	K56	DNA repair, histone deposition
	H4	K5, K8, K16	Activation
	H4	K12	Activation, histone deposition
	H4	K91	Histone deposition
Methylation	H3	K4, K79	Activation
	H3	K9, K27	Repression
	H3	R2, R8, R17, R26	Activation
	H3	K36	Elongation
	H4	R3	Activation
	H4	K20	Repression
Phosphorylation	H2A	S1, T120	Mitosis
	H2AX	S139	DNA repair
	H2B	S14	Apoptosis
	H3	T6	Activation
	H3	T3, S10, T11, S28	Mitosis, DNA repair
	H3	T45	DNA replication
	H4	S1	Mitosis, activation
Ubiquitination	H2A	K119	Repression
	H2B	K120	Elongation
	H3	K23	Maintenance of DNA methylation

K lysine, *R* arginine, *S* serine, *T* threonine

13.2.3 Histone Phosphorylation

Information is increasing rapidly on the phosphorylation of histones and their roles in cellular physiology and human diseases. Histone phosphorylation is targeted to S, T, and tyrosine (Y) residues, and its abundance can range from targeting a minute fraction of nucleosomes during the G0/G1 of the cell cycle to association with most nucleosomes of the G2/M-phase chromatin. There are eight characterized phosphorylation sites on the core canonical histones H2A, H2B, H3, and H4 (Fig. 13.1), which have been linked to specific cognate kinases. These occur in several pathways including DNA damage, mitosis/meiosis, apoptosis, and nuclear hormone signaling (Table 13.1). Among the better-characterized histone phosphorylation events are the phosphorylation of H3 at S10 and S28, which are associated with open chromatin and transcriptional activity and the phosphorylation of the H2A variant H2A.X at S139 (Table 13.1).

Serine/threonine protein kinases phosphorylate the OH group of serine or threonine, while tyrosine-specific protein kinases phosphorylate tyrosine amino acid residues. Several hundred protein kinases exist in mammals and are classified into distinct superfamilies. In contrast, protein phosphatases are the primary effectors of dephosphorylation and can be grouped into three main classes based on sequence, structure, and catalytic function. Phosphorylation of histones is diverse and complex with several different kinases being able to phosphorylate the same amino acid residues [28]. For example, phosphorylation of H3S10 in mammals is thought to occur via the actions of different kinases, including the Aurora B kinase, I-kappa-B kinase, Ribosomal protein S6 kinase A3 (a.k.a. RSK2), and AKT serine/threonine kinase 1 [28].

Probably the most important feature of histone phosphorylation is that it plays an important role in the interpretation of combinatorial posttranslational modifications [29]. An extensive cross talk exists between phosphorylation and other posttranslational modifications, which together regulate various biological processes, including gene transcription, DNA repair, and cell cycle progression [28]. For example, phosphorylation of H3S10, T11, and S28 has been clearly associated with H3 acetylation, strongly implicating these modifications in transcription activation.

13.2.4 Histone Ubiquitination

Although less well studied than the other posttranslational modifications, histone ubiquitination represents an additional opportunity for regulating the epigenome. In this context, histones can be monoubiquitinated by the addition of a single 8.5kDa (76 amino acids) ubiquitin molecule to specific lysine residues on histone tails. Histones are the most abundantly monoubiquitinated conjugates in the nucleus of mammalian cells, including the main sites of monoubiquitination at K119 on histone H2A and K120 on histone H2B (Fig. 13.1). Recently, lysine 34 was identified as a second monoubiquitination site on mammalian histone H2B [30].

The process of ubiquitination requires the sequential activities of three enzymes: ubiquitin-activating enzymes (E1), ubiquitin-conjugating enzymes (E2), and ubiquitin ligase enzymes (E3), which comprise large families, several of which specifically target histone 2A and 2B proteins [31]. Even though the ubiquitination of H2A was discovered in the 1970s, the first specific E3 ligase for this modification was not identified until 2004 [32]. Similar to other histone posttranslational modifications, H2 ubiquitination is reversible, and de-ubiquitination occurs via a class of thiol proteases known as deubiquitinating enzymes (DUBs) [31]. In contrast to polyubiquitination that marks a protein for proteasomal degradation, monoubiquitination of histones is associated with the transcriptional control of gene expression and the DNA damage response, including transcriptional reprogramming, the cell cycle checkpoint, and DNA repair.

13.3 Altered Histone Modifications in Schizophrenia

13.3.1 Human Studies

Seminal work by Mirnics and colleagues was the first to use microarray analysis to identify large-scale changes in gene expression in postmortem brains from subjects with schizophrenia [33]. Since then, a number of studies have found transcriptome-wide alterations in brain gene expression in schizophrenia (reviewed in [7, 34]). While the mechanisms(s) accounting for these gene expression changes are not clear, it has been suggested that epigenetic regulation of gene expression plays an important role in this phenomenon. Epigenetic modifications associated with repressed chromatin have been mostly implicated in schizophrenia. Such marks include increases in H3K9 and H3K27 methylation, indicative of a restrictive chromatin environment. Alternatively, lower levels of epigenetic marks associated with gene activation, i.e. H3K9 and H3K14 acetylation, would also suggest a restrictive environment. These epigenetic changes would lead to decreases in gene expression and would be consistent with the evidence for reproducible decreases in expression of a wide range of gene groups in schizophrenia, including glutamate decarboxylase 1 (GAD1), one of the most reproducible expression deficits found in schizophrenia, as well as synaptic genes, metabolic genes, and myelin-related genes. Changes in epigenetic marks have been found in both peripheral blood samples from patients with schizophrenia and from postmortem brain samples from patients with this devastation illness (see Table 13.2). One caveat of carrying out research on human subjects is that most, if not all, patients with schizophrenia have been treated with antipsychotic medications. Hence, the potential effects of drug exposure must be considered. However, the epigenetic antipsychotic drug studies published to date did not find similar changes in histone modifications due to drug treatment as those found in human patients.

Table 13.2 Altered histone modifications in patients with schizophrenia

Histone residue	Epigenetic mark	Tissue	Direction of change	Reference
Peripheral tissues:				
H3/H4	Acetylation	Lymphocytes	Decrease	[36]
H3K9/K14	Acetylation	Lymphocytes	Decrease	[37]
H3K9	Di-methylation	Lymphocytes	Increase	[39]
H3S10	Phosphorylation	PBMCs	Increase	[41, 44]
CNS tissues:				
H3R17	Methylation	PFC	Increase	[42]
H3K9	Di-methylation	Cortex	Increase	[38]
H3K4	Tri-methylation	PFC	Decrease	[43]
H3K27	Tri-methylation	PFC	Increase	[43]
H3K9	Di-methylation	Cortex	Increase	[38, 44]
H3K9/K14	Acetylation	PFC	Decrease	[73]
H3S10/H3K14	Phosphorylation/acetylation	PFC	Increase	[42]

PBMC peripheral blood mononuclear cells, *PFC* prefrontal cortex, *H* histone, *K* lysine, *R* arginine

13.3.1.1 Histone Changes in Patient Blood Cells

Over 20 years ago, the first epigenetic study using patient lymphocytes indicated an abnormal increase in heterochromatin in schizophrenia compared to normal control subjects [35]. Later studies found decreases in acetylated histones H3 and H4 in schizophrenia lymphocytes compared with lymphocytes from patients with bipolar disorder [36], as well as lower baseline levels of H3 K9/K14 acetylation in schizophrenia cells compared with those from normal control subjects [37]. Further, valproic acid, an HDAC inhibitor, elicited smaller increases in acetylated K9/K14 in schizophrenia patients compared with bipolar subjects [36]. These early studies supported the idea that schizophrenia is associated with more rigid chromatin.

Additional studies demonstrated increases in histone methylation marks associated with repressed chromatin. Significant increases in H3K9 di-methylation was demonstrated in lymphocytes from patients with schizophrenia [38, 39], in conjunction with elevated levels of three histone methyltransferase enzymes, G9a, GLP, and SETDB1, which are responsible for the majority of H3K9 methylation modifications across the genome [38]. Notably, SETDB1 is the only histone methyltransferase that specifically functions to di- and tri- methylate H3K9 [40].

Finally, histone H3S10 phosphorylation was found to be upregulated in peripheral blood mononuclear cells of schizophrenia patients in comparison to healthy controls, an effect that was negatively correlated with PANSS score [41]. Importantly, in this study, there was no significant effect of storage time or number of freeze-thaw events, age, sex, race, post-mortem interval, or history of smoking on levels of H3S10 phosphorylation [41].

13.3.1.2 Histone Modifications in Postmortem Brain from Subjects with Schizophrenia

In 2005, Akbarian and colleagues provided the first demonstration that there are histone modifications associated with schizophrenia in postmortem brain [42]. That study demonstrated that high levels of histone H3R17 methylation were associated with downregulated metabolic gene expression in the prefrontal cortex of a subset of subjects with schizophrenia [42]. In that same study, increases in histone H3S10 phosphorylation in combination with H3K14 acetylation [42] were also found in a subset of patients. Since their seminal study, they and others have provided further evidence for histone modifications in schizophrenia. With regard to histone methylation, Huang and colleagues found decreased levels of the open chromatin mark, H3K4 tri-methylation, and elevated levels of the repressive mark, H3K27 tri-methylation, in postmortem prefrontal cortex from subjects with schizophrenia, in conjunction with decreased levels of *GAD1* mRNA in patient brains [43]. Further, in a subset of clozapine-treated patients, there was an increase in H3K4 tri-methylation at the *GAD1* locus [43].

More recently, an increase in global cortical H3K9 di-methylation levels was reported in a schizophrenia cohort in the parietal cortex associated with increases of two of the enzymes that catalyze its formation, GLP and SETDB1 [38]. In a separate study, Chase and colleagues found that men with schizophrenia expressed higher levels of H3K9 di-methylation in conjunction with higher expression levels of G9α, SETDB1 methyltransferase mRNAs [44]. Additionally, higher levels of symptom presentation and an overall poorer quality of life were correlated with higher enzyme expression and H3K9 di-methylation levels [44].

Studies have also demonstrated altered acetylation levels of histone proteins. Histone H3 was found to be hypoacetylated at K9 and K14 at the promoter regions of several schizophrenia-related genes, including, *GAD1*, 5-hydroxytryptamine (serotonin) receptor 2C, and the myelin-related genes, myelin basic protein, and UDP glycosyltransferase 8 in young subjects with schizophrenia compared to age-matched controls [45]. This effect was correlated with an observed decrease in expression of these genes in young subjects with schizophrenia. In contrast, increases in acetylation of histone H3 was detected at the same promoter regions in subjects with chronic schizophrenia compared to their age-matched controls, although this effect was not directly correlated with increased expression of these genes in older subjects [45]. Interestingly, elevated levels of HDAC1 have been reported in postmortem cortical samples from subjects with schizophrenia, which may or may not be related to this effect [46].

13.3.2 Lessons from Mouse Models

Due to its heterogeneity and polygenic nature, schizophrenia has traditionally been difficult to study in animal models. Nonetheless, several diverse mouse models have been developed that replicate schizophrenia-like features. Accordingly, alterations in histone modifications have been demonstrated in different psychiatric-related mouse models.

One of the most widely studied mouse models of psychiatric disorders is maternal immune activation, which is typically achieved by injection of polyinosinic-polycytidylic acid (polyI:C), a viral mimicker, in the pregnant dam. This insult elicits a nonspecific immune response in the pregnant mouse and a range of behavioral, neuropathological, and molecular deficits in mature offspring of these mice [47]. Studies in this model have observed global hypoacetylation of histones H3 and H4 in the cortex and hippocampus of adult mice born to dams treated with polyI:C compared to sham-treated mice [48]. In addition, specific decreases in acetylation of histone H3 at K9/K14 were identified at the promoter regions of the glutamate receptor, ionotropic, AMPA 1 (*Gria1*), and Roundabout homolog 1 *(Robo1)* genes, in association with altered expression of these genes [48]. In another study, Disrupted-in-Schizophrenia 1 (*Disc1*), a schizophrenia risk gene often implicated in gene-environment interaction models, showed altered H3K4 tri-methylation after prenatal polyI:C exposure [49].

Epigenetic modifications connected with histone H3 methylation at K4 and K9 were measured in another neurodevelopmental model of schizophrenia, which is based on prenatal administration of methylazoxymethanol (MAM) at embryonic day 17. This DNA-alkylating agent produces behavioral and anatomical brain abnormalities that model some aspects of schizophrenia in the adult offspring. MAM treatment was found to alter the levels of H3K9 di-methylation in the prefrontal cortex before puberty. In contrast, H3K4 tri-methylation was noticeably decreased in adult rats [50].

Alterations in histone modifications have also been examined in response to maternal behavior. Exposure to adverse maternal conditions (i.e., stress) was associated with decreased histone H3K9/K14 acetylation at the *Bdnf* IV promoter in the brains of offspring [51]. This effect would likely lead to a decrease in expression of BDNF I in the brain. In contrast, maternal care assessed by frequency of licking the pups was associated with the opposite effect, that is, an increase in H4 acetylation in the olfactory bulb [52]. In another study, increased levels of maternal care were associated with increased levels of histone H3K9 acetylation and H3K4 tri-methylation at the metabotropic glutamate receptor gene (*mGluR1*) in hippocampus of adult offspring [53]. These posttranslational modifications were highly correlated, and both associate inversely with DNA methylation, and positively with gene transcription [53].

13.4 Therapeutic Implications

The current mainstay for treatment of schizophrenia is the use of antipsychotic medications. In general, these medications reduce symptomatology and prevent relapse in a large percentage of patients; however, it is known that many patients do not respond to traditional antipsychotic medications, which are also associated with a number of side effects. Hence, there is a critical need for improved therapeutics to treat schizophrenia and related psychiatric disorders. Given the known gene expression abnormalities associated with schizophrenia (reviewed in [7, 8]) and the evidence for a repressed chromatin state in schizophrenia, as summarized above, approaches that target relevant gene expression deficits could represent novel treatment approaches.

13.4.1 How Do Current Antipsychotics Affect the Epigenome?

The two main classes of antipsychotic medications are the "typical" and "atypical" antipsychotics. Typical antipsychotic drugs, which include haloperidol, chlorpromazine, and fluphenazine, are primarily dopamine D_2 receptor antagonists. The "atypical" antipsychotics, such as clozapine, risperidone, quetiapine, and olanzapine, have a range of affinities for several different neurotransmitter receptors in addition to those for dopamine [54–56]. Administration of antipsychotic drugs has been shown to elicit changes in gene expression. Several studies have shown increases in the expression of immediate early genes (IEGs) after acute drug exposure (reviewed in [8]). Many IEGs are known to act as transcriptional regulators, thereby linking receptor-mediated effects to changes in genomic activity. In addition, some studies have examined the expression of candidate genes in animals after chronic exposure to antipsychotic drugs [8]. Less is known about the underlying changes in chromatin structure associated with these gene expression changes. Hence, studies have begun to investigate how antipsychotic drugs might affect epigenetic mechanisms. Interestingly, both typical antipsychotics and atypical drugs have been shown to affect histone modification levels in the brains of mouse models.

Early studies showed that treatment with haloperidol rapidly induces the phosphorylation of histone H3 at S10 and the acetylation of H3 at K14 in bulk chromatin from striatum and in nuclei of striatal neurons [57]. Further, it was shown that haloperidol-induced H3 phosphoacetylation is inhibited by the NMDA receptor antagonist MK-801 [57]. Several other histone modifications, including K9 and K14 acetylation marks on histone H3, as well as K8 and K12 on histone H4, were not differentially regulated on a global level by haloperidol [57]. These results suggest that histone modifications and chromatin structure in striatal neurons are dynamically regulated by dopaminergic and glutamatergic inputs. Another study using haloperidol, demonstrated that both amphetamine, an addictive psychostimulant, and haloperidol, increase the phosphorylation of H3 at S28 and that this effect occurs in the context of H3K27 tri-methylation [58]. The increases in H3K27 tri-methylation and H3S28 phosphorylation occurred in distinct populations of neurons located in the striatum. Phosphorylation of histone H3 at S28 at genomic regions marked by tri-methylation of K27 often correlates with increased expression of genes normally repressed by polycomb group proteins, suggesting that these drugs may act by reactivating these repressed target genes [58].

Studies using chronic clozapine treatment in mice found drug-induced downregulation the transcription of metabotropic glutamate 2 receptor, in association with decreased histone H3 acetylation at its promoter [59]. This epigenetic change occurred in concert with a serotonin 5-HT2A receptor-dependent upregulation and increased binding of HDAC2 to the metabotropic glutamate 2 receptor promoter [59]. In other studies, treatment with the atypical antipsychotics, olanzapine, clozapine, as well as the HDAC inhibitor valproate, induced significant increases in acetylated histone H3 expression in the nucleus accumbens in mice [60].

13.4.2 Electroconvulsive Therapy and Histone Alterations

Electroconvulsive therapy (ECT) is one of the oldest treatments for schizophrenia, usually reserved for drug-resistant schizophrenia, although studies have shown that ECT relieves other symptoms of schizophrenia, such as delusions, hallucinations, or disorganized thinking [61]. Despite this, the molecular mechanisms underlying its clinical effect are incompletely understood. By assaying posttranslational modifications of histones at the promoter regions of several genes in rat hippocampus, it was demonstrated that electroconvulsive seizures induce histone modifications that correlate with transcriptional activation, such as global acetylation of histone H4 and acetylation of histone H3 [62]. Alterations in the expression of *c-fos*, *BDNF*, and *CREB* were proposed to play a role in the therapeutic-related effects induced after electroconvulsive seizures [62]. These data provide the first *in vivo* demonstration of the involvement of chromatin remodeling in ECT-induced regulation of gene expression in the brain and will help in understanding the mechanisms underlying the efficacy of this treatment.

13.4.3 Histone Deacetylase (HDAC) Inhibitors as Potential Novel Therapies for Schizophrenia

HDAC inhibitors have been widely implicated as a novel treatment strategy for several neurodegenerative diseases. Schizophrenia is associated with dysregulated gene expression, as evident from microarray and RNA-sequencing studies demonstrating thousands of genes altered in expression in postmortem brain tissue from patients with schizophrenia [7, 8]. Hence, epigenetic mechanisms of gene regulation are especially pertinent to this disorder due to the therapeutic potential of epigenetic drugs, which include HDAC inhibitors [63]. These drugs can act to rebalance epigenetic aberrations, thereby restoring or reversing gene expression abnormalities contributing to disease manifestation. Alternatively, HDAC inhibitors could be working to promote a relaxed state of chromatin, thereby facilitating access of immediate early genes and other transcriptional regulators to achieve proper gene transcription (Fig. 13.3). Accordingly, HDAC inhibitors have gained considerable attention as a relevant therapeutic option for psychiatric disorders [64].

Valproic acid (VPA) is a low potent HDAC inhibitor of class I and II HDACs, although it shows higher potency against class I enzymes. VPA has a long and established history of efficacy in the treatment of bipolar disorder, which is often effective as a primary medication. VPA has also been suggested to be useful in the treatment of schizophrenia. Previous reports suggest that typical and atypical antipsychotics are more potent, more efficacious, and less toxic if they are co-administered with VPA [65–67], although other studies did not report such benefit [68–70]. Conceptually, this benefit may occur by using an HDAC inhibitor to relax a restrictive chromatin environment, thereby facilitating a gene expression response

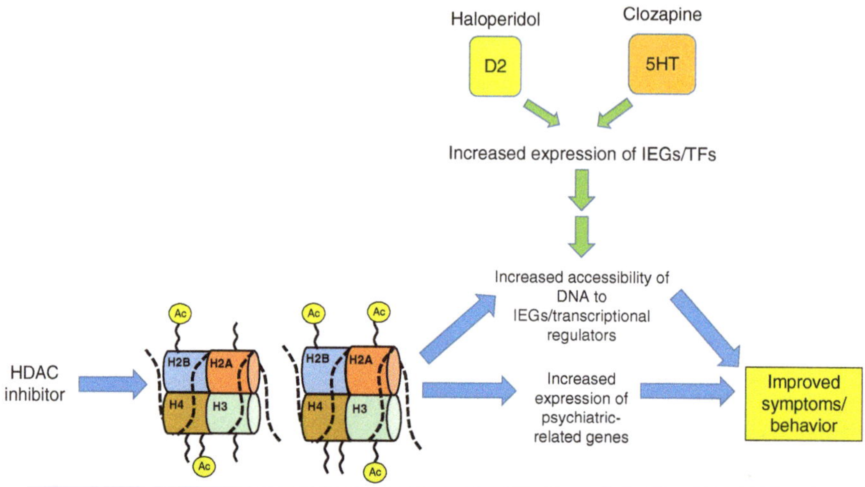

Fig. 13.3 HDAC inhibitors my act synergistically with antipsychotic drugs to promote proper gene expression. HDAC inhibitors increase acetylation of histones leading to relaxed chromatin and increased gene expression. In addition, acetylation of histones alters accessibility of chromatin and allows DNA-binding proteins to interact with exposed sites to activate gene transcription and downstream cellular functions. Immediate early gene (IEG); transcription factor (TF)

to antipsychotic medications, acting on signaling pathways via surface receptors, or by direct effects of the HDAC inhibitor. The beneficial effects obtained with VPA in the treatment of schizophrenia suggest that more potent or more highly-selective HDAC inhibitors may represent a new opportunity for pharmacological interventions for these disorders. It is believed that selective HDAC inhibitors will not show toxic properties that typically accompany treatment broad-spectrum HDAC inhibitors that are currently FDA-approved [71, 72].

Consistent with this view, studies have shown that a selective HDAC1/HDAC3 inhibitor, 4b, altered the levels of 17 schizophrenia candidate genes in mouse brain [73] Other studies have shown that another class I HDAC inhibitor, MS-275, potently activates *GAD1* gene expression in NT2 cells accompanied by decreased *GAD1* promoter methylation [74]. Further, a study using a benzamide-based inhibitor, Cpd-60, targeting HDAC1 and HDAC2, showed efficacy in mood-related behavioral assays. Cpd-60 treatment was associated with attenuated locomotor activity following acute amphetamine challenge and decreased immobility in the forced swim test [75]. Lithium has also been shown to increase the levels of histone H3 acetylation and phosphoacetylation in the central nucleus of the amygdala, in association with the induction of c-Fos [76]. Treatment with the HDAC inhibitor, sodium butyrate, was found to enhance this effect [76], providing additional support for adjunctive treatment.

13.5 Conclusion and Perspectives

In summary, studies over the past decade have demonstrated abnormalities in various histone modifications in peripheral and brain tissues from subjects with schizophrenia, as well as in animal models of the disease. The significance of these findings lies in the potential for these abnormal marks to serve as targets for novel epigenetic therapies. Much of the evidence to date supports an underlying dysfunction which would lead to an overall restrictive chromatin state and subsequent gene repression. Hence, compounds or drugs that would reverse this heterochromatin state might decrease psychiatric symptoms, whether administered as monotherapy or as an adjunct to traditional antipsychotic treatment. However, recent data indicate that the dysfunctional epigenetic mechanisms in schizophrenia may be more complex. The major challenges for epigenetic therapies are target specificity and the design of selective inhibitors against such targets. Additional epigenetic studies in schizophrenia should hopefully shed light on these issues.

References

1. Lewis DA, Lieberman JA. Catching up on schizophrenia: natural history and neurobiology. Neuron. 2000;28:325–34.
2. Meltzer HY. Suicidality in schizophrenia: a review of the evidence for risk factors and treatment options. Curr Psychiatry Rep. 2002;4:279–83.
3. Giegling I, Hartmann AM, Genius J, Benninghoff J, Moller HJ, Rujescu D. Systems biology and complex neurobehavioral traits. Pharmacopsychiatry. 2008;41(Suppl 1):S32–6.
4. Riley B, Kendler KS. Molecular genetic studies of schizophrenia. Eur J Hum Genet. 2006;14:669–80.
5. Ross CA, Margolis RL, Reading SA, Pletnikov M, Coyle JT. Neurobiology of schizophrenia. Neuron. 2006;52:139–53.
6. McDonald C, Murray RM. Early and late environmental risk factors for schizophrenia. Brain Res Brain Res Rev. 2000;31:130–7.
7. Mirnics K, Levitt P, Lewis DA. Critical appraisal of DNA microarrays in psychiatric genomics. Biol Psychiatry. 2006;60:163–76.
8. Thomas EA. Molecular profiling of antipsychotic drug function: convergent mechanisms in the pathology and treatment of psychiatric disorders. Mol Neurobiol. 2006;34:109–28.
9. Deutsch SI, Rosse RB, Mastropaolo J, Long KD, Gaskins BL. Epigenetic therapeutic strategies for the treatment of neuropsychiatric disorders: ready for prime time? Clin Neuropharmacol. 2008;31:104–19.
10. Oh G, Petronis A. Environmental studies of schizophrenia through the prism of epigenetics. Schizophr Bull. 2008;34:1122–9.
11. Roth TL, Lubin FD, Sodhi M, Kleinman JE. Epigenetic mechanisms in schizophrenia. Biochim Biophys Acta. 2009;1790:869–77.
12. Tsankova N, Renthal W, Kumar A, Nestler EJ. Epigenetic regulation in psychiatric disorders. Nat Rev Neurosci. 2007;8:355–67.
13. Grayson DR, Guidotti A. The dynamics of DNA methylation in schizophrenia and related psychiatric disorders. Neuropsychopharmacology. 2013;38:138–66.
14. Nishioka M, Bundo M, Kasai K, Iwamoto K. DNA methylation in schizophrenia: progress and challenges of epigenetic studies. Genome Med. 2012;4:96.

15. Feinberg AP. Phenotypic plasticity and the epigenetics of human disease. Nature. 2007;447:433–40.
16. Kornberg RD, Lorch Y. Twenty-five years of the nucleosome, fundamental particle of the eukaryote chromosome. Cell. 1999;98:285–94.
17. Quina AS, Buschbeck M, Di Croce L. Chromatin structure and epigenetics. Biochem Pharmacol. 2006;72:1563–9.
18. Strahl BD, Allis CD. The language of covalent histone modifications. Nature. 2000;403:41–5.
19. An W. Histone acetylation and methylation: combinatorial players for transcriptional regulation. Subcell Biochem. 2007;41:351–69.
20. Gregoretti IV, Lee YM, Goodson HV. Molecular evolution of the histone deacetylase family: functional implications of phylogenetic analysis. J Mol Biol. 2004;338:17–31.
21. Xu WS, Parmigiani RB, Marks PA. Histone deacetylase inhibitors: molecular mechanisms of action. Oncogene. 2007;26:5541–52.
22. Adcock IM, Ford P, Ito K, Barnes PJ. Epigenetics and airways disease. Respir Res. 2006;7:21.
23. Hildmann C, Riester D, Schwienhorst A. Histone deacetylases—an important class of cellular regulators with a variety of functions. Appl Microbiol Biotechnol. 2007;75:487–97.
24. Reichert N, Choukrallah MA, Matthias P. Multiple roles of class I HDACs in proliferation, differentiation, and development. Cell Mol Life Sci. 2012;69:2173–87.
25. Morales Y, Caceres T, May K, Hevel JM. Biochemistry and regulation of the protein arginine methyltransferases (PRMTs). Arch Biochem Biophys. 2016;590:138–52.
26. Morera L, Lubbert M, Jung M. Targeting histone methyltransferases and demethylases in clinical trials for cancer therapy. Clin Epigenetics. 2016;8:57.
27. Klose RJ, Yamane K, Bae Y, Zhang D, Erdjument-Bromage H, Tempst P, et al. The transcriptional repressor JHDM3A demethylates trimethyl histone H3 lysine 9 and lysine 36. Nature. 2006;442:312–6.
28. Banerjee T, Chakravarti D. A peek into the complex realm of histone phosphorylation. Mol Cell Biol. 2011;31:4858–73.
29. Sawicka A, Seiser C. Sensing core histone phosphorylation—a matter of perfect timing. Biochim Biophys Acta. 2014;1839:711–8.
30. Wu L, Zee BM, Wang Y, Garcia BA, Dou Y. The RING finger protein MSL2 in the MOF complex is an E3 ubiquitin ligase for H2B K34 and is involved in crosstalk with H3 K4 and K79 methylation. Mol Cell. 2011;43:132–44.
31. Fuchs G, Oren M. Writing and reading H2B monoubiquitylation. Biochim Biophys Acta. 2014;1839:694–701.
32. Wang H, Wang L, Erdjument-Bromage H, Vidal M, Tempst P, Jones RS, et al. Role of histone H2A ubiquitination in Polycomb silencing. Nature. 2004;431:873–8.
33. Mirnics K, Middleton FA, Marquez A, Lewis DA, Levitt P. Molecular characterization of schizophrenia viewed by microarray analysis of gene expression in prefrontal cortex. Neuron. 2000;28:53–67.
34. Horvath S, Janka Z, Mirnics K. Analyzing schizophrenia by DNA microarrays. Biol Psychiatry. 2010;69:157–62.
35. Kosower NS, Gerad L, Goldstein M, Parasol N, Zipser Y, Ragolsky M, et al. Constitutive heterochromatin of chromosome 1 and Duffy blood group alleles in schizophrenia. Am J Med Genet. 1995;60:133–8.
36. Sharma RP, Rosen C, Kartan S, Guidotti A, Costa E, Grayson DR, et al. Valproic acid and chromatin remodeling in schizophrenia and bipolar disorder: preliminary results from a clinical population. Schizophr Res. 2006;88:227–31.
37. Gavin DP, Kartan S, Chase K, Grayson DR, Sharma RP. Reduced baseline acetylated histone 3 levels, and a blunted response to HDAC inhibition in lymphocyte cultures from schizophrenia subjects. Schizophr Res. 2008;103:330–2.
38. Chase KA, Gavin DP, Guidotti A, Sharma RP. Histone methylation at H3K9: evidence for a restrictive epigenome in schizophrenia. Schizophr Res. 2013;149:15–20.

39. Gavin DP, Rosen C, Chase K, Grayson DR, Tun N, Sharma RP. Dimethylated lysine 9 of histone 3 is elevated in schizophrenia and exhibits a divergent response to histone deacetylase inhibitors in lymphocyte cultures. J Psychiatry Neurosci. 2009;34:232–7.
40. Zee BM, Levin RS, Xu B, LeRoy G, Wingreen NS, Garcia BA. In vivo residue-specific histone methylation dynamics. J Biol Chem. 2010;285:3341–50.
41. Sharma RP, Feiner B, Chase KA. Histone H3 phosphorylation is upregulated in PBMCs of schizophrenia patients in comparison to healthy controls. Schizophr Res. 2015;169:498–9.
42. Akbarian S, Ruehl MG, Bliven E, Luiz LA, Peranelli AC, Baker SP, et al. Chromatin alterations associated with down-regulated metabolic gene expression in the prefrontal cortex of subjects with schizophrenia. Arch Gen Psychiatry. 2005;62:829–40.
43. Huang HS, Matevossian A, Whittle C, Kim SY, Schumacher A, Baker SP, et al. Prefrontal dysfunction in schizophrenia involves mixed-lineage leukemia 1-regulated histone methylation at GABAergic gene promoters. J Neurosci. 2007;27:11254–62.
44. Chase KA, Rosen C, Rubin LH, Feiner B, Bodapati AS, Gin H, et al. Evidence of a sex-dependent restrictive epigenome in schizophrenia. J Psychiatr Res. 2015;65:87–94.
45. Tang B, Dean B, Thomas EA. Disease- and age-related changes in histone acetylation at gene promoters in psychiatric disorders. Transl Psychiatry. 2012;1:e64.
46. Sharma RP, Grayson DR, Gavin DP. Histone deactylase 1 expression is increased in the prefrontal cortex of schizophrenia subjects: analysis of the National Brain Databank microarray collection. Schizophr Res. 2008;98:111–7.
47. Meyer U, Feldon J. To poly(I:C) or not to poly(I:C): advancing preclinical schizophrenia research through the use of prenatal immune activation models. Neuropharmacology. 2012;62(3):1308–21.
48. Tang B, Jia H, Kast RJ, Thomas EA. Epigenetic changes at gene promoters in response to immune activation in utero. Brain Behav Immun. 2013;30:168–75.
49. Connor CM, Dincer A, Straubhaar J, Galler JR, Houston IB, Akbarian S. Maternal immune activation alters behavior in adult offspring, with subtle changes in the cortical transcriptome and epigenome. Schizophr Res. 2012;140:175–84.
50. Mackowiak M, Bator E, Latusz J, Mordalska P, Wedzony K. Prenatal MAM administration affects histone H3 methylation in postnatal life in the rat medial prefrontal cortex. Eur Neuropsychopharmacol. 2014;24:271–89.
51. Blaze J, Asok A, Roth TL. Long-term effects of early-life caregiving experiences on brain-derived neurotrophic factor histone acetylation in the adult rat mPFC. Stress. 2015;18:607–15.
52. de Moura AC, da Silva IR, Reinaldo G, Dani C, Elsner VR, Giovenardi M. Global histone H4 acetylation in the olfactory bulb of lactating rats with different patterns of maternal behavior. Cell Mol Neurobiol. 2016;36:1209–13.
53. Bagot RC, Zhang TY, Wen X, Nguyen TT, Nguyen HB, Diorio J, et al. Variations in postnatal maternal care and the epigenetic regulation of metabotropic glutamate receptor 1 expression and hippocampal function in the rat. Proc Natl Acad Sci U S A. 2012;109(Suppl 2): 17200–7.
54. Bymaster FP, Calligaro DO, Falcone JF, Marsh RD, Moore NA, Tye NC, et al. Radioreceptor binding profile of the atypical antipsychotic olanzapine. Neuropsychopharmacology. 1996;14:87–96.
55. Jann MW. Clozapine. Pharmacotherapy. 1991;11:179–95.
56. Kerwin R, Taylor D. Antipsychotics—a review of the current status and clinical potential. CNS Drugs. 1996;6:71–82.
57. Li J, Guo Y, Schroeder FA, Youngs RM, Schmidt TW, Ferris C, et al. Dopamine D2-like antagonists induce chromatin remodeling in striatal neurons through cyclic AMP-protein kinase A and NMDA receptor signaling. J Neurochem. 2004;90:1117–31.
58. Bonito-Oliva A, Sodersten E, Spigolon G, Hu X, Hellysaz A, Falconi A, et al. Differential regulation of the phosphorylation of Trimethyl-lysine27 histone H3 at serine 28 in distinct populations of striatal projection neurons. Neuropharmacology. 2016;107:89–99.

59. Kurita M, Holloway T, Garcia-Bea A, Kozlenkov A, Friedman AK, Moreno JL, et al. HDAC2 regulates atypical antipsychotic responses through the modulation of mGlu2 promoter activity. Nat Neurosci. 2012;15:1245–54.
60. Ookubo M, Kanai H, Aoki H, Yamada N. Antidepressants and mood stabilizers effects on histone deacetylase expression in C57BL/6 mice: brain region specific changes. J Psychiatr Res. 2013;47:1204–14.
61. Sommer IE, Slotema CW, Daskalakis ZJ, Derks EM, Blom JD, van der Gaag M. The treatment of hallucinations in schizophrenia spectrum disorders. Schizophr Bull. 2012;38:704–14.
62. Tsankova NM, Kumar A, Nestler EJ. Histone modifications at gene promoter regions in rat hippocampus after acute and chronic electroconvulsive seizures. J Neurosci. 2004;24:5603–10.
63. Kazantsev AG, Thompson LM. Therapeutic application of histone deacetylase inhibitors for central nervous system disorders. Nat Rev Drug Discov. 2008;7:854–68.
64. Akbarian S. Epigenetic mechanisms in schizophrenia. Dialogues Clin Neurosci. 2014;16:405–17.
65. Citrome L. Schizophrenia and valproate. Psychopharmacol Bull. 2003;37(Suppl 2):74–88.
66. Wassef AA, Dott SG, Harris A, Brown A, O'Boyle M, Meyer 3rd WJ, et al. Randomized, placebo-controlled pilot study of divalproex sodium in the treatment of acute exacerbations of chronic schizophrenia. J Clin Psychopharmacol. 2000;20:357–61.
67. Wassef AA, Hafiz NG, Hampton D, Molloy M. Divalproex sodium augmentation of haloperidol in hospitalized patients with schizophrenia: clinical and economic implications. J Clin Psychopharmacol. 2001;21:21–6.
68. Casey DE, Daniel DG, Tamminga C, Kane JM, Tran-Johnson T, Wozniak P, et al. Divalproex ER combined with olanzapine or risperidone for treatment of acute exacerbations of schizophrenia. Neuropsychopharmacology. 2009;34:1330–8.
69. Casey DE, Daniel DG, Wassef AA, Tracy KA, Wozniak P, Sommerville KW. Effect of divalproex combined with olanzapine or risperidone in patients with an acute exacerbation of schizophrenia. Neuropsychopharmacology. 2003;28:182–92.
70. Citrome L. Adjunctive lithium and anticonvulsants for the treatment of schizophrenia: what is the evidence? Expert Rev Neurother. 2009;9:55–71.
71. Balasubramanian S, Verner E, Buggy JJ. Isoform-specific histone deacetylase inhibitors: the next step? Cancer Lett. 2009;280:211–21.
72. Chateauvieux S, Morceau F, Dicato M, Diederich M. Molecular and therapeutic potential and toxicity of valproic acid. J Biomed Biotechnol. 2010;2010:18.
73. Tang B, Dean B, Thomas EA. Disease- and age-related changes in histone acetylation at gene promoters in psychiatric disorders. Transl Psychiatry. 2011;1:e64.
74. Chen Y, Dong E, Grayson DR. Analysis of the GAD1 promoter: trans-acting factors and DNA methylation converge on the 5′ untranslated region. Neuropharmacology. 2011;60:1075–87.
75. Schroeder FA, Lewis MC, Fass DM, Wagner FF, Zhang YL, Hennig KM, et al. A selective HDAC 1/2 inhibitor modulates chromatin and gene expression in brain and alters mouse behavior in two mood-related tests. PLoS One. 2013;8:e71323.
76. Kwon B, Houpt TA. Phospho-acetylation of histone H3 in the amygdala after acute lithium chloride. Brain Res. 2010;1333:36–47.

Epigenetic Mechanisms of Gene Regulation in Amyotrophic Lateral Sclerosis

14

Alba Jimenez-Pacheco, Jaime M. Franco, Soledad Lopez, Juan Miguel Gomez-Zumaquero, Maria Magdalena Leal-Lasarte, Diana E. Caballero-Hernandez, Marta Cejudo-Guillén, and David Pozo

Abstract

Despite being clinically described 150 years ago, the mechanisms underlying amyotrophic lateral sclerosis (ALS) pathogenesis have not yet been fully understood. Studies in both animal models of ALS and human patients reveal a plethora of alterations such as increased glutamate-mediated excitotoxicity, redox stress, increased apoptosis, defective axonal transport, protein-misfolding events, mitochondrial impairment and sustained unregulated immune responses. Regardless of being sporadic or familiar ALS, the final outcome at the cellular level is the death of upper and lower motor neurons, and once diagnosed, ALS is typically lethal within the next 5 years. There are neither clear biomarkers nor therapeutic or disease-modifying treatments for ALS.

Alba Jimenez-Pacheco and Jaime M. Franco contributed equally to this work.

A. Jimenez-Pacheco • J.M. Franco • S. Lopez • M. Cejudo-Guillén • D. Pozo (✉)
CABIMER, Andalusian Center for Molecular Biology and Regenerative Medicine,
Av. Americo Vespucio s/n, 41092 Seville, Spain

Department of Medical Biochemistry, Molecular Biology and Immunology, University of Seville Medical School, Seville, Spain
e-mail: david.pozo@cabimer.es

J.M. Gomez-Zumaquero
Genomic Unit, Malaga Institute of Biomedical Research (IBIMA), Malaga, Spain

M. Magdalena Leal-Lasarte
CABIMER, Andalusian Center for Molecular Biology and Regenerative Medicine,
Av. Americo Vespucio s/n, 41092 Seville, Spain

D.E. Caballero-Hernandez
CABIMER, Andalusian Center for Molecular Biology and Regenerative Medicine,
Av. Americo Vespucio s/n, 41092 Seville, Spain

School of Biological Sciences, Autonomous University of Nuevo Leon (UANL),
Nuevo Leon, Mexico

R. Delgado-Morales (ed.), *Neuroepigenomics in Aging and Disease*, Advances in Experimental Medicine and Biology, DOI 10.1007/978-3-319-53889-1_14

Accumulating evidence supports the concept that epigenetic-driven modifications, including altered chromatin remodelling events, RNA editing and non-coding RNA molecules, might shed light into the pathogenic mechanisms underlying sporadic/familiar ALS onset and/or severity to facilitate the identification of effective therapies, early diagnosis and potentially early-stage therapeutic interventions to increase the survival outcome of ALS patients.

Keywords
MicroRNAs • Frontotemporal dementia (FTP) • Neurodegeneration, DNA methyltransferases (DNMT) • Histone acetyl transferases (HAT) • Histone deacetylases (HDACs) • Neuromuscular • Neurodegeneration • Motor neurone disease (MND)

14.1 Introduction

Amyotrophic lateral sclerosis (ALS) is a fatal motor neuron disease, characterized by a progressive loss of the motor neurons at the spinal or bulbar level [1, 2]. Unfortunately, so far, there are no explicit evidences of the final causative mechanisms [1]. ALS was first described in 1874 by a French neurologist called Jean-Martin Charcot. Hence, ALS is also known as Charcot disease. However, it could be also mentioned as Lou Gehrig's disease after the name of the hall-of-fame baseball player for the New York Yankees diagnosed with ALS in 1939 [2, 3]. ALS is the most common type of motor neuron disease and the reason why it is also called motor neurone disease (MND) in UK and Australia. ALS patients (up to 50%) demonstrate mild to moderate cognitive and/or behavioural impairment due to frontotemporal degeneration (FTD) [4]. Thus, patients with FTD/ALS may first present features of either FTD or ALS with the additional symptoms developing as the disease progresses.

ALS is a rare neurodegenerative disease, being almost impossible to be diagnosed before the disease is clinically evident [1, 5]. The incidence rate of the disease remains at 2.7/100.000 [6], where the mean age of ALS onset varies from 50 to 65 years old, with less than 5% of the cases showing an onset under 30 years old [7, 8]. Once that ALS is diagnosed is typically lethal within 5 years, irrespective of the type/form of ALS [1]. ALS can be divided in two categories, sporadic and familial ALS. The most common form is sporadic ALS (90–95%) with no clear inherited component. Familial ALS is seen only in approximately 5–10% of the cases, where a genetic dominant inheritance can be associated [9, 10]. The disease has been traditionally considered as a distinctive disorder, but in the recent years, as ALS shows a wide phenotypic heterogeneity, new theories propose a common final phenotype for different disorders [11]. As we mentioned before, the motor neuron degeneration and structural and pathologic changes correlating with cognitive dysfunction have been found in ALS patients [12]. The prevalence and types of cognitive impairment associated with ALS are very variable [4]. Recent cross-sectional studies indicate that ALS patients (55–75%) also have cognitive impairment, where a subset of 15% of patients approximately exhibiting features of frontotemporal dementia (FTD), frontotemporal lobar degeneration (FTLD), as well as progressive social, behavioural and/or language dysfunctions [4, 13, 14].

The main clinical feature of ALS is a combination of upper and lower motor neuron damage where the patients experience localized muscle weakness and eventually muscle wasting [11]. Commonly, patients develop bulbar and respiratory symptoms and spasticity, affecting manual dexterity and gait [15]. The main cause of death in ALS is respiratory failure, as the result of pulmonary complications [16]. In order to prolong their lives, patients undergo tracheostomy-delivered assisted ventilation developing ultimately a state motor paralysis known as a 'totally locked-in state', involving the paralysis of all voluntary muscles and oculomotor impairment [17].

Currently, riluzole—a chemical derivative of 2-aminobeanzothiazole—is the only drug (Rilutek® as tablet form or Teglutik® as liquid form) approved for the treatment of ALS by the agencies for the evaluation of medicinal products (FDA in USA; EMA in Europe). Different clinical trials have shown a marginal increase (2–3 months) in life expectancy [18–21]. The molecular bases of action of riluzole are still unclear. Several studies have reported that riluzole disables voltage-gated sodium channels associated to damaged neurons; controversially, it has been proposed to act as an antagonist of the NMDA (*N*-methyl-D-aspartate) receptors, and, finally, it has been shown to increase the glutamate uptake from the synapse and to impair glutamate release from nerve terminals [22, 23]. Thus, to some extension, riluzole seems to act by reducing glutamate bioavailability and therefore NMDA-related excitotoxicity.

14.2 Epigenetics for ALS

14.2.1 The Genetics Context in ALS

SOD1 (superoxide dismutase 1) mutations account for a majority of cases of fALS and contribute to some of the sALS cases, where the misfolding of the mutant SOD1 forms aggregates and, through non-clearly established events, leads to motor neuron cytotoxicity in the central nervous system (CNS) [24–26]. Additional genes have been also reported to be involved in the development of fALS [1]. The TAR DNA binding protein (*TARDBP*) gene encodes a protein called transactive response DNA binding protein 43 kDa (TDP-43). Around 60 point mutations in the *TARDBP* gene have been linked to ALS. The fused-in-sarcoma (*FUS*) gene encodes a protein involved, as TDP-43, in RNA transcription, splicing, and transport [1, 6]. Recently, alterations of the frequency of a hexanucleotide repeat expansion in the non-coding region of the *C9ORF72* gene has been reported as the most common cause of fALS [27]. Other reported allelic variants of human-mutated genes have been linked to ALS [28] [1, 27, 29–31].

The recent information gathered on the human genetics of ALS indicates that the disease can be understood as the result of the complex interactions between sometimes genetically inherited factors and a given physiological context. Besides, the scenario is even more complicated as ALS shows a great deal of phenotypic variability in both sporadic and familial cases. Finally, this is particularly relevant in the context of a neurodegenerative disease like ALS, where the roles in the pathological process of ALS-associated genes in different cell types point to non-cell autonomous mechanisms for motor neuron dysfunction. Therefore, the final ALS outcome is dependent on factors external to the motor neuron, as immune imbalance or energy homeostasis impairment. A minority of ALS cases can be traced from generation to generation in a family with the identification of a gene mutation. However, most ALS cases are believed to be caused by long-lasting effects of environmental factors on the genome [32]. This situation puts emphasis on epigenetics mechanisms to better understand the pathogenesis of ALS and the smart development of therapeutic strategies.

14.2.2 Key Tenets of Epigenetics in ALS

Defined as the area of research that studies how the environmental modifications on genes could affect the normal biology of an individual and contribute to the development of different diseases, epigenetics was first described at the end of the 1930s by the British biologist Conrad H. Waddington. He defined it as the interplay between genetics and external forces during development that define the phenotype of a given organism [33]. The significance of the term has always been controversial, raising two different interpretations of the concept. On one hand, epigenetics could be described as the study of heritable changes in gene function due to modifications that maintain the so-called epigenome [34]. A second interpretation describes epigenetics as the structural changes of particular chromosomal regions in order to preserve a gene expression profile maintained by epigenetic mechanisms [35, 36].

The current working paradigm of gene expression regulation in eukaryotic cells is based on the non-covalent binding of nuclear factors to the promoter region of a gene. A fundamental concept is the degree of accessibility of the DNA sequence controlled by the chromatin structure through either binding of non-coding RNAs to specific DNA sequences or by chemical strategies for covalent modifications of the chromatin. Although epigenetic mechanisms are also highly dynamic, they usually confer some stability at mid or long term to a given transcriptional cell state [37]. Therefore, according to a more flexible interpretation, the epigenetic mechanisms are involved in long-term tissue adaptation (i.e. muscle accommodation to exercise or long-term memory in the CNS) in addition to embryogenesis and tissue differentiation. Hereafter, we will continue by analysing the different mechanisms of epigenetic gene regulation in ALS.

14.2.3 The Involvement of MicroRNAs (miRNAs) as Epigenetics Modifications in ALS

Recent evidence shows that approximately up to 60% of all protein-coding genes might be regulated by miRNAs [38]. miRNAs are short evolutionarily conserved non-coding RNA molecules involved in post-transcriptional regulation of gene expression [39]. In summary, the process is as follows: the core nuclease Drosha and the molecular anchor protein DGCR8 modify primary miRNA molecules in the nuclear 'microprocessor' complex formed after transcription in order to export them afterwards to the cytoplasm by exportin 5 [40–42] (Fig. 14.1). Once in the cytoplasm, pre-miRNAs are further processed by the Dicer complex to finally generate a ~ 22-nucleotide double-stranded mature miRNAs. Then, the protein argonaute 2 (AGO2) binds to the mature miRNAs in the RNA-induced silencing complex (RISC) promoting complementary strand separation [43]. Sequence-specific interactions between the single-stranded miRNA and the specific mRNA target mediate the silencing and the final repression of protein production either by blocking translation or by induced transcript degradation [44, 45]. The involvement and deregulation of multiple miRNAs have been reported in several neurodegenerative diseases although the implications and mechanisms associated to miRNA dysregulation in this context have not been fully elucidated. For example, whether dysregulation occurs via transcriptional, post-transcriptional or through both mechanisms still remains to be addressed [46, 47]. Some illustrations of miRNA-mediated regulation of the expression of key proteins in other neurodegenerative diseases can be found in the case of Alzheimer's disease (AD), where tau and amyloid precursor proteins have been shown to be regulated by members of the miR-20 and miR-34 families [48, 49] or parkin expression in the case of brains from Parkinson's disease (PD) patients through miR-34b/c [50]. Moreover, there are also examples of miRNA-based profiling in PD, AD and Huntington's disease (HD) [51–54]. Thus, it is not surprising to find miRNAs involved in some of the pathophysiological process linked to ALS.

In this context, the processing and biogenesis of miRNAs can be modified by the interaction with TDP-43 protein [55] as it has been suggested and supported by other authors [36, 56]. Besides the DGCR8/Drosha complex, other polypeptides such as TDP-43 and AGO2 could interact with Drosha creating an alternative complex, which facilitates the synthesis of several pre-miRNAs [56]. Thus, this facilitating role of TDP-43 in miRNAs biogenesis could be affected by a low availability of TDP-43 [56]. Furthermore, cytosolic TDP-43 through interaction with the Dicer complex also enhances the previously mentioned pre-miRNAs production [56]. Interestingly, after TDP-43 silencing, a decreased expression of several miRNAs (miR-132-5p/3p, miR-143-5p/3p and miR-574-3p) in human lymphoblast cell lines from ALS patients has been reported [57]. Furthermore, the loss of Dicer has been associated with the progression of the spinal motor neuron degeneration [58]. Taken together, the complex formed by DGCR8, Drosha and TDP43 seems to be critical in the biogenesis and processing of miRNAs in human cells being its interaction with Dicer in the cytoplasm significant for miRNA processing in ALS [13, 41].

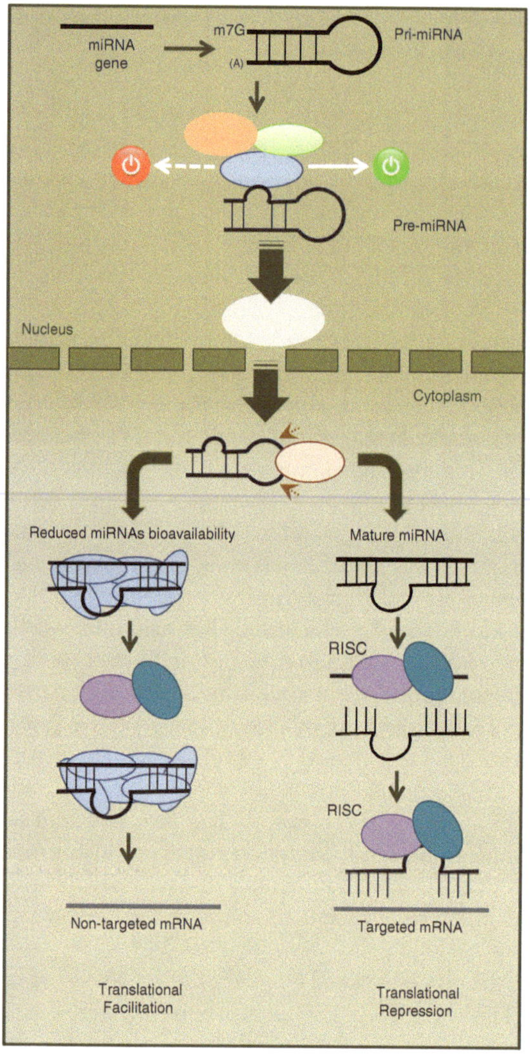

Fig. 14.1 Cytoplasmic protein aggregates disturb the correct miRNA biogenesis. A primary transcript is encoded in the nucleus by specific miRNA genes in conventional miRNA biogenesis. Further miRNA synthesis by the enzymatic complex Drosha and DGCR8 produces a stem loop pre-miRNA. TDP-43 is a component of the Drosha/DGCR8 miRNA-processing complex, and thus TDP-43 gain-of-function or loss-of-function mutations might alter miRNA processing. miRNAs correctly processed by Drosha/DGCR8/TDP43 are transported to the cytoplasm by exportin 5. Then, pre-miRNAs are cleaved by Dicer to create the 22-nucleotide double-stranded mature miRNAs. If cytosolic TDP-43 promotes the formation of aggregates with stress granules, they could bind specific miRNAs and thus alter their proper function accordingly. Mature miRNAs following a conventional processing will bind to AGO2 and TNRC6 proteins forming the RISC complex triggering the strand separation. Single-stranded miRNA complex binds to mRNA targets inducing mRNA silencing and repressing protein synthesis by blocking translation or transcript degradation. Abbreviations: AGO2, argonaute 2; DGCR8, microprocessor complex subunit DGCR8; miRNA, microRNA; RISC, RNA-induced silencing complex; TDP-43, TAR DNA binding protein 43; TNRC6, trinucleotide repeat-containing gene 6A protein

Although the precise role is still unclear, the downregulation of miRNAs as a common feature has been shown in both human ALS and in experimental ALS animal models [59].

On the opposite side, upregulation of several miRNAs have been reported in the context of ALS too. Thus, strong expression of miR-155 was reported in the SOD1-G93A animal model of ALS and further confirmed in human ALS spinal cord [60, 61]. In this sense, experiments using oligonucleotide-based miRNA inhibitors for miR-155 showed significantly prolonged survival rates in ALS mice [61]. Interestingly, miR-155 is involved in the regulation of relevant biological pathways related to ALS as TGF-β1 signalling, the stimulation of macrophage inflammatory responses and the enhancement of pro-inflammatory cytokine secretion [60]. Another example is miR-29, which has been recently reported to be highly expressed in the brain and spinal cord of the SOD1-G93A mouse model [62]. In this occasion, the knocking down of miR-29 did not affect the life expectancy of the SOD1-G93A mice [62].

The mechanisms driving microglia activation in ALS remain incompletely understood, and recent epigenetic mechanisms are gathering attention in this context. In this sense, upregulation of differentially expressed immune-related miRNAs such as miR- 22, miR-155, miR-125b, miR-146b and miR-365 was reported in activated microglia cultures of SOD1-G93A mice [63]. Particularly, miR-125b seems to be relevant in the modulation of TNF-α, while miR-365 regulates IL-6 production in the SOD1-G93A preclinical model of ALS [63, 64].

Additionally, the skeletal muscle mitochondrial dysfunction is a remarkable event in ALS, which is believed to play a central role in the progression and severity of ALS. In this context, an increased expression of specific miRNAs, including miR-23a, miR-29b, miR-206 and miR-455 in skeletal muscle of human ALS tissues, has been reported [65]. There are several functions ascribed to these miRNAs; miR-23a suppresses the mRNA and protein levels of the peroxisome proliferator-activated receptor γ coactivator 1α (PGC-1α), a transcription factor involved in thermogenesis and muscle adaptation to exercise, miR-29b is involved in muscle regeneration and miR-455 has a role in muscle wasting [65]. Moreover, in vitro and in vivo experiments have reported the involvement of miR-206 in the generation process of new neuromuscular junctions after nerve injury [66–71]. miR-206 has a role in the regeneration of neuromuscular synapses after nerve injury [71]. Thus, in SOD1-G93A mice, the lack of miR-206 accelerates disease progression [71]. In a similar way, the absence of miR-218, which is specifically expressed in motor neurons and involved in their differentiation process, has been reported to cause systemic neuromuscular failure and motor neuron cell loss [72, 73].

Interestingly, miR-218 seems to be related to glutamatergic and GABAergic signal transmission through the regulation of SLCIA2 and SLC6A1 transporters [72, 73]. Additionally, miR-9 and miR-124 have been reported to regulate the expression of the AMPA receptor, a non-NMDA-type inotropic receptor for glutamate, in the context of FTD/ALS [74, 75]. Moreover, downregulation of miR-9 expression was found in iPSC-derived neurons of FTD/ALS patients with TDP-43 mutations [76]. Besides, abnormal levels of miR-9 have been also involved in the regulation of dendritic growth associated to changes in the expression levels of neuronal cytoskeletal proteins or the repressor element 1 (RE1) silencing transcription factor REST [58, 77]. Finally, several miRNAs have been shown to be responsible for cell death and the inhibition of low molecular neurofilament (NFL) mRNA at the neuromuscular junction including miR-146a, miR-524-5p, miR-582-3p, miR-b1336 and miR-b2403 [78, 79].

Although there is still a gap in our knowledge on the precise mechanisms of action of miRNAs in the pathophysiology of ALS, their mere use as potential biomarkers for ALS is of interest for diagnosis and/or prognosis, acting as indicators for therapeutic response in clinical trials [80, 81].

14.2.4 DNA Methylation

DNA methylation is part of the epigenetic machinery. It consists in covalent modifications (addition of methyl or hydroxymethyl groups) to cytosine residues in particular DNA regions. The 5-methylcytosine (5mC) or hydroxymethylcytosine (5hmC) residues are produced by the specific actions of DNA methyltransferases (DNMTs) [82, 83]. These additions modify transcriptional protein–DNA interactions by changing chromatin structure and transcription rates, being DNA methylation usually related to transcriptional repression [84]. Recently Gaunt et al. have demonstrated that DNA methylation contains a significant heritable component, which remains consistent across the lifespan [85]. Main targets for the DNMTs are the CpG-rich sites in promoter sequences (CpG islands), where hypermethylation impairs the binding of transcription factors to these regulatory regions and therefore produces transcriptional silencing [86]. It is estimated that approximately 60–80% of the CpG islands present in the human genome are methylated [87, 88]. There are too mechanisms of passive and active reversal of CpG methylation, which mainly involves the oxidation of 5mC to 5-hmC [89–92].

Specifically related to ALS, pioneered studies reported a non-methylated pattern in the promoter sequences in some of the human genes involved in the disease, as *SOD1*, *VEGF* (vascular endothelial growth factor) and *GLT1* (glutamate type I transporter) [93, 94]. DNA methylation has recently been proposed a as marker of epigenetic dysfunction in ALS, where whole-blood DNA methylation is increased in ALS patients independently of age of the disease onset [95]. Additionally, in postmortem sporadic ALS spinal cord samples (unfortunately not in blood), Figueroa-Romero and collaborators have reported an increase in both 5mC and 5hmC levels, with opposed effects in terms of gene regulation, being transcriptional inhibitors the former and activators the latter [96]. This might be in agreement with the emerging role as a potential biomarker of 5hmC and 5mC in other neurodegenerative diseases such as AD, HD and PD [97]. Although further studies are required to assess the potential of changes in methylation status in ALS, it is worth mentioning that DNA methylation has been considered a basic mechanism for aging [98], in which it turns out to be an associated risk factor for this disease [99–101]. In this sense, by genome-wide analysis in the brain from sALS patients, Morahan and collaborators found that patients exhibited hypo- or hypermethylation at 38 methylation sites. Of these, 60% were genes involved in calcium homeostasis, neurotransmission and oxidative stress, which are relevant to ALS pathology [102].

Although there is no consensus, the functional activity of DNMTs has been associated to neurodegeneration and in particular to ALS in several studies. Chestnut and coworkers have reported that DNMT3A upregulation had a stronger proapoptotic role in neurons compared to DNMT1 in the context of ALS [82]. In the same line of evidence, Martin and collaborators have confirmed DNMT3A, but not DNMT1, as a main inducer of neuronal degeneration in cell cultures [103]. Finally, Wong and collaborators demonstrated that both mitochondrial DNA methylation and DNMT3A levels were abnormally increased in muscles and spinal cord from ALS mice [103]. The control of DNMT might have some potential in the context of cell therapies based on human bone marrow mesenchymal stromal cells in ALS patients [104, 105].

A clear example of how epigenetic changes in the promoter of a given gene modify the mRNA expression is illustrated by the ALS-linked mutations in the human *C9ORF72* gene [1]. These mutations are repeated expansion of short sequences in the promoter region of the gene and are associated to multiple deleterious mechanisms, mainly related to toxic gain of function properties of the mutated RNA produced and the subsequent formation of stable RNA-DNA hybrids (R-loops) [106].

It is recognized that the size of a pathological repeat expansion influences the severity of the symptoms of a given disease. In ALS, it is still not obvious if the clinical variability is associated to size variability of the *C9ORF72* G_4C_2 expansions while it seems to be associated to decreasing age of onset in ALS families segregating a *C9ORF72* G_4C_2 expansion [107, 108].

The methylation pattern in the CpG islands of the *C9ORF72* gene promoter has been recently addressed with contradictory conclusions. In this sense, Bauer and collaborators have suggested that an increase in the methylation level of the characteristic G_4C_2 expansion might impair the pathogenicity of *C9ORF72*-related diseases [109]. Interestingly, there is experimental evidence supporting a 'loss-of-function' hypothesis, where increased methylation of CpGs in the *C9ORF72* gene is associated to the presence of increasing G_4C_2 expansions size (i.e. number of repeat units) and ultimately to a reduction of the *C9ORF72* promoter activity [31]. However, the same authors reported a significantly delayed onset age in short expansion carriers. This illustrates a complex scenario where other alternative mechanisms of toxicity (RNA foci formation and/or aggregated dipeptide repeat (DPR) proteins through repeat-associated non-ATG translation) are likely to be involved in intermediate and long G_4C_2 expansions [31]. Thus, different groups have reported hypermethylation of the CpG islands located in the *C9ORF72* promoter region in about 10–30% of *C9ORF72* FTD-ALS patients [110–113]. However, Russ and collaborators concluded that methylation of the expanded repeats in *C9ORF72* was not significantly different between ALS and FTD patients, without predictive value in terms of age of onset, although the hypermethylation in specific tissues was associated with an extended life expectancy only in FTD [114].

In addition to DNA methylation in CpG islands in the promoter region of *C9ORF72*, a stable increase in the methylation levels of the large expansions of G_4C_2-repeats themselves (where an extra CpG island has been predicted only in expansion carriers) in ALS and FTD patients has been demonstrated very recently [115]. This opens the possibility to link methylated G_4C_2-repeats with the formation of G-quadruplex structures and related toxic events.

Beyond changes in DNA methylation, the methylation of arginine residues in particular histones is part of the epigenetic code. Interestingly, it has been recently suggested that FUS might impair the methylation of histone H4 by the human arginine methyltransferase PRMT1 as a consequence of the ALS-related aberrant accumulation of cytoplasmic FUS [116].

14.2.5 Role of Histone Modifications in ALS Epigenetic Regulation

In eukaryotic cells, DNA is packed by histones forming a nucleosome. The octameric core of the nucleosome is formed by two subunits of the histones H2A, H2B, H3 and H4. Addition of the linker histone H1 gives rise to the nucleosome. Post-translational modifications occur within arginine and/or lysine amino acids in the histone amino terminal tails protruding from the histone core. The usual modifications are addition of methyl and acetyl groups, but phosphorylation, ubiquitination or sumoylation have been also reported [117, 118].

The acetylation of the histones is one of the main epigenetic mechanisms, reducing the positive charges of the modified residues and therefore relaxing the tightly packed form of DNA in heterochromatin. Generally, increased levels of acetylation are associated to transcriptional activation (open remodelled chromatin), whereas deacetylation activity is linked to transcriptional silencing (chromatin condensation).

The overall level of histone acetylation depends on the interplay between histone acetyl transferases (HATs) and histone deacetylases (HDACs), although histones are not their exclusive substrates [119]. HDACs are grouped into four classes according to its homology to yeast deacetylases, and they differ in their subcellular localization and physiological roles. Briefly, class I HDACs (1, 2, 3 and 8) are mainly nuclear, whereas class II (IIa; 4, 5, 7 and 9; IIb 6 and 10) can shuttle between the nucleus and the cytoplasm, depending on the phosphorylation status [120]. HDAC11 is the only member of the class IV, and it can bind to HDAC6 [121]. Finally, class III HDACs are characterised by the use of nicotinamide adenine dinucleotide as acetyl group donor with no requirement of zinc. They are known as sirtuins (1 to 9) and can be localized in the nucleus, cytoplasm and mitochondria.

14.2.5.1 Changes in HDACs Expression Patterns in ALS

Pioneered work by Rouaux and collaborators reported in a SOD1 mouse model of ALS a sharp decrease of the acetylated form of H3 histone, specifically in motoneuron nuclei [122]. In ALS patients, mRNA levels of several HDACs have been determined in postmortem samples of motor cortex and spinal cord [120]. They found an upregulation of HDAC2 mRNA levels and a concomitant reduction in the HDAC11 mRNA levels in both tissues [120]. Interestingly, increased levels of HDAC2 mRNA have been described in Duchene's dystrophy, where its downregulation improved the symptoms in dystrophin-deficient mice [123, 124]. In a *Drosophila* model of Ataxia telangiectasia, a disorder characterized by progressive neurodegeneration, HDAC2, was shown to directly deacetylate the ATM (ataxia telangiectasia mutated) gene promoter, impairing its neuroprotective role [125]. In a cohort of ALS patients, those showing a rapid progressive disease exhibited an increased level of HDAC4 expression in skeletal muscle compared to patients with longer life expectancy [126].

So far, the smart manipulation of the mRNA expression of particular HDACs has not yet been addressed in the context of ALS. However, the inhibition of the HDACs activity has been explored by using small compounds that can efficiently cross the blood-brain barrier and therefore modify the epigenetic marks in the CNS. In this sense, valproic acid and butyrate can promote an increase in the acetylation status of the histones. Several of these compounds have been largely used in the treatment of some neurological diseases, such us bipolar disorders and seizures. Although the exposure of healthy cells to HDACs inhibitors might lead to apoptosis, it seems to be that the hypoacetylated status of the cells confers some kind of neuroprotective response [122].

Specifically, it has been shown in SOD1 mouse models of ALS (G93A and G86R) a state of hypoacetylation in histones from the spinal cord at the age of onset [127]. This effect is even more pronounced at the end stage of the disease in the SOD1 animal models. In this context, in vivo treatment with valproic acid induced a rapid increase in the acetylation status of histones after a few hours of administration. The chronic treatment with valproic acid induced a normalization of the histone acetylation at the end stage of the mouse model alongside an increase in total RNA transcription with a normalization of the expression of several pro-survival genes such as *Smn* and *Bcl-2* [127, 128]. This pharmacological approach did not improve the survival in the ALS mouse model but slightly delayed the onset of motor decline and muscular atrophy.

Some of the experimental strategies in preclinical studies using the SOD1 mouse model for ALS include combination regimens with HDAC pharmacological inhibition. Hence, riluzole and HDAC inhibition produced a decrease in reactive astrogliosis and motor neuron apoptosis, with an increase in the SOD1 mice survival of around a 20% compared to the separate administration of riluzole [129]. Other studies reported an enhanced neuroprotection by a combinatorial approach using mood stabilizers (lithium) in a mouse ALS model [130].

However, in terms of survival in the ALS mouse model, there are controversial reports related to the use of valproic acid or butyrate as HDAC inhibitors with positive effects [128, 131] or noncompliant results [127]. Most probably, the lack of uniformity in the starting point of the treatments might account for the differences observed. Trichostatin (TSA) is a broad HDAC inhibitor acting on class I and II HDACs, including the HDAC6 not targeted by valproic acid or butyrate. The in vivo treatment with TSA in ALS mice have shown an increase in normalized histone acetylation, neuroprotection and extended survival with partially restored normal phenotypes in microglia and motor neurons [132].

Although encouraging, the neuroprotective effects of HDACs inhibition are not particularly well understood in terms of mechanisms of action, and therefore the translation to the clinic is limited, as the clinical trials could not be properly designed. In this sense, the use of valproic acid in clinical trials—although well tolerated and regarded as safe—did not modify the overall survival rates for ALS [133].

14.2.5.2 Targeting HDACs Function Beyond Epigenetic Mechanisms

Protein acetylation is not confined to epigenetic chromatin remodelling and transcriptional regulation. So far, precise molecular targets of HDAC inhibitors are not well determined, neither the acetylation status of specific lysine residues modulating the function of other proteins [134, 135]. Thus, many other non-epigenetic-related cellular processes unregulated in the context of the pathophysiology of neurodegenerative diseases might be targets for HDAC inhibitors (e.g. microtubule dynamics or intracellular transport).

The class IIb HDAC6 represents an interesting example of this working paradigm. HDAC6 is expressed in most of the neurons; it has two catalytic domains and an additional ubiquitin-binding domain. HDAC6 in combination with sirtuin 2 is involved in the deacetylation of alpha tubulin [136], affecting the stability of microtubules, axonal transport and autophagy [137]. Remarkably, the specific inhibition of HDAC6 by tubacin has been shown to mainly affect the deacetylation of alpha tubulin, whereas the overall acetylation status of the histones is not altered [138]. In this sense, numerous works have analysed the role of both sirtuin 2 and HDAC6 inhibition in the pathogenesis of neurodegenerative diseases in the context of impairments in axonal transport [139]. HDAC6 inhibition is neuroprotective in models of HD and PD [140, 141]. In ALS, the absence of HDAC6 in the SOD1-G93A mouse model induced an increase in the levels of tubulin acetylation which was associated with an increase in the survival rates, without apparently affecting other neurological functions [142]. The role of HDAC6 in ALS beyond epigenetic mechanisms is unclear. In this sense, it has been reported that the mutated SOD1 in the SOD1-G93A mouse model promotes the formation of SOD1-tubulin-HDAC6 complexes, thus increasing alpha tubulin acetylation [143]. On the other hand, the pharmacological inhibition of sirtuin 2 through A7 has been reported to be protective in animal models of PD but not in ALS [144].

ALS is characterised for being a disease related to the imbalance of protein homeostasis too [1]. Although it has not been explored in the context of ALS therapeutic strategies, it is worth mentioning that HDAC6 reduces TDP-43 acetylation levels and, as a consequence, increases its non-aggregated status [145]. Mutations in the RNA-binding protein FUS are involved in almost 5% of the familiar ALS cases [1]. In this context, sequestration of the Prmt1 histone methylase by cytoplasmic FUS aggregates impairs its epigenetics capabilities related to the control of the histone code, and it has been recently associated to the pathogenesis of ALS [116].

Another example of the complex role of protein homeostasis in ALS, beyond well-established epigenetic mechanisms, is the sequestration of the HDAC6 mRNA molecules into protein inclusions of FUS and TDP-43 [146]. In fact, it has been reported a reduction in the levels of HDAC6 gene expression at the onset of the disease in the SOD1-G93A mouse model of ALS and further reductions at the end stages [147]. On the contrary, the CNS overexpression of HDAC6 significantly extended the survival expectancy in the SOD1-G93A mouse model of ALS [147]. The right balance of HDAC6 expression at different stages of ALS progression might be an interesting approach that deserves further consideration. Thus, the specific HDAC6 inhibition at early stages could be desirable due to its neuroprotective effects by improving axonal transport, while HDAC6 overexpression at late stages could counteract the effects of different protein aggregates in order to maintain proper autophagy functions enhancing protein aggregation clearance.

14.2.6 Environmental Factors and Epigenetic Mechanisms in ALS

The exposome can be defined as the sum of all the agents to whom an individual has been exposed to, from the time that he/she was born or even during the foetal period of pregnancy. These agents can be divided into lifestyle and occupational exposures. These interactions between the environment and the genome can lead to epigenetic changes, and, ultimately, some of these alterations can be transmitted from cell to cell [148, 149]. However, there is no a clear consensus between the so-called exposome and the onset and/or severity of ALS. Current evidences are not strong enough to show a clear association between external factors (army service, lead exposure, physical or emotional trauma, smoking, exposure to infectious agents, etc.) and ALS aetiology and survival [150–158].

14.3 Final Remarks

Although in recent years several pieces of evidence have been gathered about ALS pathogenesis and new potential therapeutic approaches, the need for an effective pharmacological option remains. A combined strategy targeting different pathogenic mechanisms may prove to be an optimal approach for preventing motor neuronal death. In this sense, the rise of new information related to the intertwined contribution of epigenetics pathways to already known mechanisms of ALS may represent a further step in developing a cure based on multi-targeted options.

Acknowledgments This work was supported by the following grants: PS09-2252 (to DP) from the Spanish Ministry of Economy (Instituto de Salud Carlos III) and co-funded by FEDER; P11-CTS8161 (to DP) from the Regional Ministry of Economy and PI13-575 (to SL) from the Regional Ministry of Health. MCG received support from an educational grant from PIF-University of Seville PhD programme. DEC received support from CONACYT postdoctoral programme (Gobierno de Mexico). We appreciate Jorge Abarca's Reto4ELA and ELA Andalucia Foundation continuous support.

References

1. Caballero-Hernandez D, Toscano MG, Cejudo-Guillen M, Garcia-Martin ML, Lopez S, Franco JM, et al. The 'Omics' of amyotrophic lateral sclerosis. Trends Mol Med. 2016;22:53–67.
2. Rowland LP, Shneider NA. Amyotrophic lateral sclerosis. N Engl J Med. 2001; 344:1688–700.
3. Leigh PN, Ray-Chaudhuri K. Motor neuron disease. J Neurol Neurosurg Psychiatry. 1994;57:886–96.
4. Ringholz GM, Appel SH, Bradshaw M, Cooke NA, Mosnik DM, Schulz PE. Prevalence and patterns of cognitive impairment in sporadic ALS. Neurology. 2005;65:586–90.
5. Zarei S, Carr K, Reiley L, Diaz K, Guerra O, Altamirano PF, et al. A comprehensive review of amyotrophic lateral sclerosis. Surg Neurol Int. 2015;6:171.
6. Kiernan MC, Vucic S, Cheah BC, Turner MR, Eisen A, Hardiman O, et al. Amyotrophic lateral sclerosis. Lancet. 2011;377:942–55.
7. Abhinav K, Stanton B, Johnston C, Hardstaff J, Orrell RW, Howard R, et al. Amyotrophic lateral sclerosis in South-East England: a population-based study. The South-East England register for amyotrophic lateral sclerosis (SEALS Registry). Neuroepidemiology. 2007;29:44–8.
8. Logroscino G, Traynor BJ, Hardiman O, Chio A, Mitchell D, Swingler RJ, et al. Incidence of amyotrophic lateral sclerosis in Europe. J Neurol Neurosurg Psychiatry. 2010;81:385–90.
9. Greenway MJ, Andersen PM, Russ C, Ennis S, Cashman S, Donaghy C, et al. ANG mutations segregate with familial and 'sporadic' amyotrophic lateral sclerosis. Nat Genet. 2006;38:411–3.
10. He J, Mangelsdorf M, Fan D, Bartlett P, Brown MA. Amyotrophic lateral sclerosis genetic studies: from genome-wide association mapping to genome sequencing. Neuroscientist. 2015;21:599–615.
11. Simon NG, Turner MR, Vucic S, Al-Chalabi A, Shefner J, Lomen-Hoerth C, et al. Quantifying disease progression in amyotrophic lateral sclerosis. Ann Neurol. 2014;76:643–57.
12. Wilson CM, Grace GM, Munoz DG, He BP, Strong MJ. Cognitive impairment in sporadic ALS: a pathologic continuum underlying a multisystem disorder. Neurology. 2001;57: 651–7.
13. Ling SC, Polymenidou M, Cleveland DW. Converging mechanisms in ALS and FTD: disrupted RNA and protein homeostasis. Neuron. 2013;79:416–38.
14. Murphy JM, Henry RG, Langmore S, Kramer JH, Miller BL, Lomen-Hoerth C. Continuum of frontal lobe impairment in amyotrophic lateral sclerosis. Arch Neurol. 2007;64:530–4.
15. Goetz CG. Amyotrophic lateral sclerosis: early contributions of Jean-Martin Charcot. Muscle Nerve. 2000;23:336–43.
16. Corcia P, Pradat PF, Salachas F, Bruneteau G, Forestier N, Seilhean D, et al. Causes of death in a post-mortem series of ALS patients. Amyotroph Lateral Scler. 2008;9:59–62.
17. Hayashi H, Kato S. Total manifestations of amyotrophic lateral sclerosis. ALS in the totally locked-in state. J Neurol Sci. 1989;93:19–35.
18. Bensimon G, Lacomblez L, Delumeau JC, Bejuit R, Truffinet P, Meininger V. A study of riluzole in the treatment of advanced stage or elderly patients with amyotrophic lateral sclerosis. J Neurol. 2002;249:609–15.
19. Bensimon G, Lacomblez L, Meininger V. A controlled trial of riluzole in amyotrophic lateral sclerosis. ALS/Riluzole Study Group. N Engl J Med. 1994;330:585–91.
20. Lacomblez L, Bensimon G, Leigh PN, Guillet P, Meininger V. Dose-ranging study of riluzole in amyotrophic lateral sclerosis. Amyotrophic Lateral Sclerosis/Riluzole Study Group II. Lancet. 1996;347:1425–31.
21. Riviere M, Meininger V, Zeisser P, Munsat T. An analysis of extended survival in patients with amyotrophic lateral sclerosis treated with riluzole. Arch Neurol. 1998;55:526–8.
22. Debono MW, Le Guern J, Canton T, Doble A, Pradier L. Inhibition by riluzole of electrophysiological responses mediated by rat kainate and NMDA receptors expressed in Xenopus oocytes. Eur J Pharmacol. 1993;235:283–9.

23. Wang SJ, Wang KY, Wang WC. Mechanisms underlying the riluzole inhibition of glutamate release from rat cerebral cortex nerve terminals (synaptosomes). Neuroscience. 2004;125:191–201.
24. Dangoumau A, Verschueren A, Hammouche E, Papon MA, Blasco H, Cherpi-Antar C, et al. Novel SOD1 mutation p.V31A identified with a slowly progressive form of amyotrophic lateral sclerosis. Neurobiol Aging. 2014;35:266 e1–4.
25. Forsberg K, Andersen PM, Marklund SL, Brannstrom T. Glial nuclear aggregates of superoxide dismutase-1 are regularly present in patients with amyotrophic lateral sclerosis. Acta Neuropathol. 2011;121:623–34.
26. Ivanova MI, Sievers SA, Guenther EL, Johnson LM, Winkler DD, Galaleldeen A, et al. Aggregation-triggering segments of SOD1 fibril formation support a common pathway for familial and sporadic ALS. Proc Natl Acad Sci U S A. 2014;111:197–201.
27. Majounie E, Renton AE, Mok K, Dopper EG, Waite A, Rollinson S, et al. Frequency of the C9orf72 hexanucleotide repeat expansion in patients with amyotrophic lateral sclerosis and frontotemporal dementia: a cross-sectional study. Lancet Neurol. 2012;11:323–30.
28. Ludolph AC, Brettschneider J, Weishaupt JH. Amyotrophic lateral sclerosis. Curr Opin Neurol. 2012;25:530–5.
29. Blair IP, Williams KL, Warraich ST, Durnall JC, Thoeng AD, Manavis J, et al. FUS mutations in amyotrophic lateral sclerosis: clinical, pathological, neurophysiological and genetic analysis. J Neurol Neurosurg Psychiatry. 2010;81:639–45.
30. Chio A, Borghero G, Pugliatti M, Ticca A, Calvo A, Moglia C, et al. Large proportion of amyotrophic lateral sclerosis cases in Sardinia due to a single founder mutation of the TARDBP gene. Arch Neurol. 2011;68:594–8.
31. Gijselinck I, Van Mossevelde S, van der Zee J, Sieben A, Engelborghs S, De Bleecker J, et al. The C9orf72 repeat size correlates with onset age of disease, DNA methylation and transcriptional downregulation of the promoter. Mol Psychiatry. 2015;21(8):1112–24.
32. Ajroud-Driss S, Siddique T. Sporadic and hereditary amyotrophic lateral sclerosis (ALS). Biochim Biophys Acta. 2015;1852:679–84.
33. Waddington CH. Preliminary notes on the development of the wings in normal and mutant strains of Drosophila. Proc Natl Acad Sci U S A. 1939;25:299–307.
34. Berger SL, Kouzarides T, Shiekhattar R, Shilatifard A. An operational definition of epigenetics. Genes Dev. 2009;23:781–3.
35. Bird A. Perceptions of epigenetics. Nature. 2007;447:396–8.
36. Paez-Colasante X, Figueroa-Romero C, Sakowski SA, Goutman SA, Feldman EL. Amyotrophic lateral sclerosis: mechanisms and therapeutics in the epigenomic era. Nat Rev Neurol. 2015;11:266–79.
37. Kaelin Jr WG, McKnight SL. Influence of metabolism on epigenetics and disease. Cell. 2013;153:56–69.
38. Friedman RC, Farh KK, Burge CB, Bartel DP. Most mammalian mRNAs are conserved targets of microRNAs. Genome Res. 2009;19:92–105.
39. Eitan C, Hornstein E. Vulnerability of microRNA biogenesis in FTD-ALS. Brain Res. 2016;1647:105–11.
40. Bohnsack MT, Czaplinski K, Gorlich D. Exportin 5 is a RanGTP-dependent dsRNA-binding protein that mediates nuclear export of pre-miRNAs. RNA. 2004;10:185–91.
41. Gregory RI, Yan KP, Amuthan G, Chendrimada T, Doratotaj B, Cooch N, et al. The Microprocessor complex mediates the genesis of microRNAs. Nature. 2004;432:235–40.
42. Yi R, Qin Y, Macara IG, Cullen BR. Exportin-5 mediates the nuclear export of pre-microRNAs and short hairpin RNAs. Genes Dev. 2003;17:3011–6.
43. Chendrimada TP, Gregory RI, Kumaraswamy E, Norman J, Cooch N, Nishikura K, et al. TRBP recruits the Dicer complex to Ago2 for microRNA processing and gene silencing. Nature. 2005;436:740–4.
44. Doench JG, Sharp PA. Specificity of microRNA target selection in translational repression. Genes Dev. 2004;18:504–11.
45. Pillai RS, Artus CG, Filipowicz W. Tethering of human Ago proteins to mRNA mimics the miRNA-mediated repression of protein synthesis. RNA. 2004;10:1518–25.

46. Greenberg DS, Soreq H. MicroRNA therapeutics in neurological disease. Curr Pharm Des. 2014;20:6022–7.
47. Wang C, Ji B, Cheng B, Chen J, Bai B. Neuroprotection of microRNA in neurological disorders. Biomed Rep. 2014;2:611–9.
48. Dickson JR, Kruse C, Montagna DR, Finsen B, Wolfe MS. Alternative polyadenylation and miR-34 family members regulate tau expression. J Neurochem. 2013;127:739–49.
49. Morais VA, Verstreken P, Roethig A, Smet J, Snellinx A, Vanbrabant M, et al. Parkinson's disease mutations in PINK1 result in decreased Complex I activity and deficient synaptic function. EMBO Mol Med. 2009;1:99–111.
50. Minones-Moyano E, Porta S, Escaramis G, Rabionet R, Iraola S, Kagerbauer B, et al. MicroRNA profiling of Parkinson's disease brains identifies early downregulation of miR-34b/c which modulate mitochondrial function. Hum Mol Genet. 2011;20:3067–78.
51. Cogswell JP, Ward J, Taylor IA, Waters M, Shi Y, Cannon B, et al. Identification of miRNA changes in Alzheimer's disease brain and CSF yields putative biomarkers and insights into disease pathways. J Alzheimers Dis. 2008;14:27–41.
52. Johnson R, Zuccato C, Belyaev ND, Guest DJ, Cattaneo E, Buckley NJ. A microRNA-based gene dysregulation pathway in Huntington's disease. Neurobiol Dis. 2008;29:438–45.
53. Margis R, Rieder CR. Identification of blood microRNAs associated to Parkinson's disease. J Biotechnol. 2011;152:96–101.
54. Packer AN, Xing Y, Harper SQ, Jones L, Davidson BL. The bifunctional microRNA miR-9/miR-9* regulates REST and CoREST and is downregulated in Huntington's disease. J Neurosci. 2008;28:14341–6.
55. Buratti E, Brindisi A, Giombi M, Tisminetzky S, Ayala YM, Baralle FE. TDP-43 binds heterogeneous nuclear ribonucleoprotein A/B through its C-terminal tail: an important region for the inhibition of cystic fibrosis transmembrane conductance regulator exon 9 splicing. J Biol Chem. 2005;280:37572–84.
56. Kawahara Y, Mieda-Sato A. TDP-43 promotes microRNA biogenesis as a component of the Drosha and Dicer complexes. Proc Natl Acad Sci U S A. 2012;109:3347–52.
57. Freischmidt A, Muller K, Ludolph AC, Weishaupt JH. Systemic dysregulation of TDP-43 binding microRNAs in amyotrophic lateral sclerosis. Acta Neuropathol Commun. 2013;1:42.
58. Haramati S, Chapnik E, Sztainberg Y, Eilam R, Zwang R, Gershoni N, et al. miRNA malfunction causes spinal motor neuron disease. Proc Natl Acad Sci U S A. 2010;107:13111–6.
59. Emde A, Eitan C, Liou LL, Libby RT, Rivkin N, Magen I, et al. Dysregulated miRNA biogenesis downstream of cellular stress and ALS-causing mutations: a new mechanism for ALS. EMBO J. 2015;34:2633–51.
60. Butovsky O, Siddiqui S, Gabriely G, Lanser AJ, Dake B, Murugaiyan G, et al. Modulating inflammatory monocytes with a unique microRNA gene signature ameliorates murine ALS. J Clin Invest. 2012;122:3063–87.
61. Koval ED, Shaner C, Zhang P, du Maine X, Fischer K, Tay J, et al. Method for widespread microRNA-155 inhibition prolongs survival in ALS-model mice. Hum Mol Genet. 2013;22:4127–35.
62. Nolan K, Mitchem MR, Jimenez-Mateos EM, Henshall DC, Concannon CG, Prehn JH. Increased expression of microRNA-29a in ALS mice: functional analysis of its inhibition. J Mol Neurosci. 2014;53:231–41.
63. Parisi C, Napoli G, Amadio S, Spalloni A, Apolloni S, Longone P, et al. MicroRNA-125b regulates microglia activation and motor neuron death in ALS. Cell Death Differ. 2016;23:531–41.
64. Parisi C, Arisi I, D'Ambrosi N, Storti AE, Brandi R, D'Onofrio M, et al. Dysregulated microRNAs in amyotrophic lateral sclerosis microglia modulate genes linked to neuroinflammation. Cell Death Dis. 2013;4:e959.
65. Russell AP, Wada S, Vergani L, Hock MB, Lamon S, Leger B, et al. Disruption of skeletal muscle mitochondrial network genes and miRNAs in amyotrophic lateral sclerosis. Neurobiol Dis. 2013;49:107–17.

66. Anderson C, Catoe H, Werner R. MIR-206 regulates connexin43 expression during skeletal muscle development. Nucleic Acids Res. 2006;34:5863–71.
67. Chen JF, Tao Y, Li J, Deng Z, Yan Z, Xiao X, et al. MicroRNA-1 and microRNA-206 regulate skeletal muscle satellite cell proliferation and differentiation by repressing Pax7. J Cell Biol. 2010;190:867–79.
68. Kim HK, Lee YS, Sivaprasad U, Malhotra A, Dutta A. Muscle-specific microRNA miR-206 promotes muscle differentiation. J Cell Biol. 2006a;174:677–87.
69. Rosenberg MI, Georges SA, Asawachaicharn A, Analau E, Tapscott SJ. MyoD inhibits Fstl1 and Utrn expression by inducing transcription of miR-206. J Cell Biol. 2006;175:77–85.
70. Valdez G, Heyer MP, Feng G, Sanes JR. The role of muscle microRNAs in repairing the neuromuscular junction. PLoS One. 2014;9:e93140.
71. Williams AH, Valdez G, Moresi V, Qi X, McAnally J, Elliott JL, et al. MicroRNA-206 delays ALS progression and promotes regeneration of neuromuscular synapses in mice. Science. 2009;326:1549–54.
72. Amin ND, Bai G, Klug JR, Bonanomi D, Pankratz MT, Gifford WD, et al. Loss of motoneuron-specific microRNA-218 causes systemic neuromuscular failure. Science. 2015;350:1525–9.
73. Thiebes KP, Nam H, Cambronne XA, Shen R, Glasgow SM, Cho HH, et al. miR-218 is essential to establish motor neuron fate as a downstream effector of Isl1-Lhx3. Nat Commun. 2015;6:7718.
74. Gascon E, Gao FB. The emerging roles of microRNAs in the pathogenesis of frontotemporal dementia-amyotrophic lateral sclerosis (FTD-ALS) spectrum disorders. J Neurogenet. 2014a;28:30–40.
75. Gascon E, Lynch K, Ruan H, Almeida S, Verheyden JM, Seeley WW, et al. Alterations in microRNA-124 and AMPA receptors contribute to social behavioral deficits in frontotemporal dementia. Nat Med. 2014b;20:1444–51.
76. Zhang Z, Almeida S, Lu Y, Nishimura AL, Peng L, Sun D, et al. Downregulation of microRNA-9 in iPSC-derived neurons of FTD/ALS patients with TDP-43 mutations. PLoS One. 2013;8:e76055.
77. Giusti SA, Vogl AM, Brockmann MM, Vercelli CA, Rein ML, Trumbach D, et al. MicroRNA-9 controls dendritic development by targeting REST. Elife. 2014; doi:10.7554/eLife.02755.
78. Campos-Melo D, Droppelmann CA, He Z, Volkening K, Strong MJ. Altered microRNA expression profile in amyotrophic lateral sclerosis: a role in the regulation of NFL mRNA levels. Mol Brain. 2013;6:26.
79. Ishtiaq M, Campos-Melo D, Volkening K, Strong MJ. Analysis of novel NEFL mRNA targeting microRNAs in amyotrophic lateral sclerosis. PLoS One. 2014;9:e85653.
80. Benigni M, Ricci C, Jones AR, Giannini F, Al-Chalabi A, Battistini S. Identification of miRNAs as potential biomarkers in cerebrospinal fluid from amyotrophic lateral sclerosis patients. Neuromol Med. 2016;18:551–60.
81. Rao P, Benito E, Fischer A. MicroRNAs as biomarkers for CNS disease. Front Mol Neurosci. 2013;6:39.
82. Chestnut BA, Chang Q, Price A, Lesuisse C, Wong M, Martin LJ. Epigenetic regulation of motor neuron cell death through DNA methylation. J Neurosci. 2011;31:16619–36.
83. Jones PA. Functions of DNA methylation: islands, start sites, gene bodies and beyond. Nat Rev Genet. 2012;13:484–92.
84. Choy MK, Movassagh M, Goh HG, Bennett MR, Down TA, Foo RS. Genome-wide conserved consensus transcription factor binding motifs are hyper-methylated. BMC Genomics. 2010;11:519.
85. Gaunt TR, Shihab HA, Hemani G, Min JL, Woodward G, Lyttleton O, et al. Systematic identification of genetic influences on methylation across the human life course. Genome Biol. 2016;17:61.
86. Kimura H, Shiota K. Methyl-CpG-binding protein, MeCP2, is a target molecule for maintenance DNA methyltransferase, Dnmt1. J Biol Chem. 2003;278:4806–12.

87. Rivera CM, Ren B. Mapping human epigenomes. Cell. 2013;155:39–55.
88. Ziller MJ, Gu H, Muller F, Donaghey J, Tsai LT, Kohlbacher O, et al. Charting a dynamic DNA methylation landscape of the human genome. Nature. 2013;500:477–81.
89. Kohli RM, Zhang Y. TET enzymes, TDG and the dynamics of DNA demethylation. Nature. 2013;502:472–9.
90. Kriaucionis S, Heintz N. The nuclear DNA base 5-hydroxymethylcytosine is present in Purkinje neurons and the brain. Science. 2009;324:929–30.
91. Penn NW, Suwalski R, O'Riley C, Bojanowski K, Yura R. The presence of 5-hydroxymethylcytosine in animal deoxyribonucleic acid. Biochem J. 1972;126:781–90.
92. Wu H, Zhang Y. Reversing DNA methylation: mechanisms, genomics, and biological functions. Cell. 2014;156:45–68.
93. Oates N, Pamphlett R. An epigenetic analysis of SOD1 and VEGF in ALS. Amyotroph Lateral Scler. 2007;8:83–6.
94. Yang Y, Gozen O, Vidensky S, Robinson MB, Rothstein JD. Epigenetic regulation of neuron-dependent induction of astroglial synaptic protein GLT1. Glia. 2010;58:277–86.
95. Tremolizzo L, Messina P, Conti E, Sala G, Cecchi M, Airoldi L, et al. Whole-blood global DNA methylation is increased in amyotrophic lateral sclerosis independently of age of onset. Amyotroph Lateral Scler Frontotemporal Degener. 2014;15:98–105.
96. Figueroa-Romero C, Hur J, Bender DE, Delaney CE, Cataldo MD, Smith AL, et al. Identification of epigenetically altered genes in sporadic amyotrophic lateral sclerosis. PLoS One. 2012;7:e52672.
97. Al-Mahdawi S, Virmouni SA, Pook MA. The emerging role of 5-hydroxymethylcytosine in neurodegenerative diseases. Front Neurosci. 2014;8:397.
98. Mazin AL. Suicidal function of DNA methylation in age-related genome disintegration. Ageing Res Rev. 2009;8:314–27.
99. Martin LJ. Mitochondrial and Cell Death Mechanisms in Neurodegenerative Diseases. Pharmaceuticals. 2010a;3:839–915.
100. Martin LJ. Olesoxime, a cholesterol-like neuroprotectant for the potential treatment of amyotrophic lateral sclerosis. IDrugs. 2010b;13:568–80.
101. Morris HR, Waite AJ, Williams NM, Neal JW, Blake DJ. Recent advances in the genetics of the ALS-FTLD complex. Curr Neurol Neurosci Rep. 2012;12:243–50.
102. Morahan JM, Yu B, Trent RJ, Pamphlett R. A genome-wide analysis of brain DNA methylation identifies new candidate genes for sporadic amyotrophic lateral sclerosis. Amyotroph Lateral Scler. 2009;10:418–29.
103. Wong M, Gertz B, Chestnut BA, Martin LJ. Mitochondrial DNMT3A and DNA methylation in skeletal muscle and CNS of transgenic mouse models of ALS. Front Cell Neurosci. 2013;7:279.
104. Oh KW, Moon C, Kim HY, Oh SI, Park J, Lee JH, et al. Phase I trial of repeated intrathecal autologous bone marrow-derived mesenchymal stromal cells in amyotrophic lateral sclerosis. Stem Cells Transl Med. 2015;4:590–7.
105. Oh YS, Kim SH, Cho GW. Functional restoration of amyotrophic lateral sclerosis patient-derived mesenchymal stromal cells through inhibition of DNA methyltransferase. Cell Mol Neurobiol. 2016;36:613–20.
106. He F, Todd PK. Epigenetics in nucleotide repeat expansion disorders. Semin Neurol. 2011;31:470–83.
107. Dols-Icardo O, Garcia-Redondo A, Rojas-Garcia R, Sanchez-Valle R, Noguera A, Gomez-Tortosa E, et al. Characterization of the repeat expansion size in C9orf72 in amyotrophic lateral sclerosis and frontotemporal dementia. Hum Mol Genet. 2014;23:749–54.
108. Van Langenhove T, van der Zee J, Gijselinck I, Engelborghs S, Vandenberghe R, Vandenbulcke M, et al. Distinct clinical characteristics of C9orf72 expansion carriers compared with GRN, MAPT, and nonmutation carriers in a Flanders-Belgian FTLD cohort. JAMA Neurol. 2013;70:365–73.
109. Bauer PO. Methylation of C9orf72 expansion reduces RNA foci formation and dipeptide-repeat proteins expression in cells. Neurosci Lett. 2016;612:204–9.

110. Belzil VV, Bauer PO, Gendron TF, Murray ME, Dickson D, Petrucelli L. Characterization of DNA hypermethylation in the cerebellum of c9FTD/ALS patients. Brain Res. 2014;1584:15–21.
111. Liu EY, Russ J, Wu K, Neal D, Suh E, McNally AG, et al. C9orf72 hypermethylation protects against repeat expansion-associated pathology in ALS/FTD. Acta Neuropathol. 2014;128:525–41.
112. Xi Z, Rainero I, Rubino E, Pinessi L, Bruni AC, Maletta RG, et al. Hypermethylation of the CpG-island near the C9orf72 G(4)C(2)-repeat expansion in FTLD patients. Hum Mol Genet. 2014;23:5630–7.
113. Xi Z, Zinman L, Moreno D, Schymick J, Liang Y, Sato C, et al. Hypermethylation of the CpG island near the G4C2 repeat in ALS with a C9orf72 expansion. Am J Hum Genet. 2013;92:981–9.
114. Russ J, Liu EY, Wu K, Neal D, Suh E, Irwin DJ, et al. Hypermethylation of repeat expanded C9orf72 is a clinical and molecular disease modifier. Acta Neuropathol. 2015;129:39–52.
115. Xi Z, Zhang M, Bruni AC, Maletta RG, Colao R, Fratta P, et al. The C9orf72 repeat expansion itself is methylated in ALS and FTLD patients. Acta Neuropathol. 2015;129:715–27.
116. Tibshirani M, Tradewell ML, Mattina KR, Minotti S, Yang W, Zhou H, et al. Cytoplasmic sequestration of FUS/TLS associated with ALS alters histone marks through loss of nuclear protein arginine methyltransferase 1. Hum Mol Genet. 2015;24:773–86.
117. Chuang DM, Leng Y, Marinova Z, Kim HJ, Chiu CT. Multiple roles of HDAC inhibition in neurodegenerative conditions. Trends Neurosci. 2009;32:591–601.
118. Khan AU, Krishnamurthy S. Histone modifications as key regulators of transcription. Front Biosci. 2005;10:866–72.
119. Sun JM, Spencer VA, Chen HY, Li L, Davie JR. Measurement of histone acetyltransferase and histone deacetylase activities and kinetics of histone acetylation. Methods. 2003;31:12–23.
120. Janssen C, Schmalbach S, Boeselt S, Sarlette A, Dengler R, Petri S. Differential histone deacetylase mRNA expression patterns in amyotrophic lateral sclerosis. J Neuropathol Exp Neurol. 2010;69:573–81.
121. Gregoretti IV, Lee YM, Goodson HV. Molecular evolution of the histone deacetylase family: functional implications of phylogenetic analysis. J Mol Biol. 2004;338:17–31.
122. Rouaux C, Jokic N, Mbebi C, Boutillier S, Loeffler JP, Boutillier AL. Critical loss of CBP/p300 histone acetylase activity by caspase-6 during neurodegeneration. EMBO J. 2003;22:6537–49.
123. Colussi C, Mozzetta C, Gurtner A, Illi B, Rosati J, Straino S, et al. HDAC2 blockade by nitric oxide and histone deacetylase inhibitors reveals a common target in Duchenne muscular dystrophy treatment. Proc Natl Acad Sci U S A. 2008;105:19183–7.
124. Minetti GC, Colussi C, Adami R, Serra C, Mozzetta C, Parente V, et al. Functional and morphological recovery of dystrophic muscles in mice treated with deacetylase inhibitors. Nat Med. 2006;12:1147–50.
125. Rimkus SA, Katzenberger RJ, Trinh AT, Dodson GE, Tibbetts RS, Wassarman DA. Mutations in String/CDC25 inhibit cell cycle re-entry and neurodegeneration in a Drosophila model of Ataxia telangiectasia. Genes Dev. 2008;22:1205–20.
126. Bruneteau G, Simonet T, Bauche S, Mandjee N, Malfatti E, Girard E, et al. Muscle histone deacetylase 4 upregulation in amyotrophic lateral sclerosis: potential role in reinnervation ability and disease progression. Brain. 2013;136:2359–68.
127. Rouaux C, Panteleeva I, Rene F, Gonzalez de Aguilar JL, Echaniz-Laguna A, Dupuis L, et al. Sodium valproate exerts neuroprotective effects in vivo through CREB-binding protein-dependent mechanisms but does not improve survival in an amyotrophic lateral sclerosis mouse model. J Neurosci. 2007;27:5535–45.
128. Petri S, Kiaei M, Kipiani K, Chen J, Calingasan NY, Crow JP, et al. Additive neuroprotective effects of a histone deacetylase inhibitor and a catalytic antioxidant in a transgenic mouse model of amyotrophic lateral sclerosis. Neurobiol Dis. 2006;22:40–9.

129. Del Signore SJ, Amante DJ, Kim J, Stack EC, Goodrich S, Cormier K, et al. Combined riluzole and sodium phenylbutyrate therapy in transgenic amyotrophic lateral sclerosis mice. Amyotroph Lateral Scler. 2009;10:85–94.
130. Feng HL, Leng Y, Ma CH, Zhang J, Ren M, Chuang DM. Combined lithium and valproate treatment delays disease onset, reduces neurological deficits and prolongs survival in an amyotrophic lateral sclerosis mouse model. Neuroscience. 2008;155:567–72.
131. Ryu H, Smith K, Camelo SI, Carreras I, Lee J, Iglesias AH, et al. Sodium phenylbutyrate prolongs survival and regulates expression of anti-apoptotic genes in transgenic amyotrophic lateral sclerosis mice. J Neurochem. 2005;93:1087–98.
132. Yoo YE, Ko CP. Treatment with trichostatin A initiated after disease onset delays disease progression and increases survival in a mouse model of amyotrophic lateral sclerosis. Exp Neurol. 2011;231:147–59.
133. Kazantsev AG, Thompson LM. Therapeutic application of histone deacetylase inhibitors for central nervous system disorders. Nat Rev Drug Discov. 2008;7:854–68.
134. Gregoire S, Xiao L, Nie J, Zhang X, Xu M, Li J, et al. Histone deacetylase 3 interacts with and deacetylates myocyte enhancer factor 2. Mol Cell Biol. 2007;27:1280–95.
135. Kim SC, Sprung R, Chen Y, Xu Y, Ball H, Pei J, et al. Substrate and functional diversity of lysine acetylation revealed by a proteomics survey. Mol Cell. 2006b;23:607–18.
136. Southwood CM, Peppi M, Dryden S, Tainsky MA, Gow A. Microtubule deacetylases, SirT2 and HDAC6, in the nervous system. Neurochem Res. 2007;32:187–95.
137. Kawaguchi Y, Kovacs JJ, McLaurin A, Vance JM, Ito A, Yao TP. The deacetylase HDAC6 regulates aggresome formation and cell viability in response to misfolded protein stress. Cell. 2003;115:727–38.
138. Haggarty SJ, Koeller KM, Wong JC, Grozinger CM, Schreiber SL. Domain-selective small-molecule inhibitor of histone deacetylase 6 (HDAC6)-mediated tubulin deacetylation. Proc Natl Acad Sci U S A. 2003;100:4389–94.
139. Luthi-Carter R, Taylor DM, Pallos J, Lambert E, Amore A, Parker A, et al. SIRT2 inhibition achieves neuroprotection by decreasing sterol biosynthesis. Proc Natl Acad Sci U S A. 2010;107:7927–32.
140. Dompierre JP, Godin JD, Charrin BC, Cordelieres FP, King SJ, Humbert S, et al. Histone deacetylase 6 inhibition compensates for the transport deficit in Huntington's disease by increasing tubulin acetylation. J Neurosci. 2007;27:3571–83.
141. Reed NA, Cai D, Blasius TL, Jih GT, Meyhofer E, Gaertig J, et al. Microtubule acetylation promotes kinesin-1 binding and transport. Curr Biol. 2006;16:2166–72.
142. Taes I, Timmers M, Hersmus N, Bento-Abreu A, Van Den Bosch L, Van Damme P, et al. Hdac6 deletion delays disease progression in the SOD1G93A mouse model of ALS. Hum Mol Genet. 2013;22:1783–90.
143. Gal J, Chen J, Barnett KR, Yang L, Brumley E, Zhu H. HDAC6 regulates mutant SOD1 aggregation through two SMIR motifs and tubulin acetylation. J Biol Chem. 2013;288:15035–45.
144. Chen X, Wales P, Quinti L, Zuo F, Moniot S, Herisson F, et al. The sirtuin-2 inhibitor AK7 is neuroprotective in models of Parkinson's disease but not amyotrophic lateral sclerosis and cerebral ischemia. PLoS One. 2015a;10:e0116919.
145. Cohen TJ, Hwang AW, Restrepo CR, Yuan CX, Trojanowski JQ, Lee VM. An acetylation switch controls TDP-43 function and aggregation propensity. Nat Commun. 2015;6:5845.
146. Kim SH, Shanware NP, Bowler MJ, Tibbetts RS. Amyotrophic lateral sclerosis-associated proteins TDP-43 and FUS/TLS function in a common biochemical complex to co-regulate HDAC6 mRNA. J Biol Chem. 2010;285:34097–105.
147. Chen S, Zhang XJ, Li LX, Wang Y, Zhong RJ, Le W. Histone deacetylase 6 delays motor neuron degeneration by ameliorating the autophagic flux defect in a transgenic mouse model of amyotrophic lateral sclerosis. Neurosci Bull. 2015b;31:459–68.
148. Belzil VV, Katzman RB, Petrucelli L. ALS and FTD: an epigenetic perspective. Acta Neuropathol. 2016;132:487–502.

149. Polsky FI, Nunn PB, Bell EA. Distribution and toxicity of alpha-amino-beta-methylaminopropionic acid. Fed Proc. 1972;31:1473–5.
150. Abel EL. Football increases the risk for Lou Gehrig's disease, amyotrophic lateral sclerosis. Percept Mot Skills. 2007;104:1251–4.
151. Armon C. Smoking may be considered an established risk factor for sporadic ALS. Neurology. 2009;73:1693–8.
152. Beard JD, Engel LS, Richardson DB, Gammon MD, Baird C, Umbach DM, et al. Military service, deployments, and exposures in relation to amyotrophic lateral sclerosis etiology. Environ Int. 2016;91:104–15.
153. Beard JD, Kamel F. Military service, deployments, and exposures in relation to amyotrophic lateral sclerosis etiology and survival. Epidemiol Rev. 2015;37:55–70.
154. Chio A, Benzi G, Dossena M, Mutani R, Mora G. Severely increased risk of amyotrophic lateral sclerosis among Italian professional football players. Brain. 2005;128:472–6.
155. Horner RD, Grambow SC, Coffman CJ, Lindquist JH, Oddone EZ, Allen KD, et al. Amyotrophic lateral sclerosis among 1991 Gulf War veterans: evidence for a time-limited outbreak. Neuroepidemiology. 2008;31:28–32.
156. Miranda ML, Alicia Overstreet Galeano M, Tassone E, Allen KD, Horner RD. Spatial analysis of the etiology of amyotrophic lateral sclerosis among 1991 Gulf War veterans. Neurotoxicology. 2008;29:964–70.
157. Oskarsson B, Horton DK, Mitsumoto H. Potential environmental factors in amyotrophic lateral sclerosis. Neurol Clin. 2015;33:877–88.
158. Vinceti M, Bottecchi I, Fan A, Finkelstein Y, Mandrioli J. Are environmental exposures to selenium, heavy metals, and pesticides risk factors for amyotrophic lateral sclerosis? Rev Environ Health. 2012;27:19–41.

Epigenetics of Huntington's Disease

15

Silvia Bassi, Takshashila Tripathi, Alan Monziani, Francesca Di Leva, and Marta Biagioli

Abstract

Huntington's disease (HD) is a genetic, fatal autosomal dominant neurodegenerative disorder typically occurring in midlife with symptoms ranging from chorea, to dementia, to personality disturbances (Philos Trans R Soc Lond Ser B Biol Sci 354:957–961, 1999). HD is inherited in a dominant fashion, and the underlying mutation in all cases is a CAG trinucleotide repeat expansion within exon 1 of the HD gene (Cell 72:971–983, 1993). The expanded CAG repeat, translated into a lengthened glutamine tract at the amino terminus of the huntingtin protein, affects its structural properties and functional activities. The effects are pleiotropic, as huntingtin is broadly expressed in different cellular compartments (i.e., cytosol, nucleus, mitochondria) as well as in all cell types of the body at all developmental stages, such that HD pathogenesis likely starts at conception and is a lifelong process (Front Neurosci 9:509, 2015). The rate-limiting mechanism(s) of neurodegeneration in HD still remains elusive: many different processes are commonly disrupted in HD cell lines and animal models, as well as in HD patient cells (Eur J Neurosci 27:2803–2820, 2008); however, epigenetic-chromatin deregulation, as determined by the analysis of DNA methylation, histone modifications, and noncoding RNAs, has now become a prevailing feature. Thus, the overarching goal of this chapter is to discuss the current status of the literature, reviewing how an aberrant epigenetic landscape can contribute to altered gene expression and neuronal dysfunction in HD.

Keywords

Huntington's disease (HD) • CAG repeat expansion • Epigenetic • Chromatin • Histone and DNA methylation/demethylation • Histone acetylation/deacetylation • Noncoding RNAs (ncRNAs) • Long noncoding RNAs (lncRNAs) • miRNAs

S. Bassi, Ph.D. • T. Tripathi, Ph.D. • A. Monziani • F. Di Leva, Ph.D. • M. Biagioli, Ph.D. (✉)
The NeuroEpigenetics Laboratory, Centre for Integrative Biology, University of Trento,
Via Sommarive 9, 38123 Povo (TN), Italy
e-mail: marta.biagioli@unitn.it

R. Delgado-Morales (ed.), *Neuroepigenomics in Aging and Disease*, Advances in Experimental Medicine and Biology, DOI 10.1007/978-3-319-53889-1_15

15.1 DNA Methylation in HD

15.1.1 DNA Methylation: General Characteristics

DNA methylation, a critical regulator of gene expression and gene silencing, is one of the earliest characterized, and most intensely studied, epigenetic modifications in mammals [1, 2]. From the structural point of view, DNA methylation involves a covalent addition of a methyl group to the 5′ position of the pyrimidine ring of cytosine bases in the nucleotide sequence 5′-CpG-3′, so called CpG island, thus creating 5-methylcytosine (5-mC) [2]. 5-mC DNA modification is prevalent in the human genome, encompassing 4–6% of all cytosines and 60–80% of all CpG dinucleotides, and it's usually associated with gene repression [3]. The 5-mC conversion is catalyzed by DNA methyltransferases (DNMTs), which transfer the methyl group to single-stranded DNA using S-adenosylmethionine as methyl donor [4, 5]. Of the five members of the DNMT family identified in mammals—DNMT1, DNMT2, DNMT3a, DNMT3b, and DNMT3L—the *de novo* methylating enzymes of DNMT3a and DNMT3b catalyze the methyl transfer onto an unmethylated DNA during the early phases of embryonic development [1, 6]. On the contrary, DNMT1—the "maintenance" DNA methylation enzyme—duplicates the DNA mark from a hemi-methylated template, thus preserving the methylation patterns in adult, mitotically active tissues [1, 7, 8]. DNA methylation patterns are not permanent; indeed, the mammalian genome is exposed to both active and passive DNA demethylation processes. Active DNA demethylation occurs when the methyl group is removed from 5-mC through a direct enzymatic process, while passive DNA demethylation refers to the failure to maintain DNA methylation pattern during DNA replication (the absence or inhibition of DNMT1) [9]. Passive demethylation is mainly caused by the conversion of 5-mC to 5-hmC (5-hydroxymethylcytosine): this modification is not recognized by DNMT1 enzyme, thus preventing the maintenance of the existing DNA methylation patterns. Instead, 5-mC to 5-hmC active conversion is catalyzed by the iron and 2-oxoglutarate-dependent TET proteins. Although 5-hmC is present in mammalian genome, it represents only the ~0.05% of all cytosines (100 times less compared to 5-mC) [10, 11]. Apart from cytosine, also guanine methylation has been reported, resulting in 7-methylguanine (7-mG) production. Similarly to 5-mC, 7-mG in CpG islands contributes to transcriptional regulation; however, increased 7-mG in CpG results in chromatin remodeling leading to increased gene expression [12, 13].

But what is the functional relevance of DNA methylation? While the presence of methyl groups in the major groove of DNA generally interferes with the binding of transcription factors, thus preventing transcriptional activation (indirect transcriptional repression) [14], a second mode of "direct transcriptional repression" involving the interaction with methyl-CpG binding domain (MBD) proteins is also well characterized [15].

15.1.2 DNA Methylation in HD

The expression of mutant huntingtin in different HD model systems and HD *post-mortem* samples has been correlated with important changes in DNA methylation [16, 1] (Fig. 15.1). Ng and colleagues used reduced representation bisulfite sequencing (RRBS) to measure DNA methylation at single base-pair resolution in mouse knock-in striatal cells carrying an expanded polyglutamine tract of 111 repeats (ST*Hdh*$^{Q111/Q111}$) [17]. This cellular model, obtained by immortalizing knock-in striatal progenitors (E14), expresses full-length mutant huntingtin under the mouse endogenous promoter, thus faithfully replicating the HD genetics [18, 19]. From the comparison between ST*Hdh*$^{Q111/Q111}$ and wild-type striatal cells, DNA methylation pattern was found to be sensibly altered at promoter, proximal, and distal regulatory regions. Importantly, promoter regions associated with neurogenesis and neuronal differentiation genes, such as *Ap-1*, *Sox2*, *Pax6*, and *Nes*, were found to be hypermethylated in mutant cells with consequently reduced expression levels [17]. Thus, mutant huntingtin protein might subtly but consistently alter chromatin regulation and transcription during neural development and specification, thus contributing to the deleterious effects observed throughout the life of an HD individual [20]. In line with the results in mouse cells, human fibroblasts obtained from HD patients showed significant alterations in 5-mC DNA methylation that can be partially ameliorated by histone deacetylase (HDAC) inhibition [21].

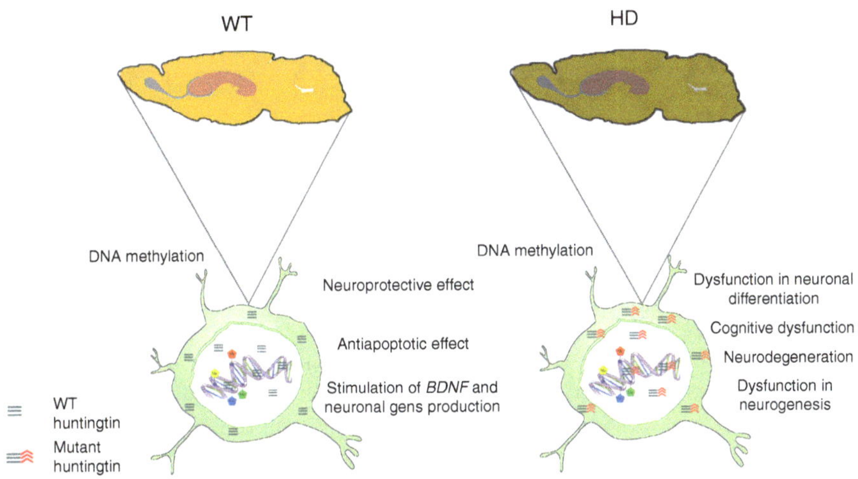

Fig. 15.1 Schematic representation of DNA methylation patterns in wild-type and HD brains. The presence of wild-type or mutant huntingtin protein differently regulates DNA methylation patterns. The consequences of these functional and dysfunctional changes in DNA methylation are also indicated for wild-type and HD neuronal cells. This figure has been modified and adapted from *Lee, J., Y. J. Hwang,* et al. *(2013)*

Independent studies have focused on 5-hmC since it's preferentially enriched in pluripotent stem cells [22, 23] and certain neuronal subtypes [24]. In physiological conditions, mouse embryonic stem cells (ESC) show high levels of 5-hmC, which transiently decrease during ESC differentiation, eventually returning to high levels during neuronal terminal differentiation [24, 25]. 5-hmC seems to be a neuronal specific epigenetic mark since it is roughly tenfold more enriched in neurons than in other tissues [17, 26]. Wang and colleagues used YAC128 (yeast artificial chromosome transgene with 128 CAG repeats) transgenic mice overexpressing full-length huntingtin protein with 128 CAG repeats [27] to characterize 5-hmC methylation in striatal and cortical regions [25]. 5-hmC markedly decreased in both the striatum and cortex of YAC128 mice as early as 6 weeks of age, suggesting an impairment of 5-hmC restoration in HD brains during postnatal development. Interestingly, and in line with the previously discussed 5-mC results, genome-wide distribution analysis of 5-hmC highlighted an aberrant epigenetic regulation impacting the correct neurogenesis, neuronal function, and survival in the HD brain [25].

Adenosine A_{2A} receptor ($A_{2A}R$), also known as ADORA2A, is a G-protein-coupled receptor that stimulates adenylyl cyclase [28] and is highly expressed in the striatum, especially in GABAergic medium-sized spiny neurons (MSNs), the most vulnerable neuronal subtype in HD [29, 30]. Many studies have correlated transcriptional alterations in $A_{2A}R$ with HD pathogenesis [29, 31–34]. More recently, Villar-Menéndez and colleagues studied a possible epigenetic regulation of $A_{2A}R$ gene in HD by measuring 5-mC and 5-hmC content in the 5′-UTR region of the gene in the putamen of HD patients as well as in the striatum of R6/1 and R6/2 mice, overexpressing the highly toxic, expanded exon-1 of the huntingtin protein, thus recapitulating the late stages of the disease [35, 36]. This study showed that $A_{2A}R$ reduction is correlated with increased levels of 5-mC in the 5′-UTR region of *ADORA2A* gene in the putamen of HD patients, while a reduction of 5-hmC correlates with reduced *ADORA2A* expression levels in the striatum of HD transgenic mice [36]. Although DNA methylation process seems to be differentially modulated between species (mice and human), the *ADORA2A* epigenetic regulation reinforces the idea that DNA methylation might be relevant for HD pathogenesis [37].

Finally, 7-methylguanine (7-mG), a DNA modification that plays important roles in transcription regulation [12, 13], has been found to be significantly changed in HD human brain samples and HD animal models, thus suggesting that DNA methylation machinery in its whole might be dysfunctional in HD [38]. These studies open potentially new avenues for HD treatment: re-establishing the native 5-mC, 5-hmC, and/or 7-mG landscape may have the potential to rescue overt transcriptional changes correlated to the disease and slow/halt the progression of HD (Fig. 15.1).

15.2 Histone Acetylation in HD

15.2.1 Histone Modifications: General Characteristics

The association of histone proteins with DNA is affected by histone modifications that modulate the dynamic nature of chromatin fibers [37, 39]. In general, histones are the central component of the nucleosome subunit, forming an octamer containing the four core histone proteins (H3, H4, H2A, H2B) around which is wrapped a 147-base-pair segment of DNA. Each of the globular histone proteins possesses a characteristic side chain which is densely populated with basic lysine and arginine residues, where most of the histone modifications reside. Gene expression is thus regulated by the concerted action of two components: on one side by the binding of transcriptional activators and repressors and, on the other side, by the chromatin structure governed by histone modification and chromatin remodeling enzymes [40, 41].

The histone tails are subject to extensive covalent posttranslational modifications (PTMs), including acetylation, methylation, phosphorylation, and ubiquitination [42, 43], collectively known as the "histone code," that directly affects the plasticity of chromatin structure [44]. Some PTMs alter the charge density between histone and DNA, impacting chromatin organization and the underlying transcriptional processes. However, they can also serve as recognition modules for specific binding proteins that, in turn, may control alterations in chromatin structure or function.

15.2.2 Histone Acetylation and Deacetylation

The process, acetylation-deacetylation, is orchestrated following the reversible transfer of an acetyl coenzyme A (Acetyl-CoA) to the NH^{3+} group on a histone lysine residue. Specifically, acetylation of lysine (K) residues on histone tails decreases the electrostatic interactions between histones and DNA, promoting a relaxed chromatin conformation that allows for the recruitment of transcription factors and the basal transcriptional machinery [45].

Gene expression, at any locus in the chromatin, is thus regulated through the interplay between histone acetyl transferases (HAT) and histone deacetylases (HDACs) in concert with other histone and DNA modifications (methylation) that facilitate or repress the recruitment of transcription factors and transcriptional machinery [46].

HAT molecules can be grouped based on sequence similarities [47]. Up to now, five different classes can be distinguished: see Table 15.1 for a complete description.

Table 15.1 Major families of histone acetyl transferases (HATs) and histone deacetylases (HDACs)

HAT[a]					HDACs[b]
Family	Subtype	Other names frequently used			**Class I (Rpd homolog)**
Cytoplasmic	KAT1	HAT1			HDAC1
	HAT4	NAA60			HDAC2
					HDAC3
GNAT	KAT2A	Gcn5			HDAC8
	KAT2B	PCAF			
					Class II (Hda 1 homolog)
MYST	KAT5	TIP60			HDAC4
	KAT6A	MOZ	MYST3		HDAC5
	KAT6B	MORF	MYST4		HDAC6
	KAT7	HBO1	MYST2		HDAC7
	KAT8	MOF	MYST1		HDAC9
					HDAC10
p300/CBP	KAT3B	p300			
	KAT3A	CBP			**Class IV**
					HDAC11
Transcription coactivators		KAT4	TAF1	TBP	
	KAT12	TIFIIIC90			**Class III (Sir2 homolog)**
					SIRT1
Steroid receptor coactivators	KAT13A	SRC1			SIRT2
	KAT13B	SCR3	AIB1	ACTR	SIRT3
	KAT13C	p600			SIRT4
	KAT13D	CLOCK			SIRT5
					SIRT6
					SIRT7

[a]HAT families, subtypes, and alternative nomenclature which are commonly used [127]
[b]HDAC families ordered in different classes [128, 129]

Similarly, four classes of HDACs have been described: class I includes HDACs 1, 2, 3, and 8; class II is divided into two subgroups, class IIA and class IIB: class IIA includes HDACs 4, 5, 7, and 9, while class IIB includes HDACs 6 and 10; class III contains sirtuins; and class IV contains only HDAC11 [48, 49] (see Table 15.1 for details).

It is thought that it is the stoichiometric balance between HAT and HDAC enzyme activities that contributes to cellular homeostasis through coordinating gene expression and repression in a spatial and temporal manner [50]. Particularly, changes in the activities of HATs and HDACs lead to dysregulated gene expression, and this perturbed acetylation homeostasis might consequently contribute to neuropathology.

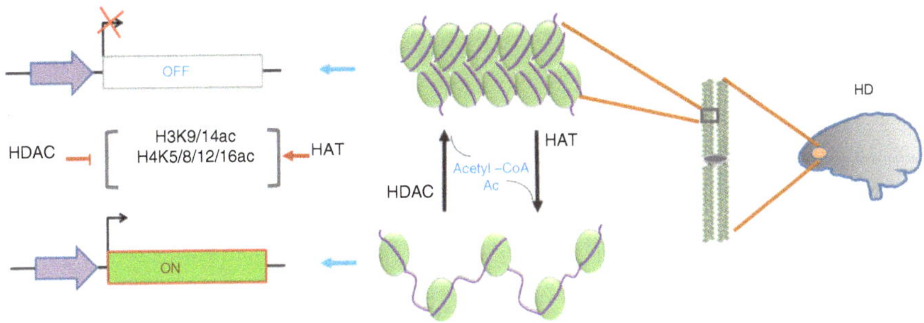

Fig. 15.2 Schematic illustration of the dynamic state of histone acetylation/deacetylation regulated by HAT and HDAC enzymes in HD. Acetylation of histones alters accessibility of chromatin (*beads on a string*) and correlates with active gene transcription, while deacetylation has been linked with transcriptional repression. Acetylation-deacetylation is orchestrated by the antagonistic action of histone acetyl transferases (HATs) and histone deacetylase (HDAC) enzymes. The *parentheses* show the modifications on histones catalyzed by HATs and HDACs. This figure has been adapted from *L.M. Valor, D. Guiretti (2014)*

15.2.3 Loss of Acetylation Homeostasis in HD

Although impairment of acetylation homeostasis has been associated to early "transcriptional dysfunction" as a contributing factor in pathogenesis of HD, the precise mechanism of this dysregulation remains unknown [51, 52] (Fig. 15.2). Mutant huntingtin was shown to interact with the acetyltransferase domain of CBP; among all the HATs, CBP is probably the most important in mediating neuronal survival response and already implicated in neurodegenerative disorders. Specifically, CBP was found in intracellular inclusions *in vitro*, in brain tissue from animal models, as well as in *post-mortem* human HD brains. Together, these studies suggested that depletion of soluble CBP can cause a global hypoacetylation of histone that might affect transcription of genes relevant for neuronal survival [45]. There have been conflicting findings linking CBP activity to soluble or aggregate forms of mutant huntingtin. This might be due to experimental confounds caused by *in vitro* overexpression of protein fragments that could force co-aggregation or variable levels of expression of mutant huntingtin. However, it was recently demonstrated that the $Hdh^{Q7/Q111}$ knock-in mice, representing a faithful replica of the human HD mutation, also exhibit impairments in spatial and recognition memory along with decreased levels of CBP and acetylated histone H3, strongly suggesting that specifically CBP may be important for the cognitive impairments seen in HD [45]. In fact, despite the high degree of similarity to CBP, p300, another HAT, is absent from intracellular mutant huntingtin inclusions, and its activity is not compromised in the presence of mutant huntingtin.

15.2.4 Histone Deacetylation Homeostasis in HD

Similar to acetylation, also deacetylation is altered in HD. The first evidences showed that the nuclear localization of the class II histone deacetylase HDAC5 was increased in the brains of HD patients without an overall change in transcript levels [53]. In contrast, a recent study in R6/2 mice demonstrated that there was an increase in HDAC1 protein levels and a concomitant decrease in class II HDACs (particularly HDAC6). However, these changes were not validated in CAG140 knock-in mice or *post-mortem* HD brain.

Apart from a direct effect on different HDACs, mutant huntingtin is able to modulate the formation of corepressor complexes through its altered binding/sequestration of "the repressor element 1 (RE1) transcription factor (REST) also called neuron-restrictive silencer factor (NRSF). In fact, polyglutamine-expanded mutant huntingtin has been shown to have a reduced interaction with REST, which is then no longer retained in the cytoplasm and can enter the nucleus causing transcriptional repression [54] through the formation of a repressor complex including HDAC, CoREST, and Sin3A. This corepressor complex is responsible for the regulation of a broad range of neuronal genes involved in neuronal development and normal neuronal functions (*BDNF*), such as ion channels, neurotransmitter receptors and their synthesizing enzymes, and synaptic vesicle proteins. [55]. In agreement with this mechanism, in both YAC transgenic and knock-in mouse models of HD, the transcription of REST/NRSF-regulated genes is positively correlated with the level of wild-type huntingtin that is present. In general, the disequilibrium caused by the loss of HAT or the increase in HDAC in HD pushes the HAT/HDAC ratio in favor of HDAC; thus HDAC inhibitors could possibly ameliorate HD-related phenotypes (Fig. 15.2). In fact, growing evidences suggest that oxidative stress-induced apoptosis is blocked in the presence of HDAC inhibitors [56], and HDAC inhibitors like TSA and SAHA arrest polyglutamine toxicity in cell lines [57] and in *Drosophila* models of HD [58]. Over the past years, a number of groups have assessed the role of HDACs in HD models via genetic knockout or knockdown approaches [45]. Particularly, recent studies have shown that HDAC inhibition ameliorates HD-polyglutamine disease. In fact, reducing the expression of HDAC6 by using RNA interference techniques or inhibiting its deacetylase activity stimulated the microtubule-dependent trafficking and BDNF release [59], while reducing HDAC4 levels delayed cytoplasmic aggregate formation and rescues neuronal and corticostriatal synaptic function in mouse models of HD, improving motor coordination and neurological phenotypes [60]. Moreover, decreasing the HDAC3 orthologue in the *C. elegans* HD model suppressed mutant huntingtin-induced degeneration of sensory neurons, while decreasing Rpd3 (orthologue to HDAC1/2/8) or Sir2 (Sirt1) was neuroprotective but did not improve survival in a *Drosophila* HD model [61]. Interestingly, Bates and colleagues have demonstrated that decreasing HDACs in the R6/2 mouse did not ameliorate the pathological phenotype associated with HD or reverse transcriptional deficit [45, 62]. In order to set apart the reasons for these controversial findings, Valor et al. (2013) conducted research on histone acetylation comparing different HD mouse models [47]. Based on this exhaustive analysis, they concluded that altered bulk histone acetylation is not a general feature of HD pathology. However, despite the absence of global changes, loci relevant to the HD pathology showed local depletions in histone acetylation marks; thus, it is not yet clear whether changes in HAT/HDAC play a causal role in HD pathogenesis, but they might still contribute to the disease progression."

15.3 Histone Methylation in HD

15.3.1 Arginine Methylation

Arginine methylation is an abundant posttranslational modification (PTM), with about 0.5% of arginine residues methylated in mammalian tissues. Three types of methylarginine species exist: monomethylarginine (MMA), asymmetric dimethylarginine (aDMA), and symmetric dimethylarginine (sDMA). The formation of MMA, aDMA, and sDMA in mammalian cells is performed by a family of nine protein arginine methyltransferases (PRMTs). PRMT1, 2, 3, 6, 8, and CARM1 (also called PRMT4) are Type I arginine methyltransferases that deposit the aDMA mark, while PRMT5 is the primary Type II arginine methyltransferase that deposits the sDMA mark [63]. The most important arginine methylations are found in histone H3 (namely, H3R2, H3R8, H3R17, H3R26) and in histone H4 (H4R3). H4R3 site is a major target for PRMT1 methylation (aDMA) and is generally associated with transcriptional activation [64, 65]. On the contrary, PRMT5 seems to be mostly associated with transcriptional repression: this enzyme symmetrically dimethylates H4R3 (H4R3sme2) and H3R8 (H3R8sme2) directly contributing to transcriptional repressive activity [66]. Interestingly, a functional interaction of wild-type huntingtin with PRMT5 has been recently uncovered [67]. Normal huntingtin stimulated PRMT5 activity *in vitro*, while mutant huntingtin showed an aberrant reduction in its ability to dimethylate endogenous substrates (histones and spliceosomal Sm proteins). Coherently, overexpression of PRMT5 or knockdown of H4R3Me2 demethylase JMJD6 reversed the toxic effects of mutant huntingtin in primary cortical neurons, suggesting that PRMT5 deficiency may mediate, at least in part, HD pathogenesis.

15.3.2 Lysine Methylation

Lysine methylation is the best known among PTMs because of its relative stability, its multivalence, and its cross talk with other modifications. Specifically, histone proteins are heavily methylated on the side chains of lysine (K) residues. Lysine methylation has been shown to be present in mono-, di-, or trimethylated states. Although Lys methylation does not alter the overall charge of the molecule, methylated H3K4, H3K36, and H3K79 are associated with open, transcriptionally active genes in euchromatin, while methylated H3K9, H3K27, and H3K20 are associated with transcriptionally inactive, heterochromatin regions. Methylation reactions are catalyzed by KTM (histone lysine methyltransferases), and they tend to be highly specific enzymes that target preferentially certain lysine residues. All KMTs (except DOT1) contain an evolutionarily conserved SET domain (usually at C-terminal) comprised of 130 amino acids. The SET domain binds the AdoMet (S-Adenosyl-L-methionine) cofactor to mediate transfer of a methyl group to the target lysine.

The removal of Lys-methylation [demethylation] is catalyzed by Jumonji demethylases [KDM5-A (JARID1A); KDM5-C(JARID1C); KDM6A (UTX)] that can demethylate mono-, di-, and trimethylated lysine via an oxidative mechanism and radical attack involving Fe(II) and a-ketoglutarate [68–70]. However, another demethylase enzyme KDM1A (LSD-lysine (K)-specific demethylase 1A) is able to demethylate only mono- and dimethyl lysines (H3K4me1/2) via an amine oxidation reaction with flavin adenine dinucleotide (FAD) as a cofactor [68].

15.3.3 Trithorax and Polycomb Complexes

Trithorax group (TrxG) proteins are evolutionarily conserved chromatin regulators that retain histone-modifying activity through SET domain-containing factors and are usually correlated with transcriptional activation. TrxG complex was found for the first time in *Drosophila*, while mammalian homologs were then identified as part of the COMPASS-like complexes, which include MLL family of KMTs methyltransferases and specifically methylates histone H3 at lysine 4. MLL3 and MLL4 specifically mediate H3K4me1 at enhancers, while, in contrast, MLL1 and MLL2 are recruited to promoters, in which they catalyze H3K4me3. MLL1 activity has been associated with active genes, whereas MLL2 seems to be specific for bivalent promoters [71].

Polycomb group (PcG) proteins are generally divided in two classes based on their biochemical characteristics and functional activity: members of the Polycomb repressive complex 1 (PRC1) and 2 (PRC2) are both required for gene expression repression, but while PRC2 has histone-modifying activity, namely, methylating H3K27 at silent genes, PRC1 can recognize and bind to this modification and induce appropriate structural change in chromatin (compaction-repression).

While all PRC1 complexes contain Ring1B (also known as Ring2/RNF2) subunit, which retains the E3 ubiquitin ligase activity [72, 73] and one member of the Polycomb group of ring finger (Pcgf1-6) protein, PRC1 complexes can be differentiated in cPRC1 and ncPRC1 complexes based on the presence of one Chromobox (Cbx) protein [cPRC1 complexes] or Ring1B, Yy1-binding protein (Rybp), or its homolog YAF2 [ncPRC1 complexes] [74, 75]. cPRC1-Cbx and ncPRC1-Rybp complexes co-occupy common as well as distinct subsets of target genes, and their distribution correlates with the levels of H3K27me3 and low transcriptional levels, with the co-occupancy of H3K27me3 by PRC1-Cbx higher than that by PRC1-Rybp. In fact, the presence of PRC1-Cbx7 (the most expressed Polycomb Cbx in embryonic stem cells) correlates with robust gene silencing, whereas genes uniquely occupied by PRC1-Rybp are moderately expressed [76].

Differently from PRC1 complexes which show a high degree of subunits heterogeneity, PRC2 contains three core proteins [Suz12 (suppressor of zeste 12); Eed (embryonic ectoderm development); and Ezh1/2 (enhancer of zeste 1 or 2 protein)], which is the SET domain methyltransferase enzyme of the complex [77]. The three components are present in a 1:1:1 stoichiometry levels [78], while the Ezh1 and Ezh2 proteins are mutually exclusive in the complex, and their expression seems to be complementary: for instance, Ezh2 is highly expressed in embryonic tissues and proliferating cells, with Ezh2 efficiently methylating H3K27, whereas Ezh1 is mostly present in adult tissues and nondividing cells with a minor methyltransferase activity [79].

PcG and TrxG groups of genes encode main effectors that transduce signals to the chromatin and participate in maintaining cellular identity, while playing a crucial role during development and tissue differentiation. The PcG and TrxG groups of proteins function, for the most part antagonistically, determining repressed and active states of gene expression, respectively. Because of the broad nature of their functions (pleiotropy), also reflected by the variety of complexes' associated subunits, deregulation of these genes contributes and even drives important disorders in mammals (such as cancer, neurological disease [71]).

Although mostly acting antagonistically, PcG and TrxG complexes are also found together to decorate several promoters mostly involved in "embryo and tissue developmental processes" and "tissue and cell differentiation" pathways in pluripotent embryonic stem cells [80, 81]. In this subset of genes, the active H3K4me3 mark (catalyzed by TrxG) coexists on the same nucleosomes together with the repressive mark H3K27me3 (catalyzed by PRC2), determining a class of promoters—BIVALENT promoters—that are poorly transcribed, but they are "primed, ready" to be activated, or repressed during tissue/cell differentiation [82].

15.3.4 Histone Methylation, Trithorax, Polycomb Complexes, and HD

A growing body of experimental evidences supports a role of wild-type and mutant huntingtin on histone methylations [83–85] (Fig. 15.3). Specifically, extending the *Htt* polyglutamine tract enhances huntingtin function as a facilitator of the chromatin repressor PRC2, while huntingtin suppression is associated with a significant impairment of PRC2 activity. These functional effects are mediated by the ability of full-length wild-type and mutant huntingtin protein to physically interact with Ezh2 and Suz12 PRC2 core components in the nucleus of embryonic stem cells and embryoid bodies [83]. Analysis of "huntingtin normal function" at genome-wide level through the dichotomous comparison between *Htt* WT and dKO genotypes demonstrated that the absence of huntingtin caused a considerable reduction in the total number of H3K27me3-marked promoters, mainly part of the "bivalent" (histone H3K27me3 and histone H3K4me3) class of promoters [85]. By contrast and in support of a "simple GAIN of FUNCTION" hypothesis for HD, "mutant huntingtin function," uncovered by the comparison of four knock-in genotypes (Q20, Q50, Q92, Q111) presenting various CAG expansion and mimicking the normal, adult, and juvenile CAG range of the human HD mutation, also showed altered pattern of genome-wide H3K27me3 distribution but not mimicking the effects caused by the absence of *Htt*. Therefore, mutant huntingtin does not replicate the loss of huntingtin function but instead highlights altered chromatin organization that leads to altered developmental potential of progenitor cells, predicting reduced fitness and an enhanced readiness for cell death [85].

The possible link between PRC2 and HD was recently further supported by the loss of neuronal PRC2 H3K27me3 sites and the upregulation of some PRC2 target genes—mainly connected with Hox gene clusters and developmentally regulated proteins—in the HD-affected human brain [86–88]. Interestingly, loss of normal regulation of PRC2 levels in adult neurons has been also associated with derepression of selected, predominantly bivalent PRC2 target genes, with detrimental effects for adult neuron functions and survival, thus further reinforcing the view according to which persistent changes in the activity or recruitment of PRC2, as well as other H3K27me3-controlling enzymes, may lead to systemic neurodegeneration [89] (Fig. 15.3). However, the presence of an expanded CAG tract does not only correlate with changes in PRC2 pattern but is also associated with a progressive change in histone H3K4me3 enrichment, leading to decreased RNA expression [85] (Fig. 15.3). This is in line with the view that histone-modifying enzymes and chromatin remodeling factors are not acting as a single entity but are part of supermolecular complexes where the regulation of transcription is accomplished by the coordinated action of complementary functions (repression and activation) [90].

Importantly, the reduced enrichment in H3K4me3, leading to reduced transcriptional levels of target genes, has been also described *in vivo*, in both cortex and striatum of R6/2 mice as well as in human HD *post-mortem* brains, supporting the role of wild-type and mutant huntingtin in chromatin regulation as an important regulatory mechanism in HD pathogenesis [84]. Specifically, H3K4me3 enrichment was reduced at the RE-1 silencing transcription factor/neuron-restrictive silencer factor (REST/NRSF) promoter II, thus suggesting that reduced transcription could be a

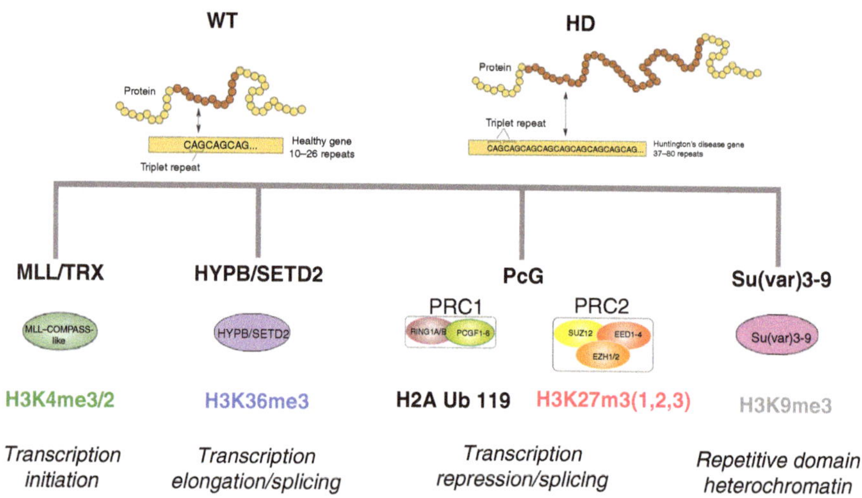

Fig. 15.3 Schematic representation of the chromatin complexes whose function is directly or indirectly regulated by wild-type or mutant huntingtin. (1) MLL/TRX specifically methylates histone H3 at lysine 4 (di- and trimethylation) and usually acts on transcription initiation; (2) HYPB/SETD2 is a histone H3 lysine 36-specific methyltransferase (trimethylation) that functions in transcription elongation and splicing; (3) PcG (Polycomb group), divided in PRC1 (the Polycomb repressive complex 1) which retains the E3 ubiquitin ligase activity and PRC2 (the Polycomb repressive complex 2) that specifically methylates (mono-, di-, or trimethylation) histone 3 at lysine 27, concertedly acts on transcription repression and splicing regulation; (4) Su(var)3-9, which selectively methylates (trimethylation) histone H3 at lysine 9, functions on repetitive domain heterochromatin. Wild-type and/or mutant huntingtin has been reported to directly or indirectly affect the activity of these four complexes

consequence of changes in chromatin structure at REST binding site and *BDNF* locus, in particular. These findings are consistent with previous report by Zuccato and Cattaneo, showing an altered REST and BDNF signaling in HD [91, 92]. Vashishtha et al. 2013 also tested the impact of reducing the expression of an H3K4me3 demethylase, JARID1C, in cell-based assays and in a *Drosophila* model of HD. In these proof-of-concept studies, the reductions of H3K4me3 demethylase restore expression of key downregulated genes (i.e., *Bdnf*) and are neuroprotective in flies, suggesting that chromatin-modulating enzymes, including JARID1C, could be rational therapeutic targets for HD treatment [84].

Importantly, other studies have recently confirmed this trend of H3K4me3 decrease at genome-wide level in HD *post-mortem* prefrontal cortical samples, thus supporting the view that the neuronal epigenome is affected in HD [86, 93]. Notably, in this latter study, reduced H3K4me3 signal is also associated with increased DNA methylation, and these epigenetic changes closely correlate with degeneration in the striatum of HD patients.

15.3.5 Other Histones Methylation in HD

In addition to the classical histone modifications correlated with transcriptional activation (H3K4me3) or repression (H3K27me3), the levels of histone H3K9me3 and the corresponding SETDB1 methyltransferase were shown to be elevated in *post-mortem* brains of HD patients as well as in HD transgenic animals [94] (Fig. 15.3). A possible explanation for this phenotype might be ascribed to CBP, since, by acting as a transcriptional cofactor, it represses SETDB1 gene expression and maintains the appropriate level of H3K9me3 [95]. Altered gene transcription in HD was also directly associated with H3K9me3-mediated chromatin remodeling, so the increase of H3K9me3 level has been correlated with the formation of large constitutive heterochromatin domains and, through this, with gene silencing [5, 37].

Because of its particular sequence structure, where a proline-rich region (PRR) directly follows the polyQ domain, the huntingtin protein has been shown to interact with WW domain-containing proteins: one of these interactors is HYPB/SETD2, the nonredundant H3K36 trimethylase [96, 97] (Fig. 15.3). While Passani and colleagues showed that the WW domain protein HYPB interacts and co-localizes with huntingtin in HD *post-mortem* brains [97], more recently, Gao and collaborators studied the structural interaction between WW domains huntingtin, discussing its possible relevance for HD pathogenesis [98]. In fact, as histone H3 lysine 36-specific methyltransferase, HYPB is associated with important cellular function such as RNAPII elongation-complex formation, minimizing aberrant, spurious transcription [99], and regulation of alternative splicing by acting on RNA transcription rate or directly regulating recruitment of splicing factors such as polypyrimidine tract binding (PTB) protein [100].

Because of this interaction pattern with both wild-type and mutant huntingtin, it's conceivable that this huntingtin-HYPB interaction might have an impact on transcription elongation and/or splicing and represent an important regulatory loop in HD pathogenesis. However, this hypothesis has not yet been tested and awaits a careful, genome-wide validation.

15.4 Noncoding RNAs (ncRNAs) in HD

Most of the mammalian genome is transcribed, but only a minimal part is translated into proteins. But what functions are correlated to this dramatically expanding repertoire of ncRNAs? For a long period of time, ncRNAs were considered "transcriptional noise," but in light of many, recent considerations—ncRNAs' inventory expanded dramatically with vertebrates evolution [101], and they can be specifically, highly expressed in the mammalian nervous system [102]—they are now increasingly associated with critical nervous system developmental and physiological functions. In particular, ncRNAs have been recently implicated in the onset of neurological disorders (i.e., Autism, Fragile-X, Parkinson's disease, and Schizophrenia) [103–106]. Moreover, a large majority of genetic risk loci for common psychiatric diseases affects noncoding regions of the genome [107].

While different classes of ncRNAs have been directly or indirectly implicated in HD, here we will focus on those ncRNAs with documented functions in chromatin regulation whose dysregulation can be associated with HD pathogenesis (Fig. 15.4).

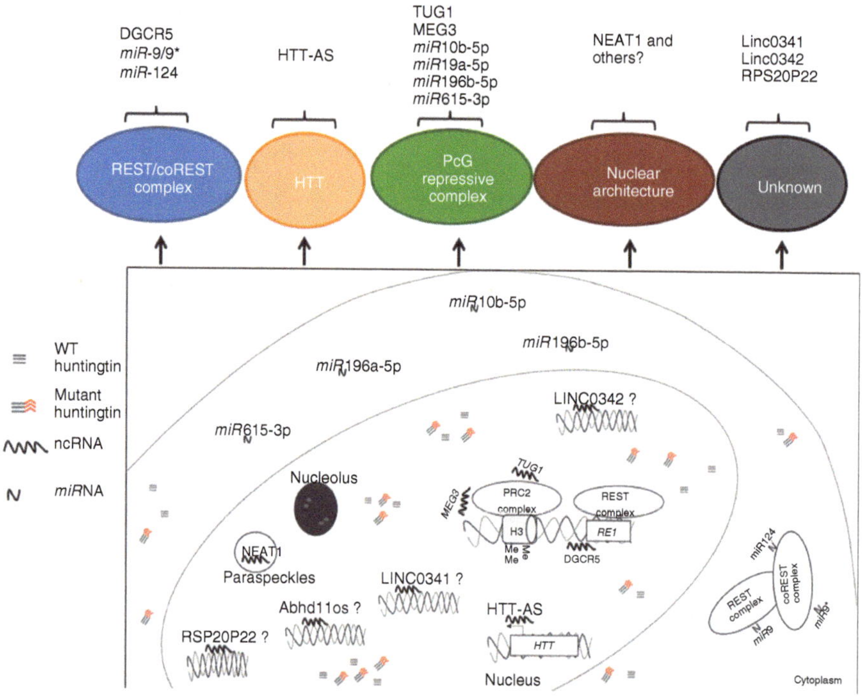

Fig. 15.4 Schematic view of ncRNAs, acting on epigenetic regulation, that are dysregulated in HD model systems and/or HD patients. The contribution of different miRNAs and lncRNAs is indicated in the cytosol or nucleus of the cell, while the epigenetic effectors are highlighted as *colored circles* at the top of the figure

15.4.1 miRNAs in HD

Mature miRNAs are single-stranded, 21–23 nucleotides long, antisense RNA that interact with target genes, downregulating their expression [108]. Although many miRNAs have been found dysregulated in HD and HD animal models [109, 110], only few of them have been shown to exert their functions by a direct or indirect regulation of chromatin complexes (Fig. 15.4). Particularly, Packer and colleagues [111] found that miR-9 and miR-9* were downregulated early in the HD cortex and that this decrease was positively correlated with the severity of the HD grade. miR-124 was also shown to be downregulated in both caudate and motor cortex of HD patient [112]; correspondingly, miR-124 target genes were highly and significantly enriched among those that are upregulated in HD. Interestingly, this bifunctional miR-9/miR-9*/miR-124 unit targets REST and CoREST chromatin repressor complexes, previously implicated in HD (see Sect. 18.4). Thus, endogenous REST protein levels are regulated by miR-9, while specific manipulation of miR-9/miR-9* directly affects REST or CoREST RNA and protein, providing functional evidence for a negative feedback loop between components of the REST complex and REST-regulated miRNAs [111]. In fact, in murine neural progenitor cells, *Rest* inhibits miR-9 and miR-124 expression [113]. Accordingly, in mature, healthy neurons, *Rest* is expressed at low levels, because of the expression of miR-9 and because it is primarily sequestered in the cytoplasm in part through interaction with wild-type huntingtin. In HD patients, instead, the levels of *REST* are aberrantly upregulated because of the altered regulation of miR-9, while altered mutant huntingtin fails to bind REST, thus enabling its nuclear translocation and the display of its full repressive activity. The opposite scenario can also be considered. In fact, elevated nuclear REST in HD is likely to contribute to gene dysregulation through aberrant repression of miRNAs [114]. In agreement, 20 mouse miRNAs with a known REST binding site were found to be downregulated in HD. Moreover, similar experiments conducted in ST*Hdh*$^{Q111/Q111}$ (*Rest* is nuclear) relative to ST*Hdh*$^{7/7}$ (*Rest* is cytoplasmatic) knock-in cell lines showed that *Rest* target miRNAs (15 out of 41) were expressed at significantly lower levels.

Interestingly, not only REST/CoREST chromatin complexes are associated with HD by miRNAs dysregulation. In fact, also an aberrant Polycomb repressive complex 2 (PRC2) regulation (see Sect. 15.3.3) seems to be correlated to a significant upregulation of five mRNAs (miR-10b-5p, miR-196a-5p, miR-196b-5p, and miR-615-3p) in prefrontal cortexes of HD brains [87]. These miRNAs, implicated in apoptosis as well as nervous system development and function, are specifically located in intergenic regions of the Hox clusters, and, conversely, in a feedback regulatory loop, 11 Hox genes are validated targets of these miRNAs. Coherently, Hox genes contiguous to the differentially expressed miRNAs are also differentially expressed in HD. Hox genes, a family of transcription factors that contribute to major morphological changes during embryonic development and are required for anterior-posterior body axis, are repressed by PcG complexes, specifically PRC2. Thus, an altered PRC2 regulation by mutant huntingtin might affect the proper regulation of specific miRNAs from the Hox clusters (and possibly elsewhere in the genome), in turn, influencing proper neuronal development and brain function [87] (Fig. 15.4).

15.4.2 LncRNAs in HD

Differently from miRNAs, lncRNAs are longer transcripts (more than 200 bp) displaying no protein-coding potential. LncRNAs undergo a canonical maturation process (they are usually spliced and can be polyadenylated); however, their sequence is shorter, and they are usually characterized by a smaller number of exons and have lower expression levels compared to protein-coding mRNAs.

Many different functions have been associated to lncRNAs through a direct regulation of proximally located [cis-regulation] or distally located genes [trans-regulation]. But, nowadays, what is known about the involvement of lncRNAs in HD pathogenesis?

The *HTT* locus itself produces a natural, head-to-head, antisense transcript *HTT*-AS which is alternatively spliced, is highly expressed in brain regions, and encompasses the CAG repeat stretch [115]. *HTT*-AS is under the control of a different promoter compared to *HTT* gene, and CAG expansion has been shown to specifically reduce its expression level. Thus, natural *HTT*-AS is downregulated by CAG repeats in *HTT* locus, but what is its effect on *HTT* transcript? *HTT*-AS overexpression (or silencing) shows a small but specific downregulation (or upregulation) of the *HTT* transcript. Thus, Chung and colleagues speculate that with normal CAG length repeats *HTT*-AS under the control of its endogenous promoter is expressed at low levels and partially inhibits *HTT* expression, representing a lncRNA-regulated feedback loop on the *HTT* transcript. However, with an expanded CAG repeat, *HTT*-AS expression is reduced, and, consequently, its regulatory effect on *HTT* transcript is abrogated (Fig. 15.4).

Apart from the cis-acting *HTT*-AS, other lncRNAs have been shown to be dysregulated in HD: microarray data obtained from *post-mortem* brains (caudate) revealed that four already described lncRNAs (*TUG1*, *NEAT1*, *MEG3*, and *DGCR5*) and three novel lncRNAs (*LINC0341*, *RPS20P22*, and *LINC00342*) were highly and significantly dysregulated in HD brains versus controls [116, 117]. Although some structural information is available for these novel lncRNAs dysregulated in HD brains, a comprehensive, full characterization of their physiological function and their role in HD pathology is still missing. More information, instead, are accessible for those previously described lncRNAs. In fact and interestingly in the context of HD, *TUG1* was initially identified to be able to bind to the PRC2 repressive complex [118], while *DGCR5*, a neural-specific lncRNA, is directly targeted by REST (Fig. 15.4). Finally, *MEG3* is a well-characterized transcript, dynamically expressed during development of the mouse nervous system and maternally imprinted [119]. *MEG3* is found in the chromatin compartment of the cell in association with PRC2 complex [120]. On another level, *NEAT1* lncRNA has been associated to the formation and maintenance of nuclear structure [121]. Thus, wild-type and/or mutant huntingtin, through a direct/indirect interaction with chromatin complexes (PcG, REST, CoREST), might be able to modulate the expression of several lncRNAs, possibly altering the constitutive nuclear architecture and contributing to disease onset or progression.

Interestingly, in a recent study, Francelle and colleagues characterized *Abhd11os* as a striatal-enriched lncRNA, displaying reduced levels in R6/2, BAC-HD, and knock-in mouse models of HD [122]. Although additional experiments will be needed to characterize the molecular mechanisms of *Abhd11os* on CAG-driven striatal phenotypes, the data are concordant in demonstrating the protective effects of abhd11os in the HD-related pathological process.

Altogether these results suggest that lncRNAs may be important epigenetic regulators not only during normal brain development but also in lifelong neurological conditions such as HD (Fig. 15.4).

15.5 Epigenetic-Based Clinical Trials in HD

15.5.1 HDAC Inhibitors in HD Clinical Trials

Because of the previously described role of acetylation in the pathological alteration of gene expression in HD, it is not surprising that several clinical trials recently focused on the use of HDAC inhibitors in HD clinical trials. Following encouraging results reporting the administration of the broad-spectrum HDAC inhibitors sodium butyrate (NaB), suberoylanilide hydroxamic acid (SAHA), and trichostatin A (TSA) as protective in HD models system [58, 123], the first report of HDAC inhibitor usage in a phase I clinical trial was associated with the oral administration of sodium phenylbutyrate and led to a moderate but significant correction of mRNA levels in a panel of blood biomarkers [124]. Despite these promising results, the use of broad-spectrum HDAC inhibitors resulted in unwanted side effects, mainly linked to the complexity of the HDAC family and the uncertainty of the rate-limiting HDAC enzymes really critical for the HD pathological process. The search for newer, selective HDAC inhibitors is currently under development for the treatment of HD. It will be important to assess their efficacy and tolerability in comparison with the results obtained by using broad-spectrum HDAC inhibitors. However, it's important to consider that HDAC inhibitor treatments (TSA mainly) seem to preferentially hyperacetylate loci already transcriptionally active, rather than reactivate transcriptionally silent genomic locations [125]. Moreover, hundreds of proteins can be acetylated and therefore altered in response to a specific HDAC inhibitor treatment, thus raising again the issue of target specificity [126].

15.5.2 Methylation Inhibitors in HD Clinical Trials

Despite the fact that extensive changes in methyltransferase(s) activities have been reported in HD, to date no compound able to alter histone methylation has reached HD clinical usage. However, there are studies reporting the identification of specific and powerful inhibitors of EZH2 activity, the PRC2-specific methyltransferase, as a therapeutic strategy for different cancers (lymphomas) [127]. Indeed, in this study,

GSK126 results a potent, highly selective, S-adenosyl-methionine-competitive, small-molecule inhibitor of EZH2 methyltransferase, leading to a global H3K27me3 reduction and reactivating transcriptionally silent PRC2 target genes [127].

These results are certainly encouraging, and similar studies meant to test the efficacy of this PRC2 inhibitor (or other equivalent molecules) on HD *in vitro* or *in vivo* model systems will shortly fill up this gap.

References

1. Thomas EA. DNA methylation in Huntington's disease: implications for transgenerational effects. Neurosci Lett. 2016;625:34–9.
2. Razin A, Riggs AD. DNA methylation and gene function. Science. 1980;210(4470): 604–10.
3. Breiling A, Lyko F. Epigenetic regulatory functions of DNA modifications: 5-methylcytosine and beyond. Epigenetics Chromatin. 2015;8:24.
4. Bird AP. CpG-rich islands and the function of DNA methylation. Nature. 1986;321(6067): 209–13.
5. Lee J, et al. Epigenetic mechanisms of neurodegeneration in Huntington's disease. Neurotherapeutics. 2013;10(4):664–76.
6. Okano M, et al. DNA methyltransferases Dnmt3a and Dnmt3b are essential for de novo methylation and mammalian development. Cell. 1999;99(3):247–57.
7. Klein CJ, et al. Mutations in DNMT1 cause hereditary sensory neuropathy with dementia and hearing loss. Nat Genet. 2011;43(6):595–600.
8. Winkelmann J, et al. Mutations in DNMT1 cause autosomal dominant cerebellar ataxia, deafness and narcolepsy. Hum Mol Genet. 2012;21(10):2205–10.
9. Wu SC, Zhang Y. Active DNA demethylation: many roads lead to Rome. Nat Rev Mol Cell Biol. 2010;11(9):607–20.
10. He YF, et al. Tet-mediated formation of 5-carboxylcytosine and its excision by TDG in mammalian DNA. Science. 2011;333(6047):1303–7.
11. Maiti A, Drohat AC. Thymine DNA glycosylase can rapidly excise 5-formylcytosine and 5-carboxylcytosine: potential implications for active demethylation of CpG sites. J Biol Chem. 2011;286(41):35334–8.
12. Herron DC, Shank RC. In vivo kinetics of O6-methylguanine and 7-methylguanine formation and persistence in DNA of rats treated with symmetrical dimethylhydrazine. Cancer Res. 1981;41(10):3967–72.
13. Watanabe S, et al. Methylated DNA-binding domain 1 and methylpurine-DNA glycosylase link transcriptional repression and DNA repair in chromatin. Proc Natl Acad Sci U S A. 2003;100(22):12859–64.
14. Watt F, Molloy PL. Cytosine methylation prevents binding to DNA of a HeLa cell transcription factor required for optimal expression of the adenovirus major late promoter. Genes Dev. 1988;2(9):1136–43.
15. Bird AP, Wolffe AP. Methylation-induced repression—belts, braces, and chromatin. Cell. 1999;99(5):451–4.
16. Gil JM, Rego AC. Mechanisms of neurodegeneration in Huntington's disease. Eur J Neurosci. 2008;27(11):2803–20.
17. Ng CW, et al. Extensive changes in DNA methylation are associated with expression of mutant huntingtin. Proc Natl Acad Sci U S A. 2013;110(6):2354–9.
18. A novel gene containing a trinucleotide repeat that is expanded and unstable on Huntington's disease chromosomes. The Huntington's Disease Collaborative Research Group. Cell. 1993; 72(6):971–83.
19. Trettel F, et al. Dominant phenotypes produced by the HD mutation in STHdh(Q111) striatal cells. Hum Mol Genet. 2000;9(19):2799–809.

20. Kerschbamer E, Biagioli M. Huntington's disease as neurodevelopmental disorder: altered chromatin regulation, coding, and non-coding RNA transcription. Front Neurosci. 2015;9:509.
21. Jia H, et al. HDAC inhibition imparts beneficial transgenerational effects in Huntington's disease mice via altered DNA and histone methylation. Proc Natl Acad Sci U S A. 2015;112(1):E56–64.
22. Pastor WA, et al. Genome-wide mapping of 5-hydroxymethylcytosine in embryonic stem cells. Nature. 2011;473(7347):394–7.
23. Tahiliani M, et al. Conversion of 5-methylcytosine to 5-hydroxymethylcytosine in mammalian DNA by MLL partner TET1. Science. 2009;324(5929):930–5.
24. Kriaucionis S, Heintz N. The nuclear DNA base 5-hydroxymethylcytosine is present in Purkinje neurons and the brain. Science. 2009;324(5929):929–30.
25. Wang FL, et al. Genome-wide loss of 5-hmC is a novel epigenetic feature of Huntingtons disease. Hum Mol Genet. 2013;22(18):3641–53.
26. Szulwach KE, et al. 5-hmC-mediated epigenetic dynamics during postnatal neurodevelopment and aging. Nat Neurosci. 2011;14(12):1607–16.
27. Slow EJ, et al. Selective striatal neuronal loss in a YAC128 mouse model of Huntington disease. Hum Mol Genet. 2003;12(13):1555–67.
28. Fredholm BB, et al. International Union of Pharmacology. XXV. Nomenclature and classification of adenosine receptors. Pharmacol Rev. 2001;53(4):527–52.
29. Harper PS. Huntington's disease: a clinical, genetic and molecular model for polyglutamine repeat disorders. Philos Trans R Soc Lond Ser B Biol Sci. 1999;354(1386):957–61.
30. Reiner A, et al. Differential loss of striatal projection neurons in Huntington disease. Proc Natl Acad Sci U S A. 1988;85(15):5733–7.
31. Zuccato C, et al. Systematic assessment of BDNF and its receptor levels in human cortices affected by Huntington's disease. Brain Pathol. 2008;18(2):225–38.
32. Chiang MC, et al. The A2A adenosine receptor rescues the urea cycle deficiency of Huntington's disease by enhancing the activity of the ubiquitin-proteasome system. Hum Mol Genet. 2009;18(16):2929–42.
33. Taherzadeh-Fard E, et al. Age at onset in Huntington's disease: replication study on the associations of ADORA2A, HAP1 and OGG1. Neurogenetics. 2010;11(4):435–9.
34. Tebano MT, et al. Role of adenosine A(2A) receptors in modulating synaptic functions and brain levels of BDNF: a possible key mechanism in the pathophysiology of Huntington's disease. Sci World J. 2010;10:1768–82.
35. Mangiarini L, et al. Exon 1 of the HD gene with an expanded CAG repeat is sufficient to cause a progressive neurological phenotype in transgenic mice. Cell. 1996;87(3):493–506.
36. Villar-Menendez I, et al. Increased 5-methylcytosine and decreased 5-hydroxymethylcytosine levels are associated with reduced striatal A2AR levels in Huntington's disease. Neruomol Med. 2013;15(2):295–309.
37. Lee J, et al. Epigenetic regulation of cholinergic receptor M1 (CHRM1) by histone H3K9me3 impairs Ca(2+) signaling in Huntington's disease. Acta Neuropathol. 2013;125(5):727–39.
38. Thomas B, et al. A novel method for detecting 7-methyl guanine reveals aberrant methylation levels in Huntington disease. Anal Biochem. 2013;436(2):112–20.
39. Hake SB, Xiao A, Allis CD. Linking the epigenetic 'language' of covalent histone modifications to cancer. Br J Cancer. 2004;90(4):761–9.
40. Marques SC, et al. Alzheimer's disease: the quest to understand complexity. J Alzheimers Dis. 2010;21(2):373–83.
41. Chouliaras L, et al. Epigenetic regulation in the pathophysiology of Alzheimer's disease. Prog Neurobiol. 2010;90(4):498–510.
42. Jenuwein T, Allis CD. Translating the histone code. Science. 2001;293(5532):1074–80.
43. Sadri-Vakili G, Cha JH. Mechanisms of disease: histone modifications in Huntington's disease. Nat Clin Pract Neurol. 2006;2(6):330–8.
44. Suzuki MM, Bird A. DNA methylation landscapes: provocative insights from epigenomics. Nat Rev Genet. 2008;9(6):465–76.
45. Glajch KE, Sadri-Vakili G. Epigenetic mechanisms involved in Huntington's disease pathogenesis. J Huntingtons Dis. 2015;4(1):1–15.

46. Verdin E, Ott M. 50 Years of protein acetylation: from gene regulation to epigenetics, metabolism and beyond. Nat Rev Mol Cell Biol. 2015;16(4):258–64.
47. Valor LM, et al. Genomic landscape of transcriptional and epigenetic dysregulation in early onset polyglutamine disease. J Neurosci. 2013;33(25):10471–82.
48. de Ruijter AJ, et al. The novel histone deacetylase inhibitor BL1521 inhibits proliferation and induces apoptosis in neuroblastoma cells. Biochem Pharmacol. 2004;68(7):1279–88.
49. Gallinari P, et al. HDACs, histone deacetylation and gene transcription: from molecular biology to cancer therapeutics. Cell Res. 2007;17(3):195–211.
50. Saha R, Pahan K. HATs and HDACs in neurodegeneration: a tale of disconcerted acetylation homeostasis. Cell Death Differ. 2006;13(4):539–50.
51. Cha JH. Transcriptional dysregulation in Huntington's disease. Trends Neurosci. 2000;23(9):387–92.
52. Sugars KL, Rubinsztein DC. Transcriptional abnormalities in Huntington disease. Trends Genet. 2003;19(5):233–8.
53. Hoshino M, et al. Histone deacetylase activity is retained in primary neurons expressing mutant huntingtin protein. J Neurochem. 2003;87(1):257–67.
54. Zuccato C, et al. Huntingtin interacts with REST/NRSF to modulate the transcription of NRSE-controlled neuronal genes. Nat Genet. 2003;35(1):76–83.
55. Cattaneo E, Zuccato C, Tartari M. Normal huntingtin function: an alternative approach to Huntington's disease. Nat Rev Neurosci. 2005;6(12):919–30.
56. Ryu H, et al. Histone deacetylase inhibitors prevent oxidative neuronal death independent of expanded polyglutamine repeats via an Sp1-dependent pathway. Proc Natl Acad Sci. 2003;100(7):4281–6.
57. McCampbell A, et al. Histone deacetylase inhibitors reduce polyglutamine toxicity. Proc Natl Acad Sci. 2001;98(26):15179–84.
58. Steffan JS, et al. Histone deacetylase inhibitors arrest polyglutamine-dependent neurodegeneration in Drosophila. Nature. 2001;413(6857):739–43.
59. Simoes-Pires C, et al. HDAC6 as a target for neurodegenerative diseases: what makes it different from the other HDACs? Mol Neurodegener. 2013;8:7.
60. Mielcarek M, et al. HDAC4 reduction: a novel therapeutic strategy to target cytoplasmic huntingtin and ameliorate neurodegeneration. PLoS Biol. 2013;11(11):e1001717.
61. Pallos J, et al. Inhibition of specific HDACs and sirtuins suppresses pathogenesis in a Drosophila model of Huntington's disease. Hum Mol Genet. 2008;17(23):3767–75.
62. Benn CL, et al. Genetic knock-down of HDAC7 does not ameliorate disease pathogenesis in the R6/2 mouse model of Huntington's disease. PLoS One. 2009;4(6):e5747.
63. Dhar S, et al. Loss of the major Type I arginine methyltransferase PRMT1 causes substrate scavenging by other PRMTs. Sci Rep. 2013;3:1311.
64. Wang H, et al. Methylation of histone H4 at arginine 3 facilitating transcriptional activation by nuclear hormone receptor. Science. 2001;293(5531):853–7.
65. Daujat S, et al. Crosstalk between CARM1 methylation and CBP acetylation on histone H3. Curr Biol. 2002;12(24):2090–7.
66. Pal S, et al. Human SWI/SNF-associated PRMT5 methylates histone H3 arginine 8 and negatively regulates expression of ST7 and NM23 tumor suppressor genes. Mol Cell Biol. 2004;24(21):9630–45.
67. Ratovitski T, et al. PRMT5-mediated symmetric arginine dimethylation is attenuated by mutant huntingtin and is impaired in Huntington's disease (HD). Cell Cycle. 2015;14(11):1716–29.
68. Shi Y, et al. Histone demethylation mediated by the nuclear amine oxidase homolog LSD1. Cell. 2004;119(7):941–53.
69. Hoffart LM, et al. Direct spectroscopic detection of a C-H-cleaving high-spin Fe(IV) complex in a prolyl-4-hydroxylase. Proc Natl Acad Sci U S A. 2006;103(40):14738–43.
70. Ozer A, Bruick RK. Non-heme dioxygenases: cellular sensors and regulators jelly rolled into one? Nat Chem Biol. 2007;3(3):144–53.

71. Schuettengruber B, et al. Trithorax group proteins: switching genes on and keeping them active. Nat Rev Mol Cell Biol. 2011;12(12):799–814.
72. Wang H, et al. Role of histone H2A ubiquitination in Polycomb silencing. Nature. 2004;431(7010):873–8.
73. Cao R, Tsukada Y, Zhang Y. Role of Bmi-1 and Ring1A in H2A ubiquitylation and Hox gene silencing. Mol Cell. 2005;20(6):845–54.
74. Gao Z, et al. PCGF homologs, CBX proteins, and RYBP define functionally distinct PRC1 family complexes. Mol Cell. 2012;45(3):344–56.
75. Tavares L, et al. RYBP-PRC1 complexes mediate H2A ubiquitylation at polycomb target sites independently of PRC2 and H3K27me3. Cell. 2012;148(4):664–78.
76. Morey L, et al. RYBP and Cbx7 define specific biological functions of polycomb complexes in mouse embryonic stem cells. Cell Rep. 2013;3(1):60–9.
77. Aranda S, Mas G, Di Croce L. Regulation of gene transcription by Polycomb proteins. Sci Adv. 2015;1(11):e1500737.
78. Smits AH, et al. Stoichiometry of chromatin-associated protein complexes revealed by label-free quantitative mass spectrometry-based proteomics. Nucleic Acids Res. 2013;41(1):e28.
79. Margueron R, et al. Ezh1 and Ezh2 maintain repressive chromatin through different mechanisms. Mol Cell. 2008;32(4):503–18.
80. Bernstein BE, et al. A bivalent chromatin structure marks key developmental genes in embryonic stem cells. Cell. 2006;125(2):315–26.
81. Mikkelsen TS, et al. Genome-wide maps of chromatin state in pluripotent and lineage-committed cells. Nature. 2007;448(7153):553–60.
82. Voigt P, et al. Asymmetrically modified nucleosomes. Cell. 2012;151(1):181–93.
83. Seong IS, et al. Huntingtin facilitates polycomb repressive complex 2. Hum Mol Genet. 2010;19(4):573–83.
84. Vashishtha M, et al. Targeting H3K4 trimethylation in Huntington disease. Proc Natl Acad Sci U S A. 2013;110(32):E3027–36.
85. Biagioli M, et al. Htt CAG repeat expansion confers pleiotropic gains of mutant huntingtin function in chromatin regulation. Hum Mol Genet. 2015;24(9):2442–57.
86. Dong X, et al. The role of H3K4me3 in transcriptional regulation is altered in Huntington's disease. PLoS One. 2015;10(12):e0144398.
87. Hoss AG, et al. MicroRNAs located in the Hox gene clusters are implicated in Huntington's disease pathogenesis. PLoS Genet. 2014;10(2):e1004188.
88. Labadorf A, et al. RNA sequence analysis of human huntington disease brain reveals an extensive increase in inflammatory and developmental gene expression. PLoS One. 2015;10(12):e0143563.
89. von Schimmelmann M, et al. Polycomb repressive complex 2 (PRC2) silences genes responsible for neurodegeneration. Nat Neurosci. 2016;19(10):1321–30.
90. Pasini D, et al. Coordinated regulation of transcriptional repression by the RBP2 H3K4 demethylase and Polycomb-Repressive Complex 2. Genes Dev. 2008;22(10):1345–55.
91. Zuccato C, et al. Loss of huntingtin-mediated BDNF gene transcription in Huntington's disease. Science. 2001;293(5529):493–8.
92. Zuccato C, Cattaneo E. Role of brain-derived neurotrophic factor in Huntington's disease. Prog Neurobiol. 2007;81(5–6):294–330.
93. Bai G, et al. Epigenetic dysregulation of hairy and enhancer of split 4 (HES4) is associated with striatal degeneration in postmortem Huntington brains. Hum Mol Genet. 2015;24(5):1441–56.
94. Ryu H, et al. ESET/SETDB1 gene expression and histone H3 (K9) trimethylation in Huntington's disease. Proc Natl Acad Sci U S A. 2006;103(50):19176–81.
95. Bannister AJ, Kouzarides T. The CBP co-activator is a histone acetyltransferase. Nature. 1996;384(6610):641–3.
96. Faber PW, et al. Huntingtin interacts with a family of WW domain proteins. Hum Mol Genet. 1998;7(9):1463–74.

97. Passani LA, et al. Huntingtin's WW domain partners in Huntington's disease post mortem brain fulfill genetic criteria for direct involvement in Huntington's disease pathogenesis. Hum Mol Genet. 2000;9(14):2175–82.
98. Gao YG, et al. Autoinhibitory structure of the WW domain of HYPB/SETD2 regulates its interaction with the proline-rich region of huntingtin. Structure. 2014;22(3):378–86.
99. Hu M, et al. Histone H3 lysine 36 methyltransferase Hypb/Setd2 is required for embryonic vascular remodeling. Proc Natl Acad Sci U S A. 2010;107(7):2956–61.
100. Luco RF, et al. Epigenetics in alternative pre-mRNA splicing. Cell. 2011;144(1):16–26.
101. Taft RJ, Pheasant M, Mattick JS. The relationship between non-protein-coding DNA and eukaryotic complexity. Bioessays. 2007;29(3):288–99.
102. Mercer TR, et al. Noncoding RNAs in long-term memory formation. Neuroscientist. 2008;14(5):434–45.
103. Ziats MN, Rennert OM. Aberrant expression of long noncoding RNAs in autistic brain. J Mol Neurosci. 2013;49(3):589–93.
104. Pastori C, et al. Comprehensive analysis of the transcriptional landscape of the human FMR1 gene reveals two new long noncoding RNAs differentially expressed in fragile X syndrome and fragile X-associated tremor/ataxia syndrome. Hum Genet. 2014;133(1):59–67.
105. Majidinia M, et al. The roles of non-coding RNAs in Parkinson's disease. Mol Biol Rep. 2016;43(11):1193–204.
106. Hu J, et al. Systematically characterizing dysfunctional long intergenic non-coding RNAs in multiple brain regions of major psychosis. Oncotarget. 2016;7(44):71087–98.
107. Hindorff LA, et al. Potential etiologic and functional implications of genome-wide association loci for human diseases and traits. Proc Natl Acad Sci U S A. 2009;106(23):9362–7.
108. Bartel DP. MicroRNAs: target recognition and regulatory functions. Cell. 2009;136(2): 215–33.
109. Molasy M, et al. MicroRNAs in glaucoma and neurodegenerative diseases. J Hum Genet. 2017;62(1):105–12.
110. Lee ST, et al. Altered microRNA regulation in Huntington's disease models. Exp Neurol. 2011;227(1):172–9.
111. Packer AN, et al. The bifunctional microRNA miR-9/miR-9* regulates REST and CoREST and is downregulated in Huntington's disease. J Neurosci. 2008;28(53):14341–6.
112. Johnson R, Buckley NJ. Gene dysregulation in Huntington's disease: REST, microRNAs and beyond. Neruomol Med. 2009;11(3):183–99.
113. Shenoy A, Blelloch RH. Regulation of microRNA function in somatic stem cell proliferation and differentiation. Nat Rev Mol Cell Biol. 2014;15(9):565–76.
114. Soldati C, et al. Dysregulation of REST-regulated coding and non-coding RNAs in a cellular model of Huntington's disease. J Neurochem. 2013;124(3):418–30.
115. Chung DW, et al. A natural antisense transcript at the Huntington's disease repeat locus regulates HTT expression. Hum Mol Genet. 2011;20(17):3467–77.
116. Hodges A, et al. Regional and cellular gene expression changes in human Huntington's disease brain. Hum Mol Genet. 2006;15(6):965–77.
117. Johnson R. Long non-coding RNAs in Huntington's disease neurodegeneration. Neurobiol Dis. 2012;46(2):245–54.
118. Khalil AM, et al. Many human large intergenic noncoding RNAs associate with chromatin-modifying complexes and affect gene expression. Proc Natl Acad Sci U S A. 2009; 106(28):11667–72.
119. Miyoshi N, et al. Identification of an imprinted gene, Meg3/Gtl2 and its human homologue MEG3, first mapped on mouse distal chromosome 12 and human chromosome 14q. Genes Cells. 2000;5(3):211–20.
120. Mondal T, et al. MEG3 long noncoding RNA regulates the TGF-beta pathway genes through formation of RNA-DNA triplex structures. Nat Commun. 2015;6:7743.
121. Sunwoo H, et al. MEN epsilon/beta nuclear-retained non-coding RNAs are up-regulated upon muscle differentiation and are essential components of paraspeckles. Genome Res. 2009;19(3):347–59.

122. Francelle et al. The striatal long noncoding RNA Abhd11os is neuroprotective against an N-terminal fragment of mutant huntingtin in vivo. Neurobiology of Aging, 2014;36(3): 1601.e7–1601.e16.
123. McCampbell A, et al. Histone deacetylase inhibitors reduce polyglutamine toxicity. Proc Natl Acad Sci U S A. 2001;98(26):15179–84.
124. Hogarth P, Lovrecic L, Krainc D. Sodium phenylbutyrate in Huntington's disease: a dose-finding study. Mov Disord. 2007;22(13):1962–4.
125. Lopez-Atalaya JP, et al. Genomic targets, and histone acetylation and gene expression profiling of neural HDAC inhibition. Nucleic Acids Res. 2013;41(17):8072–84.
126. Choudhary C, et al. Lysine acetylation targets protein complexes and co-regulates major cellular functions. Science. 2009;325(5942):834–40.
127. McCabe MT, et al. EZH2 inhibition as a therapeutic strategy for lymphoma with EZH2-activating mutations. Nature. 2012;492(7427):108–12.
128. Wapenaar H, Dekker FJ. Histone acetyltransferases: challenges in targeting bi-substrate enzymes. Clin Epigenetics. 2016;8(1):1.
129. Thiagalingam S, et al. Histone deacetylases: unique players in shaping the epigenetic histone code. Ann N Y Acad Sci. 2003;983:84–100.
130. Mortaz E, et al. Epigenetics and chromatin remodeling play a role in lung disease. Tanaffos. 2011;10(4):7–16.

Part IV

Elderly Disorders

DNA Modifications and Alzheimer's Disease

16

Rebecca G. Smith and Katie Lunnon

Abstract

Alzheimer's disease (AD) is a complex neurodegenerative disease, affecting millions of people worldwide. While a number of studies have focused on identifying genetic variants that contribute to the development and progression of late-onset AD, the majority of these only have a relatively small effect size. There are also a number of other risk factors, for example, age, gender, and other comorbidities; however, how these influence disease risk is not known. Therefore, in recent years, research has begun to investigate epigenetic mechanisms for a potential role in disease etiology. In this chapter, we discuss the current state of play for research into DNA modifications in AD, the most well studied being 5-methylcytosine (5-mC). We describe the earlier studies of candidate genes and global measures of DNA modifications in human AD samples, in addition to studies in mouse models of AD. We focus on recent epigenome-wide association studies (EWAS) in human AD, using microarray technology, examining a number of key study design issues pertinent to such studies. Finally, we discuss how new technological advances could further progress the research field.

Keywords

Alzheimer's disease • EWAS • 5-methylcytosine • 5-hydroximethylcytosine

R.G. Smith, B.Sc., Ph.D. • K. Lunnon, B.Sc., Ph.D. (✉)
University of Exeter Medical School, RILD, Barrack Road, Exeter, Devon, UK
e-mail: K.Lunnon@exeter.ac.uk

R. Delgado-Morales (ed.), *Neuroepigenomics in Aging and Disease*, Advances in Experimental Medicine and Biology, DOI 10.1007/978-3-319-53889-1_16

16.1 Genetic Contributions to Alzheimer's Disease Etiology

Alzheimer's disease (AD) is a progressive neurodegenerative disease that contributes significantly to the global disease burden, affecting in excess of 26 million people worldwide [1]. Clinically, the first signs of AD manifest as a reduction in the ability to retain new information, leading to disruptions in daily routine. This is followed by difficulty in planning and solving problems, confusion related to time and/or place, speech trouble, and mood and personality changes. AD is characterized by the accumulation of two proteins that contribute to the neuropathology of the disease: extracellular plaques of amyloid-β peptide (Aβ) and neurofibrillary tangles of hyperphosphorylated microtubule binding protein tau [2, 3]. These neuropathological changes are thought to occur perhaps decades before clinical symptoms manifest and the disease is diagnosed [4]. Moreover, given that currently prescribed medications are simply symptomatic treatments and do not modify the underlying disease process, considerable research effort is currently focused on understanding disease etiology.

While the neuropathological manifestation of AD has been well characterized in postmortem brain tissue, less is known about either the underlying risk factors for the disease or the exact mechanisms involved in disease progression. Given the high heritability estimates (60–80%) for AD derived from quantitative genetic analyses [5], the majority of etiological studies have focused predominantly on genetic contributions to the disease. Indeed autosomal dominant mutations in three genes (*APP*, *PSEN1*, and *PSEN2*), which are involved in Aβ production, can explain early-onset (<65 years) familial AD; however, these account for only 5–10% of the total disease burden. Most cases of AD are late-onset (>65 years), non-Mendelian, and highly sporadic, with susceptibility attributed to the action of common genetic variants of low penetrance. In recent years, a number of genome-wide association studies (GWAS), and a subsequent meta-analysis, have nominated around 20 common variants in a number of genes including *ABCA7*, *BIN1*, *CASS4*, *CD2AP* , *CELF1*, *CLU*, *CR1*, *FERMT2*, *FRMD4A*, *HLA-DRB5*, *INPP5D*, *MEF2C*, *MS4A4E/MS4A6A*, *NME8*, *PICALM*, *PTK2B*, *SLC24A4/RIN3*, *SORL1*, and *ZCWPW1* [6, 7]. However, collectively these only explain around a third of disease incidence [8], although polygenic risk scores based on these variants have been developed to predict disease risk [9]. The only common variants identified to date with a modest effect size exist within *APOE*, where the *APOE* ε4 variant, which arises due to single nucleotide polymorphisms (SNPs) at rs7412 and rs429358, is the largest genetic risk factor for late-onset AD, with carriers of the ε4ε4 genotype having an odds ratio of 14.9 for developing the disease [10]. Recent genome sequencing projects have identified other variants, for example, rs75932628 (R47H) within *TREM2* [11, 12]; however, these are relatively rare within the population.

16.2 A Role for Epigenetics in AD

Although many of the genomic studies to date have identified robust and reproducible findings, they do not account for all of AD incidence. Furthermore, a number of disease attributes suggest a potential epigenetic contribution to etiology, for example, the differential vulnerability of specific brain regions to disease, age of onset of disease, environmental influences on disease such as diet, and the increased risk of developing AD in individuals with obesity and type II diabetes [13]. Epigenetic processes mediate the reversible regulation of gene expression, occurring independently of DNA sequence variation, acting principally through chemical modifications to DNA and nucleosomal histone proteins, and orchestrating a diverse range of important neurobiological processes. DNA methylation is the best characterized and most stable epigenetic modification modulating the transcription of mammalian genomes and has been the focus of most human epidemiological epigenetic research to date. Standard genotyping techniques are not able to distinguish between unmodified cytosine and 5-methylcytosine (5-mC), which contains a methyl group on the five positions of the cytosine ring. Bisulfite conversion is by far the simplest method to assess the degree of DNA methylation present in a given sample as it converts unmethylated cytosine to uracil (and to thymine through subsequent PCR), while 5-mC is not converted (and thus remains as cytosine in PCR). As such, bisulfite treatment of DNA allows the differentiation between cytosine and 5-mC through downstream sequencing, amplification, or array-based techniques.

16.3 Epigenetic Studies of Mouse Models of AD

There are many available murine models that have been traditionally used for studying AD. There are, however, limitations to their utility for modeling human AD, for example, mice do not naturally get AD symptoms or produce amyloid plaques. Therefore, in mouse models of amyloid pathology, the amyloid is derived from human transgenes and so produces human amyloid. Currently, no mouse model has all the features of AD seen in humans, but they have differing combinations of the disease characteristics, such as behavioral changes, neurodegeneration, neuropathology, and cognitive deficits at different ages [14]. Although most AD cases are sporadic, transgenic animal models have relied on the utilization of genetic mutations associated with familial AD and are thus a model of the effects of the accumulation of amyloid-β (Aβ) plaques and/or neurofibrillary tangles, rather than a model of sporadic AD, where the causes are unknown. Mouse models have been made for mutations in *APP* [15–17] and knock-out and knock-in mouse models are also available for the *APP* secretases (*BACE* [18], *PSEN1*, *PSEN2* [19], *ADAM10*, and *ADAM17* [20]). The 3xTg-AD mice, which contain human *APP*, *PSEN1*, and tau mutant transgenes [21], are useful as they exhibit both plaque and tangle pathology. Nonetheless, despite this issue of translation from murine models to human sporadic patients, there are many advantages to using them to study epigenetic changes beyond the advantages common of most mouse models, such as having experimental control and a short life span. For example, one can assess longitudinal changes in the epigenome across specific regions of the brain in genetically identical mice. Furthermore, with specific murine models available, which have already been well characterized, one can accurately predict when pathology will start to develop and easily look for epigenetic alterations associated with behavioral, cognitive, and physiological changes at different stages of pathology [14].

To date, three epigenome-wide association studies (EWAS) of DNA methylation in AD transgenic mice have been published. The first by Sanchez-Mut et al. analyzed 12 brain regions from C57BL/6J on a genome-wide promoter DNA methylation array focusing on 762 genes associated with sensory perception, cognition, neuroplasticity, brain physiology, and mental disease. The authors limited their investigation to two transgenic mouse models, *APP/PSEN1* (double-transgenic mice carrying the APPswe/PS1dE9 mutations) and 3xTg-AD, and their analysis of their non-transgenic littermates to the prefrontal cortex used only seven genes which had the largest degree of differential methylation between cerebral cortex and the rest of the brain. They observed that *Tbxa2r*, *F2rl2*, *Sorbs3*, and *Spnb4* were hypermethylated in the frontal cortex of *APP/PSEN1* and 3xTg-AD mice and were replicated in independent samples by pyrosequencing. Further, they also found *TBXA2R*, *SORBS3*, and *SPTBN4* to be hypermethylated in AD Braak stage V–VI cases compared to controls [22].

Cong and colleagues performed MeDIP-chip analysis on cortex samples from *APP/PSEN1* transgenic mice. They identified 2346 hypermethylated CpG sites in 485 unique genes associated with AD compared to non-transgenic littermates. Subsequent pathway analyses showed differentially methylated genes were enriched in inflammatory response and disease, organismal injury and abnormalities, respiratory disease and cancer pathways [23]. Another MeDIP-chip study by Agbemenyah et al. using hippocampal tissue from APPPS1-21 mice, which have Thy1-APP and Thy1-PS1 transgenes, found hypomethylation at the *Igfbp7* promoter was lower in transgenic mice when compared to wild type. They also demonstrated that *Igfbp7* gene expression and IGFBP7 protein levels were also increased [24]. The majority of DNA methylomic studies of AD have however focused on studying the human disease in postmortem brain samples; however, there are some specific issues when performing EWAS in AD brain samples, which require careful consideration.

16.4 The Importance of Study Design for EWAS in Human Tissues

Although assessing epigenomic variation is relatively straightforward, there are a number of caveats when compared to genomic studies. First, and foremost, epigenomic variation is tissue specific, and as such it is important to specifically examine the tissue of interest [25]. Given that AD is a progressive neurodegenerative disorder, with the spread of neurofibrillary tangles throughout the brain being well documented, from the transentorhinal region through the cortex, one would expect different regions of the brain to show disease-specific alterations at different stages of the disease process. This thus poses the question as to where would be the ideal brain region to profile; areas of the brain that are affected early in AD would have large amounts of neuronal loss, while other regions may not exhibit disease pathology. As such, the profiling of multiple brain regions, representing the spectrum of pathology, is optimal as it would allow spatiotemporal mapping of disease-related changes. Further by profiling multiple brain regions, one could look for patterns of epigenetic changes prior to neuropathology, to attempt to assess causality.

Second is the issue of differences in cell abundance when assessing epigenetic variation in heterogeneous tissue such as the brain [26]. This is particularly pertinent for diseases such as AD, which are characterized by neuronal cell loss and gliosis. Given that distinct cell types have potentially different epigenomes, it is important to acknowledge this in analyses. By comparing epigenetic changes at a population level in DNA extracted from whole tissue, which is a collection of cell types with potentially different methylomes, one will be assessing the percentage of cells which do, or do not, have a methylated cytosine at a specific position. This means that cell-specific changes in heterogeneous cell populations could be diluted by unaffected cell types or could be a combination of small changes in many cell types. Some studies have used bioinformatic approaches to provide a proxy measure of neuron/glia proportions [27] and include this as a covariate in analyses; however, the optimal study would use a method such as fluorescence-activated cell sorting (FACS) or laser capture microdissection (LCM) to yield pure populations of different cell types prior to epigenomic profiling. However, such methods are labor intensive, slow, and expensive and are thus not generally feasible for large cohort studies. Third, sample size is another important consideration for EWAS. Although it is generally appreciated that numbers required for EWAS are considerably smaller than for a standard GWAS, with ~75 samples per group giving sufficient power to detect modest changes (~5%) in DNA methylation. It is however imperative that EWAS are tissue specific, and as such it can be challenging to access sufficient numbers of highly characterized brain samples from specific brain regions to ensure adequate power. Related to this issue are co-diagnoses of AD with other dementias. Many postmortem diagnoses of AD are made in combination with other dementias such as Lewy body dementia (LBD) or vascular dementia (VD). As such getting sufficient numbers of donor AD samples for analysis, in the absence of other dementias, can be difficult. Although it is of interest to identify molecular mechanisms associated with dementia, it is also important to identify disease-specific signatures when looking for new pharmacological targets. Finally, many dementia sufferers die with a systemic infection; in fact one study of postmortem records showed that 80% of AD patients had an infection at the time of death [28]. This could also be a confounder in the analysis of data as infections could elicit the activation of pro-inflammatory pathways within the brain. There are thus many issues to consider when planning, designing, and performing an epigenomic study on any disease but particularly in age-related neurodegenerative diseases such as AD.

Aside from identifying novel mechanistic pathways involved in the etiology of AD in the brain, epigenomic analyses could allow the development of novel translational clinical tools for AD. Although there is growing interest in the identification of novel epigenetic biomarkers for the disease in blood, there are other important considerations for these types of studies. A number of environmental exposures have been associated with DNA methylation changes in blood cells, for example, smoking [29], exposure to environmental chemicals [30], and diet [31], which may not have the same effect in disease-relevant tissues and blood and may be more susceptible to changes due to the environment. Furthermore, the timing of sampling could be important, for example, normal aging can alter the epigenome [32] and may have differing effects in different tissues. This can be somewhat adjusted for with the use of "epigenetic age" tools, which allow one to estimate biological age as a result of age-sensitive DNA methylation marks and that can then be used for analysis adjustment [32]. Using peripheral tissues to find detectable disease-associated differences is a goal of most studies due to the fact that neurodegeneration starts decades before clinical diagnosis. To date, however, no robust epigenetic biomarkers have been identified in blood even though accessing larger sample numbers is easier than for brain and relatively noninvasive. The potential of using blood or peripheral tissues to develop epigenetic biomarkers is still feasible given the correlation of DNA methylation between blood and brain for some, but not all, genetic loci [33]. Finally, longitudinal studies would be useful in being able to identify epigenetic biomarkers of disease progression and neuropathology.

16.5 DNA Methylomic Studies of Human AD

Until relatively recently, published literature examining a role for epigenetic modifications in AD development had been largely limited to either speculative reviews or a limited amount of empirical research focused on candidate genes or global changes. While some global methylomic studies using antibodies to detect DNA methylation have shown reductions in DNA methylation in the entorhinal cortex [34], temporal neocortex [35], and hippocampus [36] of AD sufferers postmortem, other studies have reported conflicting results [37–39]. A number of candidate gene studies have also been carried out in human tissue to try and identify AD-associated methylation changes. There have been many candidate gene studies on blood on *5-LOX* [40], repetitive elements Alu, LINE-1, and SAT-α [41, 42], *FAAH* [43], *PIN1* [44], *SNAP25* [45], *SORL1*, and *SIRT1* [46] in small number of samples, but some of these were variable in their results. Other candidate gene studies have used brain tissue samples from AD cases and controls to assess DNA methylation. These have used assays for *HSPA8* and *HSPA9* [47]; *ACE*, *APOE*, *APP*, *BACE1*, *GSK3B*, *MAPT*, and *PSEN1* [48–51]; *CNP* and *DPYSL2* [52]; *PP2AC* [53]; and *RAGE*, *ADORA2A*, and *UCHL1* [51] with associations with AD being found in *APP*, *GSK3B*, *MAPT*, *PP2AC*, *APOE*, *DNMT1*, *MTHFR*, *5-LOX*, *FAAH*, and *PIN1*. Meanwhile, two MethyLight PCR studies assessed DNA methylation in AD, the first in 50 candidate genes and the second accessing promoter methylation for a small selection of genes (*COX-2*, *BDNF*, *NF-κβ*, *CREB*, *DBNL*, *SYP*, *ALOX12*, and genes associated with p450 epoxygenase), in a limited number of AD samples [54, 55].

However, in recent years, advances in epigenomic technology have allowed the quantification of DNA methylomic variation in a number of complex disease phenotypes, including AD (Table 16.1). The workhorses for epigenome-wide association studies (EWAS) have been the Illumina Infinium 27 K array, the 450K BeadArray, and their recent successor, the EPIC 850 K array, which are

Table 16.1 Summary of current EWAS on DNA methylation in AD. A number of studies have examined a role for DNA methylation in Alzheimer's disease, utilizing the Illumina Infinium 27K or 450K BeadArrays

Study	Year	Number of samples	Tissue(s)	Brain bank	Method	Analysis	Genes identified
Bakulski et al. [56]	2012	24	Prefrontal cortex	Michigan Alzheimer's Disease Center	27 K	Disease status	948 CpG sites representing 918 genes
Lunnon et al. [59]	2014	117	Prefrontal cortex	MRC London Neurodegenerative Disease Brain Bank	450 K	Braak	*ANK1, MIR486, PCBD1, SLC15A4, SIRT6, MEST, MLST8, ZNF512, TMX4*
			Superior temporal gyrus				
			Entorhinal cortex				
			Cerebellum				
			Premortem blood				
De Jager et al. [58]	2014	708	Prefrontal cortex	Religious Order Study (ROS) or the Memory and Aging Project (MAP)	450 K	Amyloid	*ANK1, FOXK1, RHBDF2, CDH23, SPG7, RHBDF2, KDM2B, WDR81, HMHA1, C10orf54, ITPRIPL2, PCNT, HOXA3*
Sanchez-Mut et al. [57]	2014	20	Hippocampus	Institute of Neuropathology Brain Bank	27 K	Disease status	*DUSP22, CLDN15, and QSCN6*
Watson et al. [64]	2016	68	Superior temporal gyrus	Mount Sinai Alzheimer's Disease and Schizophrenia Brain Bank	450 K	Disease status	DMRs identified in 475 genes, enriched for neuron function, development, and cellular metabolism

cost-effective approaches to screen methylomic variation at ~27,000, ~450,000, and ~850,000 methylation sites in the human genome, respectively. Other methods, such as methylated DNA immunoprecipitation sequencing (MeDIP-seq), whole-genome bisulfite sequencing (WGBS), and reduced representation bisulfite sequencing (RRBS), are also available; however, due to the prohibitive cost of sequencing, the human genome at sufficient depth has not been widely utilized in epigenetic epidemiological studies of AD. The first empirical EWAS, by Bakulski and colleagues, used the Illumina Infinium 27K BeadArray to quantify DNA methylation at ~27,000 CpG sites in the frontal cortex of 12 late-onset AD donors and 12 cognitively normal matched control subjects. They identified 948 nominally significant DNA methylation differences mapped to 918 unique genes with an average methylation difference between AD cases and controls of 2.9%. Interestingly, their most significant loci, with a 7.3% reduction in AD, resided in the *TMEM59* gene, which is believed to be involved in APP posttranslational glycolytic processing [56]. Using the same technology, Sanchez-Mut and colleagues examined hippocampal samples from five control donors to those with early-stage AD (Braak I–II), mid-stage AD (Braak III–IV), and late-stage AD (Braak V–VI) [57], demonstrating a >25% methylation difference between controls and Braak stages V–VI at single loci in *CLDN15* and *QSCN6* and two loci in *DUSP22*.

The advent of the Illumina Infinium 450K BeadArray has since allowed more in-depth studies of DNA methylomic differences in AD. Two independent back-to-back publications both demonstrated highly robust and reproducible alterations in four genes not previously associated with AD, namely, *ANK1*, *RHBDF2*, *RPL13*, and *CDH23* [58–60]. De Jager et al. used a large cohort of 708 prefrontal cortex samples to examine DNA methylomic differences associated with neuritic plaque burden. They identified 71 differentially methylated probes, 11 of which were replicated by Lunnon et al. Differentially methylated loci associated with neuropathology included genes such as *RNF34*, *CDH23*, *SLC2A1*, *COQ7*, and the *HOXA* gene cluster, and each of the 71 CpGs explained on average 5% of the variance in neuritic amyloid plaque burden, with a range of 3.7–9.7% [58]. In this study, they also attempted to look for altered gene expression with AD pathology in the replicated differentially methylated genes in independent temporal cortical samples. They found that *ANK1*, *CDH23*, *DIP2A*, *RHBDF2*, *RPL13*, *SERPINF1*, and *SERPINF2* all showed differential gene expression with amyloid burden providing some evidence of further reaching consequences as a result of these DNA methylation differences.

Meanwhile Lunnon and colleagues used a cross tissue approach to assess DNA methylomic changes in AD in a range of brain regions representing the spectrum of AD pathology in a discovery cohort of 117 individuals [59]. They initially focused on the entorhinal cortex, as it shows neuropathology in the early stages of disease, to identify a number of differentially methylated loci associated with Braak stage, a standardized measure of neurofibrillary tangle deposition. They then examined other matched brain regions from the same donors, namely, the prefrontal cortex and superior temporal gyrus, to identify cross cortex differences in the identified top loci from the entorhinal cortex. Probes in *ANK1*, *SLC15A4*, *MEST*, and *TMX4* were only seen to replicate in the prefrontal cortex, superior temporal gyrus, or both [59], with probes in *PCBD1*, *MLST8, and ZNF512* not being significant in other cortical tissues and probes in *SIRT6* and near *CLYBL* being significant in the cerebellum, a region of the brain largely protected from neurodegeneration. They utilized two independent replication cohorts to validate their findings, one again utilizing the Illumina Infinium 450K BeadArray to profile genome-wide methylomic differences in the prefrontal cortex and superior temporal gyrus in a cohort of 147 individuals and the other utilizing pyrosequencing in the entorhinal cortex, prefrontal cortex, and superior temporal gyrus of 62 individuals. The authors showed an extended region of hypermethylation in the *ANK1* gene in AD cortex that spanned at least six CpG sites.

Interestingly, *ANK1* encodes a brain-expressed protein involved in compartmentalization of the neuronal plasma membrane but has not previously implicated in AD. *ANK1* is primarily expressed in red blood cells but is also expressed in brain and muscle and is thought to play a role in cell-surface protein binding to the underlying spectrin-actin cytoskeleton and is used in cell motility, activation, proliferation, and contact. To date, *ANK1* is the most robust AD-associated DNA methylation difference observed in the brain [61] (Fig. 16.1). Further studies from the same groups that have since built on these are now publically available EWAS datasets.

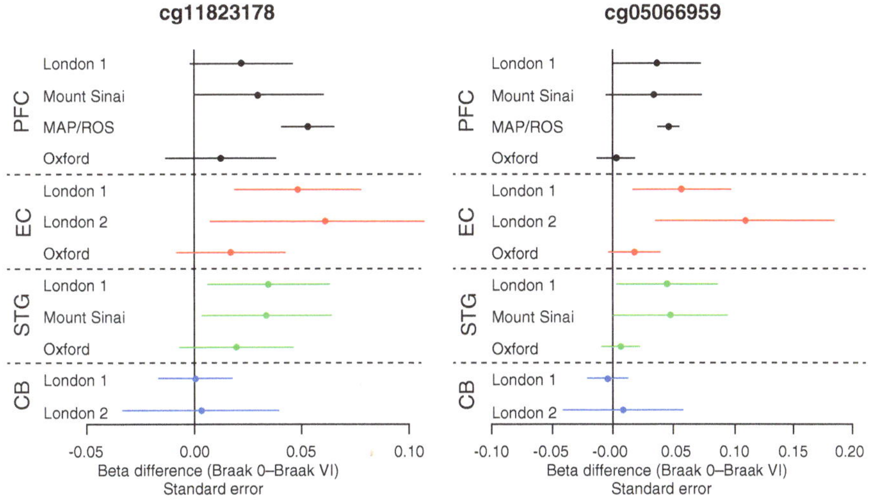

Fig. 16.1 *ANK1* shows hypermethylation associated with Braak stage across multiple studies. DNA methylation (beta) difference associated with Braak stages at two probes in the *ANK1* gene (cg11823178 and cg05066959) was observed in the prefrontal cortex (PFC), entorhinal cortex (EC), and superior temporal gyrus (STG), but not in the cerebellum in four cohorts. London 1 ($N = 117$), Oxford ($N = 62$) and Mount Sinai ($N = 147$) cohort data were taken from Lunnon et al. [59], MAP/ROS cohort data ($N = 708$) was taken from De Jager et al. [58]. London 2 cohort data ($N = 94$) is currently unpublished data from our group. Consistent AD-associated hypermethylation was observed across all cortical tissues but not in the cerebellum

Chibnik et al. have examined DNA methylomic variation in AD loci nominated from GWAS, demonstrating that DNA methylation at 17 CpG sites spanning six AD-risk genes (*BIN1, CLU, ABCA7, MS4A6A, CD2AP*, and *APOE*) shows an association with amyloid burden, independent of genotype, and collectively explain 16.8% of variability in neuritic plaques [62]. Smith and colleagues examined DNA methylation at a locus within the *TREM2* gene, showing consistent hypermethylation in three different cohorts in the superior temporal gyrus, which appeared to be independent of the SNP previously implicated in the disease [63]. Finally, Watson et al. used the Illumina Infinium 450K BeadArray to identify AD-related DNA methylation changes in the superior temporal gyrus in 34 AD cases and 34 matched controls [64]. They identified 479 differentially methylated regions (DMRs, clusters of significantly differentially methylated positions) encompassing 4565 CpG sites, with the majority of differentially methylated positions being hypermethylated. They also showed overlap between their most significant DMRs and the Lunnon et al. and De Jager et al. studies, with eight of their top 25 DMRs containing genes having differentially methylated positions in the previously published studies (*LOC100507547, PRDM16, PRRT1, C10orf105, CDH23, PPT2, PPT2-EGFL8*, and *RNF39*) [58, 59].

16.6 A Role for Other DNA Modifications in AD?

Although DNA methylation has been the focus of published research to date, a number of additional DNA modifications are now starting to receive considerable attention. 5-Hydroxymethylcytosine (5-hmC), 5-formylcytosine (5-fC), and 5-carboxylcytosine (5-caC) were originally thought to represent intermediates in the demethylation of 5-mC to unmodified cytosine [65] (Fig. 16.2). However, recent evidence suggests that they may represent independent epigenetic marks. There has been particular interest in 5-hmC in the context of brain disorders as it appears to be found at relatively high levels in the brain compared to other tissues [66, 67] and is particularly enriched in the vicinity of genes with synapse-related functions [68]. Until recently, studies of 5-hmC in AD brain have been limited to global profiling methods, two of which have shown global decreases in the hippocampus [36], entorhinal cortex, and cerebellum [69], although another study showed increased 5-hmC in the middle frontal gyrus and middle temporal gyrus [38], while another revealed no difference in the entorhinal cortex [37]. Given that we know that DNA methylomic differences in AD are loci and tissue specific and that global studies of alterations in DNA methylation levels were inconclusive, this demonstrates the importance of assessing DNA hydroxymethylation levels at single-base resolution. Interestingly, all of the published EWAS of DNA methylation in AD to date have utilized bisulfite converted DNA; however, this treatment is unable to convert either 5-mC or 5-hmC to uracil [70]. As such, all of the published EWAS of AD actually represent a summative measure of these two modifications. A recently published adaptation to bisulfite treatment has allowed the

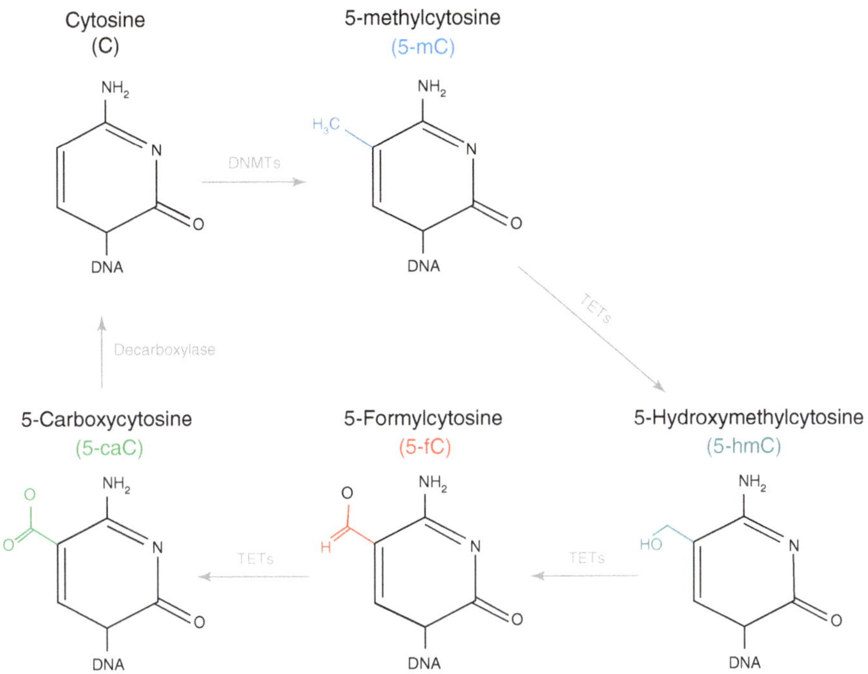

Fig. 16.2 Epigenetic DNA modifications. DNA methyltransferases (DNMTs) can add and maintain methyl groups on cytosine to create 5-methylcytosine (5-mC). Through the action of the ten-eleven translocation (TET) family of DNA hydroxylases, the process of active DNA demethylation occurs. This occurs via 5-hydroxymethylcytosine (5-hmC) and into 5-formylcytosine (5-fC) and 5-carboxycytosine (5-caC)

simultaneous measurement of 5-mC and 5-hmC in a sample. Oxidative bisulfite technology uses a selective chemical oxidation to accurately distinguish between 5-mC and 5-hmC by first converting 5-hmC into 5-fC, and through bisulfite conversion, this is converted to uracil. As such, this allows the quantification of 5-mC in the absence of confounding by 5-hmC. Further, by profiling bisulfite and oxidative bisulfite-treated DNA in parallel, one can generate a quantitative measurement of 5-hmC by subtracting the oxidative bisulfite data from the bisulfite data. This method has been utilized together with the Illumina Infinium 450K BeadArray to assess 5-mC and 5-hmC across different regions of the human brain [71–73]; however, there are nuances to the method, for example, technical artefacts can sometimes result in negative 5-hmC values. Two current studies in AD that are expected to be published soon are using this approach to quantify true 5-mC and 5-hmC measures in AD brain. Smith et al. have profiled both modifications in the entorhinal cortex and cerebellum of 96 individuals ranging from Braak 0 to Braak VI, while Roubroeks et al. have assessed the middle temporal gyrus in a similar number of individuals.

16.7 Looking to the Future

Although the published EWAS in AD have shown a number of robust and reproducible DNA modification changes in disease, it is to be expected that technological advances will allow more in depth assessments. The recent release of the of Illumina Infinium Methylation EPIC 850K BeadArray now allows the simultaneous assessment of ~850,000 methylation sites in the human genome. In addition, the falling cost of next-generation sequencing means that methods such as RRBS and WGBS are becoming more affordable for cohort studies. Similarly, the advent of third-generation sequencing technologies such as the PacBio RS II from Pacific Biosciences allows for whole-genome sequencing and targeted sequencing. These methods have the potential to detect different DNA modifications during standard sequencing as well as allowing for single-base and DNA-strand resolution. Targeted and whole-genome sequencing approaches could also be utilized to assess epigenetic variation in mitochondrial DNA (mtDNA). Mitochondrial dysfunction has been proposed to be a potential mechanism in the development of AD, which has been reported in various studies [74, 75]. Interestingly, mitochondria possess their own circular genome of 16.6 Kb, which is separate from the nuclear genome and contains 37 genes [76]. However, with no coverage of the mitochondrial genome on the Illumina Infinium BeadArrays, a potential role for mtDNA modifications has not been examined in the AD EWAS published to date [77, 78].

By far, the major criticism of epigenetic studies in various diseases relates to the issue of causality. Unlike genetic variation, it is not known whether disease-associated epigenetic changes represent a cause or a consequence of disease. Methods such as Mendelian randomization (MR) with existing "omics" datasets could provide some evidence for the direction of effect of the epigenetic changes observed, but, more recently, there have been suggestions of using genetic editing techniques to determine causality. New technologies such as clustered regularly interspaced short palindromic repeats (CRISPR)/Cas9 allow researchers to impose genetic modifications to DNA and observe the results in cell lines and model organisms [79]. This technology could be lent to loci-specific epigenetic editing via the use of targeting proteins to methylate of demethylate target sites and then testing whether this accelerates or reverses disease pathology. Previously, CRISPR technology has been used to target nuclease activity to introduce single or double strand breaks in DNA, target transcriptional transactivation, and regulate gene expression. Genome editing with CRISPR targets enzyme activity to specific target DNA sequences depending on the specificity of guide RNAs in the CRISPR complex. These methods could be used to alter DNA methylation levels in model organisms or cell lines to replicate the differences seen in human EWAS studies to attempt to establish causality and the effects of disease-associated changes. Ultimately, even if such studies prove that a nominated locus is not causal in disease, it does not make EWAS any less worthwhile, as even identifying consequences of disease will teach us more about the disease process.

Conclusion

The role of epigenetic mechanisms in AD is still a research field in its infancy, and particularly how epigenetic DNA modifications could contribute to the cause and progression of disease is still yet to be explored. Currently, epigenetics has not been well studied in regard to AD, and there are only a handful of studies which provide any empirical data. Due to the relative ease and affordability of EWAS however, the amount of data being generated is increasing for both 5-mC and 5-hmC. Replication is integral to finding robust epigenetic changes, and so far, a few replicable differences have been observed in relevant brain tissues, such as in *ANK1* and genes in the *HOXA* gene cluster. More work is needed, and combining multiple EWAS datasets into meta-analyses to provide strong evidence for the contribution of DNA methylation to disease progression is warranted. Although EWAS using bisulfite-treated DNA are a combination of measures of both 5-mC and 5-hmC, using oxidative bisulfite methods give a truer measure of 5-mC as well as allowing the quantification of 5-hmC. As of now, there are no AD epigenetic peripheral tissue biomarkers for AD, which is a major goal for dementia research. As epigenetic mechanisms are malleable and changeable over the course of development, life, exposure to environmental influences, and normal aging, it provides an attractive target for a proxy of disease progression and a target for drugs. By combining epigenetic measurements in peripheral tissues, such as blood, with neuroimaging and clinical assessments, we can associate what is happening in the brain to blood. As with many diseases that are believed to have both genetic and environmental components, it is important to integrate different data modalities to generate a full picture of AD risk. Integrating genomic, epigenomic, and transcriptomic data will allow the identification of methylation and expression quantitative trait loci (mQTLs/eQTLs), showing how genetic variation may influence methylation and expression in a tissue and disease-specific manner.

References

1. Brookmeyer R, Johnson E, Ziegler-Graham K, Arrighi HM. Forecasting the global burden of Alzheimer's disease. Alzheimers Dement. 2007;3(3):186–91.
2. Hardy J, Selkoe DJ. The amyloid hypothesis of Alzheimer's disease: progress and problems on the road to therapeutics. Science. 2002;297(5580):353–6.
3. Hardy JA, Higgins GA. Alzheimer's disease: the amyloid cascade hypothesis. Science. 1992;256(5054):184–5.
4. Mortimer JA, Borenstein AR, Gosche KM, Snowdon DA. Very early detection of Alzheimer neuropathology and the role of brain reserve in modifying its clinical expression. J Geriatr Psychiatry Neurol. 2005;18(4):218–23.
5. Gatz M, Reynolds CA, Fratiglioni L, Johansson B, Mortimer JA, Berg S, et al. Role of genes and environments for explaining Alzheimer disease. Arch Gen Psychiatry. 2006;63(2):168–74.

6. Lambert JC, Heath S, Even G, Campion D, Sleegers K, Hiltunen M, et al. Genome-wide association study identifies variants at CLU and CR1 associated with Alzheimer's disease. Nat Genet. 2009;41(10):1094–9.
7. Lambert JC, Ibrahim-Verbaas CA, Harold D, Naj AC, Sims R, Bellenguez C, et al. Meta-analysis of 74,046 individuals identifies 11 new susceptibility loci for Alzheimer's disease. Nat Genet. 2013;45(12):1452–8.
8. Ridge PG, Mukherjee S, Crane PK, Kauwe JS, Alzheimer's Disease Genetics Consortium. Alzheimer's disease: analyzing the missing heritability. PLoS One. 2013;8(11):e79771.
9. Escott-Price V, Sims R, Bannister C, Harold D, Vronskaya M, Majounie E, et al. Common polygenic variation enhances risk prediction for Alzheimer's disease. Brain J Neurol. 2015;138(Pt 12):3673–84.
10. Farrer LA, Cupples LA, Haines JL, Hyman B, Kukull WA, Mayeux R, et al. Effects of age, sex, and ethnicity on the association between apolipoprotein E genotype and Alzheimer disease. A meta-analysis. APOE and Alzheimer Disease Meta Analysis Consortium. JAMA. 1997;278(16):1349–56.
11. Jonsson T, Stefansson H, Steinberg S, Jonsdottir I, Jonsson PV, Snaedal J, et al. Variant of TREM2 associated with the risk of Alzheimer's disease. N Engl J Med. 2013;368(2):107–16.
12. Guerreiro R, Wojtas A, Bras J, Carrasquillo M, Rogaeva E, Majounie E, et al. TREM2 variants in Alzheimer's disease. N Engl J Med. 2013;368(2):117–27.
13. Profenno LA, Porsteinsson AP, Faraone SV. Meta-analysis of Alzheimer's disease risk with obesity, diabetes, and related disorders. Biol Psychiatry. 2010;67(6):505–12.
14. Hall AM, Roberson ED. Mouse models of Alzheimer's disease. Brain Res Bull. 2012;88(1):3–12.
15. Games D, Adams D, Alessandrini R, Barbour R, Berthelette P, Blackwell C, et al. Alzheimer-type neuropathology in transgenic mice overexpressing V717F beta-amyloid precursor protein. Nature. 1995;373(6514):523–7.
16. Hsiao K, Chapman P, Nilsen S, Eckman C, Harigaya Y, Younkin S, et al. Correlative memory deficits, Abeta elevation, and amyloid plaques in transgenic mice. Science. 1996;274(5284):99–102.
17. Calhoun ME, Burgermeister P, Phinney AL, Stalder M, Tolnay M, Wiederhold KH, et al. Neuronal overexpression of mutant amyloid precursor protein results in prominent deposition of cerebrovascular amyloid. Proc Natl Acad Sci U S A. 1999;96(24):14088–93.
18. Roberds SL, Anderson J, Basi G, Bienkowski MJ, Branstetter DG, Chen KS, et al. BACE knockout mice are healthy despite lacking the primary beta-secretase activity in brain: implications for Alzheimer's disease therapeutics. Hum Mol Genet. 2001;10(12):1317–24.
19. Saura CA, Choi SY, Beglopoulos V, Malkani S, Zhang D, Shankaranarayana Rao BS, et al. Loss of presenilin function causes impairments of memory and synaptic plasticity followed by age-dependent neurodegeneration. Neuron. 2004;42(1):23–36.
20. Black RA, Rauch CT, Kozlosky CJ, Peschon JJ, Slack JL, Wolfson MF, et al. A metalloproteinase disintegrin that releases tumour-necrosis factor-alpha from cells. Nature. 1997;385(6618):729–33.
21. Oddo S, Caccamo A, Shepherd JD, Murphy MP, Golde TE, Kayed R, et al. Triple-transgenic model of Alzheimer's disease with plaques and tangles: intracellular Abeta and synaptic dysfunction. Neuron. 2003;39(3):409–21.
22. Sanchez-Mut JV, Aso E, Panayotis N, Lott I, Dierssen M, Rabano A, et al. DNA methylation map of mouse and human brain identifies target genes in Alzheimer's disease. Brain. 2013;136(Pt 10):3018–27.
23. Cong L, Jia J, Qin W, Ren Y, Sun Y. Genome-wide analysis of DNA methylation in an APP/PS1 mouse model of Alzheimer's disease. Acta Neurol Belg. 2014;114(3):195–206.
24. Agbemenyah HY, Agis-Balboa RC, Burkhardt S, Delalle I, Fischer A. Insulin growth factor binding protein 7 is a novel target to treat dementia. Neurobiol Dis. 2014;62:135–43.
25. Thompson RF, Atzmon G, Gheorghe C, Liang HQ, Lowes C, Greally JM, et al. Tissue-specific dysregulation of DNA methylation in aging. Aging Cell. 2010;9(4):506–18.

26. Heijmans BT, Mill J. Commentary: The seven plagues of epigenetic epidemiology. Int J Epidemiol. 2012;41(1):74–8.
27. Guintivano J, Aryee MJ, Kaminsky ZA. A cell epigenotype specific model for the correction of brain cellular heterogeneity bias and its application to age, brain region and major depression. Epigenetics. 2013;8(3):290–302.
28. Burns A, Jacoby R, Luthert P, Levy R. Cause of death in Alzheimer's disease. Age Ageing. 1990;19(5):341–4.
29. Breitling LP, Yang R, Korn B, Burwinkel B, Brenner H. Tobacco-smoking-related differential DNA methylation: 27 K discovery and replication. Am J Hum Genet. 2011;88(4):450–7.
30. Ruiz-Hernandez A, Kuo CC, Rentero-Garrido P, Tang WY, Redon J, Ordovas JM, et al. Environmental chemicals and DNA methylation in adults: a systematic review of the epidemiologic evidence. Clin Epigenetics. 2015;7:55.
31. Crider KS, Yang TP, Berry RJ, Bailey LB. Folate and DNA methylation: a review of molecular mechanisms and the evidence for folate's role. Adv Nutr. 2012;3(1):21–38.
32. Horvath S. DNA methylation age of human tissues and cell types. Genome Biol. 2013;14(10):R115.
33. Hannon E, Lunnon K, Schalkwyk L, Mill J. Interindividual methylomic variation across blood, cortex, and cerebellum: implications for epigenetic studies of neurological and neuropsychiatric phenotypes. Epigenetics. 2015;10(11):1024–32.
34. Mastroeni D, Grover A, Delvaux E, Whiteside C, Coleman PD, Rogers J. Epigenetic changes in Alzheimer's disease: decrements in DNA methylation. Neurobiol Aging. 2010;31(12):2025–37.
35. Mastroeni D, McKee A, Grover A, Rogers J, Coleman PD. Epigenetic differences in cortical neurons from a pair of monozygotic twins discordant for Alzheimer's disease. PLoS One. 2009;4(8):e6617.
36. Chouliaras L, Mastroeni D, Delvaux E, Grover A, Kenis G, Hof PR, et al. Consistent decrease in global DNA methylation and hydroxymethylation in the hippocampus of Alzheimer's disease patients. Neurobiol Aging. 2013;34(9):2091–9.
37. Lashley T, Gami P, Valizadeh N, Li A, Revesz T, Balazs R. Alterations in global DNA methylation and hydroxymethylation are not detected in Alzheimer's disease. Neuropathol Appl Neurobiol. 2015;41(4):497–506.
38. Coppieters N, Dieriks BV, Lill C, Faull RL, Curtis MA, Dragunow M. Global changes in DNA methylation and hydroxymethylation in Alzheimer's disease human brain. Neurobiol Aging. 2014;35(6):1334–44.
39. Bradley-Whitman MA, Lovell MA. Epigenetic changes in the progression of Alzheimer's disease. Mech Ageing Dev. 2013;134(10):486–95.
40. Di Francesco A, Arosio B, Gussago C, Dainese E, Mari D, D'Addario C, et al. Involvement of 5-lipoxygenase in Alzheimer's disease: a role for DNA methylation. J Alzheimers Dis. 2013;37(1):3–8.
41. Bollati V, Galimberti D, Pergoli L, Dalla Valle E, Barretta F, Cortini F, et al. DNA methylation in repetitive elements and Alzheimer disease. Brain Behav Immun. 2011;25(6):1078–83.
42. Hernandez HG, Mahecha MF, Mejia A, Arboleda H, Forero DA. Global long interspersed nuclear element 1 DNA methylation in a Colombian sample of patients with late-onset Alzheimer's disease. Am J Alzheimers Dis Other Demen. 2014;29(1):50–3.
43. D'Addario C, Di Francesco A, Arosio B, Gussago C, Dell'Osso B, Bari M, et al. Epigenetic regulation of fatty acid amide hydrolase in Alzheimer disease. PLoS One. 2012;7(6):e39186.
44. Arosio B, Bulbarelli A, Bastias Candia S, Lonati E, Mastronardi L, Romualdi P, et al. Pin1 contribution to Alzheimer's disease: transcriptional and epigenetic mechanisms in patients with late-onset Alzheimer's disease. Neurodegener Dis. 2012;10(1–4):207–11.
45. Furuya TK, Silva PN, Payao SL, Bertolucci PH, Rasmussen LT, De Labio RW, et al. Analysis of SNAP25 mRNA expression and promoter DNA methylation in brain areas of Alzheimer's disease patients. Neuroscience. 2012;220:41–6.

46. Furuya TK, da Silva PN, Payao SL, Rasmussen LT, de Labio RW, Bertolucci PH, et al. SORL1 and SIRT1 mRNA expression and promoter methylation levels in aging and Alzheimer's disease. Neurochem Int. 2012;61(7):973–5.
47. Silva PN, Furuya TK, Braga IL, Rasmussen LT, Labio RW, Bertolucci PH, et al. Analysis of HSPA8 and HSPA9 mRNA expression and promoter methylation in the brain and blood of Alzheimer's disease patients. J Alzheimers Dis. 2014;38(1):165–70.
48. Iwata A, Nagata K, Hatsuta H, Takuma H, Bundo M, Iwamoto K, et al. Altered CpG methylation in sporadic Alzheimer's disease is associated with APP and MAPT dysregulation. Hum Mol Genet. 2014;23(3):648–56.
49. Brohede J, Rinde M, Winblad B, Graff C. A DNA methylation study of the amyloid precursor protein gene in several brain regions from patients with familial Alzheimer disease. J Neurogenet. 2010;24(4):179–81.
50. Wang SC, Oelze B, Schumacher A. Age-specific epigenetic drift in late-onset Alzheimer's disease. PLoS One. 2008;3(7):e2698.
51. Barrachina M, Ferrer I. DNA methylation of Alzheimer disease and tauopathy-related genes in postmortem brain. J Neuropathol Exp Neurol. 2009;68(8):880–91.
52. Silva PN, Furuya TK, Sampaio Braga I, Rasmussen LT, de Labio RW, Bertolucci PH, et al. CNP and DPYSL2 mRNA expression and promoter methylation levels in brain of Alzheimer's disease patients. J Alzheimers Dis. 2013;33(2):349–55.
53. Sontag E, Hladik C, Montgomery L, Luangpirom A, Mudrak I, Ogris E, et al. Downregulation of protein phosphatase 2A carboxyl methylation and methyltransferase may contribute to Alzheimer disease pathogenesis. J Neuropathol Exp Neurol. 2004;63(10):1080–91.
54. Siegmund KD, Connor CM, Campan M, Long TI, Weisenberger DJ, Biniszkiewicz D, et al. DNA methylation in the human cerebral cortex is dynamically regulated throughout the life span and involves differentiated neurons. PLoS One. 2007;2(9):e895.
55. Rao JS, Keleshian VL, Klein S, Rapoport SI. Epigenetic modifications in frontal cortex from Alzheimer's disease and bipolar disorder patients. Transl Psychiatry. 2012;2:e132.
56. Bakulski KM, Dolinoy DC, Sartor MA, Paulson HL, Konen JR, Lieberman AP, et al. Genome-wide DNA methylation differences between late-onset Alzheimer's disease and cognitively normal controls in human frontal cortex. J Alzheimers Dis. 2012;29(3):571–88.
57. Sanchez-Mut JV, Aso E, Heyn H, Matsuda T, Bock C, Ferrer I, et al. Promoter hypermethylation of the phosphatase DUSP22 mediates PKA-dependent TAU phosphorylation and CREB activation in Alzheimer's disease. Hippocampus. 2014;24(4):363–8.
58. De Jager PL, Srivastava G, Lunnon K, Burgess J, Schalkwyk LC, Yu L, et al. Alzheimer's disease: early alterations in brain DNA methylation at ANK1, BIN1, RHBDF2 and other loci. Nat Neurosci. 2014;17(9):1156–63.
59. Lunnon K, Smith R, Hannon E, De Jager PL, Srivastava G, Volta M, et al. Methylomic profiling implicates cortical deregulation of ANK1 in Alzheimer's disease. Nat Neurosci. 2014;17(9):1164–70.
60. Lord J, Cruchaga C. The epigenetic landscape of Alzheimer's disease. Nat Neurosci. 2014;17(9):1138–40.
61. Smith AR, Mill J, Smith RG, Lunnon K. Elucidating novel dysfunctional pathways in Alzheimer's disease by integrating loci identified in genetic and epigenetic studies. Neuroepigenetics. 2016;6:32–50.
62. Chibnik LB, Yu L, Eaton ML, Srivastava G, Schneider JA, Kellis M, et al. Alzheimer's loci: epigenetic associations and interaction with genetic factors. Ann Clin Transl Neurol. 2015;2(6):636–47.
63. Smith AR, Smith RG, Condliffe D, Hannon E, Schalkwyk L, Mill J, et al. Increased DNA methylation near TREM2 is consistently seen in the superior temporal gyrus in Alzheimer's disease brain. Neurobiol Aging. 2016;47:35–40.

64. Watson CT, Roussos P, Garg P, Ho DJ, Azam N, Katsel PL, et al. Genome-wide DNA methylation profiling in the superior temporal gyrus reveals epigenetic signatures associated with Alzheimer's disease. Genome Med. 2016;8(1):5.
65. Song CX, Yi C, He C. Mapping recently identified nucleotide variants in the genome and transcriptome. Nat Biotechnol. 2012;30(11):1107–16.
66. Song CX, Szulwach KE, Fu Y, Dai Q, Yi C, Li X, et al. Selective chemical labeling reveals the genome-wide distribution of 5-hydroxymethylcytosine. Nat Biotechnol. 2011;29(1):68–72.
67. Nestor CE, Ottaviano R, Reddington J, Sproul D, Reinhardt D, Dunican D, et al. Tissue type is a major modifier of the 5-hydroxymethylcytosine content of human genes. Genome Res. 2012;22(3):467–77.
68. Khare T, Pai S, Koncevicius K, Pal M, Kriukiene E, Liutkeviciute Z, et al. 5-hmC in the brain is abundant in synaptic genes and shows differences at the exon-intron boundary. Nat Struct Mol Biol. 2012;19(10):1037–43.
69. Condliffe D, Wong A, Troakes C, Proitsi P, Patel Y, Chouliaras L, et al. Cross-region reduction in 5-hydroxymethylcytosine in Alzheimer's disease brain. Neurobiol Aging. 2014;35(8):1850–4.
70. Nestor C, Ruzov A, Meehan R, Dunican D. Enzymatic approaches and bisulfite sequencing cannot distinguish between 5-methylcytosine and 5-hydroxymethylcytosine in DNA. Biotechniques. 2010;48(4):317–9.
71. Stewart SK, Morris TJ, Guilhamon P, Bulstrode H, Bachman M, Balasubramanian S, et al. oxBS-450 K: a method for analysing hydroxymethylation using 450 K BeadChips. Methods. 2015;72:9–15.
72. Lunnon K, Hannon E, Smith RG, Dempster E, Wong C, Burrage J, et al. Variation in 5-hydroxymethylcytosine across human cortex and cerebellum. Genome Biol. 2016;17:27.
73. Field SF, Beraldi D, Bachman M, Stewart SK, Beck S, Balasubramanian S. Accurate measurement of 5-methylcytosine and 5-hydroxymethylcytosine in human cerebellum DNA by oxidative bisulfite on an array (OxBS-array). PLoS One. 2015;10(2):e0118202.
74. Manczak M, Park BS, Jung Y, Reddy PH. Differential expression of oxidative phosphorylation genes in patients with Alzheimer's disease: implications for early mitochondrial dysfunction and oxidative damage. Neruomol Med. 2004;5(2):147–62.
75. Lunnon K, Ibrahim Z, Proitsi P, Lourdusamy A, Newhouse S, Sattlecker M, et al. Mitochondrial dysfunction and immune activation are detectable in early Alzheimer's disease blood. J Alzheimers Dis. 2012;30(3):685–710.
76. Anderson S, Bankier AT, Barrell BG, de Bruijn MH, Coulson AR, Drouin J, et al. Sequence and organization of the human mitochondrial genome. Nature. 1981;290(5806):457–65.
77. Devall M, Mill J, Lunnon K. The mitochondrial epigenome: a role in Alzheimer's disease? Epigenomics. 2014;6(6):665–75.
78. Devall M, Roubroeks J, Mill J, Weedon M, Lunnon K. Epigenetic regulation of mitochondrial function in neurodegenerative disease: new insights from advances in genomic technologies. Neurosci Lett. 2016;625:47–55.
79. Doudna JA, Charpentier E. Genome editing. The new frontier of genome engineering with CRISPR-Cas9. Science. 2014;346(6213):1258096.

Alzheimer's Disease and Histone Code Alterations

17

Pritika Narayan and Mike Dragunow

Abstract

Substantial progress has been made in identifying Alzheimer's disease (AD) risk-associated variants using genome-wide association studies (GWAS). The majority of these risk variants reside in noncoding regions of the genome making their functional evaluation difficult; however, they also infer the presence of unconventional regulatory regions that may reside at these locations. We know from these studies that rare familial cases of AD account for less than 5% of all AD cases and autosomal dominant mutations in APP, PSEN1 and PSEN2 account for less than 10% of the genetic basis of these familial cases [1]. The sporadic form of AD, while more complex, still has a substantial genetic component evidenced by observational studies where 30–48% of AD patients have a first degree relative who is also affected [2]. In addition, the strongest risk factor after age is the APOE E4 polymorphism, and more than 20 other risk variants have been identified to date, reviewed in two recent papers [3, 4]. Monozygotic twin studies have revealed a discordance for AD, implicating that a combination of epigenetic and genetic factors are likely involved in the development of AD [5].

Keywords

Epigenetics • Histone modifications • Human brain

P. Narayan, B.M.Sc., P.G.Dip.Sci., Ph.D.
School of Biological Sciences and Centre for Brain Research, University of Auckland, Auckland, New Zealand

M. Dragunow, Ph.D. (✉)
Department of Pharmacology and Centre for Brain Research, University of Auckland, Auckland, New Zealand
e-mail: m.dragunow@auckland.ac.nz

R. Delgado-Morales (ed.), *Neuroepigenomics in Aging and Disease*, Advances in Experimental Medicine and Biology, DOI 10.1007/978-3-319-53889-1_17

17.1 Introduction

Substantial progress has been made in identifying Alzheimer's disease (AD) risk-associated variants using genome-wide association studies (GWAS). The majority of these risk variants reside in noncoding regions of the genome making their functional evaluation difficult; however, they also infer the presence of unconventional regulatory regions that may reside at these locations. We know from these studies that rare familial cases of AD account for less than 5% of all AD cases and autosomal dominant mutations in APP, PSEN1 and PSEN2 account for less than 10% of the genetic basis of these familial cases [1]. The sporadic form of AD, while more complex, still has a substantial genetic component evidenced by observational studies where 30–48% of AD patients have a first degree relative who is also affected [2]. In addition, the strongest risk factor after age is the APOE E4 polymorphism, and more than 20 other risk variants have been identified to date, reviewed in two recent papers [3, 4]. Monozygotic twin studies have revealed a discordance for AD, implicating that a combination of epigenetic and genetic factors are likely involved in the development of AD [5].

Epigenomic regulation encompasses DNA and histone modifications and the higher-order architecture of DNA associating with histones, alongside a plethora of transcription factors/proteins. These associations are plastic and responsive to environmental stimuli and dictate whether a specific region of DNA will be repressed, transcribed or involved in controlling the expression of other gene segments over time. In the context of Alzheimer's disease, there has been a significant shift in literature, from searching for common disease-associated variants to epigenome-wide exploration of these complex interactions, structures and modifications. Techniques for genome-wide profiling of peaks for different histone modifications, have facilitated a rapid increase in the characterisation of these modifications and the identification of specific genes they regulate across the genome in relation to specific diseases, such as AD.

This chapter will review histone modifications in the context of AD disease with a focus on studies of post-mortem human brain as well as pharmacological intervention strategies that have been tested in vivo/in clinic.

17.2 Histone Modifications

17.2.1 Enzymes That Regulate Histone Modifications

Histone modifications include the addition of methyl, acetyl, phospho and other groups to specific amino acid residues on the N-terminal tails of histones H2A, H2B, H3 and H4. The different modifications and/or combinations of modifications on a given tail determine the availability of bound DNA for transcription. Each modification is maintained by the balanced and opposing actions of enzymes: acetyltransferases add acetyl groups and deacetylases remove, methyl transferases add methyl groups (mono-, di- or trimethyl groups can be added) and demethylases remove, kinases add phosphate groups and phosphatases remove, reviewed previously [6].

It is important to note that each enzyme group contains multiple proteins/enzymes. For example, there are at least 18 known histone acetyltransferases (HATs), more aptly renamed as lysine acetyltransferases (KATs), and the major subgroups include Gcn5 *N*-acetyltransferases (GNATs), MYST (MOZ, ybf2, Sas3, Sas2, Tip60) and p300 and CBP subfamily (p300/KAT3b, CBP/KAT3a, PCAF/KAT2a) [7]. Similarly HDACs are divided into 11 main subclasses, which can be further divided into 38 different sequence variants of the canonical sequence [8].

Early studies investigating the role of these enzymes in development have highlighted their significant roles in learning and memory. For example, mice lacking the HAT, CBP develop impaired memory function [9, 10]. Mice lacking some isoforms of HDACs such as HDAC2 and HDAC3 show improved learning [11, 12], while loss of HDAC4 and HDAC5 has been shown to impair memory function [13–15].

A recent study by Anderson and colleagues demonstrated that in comparison to mouse models of AD, where relatively high concentrations of HDAC3 and HDAC4 were observed in the brain, the same isoforms were undetectable in the human AD prefrontal cortex [16]. A previous study has shown that HDAC4 is undetectable and low levels of HDAC3 were measured in human brain [17]. Interestingly reduced HDAC4 levels in humans has been linked to mental retardation [18], emphasising the need for isoform selectivity if HDACs were to be targeted therapeutically in AD. Anderson and colleagues also showed that HDAC1 and HDAC2 were decreased and HDAC5 and HDAC6 were significantly increased in AD compared to control cases [16]. HDAC6 overexpression in AD has been observed previously [19].

It is also important to note that enzymes that add or remove different chemical groups on histone tails do not only act on histone molecules but a range of different proteins within the cell. For example, investigations of the acetylome in three different human cell lines have revealed 3600 acetylation sites across 1750 different proteins [20], while in human liver samples, another study found 1300 acetylation sites spread across 1042 proteins [21]. If all the unique proteins are combined between the two studies, then the human acetylome contains at least 2500 proteins that can be acetylated [22]. The sheer magnitude of molecules therefore regulated by HATs and HDACs raises the need for caution and highlights the need for molecular specificity of therapeutics targeting histone acetylation.

17.2.2 APP Processing and Histone Modifications

Tip60 is an acetyltransferase that interacts with the APP intracellular domain (AICD) [23] and Fe65, an adapter protein, resulting in translocation of this 'AFT' complex to the nucleus to alter gene expression [24]. The AFT complex has been shown to regulate APP itself [25], along with stathmin, a molecule involved in Tau pathology [26]. A recent study [27] demonstrated that RanBP9 modulates the interaction of the AFT complex and can regulate whether this complex localises to nuclear spots [28] where transcription factories reside or nuclear speckles where RanBP9 may relocate AICD away from transcription factories [27]. AICD has been shown to compete with HDACs 1 and 3 for binding at the promoters of AB degrading enzymes neprilysin and transthyretin [29, 30].

Acetyltransferases such as Tip60 are promising therapeutic targets in comparison to deacetylases because specific acetyltransferases have less redundant targets within the genome and could be used to upregulate specific neuroprotective pathways [31]. For example, Tip60 overexpression can rescue AD drosophila from APP-induced learning and memory deficits [32, 33] and can also regulate the transcription of genes involved in a variety of neuronal processes [34]. A loss of Tip60 leads to axonopathy and aberrant histone acetylation-mediated expression of axonal transport genes [35]. Other acetyl transferases that have been shown to play a significant role in neuroprotection include p300 and CBP [36–38], and a critical role in mediating memory consolidation [39]. However, another study has suggested that CBP and p300 knockout mice are resistant to amyloid beta-mediated toxicity [40].

17.2.3 Global Histone Modifications Observed in PM Tissue to Date

Literature on global histone modifications in post-mortem AD brain is limited and varies significantly with regard to the methodology used, brain regions studied and sample sizes. Table 17.1 and Fig. 17.1 highlight the relevant studies published to date.

Increases in global histone modifications that would result in transcriptional activation in AD brain include trimethylation at histone H3 lysine 4 (H3K4Me3); acetylation at histones H3K9, H3K14, H3K18, and H3K23 and histone H4 lysines 5, 8, 12 and 16; as well as phosphorylation at histone H3 serine 10. In contrast, an increase in di-methylation on histone H3 lysine 9 is a signature of heterochromatin and results in transcriptional repression. Transcriptionally activating global histone changes in post-mortem AD brain have been observed in different regions of the brain, frontal cortex [41, 42], hippocampus [43–45], middle temporal gyrus [44, 46], inferior temporal gyrus [46] and occipital cortex [47], while transcriptionally repressive changes for AD brain have also been observed in temporal lobe [48], occipital lobe [47] and hippocampus [43]. It is difficult to reach a consensus from this data and begs the development of universal standards for tissue preparation and fixation, minimum sample size, developing robust techniques for addressing cellular heterogeneity, standardising imaging methods and the equipment/software used for quantification to enable more robust comparisons to be made between independent groups. Given the range of different methods and sample sizes used to study global histone changes,

Table 17.1 An overview of the studies to date that have investigated global histone modifications in PM-AD brain

First author	Title	Sample size		Methods	Regions	Overall change in AD relative to control
		Control	AD			
Rao et al. [41]	Epigenetic modifications in frontal cortex from Alzheimer's disease and bipolar disorder patients	10	10	Epigenetic global H3 acetylation and global H3 phosphorylation ELISA kits	Frontal cortex	No change in AcH3, increase in H3 phosphorylation (specific residues not specified)
Zhang et al. [48]	Targeted proteomics for quantification of histone acetylation in Alzheimer's disease	4	6	Selected reaction monitoring-based proteomics and Western blots to study H3K18/K23 acetylation	Temporal lobe	Significant decrease in H3K18/K23 acetylation
Chaput et al. [98]	Potential role of PCTAIRE-2, PCTAIRE-3 and P-histone H4 in amyloid precursor protein-dependent Alzheimer pathology	4	5	Western blots	Undefined	Significant increase in H4Ser47 phosphorylation
Mastroeni et al. [44]	Aberrant intracellular localisation of H3k4me3 demonstrates an early epigenetic phenomenon in Alzheimer's disease	19	18	Global IHC, WB and ELISA	Hp and MTG	H3K4Me3 colocalisation with NFTs significantly increased, nuclear labelling reduced

(continued)

Table 17.1 (continued)

First author	Title	Sample size Control	AD	Methods	Regions	Overall change in AD relative to control
Narayan et al. [46]	Increased acetyl and total histone levels in post-mortem Alzheimer's disease brain	28 (MTG); 17 (ITG)	29 (MTG); 14(ITG)	Global IHC-free floating sections for ITG, IHC-P for tissue microarray of MTG and WB	ITG and MTG	Significant increase in AcH3 (K9/K14) and AcH4 (K5/K8/K12/K16) and proportional increase in respective total histone protein also
Hernández-Ortega et al. [43]	Altered machinery of protein synthesis in Alzheimer's: from the nucleolus to the ribosome	18	49	Global IHC-P	Hp	Observed decrease in H3K9Me2 and H4K12Ac
Lithner et al. [47]	Disruption of neocortical histone H3 homeostasis by soluble Aβ: implications for Alzheimer's disease	3	7	Histones isolated with nuclear extraction kit (Pierce Biotechnology) and analysed with WB	Occipital Ctx	Significant increase in AcH3K14 and H3K9Me2
Ogawa et al. [45]	Ectopic localisation of phosphorylated histone H3 in Alzheimer's disease: a mitotic catastrophe?	9	17	IHC-P and WB	Hp (IHC-P) and cortical grey matter for WB	Significant increase in H3Ser10 phosphorylation, staining appeared to colocalise with tangles
Lu et al. [42]	REST and stress resistance in ageing and Alzheimer's disease	11	8	IHC in isolated neuronal nuclei	PFC	Significant increase in AcH3K9 in AD compared to control cases

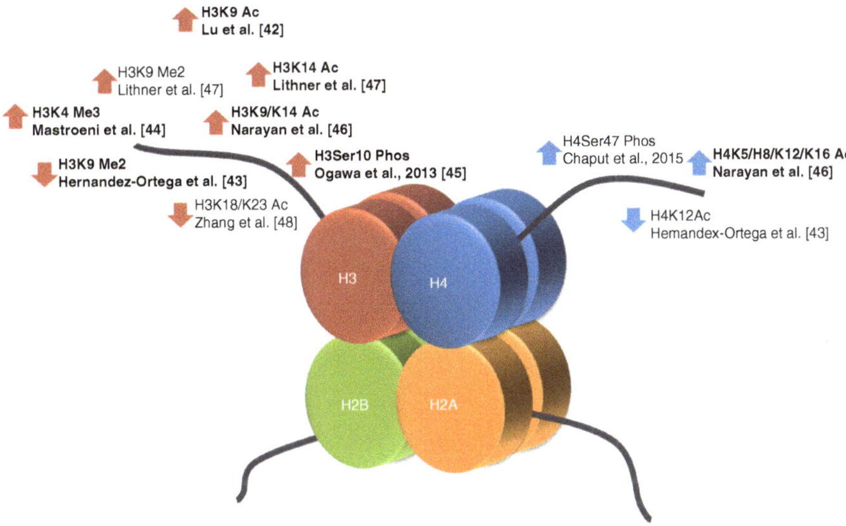

Fig. 17.1 A schematic overview of studies that have examined global histone modifications in PM-AD brain. The modifications that appear in *bold* delineate modifications that would result in transcriptional activation in AD cases

to be able to draw conclusions from similar studies, we focused below on reviewing studies that used immunohistochemical techniques and a sample size of ten or more cases in each control and AD group.

Mastroeni and colleagues [44] performed immunohistochemical analysis of hippocampal tissue from control (n = 19) and AD (n = 18) cases to analyse global changes in histone H3 lysine 4 trimethylation (H3K4Me3). H3K4Me3 is widely accepted as an epigenetic signature for actively transcribed or poised (waiting for transcriptional activation) genomic regions. They showed a reduction in nuclear labelling and an increase in cytoplasmic labelling—significantly colocalising with hyperphosphorylated tau tangles, in AD brains compared to non-demented controls. When cases were analysed by Braak stage, they showed that cytoplasmic localisation of H3K4Me3 preceded the earliest observations of tau hyperphosphorylation at epitopes known to be early markers of AD (PG5 and MG1 which detect phosphorylation at serines 409 and 312–322, respectively) [49], suggesting that intracellular localisation of this histone marker may be important in altering transcription in AD. They also demonstrated that the cytoplasmic staining increased in a manner corresponding to neuropathology (increasing Braak stage).

Narayan et al. [46] used immunolabelling of free-floating AD (n = 14) and control (n = 17) ITG sections and of tissue microarrays containing paraffin-embedded MTG from 28 control and 29 AD cases. Their results showed a significant increase in not only histone H3 and H4 acetylation but also corresponding increases in the total protein loads of histone H3 and H4. They found that each marker correlated significantly with levels of gliosis (HLA for microglia and GFAP for astrocytes) and with neuropathological hallmarks (tau and amyloid load) in AD but not control cases. Significant correlations with ubiquitin load and each of the histone markers reinforced their hypothesis that protein degradation is compromised in AD and may cause the observed changes in histone markers, bringing into question the therapeutic efficacy of drugs that target the epigenome alone in AD.

Hernández-Ortega and colleagues [43] investigated whether major nucleolar proteins (which act as histone binding chaperones) were altered in AD in relation to histone markers H3K9Me2 and AcH4K12. Using immunohistochemical techniques on hippocampal tissue from 18 control and 49 AD cases (of mixed Braak stages), they found that decreases in the nucleolar proteins corresponded to decreases in H3K9Me2 and AcH4K12 levels in the hippocampus. They unfortunately did not show correlations of this relationship; however, they did show that the loss of nucleolar proteins increased with Braak stage.

17.2.4 ChIP Sequencing for Histone Markers in AD

Gjoneska and colleagues [50] studied a CK-p25 mouse model of AD compared to CK wild-type littermates to conduct chromatin immunoprecipitation sequencing (ChIP-Seq) experiments of seven different histone markers. They included markers associated with active promoters (H3K4Me3), those associated with enhancers (H3K4Me1 and H3K27Ac), or associated with repression (H3K27Me3 and H3K9Me3), markers associated with gene bodies (H3K36Me3 and H4K20Me1). Upregulated genes (3667) corresponding to H3K4Me3 peaks (relative to controls) were found to be enriched for immune and stimulus response functions, while downregulated genes (5056), corresponding to H3K4Me3 decreased-level peaks, were enriched for synaptic function and learning. Immune regulators that were identified to have an increased level of H3K4Me3 peaks included NFkB and PU.1 consistent with previous findings where PU.1 has been shown to regulate microglial activation and proliferation in AD [51].

This study however is the first of its kind in the context of AD, and this area begs more research particularly using neuronal specific ChIP-seq data generated from control and AD post-mortem human brain samples.

17.3 Therapeutic Implications

Over the past few years, the results of a number of animal studies have shown that in models of AD there is a consistent **reduction in histone acetylation**. This reduced histone acetylation is associated with cognitive changes, and both the histone 'defect' and the cognitive changes can be reversed using common inhibitors of histone deacetylases. We will review this literature and then compare these animal studies to results in human brain to determine whether they can be translated into effective therapies for AD.

Using an AD transgenic mouse expressing the Swedish double mutation of APP, Ricobaraza et al. [52] found a dramatic reduction in acetylation of histone H4 in lysates of frontal cortex and in the hippocampus using immunohistochemistry. This reduced H4 acetylation was almost completely reversed by the HDAC inhibitor phenylbutyrate, which also reversed memory deficits in these mice. This reduced acetylation was only present in primary neuronal cultures grown from AD transgenic mice and reversed in vitro by phenylbutyrate. These authors also found that phenylbutyrate reduced tau phosphorylation in AD transgenic mice, but did not modify amyloid load. These results support the hypothesis of a hypoacetylation mechanism underlying AD, at least in a mouse model. Similar overall results were seen in a study by Francis et al. [53] using a different AD transgenic model in mice. Using an associative learning model, they found that wild-type mice showed increased acetylation of histone H4 24 h after learning, whereas AD mice showed reduced learning-associated H4 histone acetylation. Trichostatin A, a HDAC inhibitor, rescued both the histone acetylation defect and the memory performance of the AD mice. Using another HDAC inhibitor, sodium butyrate, Govindarajan et al. [54] also found that increasing histone acetylation in an AD mouse model alleviated both the hypoacetylation and memory function. Graff et al. [55] using the CK-p25 mouse model of neurodegeneration showed elevated HDAC2 levels correlated with reduced cognition and histone acetylation. They reversed both with a siRNA to HDAC2. Yao et al. [56] found that the HDAC inhibitor valproic acid reversed hypoacetylation of histones H3 and H4 and improved memory performance in AD transgenic mice. Qing et al. [57] also found VPA treatment rescued amyloid pathology and memory deficits in APP23 transgenic mice.

Using a high fat diet model of insulin resistance associated cognitive deficits, Sharma and Taliyan [58] found that the cognitive deficits in these mice were accompanied by histone H3 hypoacetylation (and reduced BDNF levels). The HDAC inhibitor SAHA reversed both the hypoacetylation (and elevated BDNF) and improved cognition. SAHA (vorinostat) was also used by Benito et al. [59] who showed that this drug improved cognition, reversed hypoacetylation of histones and had an anti-inflammatory effect on AD transgenic mice.

Cuadrado-Tejedor et al. [60] found that a combination of SAHA and phosphodiesterase 5 inhibitor tadalafil synergistically increased acetylation of histone H3 on the lysine 9 residue in APP transgenic mice. They speculated that the tadalafil might have augmented histone acetylation by driving the CBP histone acetyl transferase. This drug combination also enhanced LTP in hippocampal slices and improved memory in vivo. Pavlopoulos and colleagues [61] have also demonstrated a critical role for CBP-driven histone acetylation in maintaining memory in AD. They identified a histone binding protein RBAp48 to be significantly depleted in the entorhinal cortex and dentate gyrus of post-mortem AD cases. RbAp48 is an important regulator of CBP and plays a key role in modifying histone acetylation patterns in the brain. They developed a transgenic mouse model that expressed a mutant form of RbAp48 (lacking 54 N-terminal amino acids which are critical for its interaction with histone H4). The mutant mice displayed impaired memory and cognitive performance which correlated with loss of CBP-mediated histone acetylation, which was rescued when RbAp48 was reintroduced in the dentate gyrus via viral vector [61].

Thus, there is consistent and compelling data to suggest that in mouse transgenic models of AD, as well as in other mouse models of memory impairment, histones H3 and/or H4 are hypoacetylated. This leads to reduced expression of plasticity associated genes and may be responsible for memory impairment in these models. A range of non-specific HDAC inhibitors such as valproate, SAHA, sodium butyrate and phenylbutyrate reverse the hypoacetylation and improve memory processes.

However, there is some data showing hyperacetylation of histones in models of AD. Walker et al. [62] found that neurons grown from AD transgenic mice showed increased H3 and H4 histone acetylation compared to neurons from non-transgenic mice. They further showed that amyloid could increase acetylation in neurons in vitro from non-transgenic mice indicating that amyloid load may have been responsible for these effects. Guo et al. [63] used an indirect in vitro model of cellular stress-induced amyloid production in human neuroblastoma cells to show that increased amyloid production correlated with hyperacetylation of histones, perhaps mediated by decreased HDACs and increased HATs (CBP). Using another indirect in vitro model of hydrogen peroxide-induced amyloid production in human neuroblastoma cells, Gu et al. [64] found similar results. Finally, transfection of rodent neuroblastoma cells with a mutant APP also generated increases in gene-specific (PS1, BACE1) histone acetylation, possibly by enhancing the HAT p300 [65]. This in vitro data showing that elevating amyloid levels increases histone acetylation is supported by the work of Narayan et al. [46] who found that increased histone H3 and H4 expression and acetylation was correlated with amyloid load in AD brains.

Thus, there is evidence for both hyper- and hypo-acetylation in cellular and in vivo models of AD, although the overwhelming results of in vivo rodent AD models is clearly hypoacetylation. Furthermore, HDAC inhibitors in animal models of AD reverse the hypoacetylation defect and improve cognition.

Does this work using in vivo rodent models of AD translate to the clinic? A search of the clinical literature where HDAC inhibitors such as valproate have been used to treat AD is rather disappointing [66–74]. Of the most recent publications in this area, dose appears to be a significant factor; higher doses were not tolerated well, causing significant adverse side effects in AD patients, particularly in patients displaying symptoms of aggression and agitation [75–77].

Why then does it appear that this very convincing rodent work does not translate, at least based on current clinical data, to humans with AD? This question has plagued neuroscience research in general for many decades now, and this is not the place to discuss this in depth. The reader is referred to a number of publications discussing these issues [78–82]. However, there is data showing that one widely used HDAC inhibitor valproate shows species-specific pharmacological activity. Many years ago, we showed that valproate when added to rodent microglia induces caspase 3-mediated apoptosis [83] and also in surviving microglia stimulates their phagocytic activity against amyloid peptides [84]. However, when human microglia are exposed to valproate, there is no evidence of apoptosis, and phagocytosis of amyloid peptides is inhibited by valproate [85], at concentrations that greatly enhance acetylation of histone H3 and H4. These results suggest that perhaps HDAC inhibitors have species-specific actions. Many more studies are required using different inhibitors to confirm this hypothesis, but this might explain the lack of effects of valproate on AD in the clinic.

As discussed above, Narayan et al. [46] found that increased histone H3 and H4 expression and acetylation in AD brains was correlated with amyloid load. Furthermore, histone increases also strongly correlated with ubiquitin load, suggesting that compromised protein degradation in AD brains might also contribute to increases in histones. Indeed, Narayan et al. [46] found that the proteasome inhibitor MG132 elevated ubiquitin levels and acetylation of histone H3 in human neuroblastoma cells. Interestingly, when valproate was combined with MG132, as expected, there was an increase in histone acetylation, but unexpectedly there was also an increase in ubiquitination and cell death. This result, if applicable to humans in the clinic, would suggest caution when using HDAC inhibitors to 'treat' AD.

17.4 Concluding Remarks

There has been some progress made in literature in moving from reductionist or global approaches of studying histone modifications, in the context of AD, towards methods that allow visualisation of epigenetic marker peaks scattered throughout the genome. However, there is significant scope for improvement. Moving forward, it will be essential to utilise international coordinated efforts such as the ENCODE project [86, 87], GENCODE [88], the National Institutes of health Roadmap Epigenomics Project [89] and the Broad Institute Reference Epigenomic Mapping Centre, in understanding and deciphering the epigenomes role in transcriptional regulation/dysregulation and AD. A major area of improvement in these databases will be not only brain region-specific mapping but shifting to purified cell populations and subpopulations. Using cell sorting or nuclei sorting methods prior to sample isolation will be essential. Also, sample sizes are small, and while efforts are being made to analyse multicentre cohorts of control and AD brain, there is room for improvement in this area.

In addition, utilising tools that will allow the analysis of crosstalk between histone and DNA modifications will be really important, and study designs should consider the use of sequential ChIP BS, ox-BS sequencing methods [90, 91]. Also incorporating methods that will address the contributions of non-CpG methylation [92–94], methylation of RNA (c5) [95], N1-methyladenosine [96] and hydroxymethyl RNA [97] in the context of AD will be of great interest.

References

1. Brouwers N, Sleegers K, Van Broeckhoven C. Molecular genetics of Alzheimer's disease: an update. Ann Med. 2008;40:562–83.
2. Liddell MB, Lovestone S, Owen MJ. Genetic risk of Alzheimer's disease: advising relatives. Br J Psychiatry. 2001;178:7–11.

3. Cuyvers E, Sleegers K. Genetic variations underlying Alzheimer's disease: evidence from genome-wide association studies and beyond. Lancet Neurol. 2016;15:857–68.
4. Shen L, Jia J. An overview of genome-wide association studies in Alzheimer's disease. Neurosci Bull. 2016;32:183–90.
5. Gatz M, Reynolds CA, Fratiglioni L, Johansson B, Mortimer JA, Berg S, et al. Role of genes and environments for explaining Alzheimer disease. Arch Gen Psychiatry. 2006;63:168–74.
6. Narayan P, Dragunow M. Pharmacology of epigenetics in brain disorders. Br J Pharmacol. 2010;159:285–303.
7. Allis CD, Berger SL, Cote J, Dent S, Jenuwien T, Kouzarides T, et al. New nomenclature for chromatin-modifying enzymes. Cell. 2007;131:633–6.
8. Apweiler R, Bateman A, Martin MJ, O'Donovan C, Magrane M, Alam-Faruque Y, et al. Activities at the Universal Protein Resource (UniProt). Nucleic Acids Res. 2014;42:D191–D8.
9. Alarcón JM, Malleret G, Touzani K, Vronskaya S, Ishii S, Kandel ER, et al. Chromatin acetylation, memory, and LTP are impaired in CBP+/− mice: a model for the cognitive deficit in Rubinstein-Taybi syndrome and its amelioration. Neuron. 2004;42:947–59.
10. Korzus E, Rosenfeld MG, Mayford M. CBP histone acetyltransferase activity is a critical component of memory consolidation. Neuron. 2004;42:961–72.
11. Guan JS, Haggarty SJ, Giacometti E, Dannenberg JH, Joseph N, Gao J, et al. HDAC2 negatively regulates memory formation and synaptic plasticity. Nature. 2009;459:55–60.
12. McQuown SC, Barrett RM, Matheos DP, Post RJ, Rogge GA, Alenghat T, et al. HDAC3 is a critical negative regulator of long-term memory formation. J Neurosci. 2011;31:764–74.
13. Kim MS, Akhtar MW, Adachi M, Mahgoub M, Bassel-Duby R, Kavalali ET, et al. An essential role for histone deacetylase 4 in synaptic plasticity and memory formation. J Neurosci. 2012;32:10879–86.
14. Sando Iii R, Gounko N, Pieraut S, Liao L, Yates Iii J, Maximov A. HDAC4 governs a transcriptional program essential for synaptic plasticity and memory. Cell. 2012;151:821–34.
15. Agis-Balboa RC, Pavelka Z, Kerimoglu C, Fischer A. Loss of HDAC5 impairs memory function: implications for Alzheimer's disease. J Alzheimers Dis. 2013;33:35–44.
16. Anderson KW, Chen J, Wang M, Mast N, Pikuleva IA, Turko IV. Quantification of histone deacetylase isoforms in human frontal cortex, human retina, and mouse brain. PLoS One. 2015;10:e0126592.
17. de Ruijter AJ, van Gennip AH, Caron HN, Kemp S, van Kuilenburg AB. Histone deacetylases (HDACs): characterization of the classical HDAC family. Biochem J. 2003;370:737–49.
18. Williams SR, Aldred MA, Der Kaloustian VM, Halal F, Gowans G, McLeod DR, et al. Haploinsufficiency of HDAC4 causes brachydactyly mental retardation syndrome, with brachydactyly type E, developmental delays, and behavioral problems. Am J Hum Genet. 2010;87:219–28.
19. Zhang L, Sheng S, Qin C. The role of HDAC6 in Alzheimer's disease. J Alzheimers Dis. 2013;33:283–95.
20. Choudhary C, Kumar C, Gnad F, Nielsen ML, Rehman M, Walther TC, et al. Lysine acetylation targets protein complexes and co-regulates major cellular functions. Science. 2009;325:834–40.
21. Zhao S, Xu W, Jiang W, Yu W, Lin Y, Zhang T, et al. Regulation of cellular metabolism by protein lysine acetylation. Science. 2010;327:1000–4.
22. Kim GW, Yang XJ. Comprehensive lysine acetylomes emerging from bacteria to humans. Trends Biochem Sci. 2011;36:211–20.
23. Cao X, Sudhof TC. A transcriptionally [correction of transcriptively] active complex of APP with Fe65 and histone acetyltransferase Tip60. Science. 2001;293:115–20.
24. Muller T, Concannon CG, Ward MW, Walsh CM, Tirniceriu AL, Tribl F, et al. Modulation of gene expression and cytoskeletal dynamics by the amyloid precursor protein intracellular domain (AICD). Mol Biol Cell. 2007;18:201–10.

25. Konietzko U, Goodger ZV, Meyer M, Kohli BM, Bosset J, Lahiri DK, et al. Co-localization of the amyloid precursor protein and Notch intracellular domains in nuclear transcription factories. Neurobiol Aging. 2010;31:58–73.
26. Müller T, Schrötter A, Loosse C, Pfeiffer K, Theiss C, Kauth M, et al. A ternary complex consisting of AICD, FE65, and TIP60 down-regulates Stathmin1. Biochim Biophys Acta. 1834;2013:387–94.
27. Domingues SC, Konietzko U, Henriques AG, Rebelo S, Fardilha M, Nishitani H, et al. RanBP9 modulates AICD localization and transcriptional activity via direct interaction with Tip60. J Alzheimers Dis. 2014;42:1415–33.
28. von Rotz RC, Kohli BM, Bosset J, Meier M, Suzuki T, Nitsch RM, et al. The APP intracellular domain forms nuclear multiprotein complexes and regulates the transcription of its own precursor. J Cell Sci. 2004;117:4435–48.
29. Kerridge C, Belyaev ND, Nalivaeva NN, Turner AJ. The Aβ-clearance protein transthyretin, like neprilysin, is epigenetically regulated by the amyloid precursor protein intracellular domain. J Neurochem. 2014;130:419–31.
30. Nalivaeva NN, Belyaev ND, Turner AJ. New insights into epigenetic and pharmacological regulation of amyloid-degrading enzymes. Neurochem Res. 2016;41:620–30.
31. Meaney MJ, Ferguson-Smith AC. Epigenetic regulation of the neural transcriptome: the meaning of the marks. Nat Neurosci. 2010;13:1313–8.
32. Xu S, Wilf R, Menon T, Panikker P, Sarthi J, Elefant F. Epigenetic control of learning and memory in drosophila by Tip60 HAT action. Genetics. 2014;198:1571–86.
33. Xu S, Elefant F. Tip off the HAT—epigenetic control of learning and memory by drosophila tip60. Fly. 2015;9:22–8.
34. Lorbeck M, Pirooznia K, Sarthi J, Zhu X, Elefant F. Microarray analysis uncovers a role for Tip60 in nervous system function and general metabolism. PLoS One. 2011;6:e18412.
35. Johnson AA, Sarthi J, Pirooznia SK, Reube W, Elefant F. Increasing Tip60 HAT levels rescues axonal transport defects and associated behavioral phenotypes in a Drosophila Alzheimer's disease model. J Neurosci. 2013;33:7535–47.
36. Marek R, Coelho CM, Sullivan RK, Baker-Andresen D, Li X, Ratnu V, et al. Paradoxical enhancement of fear extinction memory and synaptic plasticity by inhibition of the histone acetyltransferase p300. J Neurosci. 2011;31:7486–91.
37. Caccamo A, Maldonado MA, Bokov AF, Majumder S, Oddo S. CBP gene transfer increases BDNF levels and ameliorates learning and memory deficits in a mouse model of Alzheimer's disease. Proc Natl Acad Sci U S A. 2010;107:22687–92.
38. Rouaux C, Jokic N, Mbebi C, Boutillier S, Loeffler JP, Boutillier AL. Critical loss of CBP/p300 histone acetylase activity by caspase-6 during neurodegeneration. EMBO J. 2003;22:6537–49.
39. Bousiges O, Vasconcelos APD, Neidl R, Cosquer B, Herbeaux K, Panteleeva I, et al. Spatial memory consolidation is associated with induction of several lysine-acetyltransferase (histone acetyltransferase) expression levels and H2B/H4 acetylation-dependent transcriptional events in the rat hippocampus. Neuropsychopharmacology. 2010;35:2521–37.
40. Duclot F, Meffre J, Jacquet C, Gongora C, Maurice T. Mice knock out for the histone acetyltransferase p300/CREB binding protein-associated factor develop a resistance to amyloid toxicity. Neuroscience. 2010;167:850–63.
41. Rao JS, Keleshian VL, Klein S, Rapoport SI. Epigenetic modifications in frontal cortex from Alzheimer's disease and bipolar disorder patients. Transl Psychiatry. 2012;2:e132.
42. Lu T, Aron L, Zullo J, Pan Y, Kim H, Chen Y, et al. REST and stress resistance in ageing and Alzheimer's disease. Nature. 2014;507:448–54.
43. Hernández-Ortega K, Garcia-Esparcia P, Gil L, Lucas JJ, Ferrer I. Altered machinery of protein synthesis in Alzheimer's: from the nucleolus to the ribosome. Brain Pathol. 2015;26:593–605.

44. Mastroeni D, Delvaux E, Nolz J, Tan Y, Grover A, Oddo S, et al. Aberrant intracellular localization of H3k4me3 demonstrates an early epigenetic phenomenon in Alzheimer's disease. Neurobiol Aging. 2015;36:3121–9.
45. Ogawa O, Zhu X, Lee HG, Raina A, Obrenovich ME, Bowser R, et al. Ectopic localization of phosphorylated histone H3 in Alzheimer's disease: a mitotic catastrophe? Acta Neuropathol. 2003;105:524–8.
46. Narayan PJ, Lill C, Faull R, Curtis MA, Dragunow M. Increased acetyl and total histone levels in post-mortem Alzheimer's disease brain. Neurobiol Dis. 2015;74:281–94.
47. Lithner CU, Lacor PN, Zhao WQ, Mustafiz T, Klein WL, Sweatt JD, et al. Disruption of neocortical histone H3 homeostasis by soluble Aβ: implications for Alzheimer's disease. Neurobiol Aging. 2013;34:2081–90.
48. Zhang K, Schrag M, Crofton A, Trivedi R, Vinters H, Kirsch W. Targeted proteomics for quantification of histone acetylation in Alzheimer's disease. Proteomics. 2012;12:1261–8.
49. Davies P. Characterization and use of monoclonal antibodies to tau and paired helical filament tau. Methods Mol Med. 2000;32:361–73.
50. Gjoneska E, Pfenning AR, Mathys H, Quon G, Kundaje A, Tsai LH, et al. Conserved epigenomic signals in mice and humans reveal immune basis of Alzheimer's disease. Nature. 2015;518:365–9.
51. Satoh JI, Asahina N, Kitano S, Kino Y. A comprehensive profile of ChIp-Seq-Based PU.1/Spi1 target genes in microglia. Gene Regul Syst Biol. 2014;8:127–39.
52. Ricobaraza A, Cuadrado-Tejedor M, Perez-Mediavilla A, Frechilla D, Del Rio J, Garcia-Osta A. Phenylbutyrate ameliorates cognitive deficit and reduces tau pathology in an Alzheimer's disease mouse model. Neuropsychopharmacology. 2009;34:1721–32.
53. Francis YI, Fa M, Ashraf H, Zhang H, Staniszewski A, Latchman DS, et al. Dysregulation of histone acetylation in the APP/PS1 mouse model of Alzheimer's disease. J Alzheimers Dis. 2009;18:131–9.
54. Govindarajan N, Agis-Balboa RC, Walter J, Sananbenesi F, Fischer A. Sodium butyrate improves memory function in an Alzheimer's disease mouse model when administered at an advanced stage of disease progression. J Alzheimers Dis. 2011;26:187–97.
55. Graff J, Rei D, Guan JS, Wang WY, Seo J, Hennig KM, et al. An epigenetic blockade of cognitive functions in the neurodegenerating brain. Nature. 2012;483:222–6.
56. Yao ZG, Liang L, Liu Y, Zhang L, Zhu H, Huang L, et al. Valproate improves memory deficits in an Alzheimer's disease mouse model: investigation of possible mechanisms of action. Cell Mol Neurobiol. 2014;34:805–12.
57. Qing H, He G, Ly PTT, Fox CJ, Staufenbiel M, Cai F, et al. Valproic acid inhibits aβ production, neuritic plaque formation, and behavioral deficits in Alzheimer's disease mouse models. J Exp Med. 2008;205:2781–9.
58. Sharma S, Taliyan R. Epigenetic modifications by inhibiting histone deacetylases reverse memory impairment in insulin resistance induced cognitive deficit in mice. Neuropharmacology. 2016;105:285–97.
59. Benito E, Urbanke H, Ramachandran B, Barth J, Halder R, Awasthi A, et al. HDAC inhibitor-dependent transcriptome and memory reinstatement in cognitive decline models. J Clin Invest. 2015;125:3572–84.
60. Cuadrado-Tejedor M, Garcia-Barroso C, Sanzhez-Arias J, Mederos S, Rabal O, Ugarte A, et al. Concomitant histone deacetylase and phosphodiesterase 5 inhibition synergistically prevents the disruption in synaptic plasticity and it reverses cognitive impairment in a mouse model of Alzheimer's disease. Clin Epigenetics. 2015;7:108.
61. Pavlopoulos E, Jones S, Kosmidis S, Close M, Kim C, Kovalerchik O, et al. Molecular mechanism for age-related memory loss: the histone-binding protein RbAp48. Sci Transl Med. 2013;5:1–12.
62. Walker MP, Laferla FM, Oddo SS, Brewer GJ. Reversible epigenetic histone modifications and Bdnf expression in neurons with aging and from a mouse model of Alzheimer's disease. Age. 2013;35:519–31.

63. Guo X, Wu X, Ren L, Liu G, Li L. Epigenetic mechanisms of amyloid-beta production in anisomycin-treated SH-SY5Y cells. Neuroscience. 2011;194:272–81.
64. Gu X, Sun J, Li S, Wu X, Li L. Oxidative stress induces DNA demethylation and histone acetylation in SH-SY5Y cells: potential epigenetic mechanisms in gene transcription in Aβ production. Neurobiol Aging. 2013;34:1069–79.
65. Lu X, Deng Y, Yu D, Cao H, Wang L, Liu L, et al. Histone acetyltransferase p300 mediates histone acetylation of PS1 and BACE1 in a cellular model of Alzheimer's disease. PLoS One. 2014;9:e103067.
66. Porsteinsson AP. Divalproex sodium for the treatment of behavioural problems associated with dementia in the elderly. Drugs Aging. 2006;23:877–86.
67. Porsteinsson AP, Tariot PN, Erb R, Cox C, Smith E, Jakimovich L, et al. Placebo-controlled study of divalproex sodium for agitation in dementia. Am J Geriatr Psychiatry. 2001;9:58–66.
68. Porsteinsson AP, Tariot PN, Erb R, Gaile S. An open trial of valproate for agitation in geriatric neuropsychiatric disorders. Am J Geriatr Psychiatry. 1997;5:344–51.
69. Porsteinsson AP, Tariot PN, Jakimovich LJ, Kowalski N, Holt C, Erb R, et al. Valproate therapy for agitation in dementia: open-label extension of a double-blind trial. Am J Geriatr Psychiatry. 2003;11:434–40.
70. Tariot PN, Raman R, Jakimovich L, Schneider L, Porsteinsson A, Thomas R, et al. Divalproex sodium in nursing home residents with possible or probable Alzheimer disease complicated by agitation: a randomized, controlled trial. Am J Geriatr Psychiatr. 2005;13:942–9.
71. Tariot PN, Schneider LS, Mintzer JE, Cutler AJ, Cunningham MR, Thomas JW, et al. Safety and tolerability of divalproex sodium in the treatment of signs and symptoms of mania in elderly patients with dementia: results of a double-blind, placebo-controlled trial. Curr Ther Res. 2001;62:51–67.
72. Herrmann N. Valproic acid treatment of agitation in dementia. Can J Psychiatr. 1998;43:69–72.
73. Herrmann N, Lanctot KL, Rothenburg LS, Eryavec G. A placebo-controlled trial of valproate for agitation and aggression in Alzheimer's disease. Dement Geriatr Cogn Disord. 2007;23:116–9.
74. Kasckow JW, McElroy SL, Cameron RL, Mahler LL, Fudala SJ. A pilot study on the use of divalproex sodium in the treatment of behavioral agitation in elderly patients with dementia: assessment with the BEHAVE-AD and CGI rating scales. Curr Ther Res. 1997;58:981–9.
75. Dolder C, McKinsey J. Low-dose divalproex in agitated patients with Alzheimer's disease. J Psychiatr Pract. 2010;16:63–7.
76. Fleisher AS, Truran D, Mai JT, Langbaum JBS, Aisen PS, Cummings JL, et al. Chronic divalproex sodium use and brain atrophy in Alzheimer disease. Neurology. 2011;77:1263–71.
77. Tariot PN, Schneider LS, Cummings J, Thomas RG, Raman R, Jakimovich LJ, et al. Chronic divalproex sodium to attenuate agitation and clinical progression of Alzheimer disease. Arch Gen Psychiatry. 2011;68:853–61.
78. Chico LK, Watterson DM. The 6th drug discovery for neurodegeneration conference: an intensive course on translating research into drugs. Expert Opin Drug Discovery. 2012;7:1225–8.
79. Beach TG. Alzheimer's disease and the "Valley Of Death": not enough guidance from human brain tissue? J Alzheimers Dis. 2013;33:S219–33.
80. Frye SV, Arkin MR, Arrowsmith CH, Conn PJ, Glicksman MA, Hull-Ryde EA, et al. Tackling reproducibility in academic preclinical drug discovery. Nat Rev Drug Discov. 2015;14:733–4.
81. Dragunow M. The adult human brain in preclinical drug development. Nat Rev Drug Discov. 2008;7:659–66.
82. Smith AM, Dragunow M. The human side of microglia. Trends Neurosci. 2014;37:125–35.
83. Dragunow M, Greenwood JM, Cameron RE, Narayan PJ, O'Carroll SJ, Pearson AG, et al. Valproic acid induces caspase 3-mediated apoptosis in microglial cells. Neuroscience. 2006;140:1149–56.

84. Smith AM, Gibbons HM, Dragunow M. Valproic acid enhances microglial phagocytosis of amyloid-beta(1-42). Neuroscience. 2010;169:505–15.
85. Gibbons HM, Smith AM, Teoh HH, Bergin PM, Mee EW, Faull RL, et al. Valproic acid induces microglial dysfunction, not apoptosis, in human glial cultures. Neurobiol Dis. 2011;41:96–103.
86. Feingold EA, Good PJ, Guyer MS, Kamholz S, Liefer L, Wetterstrand K, et al. The ENCODE (ENCyclopedia of DNA Elements) Project. Science. 2004;306:636–40.
87. Myers RM, Stamatoyannopoulos J, Snyder M, Dunham I, Hardison RC, Bernstein BE, et al. A user's guide to the Encyclopedia of DNA elements (ENCODE). PLoS Biol. 2011;9:1–21.
88. Harrow J, Frankish A, Gonzalez JM, Tapanari E, Diekhans M, Kokocinski F, et al. GENCODE: the reference human genome annotation for the ENCODE project. Genome Res. 2012;22:1760–74.
89. Bernstein BE, Stamatoyannopoulos JA, Costello JF, Ren B, Milosavljevic A, Meissner A, et al. The NIH Roadmap Epigenomics Mapping Consortium. Nat Biotechnol. 2010;28:1045–8.
90. Brinkman AB, Gu H, Bartels SJJ, Zhang Y, Matarese F, Simmer F, et al. Sequential ChIP-bisulfite sequencing enables direct genome-scale investigation of chromatin and DNA methylation cross-talk. Genome Res. 2012;22:1128–38.
91. Statham AL, Robinson MD, Song JZ, Coolen MW, Stirzaker C, Clark SJ. Bisulfite sequencing of chromatin immunoprecipitated DNA (BisChIP-seq) directly informs methylation status of histone-modified DNA. Genome Res. 2012;22:1120–7.
92. Yan J, Zierath JR, Barres R. Evidence for non-CpG methylation in mammals. Exp Cell Res. 2011;317:2555–61.
93. Guo JU, Su Y, Shin JH, Shin J, Li H, Xie B, et al. Distribution, recognition and regulation of non-CpG methylation in the adult mammalian brain. Nat Neurosci. 2014;17:215–22.
94. Guo W, Chung WY, Qian M, Pellegrini M, Zhang MQ. Characterizing the strand-specific distribution of non-CpG methylation in human pluripotent cells. Nucleic Acids Res. 2014;42:3009–16.
95. Squires JE, Patel HR, Nousch M, Sibbritt T, Humphreys DT, Parker BJ, et al. Widespread occurrence of 5-methylcytosine in human coding and non-coding RNA. Nucleic Acids Res. 2012;40:5023–33.
96. Dominissini D, Nachtergaele S, Moshitch-Moshkovitz S, Peer E, Kol N, Ben-Haim MS, et al. The dynamic N(1)-methyladenosine methylome in eukaryotic messenger RNA. Nature. 2016;530:441–6.
97. Delatte B, Wang F, Ngoc LV, Collignon E, Bonvin E, Deplus R, et al. RNA biochemistry. Transcriptome-wide distribution and function of RNA hydroxymethylcytosine. Science. 2016;351:282–5.
98. Chaput D, Kirouac L, Stevens SM, Padmanabhan J. Potential role of PCTAIRE-2, PCTAIRE-3 and P-Histone H4 in amyloid precursor protein dependent Alzheimer pathology. Oncotarget. 2016;7:8481–97.

Alzheimer's Disease and ncRNAs

18

Rotem Maoz, Benjamin P. Garfinkel, and Hermona Soreq

Abstract

Alzheimer's disease is a devastating neurodegenerative disorder affecting a significant portion of the world's rapidly growing aging population. In spite of its prevalence, the etiology of the disease is still poorly understood, and effective therapy is all but unavailable. Over the past decade, noncoding RNA, including microRNA (miRNA), has emerged as a major class of regulatory molecules involved in virtually all physiological and disease states. The specificity provided by miRNA sequence complementarity, together with the ability of these molecules to regulate complex networks of genes, has made them exciting novel targets for therapeutic agents. In this chapter, we review recent progress on understanding the role of noncoding RNA in Alzheimer's disease (AD). The majority of available work has focused on miRNA, and we review the many studies implicating specific miRNAs in the development of the disease. More recently, several studies have tied other RNA classes to the disorder, including long noncoding RNA, circular RNA, and Y RNAs, and we review this fascinating field as well. Finally, we explore the potential promise of these findings for future therapeutic applications.

Keywords

Alzheimer's disease • Amyloid beta • Lipids • Long noncoding RNA • MicroRNA • Tau

---

Rotem Maoz and Benjamin P. Garfinkel are contributed equally.

R. Maoz, B.Sc. • B.P. Garfinkel, Ph.D. • H. Soreq, Ph.D. (✉)
The Edmond and Lily Safra Center of Brain Science, The Hebrew University of Jerusalem, Jerusalem 9190401, Israel
e-mail: rotem.maoz@mail.huji.ac.il; benny.garfinkel@mail.huji.ac.il; Hermona.soreq@mail.huji.ac.il

R. Delgado-Morales (ed.), *Neuroepigenomics in Aging and Disease*, Advances in Experimental Medicine and Biology, DOI 10.1007/978-3-319-53889-1_18

18.1 Alzheimer's Disease

AD is an irreversible progressive neurodegenerative disorder affecting the central nervous system, which accounts for >80% of dementia cases in people over the age of 65 [1]. The major physical hallmarks of the disease are the presence of amyloid beta plaques and hyper-phosphorylated paired helical filaments of tau protein-rich neurofibrillary tangles (NFTs) [2]. Although the disease etiology is incompletely understood, both types of lesions have been linked to AD dementia, and their presence among those afflicted with AD is typically associated with much steeper trajectories of cognitive deficit accumulation with age [3]. Furthermore, families with mutations in three proteins directly involved in the Aβ processing pathway present inherited early-onset AD [4], suggesting a key role for Aβ in disease pathogenesis. The second hallmark of AD pathology is the accumulation of intracellular NFTs, made up primarily of aggregated tau bearing abnormal posttranslational modifications, including increased phosphorylation and acetylation. The exact role of tau in AD progression remains unclear, but recent work has suggested that abnormal modifications of the protein can lead to enrichment of misfolded tau in dendritic spines where it can interfere with neurotransmission [5]. The extracellular Aβ and intracellular NFTs are intricately linked, and Aβ oligomers have been implicated in the postsynaptic enrichment of tau [6], as well as in altered tau phosphorylation [7], proteasomal degradation [8], and nucleation [9].

While the Aβ and NFT structures are well-accepted hallmarks of the disease, AD is a multifactorial disorder, involving numerous cell types and pathways. Interestingly, Aβ peptide aggregates have been implicated in many of these pathways, including mitochondrial dysfunction [10], oxidative damage [11], excessive calcium influx [12], lipid dysregulation [13], synaptic dysfunction [14], apoptosis [15], aberrant neurogenesis [16], and neuroinflammation [17]. Furthermore, Aβ has been shown to have complex interactions with nonneuronal cell types. For example, microglia have been shown to contribute to Aβ clearance at early stages of AD, while the same cells release pro-inflammatory cytokines that exacerbate symptoms at late stages of the disease [18, 19]. Collectively, these results indicate that AD is a complex disease involving multiple interlinked pathways and cell types, hinting at the potential role for co-regulatory mechanisms in disease development and in future therapeutics.

18.2 The Noncoding RNA Revolution

For many years, RNA was seen as no more than an inert participant in the process of protein synthesis, with mRNA providing the template and tRNA the building blocks and rRNA acting as a scaffold [20]. It was not until the early 1980s, with the discovery of the catalytic activity of small nuclear (sn)RNAs in the excision of introns, that noncoding RNA (ncRNA) began to be seen as a more active agent within the cell. A major turning point in our understanding of the diverse roles of

RNA came in the early 2000s with the discovery of miRNAs and their many relatives, underscoring the importance of posttranscriptional events in gene expression, particularly in eukaryotic organisms [21, 22]. Now, with the advent of RNA sequencing, the field has seen an explosion in the depth and breadth of discovery, with new classes of RNA being described on a regular basis [23]. Thousands of novel ncRNAs have been catalogued, and their many roles in regulating gene expression and organizing the genome have only begun to be explored.

Of all classes of ncRNA, miRNAs are by far the most studied, and their functional regulatory relevance is clearly evident and well established [24]. In human diseases, particularly cancer, it has been shown that epigenetic and genetic defects in miRNAs and their processing machinery are a common hallmark of disease [25–27]. The informational bias toward miRNA holds true for nervous system diseases as well, including AD [28], and most of the work described herein relates to this fascinating class of regulatory RNA. However, miRNAs are just the tip of the iceberg, and other ncRNAs such as small nucleolar RNAs (snoRNAs), PIWI-interacting RNAs (piRNAs), Y RNAs, the large heterogeneous group of long noncoding RNAs (lncRNAs), and circular RNAs are all emerging as players in the development of many different human disorders [29–31]. In fact, reports regarding potential involvement of some of these novel classes of ncRNA in the pathophysiology of AD are beginning to appear, and these will be explored below as well.

18.3 MicroRNA in AD

miRNAs are endogenous 20–24 nucleotide noncoding RNAs that bind to target motifs in mRNAs of protein-coding genes to direct posttranscriptional silencing either through transcript degradation or by translational repression. So far more than 2000 miRNAs have been denoted in the human genome, and the number is ever increasing, illustrating the potential of miRNAs as important players in gene regulation [32]. Recent studies have indicated that miRNAs play a pivotal role in most critical biological events, including development, proliferation, differentiation, cell fate determination, apoptosis, signal transduction, organ development, hematopoietic lineage differentiation, host-virus interactions, tumor genesis, and more [33]. Specifically, in the brain, miRNAs are highly expressed in neurons where they play key roles during neuronal differentiation, synaptogenesis, and plasticity. Correspondingly, it is becoming increasingly evident that miRNAs have a profound impact on higher cognitive functions and that impairments in their functioning are involved in the etiology of several neurological diseases and disorders, including AD, Parkinson's disease, amyotrophic lateral sclerosis (ALS), and Huntington's disease [34, 35]. Over the past decade, accumulating evidence in AD research suggests that alterations in the miRNA network could actively contribute to the disease process. In the following sections, we will review recent progress in the field.

18.4 miRNA Regulation of Amyloid Beta

Aβ peptides accumulating in the brain of AD patients result from proteolytic cleavage of APP by β-secretase (BACE1) and γ-secretase [36, 37]. AD pathology is associated with increased APP expression, polymorphisms in the APP promoter, abnormal APP processing, and altered Aβ clearance [10]. miRNAs have been shown to be involved in Aβ pathology through regulation of APP expression and of other enzymes involved in the Aβ processing, mostly BACE1. The following section describes the main findings in this field.

18.4.1 Regulation of Aβ by miRNA Control of the APP Transcript

The first studies showing miRNA regulation of APP mRNA came in 2008 from *C. elegans*, where the worm APP homolog, APL-1, was shown to be developmentally regulated by miRNA let-7 [38]. Later that same year, Patel et al. showed that APP is regulated by miRNA in humans as well, as overexpression of miR-106a and miR-520c results in translational repression of APP mRNA and significantly reduces APP levels [39]. Since then, a large number of other miRNAs have been shown to directly regulate APP mRNA in human cells in vitro. These include miR-20a, miR-19, and miR-106a/b, all from the miR-20a family [40, 41] and miR-101 [42, 43], among others. In light of these findings, it was speculated that SNPs in miRNA binding sites in the 3′-UTR of APP could affect Aβ pathology and AD risk, similar to what has been observed for several neurological diseases [44]. Indeed, in 2011, Delay et al. investigated miRNAs that have potential binding sites in or near a polymorphism located on the 3′-untranslated region (3′-UTR) of human APP. In their study, they used luciferase assays to confirm that miR-20a, miR-17, miR-147, miR-323-3p, miR-644, and miR-153 can all regulate APP in vitro and, furthermore, that miR-147 and miR-20a are affected by AD-associated 3′-UTR single nucleotide polymorphism (SNP) variants in the APP gene [45].

miRNAs have also been implicated in the regulation of APP's alternative splicing. Neuronal APP undergoes alternative splicing of exons 7, 8, and 15, and there is substantial evidence for increased levels of exon 7 and/or 8 isoforms of APP in AD brains [46–50]. Furthermore, abnormal neuronal splicing of APP was found to be associated with increased Aβ production [51]. Clear evidence of a role for miRNA in APP splicing came from a study by Smith et al. who detected abnormal splicing of exons 7 and 8 of APP in the cortex of dicer knockout (KO) mice, with no change in the overall APP mRNA levels. In the same study, knockdown of polypyrimidine tract-binding protein 1 (PTB1) in neuronal cell lines altered APP's exon 7 and 8 splicing, and PTB1 itself was regulated by miR-124. Furthermore, ectopic expression of this abundant neuronal-specific miRNA in cultured neurons reversed the above effects on APP splicing, and miR-124 was found to be downregulated in AD brains [52–55].

18.4.2 miRNA Regulation of BACE1

The activity of BACE1 is a key factor in AD, since the cleavage of APP by BACE1 is the first and rate-limiting step in the formation of Aβ. Indeed, increased levels of BACE1 expression and enzymatic activity were found in sporadic AD brains [56]. The best-studied miRNAs in the context of BACE1 belong to the miR-29 family, and several studies have linked this family to regulation of BACE1 activity both in vitro and in vivo. Members of the miR-29 family, which are processed from different precursors, include three main mature miRNAs, known as hsa-miR-29a, hsa-miR-29b, and hsa-miR-29c. miR-29c was shown to regulate the expression of BACE1 directly by targeting its 3′-UTR both in human and mouse cell lines. In addition, this miRNA is downregulated in AD brains, and its upregulation in the hippocampus of SAMP8 mice, a naturally occurring mouse line that displays accelerated aging, improved spatial memory [57–59]. The other two members of the miR-29 family, miR-29a and miR-29b, were significantly decreased in AD brains, in a manner specific to AD dementia. Their decrease associates with abnormally high BACE1 protein, and loss of the suppressing activity of the miR-29a/b-1 cluster in human cell cultures led to an increase in Aβ production [60]. Other miRNAs that are deregulated in AD brains and can directly target BACE1 in vitro include miR-339-5p, miR-195 [61], and miR-107 [62].

Other miRNAs that were found to be involved in Aβ regulation are miR-9 that was shown to attenuate Aβ-induced synaptic toxicity by targeting CAMKK2 (calcium-/calmodulin-dependent protein kinase kinase 2) in human cells [63]. Interestingly, miR-9 has also been implicated in regulation of insulin signaling [64], potentially providing a link to the increased risk of AD patients for diabetes [65]. Finally, miR-144/miR-451 was found to regulate α-secretase ADAM10 (a disintegrin and metalloprotease 10), a protein that protects the brain from Aβ production [66].

18.5 miRNA Regulation of Tau

Tau is a microtubule-associated protein normally located in neuronal axons, where it stabilizes microtubule structures and facilitates axonal transport. In AD, tau is abnormally translocated to the somato-dendritic compartment and further undergoes hyper-phosphorylation and misfolding resulting in the generation of intracellular aggregates (neurofibrillary tangles) that are toxic to neurons [67]. Tau pathology can potentially be caused by disruption in any of the processes regulating its metabolism, including expression, localization, transcriptional and posttranslational modifications, and clearance. In particular, tau hyper-phosphorylation may result from upregulation or aberrant expression of tau kinases, downregulation of phosphatases, mutations, covalent modifications of tau, and others [68, 69]. The following two sections describe the involvement of miRNAs in the regulation of tau metabolism under pathological conditions.

18.5.1 miRNA Regulation of Tau mRNA Expression and Metabolism

Several studies have shown direct regulation of tau expression by miRNAs. One of these is miR-34a, which was found to regulate tau expression through specifically targeting the longer variant of tau alternative polyadenylation (APA). Both human and rodent tau 3′-UTR contain two polyadenylation signals in tandem and can undergo APA to produce transcripts of approximately 2 or 6 kb [70]. Dickson et al. found that the expression of the two human tau 3′-UTR isoforms is differentially regulated, affecting both protein and mRNA levels. In the same study, they showed that miR-34a has a binding site in the long tau 3′-UTR isoform and demonstrated that miR-34a can inhibit the expression of endogenous tau [71].

The miR-132/miR-212 cluster has also been linked to regulation of tau expression. Smith et al. showed that miR-132/miR-212 deficiency in mice leads to increased tau expression, phosphorylation, and aggregation. They demonstrated that deletion of miR-132/miR-212 induced tau aggregation in mice expressing endogenous or human mutant tau, an effect associated with autophagy dysfunction. And conversely, treatment of AD mice with miR-132 mimics restored in part memory function and tau [72]. At the same time, deletion of this miRNA in another study impaired learning and memory [73]. These results are of particular interest as miR-132 levels were shown to be decreased in the brains of advanced stage patients in more than one study [28, 74].

More recently, Santa-Maria et al. used experimental evidence from *Drosophila*, postmortem human samples, and mammalian cells to show that the highly conserved miR-219 directly regulates tau and that this is an ancient mechanism. They demonstrated that reduction of miR-219 in a *Drosophila* model, which produces human tau in vivo, exacerbated tau toxicity, while overexpression of the miRNA partially abrogated toxic effects. They further found that miR-219 is downregulated in brain tissue taken at autopsy from patients with AD and from those with severe primary age-related tauopathy. Finally, they showed in mammalian cellular models that miR-219 binds directly to the 3′-UTR of the tau mRNA and represses tau synthesis at the posttranscriptional level [75]. Another mechanism impacting the levels of tau mRNA involves disruption of the degradation process. Carrettiero et al. showed that miR-128a modulates the expression of BAG2, a co-chaperone potentially involved in tau degradation and aggregation in cultured COS-7 cells and in primary neurons [76]. Thus, both the birth and destruction of tau appear to be causally involved in AD progression.

18.5.1.1 miRNA-Mediated Mis-splicing of Tau Is causally Linked to Dementia and Neurodegeneration

The tau primary transcript contains 16 exons, of which exons 2, 3, and 10 are alternatively spliced and are adult brain specific [77]. Exclusion of exon 10 (E10) results in a protein with three microtubule-binding repeats (3R tau), whereas E10 inclusion produces a protein with an additional microtubule-binding domain (4R tau). In the normal human adult brain, the 3R and 4R tau isoforms are expressed in a one-to-one ratio [78–80]. Clinical and biochemical evidence suggests that deviation from the 4R:3R ratio due to altered tau exon 10 splicing is causally linked to neurodegeneration and dementia. For example, mutations in or surrounding tau E10 were identified in patients suffering from rare forms of familial frontotemporal dementia and parkinsonism. Additional changes in tau isoforms have also been observed in other tauopathies, including PSP and Pick's disease or myotonic dystrophy [69, 80–82].

Interestingly, Dicer deficiency in the adult brain is associated with changes in tau splicing, and Smith et al. [52] showed that Dicer mutant mice display changes in tau E10 splicing. In the same study, a number of brain miRNAs were identified, including miR-124, miR-9, miR-132, and miR-137, which regulate 4R:3R tau ratios in neuronal cells by targeting particular regulatory or splicing factors. Specifically, they demonstrated that miR-132 directly targets the neuronal splicing factor polypyrimidine tract-binding protein 2 (PTBP2). This result is of particular interest, since these authors also found that PTBP2 protein levels were increased in patients with PSP, a major 4R–tau tauopathy [52].

18.5.2 Tau Phosphorylation

While the absolute levels of tau protein are important, aberrant posttranslational processes play a major role in tau-mediated pathology. In this regard, hyper-phosphorylated microtubule-associated tau proteins are the main components of NFTs in AD, and molecular analysis has revealed that abnormal phosphorylation might be one of the important events in the process leading to their aggregation [77]. Tau phosphorylation is regulated by a balanced interplay of kinases and phosphatases, and recent studies have found several miRNAs that regulate these processes.

In 2010, Hebert et al. showed that hyper-phosphorylation of endogenous tau at pathological sites coincided with an increase in the protein and phosphorylation of mitogen-activated protein kinase 3 (MAPK3/ERK1). Furthermore, they showed in mouse neuronal cells that ERK1 is regulated by several miR-15 family members, including miR-15a, and that miR-15a is decreased in AD brains [83]. GSK-3b, another direct tau kinase, has a critical role in Aβ production and NFT formation, and it was shown to be regulated by miR-26a in smooth muscle [84, 85]. This miRNA, together with its family member miR-26b, was shown to down-regulate brain-derived neurotrophic factor (BDNF) expression, a neurotrophin that plays an essential role in neuronal development and plasticity [86]. Recently, Absalon et al. found that miR-26b is significantly elevated in the temporal gyrus of human postmortem brains, starting from early stages of AD neuropathology (Braak III). They also showed that ectopic overexpression of this miRNA in rat primary postmitotic neurons led to DNA replication and aberrant cell cycle entry and, in parallel, increased tau phosphorylation, which culminated in the apoptotic cell death of neurons [87]. In a different study, miR-26a was also found to have altered expression in AD [88].

Another miRNA that was found to increase tau phosphorylation is miR-922. Zhao et al. [89] showed in human cell lines that miR-922 increases tau phosphorylation by downregulating UCHL1, a member of deubiquitinating enzymes (DUBs) that is decreased in the brains of AD patients, and its levels were found to be inversely proportional to tangle numbers [90]. Correspondingly, miR-138, that was found to be increased in AD models including N2a/APP and HEK293/tau cell lines, was shown to promote tau phosphorylation by targeting the RARA/GSK-3β pathway in HEK293/tau cells [91]. Additional works found other miRNAs that effect tau phosphorylation, including miR-126b [87] and miR-125b [79]. Thus, tau regulation by miRNAs emerges as a complex and context-dependent process.

18.6 miRNA Regulation of Lipid Metabolism

A clear link between lipid metabolism and AD was established in 1993 when the ε4 allele of the apolipoprotein E (*APOE*) gene was identified as a risk factor for AD, and it remains to this day the strongest known genetic risk factor for AD [92]. Over the course of the following two decades, altered metabolism of various lipids has been linked to AD in a large number of studies [93–95]. Cholesterol metabolism in particular was identified as a key player, and dysregulation of genes involved in cholesterol biosynthesis and cholesterol efflux has been associated with developing AD [96]. Studies of miRNA regulation of lipid homeostasis in AD begin to emerge, with most current studies investigating miRNA regulation of cholesterol metabolism.

Several studies have linked miR-33 to AD through its regulation of lipid metabolism, mostly via inhibition of the ATP-binding cassette transporter A1 (ABCA1). ABCA1 is a membrane-bound protein that mediates the efflux cholesterol and phospholipids onto lipid-poor apolipoproteins during HDL biogenesis, and it was shown to be relevant to AD by decreasing the levels of Aβ [97–99]. Specifically, miR-33 was shown to directly regulate ABCA1 in vitro, both in human neuronal cell lines and in mouse neurons and primary astrocytes [100–103]. This regulation has functional implications relevant to AD, and Kim et al. have recently shown that downregulation of ABCA1 by miR-33 affected Aβ levels. In in vitro conditions, they showed that overexpression of miR-33 impaired cellular cholesterol efflux and dramatically increased extracellular Aβ levels by promoting Aβ secretion and impairing Aβ clearance. In vivo, they found that the cortex of miR-33−/− mice shows increased ABCA1 levels and ApoE-mediated lipid production, accompanied by decreased endogenous Aβ levels. Furthermore, they showed that chronic treatment with a miR-33 antagonist caused a significant decrease in Aβ levels in the cortex of *APP/PS1* mice [101]. A very recent review by Jaouen and Gascon summarizes the involvement of miR-33 in brain lipid metabolism and the implication for AD [104]. It should be mentioned, however, that this miRNA has not been observed to be deregulated in brains of AD patients in any study to date.

Another miRNA that regulates ABCA1 is miR-106b, overexpression of which in cultured mouse neuronal cells caused a significant increase in secreted Aβ. This increase was caused by both increased Aβ production and prevention of Aβ clearance that were both mediated by ABCA1. This finding was supported by over-expressing an ABCA1 construct that lacked the 3′-UTR sequence in the same miR-106b-5p overexpressing cells, which resulted in rescue of the impaired Aβ production and clearance processes [105]. ABCA1 and cholesterol metabolism have also been shown to be regulated by miR-758, and the Fernández-Hernando group suggested that this miRNA could have important implications for the regulation of AD development [96, 106]. A number of additional miRNAs have also been linked to impaired lipid metabolism in AD. Among these are miR-137, miR-181c, miR-9, and miR29a/b that were shown to target serine palmitoyltransferase (SPT), a regulator of ceramide levels that is upregulated in AD brains and directly mediates amyloid β (Aβ) levels [107, 108]. Another relevant miRNA is miR-188-3p that was implicated in 2-arachidonoylglycerol (2-AG)-mediated regulation of BACE1, affecting synaptic and cognitive functions in an AD mouse model [109]. Other miRNAs that are involved in cholesterol metabolism and neurodegeneration are covered by Goedeke and Fernández-Hernando [96].

18.7 miRNA Regulation of Neuroinflammation

AD pathology is intricately connected to immunological mechanisms in the brain. Binding of misfolded and aggregated proteins to receptors on glia initiates an innate immune response that contributes to disease progression. Furthermore, genes which regulate glial clearance of misfolded proteins and inflammatory reactions have been associated with increased risk for AD [110]. The transcription of several miRNAs known to be involved in neuroinflammation is regulated by NF-κB, an immune and stress-induced transcription factor. Under normal conditions, inflammation-inducing NF-κB activity is blocked by acetylcholine [111]. However, acetylcholine production is severely impaired in AD [112]. Consequently, one of the NF-κB-controlled miRNAs, miR-34a, was shown by Zaho et al. to be upregulated in the hippocampal CA1 region of AD patients. In their study, they used murine microglia to show that miR-34a regulates the microglial receptor TREM2 (triggering receptor expressed on myeloid cells-2), a critical component of Aβ42 peptide clearance that was shown to be downregulated in the CA1 of AD patients [113–115]. This link is lent strong support by the finding that rare heterozygous variants of TREM2 are associated with a significant increase in the risk of AD [116].

Another NF-κB-dependent miRNA, miR-146a, targets key members of the innate immune system with known relevance to AD. These include the sialic acid containing glycoprotein immune repressor complement factor H (CFH), the membrane-spanning beta-amyloid precursor protein (βAPP)-associated TSPAN12, and the inflammation mediator interleukin receptor-associated kinase IRAK-1 [117]. miR-155, which is also a known target of transcriptional regulation by NF-κB, was shown by Guedes et al. to be upregulated in the brain of 12-month-old 3 × Tg-AD animals. Surprisingly, this upregulation was not dependent on NF-κB, but rather on the transcription factor c-Jun, and it occurred before the appearance of extracellular Aβ aggregates and was accompanied by hyper-activation of microglia and astrocytes [118]. Interestingly, another study also identified miR-155 as an immune-related miRNA relevant to AD, this time through regulation of T lymphocyte function [119].

Yet another miRNA family involved in neuroinflammatory responses and showing relevance to AD is the miR-181 family [120]. Rodriguez-Ortiz et al. showed that 12-month-old 3 × Tg-AD mice with plaques and tangles presented a significant upregulation of miR-181a in the hippocampus compared to age-matched wild-type mice. They further showed that overexpression of this miRNA in SH-SY5Y cells significantly decreased its validated target SIRT1 and its predicted target c-Fos, hinting at its potential mechanism of action [121, 122]. Interestingly, the other miRNAs in miR-181 family were shown to be related to stress response. Remarkably, miR-181d is one of the most stress-responsive miRNAs identified in the thymus [123], together indicating a link between stress responses and neuroinflammation in the AD brain.

18.8 Global Assays

The study of miRNAs in AD is still in the discovery phase, and a growing number of groups are attempting to harness the power of unbiased high-throughput assays to uncover novel miRNAs implicated in the disease. The majority of these studies used qPCR or microarrays, and only with the recent progress in sequencing technologies have RNA-sequencing studies begun to appear. This is of great importance, as RNA-seq offers clear advantages over other methods, including the detection of absolute values rather than fold changes, high accuracy, and the ability to detect novel miRNA and miRNA editing.

Muller et al. [124] and Cogswell et al. [88] both used qPCR on human brain samples to assess global changes in miRNA expression. While studies using qPCR are considered to be relatively reliable and specific, they are prone to significant bias, as only a limited number of known miRNAs can be tested [125]. Before the advent of high-throughput RNA sequencing, microarrays allowed researchers to overcome the size limitation of qPCR [126, 127]. Two recent studies of relatively large cohorts ($n = 42$ and $n = 32$) used microarrays accompanied by qPCR validations [128, 129] to assess miRNA expression in postmortem brains. It is important to note that microarrays suffer from several drawbacks: They have lower specificity than qPCR or RNA sequencing, are difficult to use for absolute quantification, ignore ncRNA contributions, and cannot typically identify novel miRNAs [125].

The most comprehensive discovery study performed so far is from 2013, by Lau et al. In this project, an international collaboration between several groups allowed for an elaborate joint study combining three different methods. They interrogated data from three different brain areas: hippocampus (41 AD, 23 control), prefrontal cortex ($n = 49$ in different Braak stages, 7 control), and temporal gyrus (8 AD, 8 control), taken from three large cohorts. The nCounter method [130], which resembles microarray analysis, was performed on most of the samples and RNA sequencing on 12 prefrontal cortex samples, and qPCR was used for validation. These methods produced lists of differentially expressed miRNAs for each tissue. Remarkably, they found deregulation of miR-132-3p in all the three brain regions that were tested using all technologies. In contrast, the rest of the differentially expressed miRNAs were only identified in one or more tissue sets [28].

Several novel and interesting directions have also been attempted in the field of global miRNA assays. Roy et al. searched for miRNAs that might have significant involvement in AD through the identification of differential methylation patterns between AD and control brains [7]. Another evolving field is the analysis of altered miRNA editing based on high-throughput sequencing, and a number of groups have observed indications for this phenomenon in AD [8, 9].

18.9 Long Noncoding RNA (lncRNA) in AD

Recent genomic studies have revealed tens of thousands of lncRNAs in mammalian genomes [131]. lncRNA genes give rise to long (>200 nt) and often multi-exonic transcripts that are supposedly not translated to protein, as assessed by means of in silico prediction algorithms. Akin to mRNAs, lncRNAs are transcribed as precursor transcripts and are subject to splicing and maturation in the nucleus, as well as to cytoplasmic export, editing, transport, and decay. In comparison with their protein-coding counterparts, lncRNA genes are poorly conserved [132] and are more numerous in biologically complex species [133]. lncRNAs are particularly difficult to classify and categorize [134]. One relatively well-accepted categorization is the broad division into two classes: natural antisense (AS) RNAs which are transcribed from the opposite DNA strand of other RNA transcripts and long intergenic RNAs (lincRNAs) which are transcribed from intergenic regions [135]. Through their impact on gene expression patterns, lncRNAs are emerging as key regulators of cellular processes (e.g., proliferation, apoptosis, stress response, differentiation, senescence) as well as physiologic and pathologic processes (such as immune adaptation, cancer, neurodegeneration, cardiovascular disease, and aging) [31, 136–138]. In both the nucleus and the cytoplasm, lncRNAs are believed to control gene expression by interacting with chromatin regulators, transcriptional activators and repressors, chromosomal DNA, miRNAs, RBPs, and mRNAs [139]. However, the full spectrum of functions for the vast class of lncRNAs is poorly understood.

The roles of lncRNAs in AD have only recently begun to be explored, and very few examples existed prior to the recent advance in RNA sequencing. Interestingly, most of these examples are of AS lncRNAs. Thus, one lncRNA with a relatively well-defined involvement in AD is the conserved antisense lncRNA BACE1-AS which is transcribed from the complementary strand of beta-secretase-1 [140]. Expression of BACE1-AS drives feed-forward regulation of beta-secretase and is directly implicated in the increased abundance of Aβ 1–42 in AD. Several other antisense lncRNAs with a potential involvement in amyloid pathology have been described, including NAT-Rad18 which was found to be upregulated in rat neurons in response to Aβ peptide [141] and 51A, which overlaps with SORL1 [142] and was shown to affect Aβ formation and to be upregulated in AD. Finally, 17A is an antisense lncRNA complementary to an intronic region of the GABA receptor gene, and its expression leads to the production of alternative splicing transcripts of this receptor [143]. 17A is also upregulated in AD, and its expression in neuroblastoma cells led to increased Aβ secretion. In contrast to the AS lncRNAs, lincRNAs are apparently more common in the genome, but far less understood. One example of a lincRNA potentially involved in AD is the primate-specific BC200 RNA (BCYRN1), which was found to be expressed in dendritic domains of neurons and is downregulated during aging [144].

More recently, two groups have aimed to systematically assess lncRNAs involved in AD. Zhou et al. used a novel algorithm to reanalyze microarray data from postmortem brains and found ~100 lncRNAs with greater than twofold change in AD [145]. Interestingly, most of their identified lncRNAs were intergenic and brain specific. Furthermore, the altered expression signatures of lncRNAs predicted AD with equal accuracy as altered protein-coding genes, but the number of lncRNAs required for optimal prediction was less than that of protein-coding genes, suggesting a potential use as a biomarker. Magistri et al. used RNA sequencing to identify novel lncRNAs in AD [135] and also observed significant alterations in the lncRNA expression profile in AD brains. This study similarly found the majority of the altered lncRNAs to be intergenic. They further focused on two AS and two intergenic novel lncRNAs that were upregulated in AD and found that all four localized to chromatin, suggesting regulatory roles. Interestingly, three of these lncRNAs were neuronal activity dependent, and the fourth was upregulated in response to Aβ. These studies suggest that we have only begun to scratch the surface of the involvement of lncRNA in AD, and future follow-up studies will expand this exciting arena.

18.10 Other ncRNAs

Relative to miRNA and lncRNA, even less is known regarding the roles of other classes of ncRNA in AD. circRNA is a novel and intriguing form of RNA (reviewed in [147]). circRNAs are particularly abundant in mammalian brains [147], and their expression in tissues such as skeletal muscle was found to be age dependent [148]. Thus, through their predicted interaction with miRNA, circRNAs are very likely to play a role in neurodegenerative disease and particularly in AD. Pioneering work in this field by Lukiw and colleagues [149] found that the circRNA CIRS-7, a well-known sponge for miR-7 [150], is dramatically reduced in brains with AD. This could potentially lead to a significant increase in the activity of miR-7 and down-regulation of its targets, including such relevant genes as alpha-synuclein, mTOR, and the IGF receptor.

Another mysterious and underexplored class of ncRNA is Y RNA, a family of highly expressed 100 nt long structured noncoding RNAs usually found in complex with the protein RO60 [151]. Four canonical Y RNAs, Y1/3/4/5, have been characterized in humans, but numerous slightly divergent copies of these Y RNAs, especially Y1 and Y3, are distributed throughout the human genome [152]. In a fascinating recent work, Scheckel et al. [153] describe a massive shift in the RNA-binding features of ELAV-like proteins, especially with regard to their preference from mRNA to Y RNA in brains of AD patients, even though Y RNA levels remained largely unchanged. ELAVL proteins have been shown to regulate several aspects of RNA metabolism, and they have been implicated in the control over the stabilization and/or translation of specific mRNAs, as well as in the regulation of splicing and polyadenylation of select transcripts. Thus, Y RNA appears to play an important pathological role in the altered mRNA landscape of AD.

An interesting and unique case of ncRNA involvement in AD is the small NRSE dsRNA which is a ~20 bp double-stranded RNA corresponding to the NRSE/RE1 sequence. This motif is usually localized within promoter regions of neuron-specific genes and is recognized by the repressor element 1-silencing transcription factor (REST) protein to restrict neuron-specific gene expression. NRSE dsRNA was shown to interact with and convert the REST complex from repressor to activator [154]. This is of particular interest in light of the recent description of the crucial role played by REST in protecting neurons from oxidative stress and amyloid β-protein toxicity, a function that is lost in the brains of AD patients [155]. Finally, there is increasing evidence for the involvement of the U1 snRNP splicing factor in AD. Aberrant complex formation of U1 snRNP with the DNA-binding protein HMGA1a leads to altered exon skipping in AD [156]. Furthermore, U1 snRNP was found to be highly enriched in neurofibrillary tangles in both sporadic [157] and early-onset [158] AD. The neurodegeneration characteristic impairments in RNA metabolism are hence intimately related with ncRNA alterations.

18.11 Using MicroRNA to Treat AD

At present, Alzheimer's remains an incurable disease. In fact, no new drugs have been approved for treatment of AD since 2003, and the treatment currently available is essentially palliative with only modest and short-term effects [159]. A broad range of agents targeting different aspects of the disease are currently under development and clinical testing, but so far, no drug has proven to be both safe and effective. To a large extent, this therapeutic deficit stems from our ongoing difficulty in understanding the pathogenesis of the disease. Despite continuing debate about the Aβ hypothesis, imbalance between production and clearance of Aβ42 and related Aβ peptides remains one of the more compelling explanations for disease progression [160], and many of the developing clinical approaches target the accumulation of extracellular Aβ. The potential efficacy of such an approach has just recently been demonstrated in a study where immunotherapy against Aβ was safe, and preliminary results indicate simultaneous slowing down of cognitive decline together with Aβ clearance [161]. In addition to enhancing its clearance, another potential method of reducing Aβ levels is slowing its accumulation by targeting the amyloid protein processing enzymes γ-secretase and β-secretase (BACE1). Reducing the activity of these enzymes should dramatically impact the buildup of extracellular

Aβ, and a number of clinical trials with small molecule inhibitors of these proteins are ongoing, although some of these were discontinued due to worrying side effects. Instead of targeting their activity, it is also possible to reduce the actual amount of the enzymes themselves, and in this regard, several miRNAs have been shown to regulate BACE1, suggesting a potential for therapeutic intervention [162].

MicroRNAs present an exciting novel avenue for therapeutic applications in general, and as of 2016, the annual number of US and European published patent applications and issued patents related to miRNAs is close to 500 [163]. The aberrant expression of miRNAs in many human diseases and their involvement in key biological pathways has made them attractive drug targets, and miRNA therapeutics are being devised for various diseases which either downregulate or upregulate the expression or function of miRNAs. One major advantage of miRNA as a therapeutic agent is that these target not only one but multiple genes, allowing potential effects on whole disease networks and pathways. This could be of great significance for a complex and multifactorial disease like AD. Proof of concept for this notion was recently shown for miR-16, which regulated levels of APP, BACE1, and tau in culture and also when administered to mice [164]. In the long run, this could provide the rationale for the use of a small number of miRNA mimics or miRNA blockers to achieve an orchestrated and broad therapeutic effect. It is important to note, however, that this same feature also entails considerable risk of off-target effects. Thus, the study of miRNA therapy in AD and in general is still in early stages. The complex nature of miRNA target recognition means that a much deeper knowledge of miRNA targets is necessary before these can effectively be considered for therapy. Another major hurdle for successful miRNA-based therapy is drug delivery to the brain, which remains a mostly unsolved issue (for a recent review, see [166]).

18.12 MicroRNAs as Biomarkers for AD

AD is notoriously difficult to diagnose early, and timely detection of the disease could potentially offer the opportunities of early intervention, implementation of coordinated care plans, better management of symptoms, patient safety, cost savings, and postponement of institutionalization [166]. The search for reliable and effective biomarkers for early stages of AD is ongoing, and microRNAs have been suggested as good candidates due to their presence in biofluids and their high stability under storage and handling conditions [167]. Over the past several years, a large number of groups have tried to identify circulating miRNAs in serum, plasma, and CSF as biomarkers for AD. Many of these studies have been reviewed previously [168, 169], and here we focus on studies from 2014 and on. Most of these were performed on dozens of subjects and produced a list of miRNAs that are differentially expressed between AD and control groups. The most prevalent profiling method in these studies was qPCR, but the use of RNA-seq is on the rise.

The recent studies analyzing blood samples are summarized in Table 18.1, and those analyzing CSF in Table 18.2. Unfortunately, the consistency between the

Table 18.1 Biomarker assays in blood and serum

Study	Cohort	Technology	Main results
Blood assays			
Keller et al. [176]	Two cohorts: 54 AD, 22 control + 49 AD, 55 control, 20 MCI, 90 multiple sclerosis	Next-generation sequencing	Overlap of 68 dysregulated miRNAs in two cohorts
Sørensen and Nygaard [177]	10 AD, 10 control	qPCR	Up: miR-590-5p, miR-142-5p
			Down: miR-194-5p
Yang et al. [57]	30 AD, 30 control	qPCR	miR-29c was downregulated in AD patients, in a negative correlation with BACE1 protein levels
Bhatnagar et al. [178]	100 AD, 123 control	qPCR	Up: miR-34c
			No change: miR-34b
Serum assays			
Jia and Liu [179]	84 AD, 62 control	qPCR	Up: miR-519
			Down: miR-29, miR-125b, miR-223
			miR-223 expression was also correlated with the severity of AD. miR-223 and the combination of miR-223 and miR-125 gave the greatest sensitivity/specificity for AD
Tan et al. [170]	105 AD, 150 control	qPCR	Up: miR-9
			Down: miR-125b, miR-81c
			No change: miR-29a, miR-29b, miR-10
			miR-125b gave the best results in the operating characteristic (ROC) and was correlated with the Mini-Mental State Examination (MMSE) in AD patients
Liu et al. [180]	38 AD, 30 control	qPCR	Down: miR-135a, miR-200b

Table 18.2 Biomarker studies in CSF

Study	Cohort	Technology	Main results
Liu et al. [180]	5 AD, 5 control	qPCR	Down: miR-135a, miR-200b
Sørensen and Nygaard [177]	10 AD, 10 control	qPCR	Down: miR-29c-3p
			Up: let-7i-5p, miR-15a-5p
Denk et al. [171]	22 AD, 28 control	Open array	Up: 74 miRNAs
			Down: 74 miRNAs
			Discrimination analysis using a combination of miR-100, miR-103, and miR-375 was able to detect AD in CSF with 96.4 and 95.5% accuracy, respectively
Muller et al. [181]	18 AD, 20 control	qPCR	Up: miR-29a
			No change: miR-27a, miR-29b, miR-125b

studies is very small, and most miRNAs were not identified by more than one group. One of the few miRNAs identified in more than one study is miR-181c, which was downregulated in the serum and CSF of AD patients [88, 170], and interestingly, a parallel downregulation has also been shown in AD brains in the context of lipid metabolism, as discussed above [108]. Another notable miRNA is miR-146a which is upregulated in AD brains and is involved in the innate immune response and was found to be upregulated in the CSF of AD patients [117, 171, 172].

18.13 lncRNA Therapeutics and Future Perspectives

Relative to miRNA, the study of lncRNA as therapeutic targets is at an even earlier stage. However, these transcripts could be exciting targets since they promise very high specificity at two important levels. First, expression of lncRNAs appears to be highly cell and tissue specific [151], which could provide a unique opportunity for specific regulation by lncRNA-targeting therapeutics. Second, lncRNA function is highly sequence specific, and this can be advantageous in the design of specific therapies. Many lncRNAs function through epigenetic regulation of chromatin in cis, and targeting these lncRNAs can reverse or activate epigenetic modifications in a very specific manner. The potential for this type of gene-specific epigenetic toggling was demonstrated exquisitely in a recent study on Angelman syndrome. This devastating neurodevelopmental disorder is caused by maternal deficiency of the imprinted gene UBE3A [134, 154] and could potentially be rescued by activating the paternal allele. This allele is constitutively silenced by the lncRNA UBE3A-ATS [155], and in their work, Meng et al. show that targeting the lncRNA with DNA antisense oligonucleotides in mice led to degradation of UBE3A-ATS, partial restoration of UBE3A protein, and amelioration of some cognitive deficits [173]. Another recent example is the targeting of the brain-derived neurotrophic factor antisense (BDNF-AS) transcript which efficiently inhibited BDNF-AS function and thus increased transcription of the sense BDNF mRNA both in vitro and in vivo [174]. As described above, the field of lncRNA function in AD is at a very nascent stage. Still, therapies targeting the BACE1-AS transcript and leading to reduced abundance of Aβ 1–42 can already be envisioned [175], and other interesting opportunities are sure to arise as our knowledge of the roles of lncRNA in AD increases.

Acknowledgments This work was supported by the Legacy Heritage Science Initiative (LHSI) of the Israel Science Foundation Grant No. 378/11 and the Nofar-Teva grant of the Israel Innovation Authority of the Ministry of Economics and Industry (to H. S.). B. P. G. and R. M. were supported by postdoctoral and predoctoral fellowships by the Edmond and Lily Safra Center for Brain Sciences.

References

1. Anand R, Gill KD, Mahdi AA. Therapeutics of Alzheimer's disease: past, present and future. Neuropharmacology. 2014;76(Pt A):27–50.
2. Huang Y, Mucke L. Alzheimer mechanisms and therapeutic strategies. Cell. 2012;148(6):1204–22.
3. Bennett DA, Wilson RS, Arvanitakis Z, Boyle PA, de Toledo-Morrell L, Schneider JA. Selected findings from the religious orders study and rush memory and aging project. J Alzheimers Dis. 2013;33(Suppl 1):S397–403.
4. Bertram L, Lill CM, Tanzi RE. The genetics of Alzheimer disease: back to the future. Neuron. 2010;68(2):270–81.
5. Hoover BR, Reed MN, Su J, Penrod RD, Kotilinek LA, Grant MK, et al. Tau mislocalization to dendritic spines mediates synaptic dysfunction independently of neurodegeneration. Neuron. 2010;68(6):1067–81.
6. Zempel H, Thies E, Mandelkow E, Mandelkow EM. Abeta oligomers cause localized Ca(2+) elevation, missorting of endogenous Tau into dendrites, Tau phosphorylation, and destruction of microtubules and spines. J Neurosci. 2010;30(36):11938–50.
7. Llorens-Martin M, Jurado J, Hernandez F, Avila J. GSK-3beta, a pivotal kinase in Alzheimer disease. Front Mol Neurosci. 2014;7:46.
8. Oddo S, Caccamo A, Cheng D, LaFerla FM. Genetically altering Abeta distribution from the brain to the vasculature ameliorates tau pathology. Brain Pathol. 2009;19(3):421–30.
9. Manassero G, Guglielmotto M, Zamfir R, Borghi R, Colombo L, Salmona M, et al. Beta-amyloid 1-42 monomers, but not oligomers, produce PHF-like conformation of Tau protein. Aging Cell. 2016;15(5):914–23.
10. Abramov AY, Canevari L, Duchen MR. β-amyloid peptides induce mitochondrial dysfunction and oxidative stress in astrocytes and death of neurons through activation of NADPH oxidase. J Neurosci. 2004;24(2):565–75.
11. Butterfield DA, Drake J, Pocernich C, Castegna A. Evidence of oxidative damage in Alzheimer's disease brain: central role for amyloid β-peptide. Trends Mol Med. 2001;7(12):548–54.
12. Mattson MP, Cheng B, Davis D, Bryant K, Lieberburg I, Rydel RE. beta-Amyloid peptides destabilize calcium homeostasis and render human cortical neurons vulnerable to excitotoxicity. J Neurosci. 1992;12(2):376–89.
13. Grimm MO, Grimm HS, Pätzold AJ, Zinser EG, Halonen R, Duering M, et al. Regulation of cholesterol and sphingomyelin metabolism by amyloid-β and presenilin. Nat Cell Biol. 2005;7(11):1118–23.
14. Shankar GM, Li S, Mehta TH, Garcia-Munoz A, Shepardson NE, Smith I, et al. Amyloid-β protein dimers isolated directly from Alzheimer's brains impair synaptic plasticity and memory. Nat Med. 2008;14(8):837–42.

15. Keil U, Bonert A, Marques CA, Scherping I, Weyermann J, Strosznajder JB, et al. Amyloid β-induced changes in nitric oxide production and mitochondrial activity lead to apoptosis. J Biol Chem. 2004;279(48):50310–20.
16. Feng R, Rampon C, Tang Y-P, Shrom D, Jin J, Kyin M, et al. Deficient neurogenesis in forebrain-specific presenilin-1 knockout mice is associated with reduced clearance of hippocampal memory traces. Neuron. 2001;32(5):911–26.
17. Liu S, Liu Y, Hao W, Wolf L, Kiliaan AJ, Penke B, et al. TLR2 is a primary receptor for Alzheimer's amyloid β peptide to trigger neuroinflammatory activation. J Immunol. 2012;188(3):1098–107.
18. El Khoury J, Toft M, Hickman SE, Means TK, Terada K, Geula C, et al. Ccr2 deficiency impairs microglial accumulation and accelerates progression of Alzheimer-like disease. Nat Med. 2007;13:432–8.
19. Hickman SE, Allison EK, El Khoury J. Microglial dysfunction and defective -amyloid clearance pathways in aging Alzheimer's disease mice. J Neurosci. 2008;28:8354–60.
20. Cech TR, Steitz JA. The noncoding RNA revolution—trashing old rules to forge new ones. Cell. 2014;157(1):77–94.
21. Lagos-Quintana M, Rauhut R, Lendeckel W, Tuschl T. Identification of novel genes coding for small expressed RNAs. Science. 2001;294(5543):853–8.
22. Lee RC, Ambros V. An extensive class of small RNAs in *Caenorhabditis elegans*. Science. 2001;294(5543):862–4.
23. Mattick JS, Rinn JL. Discovery and annotation of long noncoding RNAs. Nat Struct Mol Biol. 2015;22(1):5–7.
24. He L, Hannon GJ. MicroRNAs: small RNAs with a big role in gene regulation. Nat Rev Genet. 2004;5(7):522–31.
25. Lu J, Getz G, Miska EA, Alvarez-Saavedra E, Lamb J, Peck D, et al. MicroRNA expression profiles classify human cancers nature. Nature. 2005;435(7043):834–8.
26. Ventura A, Jacks T. MicroRNAs and cancer: short RNAs go a long way. Cell. 2009;136(4):586–91.
27. Iorio MV, Croce CM. MicroRNA dysregulation in cancer: diagnostics, monitoring and therapeutics. A comprehensive review. EMBO Mol Med. 2012;4(3):143–59.
28. Lau P, Bossers K, Janky R, Salta E, Frigerio CS, Barbash S, et al. Alteration of the microRNA network during the progression of Alzheimer's disease. EMBO Mol Med. 2013;5(10):1613–34.
29. Shi X, Sun M, Liu H, Yao Y, Song Y. Long non-coding RNAs: a new frontier in the study of human diseases. Cancer Lett. 2013;339(2):159–66.
30. Batista PJ, Chang HY. Long noncoding RNAs: cellular address codes in development and disease. Cell. 2013;152(6):1298–307.
31. Esteller M. Non-coding RNAs in human disease. Nat Rev Genet. 2011;12(12):861–74.
32. Sun AX, Crabtree GR, Yoo AS. MicroRNAs: regulators of neuronal fate. Curr Opin Cell Biol. 2013;25(2):215–21.
33. Huang Y, Shen XJ, Zou Q, Wang SP, Tang SM, Zhang GZ. Biological functions of microRNAs: a review. J Physiol Biochem. 2011;67(1):129–39.
34. Fiore R, Khudayberdiev S, Saba R, Schratt G. MicroRNA function in the nervous system. Prog Mol Biol Transl Sci. 2011;102:47–100.
35. Goodall EF, Heath PR, Bandmann O, Kirby J, Shaw PJ. Neuronal dark matter: the emerging role of microRNAs in neurodegeneration. Front Cell Neurosci. 2013;7:178.
36. Hardy JA, Higgins GA. Alzheimer's disease: the amyloid cascade hypothesis. Science. 1992;256(5054):184.
37. Karran E, Mercken M, De Strooper B. The amyloid cascade hypothesis for Alzheimer's disease: an appraisal for the development of therapeutics. Nat Rev Drug Discov. 2011;10(9):698–712.
38. Niwa R, Zhou F, Li C, Slack FJ. The expression of the Alzheimer's amyloid precursor protein-like gene is regulated by developmental timing microRNAs and their targets in *Caenorhabditis elegans*. Dev Biol. 2008;315:418–25.
39. Patel N, Hoang D, Miller N, Ansaloni S, Huang Q, Rogers JT, et al. MicroRNAs can regulate human APP levels. Mol Neurodegener. 2008;3:10.

40. Fan X, Liu Y, Jiang J, Ma Z, Wu H, Liu T, et al. miR-20a promotes proliferation and invasion by targeting APP in human ovarian cancer cells. Acta Biochim Biophys Sin. 2010;42:318–24.
41. Hebert SS, Horre K, Nicolai L, Bergmans B, Papadopoulou AS, Delacourte A, et al. MicroRNA regulation of Alzheimer's amyloid precursor protein expression. Neurobiol Dis. 2009;33(3):422–8.
42. Vilardo E, Barbato C, Ciotti M, Cogoni C, Ruberti F. MicroRNA-101 regulates amyloid precursor protein expression in hippocampal neurons. J Biol Chem. 2010;285:18344–51.
43. Long JM, Ray B, Lahiri DK. MicroRNA-339-5p down-regulates protein expression of beta-site amyloid precursor protein-cleaving enzyme 1 (BACE1) in human primary brain cultures and is reduced in brain tissue specimens of Alzheimer disease subjects. J Biol Chem. 2014;289(8):5184–98.
44. Glinsky GV. An SNP-guided microRNA map of fifteen common human disorders identifies a consensus disease phenocode aiming at principal components of the nuclear import pathway. Cell Cycle. 2008;7:2570–83. doi:10.4161/cc.7.16.6524.
45. Delay C, Calon F, Mathews P, Hébert SS, Strooper BD, Mucke L, et al. Alzheimer-specific variants in the 3'UTR of Amyloid precursor protein affect microRNA function. Mol Neurodegener. 2011;6:70.
46. Golde TE, Estus S, Usiak M, Younkin LH, Younkin SG, Angerer LM, et al. Expression of beta amyloid protein precursor mRNAs: recognition of a novel alternatively spliced form and quantitation in Alzheimer's disease using PCR. Neuron. 1990;4:253–67.
47. Neve RL, Rogers J, Higgins GA, Ball MJ, Benowitz LI, Rodriguez W, et al. The Alzheimer amyloid precursor-related transcript lacking the beta/A4 sequence is specifically increased in Alzheimer's disease brain. Neuron. 1990;5:329–38.
48. Jacobsen JS, Blume AJ, Vitek MP. Quantitative measurement of alternatively spliced amyloid precursor protein mRNA expression in Alzheimer's disease and normal brain by S1 nuclease protection analysis. Neurobiol Aging. 1991;12:585–92.
49. Tanzi RE, Wenniger JJ, Hyman BT. Cellular specificity and regional distribution of amyloid beta protein precursor alternative transcripts are unaltered in Alzheimer hippocampal formation. Brain Res Mol Brain Res. 1993;18:246–52.
50. Rockenstein EM, McConlogue L, Tan H, Power M, Masliah E, Mucke L. Levels and alternative splicing of amyloid beta protein precursor (APP) transcripts in brains of APP transgenic mice and humans with Alzheimer's disease. J Biol Chem. 1995;270:28257–67.
51. Donev R, Newall A, Thome J, Sheer D. A role for SC35 and hnRNPA1 in the determination of amyloid precursor protein isoforms. Mol Psychiatry. 2007;12:681–90.
52. Smith P, Al Hashimi A, Girard J, Delay C, Hébert SS. In vivo regulation of amyloid precursor protein neuronal splicing by microRNAs. J Neurochem. 2011;116:240–7.
53. Lukiw WJ. Micro-RNA speciation in fetal, adult and Alzheimer's disease hippocampus. Neuroreport. 2007;18:297–300.
54. Kong Y, Wu J, Zhang D, Wan C, Yuan L. The role of miR-124 in Drosophila Alzheimer's disease model by targeting delta in notch signaling pathway. Curr Mol Med. 2015;15(10):980–9.
55. Schonrock N, Matamales M, Ittner LM, Götz J. MicroRNA networks surrounding APP and amyloid-β metabolism—implications for Alzheimer's disease. Exp Neurol. 2012;235:447–54.
56. Yang L-B, Lindholm K, Yan R, Citron M, Xia W, Yang X-L, et al. Elevated β-secretase expression and enzymatic activity detected in sporadic Alzheimer disease. Nat Med. 2003;9:3–4.
57. Yang G, Song Y, Zhou X, Deng Y, Liu T, Weng G, et al. MicroRNA-29c targets beta-site amyloid precursor protein-cleaving enzyme 1 and has a neuroprotective role in vitro and in vivo. Mol Med Rep. 2015;12(2):3081–8.
58. Lei X, Lei L, Zhang Z, Zhang Z, Cheng Y. Downregulated miR-29c correlates with increased BACE1 expression in sporadic Alzheimer's disease. Int J Clin Exp Pathol. 2015;8(2):1565–74.
59. Zong Y, Wang H, Dong W, Quan X, Zhu H, Xu Y, et al. miR-29c regulates BACE1 protein expression. Brain Res. 2011;1395:108–15.
60. Hébert SS, Horré K, Nicolaï L, Papadopoulou AS, Mandemakers W, Silahtaroglu AN, et al. Loss of microRNA cluster miR-29a/b-1 in sporadic Alzheimer's disease correlates with increased BACE1/beta-secretase expression. Proc Natl Acad Sci U S A. 2008;105:6415–20.

61. Zhu HC, Wang LM, Wang M, Song B, Tan S, Teng JF, et al. MicroRNA-195 downregulates Alzheimer's disease amyloid-beta production by targeting BACE1. Brain Res Bull. 2012;88(6):596–601.
62. Wang W-X, Rajeev BW, Stromberg AJ, Ren N, Tang G, Huang Q, et al. The expression of MicroRNA miR-107 decreases early in Alzheimer's disease and may accelerate disease progression through Regulation of -site amyloid precursor protein-cleaving enzyme 1. J Neurosci. 2008;28:1213–23.
63. Chang F, Zhang LH, Xu WP, Jing P, Zhan PY. microRNA-9 attenuates amyloidbeta-induced synaptotoxicity by targeting calcium/calmodulin-dependent protein kinase kinase 2. Mol Med Rep. 2014;9(5):1917–22.
64. Yan C, Chen J, Li M, Xuan W, Su D, You H, et al. A decrease in hepatic microRNA-9 expression impairs gluconeogenesis by targeting FOXO1 in obese mice. Diabetologia. 2016;59(7):1524–32.
65. Janson J, Laedtke T, Parisi JE, O'Brien P, Petersen RC, Butler PC. Increased risk of type 2 diabetes in Alzheimer disease. Diabetes. 2004;53(2):474–81.
66. Cheng C, Li W, Zhang Z, Yoshimura S, Hao Q, Zhang C, et al. MicroRNA-144 is regulated by activator protein-1 (AP-1) and decreases expression of Alzheimer disease-related a disintegrin and metalloprotease 10 (ADAM10). J Biol Chem. 2013;288(19):13748–61.
67. Serrano-Pozo A, Frosch MP, Masliah E, Hyman BT. Neuropathological alterations in Alzheimer disease. Cold Spring Harb Perspect Med. 2011;1(1):a006189.
68. Ballatore C, Lee VM-Y, Trojanowski JQ. Tau-mediated neurodegeneration in Alzheimer's disease and related disorders. Nat Rev Neurosci. 2007;8:663–72.
69. Schonrock N, Gotz J. Decoding the non-coding RNAs in Alzheimer's disease. Cell Mol Life Sci. 2012;69(21):3543–59.
70. Poorkaj P, Kas A, D'Souza I, Zhou Y, Pham Q, Stone M, et al. A genomic sequence analysis of the mouse and human microtubule-associated protein tau. Mamm Genome. 2001;12(9):700–12.
71. Dickson JR, Kruse C, Montagna DR, Finsen B, Wolfe MS. Alternative polyadenylation and miR-34 family members regulate tau expression. J Neurochem. 2013;127(6):739–49.
72. Smith PY, Hernandez-Rapp J, Jolivette F, Lecours C, Bisht K, Goupil C, et al. miR-132/212 deficiency impairs tau metabolism and promotes pathological aggregation in vivo. Hum Mol Genet. 2015;24(23):6721–35.
73. Hansen KF, Sakamoto K, Aten S, Snider KH, Loeser J, Hesse AM, et al. Targeted deletion of miR-132/−212 impairs memory and alters the hippocampal transcriptome. Learn Mem. 2016;23(2):61–71.
74. Zhu Q-B, Unmehopa U, Bossers K, Hu Y-T, Verwer R, Balesar R, et al. MicroRNA-132 and early growth response-1 in nucleus basalis of Meynert during the course of Alzheimer's disease. Brain. 2016;139:908–21.
75. Santa-Maria I, Alaniz ME, Renwick N, Cela C, Fulga TA, Van Vactor D, et al. Dysregulation of microRNA-219 promotes neurodegeneration through post-transcriptional regulation of tau. J Clin Invest. 2015;125(2):681–6.
76. Carrettiero DC, Hernandez I, Neveu P, Papagiannakopoulos T, Kosik KS. The cochaperone BAG2 sweeps paired helical filament- insoluble tau from the microtubule. J Neurosci. 2009;29:2151–61.
77. Buée L, Bussière T, Buée-Scherrer V, Delacourte A, Hof PR. Tau protein isoforms, phosphorylation and role in neurodegenerative disorders. Brain Res Rev. 2000;33:95–130.
78. Hong M, Zhukareva V, Vogelsberg-Ragaglia V, Wszolek Z, Reed L, Miller BI, et al. Mutation-specific functional impairments in distinct tau isoforms of hereditary FTDP-17. Science. 1998;282:1914–7.
79. Banzhaf-Strathmann J, Benito E, May S, Arzberger T, Tahirovic S, Kretzschmar H, et al. MicroRNA-125b induces tau hyperphosphorylation and cognitive deficits in Alzheimer's disease. EMBO J. 2014;33(15):1667–80.
80. D'Souza I, Schellenberg GD. Regulation of tau isoform expression and dementia. Biochim Biophys Acta. 2005;1739:104–15.

81. Sergeant N, Bretteville A, Hamdane M, Caillet-Boudin M-L, Grognet P, Bombois S, et al. Biochemistry of Tau in Alzheimer's disease and related neurological disorders. Expert Rev Proteomics. 2008;5:207–24.
82. Hebert SS, Sergeant N, Buee L. MicroRNAs and the regulation of tau metabolism. Int J Alzheimers Dis. 2012;2012:406561.
83. Hebert SS, Papadopoulou AS, Smith P, Galas MC, Planel E, Silahtaroglu AN, et al. Genetic ablation of Dicer in adult forebrain neurons results in abnormal tau hyperphosphorylation and neurodegeneration. Hum Mol Genet. 2010;19(20):3959–69.
84. Cai Z, Zhao Y, Zhao B. Roles of glycogen synthase kinase 3 in Alzheimer's disease. Curr Alzheimer Res. 2012;9:864–79.
85. Mohamed JS, Lopez MA, Boriek AM. Mechanical stretch up-regulates MicroRNA-26a and induces human airway smooth muscle hypertrophy by suppressing glycogen synthase kinase-3β. J Biol Chem. 2010;285:29336–47.
86. Caputo V, Sinibaldi L, Fiorentino A, Parisi C, Catalanotto C, Pasini A, et al. Brain derived neurotrophic factor (BDNF) expression is regulated by MicroRNAs miR-26a and miR-26b allele-specific binding. PLoS One. 2011;6:e28656.
87. Absalon S, Kochanek DM, Raghavan V, Krichevsky AM. MiR-26b, upregulated in Alzheimer's disease, activates cell cycle entry, tau-phosphorylation, and apoptosis in postmitotic neurons. J Neurosci. 2013;33(37):14645–59.
88. Cogswell JP, Ward J, Taylor IA, Waters M, Shi Y, Cannon B, et al. Identification of miRNA changes in Alzheimer's disease brain and CSF yields putative biomarkers and insights into disease pathways. J Alzheimers Dis. 2008;14(1):27–41.
89. Zhao ZB, Wu L, Xiong R, Wang LL, Zhang B, Wang C, et al. MicroRNA-922 promotes tau phosphorylation by downregulating ubiquitin carboxy-terminal hydrolase L1 (UCHL1) expression in the pathogenesis of Alzheimer's disease. Neuroscience. 2014;275:232–7.
90. Choi J, Levey AI, Weintraub ST, Rees HD, Gearing M, Chin L-S, et al. Oxidative modifications and down-regulation of ubiquitin carboxyl-terminal hydrolase L1 associated with idiopathic Parkinson's and Alzheimer's diseases. J Biol Chem. 2004;279:13256–64.
91. Wang X, Tan L, Lu Y, Peng J, Zhu Y, Zhang Y, et al. MicroRNA-138 promotes tau phosphorylation by targeting retinoic acid receptor alpha. FEBS Lett. 2015;589(6):726–9.
92. Corder EH, Saunders AM, Strittmatter WJ, Schmechel DE, Gaskell PC, Small GW, et al. Gene dose of apolipoprotein E type 4 allele and the risk of Alzheimer's disease in late onset families. Science. 1993;261:921–3.
93. Han X. Multi-dimensional mass spectrometry-based shotgun lipidomics and the altered lipids at the mild cognitive impairment stage of Alzheimer's disease. Biochim Biophys Acta. 1801;2010:774–83.
94. Frisardi V, Panza F, Seripa D, Farooqui T, Farooqui AA. Glycerophospholipids and glycerophospholipid-derived lipid mediators: a complex meshwork in Alzheimerâ€™s disease pathology. Prog Lipid Res. 2011;50:313–30.
95. Kosicek M, Hecimovic S. Phospholipids and Alzheimer's disease: alterations, mechanisms and potential biomarkers. Int J Mol Sci. 2013;14:1310–22.
96. Goedeke L, Fernandez-Hernando C. MicroRNAs: a connection between cholesterol metabolism and neurodegeneration. Neurobiol Dis. 2014;72(Pt A):48–53.
97. Koldamova R, Staufenbiel M, Lefterov I. Lack of ABCA1 considerably decreases brain ApoE level and increases amyloid deposition in APP23 mice. J Biol Chem. 2005;280:43224–35.
98. Fitz NF, Cronican AA, Saleem M, Fauq AH, Chapman R, Lefterov I, et al. Abca1 deficiency affects Alzheimer's disease-like phenotype in human ApoE4 but not in ApoE3-targeted replacement mice. J Neurosci. 2012;32:13125–36.
99. Wahrle SE, Jiang H, Parsadanian M, Hartman RE, Bales KR, Paul SM, et al. Deletion of Abca1 increases Abeta deposition in the PDAPP transgenic mouse model of Alzheimer disease. J Biol Chem. 2005;280(52):43236–42.
100. Horie T, Ono K, Horiguchi M, Nishi H, Nakamura T, Nagao K, et al. MicroRNA-33 encoded by an intron of sterol regulatory element-binding protein 2 (Srebp2) regulates HDL in vivo. Proc Natl Acad Sci. 2010;107:17321–6.

101. Kim J, Yoon H, Horie T, Burchett JM, Restivo JL, Rotllan N, et al. microRNA-33 regulates ApoE lipidation and amyloid-β metabolism in the brain. J Neurosci. 2015;35(44):14717–26.
102. Jan A, Karasinska JM, Kang MH, de Haan W, Ruddle P, Kaur A, et al. Direct intracerebral delivery of a miR-33 antisense oligonucleotide into mouse brain increases brain ABCA1 expression. [Corrected]. Neurosci Lett. 2015;598:66–72.
103. Rayner KJ, Suárez Y, Dávalos A, Parathath S, Fitzgerald ML, Tamehiro N, et al. MiR-33 contributes to the regulation of cholesterol homeostasis. Science. 2010;328:1570–3.
104. Jaouen F, Gascon E. Understanding the role of miR-33 in brain lipid metabolism: implications for Alzheimer's disease. J Neurosci. 2016;36:2558–60.
105. Kim J, Yoon H, Ramirez CM, Lee SM, Hoe HS, Fernandez-Hernando C, et al. MiR-106b impairs cholesterol efflux and increases Abeta levels by repressing ABCA1 expression. Exp Neurol. 2012;235(2):476–83.
106. Ramirez CM, Dávalos A, Goedeke L, Salerno AG, Warrier N, Cirera-Salinas D, et al. MicroRNA-758 regulates cholesterol efflux through posttranscriptional repression of ATP-binding cassette transporter A1. Arterioscler Thromb Vasc Biol. 2011;31:2707–14.
107. Geekiyanage H, Upadhye A, Chan C. Inhibition of serine palmitoyltransferase reduces Abeta and tau hyperphosphorylation in a murine model: a safe therapeutic strategy for Alzheimer's disease. Neurobiol Aging. 2013;34(8):2037–51.
108. Geekiyanage H, Chan C. MicroRNA-137/181c regulates serine palmitoyltransferase and in turn amyloid β, novel targets in sporadic Alzheimer's disease. J Neurosci. 2011;31:14820–30.
109. Zhang J, Hu M, Teng Z, Tang Y-P, Chen C. Synaptic and cognitive improvements by inhibition of 2-AG metabolism are through upregulation of microRNA-188-3p in a mouse model of Alzheimer's disease. J Neurosci. 2014;34(45):14919–33.
110. Heneka MT, Carson MJ, Khoury JE, Landreth GE, Brosseron F, Feinstein DL, et al. Neuroinflammation in Alzheimer's disease. Lancet Neurol. 2015;14:388–405.
111. Altavilla D, Guarini S, Bitto A, Mioni C, Giuliani D, Bigiani A, et al. Activation of the cholinergic anti-inflammatory pathway reduces NF-kappab activation, blunts TNF-alpha production, and protects against splanchic artery occlusion shock. Shock. 2006;25(5):500–6.
112. Coyle JT, Price DL, DeLong MR. Alzheimer's disease: a disorder of cortical cholinergic innervation. Science. 1983;219(4589):1184–90.
113. Zhao Y, Bhattacharjee S, Jones BM, Dua P, Alexandrov PN, Hill JM, et al. Regulation of TREM2 expression by an NF-small ka, CyrillicB-sensitive miRNA-34a. Neuroreport. 2013;24(6):318–23.
114. Bhattacharjee S, Zhao Y, Lukiw WJ. Deficits in the miRNA-34a-regulated endogenous TREM2 phagocytosis sensor-receptor in Alzheimer's disease (AD); an update. Front Aging Neurosci. 2014;6:116.
115. Bhattacharjee S, Zhao Y, Dua P, Rogaev EI, Lukiw WJ. microRNA-34a-mediated down-regulation of the microglial-enriched triggering receptor and phagocytosis-sensor TREM2 in age-related macular degeneration. PLoS One. 2016;11(3):e0150211.
116. Guerreiro R, Wojtas A, Bras J, Carrasquillo M, Rogaeva E, Majounie E, et al. *TREM2* variants in Alzheimer's disease. N Engl J Med. 2013;368:117–27.
117. Alexandrov PN, Dua P, Lukiw WJ. Up-REGULATION of miRNA-146a in progressive age-related inflammatory neurodegenerative disorders of the human CNS. Front Neurol. 2014;5:181.
118. Guedes JR, Custodia CM, Silva RJ, de Almeida LP, Pedroso de Lima MC, Cardoso AL. Early miR-155 upregulation contributes to neuroinflammation in Alzheimer's disease triple transgenic mouse model. Hum Mol Genet. 2014;23(23):6286–301.
119. Song J, Lee JE. miR-155 is involved in Alzheimer's disease by regulating T lymphocyte function. Front Aging Neurosci. 2015;7:61.
120. Hutchison ER, Kawamoto EM, Taub DD, Lal A, Abdelmohsen K, Zhang Y, et al. Evidence for miR-181 involvement in neuroinflammatory responses of astrocytes. Glia. 2013;61(7):1018–28.
121. Rodriguez-Ortiz CJ, Baglietto-Vargas D, Martinez-Coria H, LaFerla FM, Kitazawa M. Upregulation of miR-181 decreases c-Fos and SIRT-1 in the hippocampus of 3xTg-AD mice. J Alzheimers Dis. 2014;42(4):1229–38.

122. Rivetti di Val Cervo P, Lena AM, Nicoloso M, Rossi S, Mancini M, Zhou H, et al. p63-microRNA feedback in keratinocyte senescence. Proc Natl Acad Sci. 2012;109:1133–8.
123. Belkaya S, van Oers NSC, Ageev A, Sidorin V, Rogachev M, Timofeev I, et al. Transgenic expression of MicroRNA-181d augments the stress-sensitivity of CD4+CD8+thymocytes. PLoS One. 2014;9:e85274.
124. Muller M, Kuiperij HB, Claassen JA, Kusters B, Verbeek MM. MicroRNAs in Alzheimer's disease: differential expression in hippocampus and cell-free cerebrospinal fluid. Neurobiol Aging. 2014;35(1):152–8.
125. Pritchard CC, Cheng HH, Tewari M. MicroRNA profiling: approaches and considerations. Nat Rev Genet. 2012;13(5):358–69.
126. Wang W-X, Huang Q, Hu Y, Stromberg AJ, Nelson PT. Patterns of microRNA expression in normal and early Alzheimer's disease human temporal cortex: white matter versus gray matter. Acta Neuropathol. 2011;121:193–205.
127. Nunez-Iglesias J, Liu CC, Morgan TE, Finch CE, Zhou XJ. Joint genome-wide profiling of miRNA and mRNA expression in Alzheimer's disease cortex reveals altered miRNA regulation. PLoS One. 2010;5(2):e8898.
128. Bekris LM, Lutz F, Montine TJ, Yu CE, Tsuang D, Peskind ER, et al. MicroRNA in Alzheimer's disease: an exploratory study in brain, cerebrospinal fluid and plasma. Biomarkers. 2013;18(5):455–66.
129. Weinberg RB, Mufson EJ, Counts SE. Evidence for a neuroprotective microRNA pathway in amnestic mild cognitive impairment. Front Neurosci. 2015;9:430.
130. Wyman SK, Knouf EC, Parkin RK, Fritz BR, Lin DW, Dennis LM, et al. Post-transcriptional generation of miRNA variants by multiple nucleotidyl transferases contributes to miRNA transcriptome complexity. Genome Res. 2011;21(9):1450–61.
131. Gellert P, Ponomareva Y, Braun T, Uchida S. Noncoder: a web interface for exon array-based detection of long non-coding RNAs. Nucleic Acids Res. 2012;41:e20.
132. Cabili MN, Trapnell C, Goff L, Koziol M, Tazon-Vega B, Regev A, et al. Integrative annotation of human large intergenic noncoding RNAs reveals global properties and specific subclasses. Genes Dev. 2011;25(18):1915–27.
133. Taft RJ, Mattick JS. Increasing biological complexity is positively correlated with the relative genome-wide expansion of non-protein-coding DNA sequences. Genome Biol. 2004;5(1):P1.
134. Kishino T, Lalande M, Wagstaff J. UBE3A/E6-AP mutations cause Angelman syndrome. Nat Genet. 1997;15(1):70–3.
135. Magistri M, Velmeshev D, Makhmutova M, Faghihi MA. Transcriptomics profiling of Alzheimer's disease reveal Neurovascular defects, altered amyloid-beta homeostasis, and deregulated expression of long noncoding RNAs. J Alzheimers Dis. 2015;48(3):647–65.
136. Qureshi IA, Mattick JS, Mehler MF. Long non-coding RNAs in nervous system function and disease. Brain Res. 2010;1338:20–35.
137. Gutschner T, Diederichs S. The hallmarks of cancer: a long non-coding RNA point of view. RNA Biol. 2012;9(6):703–19.
138. Briggs JA, Wolvetang EJ, Mattick JS, Rinn JL, Barry G. Mechanisms of long non-coding RNAs in mammalian nervous system development, Plasticity, disease, and evolution. Neuron. 2015;88(5):861–77.
139. Yoon J-H, Abdelmohsen K, Gorospe M. Posttranscriptional gene regulation by long noncoding RNA. J Mol Biol. 2013;425(19):3723–30.
140. Faghihi MA, Modarresi F, Khalil AM, Wood DE, Sahagan BG, Morgan TE, et al. Expression of a noncoding RNA is elevated in Alzheimer's disease and drives rapid feed-forward regulation of [beta]-secretase. Nat Med. 2008;14(7):723–30.
141. Parenti R, Paratore S, Torrisi A, Cavallaro S. A natural antisense transcript against Rad18, specifically expressed in neurons and upregulated during β-amyloid-induced apoptosis. Eur J Neurosci. 2007;26(9):2444–57.
142. Ciarlo E, Massone S, Penna I, Nizzari M, Gigoni A, Dieci G, et al. An intronic ncRNA-dependent regulation of SORL1 expression affecting Abeta formation is upregulated in post-mortem Alzheimer's disease brain samples. Dis Model Mech. 2013;6(2):424–33.

143. Massone S, Vassallo I, Fiorino G, Castelnuovo M, Barbieri F, Borghi R, et al. 17A, a novel non-coding RNA, regulates GABA B alternative splicing and signaling in response to inflammatory stimuli and in Alzheimer disease. Neurobiol Dis. 2011;41(2):308–17.
144. Mus E, Hof PR, Tiedge H. Dendritic BC200 RNA in aging and in Alzheimer's disease. Proc Natl Acad Sci U S A. 2007;104(25):10679–84.
145. Zhou X, Xu J. Identification of Alzheimer's disease-associated long noncoding RNAs. Neurobiol Aging. 2015;36(11):2925–31.
146. Chen L-L. The biogenesis and emerging roles of circular RNAs. Nat Rev Mol Cell Biol. 2016;17(4):205–11.
147. Rybak-Wolf A, Stottmeister C, Glažar P, Jens M, Pino N, Giusti S, et al. Circular RNAs in the mammalian brain are highly abundant, conserved, and dynamically expressed. Mol Cell. 2015;58(5):870–85.
148. Abdelmohsen K, Panda AC, De S, Grammatikakis I, Kim J, Ding J, et al. Circular RNAs in monkey muscle: age-dependent changes. Aging. 2015;7(11):903.
149. Lukiw WJ. Circular RNA (circRNA) in Alzheimer's disease (AD). Front Genet. 2013;4:307.
150. Hansen TB, Jensen TI, Clausen BH, Bramsen JB, Finsen B, Damgaard CK, et al. Natural RNA circles function as efficient microRNA sponges. Nature. 2013;495(7441):384–8.
151. Djebali S, Davis CA, Merkel A, Dobin A, Lassmann T, Mortazavi A, et al. Landscape of transcription in human cells. Nature. 2012;489(7414):101–8.
152. Neuner SM, Garfinkel BP, Wilmott LA, Ignatowska-Jankowska BM, Citri A, Orly J, et al. Systems genetics identifies Hp1bp3 as a novel modulator of cognitive aging. Neurobiol Aging. 2016;46:58–67.
153. Champagne F, Diorio J, Sharma S, Meaney MJ. Naturally occurring variations in maternal behavior in the rat are associated with differences in estrogen-inducible central oxytocin receptors. Proc Natl Acad Sci. 2001;98(22):12736–41.
154. Albrecht U, Sutcliffe JS, Cattanach BM, Beechey CV, Armstrong D, Eichele G, et al. Imprinted expression of the murine Angelman syndrome gene, Ube3a, in hippocampal and Purkinje neurons. Nat Genet. 1997;17(1):75–8.
155. Rougeulle C, Cardoso C, Fontés M, Colleaux L, Lalande M. An imprinted antisense RNA overlaps UBE3A and a second maternally expressed transcript. Nat Genet. 1998;19(1):15.
156. Ohe K, Mayeda A. HMGA1a trapping of U1 snRNP at an authentic 5′ splice site induces aberrant exon skipping in sporadic Alzheimer's disease. Mol Cell Biol. 2010;30(9):2220–8.
157. Hales CM, Dammer EB, Diner I, Yi H, Seyfried NT, Gearing M, et al. Aggregates of small nuclear ribonucleic acids (snRNAs) in Alzheimer's disease. Brain Pathol. 2014;24(4):344–51.
158. Hales CM, Seyfried NT, Dammer EB, Duong D, Yi H, Gearing M, et al. U1 small nuclear ribonucleoproteins (snRNPs) aggregate in Alzheimer's disease due to autosomal dominant genetic mutations and trisomy 21. Mol Neurodegener. 2014;9:15.
159. Godyń J, Jończyk J, Panek D, Malawska B. Therapeutic strategies for Alzheimer's disease in clinical trials. Pharmacol Rep. 2016;68(1):127–38.
160. Selkoe DJ, Hardy J. The amyloid hypothesis of Alzheimer's disease at 25 years. EMBO Mol Med. 2016;8(6):595–608.
161. Sevigny J, Chiao P, Bussiere T, Weinreb PH, Williams L, Maier M, et al. The antibody aducanumab reduces Abeta plaques in Alzheimer's disease. Nature. 2016;537(7618):50–6.
162. Pereira PA, Tomas JF, Queiroz JA, Figueiras AR, Sousa F. Recombinant pre-miR-29b for Alzheimer's disease therapeutics. Sci Rep. 2016;6:19946.
163. Christopher AF, Kaur RP, Kaur G, Kaur A, Gupta V, Bansal P. MicroRNA therapeutics: discovering novel targets and developing specific therapy. Perspect Clin Res. 2016;7(2):68.
164. Parsi S, Smith PY, Goupil C, Dorval V, Hebert SS. Preclinical evaluation of miR-15/107 family members as multifactorial drug targets for Alzheimer's disease. Mol Therapy Nucl Acids. 2015;4:e256.
165. Wang H, Jiang Y, Peng H, Chen Y, Zhu P, Huang Y. Recent progress in microRNA delivery for cancer therapy by non-viral synthetic vectors. Adv Drug Deliv Rev. 2015;81:142–60.
166. Dubois B, Padovani A, Scheltens P, Rossi A, Dell'Agnello G. Timely diagnosis for Alzheimer's disease: a literature review on benefits and challenges. J Alzheimers Dis. 2015;49(3):617–31.

167. Schwarzenbach H, Nishida N, Calin GA, Pantel K. Clinical relevance of circulating cell-free microRNAs in cancer. Nat Rev Clin Oncol. 2014;11:145–56.
168. Lau P, Sala Frigerio C, De Strooper B. Variance in the identification of microRNAs deregulated in Alzheimer's disease and possible role of lincRNAs in the pathology: the need of larger datasets. Ageing Res Rev. 2014;17:43–53.
169. Bekris LM, Leverenz JB. The biomarker and therapeutic potential of miRNA in Alzheimer's disease. Neurodegener Dis Manag. 2015;5(1):61–74.
170. Tan L, Yu JT, Liu QY, Tan MS, Zhang W, Hu N, et al. Circulating miR-125b as a biomarker of Alzheimer's disease. J Neurol Sci. 2014;336(1–2):52–6.
171. Denk J, Boelmans K, Siegismund C, Lassner D, Arlt S, Jahn H. MicroRNA profiling of CSF reveals potential biomarkers to detect Alzheimer's disease. PLoS One. 2015;10(5):e0126423.
172. Lukiw WJ, Alexandrov PN, Zhao Y, Hill JM, Bhattacharjee S. Spreading of Alzheimer's disease inflammatory signaling through soluble micro-RNA. Neuroreport. 2012;23(10):621–6.
173. Meng L, Ward AJ, Chun S, Bennett CF, Beaudet AL, Rigo F. Towards a therapy for Angelman syndrome by targeting a long non-coding RNA. Nature. 2015;518(7539):409–12.
174. Modarresi F, Faghihi MA, Lopez-Toledano MA, Fatemi RP, Magistri M, Brothers SP, et al. Inhibition of natural antisense transcripts in vivo results in gene-specific transcriptional upregulation. Nat Biotechnol. 2012;30(5):453–9.
175. Modarresi F, Faghihi MA, Patel NS, Sahagan BG, Wahlestedt C, Lopez-Toledano MA. Knockdown of BACE1-AS nonprotein-coding transcript modulates beta-amyloid-related hippocampal neurogenesis. Int J Alzheimers Dis. 2011;2011:929042.
176. Keller A, Backes C, Haas J, Leidinger P, Maetzler W, Deuschle C, et al. Validating Alzheimer's disease micro RNAs using next-generation sequencing. Alzheimers Dement. 2016;12(5):565–76.
177. Sørensen SS, Nygaard A-B, Christensen T. miRNA expression profiles in cerebrospinal fluid and blood of patients with Alzheimer's disease and other types of dementia—an exploratory study. Transl Neurodegener. 2016;5:6.
178. Bhatnagar S, Chertkow H, Schipper HM, Yuan Z, Shetty V, Jenkins S, et al. Increased microRNA-34c abundance in Alzheimer's disease circulating blood plasma. Front Mol Neurosci. 2014;7:2.
179. Jia LH, Liu YN. Downregulated serum miR-223 servers as biomarker in Alzheimer's disease. Cell Biochem Funct. 2016;34(4):233–7.
180. Liu CG, Wang JL, Li L, Xue LX, Zhang YQ, Wang PC. MicroRNA-135a and -200b, potential biomarkers for Alzheimer's disease, regulate beta secretase and amyloid precursor protein. Brain Res. 2014;1583:55–64.
181. Muller M, Jakel L, Bruinsma IB, Claassen JA, Kuiperij HB, Verbeek MM. MicroRNA-29a is a candidate biomarker for Alzheimer's disease in cell-free cerebrospinal fluid. Mol Neurobiol. 2015;53(5):2894–9.

Epigenetics in Parkinson's Disease

19

Maria Angeliki S. Pavlou and Tiago Fleming Outeiro

Abstract

Parkinson's disease (PD) is a highly complex neurodegenerative disorder with a multifactorial origin. Although several cellular mechanisms and genes have been implicated in the onset and progression of the disease, the precise molecular underpinnings of the disease remain unclear. In this context, epigenetic modulation of gene expression by environmental factors is emerging as an important mechanism in PD and in other neurodegenerative disorders. Thus, epigenetic mechanisms, such as DNA methylation, histone modifications and altered microRNA expression, have been under intense investigation due to their possible involvement in PD. Epigenetic modulation is responsible for inducing differential gene expression, a phenomenon which is essential throughout life in order to regulate multiple cellular responses such as development, cellular fate commitment and adaptation to the environment. Disturbances of a balanced gene expression can, therefore, have detrimental effects. Environmental factors can

M.A.S. Pavlou, B.Sc., M.Sc., Ph.D.
Department of NeuroDegeneration and Restorative Research, Center for Nanoscale Microscopy and Molecular Physiology of the Brain, University Medical Center Göttingen, Göttingen, Lower Saxony, Germany

Department of NeuroDegeneration and Restorative Research, University Medical Center Göttingen, Waldweg 33, 37073 Göttingen, Lower Saxony, Germany
e-mail: mpavlou@gwdg.de

T.F. Outeiro, B.Sc., M.Sc., Ph.D. (✉)
Department of NeuroDegeneration and Restorative Research, Center for Nanoscale Microscopy and Molecular Physiology of the Brain, University Medical Center Göttingen, Göttingen, Lower Saxony, Germany

Department of NeuroDegeneration and Restorative Research, University Medical Center Göttingen, Waldweg 33, 37073 Göttingen, Lower Saxony, Germany

Max Planck Institute for Experimental Medicine, Goettingen, Germany
e-mail: touteir@gwdg.de

R. Delgado-Morales (ed.), *Neuroepigenomics in Aging and Disease*, Advances in Experimental Medicine and Biology, DOI 10.1007/978-3-319-53889-1_19

challenge the establishment and maintenance of epigenetic modifications and could thereby fill the gap in our further understanding of origin and/or progression of neurodegenerative diseases. In this chapter, we focus on the role of epigenetics in PD.

Keywords
Parkinson's disease • Epigenetics • DNA methylation • Histone modifications • miRNA • Alpha-synuclein

Abbreviations

3′ UTR	3′ untranslated region
5-hmC	5-hydroxymethylcytosine
5-mC	5-methylcytosine is formed
Ago	Argonaute
aSyn	Alpha-synuclein protein
DLB	dementia with Lewy bodies
DNMT1	DNA methyltransferase 1
DNMT3a	DNA methyltransferase 3a
DNMT3b	DNA methyltransferase 3b
DNMTs	DNA methyltransferases
FGF20	Fibroblast growth factor 20
HAT	Histone acetyltransferase
HDACis	Histone deacetylase inhibitors
HDACs	Histone deacetylases
Hsp70	Heat sock protein 70
LB	Lewy bodies
LDID	Levodopa-induced dyskinesia
L-DOPA	Levodopa
LN	Lewy neurites
MBDs	methyl-CpG-binding domain proteins
mDNMT	Mitochondrial DNMT
miRNAs	microRNAs
MPP^+	1-methyl-4-phenylpyridinium
MPTP	1-methyl-4-phenyl-1,2,3,6-tetrahydropyridine
NACP-Rep1	polymorphic microsatellite repeat region
Nurr1	Nuclear receptor-related 1 protein
PBMCs	Peripheral blood mononuclear cells
PD	Parkinson's disease
PKC δ	Protein kinase C δ
SAH	S-adenosylhomocysteine
SAM	S-adenosyl methionine

SN	*substantia nigra pars compacta*
SNCA	alpha-synuclein gene
SNPs	Single nucleotide polymorphisms
TSA	Trichostatin
VPA	Valproic acid
VSP35	vacuolar protein sorting 35

19.1 Introduction

The inability of the brain to replenish certain cell types, upon their death, is associated with the development of specific conditions, known as neurodegenerative disorders. In Parkinson's disease (PD), one of those conditions that usually manifests after 60 years of age, the demise of dopaminergic neurons from the substantia nigra, explains the typical motor symptoms of the disease. Due to the increase in life expectancy, the number of individuals affected by PD has also drastically increased, resulting in extensive socioeconomic challenges. In the USA alone, it has been estimated that the annual costs of PD rise up to $23 billion [1]. Several therapeutic options are presently available to treat some of the symptoms associated with PD. However, there is currently no cure or preventive strategy. The majority of PD cases is sporadic, with no known cause, and is thought to occur due to the interplay between susceptibility genes and the environment, in ways that are poorly understood.

The term epigenetics refers to alterations in gene expression, usually reversible, which can be inherited but are not engraved in the DNA sequence. These modifications can be implemented via methylation of the DNA, histone modifications or microRNAs (miRNAs). Chemical pollutants, nutrition, temperature changes and other environmental stresses can influence gene expression via changes in epigenetic modifications. Although no solid relationship has been yet identified, epigenetic deregulation is thought to play an important and poorly understood role in the aetiopathogenesis of various neurodegenerative disorders, including PD.

19.2 Parkinson's Disease

19.2.1 Pathology and Clinical Features

PD, named after Dr. James Parkinson who first documented it in 1817, constitutes the second most prevalent neurodegenerative disorder today. With a prevalence of 1–2% over the age of 65 [2, 3] and of 4–5% over the age of 85 [4], it is estimated that this progressive disorder affects approximately 6.3 million individuals worldwide, with the number expected to increase to 8.3 million by 2030 [5].

The typical neuropathological hallmarks of PD are the loss of dopaminergic neurons from the *substantia nigra pars compacta* (SN) and the accumulation of intracellular protein inclusions termed Lewy bodies (LBs), mainly composed of alpha-synuclein (aSyn) [6, 7]. Dopaminergic neurons extend their fibres from the SN towards the striatum, where they release dopamine, the neurotransmitter responsible for the learning and execution of motor functions [8, 9]. Due to decreased levels of dopamine, PD patients present characteristic motor dysfunctions such as bradykinesia, muscle rigidity, resting tremor and postural instability [10, 11]. Nonmotor symptoms, including anxiety, depression, dementia, sleep disturbances, constipation, hyposmia and anosmia, are also apparent and limit the quality of life of patients even further [8, 12]. Motor features remain the principal criteria for the clinical diagnosis of PD, although some nonmotor impairments are now valued as predictive markers for the disorder since they tend to appear prior to the onset of motor symptoms [13, 14]. Indeed, according to the Braak staging hypothesis, Lewy body pathology is quite dispersed not only throughout the brain but also in other tissues, such as the gut. According to this hypothesis, the progression of PD is classified into six stages. Stages 1–2 are linked with the presymptomatic phase where Lewy bodies appear in the enteric and peripheral autonomic nervous system and also spread from the olfactory bulb and vagus nerve to the lower brainstem. The symptomatic period starts on stage 3, when the midbrain, including the SN, starts to be affected. Finally, pathological changes involve the mesocortex in stage 4 and the neocortex in stages 5 and 6 [15]. Although this staging system has been confirmed by other groups and applies for the majority of the cases, deviations from this model can be observed, raising questions about the overall validity of the hypothesis [16, 17].

19.2.2 Genetic Forms of PD

Familial forms of PD account for only about 10–15% of all the cases [18]. However, it is possible that additional cases might be associated with yet unidentified genes, as additional genetic studies are conducted [19]. Thus, the list of genes implicated in the onset of PD (PARK genes) is expanding. The PARK gene family currently comprises 20 genes (Table 19.1) which are responsible for autosomal recessive, dominant or X-linked modes of inheritance. Moreover, PD-related genes can present point mutations, duplications or triplications and account for both early- or late-onset forms PD [20, 21]. Interestingly, over 500 DNA variants have been described in only five of the PD-associated genes [22].

A mutation in gene encoding for alpha-synuclein (*SNCA*) was the first to be associated with familial PD. Presently, six point mutations leading to amino acid substitutions have been linked with autosomal dominant forms of PD. In addition, duplications and triplications of the *SNCA* locus have also been associated with autosomal dominant forms of PD [23–25]. Although *SNCA* is an extensively studied gene, the precise function of alpha-synuclein (aSyn) and how it causes disease remain elusive. aSyn is typically described as a presynaptic protein participating in the regulation of the synaptic vesicle pool and in neurotransmitter release. However,

Table 19.1 Genes associated with familial forms of PD

Locus	Gene	Gene product	Inheritance/ PD onset	Chromosomal locus	References
PARK1/PARK4	*SNCA*	Alpha-synuclein (aSyn)	AD/EO	4q21.3-q22	[149]
PARK2	*PARKIN*	Parkin RBR E3 ubiquitin protein ligase	AR/EO	6q25.2-q27	[150, 151]
PARK3	Unknown	Unknown	AD	2p13	[152]
PARK5	*UCHL1*	Ubiquitin C-terminal hydrolase L1	AD	4p13	[153, 154]
PARK6	*PINK1*	PTEN induced putative kinase 1	AR/EO	1p36.12	[155]
PARK7	*DJ-1*	DJ-1	AR/EO	1p36.23	[156–158]
PARK8	*LRRK2*	Leucine-rich repeat kinase 2 (LRRK2)	AD/EO and LO cases	12q12	[159–161]
PARK9	*ATP13A2*	ATPase type 13A2 (ATP13A2)	AR/EO	1p36	[162–164]
PARK10	Unknown	AAOPD	Susceptibility	1p32	[165, 166]
PARK11	Unknown	GIGYF2 (GRB10 interacting GYF protein 2)	AR/EO	2q36-q37	[167–169]
PARK12	Unknown	Unknown	Susceptibility	Xq21-q25	[166, 168–170]
PARK13	*HTRA2*	HtrA serine peptidase 2	AD	2p13.1	[171, 172]
PARK14	*PLA2G6*	Phospholipase A2 group VI	AR/LO	22q13.1	[173–175]
PARK15	*FBXO7*	F-box protein 7 (FBXO7)	AR/EO	22q12.3	[176, 177]
PARK16	Unknown	Unknown	Susceptibility	1q32	[178, 179]
PARK17	*VPS35*	VPS35 retromer complex component	AD/LO	16q12	[32, 180]
PARK18	*EIF4G1*	Eukaryotic translation initiation factor 4 gamma 1	AD/LO	3q27.1	[181–183]
PARK19	*DNAJC6*	Auxilin	AR/EO	1p31.3	[184, 185]
PARK20	*SYNJ1*	Synaptojanin-1	AR/EO	21q22.11	[186, 187]

AD autosomal dominant, *AR* autosomal recessive, *EO* early onset, *LO* late onset

other studies reported aSyn binds mitochondria and is present in the interconnection of mitochondrial membranes and ER or in the nucleus [26–29].

LRRK2 mutations are the most common cause of autosomal dominant PD [30]. Some *LRRK2* mutations are more prevalent in certain ethnic groups [22]. The majority of patients carrying *LRRK2* mutations present the classical pathological features of PD, including the presence of LBs, but the age of onset of the symptoms can vary appearing either earlier or later than idiopathic forms of the disease [30].

The *VPS35* gene codes for the vacuolar protein sorting 35 (VPS35). VPS35 is one of the central components of the retromer cargo-recognition complex which is involved in the trafficking and recycling of synaptic vesicles and proteins [30]. The p.D620N mutation was recognised as a novel cause of autosomal dominant, late-onset PD [31, 32], displaying a dominant negative protein sorting phenotype [33, 34].

The lysosomal enzyme β-glucocerebrosidase, encoded by *GBA*, plays an important role in glycolipid metabolism [35]. Mutations in this gene are known to cause Gaucher disease, one of a growing list of lysosomal storage disorders. However, *GBA* mutations have been described to increase the risk of developing PD and are quite common in PD patients [36–39].

On the other hand, mutations in *PARK2*, *PINK1* and *PARK7* can cause autosomal recessive forms of early-onset PD. All three genes share identical clinical phenotypes, but LB pathology appears to be more variable [35]. *PARK3*, *PARK10* and *PARK12* loci have been implicated in PD, but the genes have not yet been identified. Thus, further analyses will be necessary in order to elucidate the role these loci play in PD pathogenesis [40].

19.2.3 Sporadic Forms of PD

Most PD cases have no known cause, suggesting environmental and lifestyle factors play important and poorly understood roles in the disease. Although these factors are indeed valid and important, it is now estimated that genetics may explain up to 60% of PD cases, underscoring the complexity of the disorder [12]. Toxins, such as methyl-phenyl-tetrahydropyridine (MPTP) [41], 6-hydroxydopamine [42], the herbicide paraquat [43] and the pesticide rotenone [44], have been shown to cause loss of dopaminergic cells in the substantia nigra. In addition, exposure to heavy metals or electromagnetic radiation, head trauma and viral infections are also known risk factors in PD [12, 45]. On the contrary, caffeine [46], uric acid levels [47], nicotine [48] and antagonists of the A_{2A} receptor [49] have been suggested to act as neuroprotectors.

19.3 Epigenetics in PD

PD, as other neurodegenerative diseases, is a complex disorder occurring from the interplay between genetic, environmental, nutritional and other factors, together with ageing. As epigenetics may be altered in response to, at least, some of these factors, it is becoming increasingly accepted; it may also play an important role in the aetiology and pathogenesis of PD.

19.3.1 The Role of DNA Methylation

DNA methylation involves the covalent addition of a methyl group from S-adenosyl methionine (SAM) to the 5′ position of cytosines. In this way, 5-methylcytosine is formed (5-mC), with the concomitant conversion of SAM to S-adenosylhomocysteine (SAH) [50–52]. Methylation is a dynamic process that is apparent in multiple genomic sites, although it is mainly described to occur in repeats of CG dinucleotides [53]. In the human genome, these dinucleotides cluster in areas known as CpG islands which are associated with promoter regions, at least for about 60% of human genes [54]. Functionally, DNA methylation is associated with transcriptional inhibition. This can be executed either directly, by hindering the association of the DNA machinery with chromatin, or indirectly, with the recruitment of methyl-CpG-binding domain proteins (MBDs) [55, 56]. MBDs, in turn, attract histone-modifying and chromatin-remodelling complexes to the methylated sites. DNA methyltransferases (DNMTs) are the enzymes responsible for mediating DNA methylation. In mammals, DNMT1 is able to maintain DNA methylation following replication, while DNMT3a and DNMT3b exert de novo methylation [50].

Genome-wide DNA methylation analysis in blood and brain samples of healthy individuals and PD patients revealed a significant dysregulation of CpG island methylation in the group of patients. Many genes were found to be either hypo- or hypermethylated, including PD risk genes [57]. Another study identified 20 genes that were differentially methylated in blood samples obtained from PD patients in comparison to controls [58].

The observation that the *SNCA* promoter is hypermethylated in patients with alcoholism [59] or anorexia [60] suggested that epigenetics, perhaps through metabolic alterations, may also play a role in PD. Indeed, it was described that *SNCA* expression was upregulated upon methylation-mediated inhibition of *SNCA* intron 1 and that the SN, putamen and cortex of PD patients exhibited a significant hypomethylation pattern compared to healthy controls (Fig. 19.1) [61]. Another study was not able to detect methylation differences in the anterior cingulate or putamen of PD patients when examined a CpG region of the promoter of *SNCA*. However, substantial methylation reduction was apparent in the SN of these patients [62]. A reduction in the nuclear levels of DNMT1 was reported in postmortem brain tissue from dementia with Lewy bodies (DLB) or PD patients, as well as in brains from transgenic mice overexpressing *SNCA*. This alteration in the subcellular localisation of DNMT1 resulted in a global hypomethylation, including CpG islands upstream of *SNCA* and other genes, while aSyn was identified as the sequester of DNMT1 from the nucleus to the cytoplasm (Fig. 19.1) [63]. On the other hand, when the promoter and a CpG-rich region of *SNCA* intron 1 were analysed in patients with PD versus healthy individuals, hypermethylation at various positions in different brain regions was detected [64].

The methylation status of *SNCA* intron 1 was further investigated in blood samples [65], peripheral blood mononuclear cells (PBMCs) [66] or leukocytes of PD patients [67]. In agreement with results in brain tissue, these studies reported a significant decrease in methylation of the *SNCA* promoter. Nevertheless, a correlation between *SNCA* mRNA levels and the methylation pattern of its promoter could not

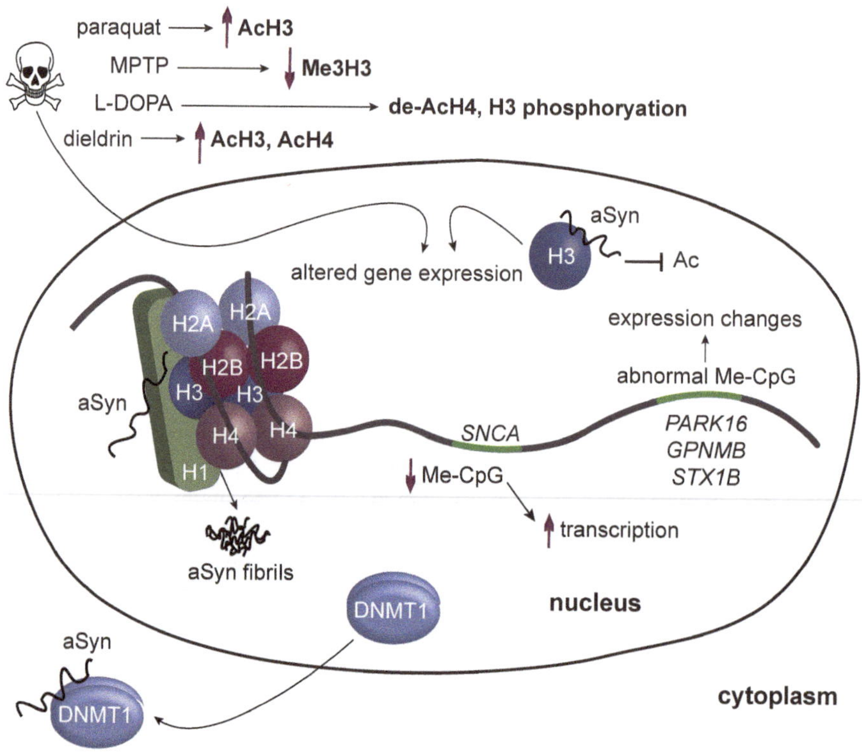

Fig. 19.1 Epigenetic modifications in dopaminergic neurons. Certain toxins enter the neuronal cells and cause histone modifications, thereby influencing the expression of several genes. In the nucleus, aSyn interacts with H1 forming a tight complex and also with H3 inhibiting its acetylation. In turn, histones trigger the aggregation of aSyn. Several PD-associated genes, such as *PARK16*, *GPNMB* and *STX1B*, show altered expression as a result of aberrant DNA methylation. The promoter of *SNCA* is usually found hypomethylated in PD, leading to increased levels of aSyn. aSyn is able to sequester DNMT1 from the cytoplasm in the nucleus resulting in a general reduction of the methylation pattern. *Ac* acetylation, *de-Ac* deacetylation, ↑ increase, ↓ decrease, ⊣ inhibition

be firmly established [65, 66]. Another study in leukocytes from PD patients and healthy individuals revealed no alterations in the levels of methylation in any of the investigated regions [68].

Additional genes, namely, *PARK16*, *GPNMB* and *STX1B*, have been found to present aberrant methylation in postmortem PD brain samples (Fig. 19.1) [69]. The methylation status of the *TNF* promoter was significantly diminished in the SN compared to the cortex of both PD patients and healthy individuals, suggesting that a possible overexpression of TNF may trigger inflammatory reactions compromising the vulnerability of the dopaminergic neurons [70]. Postmortem samples obtained from the cortex and putamen showed decreased CpG methylation and increased mRNA levels of the *CYP2E1* gene in PD patients [71]. Interestingly, a single nucleotide polymorphism (SNP) in this gene has been associated with PD [72], and its protein product, cytochrome P450 2E1, is implicated in the production of toxic metabolites that influence degeneration of dopaminergic neurons [50]. Although mutations in *PARK2* have been associated with autosomal recessive juvenile parkinsonism, abnormal methylation levels of *PARK2* promoter have been described in acute lymphoblastic and in chronic myeloid leukaemia [73], but not in PD cases [74]. In a similar manner, increased methylation of the *UCHL1* promoter was reported in diverse types of cancer [75, 76], while no significant alterations in CpG methylation was observed in the hippocampus and frontal cortex from PD brains [77]. Similar results were obtained for *ATP13A2* gene. DNA methylation of the promoter revealed an association with the progression of Kufor-Rakeb syndrome, although no such link has been made for PD so far [78].

DNA methylation in mitochondria might also be a relevant phenomenon in the context of PD. Recently, the mammalian mitochondrial DNMT (mtDNMT) was discovered [79]. Despite some controversy regarding CpG methylation in the genome of human mitochondria [80], some studies claim this can occur [81, 82]. Moreover, alterations in mitochondrial DNA methylation have been associated with cancer [83] and liver disease [84]. Finally, it was suggested that age-related changes in the DNA methylation of mitochondria may influence gene expression, alter mitochondrial metabolism and increase ROS production [85]. On the other hand, both *PARK2* and *PINK1* genes are essential for physiological mitochondrial function, and, when either of them is mutated, they can lead to mitochondrial impairment [12]. Considering the involvement of mitochondria in PD, further investigation will unravel possible implication of mitochondrial DNA methylation in PD pathogenesis.

19.3.2 Hydroxymethylation

Recently, the enzyme ten-eleven translocation1 (Tet1) was found to catalyse the oxidation of 5-mC to 5-hydroxymethylcytosine (5-hmC) [86]. Following studies have associated 5-hmC with euchromatin, indicating its relation with promoter regions and increased transcriptional levels [87, 88]. This intriguing, novel epigenetic modification is essentially unexplored in the context of neurodegeneration.

A detailed study revealed that 5-hmC levels increase in the mouse cerebellum in an age-dependent manner. In addition, an intragenic and proximal (to transcription start or termination sites regions) enrichment of 5-hmC was identified and associated with elevated gene expression. Gene ontology pathway analysis of the differentially expressed genes pointed towards pathways which are associated with neurodegenerative diseases such as Alzheimer's disease, Huntington's disease and PD [89], but additional studies are necessary in order to establish whether this type of DNA alteration is relevant in neurodegeneration.

19.3.3 Histone Modifications

The N-terminal tails of the histones are around 25–40 amino acid residues long and constitute a suitable region where chromatin-modifying enzymes can execute their function [90]. Histone modifications include methylation of lysine or arginine residues, acetylation, phosphorylation, ubiquitination, SUMOylation, ADP-ribosylation, crotonylation, hydroxylation and proline isomerisation [52, 81]. Histone modifications have been described to play pivotal roles in the development, differentiation and maintenance of dopaminergic neurons [91]. However, little is known concerning alterations in the physiological pattern histone modifications and their implications in PD pathogenesis.

In a recent study, the use of isolated dopaminergic neurons from brain tissue from PD patients revealed increased acetylation levels of histone H2A, H3 and H4 compared to age-matched control individuals. Furthermore, the levels of various histone deacetylases (HDACs) are reduced in 1-methyl-4-phenylpyridinium (MPP^+)-treated cells and in MPTP-treated mouse brains and also in midbrain samples from PD patients [92]. These findings highlighted the presence of histone modifications suggesting that chromatin remodelling may be highly implicated in the pathogenesis of PD. Exposure to additional toxins also induces alterations into histones. For instance, when the pesticide dieldrin was administered in mice, elevated acetylation of histones H3 and H4 occurred in mesencephalic dopaminergic neurons due to proteasomal dysfunction (Fig. 19.1). Subsequently, the cAMP response element-binding protein, a histone acetyltransferase (HAT), was found to accumulate in the cells [93]. Another neurotoxic agent, paraquat, induces acetylation of histone H3 in dopaminergic cells in vitro (Fig. 19.1) [94].

In murine and primate models of levodopa-induced dyskinesia (LDID), dopamine depletion via MPTP administration was associated with a reduction in histone H3 trimethylation at Lys4 (Fig. 19.1). Chronic levodopa (or L-DOPA) therapy of these models was accompanied by deacetylation of striatal histone H4 at Lys5, 8, 12 and 16 (Fig. 19.1). The presence of histone modifications is evident, suggesting they may contribute to the development and maintenance of LDID in PD [95]. LDID has been associated with abnormal dopamine D1 receptor transmission. Histone H3 phosphoacetylation is blocked by D1 receptor inactivation, suggesting that inhibition of histone H3 acetylation and/or phosphorylation may be used for the prevention or reversion of dyskinesia [96]. In a mouse model of PD, it was shown that administration of L-DOPA induced phosphorylation of histone H3 on Ser28 in regions marked by trimethylation of the adjacent Lys27 (Fig. 19.1). This phenomenon was specifically observed in neurons expressing the D1 receptor and correlated with aberrant expression of genes that may be accountable for motor complications or dyskinesia [97].

Dopaminergic neurons of paraquat-treated mice displayed accumulation of aSyn in the nucleus, where it co-localises with acetylated histone H3. Further investigation revealed that aSyn binds directly to histone H1 and forms a tight 2:1 complex (Fig. 19.1). On the other hand, histone H1, together with the core histones, was able to boost the formation of aSyn fibrils (Fig. 19.1) [98]. Another study reported both in vitro and in *Drosophila* that nuclear aSyn associated with histone H3 reduces its acetylation (Fig. 19.1) [99]. Similar results were also described in PC12 cells expressing monoamine oxidase B. aSyn co-localised with histone H3 and once more was able to decrease its acetylation [27]. Finally, overexpression of dHDAC6 in a *Drosophila* model of PD ectopically expressing *SNCA* promoted aSyn inclusion formation and reduced aSyn oligomerisation. On the other hand, depletion of dHDAC6 enhanced the detrimental effects of aSyn overexpression, including the loss of dopaminergic neurons and locomotor dysfunction [100].

In *C. elegans* overexpressing human wt or A53T *SNCA*, nine histone genes coding for linker H1 and two core histones, H2B and H4, were downregulated [101].

19.3.4 miRNAs in PD

miRNAs bind to the 3′ untranslated region (UTR) of mRNA targets and modulate protein translation [102]. Thus, given their pleiotropic effects in cell biology, miRNAs are also emerging as relevant contributors to neurodegeneration in PD. Recently, an overall downregulation of miRNAs was found in tissue samples isolated from the SN of PD patients when compared to samples from healthy individuals [103].

Transgenic mice lacking Dicer in their dopaminergic neurons display neuronal cell death in the SN [104], suggesting overall miRNA processing is detrimental for dopaminergic cell function. Interestingly, studies in PD patients revealed that miR-133b, which is specifically expressed in midbrain dopaminergic neurons, is deficient in midbrain tissue. miR-133b is involved in a negative feedback circuit that contains the paired-like homeodomain transcription factor Pitx3, having a regulatory role in the maturation and function of midbrain dopaminergic neurons [104]. miR-132 has also been linked to midbrain dopaminergic neuronal differentiation. In a rat model of PD, miR-132 was significantly increased, and, in turn, the levels of its target protein, nuclear receptor-related 1 protein (Nurr1), were reduced [105, 106].

In a study using the MPTP-induced mouse model of PD, miR-124 was found to be downregulated in the SN of the mice, along with an increase in the levels of calpain/CDK5 proteins [107]. Interestingly, activation of calpains has been associated with dopaminergic cell death in the MPTP-induced mouse model and in postmortem nigral tissue from PD brains [108]. Another study reported a functional role of elevated miR-126 in SN dopaminergic neurons of PD patients through the inhibition of IGF-1/PI3K signalling pathway, contributing to neurotoxicity [109].

The levels of miR-1, miR-22* and miR-29 are reduced in blood samples of PD patients. Interestingly, the levels of miR-16-2*, miR-26a2* and miR30a enabled the distinction between treated from non-treated PD patients [110]. On the other hand, miR-1826/miR-450b-3p, miR-505 and miR-626 are upregulated in the plasma of PD patients and may be useful as PD biomarkers [111].

LRRK2 was found to influence the miRNA pathway, possibly by associating with Argonaute (Ago), in both human and *Drosophila* samples. Furthermore, in a *Drosophila* model of PD, it was observed that mutant LRRK2 suppresses the function of let-7 and miR-184* which normally regulate the translation of E2F1/DP complex, involved in cell cycle and survival control (Fig. 19.2) [112]. Furthermore, frontal cortex samples from PD patients contain high levels of LRRK2 and reduced levels of miR-205. It was then found that miR-205 is able to bind to the 3′ UTR of *LRRK2* mRNA and suppress its expression. Further in vitro studies included the introduction of miR-205 in neurons carrying the R1441G LRRK2 mutation, which prevented outgrowth defects [113]. These findings suggested the regulatory role of miR-205 on *LRRK2* expression and, therefore, a possible role in PD pathogenesis (Fig. 19.2).

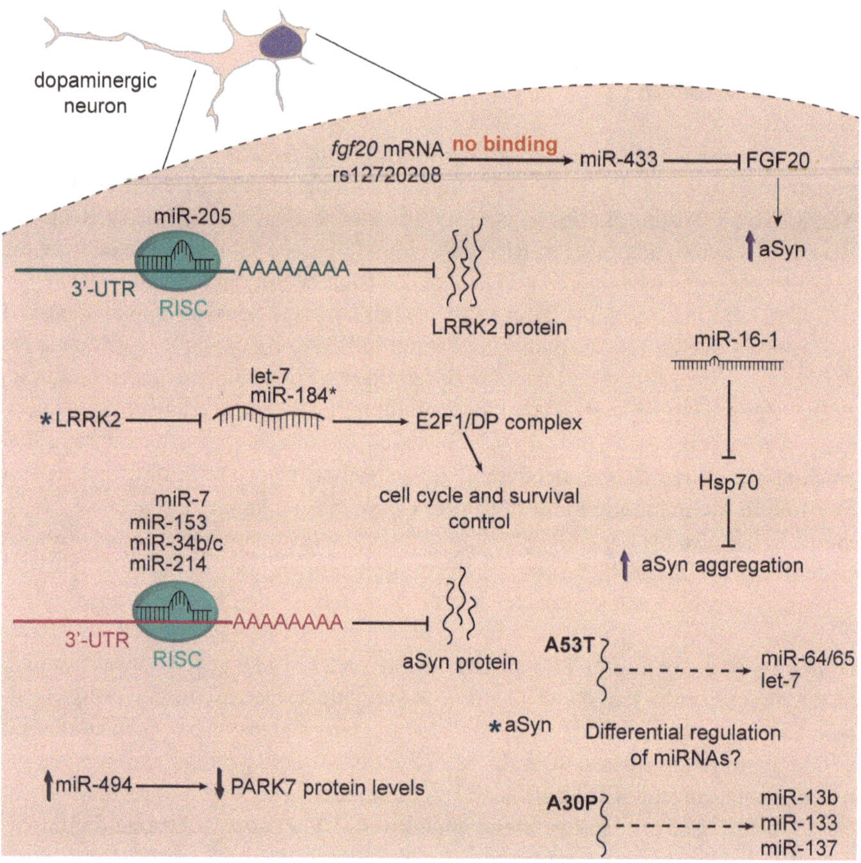

Fig. 19.2 The impact of miRNAs on TH⁺ neurons. miR-205 is able to suppress the expression of LRRK2 protein by binding to its 3′ UTR mRNA region. On the contrary, mutant LRRK2 inhibits let-7 and miR-184* which participate in cell survival. Overexpression of miR-494 reduces the levels of PARK7. Furthermore, several miRNAs bind to *snca* mRNA sequence and prevent its translation. On the other hand, the levels and aggregation of aSyn are indirectly increased due to increased FGF20 or decreased Hsp70 protein levels. Finally, mutant aSyn is thought to affect the production of certain miRNAs. **LRRK2* mutant LRRK2, **aSyn* mutant aSyn, ↑ increase, ↓ decrease, ⊣ inhibition

DJ-1, the product of *PARK7*, is thought to be an oxidative sensor that protects cells from oxidative stress. Decreased levels of DJ-1 have been detected in the SN of sporadic PD patients suggesting a connection with PD. miR-494 was found to bind to the 3′ UTR of *PARK7* mRNA and, when overexpressed, was able to significantly reduce DJ-1 protein levels in vitro and in an MPTP mouse model, while concomitantly rendering the cells more susceptible to oxidative stress and leading to dopaminergic cell death (Fig. 19.2) [114].

A global miRNA expression profiling in *C. elegans* showed that three members of the let-7 family (cel-miR-241, 230 and 48) were deregulated in animals mutated for *PARK2*. Similarly, 12 differentially regulated miRNAs from the miR-64/miR-65 and let-7 families were identified in animals overexpressing human A53T *SNCA* (Fig. 19.2) [115].

The levels of miR-34b and c were found significantly reduced in the amygdala, frontal cortex, cerebellum and SN of PD patients, accompanied by a decrease in the expression of *PARK2* and *PARK7*. In addition, depletion of miR34-b and c in in vitro differentiated dopaminergic neurons caused an alteration of mitochondrial function and oxidative stress [116, 117]. In addition, both miRNAs appear to repress *SNCA* expression. Overexpression of miR-34b and c in SH-SY5Y cells resulted in a substantial reduction of aSyn protein levels via targeting the 3′ UTR of *SNCA* mRNA (Fig. 19.2), while inhibition, using anti-miRs, increased both the levels and the aggregation of the protein. Finally, a polymorphic variation in the 3′ UTR of human *SNCA* mRNA was associated with resistance to miR-34b binding and therefore to increased aSyn [118].

Two other abundant brain miRNAs, miR-7 and miR-153, bind to the 3′ UTR of *SNCA* mRNA and inhibit its translation (Fig. 19.2). More precisely, miR-7, a neuron-specific miRNA, was found to downregulate the expression of *SNCA* in HEK293T cells, protecting against oxidative stress. On the other hand, a specific miR-7 inhibitor caused a significant increase of aSyn protein levels in SH-SY5Y cells. Results obtained from MPTP-treated mice were in agreement with those obtained in the in vitro models, showing a substantial reduction of miR-7 levels and suggesting that elevated *SNCA* expression may be attributed to this downregulation [119]. Furthermore, treatment of primary cortical neurons with MPP^+ followed by miR-7 overexpression resulted in neuronal protection from MPP^+-induced toxicity and restored neuronal viability [120]. This protection from cell death was achieved via preservation of active mTOR signalling, possibly promoting aSyn clearance [120, 121].

miR-153 is another brain predominant miRNA that binds to the 3′ UTR of *SNCA* mRNA resulting in a significant decrease of its mRNA and protein levels [122]. The miR-153 binding site is predicted to be located within nucleotides 459–465. A variation identified in one male PD patient (464 C > A) was never encountered in healthy individuals or in patients with familial PD that were involved in the study and was suggested to be a rare cause of PD [123]. Interestingly, it seems that miR-7 and miR-153 have a synergistic effect on reducing aSyn levels [122].

In contrast, it was reported that SH-SY5Y cells treated with miR-106a* significantly increased their aSyn protein levels [124]. Moreover, other miRNAs such as miR-301b, miR-26b, miR-373* and miR-21 which regulate the levels of chaperone-mediated autophagy proteins were significantly increased in the SN of human PD brain tissues [124].

Administration of MPP^+ or MPTP to cell or mouse models, respectively, resulted in a decline of miR-214 levels and in an increase in aSyn levels. In particular, a miR-214 inhibitor caused a reduction in the amount of TH^+ cells when administered in vivo. Thus, as a result, miR-214 may contribute to the upregulation of *SNCA* and, therefore, to the toxic effects of aSyn in dopaminergic neurons (Fig. 19.2) [125].

Alterations in synaptosomal proteins were investigated in early symptomatic A30P *SNCA* transgenic mice, indicating that several proteins related to mitochondrial function were differentially expressed. Moreover, miRNA expression profiling revealed that the levels of miR-10a, 10b, 212, 132 and 495 were altered in brainstem samples when compared those from wild-type control animals [126]. In a *Drosophila* A30P *SNCA* model, high-throughput sequencing of small RNAs revealed that five miRNAs were upregulated. Among them, miR-13b, miR-133 and miR-137 are enriched in the brain and highly conserved from *Drosophila* to humans. miR-137 was shown to target the 3′ UTR mRNA of the dopamine D2 receptor. Therefore, it was suggested that mutant aSyn may be responsible for the dysregulation of miRNAs which are implicated in neuroactive-ligand receptor pathways (Fig. 19.2) [127].

Heat sock protein 70 (Hsp70) is capable of inhibiting cellular toxicity caused by aSyn via reduction of aSyn misfolding and aggregation [128–132]. Chemical blockade of Hsp70 in a cellular model (SH-SY5Y cells) overexpressing *SNCA* promotes aSyn aggregation. Interestingly, administration of miR-16-1 mimics those results given that miR-16-1 targets *HSP70* mRNA and downregulates both its mRNA and protein levels (Fig. 19.2) [133]. Therefore, aSyn toxicity and the protective effects of Hsp70 are corroborated via this novel mechanism, opening new perspectives for intervention in PD.

A polymorphism (rs1989754) in the *FGF20* gene was reported to be associated with increased risk of developing PD [134]. Another *FGF20* polymorphism that was identified a few years later (rs12720208) was suggested to obstruct the binding of miR-433 to the *FGF20* mRNA both in vitro and in vivo and, therefore, lead to increased FGF20 protein levels. Interestingly, elevated FGF20 protein levels have been linked to the subsequent increase of aSyn levels, observed both in SH-SY5Y cells and in human brain samples. In this way, elevated FGF20 levels may account for susceptibility towards developing PD through the increase of aSyn (Fig. 19.2) [135].

19.4 Epigenetic-Based Therapeutic Approaches for PD

HDAC inhibitors (HDACis) are commonly used as anticancer molecules. However, they have also emerged in the field of neurodegenerative disorders, in models of PD and AD, due to their effects on different members of the histone deacetylase family of proteins [136–139]. Valproic acid (VPA) has been shown to protect against rotenone [140], aSyn [140] and MPTP toxicity [141]. The responses triggered by VPA were mediated by decreasing the levels of pro-inflammatory factors and inducing microglia apoptosis [142, 143]. Finally, trichostatin A (TSA) has been described to increase the expression of *HSP70,* thereby having neuroprotective and anti-inflammatory properties [144], and to induce microglia apoptosis accompanied by increased histone H3 acetylation [143]. Nevertheless, the positive effects of these compounds conceal certain drawbacks. For example, in one study, it was shown that hyperacetylation of histone H4 via the administration of sodium butyrate, an HDACi, induces the expression of the protein kinase C δ (PKCδ) in the striatum and SN of mice. This upregulation was responsible for increasing the sensitivity of the cells to oxidative stress, rendering the dopaminergic neurons more prone to cell death and potentially contributing to PD [145]. TSA was also found to induce neuronal cell death and activate pro-apoptotic genes, likely contributing to PD pathogenesis [146, 147]. In addition, it was described that TSA potentiated pro-inflammatory responses in microglial cells, a process that is associated with several degenerative conditions [148]. The balance between HAT and HDAC activities is vital for normal cellular function, and, although many studies are evaluating the therapeutic potential of HDACis in PD, it should also be noted that they may cause undesired side effects and responses not only in neurons but also in other cell types, due to putative effects in nonhistone protein targets. Thus, despite current hopes and potential, additional work is still necessary in order to improve the applicability of these approaches.

Acknowledgements TFO is supported by the DFG Center for Nanoscale Microscopy and Molecular Physiology of the Brain and from BMBF Grant DecipherPD (01KU1503B).

References

1. Huse DM, Schulman K, Orsini L, Castelli-Haley J, Kennedy S, Lenhart G. Burden of illness in Parkinson's disease. Mov Disord [Internet]. 2005;20(11):1449–54. http://www.ncbi.nlm.nih.gov/pubmed/16007641
2. de Rijk MC, Launer LJ, Berger K, Breteler MM, Dartigues JF, Baldereschi M, et al. Prevalence of Parkinson's disease in Europe: A collaborative study of population-based cohorts. Neurologic Diseases in the Elderly Research Group. Neurology [Internet]. Department of Epidemiology & Biostatistics, Erasmus Medical Center, Rotterdam, The Netherlands.; 2000;54:S21–3. http://www.ncbi.nlm.nih.gov/pubmed/10854357
3. Van Den Eeden SK, Tanner CM, Bernstein AL, Fross RD, Leimpeter A, Bloch DA, et al. Incidence of Parkinson's disease: variation by age, gender, and race/ethnicity. Am J Epidemiol [Internet]. Division of Research, Kaiser Permanente, Oakland, CA 94612, USA. skv@dor.kaiser.org; 2003;157:1015–22. http://www.ncbi.nlm.nih.gov/pubmed/12777365
4. de Lau LM, Breteler MM. Epidemiology of Parkinson's disease. Lancet Neurol [Internet]. Department of Epidemiology & Biostatistics, Erasmus Medical Centre Rotterdam, Netherlands.; 2006;5:525–35. http://www.ncbi.nlm.nih.gov/pubmed/16713924
5. Dorsey ER, Constantinescu R, Thompson JP, Biglan KM, Holloway RG, Kieburtz K, et al. Projected number of people with Parkinson disease in the most populous nations, 2005 through 2030. Neurology. 2007;68(5):384–6.

6. Dickson DW, Braak H, Duda JE, Duyckaerts C, Gasser T, Halliday GM, et al. Neuropathological assessment of Parkinson's disease: refining the diagnostic criteria. Lancet Neurol [Internet]. Mayo Clinic, Jacksonville, FL, USA.; 2009;8:1150–7. http://www.ncbi.nlm.nih.gov/pubmed/19909913
7. Spillantini MG, Schmidt ML, Lee VM, Trojanowski JQ, Jakes R, Goedert M. Alpha-synuclein in Lewy bodies. Nature [Internet]. 1997;388:839–40. http://www.ncbi.nlm.nih.gov/pubmed/9278044
8. Winner B, Kohl Z, Gage FH. Neurodegenerative disease and adult neurogenesis. Eur J Neurosci. 2011;33(6):1139–51.
9. Goedert M. Alpha-synuclein and neurodegenerative diseases. Nat Rev Neurosci. 2001;2(7):492–501.
10. Fahn S. Description of Parkinson's disease as a clinical syndrome. Ann N Y Acad Sci [Internet]. Department of Neurology, Columbia University College of Physicians Surgeons, New York, New York 10032, USA. fahn@neuro.columbia.edu; 2003;991:1–14. http://www.ncbi.nlm.nih.gov/pubmed/12846969
11. Lang AE, Lozano AM. Parkinson's disease. Second of two parts. N Engl J Med [Internet]. Department of Medicine, University of Toronto and the Toronto Hospital, Canada.; 1998;339:1130–43. http://www.ncbi.nlm.nih.gov/pubmed/9770561
12. Mhyre TR, Boyd JT, Hamill RW, Maguire-Zeiss KA. Parkinson's disease. Subcell Biochem [Internet]. Department of Neuroscience, Georgetown University Medical Center, 3970 Reservoir Road, NW NRB WP-24A, 20057, Washington, DC, USA, trm36@georgetown.edu.; 2012;65:389–455. http://www.ncbi.nlm.nih.gov/pubmed/23225012
13. Savica R, Rocca W a, Ahlskog JE. When does Parkinson disease start? Arch Neurol. 2010;67(7):798–801.
14. Hawkes CH. The prodromal phase of sporadic Parkinson's disease: Does it exist and if so how long is it? Movement Disorders. 2008. p. 1799–807.
15. Braak H, Del Tredici K, Rüb U, De Vos RAI, Jansen Steur ENH, Braak E. Staging of brain pathology related to sporadic Parkinson's disease. Neurobiol Aging. 2003;24(2):197–211.
16. Halliday G, Hely M, Reid W, Morris J. The progression of pathology in longitudinally followed patients with Parkinson's disease. Acta Neuropathol. 2008;115(4):409–15.
17. Zaccai J, Brayne C, McKeith I, Matthews F, Ince PG. Patterns and stages of alpha-synucleinopathy: Relevance in a population-based cohort. Neurology. 2008;70(13):1042–8.
18. Verstraeten A, Theuns J, Van Broeckhoven C. Progress in unraveling the genetic etiology of Parkinson disease in a genomic era. Trends Genet [Internet]. Elsevier Ltd; 2015;31(3):140–9. http://dx.doi.org/10.1016/j.tig.2015.01.004
19. Hamza TH, Payami H. The heritability of risk and age at onset of Parkinson's disease after accounting for known genetic risk factors. J Hum Genet [Internet]. Division of Genetics, New York State Department of Health, Wadsworth Center, Albany, NY 12201–2002, USA.; 2010;55:241–3. http://www.ncbi.nlm.nih.gov/pubmed/20203693
20. Martin I, Dawson VL, Dawson TM. Recent advances in the genetics of Parkinson's disease. Annu Rev Genomics Hum Genet [Internet]. 2011;12:301–25. http://www.annualreviews.org/doi/abs/10.1146/annurev-genom-082410-101440
21. Hardy J, Cai H, Cookson MR, Gwinn-Hardy K, Singleton A. Genetics of Parkinson's disease and parkinsonism. Annals of Neurology. 2006. p. 389–98.
22. Nuytemans K, Theuns J, Cruts M, Van Broeckhoven C. Genetic etiology of Parkinson disease associated with mutations in the SNCA, PARK2, PINK1, PARK7, and LRRK2 genes: A mutation update. Hum Mutat. 2010;31(7):763–80.
23. Chartier-Harlin MC, Kachergus J, Roumier C, Mouroux V, Douay X, Lincoln S, et al. α-synuclein locus duplication as a cause of familial Parkinson's disease. Lancet. 2004;364(9440):1167–9.
24. Singleton a B, Farrer M, Johnson J, Singleton a, Hague S, Kachergus J, et al. alpha-Synuclein locus triplication causes Parkinson's disease. Science. 2003;302(5646):841.

25. Ibáñez P, Bonnet A-M, Débarges B, Lohmann E, Tison F, Pollak P, et al. Causal relation between alpha-synuclein gene duplication and familial Parkinson's disease. Lancet (London, England) [Internet]. 2004;364(9440):1169–71. http://www.sciencedirect.com/science/article/pii/S0140673604171043
26. Guardia-Laguarta C, Area-Gomez E, Schon E a, Przedborski S. Novel subcellular localization for α-synuclein: possible functional consequences. Front Neuroanat [Internet]. 2015;9(February):17. http://journal.frontiersin.org.ezproxy.nihlibrary.nih.gov/article/10.3389/fnana.2015.00017/abstract
27. Siddiqui A, Chinta SJ, Mallajosyula JK, Rajagopolan S, Hanson I, Rane A, et al. Selective binding of nuclear alpha-synuclein to the PGC1alpha promoter under conditions of oxidative stress may contribute to losses in mitochondrial function: implications for Parkinson's disease. Free Radic Biol Med [Internet]. Buck Institute for Age Research, 8001 Redwood Boulevard, Novato, CA 94945, USA.; 2012;53:993–1003. http://www.ncbi.nlm.nih.gov/pubmed/22705949
28. Yu S, Zuo X, Li Y, Zhang C, Zhou M, Zhang YA, et al. Inhibition of tyrosine hydroxylase expression in alpha-synuclein-transfected dopaminergic neuronal cells. Neurosci Lett [Internet]. Department of Neurobiology and the Sino-Japan Joint Laboratory on Neurodegenerative Diseases, Beijing Institute of Geriatrics, Xuanwu Hospital of the Capital University of Medical Sciences, 45 Changchun Street, Beijing 100053, China.; 2004;367:34–9. http://www.ncbi.nlm.nih.gov/pubmed/15308292
29. Specht CG, Tigaret CM, Rast GF, Thalhammer A, Rudhard Y, Schoepfer R. Subcellular localisation of recombinant alpha- and gamma-synuclein. Mol Cell Neurosci [Internet]. Laboratory for Molecular Pharmacology, Department of Pharmacology, University College London, UCL, London WC1E 6BT, UK.; 2005;28:326–34. http://www.ncbi.nlm.nih.gov/pubmed/15691713
30. Bonifati V. Genetics of Parkinson's disease – state of the art, 2013. Parkinsonism Relat Disord [Internet]. Elsevier Ltd; 2014;20:S23–8. http://linkinghub.elsevier.com/retrieve/pii/S1353802013700099
31. Zimprich A, Benet-Pagès A, Struhal W, Graf E, Eck SH, Offman MN, et al. A mutation in VPS35, encoding a subunit of the retromer complex, causes late-onset parkinson disease. Am J Hum Genet. 2011;89(1):168–75.
32. Vilariño-Güell C, Wider C, Ross OA, Dachsel JC, Kachergus JM, Lincoln SJ, et al. VPS35 mutations in parkinson disease. Am J Hum Genet. 2011;89(1):162–7.
33. Braschi E, Goyon V, Zunino R, Mohanty A, Xu L, McBride HM. Vps35 mediates vesicle transport between the mitochondria and peroxisomes. Curr Biol. 2010;20(14):1310–5.
34. MacLeod DA, Rhinn H, Kuwahara T, Zolin A, Di Paolo G, MacCabe BD, et al. RAB7L1 Interacts with LRRK2 to Modify Intraneuronal Protein Sorting and Parkinson's Disease Risk. Neuron. 2013;77(3):425–39.
35. Klein C, Westenberger A. Genetics of Parkinson's Disease. Cold Spring Harb Perspect Med [Internet]. Cold Spring Harbor Laboratory Press; 2012 Jan;2(1):a008888. http://www.ncbi.nlm.nih.gov/pmc/articles/PMC3253033/
36. The Glucocerebrosidase Gene and Parkinson's Disease in Ashkenazi Jews. N Engl J Med [Internet]. Massachusetts Medical Society; 2005 Feb 17;352(7):728–31. http://dx.doi.org/10.1056/NEJM200502173520719
37. Goker-Alpan O, Schiffmann R, LaMarca ME, Nussbaum RL, McInerney-Leo A, Sidransky E. Parkinsonism among Gaucher disease carriers. J Med Genet [Internet]. 2004;41(12):937–40. http://jmg.bmj.com/content/41/12/937.full
38. Lwin A, Orvisky E, Goker-Alpan O, LaMarca ME, Sidransky E. Glucocerebrosidase mutations in subjects with parkinsonism. Mol Genet Metab. 2004;81(1):70–3.
39. Sidransky E, Nalls MA, Aasly JO, Aharon-Peretz J, Annesi G, Barbosa ER, et al. Multicenter analysis of glucocerebrosidase mutations in Parkinson's disease. N Engl J Med [Internet]. 2009;361(17):1651–61. http://www.pubmedcentral.nih.gov/articlerender.fcgi?artid=2856322&tool=pmcentrez&rendertype=abstract

40. Schulte C, Gasser T. Genetic basis of Parkinson's disease: Inheritance, penetrance, and expression. Application of Clinical Genetics. 2011. p. 67–80.
41. Langston JW, Ballard P, Tetrud JW, Irwin I. Chronic Parkinsonism in humans due to a product of meperidine-analog synthesis. Science. 1983;219(4587):979–80.
42. Ungerstedt U. 6-Hydroxy-dopamine induced degeneration of central monoamine neurons. Eur J Pharmacol. 1968;5:107–10.
43. Dauer W, Przedborski S. Parkinson's disease: Mechanisms and models. Neuron. 2003. p. 889–909.
44. Betarbet R, Sherer TB, MacKenzie G, Garcia-Osuna M, Panov a V, Greenamyre JT. Chronic systemic pesticide exposure reproduces features of Parkinson's disease. Nat Neurosci. 2000;3(12):1301–6.
45. Shulman JM, De Jager PL, Feany MB. Parkinson's Disease: Genetics and Pathogenesis. Annu Rev Pathol Mech Dis [Internet]. 2011;6(1):193–222. http://www.annualreviews.org/doi/abs/10.1146/annurev-pathol-011110-130242
46. Kalda A, Yu L, Oztas E, Chen JF. Novel neuroprotection by caffeine and adenosine A(2A) receptor antagonists in animal models of Parkinson's disease. J Neurol Sci [Internet]. 2006;248(1–2):9–15. http://www.ncbi.nlm.nih.gov/pubmed/16806272
47. Wirdefeldt K, Adami H-O, Cole P, Trichopoulos D, Mandel J. Epidemiology and etiology of Parkinson's disease: a review of the evidence [Internet]. Eur J Epidemiol 2011 p. 1–58. http://link.springer.com/article/10.1007/s10654-011-9581-6/fulltext.html
48. Quik M. Smoking, nicotine and Parkinson's disease. Trends in Neurosciences. 2004. p. 561–8.
49. Ferreira DG, Batalha VL, Vicente Miranda H, Coelho JE, Gomes R, Gonçalves FQ, et al. Adenosine A2A Receptors Modulate α-Synuclein Aggregation and Toxicity. Cereb Cortex [Internet]. 2015;bhv268. http://cercor.oxfordjournals.org/content/early/2015/11/02/cercor.bhv268.abstract
50. Feng Y, Jankovic J, Wu YC. Epigenetic mechanisms in Parkinson's disease. J Neurol Sci [Internet]. Department of Neurology, Shanghai First People's Hospital, Shanghai Jiao Tong University School of Medicine, Shanghai 200080, PR China. Parkinson's Disease Center and Movement Disorders Clinic, Department of Neurology, Baylor College of Medicine, Houston,; 2015;349:3–9. http://www.ncbi.nlm.nih.gov/pubmed/25553963
51. Coppede F. The potential of epigenetic therapies in neurodegenerative diseases. Front Genet [Internet]. Department of Translational Research and New Technologies in Medicine and Surgery, University of Pisa Pisa, Italy.; 2014;5:220. http://www.ncbi.nlm.nih.gov/pubmed/25071843
52. Urdinguio RG, Sanchez-Mut J V., Esteller M. Epigenetic mechanisms in neurological diseases: genes, syndromes, and therapies. Lancet Neurol [Internet]. Elsevier Ltd; 2009;8(11):1056–72. http://dx.doi.org/10.1016/S1474-4422(09)70262-5
53. Georgel PT. The danger of epigenetics misconceptions (epigenetics and stuff…). Biochem Cell Biol [Internet]. 2015;4(August):1–4. http://www.nrcresearchpress.com/doi/abs/10.1139/bcb-2015-0091?url_ver=Z39.88-2003&rfr_id=ori:rid:crossref.org&rfr_dat=cr_pub=pubmed
54. Portela A, Esteller M. Epigenetic modifications and human disease. Nat Biotechnol [Internet]. Nature Publishing Group; 2010;28(10):1057–68. http://www.ncbi.nlm.nih.gov/pubmed/20944598
55. Lopez-Serra L, Esteller M. Proteins that bind methylated DNA and human cancer: reading the wrong words. Br J Cancer [Internet]. 2008;98(12):1881–5. http://www.pubmedcentral.nih.gov/articlerender.fcgi?artid=2441952&tool=pmcentrez&rendertype=abstract
56. Qureshi IA, Mehler MF. Epigenetic mechanisms governing the process of neurodegeneration. Mol Asp Med [Internet]. Roslyn and Leslie Goldstein Laboratory for Stem Cell Biology and Regenerative Medicine, Albert Einstein College of Medicine, Bronx, New York, NY 10461, USA. irfan@jhu.edu; 2013; 34:875–82. http://www.ncbi.nlm.nih.gov/pubmed/22782013
57. Masliah E, Dumaop W, Galasko D, Desplats P. Distinctive patterns of DNA methylation associated with Parkinson disease: identification of concordant epigenetic changes in brain

and peripheral blood leukocytes. Epigenetics [Internet]. Department of Neuroscience; University of California San Diego; La Jolla, CA USA; Department of Pathology; University of California San Diego; La Jolla, CA USA. Department of Pathology; University of California San Diego; La Jolla, CA USA. Department of N; 2013;8:1030–8. http://www.ncbi.nlm.nih.gov/pubmed/23907097
58. Moore K, McKnight AJ, Craig D, O'Neill F. Epigenome-wide association study for Parkinson's disease. Neuromolecular Med [Internet]. Queens University Belfast, Belfast, UK, kerry.moore@btinternet.com.; 2014;16:845–55. http://www.ncbi.nlm.nih.gov/pubmed/25304910
59. Bonsch D, Lenz B, Kornhuber J, Bleich S. DNA hypermethylation of the alpha synuclein promoter in patients with alcoholism. Neuroreport [Internet]. Department of Psychiatry and Psychotherapy, Friedrich-Alexander-University of Erlangen-Nuremberg, Schwabachanlage 6–10, D-91054 Erlangen, Germany.; 2005;16:167–70. http://www.ncbi.nlm.nih.gov/pubmed/15671870
60. Frieling H, Gozner A, Romer KD, Lenz B, Bonsch D, Wilhelm J, et al. Global DNA hypomethylation and DNA hypermethylation of the alpha synuclein promoter in females with anorexia nervosa. Mol Psychiatry [Internet]. 2007;12:229–30. http://www.ncbi.nlm.nih.gov/pubmed/17325715
61. Jowaed A, Schmitt I, Kaut O, Wullner U. Methylation regulates alpha-synuclein expression and is decreased in Parkinson's disease patients' brains. J Neurosci [Internet]. Department of Neurology, Rheinische Friedrich-Wilhelms-Universitat, Universitatsklinikum Bonn, D-53105 Bonn, Germany.; 2010;30:6355–9. http://www.ncbi.nlm.nih.gov/pubmed/20445061
62. Matsumoto L, Takuma H, Tamaoka A, Kurisaki H, Date H, Tsuji S, et al. CpG demethylation enhances alpha-synuclein expression and affects the pathogenesis of Parkinson's disease. PLoS One [Internet]. Division of Neuroscience, Department of Neurology, Graduate School of Medicine, The University of Tokyo, Bunkyo, Tokyo, Japan.; 2010;5:e15522. http://www.ncbi.nlm.nih.gov/pubmed/21124796
63. Desplats P, Spencer B, Coffee E, Patel P, Michael S, Patrick C, et al. Alpha-synuclein sequesters Dnmt1 from the nucleus: a novel mechanism for epigenetic alterations in Lewy body diseases. J Biol Chem [Internet]. Department of Neurosciences, School of Medicine, University of California at San Diego, La Jolla, California 92093, USA. pdesplat@ucsd.edu; 2011;286:9031–7. http://www.ncbi.nlm.nih.gov/pubmed/21296890
64. de Boni L, Tierling S, Roeber S, Walter J, Giese A, Kretzschmar HA. Next-generation sequencing reveals regional differences of the alpha-synuclein methylation state independent of Lewy body disease. Neuromolecular Med [Internet]. The Center for Neuropathology and Prion Research, Ludwig-Maximilians-University Munich, Feodor-Lynen-Str. 23, 81377, Munich, Germany. Laura.de_Boni@med.uni.muenchen.de; 2011;13:310–20. http://www.ncbi.nlm.nih.gov/pubmed/22042430
65. Pihlstrom L, Berge V, Rengmark A, Toft M. Parkinson's disease correlates with promoter methylation in the alpha-synuclein gene. Mov Disord [Internet]. Department of Neurology, Oslo University Hospital, Oslo, Norway.; 2015;30:577–80. http://www.ncbi.nlm.nih.gov/pubmed/25545759
66. Ai SX, Xu Q, Hu YC, Song CY, Guo JF, Shen L, et al. Hypomethylation of SNCA in blood of patients with sporadic Parkinson's disease. J Neurol Sci [Internet]. Department of Neurology, Xiangya Hospital, Central South University, Changsha, China. Department of Geriatric Neurology, Xiangya Hospital, Central South University, Changsha, China. Department of Neurology, Xiangya Hospital, Central South University, Chan; 2014;337:123–8. http://www.ncbi.nlm.nih.gov/pubmed/24326201
67. Tan YY, Wu L, Zhao ZB, Wang Y, Xiao Q, Liu J, et al. Methylation of alpha-synuclein and leucine-rich repeat kinase 2 in leukocyte DNA of Parkinson's disease patients. Park Relat Disord [Internet]. Department of Neurology and Institute of Neurology, Ruijin Hospital Affiliated to Shanghai Jiao Tong University School of Medicine, Shanghai 200025, China.

Electronic address: rabbit82@gmail.com. Department of Neurology and Institute of Neurology, Ruijin; 2014;20:308–13. http://www.ncbi.nlm.nih.gov/pubmed/24398085
68. Song Y, Ding H, Yang J, Lin Q, Xue J, Zhang Y, et al. Pyrosequencing analysis of SNCA methylation levels in leukocytes from Parkinson's disease patients. Neurosci Lett [Internet]. Department of Neurology, Xuanwu Hospital of Capital Medical University, Key Laboratory for Neurodegenerative Diseases of Ministry of Education, Beijing 100053, PR China. Department of Neurology, Xuanwu Hospital of Capital Medical University, Key Laborator; 2014;569:85–8. http://www.ncbi.nlm.nih.gov/pubmed/24721670
69. Plagnol V, Nalls MA, Bras JM, Hernandez DG, Sharma M, Sheerin U-M, et al. A Two-Stage Meta-Analysis Identifies Several New Loci for Parkinson's Disease. Gibson G, editor. PLoS Genet [Internet]. 2011 Jun 30;7(6):e1002142. http://dx.plos.org/10.1371/journal.pgen.1002142
70. Pieper HC, Evert BO, Kaut O, Riederer PF, Waha A, Wüllner U. Different methylation of the TNF-alpha promoter in cortex and substantia nigra: Implications for selective neuronal vulnerability. Neurobiol Dis. 2008;32(3):521–7.
71. Kaut O, Schmitt I, Wüllner U. Genome-scale methylation analysis of Parkinson's disease patients' brains reveals DNA hypomethylation and increased mRNA expression of cytochrome P450 2E1. Neurogenetics. 2012;13(1):87–91.
72. Shahabi HN, Westberg L, Melke J, Håkansson A, Belin AC, Sydow O, et al. Cytochrome P450 2E1 gene polymorphisms/haplotypes and Parkinson's disease in a Swedish population. J Neural Transm. 2009;116(5):567–73.
73. Agirre X, Román-Gómez J, Vázquez I, Jiménez-Velasco A, Garate L, Montiel-Duarte C, et al. Abnormal methylation of the common PARK2 and PACRG promoter is associated with downregulation of gene expression in acute lymphoblastic leukemia and chronic myeloid leukemia. Int J Cancer. 2006;118(8):1945–53.
74. Cai M, Tian J, Zhao G, Luo W, Zhang B. Study of methylation levels of parkin gene promoter in Parkinson's disease patients. Int J Neurosci [Internet]. 2011;121(9):497–502. http://www.ncbi.nlm.nih.gov/pubmed/21663383
75. Yu J, Tao Q, Cheung KF, Jin H, Poon FF, Wang X, et al. Epigenetic identification of ubiquitin carboxyl-terminal hydrolase L1 as a functional tumor suppressor and biomarker for hepatocellular carcinoma and other digestive tumors. Hepatology. 2008;48(2):508–18.
76. Kagara I, Enokida H, Kawakami K, Matsuda R, Toki K, Nishimura H, et al. CpG Hypermethylation of the UCHL1 Gene Promoter is Associated With Pathogenesis and Poor Prognosis in Renal Cell Carcinoma. J Urol. 2008;180(1):343–51.
77. Barrachina M, Ferrer I. DNA methylation of Alzheimer disease and tauopathy-related genes in postmortem brain. J Neuropathol Exp Neurol. 2009;68(8):880–91.
78. Behrens MI, Brüggemann N, Chana P, Venegas P, Kägi M, Parrao T, et al. Clinical spectrum of Kufor-Rakeb syndrome in the Chilean kindred with ATP13A2 mutations. Mov Disord. 2010;25(12):1929–37.
79. Shock LS, Thakkar P V, Peterson EJ, Moran RG, Taylor SM. DNA methyltransferase 1, cytosine methylation, and cytosine hydroxymethylation in mammalian mitochondria. Proc Natl Acad Sci U S A [Internet]. 2011;108(9):3630–5. http://www.pubmedcentral.nih.gov/articlerender.fcgi?artid=3048134&tool=pmcentrez&rendertype=abstract
80. Hong EE, Okitsu CY, Smith AD, Hsieh C-L. Regionally specific and genome-wide analyses conclusively demonstrate the absence of CpG methylation in human mitochondrial DNA. Mol Cell Biol [Internet]. 2013;33(14):2683–90. http://www.pubmedcentral.nih.gov/articlerender.fcgi?artid=3700126&tool=pmcentrez&rendertype=abstract
81. Lardenoije R, Iatrou A, Kenis G, Kompotis K, Steinbusch HW, Mastroeni D, et al. The epigenetics of aging and neurodegeneration. Prog Neurobiol [Internet]. School for Mental Health and Neuroscience (MHeNS), Department of Psychiatry and Neuropsychology, Maastricht University, Universiteitssingel 50, 6200 MD Maastricht, The Netherlands. Center for Integrative Genomics, University of Lausanne, Genopode Building; 2015;131:21–64. http://www.ncbi.nlm.nih.gov/pubmed/26072273

82. Iacobazzi V, Castegna A, Infantino V, Andria G. Mitochondrial DNA methylation as a next-generation biomarker and diagnostic tool. Molecular Genetics and Metabolism. 2013. p. 25–34.
83. Feng S, Xiong L, Ji Z, Cheng W, Yang H. Correlation between increased ND2 expression and demethylated displacement loop of mtDNA in colorectal cancer. Mol Med Rep. 2012;6(1):125–30.
84. Pirola CJ, Fernandez Gianotti T, Burgueno a. L, Rey-Funes M, Loidl CF, Mallardi P, et al. Epigenetic modification of liver mitochondrial DNA is associated with histological severity of nonalcoholic fatty liver disease. Gut. 2012;1356–63.
85. Zinovkina LA, Zinovkin RA. DNA Methylation, Mitochondria, and Programmed Aging. Biochemistry [Internet]. 2015;80(12):1571–7. http://www.ncbi.nlm.nih.gov/pubmed/26638681
86. Tahiliani M, Koh KP, Shen Y, Pastor WA, Bandukwala H, Brudno Y, et al. Conversion of 5-methylcytosine to 5-hydroxymethylcytosine in mammalian DNA by MLL partner TET1. Science [Internet]. 2009;324(5929):930–5. http://www.pubmedcentral.nih.gov/articlerender.fcgi?artid=2715015&tool=pmcentrez&rendertype=abstract
87. Ficz G, Branco MR, Seisenberger S, Santos F, Krueger F, Hore T a, et al. Dynamic regulation of 5-hydroxymethylcytosine in mouse ES cells and during differentiation. Nature [Internet]. 2011;473(7347):398–402. http://www.ncbi.nlm.nih.gov/pubmed/21460836
88. Branco MR, Ficz G, Reik W. Uncovering the role of 5-hydroxymethylcytosine in the epigenome. Nat Rev Genet [Internet]. 2011;13(1):7–13. http://dx.doi.org/10.1038/nrg3080
89. Song C-X, Szulwach KE, Fu Y, Dai Q, Yi C, Li X, et al. Selective chemical labeling reveals the genome-wide distribution of 5-hydroxymethylcytosine. Nat Biotechnol. 2011;29(1):68–72.
90. Habibi E, Masoudi-Nejad A, Abdolmaleky HM, Haggarty SJ. Emerging roles of epigenetic mechanisms in Parkinson's disease. Funct Integr Genomics [Internet]. Laboratory of Systems Biology and Bioinformatics (LBB), Institute of Biochemistry and Biophysics and Center of Excellence in Biomathematics, University of Tehran, Tehran, Iran.; 2011;11:523–37. http://www.ncbi.nlm.nih.gov/pubmed/21892731
91. van Heesbeen HJ, Mesman S, Veenvliet J V, Smidt MP. Epigenetic mechanisms in the development and maintenance of dopaminergic neurons. Development [Internet]. 2013;140(6):1159–69. http://www.ncbi.nlm.nih.gov/pubmed/23444349
92. Park G, Tan J, Garcia G, Kang Y, Salvesen G, Zhang Z. Regulation of Histone Acetylation by Autophagy in Parkinson Disease. J Biol Chem [Internet]. Sanford-Burnham Medical Research Institute, United States; Central South University, China. Central South University, China zhangzhuohua@sklmg.edu.cn.; 2015; http://www.ncbi.nlm.nih.gov/pubmed/26699403
93. Song C, Kanthasamy A, Anantharam V, Sun F, Kanthasamy AG. Environmental neurotoxic pesticide increases histone acetylation to promote apoptosis in dopaminergic neuronal cells: relevance to epigenetic mechanisms of neurodegeneration. Mol Pharmacol [Internet]. Iowa Center for Advanced Neurotoxicology, Department of Biomedical Sciences, Iowa State University, Ames, IA 50011, USA.; 2010;77:621–32. http://www.ncbi.nlm.nih.gov/pubmed/20097775
94. Song C, Kanthasamy A, Jin H, Anantharam V, Kanthasamy AG. Paraquat induces epigenetic changes by promoting histone acetylation in cell culture models of dopaminergic degeneration. Neurotoxicology [Internet]. Department of Biomedical Sciences, Iowa Center for Advanced Neurotoxicology, Iowa State University, Ames, IA 50011, USA.; 2011;32:586–95. http://www.ncbi.nlm.nih.gov/pubmed/21777615
95. Nicholas AP, Lubin FD, Hallett PJ, Vattem P, Ravenscroft P, Bezard E, et al. Striatal histone modifications in models of levodopa-induced dyskinesia. J Neurochem [Internet]. Center for Neurodegeneration and Experimental Therapeutics, Department of Neurology, University of Alabama at Birmingham, AL 35294–0017, USA. nicholas@uab.edu; 2008;106:486–94. http://www.ncbi.nlm.nih.gov/pubmed/18410512

96. Darmopil S, Martín AB, De Diego IR, Ares S, Moratalla R. Genetic Inactivation of Dopamine D1 but Not D2 Receptors Inhibits L-DOPA-Induced Dyskinesia and Histone Activation. Biol Psychiatry. 2009;66(6):603–13.
97. Södersten E, Feyder M, Lerdrup M, Gomes AL, Kryh H, Spigolon G, et al. Dopamine Signaling Leads to Loss of Polycomb Repression and Aberrant Gene Activation in Experimental Parkinsonism. PLoS Genet. 2014;10(9).
98. Goers J, Manning-Bog AB, McCormack AL, Millett IS, Doniach S, Di Monte DA, et al. Nuclear localization of alpha-synuclein and its interaction with histones. Biochemistry [Internet]. Department of Chemistry and Biochemistry, University of California, Santa Cruz, California 95064, USA.; 2003;42:8465–71. http://www.ncbi.nlm.nih.gov/pubmed/12859192
99. Kontopoulos E, Parvin JD, Feany MB. Alpha-synuclein acts in the nucleus to inhibit histone acetylation and promote neurotoxicity. Hum Mol Genet [Internet]. Department of Pathology, Brigham and Women's Hospital, Program in Neuroscience, Harvard Medical School, 77 Avenue Louis Pasteur, Boston, MA 02115, USA.; 2006;15:3012–23. http://www.ncbi.nlm.nih.gov/pubmed/16959795
100. Du G, Liu X, Chen X, Song M, Yan Y, Jiao R, et al. Drosophila Histone Deacetylase 6 Protects Dopaminergic Neurons against -Synuclein Toxicity by Promoting Inclusion Formation. Mol Biol Cell [Internet]. 2010;21(13):2128–37. http://www.molbiolcell.org/content/21/13/2128.abstract
101. Vartiainen S, Pehkonen P, Lakso M, Nass R, Wong G. Identification of gene expression changes in transgenic *C. elegans* overexpressing human α-synuclein. Neurobiol Dis. 2006;22(3):477–86.
102. Bartel DP. MicroRNAs: target recognition and regulatory functions. Cell. 2009/01/27 ed. Howard Hughes Medical Institute, Massachusetts Institute of Technology, Cambridge, MA 02139, USA. dbartel@wi.mit.edu; 2009;136:215–33.
103. Cardo LF, Coto E, Ribacoba R, Menendez M, Moris G, Suarez E, et al. MiRNA profile in the substantia nigra of Parkinson's disease and healthy subjects. J Mol Neurosci [Internet]. Genetica Molecular-Laboratorio de Medicina, Hospital Universitario Central de Asturias, 33006, Oviedo, Spain.; 2014;54:830–6. http://www.ncbi.nlm.nih.gov/pubmed/25284245
104. Kim J, Inoue K, Ishii J, Vanti WB, Voronov S V, Murchison E, et al. A MicroRNA feedback circuit in midbrain dopamine neurons. Science (80-) [Internet]. Departments of Pathology and Neurology, Center for Neurobiology and Behavior, and Taub Institute, Columbia University, College of Physicians and Surgeons 15–403, 630 West 168th Street, New York, NY 10032, USA.; 2007;317:1220–4. http://www.ncbi.nlm.nih.gov/pubmed/17761882
105. Jankovic J, Chen S, Le WD. The role of Nurr1 in the development of dopaminergic neurons and Parkinson's disease. Prog Neurobiol. 2005;77(1–2):128–38.
106. Lungu G, Stoica G, Ambrus A. MicroRNA profiling and the role of microRNA-132 in neurodegeneration using a rat model. Neurosci Lett. 2013;553:153–8.
107. Kanagaraj N, Beiping H, Dheen ST, Tay SS. Downregulation of miR-124 in MPTP-treated mouse model of Parkinson's disease and MPP iodide-treated MN9D cells modulates the expression of the calpain/cdk5 pathway proteins. Neuroscience [Internet]. Department of Anatomy, Yong Loo Lin School of Medicine, National University Health System, National University of Singapore, Singapore 117597, Singapore. Department of Anatomy, Yong Loo Lin School of Medicine, National University Health System, National U; 2014;272:167–79. http://www.ncbi.nlm.nih.gov/pubmed/24792712
108. Crocker SJ, Smith PD, Jackson-Lewis V, Lamba WR, Hayley SP, Grimm E, et al. Inhibition of calpains prevents neuronal and behavioral deficits in an MPTP mouse model of Parkinson's disease. J Neurosci. 2003;23(10):4081–91.
109. Kim W, Lee Y, McKenna ND, Yi M, Simunovic F, Wang Y, et al. miR-126 contributes to Parkinson's disease by dysregulating the insulin-like growth factor/phosphoinositide 3-kinase signaling. Neurobiol Aging [Internet]. Department of Psychiatry, McLean Hospital, Harvard Medical School, Belmont, MA, USA; Department of Psychiatry, McLean Hospital, Harvard

Medical School, Belmont, MA, USA. Department of Psychiatry, McLean Hospital, Harvard Medical School, Belmont, MA, USA.; 2014;35:1712–21. http://www.ncbi.nlm.nih.gov/pubmed/24559646

110. Margis R, Margis R, Rieder CR. Identification of blood microRNAs associated to Parkinsonis disease. J Biotechnol [Internet]. Neurology Section, Movement Disorders Unit, Hospital de Clinicas de Porto Alegre, Rua Ramiro Barcelos 2350, 90035–00 Porto Alegre, RS, Brazil.; 2011;152:96–101. http://www.ncbi.nlm.nih.gov/pubmed/21295623
111. Khoo SK, Petillo D, Kang UJ, Resau JH, Berryhill B, Linder J, et al. Plasma-based circulating MicroRNA biomarkers for Parkinson's disease. J Park Dis [Internet]. Center for Neurodegenerative Science, Van Andel Institute, Grand Rapids, MI 49503, USA. Kean.Khoo@vai.org; 2012;2:321–31. http://www.ncbi.nlm.nih.gov/pubmed/23938262
112. Gehrke S, Imai Y, Sokol N, Lu B. Pathogenic LRRK2 negatively regulates microRNA-mediated translational repression. Nature [Internet]. 2010;466(7306):637–41. http://han.sub.uni-goettingen.de/han/GoogleScholar/www.nature.com/nature/journal/v466/n7306/full/nature09191.html
113. Cho HJ, Liu G, Jin SM, Parisiadou L, Xie C, Yu J, et al. Microrna-205 regulates the expression of parkinson's disease-related leucine-rich repeat kinase 2 protein. Hum Mol Genet. 2013;22(3):608–20.
114. Xiong R, Wang Z, Zhao Z, Li H, Chen W, Zhang B, et al. MicroRNA-494 reduces DJ-1 expression and exacerbates neurodegeneration. Neurobiol Aging. 2014;35(3):705–14.
115. Asikainen S, Rudgalvyte M, Heikkinen L, Louhiranta K, Lakso M, Wong G, et al. Global microRNA expression profiling of *Caenorhabditis elegans* Parkinson's disease models. J Mol Neurosci. 2010/01/22 ed. Department of Biosciences, Kuopio University, Kuopio, Finland.; 2010;41:210–8.
116. Minones-Moyano E, Porta S, Escaramis G, Rabionet R, Iraola S, Kagerbauer B, et al. MicroRNA profiling of Parkinson's disease brains identifies early downregulation of miR-34b/c which modulate mitochondrial function. Hum Mol Genet. 2011/05/12 ed. Genetic Causes of Disease Group, Genes and Disease Program, Centre for Genomic Regulation, Barcelona, Catalonia, Spain.; 2011;20:3067–78.
117. Villar-Menendez I, Porta S, Buira SP, Pereira-Veiga T, Diaz-Sanchez S, Albasanz JL, et al. Increased striatal adenosine A2A receptor levels is an early event in Parkinson's disease-related pathology and it is potentially regulated by miR-34b. Neurobiol Dis. 2014/06/04 ed. Institute of Neuropathology, Bellvitge University Hospital-ICS, [Bellvitge Biomedical Research Institute-] IDIBELL, L'Hospitalet de Llobregat, Spain. Departamento de Quimica Inorganica, Organica y Bioquimica, Facultad de Ciencias y Tecnologias Quimicas, C; 2014;69:206–14.
118. Kabaria S, Choi DC, Chaudhuri AD, Mouradian MM, Junn E. Inhibition of miR-34b and miR-34c enhances alpha-synuclein expression in Parkinson's disease. FEBS Lett. 2014/12/30 ed. Center for Neurodegenerative and Neuroimmunologic Diseases, Department of Neurology, Rutgers - Robert Wood Johnson Medical School, Piscataway, NJ 08854, USA. Center for Neurodegenerative and Neuroimmunologic Diseases, Department of Neurology, Rutgers - Ro; 2015;589:319–25.
119. Junn E, Lee KW, Jeong BS, Chan TW, Im JY, Mouradian MM. Repression of alpha-synuclein expression and toxicity by microRNA-7. Proc Natl Acad Sci U S A. 2009/07/25 ed. Center for Neurodegenerative and Neuroimmunologic Diseases, Department of Neurology, University of Medicine and Dentistry of New Jersey-Robert Wood Johnson Medical School, Piscataway, NJ 08854, USA. junneu@umdnj.edu; 2009;106:13052–7.
120. Fragkouli A, Doxakis E. miR-7 and miR-153 protect neurons against MPP(+)-induced cell death via upregulation of mTOR pathway. Front Cell Neurosci. 2014/07/30 ed. Lab of Molecular and Cellular Neuroscience, Center for Basic Research, Biomedical Research Foundation of the Academy of Athens Athens, Greece.; 2014;8:182.
121. Chen LL, Song JX, Lu JH, Yuan ZW, Liu LF, Durairajan SS, et al. Corynoxine, a natural autophagy enhancer, promotes the clearance of alpha-synuclein via Akt/mTOR pathway.

J Neuroimmune Pharmacol. 2014/02/14 ed. School of Chinese Medicine, Hong Kong Baptist University, Kowloon, Hong Kong.; 2014;9:380–7.
122. Doxakis E. Post-transcriptional regulation of alpha-synuclein expression by mir-7 and mir-153. J Biol Chem. 2010/01/29 ed. Basic Neurosciences Division, Biomedical Research Foundation of the Academy of Athens, Soranou Efesiou 4, Athens 11527, Greece. edoxakis@bioacademy.gr; 2010;285:12726–34.
123. Kim HJ, Park G, Jeon BS, Park WY, Kim YE. A mir-153 binding site variation in SNCA in a patient with Parkinson's disease. Mov Disord. 2013/05/16 ed. Departments of Neurology and Movement Disorder Center, Parkinson Study Group, and Neuroscience Research Institute, Seoul National University College of Medicine, Seoul, Korea.; 2013;28:1755–6.
124. Alvarez-Erviti L, Seow Y, Schapira AH V, Rodriguez-Oroz MC, Obeso JA, Cooper JM. Influence of microRNA deregulation on chaperone-mediated autophagy and α-synuclein pathology in Parkinson's disease. Cell Death Dis [Internet]. 2013;4:e545. http://www.pubmedcentral.nih.gov/articlerender.fcgi?artid=3615743&tool=pmcentrez&rendertype=abstract
125. Wang ZH, Zhang JL, Duan YL, Zhang QS, Li GF, Zheng DL. MicroRNA-214 participates in the neuroprotective effect of Resveratrol via inhibiting alpha-synuclein expression in MPTP-induced Parkinson's disease mouse. Biomed Pharmacother. 2015/09/10 ed. Department of Internal Neurology, Huaihe Hospital of Henan University, Kaifeng 475000, China. Electronic address: wzhdyl0526@163.com. Department of Internal Neurology, Huaihe Hospital of Henan University, Kaifeng 475000, China. Department of Ultrasound, K; 2015;74:252–6.
126. Gillardon F, Mack M, Rist W, Schnack C, Lenter M, Hildebrandt T, et al. MicroRNA and proteome expression profiling in early-symptomatic α-synuclein(A30P)-transgenic mice. Proteomics - Clin Appl. 2008;2(5):697–705.
127. Kong Y, Liang X, Liu L, Zhang D, Wan C, Gan Z, et al. High Throughput Sequencing Identifies MicroRNAs Mediating α-Synuclein Toxicity by Targeting Neuroactive-Ligand Receptor Interaction Pathway in Early Stage of Drosophila Parkinson's Disease Model. PLoS One [Internet]. 2015;10:e0137432. http://www.ncbi.nlm.nih.gov/pmc/articles/PMC4567341/
128. Outeiro TF, Klucken J, Strathearn KE, Liu F, Nguyen P, Rochet JC, et al. Small heat shock proteins protect against alpha-synuclein-induced toxicity and aggregation. Biochem Biophys Res Commun [Internet]. Alzheimer's Research Unit, MassGeneral Institute for Neurodegenerative Disease, MGH, Harvard Medical School, Charlestown, MA 02129, USA. touteir@partners.org; 2006;351:631–8. http://www.ncbi.nlm.nih.gov/pubmed/17081499
129. Klucken J, Outeiro TF, Nguyen P, McLean PJ, Hyman BT. Detection of novel intracellular alpha-synuclein oligomeric species by fluorescence lifetime imaging. FASEB J [Internet]. MassGeneral Institute for Neurodegenerative Disease, Alzheimer's Disease Research Unit, Massachusetts General Hospital, 114 16 St., Charlestown, MA 02129, USA.; 2006;20:2050–7. http://www.ncbi.nlm.nih.gov/pubmed/17012257
130. Klucken J, Shin Y, Masliah E, Hyman BT, McLean PJ. Hsp70 Reduces alpha-Synuclein Aggregation and Toxicity. J Biol Chem [Internet]. Alzheimer's Disease Research Laboratory, Harvard Medical School, Massachusetts General Hospital, Charlestown, Massachusetts 02129, USA.; 2004;279:25497–502. http://www.ncbi.nlm.nih.gov/pubmed/15044495
131. Outeiro TF, Putcha P, Tetzlaff JE, Spoelgen R, Koker M, Carvalho F, et al. Formation of toxic oligomeric alpha-synuclein species in living cells. PLoS One [Internet]. Alzheimer's Research Unit, MassGeneral Institute for Neurodegenerative Disease, MGH Harvard Medical School, Charlestown, Massachusetts, United States of America. touteiro@fm.ul.pt; 2008;3:e1867. http://www.ncbi.nlm.nih.gov/pubmed/18382657
132. Shin Y, Klucken J, Patterson C, Hyman BT, McLean PJ. The co-chaperone carboxyl terminus of Hsp70-interacting protein (CHIP) mediates alpha-synuclein degradation decisions between proteasomal and lysosomal pathways. J Biol Chem [Internet]. Alzheimer Disease Research Unit, Department of Neurology, Massachusetts General Hospital, Charlestown, Massachusetts 02129, USA.; 2005;280:23727–34. http://www.ncbi.nlm.nih.gov/pubmed/15845543

133. Zhang Z, Cheng Y. miR-16-1 Promotes the Aberrant α-Synuclein Accumulation in Parkinson Disease via Targeting Heat Shock Protein 70. Sci World J [Internet]. 2014;2014:8. http://dx.doi.org/10.1155/2014/938348
134. van der Walt JM, Noureddine MA, Kittappa R, Hauser MA, Scott WK, McKay R, et al. Fibroblast growth factor 20 polymorphisms and haplotypes strongly influence risk of Parkinson disease. Am J Hum Genet. 2004/05/04 ed. Department of Medicine and Center for Human Genetics, Duke University Medical Center, Durham, NC 27710, USA.; 2004;74:1121–7.
135. Wang G, van der Walt JM, Mayhew G, Li YJ, Zuchner S, Scott WK, et al. Variation in the miRNA-433 binding site of FGF20 confers risk for Parkinson disease by overexpression of alpha-synuclein. Am J Hum Genet. 2008/02/07 ed. Center for Human Genetics, Duke University Medical Center, Durham, NC 27710, USA.; 2008;82:283–9.
136. Bahari-Javan S, Sananbenesi F, Fischer A. Histone-acetylation: a link between Alzheimer's disease and post-traumatic stress disorder? Front Neurosci [Internet]. Department of Psychiatry and Psychotherapy, University Medical Center Gottingen Gottingen, Germany ; Research Group for Epigenetics in Neurodegenerative Diseases, German Center for Neurodegenerative Diseases (DZNE) Gottingen Germany. Research Group for Ep; 2014;8:160. http://www.ncbi.nlm.nih.gov/pubmed/25009454
137. Benito E, Urbanke H, Ramachandran B, Barth J, Halder R, Awasthi A, et al. HDAC inhibitor-dependent transcriptome and memory reinstatement in cognitive decline models. J Clin Invest [Internet]. 2015;125:3572–84. http://www.ncbi.nlm.nih.gov/pubmed/26280576
138. Fischer A, Sananbenesi F, Mungenast A, Tsai LH. Targeting the correct HDAC(s) to treat cognitive disorders. Trends Pharmacol Sci [Internet]. Laboratory for Aging and Cognitive Diseases, European Neuroscience Institute, Grisebach Str. 5, D-37077 Goettingen, Germany. afische2@gwdg.de; 2010;31:605–17. http://www.ncbi.nlm.nih.gov/pubmed/20980063
139. Outeiro TF, Kontopoulos E, Altmann SM, Kufareva I, Strathearn KE, Amore AM, et al. Sirtuin 2 inhibitors rescue alpha-synuclein-mediated toxicity in models of Parkinson's disease. Science (80-) [Internet]. Alzheimer's Research Unit, MGH, Harvard Medical School, CNY 114, 16th Street, Charlestown, MA 02129, USA.; 2007;317:516–9. http://www.ncbi.nlm.nih.gov/pubmed/17588900
140. Monti B, Gatta V, Piretti F, Raffaelli SS, Virgili M, Contestabile A. Valproic acid is neuroprotective in the rotenone rat model of Parkinson's disease: involvement of alpha-synuclein. Neurotox Res [Internet]. Department of Biology, University of Bologna, Bologna, Italy.; 2010;17:130–41. http://www.ncbi.nlm.nih.gov/pubmed/19626387
141. Kidd SK, Schneider JS. Protective effects of valproic acid on the nigrostriatal dopamine system in a 1-methyl-4-phenyl-1,2,3,6-tetrahydropyridine mouse model of Parkinson's disease. Neuroscience [Internet]. Department of Pathology, Anatomy and Cell Biology, Thomas Jefferson University, 1020 Locust Street, JAH 521, Philadelphia, PA 19107, USA.; 2011;194:189–94. http://www.ncbi.nlm.nih.gov/pubmed/21846494
142. Peng GS, Li G, Tzeng NS, Chen PS, Chuang DM, Hsu YD, et al. Valproate pretreatment protects dopaminergic neurons from LPS-induced neurotoxicity in rat primary midbrain cultures: role of microglia. Brain Res Mol Brain Res [Internet]. Department of Neurology, Tri-Service General Hospital, National Defense Medical Center, Taipei, Taiwan.; 2005;134:162–9. http://www.ncbi.nlm.nih.gov/pubmed/15790540
143. Chen PS, Wang CC, Bortner CD, Peng GS, Wu X, Pang H, et al. Valproic acid and other histone deacetylase inhibitors induce microglial apoptosis and attenuate lipopolysaccharide-induced dopaminergic neurotoxicity. Neuroscience [Internet]. Laboratory of Pharmacology and Chemistry, National Institute of Environmental Health Sciences, National Institutes of Health, Research Triangle Park, NC 27709, USA.; 2007;149:203–12. http://www.ncbi.nlm.nih.gov/pubmed/17850978
144. Marinova Z, Ren M, Wendland JR, Leng Y, Liang MH, Yasuda S, et al. Valproic acid induces functional heat-shock protein 70 via Class I histone deacetylase inhibition in cortical neurons: a potential role of Sp1 acetylation. J Neurochem [Internet]. Molecular Neurobiology

Section, National Institute of Mental Health, National Institutes of Health, Bethesda, Maryland 20892–1363, USA.; 2009;111:976–87. http://www.ncbi.nlm.nih.gov/pubmed/19765194
145. Jin H, Kanthasamy A, Harischandra DS, Kondru N, Ghosh A, Panicker N, et al. Histone hyperacetylation up-regulates protein kinase Cdelta in dopaminergic neurons to induce cell death: relevance to epigenetic mechanisms of neurodegeneration in Parkinson disease. J Biol Chem [Internet]. From the Department of Biomedical Sciences, Iowa Center for Advanced Neurotoxicology, Iowa State University, Ames, Iowa 50011. the Department of Molecular Pharmacology and Therapeutics, Stritch School of Medicine, Loyola University Chicago, Maywood, Illin; 2014;289:34743–67. http://www.ncbi.nlm.nih.gov/pubmed/25342743
146. Boutillier AL, Trinh E, Loeffler JP. Selective E2F-dependent gene transcription is controlled by histone deacetylase activity during neuronal apoptosis. J Neurochem [Internet]. Laboratoire de Signalisations Moleculaires et Neurodegenerescence, EA no. 3433, 11 rue Humann, 67085 Strasbourg cedex, France. laurette@neurochem.u-strasbg.fr; 2003;84:814–28. http://www.ncbi.nlm.nih.gov/pubmed/12562525
147. Wang Y, Wang X, Liu L, Wang X. HDAC inhibitor trichostatin A-inhibited survival of dopaminergic neuronal cells. Neurosci Lett [Internet]. Department of Physiology and Key Laboratory of the Neurodegenerative Disorders of the Chinese Ministry of Education, Capital Medical University, Youanmen, Beijing 100069, China.; 2009;467:212–6. http://www.ncbi.nlm.nih.gov/pubmed/19835929
148. Suuronen T, Huuskonen J, Pihlaja R, Kyrylenko S, Salminen A. Regulation of microglial inflammatory response by histone deacetylase inhibitors. J Neurochem [Internet]. Department of Neuroscience and Neurology, University of Kuopio, Kuopio, Finland.; 2003;87:407–16. http://www.ncbi.nlm.nih.gov/pubmed/14511118
149. Polymeropoulos MH, Higgins JJ, Golbe LI, Johnson WG, Ide SE, Di Iorio G, et al. Mapping of a gene for Parkinson's disease to chromosome 4q21-q23. Science. 1996;274(5290):1197–9.
150. Takahashi H, Ohama E, Suzuki S, Horikawa Y, Ishikawa a, Morita T, et al. Familial juvenile parkinsonism: clinical and pathologic study in a family. Neurology [Internet]. 1994;44(3 Pt 1):437–41. http://www.ncbi.nlm.nih.gov/pubmed/8145912
151. Matsumine H, Saito M, Shimoda-Matsubayashi S, Tanaka H, Ishikawa a, Nakagawa-Hattori Y, et al. Localization of a gene for an autosomal recessive form of juvenile Parkinsonism to chromosome 6q25.2-27. Am J Hum Genet. 1997;60:588–96.
152. Gasser T, Müller-Myhsok B, Wszolek ZK, Oehlmann R, Calne DB, Bonifati V, et al. A susceptibility locus for Parkinson's disease maps to chromosome 2p13. Nat Genet [Internet]. 1998;18(3):262–5. http://www.ncbi.nlm.nih.gov/pubmed/9500549, http://www.nature.com/ng/journal/v18/n3/pdf/ng0398-262.pdf
153. Leroy E, Boyer R, Auburger G, Leube B, Ulm G, Mezey E, et al. The ubiquitin pathway in Parkinson's disease. Nature. 1998;395(6701):451–2.
154. Leroy E. Intron-exon Structure of Ubiquitin C-terminal Hydrolase-L1. DNA Res [Internet]. Oxford University Press; 1998 Jan 1 [cited 2016 Apr 19];5(6):397–400. http://dnaresearch.oxfordjournals.org/cgi/content/long/5/6/397
155. Valente EM, Bentivoglio AR, Dixon PH, Ferraris A, Ialongo T, Frontali M, et al. Localization of a novel locus for autosomal recessive early-onset parkinsonism, PARK6, on human chromosome 1p35-p36. Am J Hum Genet [Internet]. 2001;68(4):895–900. http://www.pubmedcentral.nih.gov/articlerender.fcgi?artid=1275643&tool=pmcentrez&rendertype=abstract
156. van Duijn CM, Dekker MC, Bonifati V, Galjaard RJ, Houwing-Duistermaat JJ, Snijders PJ, et al. Park7, a novel locus for autosomal recessive early-onset parkinsonism, on chromosome 1p36. Am J Hum Genet. 2001;69(3):629–34.
157. Annesi G, Savettieri G, Pugliese P, D'Amelio M, Tarantino P, Ragonese P, et al. DJ-1 mutations and parkinsonism-dementia-amyotrophic lateral sclerosis complex. Ann Neurol. 2005;58(5):803–7.

158. Abou-Sleiman PM, Healy DG, Quinn N, Lees AJ, Wood NW. The role of pathogenic DJ-1 mutations in Parkinson's disease. Ann Neurol. 2003;54(3):283–6.
159. Hasegawa K, Kowa H. Autosomal dominant familial Parkinson disease: Older onset of age, and good response to Levodopa therapy. Eur Neurol. 1997;38(suppl1):39–43.
160. Wszolek ZK, Pfeiffer B, Fulgham JR, Parisi JE, Thompson BM, Uitti RJ, et al. Western Nebraska family (family D) with autosomal dominant parkinsonism. Neurology. 1995;45(3 Pt 1):502–5.
161. Paisán-Ruíz C, Jain S, Evans EW, Gilks WP, Simón J, Van Der Brug M, et al. Cloning of the gene containing mutations that cause PARK8-linked Parkinson's disease. Neuron. 2004;44(4):595–600.
162. Lin CH, Tan EK, Chen ML, Tan LC, Lim HQ, Chen GS, et al. Novel ATP13A2 variant associated with Parkinson disease in Taiwan and Singapore. Neurology. 2008;71(21):1727–32.
163. Di Fonzo A, Chien HF, Socal M, Giraudo S, Tassorelli C, Iliceto G, et al. ATP13A2 missense mutations in juvenile parkinsonism and young onset Parkinson disease. Neurology. 2007;68(19):1557–62.
164. Chan AYY, Baum L, Tang NLS, Lau CYK, Ng PW, Hui KF, et al. The role of the Ala746Thr variant in the ATP13A2 gene among Chinese patients with Parkinson's disease. J Clin Neurosci. 2013;20(5):761–2.
165. Li YJ, Deng J, Mayhew GM, Grimsley JW, Huo X, Vance JM. Investigation of the PARK10 gene in Parkinson disease. Ann Hum Genet. 2007;71(5):639–47.
166. Hicks AA, Petursson H, Jonsson T, Stefansson H, Johannsdottir HS, Sainz J, et al. A susceptibility gene for late-onset idiopathic Parkinson's disease. Ann Neurol. 2002;52(5):549–55.
167. Pankratz N, Nichols WC, Uniacke SK, Halter C, Rudolph A, Shults C, et al. Significant linkage of Parkinson disease to chromosome 2q36-37. Am J Hum Genet. 2003;72(4):1053–7.
168. Pankratz N, Nichols WC, Uniacke SK, Halter C, Murrell J, Rudolph A, et al. Genome-wide linkage analysis and evidence of gene-by-gene interactions in a sample of 362 multiplex Parkinson disease families. Hum Mol Genet. 2003;12(20):2599–608.
169. Pankratz N, Nichols WC, Uniacke SK, Halter C, Rudolph A, Shults C, et al. Genome screen to identify susceptibility genes for Parkinson disease in a sample without parkin mutations. Am J Hum Genet. 2002;71:124–35.
170. Scott WK, Nance MA, Watts RL, Hubble JP, Koller WC, Lyons K, et al. Complete genomic screen in Parkinson disease: evidence for multiple genes. JAMA [Internet]. 2001;286(18):2239–44. http://www.ncbi.nlm.nih.gov/pubmed/11710888
171. Strauss KM, Martins LM, Plun-Favreau H, Marx FP, Kautzmann S, Berg D, et al. Loss of function mutations in the gene encoding Omi/HtrA2 in Parkinson's disease. Hum Mol Genet. 2005;14(15):2099–111.
172. Simón-Sánchez J, Singleton AB. Sequencing analysis of OMI/HTRA2 shows previously reported pathogenic mutations in neurologically normal controls. Hum Mol Genet. 2008;17(13):1988–93.
173. Shi CH, Tang BS, Wang L, Lv ZY, Wang J, Luo LZ, et al. PLA2G6 gene mutation in autosomal recessive early-onset parkinsonism in a Chinese cohort. Neurology. 2011;77(1):75–81.
174. Yoshino H, Tomiyama H, Tachibana N, Ogaki K, Li Y, Funayama M, et al. Phenotypic spectrum of patients with PLA2G6 mutation and PARK14-linked parkinsonism. Neurology. 2010;75(15):1356–61.
175. Paisan-Ruiz C, Bhatia KP, Li A, Hernandez D, Davis M, Wood NW, et al. Characterization of PLA2G6 as a locus for dystonia-parkinsonism. Ann Neurol. 2009;65(1):19–23.
176. Di Fonzo A, Dekker MC, Montagna P, Baruzzi A, Yonova EH, Correia Guedes L, et al. FBXO7 mutations cause autosomal recessive, early-onset parkinsonian-pyramidal syndrome. Neurology [Internet]. 2009;72(3):240–5. http://www.ncbi.nlm.nih.gov/pubmed/19038853, http://www.neurology.org/content/72/3/240.full.pdf
177. Shojaee S, Sina F, Banihosseini SS, Kazemi MH, Kalhor R, Shahidi GA, et al. Genome-wide Linkage Analysis of a Parkinsonian-Pyramidal Syndrome Pedigree by 500 K SNP Arrays. Am J Hum Genet. 2008;82(6):1375–84.

178. Satake W, Nakabayashi Y, Mizuta I, Hirota Y, Ito C, Kubo M, et al. Genome-wide association study identifies common variants at four loci as genetic risk factors for Parkinson's disease. Nat Genet [Internet]. 2009;41(12):1303–7. http://www.ncbi.nlm.nih.gov/pubmed/19915576
179. Simón-Sánchez J, Schulte C, Bras JM, Sharma M, Gibbs JR, Berg D, et al. Genome-wide association study reveals genetic risk underlying Parkinson's disease. Nat Genet [Internet]. 2009;41(12):1308–12. http://www.pubmedcentral.nih.gov/articlerender.fcgi?artid=2787725&tool=pmcentrez&rendertype=abstract
180. Wider C, Skipper L, Solida A, Brown L, Farrer M, Dickson D, et al. Autosomal dominant dopa-responsive parkinsonism in a multigenerational Swiss family. Park Relat Disord. 2008;14(6):465–70.
181. Chartier-Harlin MC, Dachsel JC, Vilariño-Güell C, Lincoln SJ, Leprêtre F, Hulihan MM, et al. Translation initiator EIF4G1 mutations in familial parkinson disease. Am J Hum Genet. 2011;89(3):398–406.
182. Hamza TH, Zabetian CP, Tenesa A, Laederach A, Montimurro J, Yearout D, et al. Common genetic variation in the HLA region is associated with late-onset sporadic Parkinson's disease. Nat Genet [Internet]. 2010;42(9):781–5. http://dx.doi.org/10.1038/ng.642
183. Nuytemans K, Bademci G, Inchausti V, Dressen A, Kinnamon DD, Mehta A, et al. Whole exome sequencing of rare variants in EIF4G1 and VPS35 in Parkinson disease. Neurology. 2013;80(11):982–9.
184. Edvardson S, Cinnamon Y, Ta-Shma A, Shaag A, Yim YI, Zenvirt S, et al. A deleterious mutation in DNAJC6 encoding the neuronal-specific clathrin-uncoating Co-chaperone auxilin, is associated with juvenile parkinsonism. PLoS One. 2012;7(5).
185. Köroğlu Ç, Baysal L, Cetinkaya M, Karasoy H, Tolun A. DNAJC6 is responsible for juvenile parkinsonism with phenotypic variability. Park Relat Disord. 2013;19(3):320–4.
186. Krebs CE, Karkheiran S, Powell JC, Cao M, Makarov V, Darvish H, et al. The sac1 domain of SYNJ1 identified mutated in a family with early-onset progressive parkinsonism with generalized seizures. Hum Mutat. 2013;34(9):1200–7.
187. Quadri M, Fang M, Picillo M, Olgiati S, Breedveld GJ, Graafland J, et al. Mutation in the SYNJ1 gene associated with autosomal recessive, early-onset parkinsonism. Hum Mutat. 2013;34(9):1208–15.

Part V

New Approaches for Neuroepigenomic Studies

Single-Cell Genomics Unravels Brain Cell-Type Complexity

20

Amy Guillaumet-Adkins and Holger Heyn

Abstract

The brain is the most complex tissue in terms of cell types that it comprises, to the extent that it is still poorly understood. Single cell genome and transcriptome profiling allow to disentangle the neuronal heterogeneity, enabling the categorization of individual neurons into groups with similar molecular signatures. Herein, we unravel the current state of knowledge in single cell neurogenomics. We describe the molecular understanding of the cellular architecture of the mammalian nervous system in health and in disease; from the discovery of unrecognized cell types to the validation of known ones, applying these state-of-the-art technologies.

Keywords

Single cell genomics • single cell transcriptomics • neurobiology • cortex • hippocampus • midbrain • somatosensory nervous system • oligodendrocyte • neurodegenerative diseases and neuro-oncology

---

A. Guillaumet-Adkins, Ph.D.
CNAG-CRG, Centre for Genomic Regulation (CRG), Barcelona Institute of Science and Technology (BIST), Universitat Pompeu Fabra (UPF), Parc Científic de Barcelona – Torre I, Baldiri Reixac, 4, Barcelona, Catalonia 08028, Spain
e-mail: amy.guillaumet@cnag.crg.eu

H. Heyn, Ph.D. (✉)
CNAG-CRG, Centre for Genomic Regulation (CRG), Barcelona Institute of Science and Technology (BIST), National Centre for Genomic Analysis (CNAG), Parc Científic de Barcelona – Torre I, Baldiri Reixac, 4, Barcelona, Catalonia 08028, Spain
e-mail: holger.heyn@cnag.crg.eu

R. Delgado-Morales (ed.), *Neuroepigenomics in Aging and Disease*, Advances in Experimental Medicine and Biology, DOI 10.1007/978-3-319-53889-1_20

20.1 Introduction

Single cells are the fundamental units of life. It has not been until recently that single-cell analysis has enabled us to interrogate the heterogeneity of complex cellular populations at ultrahigh resolution. The development of powerful single-cell genomics techniques and advances in next-generation sequencing technologies made the sequencing of thousands of cells feasible and affordable. Single-cell studies reveal information that work performed on bulk populations could not address. They provide fine-grained resolution to define cell-type heterogeneity of complex tissues and define cell states in dynamic processes with high sensitivity. The mammalian brain is the most complex organ in our body, which is thought to contain thousands of cell types [1]. However, the function of most cell types in the nervous system is still unknown. Cells have historically been classified according to location, morphology, and electrophysiological characteristics, combined with molecular markers. Cell identities and function have been assigned through the expression of marker genes [2]. Large-scale projects, such as Allen Brain Atlas (ABA) or Gene Expression Nervous System Atlas (GENSAT) provided gene expression profiles of different brain regions and cell types [3, 4]. Single-cell technologies now allow the analysis of single-cell units that form this complex tissue and to identify cell types in an unbiased manner without prior knowledge about phenotype or function. In this chapter, we review how the single-cell genomics approaches are providing novel insights into the neural cell type heterogeneity and classification. We conclude by summarizing future applications of single-cell technologies in neuroscience.

20.2 Single-Cell Capture and Isolation

Isolation of single cells from complex tissues is a critical step in single-cell sequencing. It is very important to preserve cellular integrity to produce a representative image of the transcriptome for phenotype inference. For solid tissues, a proteolytic enzymatic treatment, such as collagenase, papain, and trypsin, is applied to obtain single-cell solutions. Caution has to be taken as such treatments can affect cell viability, which can later impact on transcriptional profiles or causes underrepresentation of certain cell types. The main approaches in isolation of cells from tissues or cell cultures are manual or automated micropipetting, laser capture microdissection (LCM), fluorescence-activated cell sorting (FACS), microfluidics, and droplet-based separation. In the following we summarize mechanisms, applications, and their pros and cons for each technique. Manual or automated micropipette and laser capture microdissection are low-throughput methods. Isolation is performed under a microscope based on visual inspection of cellular morphology. It enables specific cell selection but requires a trained experimentalist and very time-consuming. Micromanipulation is mostly used on early embryos and LCM on tissue sections [5]. FACS, microfluidics, and droplet techniques are high-throughput methods that are based on isolating cells based on cell-specific characteristics of fluorescence markers, light scattering, and size. FACS requires a large number of cells in suspension to set up the instrument, being a downfall on samples with low quantity. FACS instrument can accurately sort cells into the center of a well of microtiter plates, ensuring cells are immersed in the lysis buffer. FACS-based systems require a minimum reaction volume increasing the costs per cell. In contrary, microfluidic systems allow the isolation of single cells in microfluidic chips with individual micro-reaction chambers and micro-mechanical valves that allow automating downstream of biochemical reactions [6]. The most popular and commercial microfluidic systems in the market is the Fluidigm C1, which can capture up to 800 cells per chip, also allowing the visualization of cells under the microscope to identify potential doublets. Afterward, the reactions in nanoliter volumes are monitored, reducing the quantity of reagents required. Microdroplet systems, an emerging new technology, such as DROP-seq [7] or commercial 10× Genomics devices, encapsulate cells in aqueous droplets in flowing oil that included cell-specific barcoding. These techniques enable the processing of thousands of cells in parallel with relative low-sequencing library preparation cost per cell.

20.3 Single-Cell DNA Sequencing Methods

Only 6 pg of DNA is contained in a human diploid cell, being an insufficient material for standard DNA sequencing applications [8]. Therefore, single-cell genome sequencing requires whole-genome amplification (WGA) prior to library preparation. However, there are major challenges while generating high fidelity and unbiased WGA for adequate single-cell applications. Different approaches exist, but all have their advantages and limitations. Current WGA methodologies are based on multiple displacement amplification (MDA), polymerase chain reaction (PCR), or a combination of both methods.

MDA is based on the use of random hexamers binding to denatured DNA, where the phi29 polymerase catalyzes a strand displacement synthesis at a constant temperature [9]. The polymerase produces high DNA yield with high amplification fidelity [10]. However, compared to PCR-based techniques, it shows a significant amplification bias resulting in less evenly amplified genomes. The phi29 presents unique molecular properties, due to the ability of proofreading activity and high replication fidelity ($3' \rightarrow 5'$ exonuclease activity) [11]. It generates DNA amplicons of up to 10 kb in length.

PCR-based WGA methods rely on primer extension pre-amplification PCR (PEP-PCR), degenerated oligonucleotide-primed PCR (DOP-PCR), or linker-adaptor PCR (LA-PCR). During PEP-PCR, DNA is amplified with oligonucleotides of degenerate sequences using permissive thermocycling with increasing annealing temperatures [12]. In DOP-PCR, DNA is amplified with hybrid oligonucleotides containing the degenerate and unique sequences, starting with thermocycling at low annealing temperature (semi-random priming), followed by PCR at high annealing temperature (nonrandom priming) [13]. In contrary to the random priming-based methods, LA-PCR utilizes sheared or digested DNA and adaptors with universal sequences ligated to the DNA ends. The unique sequences are then used for subsequent PCR amplification [14].

A methodology that combines both MDA and PCR-based method is known as multiple annealing and looping-based amplification cycles (MALBAC). Its unique feature is the quasi-linear amplification to reduce the bias associated with nonlinear amplification methods [15]. The primers that anneal randomly to the genome contain specific sequences that allow the amplicons to form looped pre-amplification products. This looping protects previously copied segments to be further pre-amplified, therefore avoiding sequence-dependent biases by exponential amplification.

Several studies have been performed to compare WGA amplification techniques [16, 17]. They conclude that WGA from single cells presents a suitable tool for profiling copy number and structural variants or the detection of small-scale alteration, such as point mutations. Nevertheless, whole-genome DNA sequencing remains challenging due to the loss of material that causes dropout events or the introduction of sequencing errors that complicate variant calling.

20.4 Single-Cell RNA Sequencing Methods

The vast majority of single-cell RNA sequencing methodologies follow a similar strategy as DNA-based methods, since the typical mammalian cell contains 10 pg of RNA, but only 0.1 pg of messenger RNA. Thus profound amplification is required before sequencing libraries can be prepared. In initial steps, single cells are captured and lysed, and reverse transcription converts polyA-tailed RNA into cDNA. Then, the minute amounts of cDNA are amplified by PCR or by in vitro transcription before sequencing library preparation. The first single-cell whole-transcriptome method for mammalian cells was described by Tang [18]. Since then, many new methods have been developed that tackle different challenges and range from full-length approaches (SMART-seq [19] and SMART-seq2 [20]) to 5′-end- (STRT and STRT-C1 [21]) or 3′-end-focused (CEL-seq [22], CEL-seq2 [23], MARS-seq [24], Quartz-seq [25], and DROP-seq [7]) methods. Techniques that focus on the full-length approach rely on the Moloney murine leukemia virus (MMLV) reverse transcriptase. This enzyme presents unique properties that enable both template-switching and terminal transferase activity, resulting in the addition of a non-templated cytosine residue to the 5′ end of the cDNA. By adding a poly(G) template primer with an adapter sequence to the reaction, the enzyme can switch templates and transcribe the other strand. The resulting full-length cDNA can be amplified by PCR [19]. It is of note that all single-cell RNA sequencing methods depend on the amplification of the minute amount of starting material which can introduce technical variability and amplification bias. To correct such errors, methodologies focused on 5′- or 3′-end amplification include unique molecular indexes (UMIs) in the reverse transcription primer to label the original pool of RNA molecules [26]. Another approach to controlling this technical variability is the addition of external spike-in RNAs of known concentrations, such as the External RNA Controls Consortium (ERCC) mix [27].

20.5 Single-Cell Epigenomic Sequencing Methods

DNA methylome can be profiled at a single-cell level; however, technical peculiarities, such as the DNA degradation caused by the bisulfite conversion, challenge the preparation procedures. The first single-cell method to measure genome-wide 5-methylcytosine (5mC) levels utilized reduced single-cell representation bisulfite sequencing (scRBS). This technique digests the genomic DNA with restriction enzyme, prior to treatment with bisulfite. Resulting sequencing libraries are enriched in CpG methylation and, however, present limited genome coverage and a bias toward CpG-dense regions [28]. On the other hand, genome-wide DNA methylation profiling techniques provide a reasonable representation of each cell's DNA methylome [29, 30] but increasing sequencing cost per single cell. Subsequent comparative analysis enables the modeling of epigenetic dynamics and variability in contexts, such as development or differentiation.

Histones are subjected to a wide variety of posttranslational modifications. Technically, the profiling of the histone marks at a given genomic location is performed by chromatin immunoprecipitation followed by sequencing (ChIP-seq). On single-cell level, this technique is extremely challenging due to the background noise caused by nonspecific antibody binding. To overcome this problem, immunoprecipitation was performed on a pool of single-cell chromatin that underwent prior barcoding in droplet-based systems [31]. Another layer of epigenetic regulation is chromatin structure that can also be evaluated at a single-cell level. Two independent approaches have been developed to evaluate open chromatin regions that indicate regulatory activity. Buenrostro et al. used on a microfluidic device that implements an assay for transposase-accessible chromatin sequencing (ATAC-seq) [32]. Here, hyperactive prokaryotic Tn5 transposase inserts into accessible chromatin and tags the sites with sequencing adaptors. Cusanovich et al. used a combinatorial indexing strategy, where two-level tagmentation is carried out which introduces a unique barcode to each pool [33]. This strategy allows reactions with multiple cells to increase tagmentation efficiencies, while barcode combination allows the subsequent deconvolution to single-cell level. Chromosome conformation, a higher order of epigenome regulation, can be assessed by 3C-based method that can profile genome-wide chromosome interactions (e.g., HiC methods). At a single-cell level, HiC analysis revealed cell-to-cell variability in chromosome structure and compartmentalization [34, 35].

20.6 Linking Different Single-Cell Genomics Strategies

The combination of different single-cell genomics methods is an actively pursued issue in the field. Certain combinations of techniques measuring two modalities from a single cell are now possible. Macaulay et al. developed "genome and transcriptome" sequencing (scG&T) enabling the assessment of gene expression level and genetic variant from the same single cell [36]. Specifically, G&T-seq allows whole-genome and whole-transcriptome amplification following the physical separation of nucleic acids (DNA and mRNA) from a single cell. G&T-seq further modified the technique to allow the simultaneous assessment of DNA methylation and gene transcription (scM&T-seq) [37]. Both methodologies will link phenotypes (defined by gene expression or epigenetic states) to their genotypes. Combined single-cell methods enable the clear assessment of cellular relationships to better understand tissue heterogeneity in normal diseased states.

20.6.1 Studying Neurobiology Systems Using Single-Cell Sequencing Approaches

To date single-cell studies have been used to characterize cells from complex tissues, such as lung epithelium [38], spleen [24], or the pancreas [39]. In neuronal systems, single-cell genomics identified different neural types in various regions of the mouse and human nervous system. Here sample preparation and computational analysis face specific challenges due to the complex cell morphology and strongly interconnected expression profiles. Nevertheless, single-cell analysis allowed the identification of novel cell types with unique biological properties, the inference of neural connectivity, and the association to neurological diseases.

20.6.2 Transcriptome of the Mouse Cortex and Hippocampus at Single-Cell Resolution

The mammalian cerebral cortex is involved in cognitive functions such as sensorimotor integration, memory, and social behaviors. To elucidate the transcriptome of the primary somatosensory cortex (S1) and hippocampal (CA1) region, single-cell RNA-seq was performed on 3005 single cells, applying the 5′-end-focused STRT technology [40]. The resulting fine-grained characterization of cell type composition was confirmed by single-molecule RNA fluorescence in situ hybridization (RNA-FISH). The study identified nine molecularly distinct classes of cells by computational clustering, which were confirmed by the presence of specific markers, which play a functional role in the cell types. The nine clusters represented cortical and hippocampal pyramidal neurons, interneurons, oligodendrocytes, astrocytes, microglia, vascular endothelial cells, mural cells (pericytes and vascular smooth muscle cells), and ependymal cells. Exemplarily, S1 pyramidal cells were identified by the expression of *Tbr1* (*T-Box Brain 1*), a transcription factor implicated in neuronal migration and axonal projection; and oligodendrocytes were marked by *Mbp* (*Myelin basic protein*), a major constituent of the myelin sheath. Subsequently, biclustering analysis was performed on the nine major classes, identifying a total of 47 molecularly distinct subclasses of cell types. The authors observed that the RNA content was different among cells, with neurons containing more molecules than glia and vascular cells, correlating with a higher number of detected genes. Within the cortical pyramidal neurons, the work identified seven subclasses with layer specificity. Two subtypes of hippocampal pyramidal CA1 neurons were identified, associated to high mitochondrial function, plus cells derived from the adjacent CA2 and subiculum. Sixteen subclasses of interneurons were described, with regions both in the cortex and hippocampus containing these closely related subclasses. As for non-neural diversity, two major subtypes were identified for astrocytes and immune cells. In oligodendrocytes, six populations were found, representing different stages of maturation [40]. Ependymal cells expressed a largest set of subclass-specific genes. Altogether, the single-cell transcriptome study of primary somatosensory cortex and hippocampus indicates an extensive functional specialization between cellular subclasses reflected by specific gene expression patterns.

Another study profiling the cortex by single-cell RNA-seq utilized full-length transcriptome sequencing (SMART-seq technology). The work focused on a single cortical region, the primary visual mouse cortex [41], and used a transgenic mouse line with specifically labeled cortical cells [42]. By analyzing more than 1600 single cells, the authors identified 49 transcriptomic cortical cell types assigned by cell location and marker gene expression. These included 23 GABAergic, 19 glutamatergic, and 7 nonneuronal subtypes, with the majority of marker genes also detected in the RNA-ISH data from the Allen Brain Atlas [3]. Together, both studies provide a comprehensive overview of the transcriptomic landscape in the cortex and hippocampus areas of the mouse. Importantly, the unbiased analysis of single-cell transcriptomes allowed the identification of novel cell subtypes with likely highly specialized functions in these areas.

20.7 Transcriptome on the Human Cortex at Single-Cell Resolution

The human brain is a highly complex tissue, and study designs are hindered by the accessibility of material mostly being postmortem. Consequently, seminal studies on neuronal tissues have been performed on nonhuman model organisms, which may not recapitulate the full molecular complexity of the human brain. To tackle this problem, the use of human cerebral organoids has proven to reflect gene expression programs of the fetal neocortex development at a single-cell level [43].

To interrogate the heterogeneity of the human cortex, Darmanis et al. evaluated single-cell transcriptomes of normal primary fetal and adult brain tissue subjected to surgery using the SMART-seq technology. Their findings demonstrate that the transcriptomic profile obtained from single cells can successfully identify all major neuronal, glial, and vascular cell types in the human cerebral cortex. Moreover, the work supplemented traditional neuronal classifications based on marker genes with the underlying transcriptome. Also, subclustering neurons revealed two distinct groups, excitatory and inhibitory neurons. These cells represent a neuronal community with unique expression signatures and a specific role in the network niche. Gene expression analysis between pre- and postnatal neurons determined considerably distinct expression patterns. For example, adult neurons displayed high expression of *SNAP25* and *GAD1*; fetal neuronal progenitors expressed *MKI67* and *PAX6*;

and quiescent neurons revealed specific activation of *DCX* and *TUBB3*. Moreover, the work interrogated HLA expression in fetal and adult neurons, since the central nervous system (CNS) is believed to be immunologically inert. Although MCHI proteins are expressed in adult mouse brain [44], the expression of these genes in the human brain has been subject of controversial discussion, since their identification in a subpopulation of adult neurons.

A second study analyzed the transcriptome of a postmortem brain using neuronal nuclear antigen (NeuN) to isolate single neuronal nuclei and applying the SMART-seq technology for single-cell transcriptome generation. The study focused on six classically defined Brodmann areas (BAs) with well-documented anatomical and electrophysiological properties. Single-cell transcriptome profiles identified two major classifications within the cerebral cortex, inhibitory neurons that encompass interneurons, and excitatory including pyramidal or projection neurons. Each class and their associated subtypes revealed significant cell heterogeneity among the BAs, indicating that neural composition varies profoundly among regions in the brain. Another study on the developmental cerebral cortex identified different cell types of dividing neural progenitor, radial glia, and newborn neurons to maturing neurons [45].

Taking together single-cell transcriptome studies laid the groundwork for the construction of a cellular atlas of the human cortex, being a stepping stone toward elucidating the full cellular complexity of the human brain [46].

20.8 Midbrain in Mouse and Human at Single-Cell Resolution

The midbrain is a portion of the central nervous system associated with vision, hearing, motor control, alertness, and temperature regulation. An initial single-cell study was performed on the midbrain dopaminergic (DA) system, to assess neuron diversity. The classical anatomical classification of the midbrain DA neurons are substantia nigra pars compacta (SNc), the ventral tegmental area (VTA), and the retrorubral area (RR). This area has its clinical importance due to its implication in Parkinson's disease (PD), schizophrenia, attention deficit hyperactivity disorders (ADHD), obsessive-compulsive disorder, addiction, and depression. To identify molecular distinct DA neurons, early postnatal brain from mice was analyzed. Specifically, FACS separation combined with the expression analysis of 96 genes (using Fluidigm Biomark system) discovered six different types of cells (DA^{1A}, DA^{1B}, DA^{2A}, DA^{2B}, DA^{3A}, and DA^{3B}), which could be validated by RNA-ISH [47]. A more recent study focused on the development of the midbrain in human and mouse [48]. Although it is thought that the development of the organ in humans follows similar sequence of events as in rodent, the degree of conservation is unclear. Single-cell RNA-seq was performed on the developmental ventral midbrain and 1907 mouse cells from 271 embryos and in 1977 human cells from 10 human embryos. The analysis at different time points uncovered dopaminergic progenitor specification, neurogenesis, and differentiation. Moreover, it identified that gene expression profiles were conserved across species, which, however, showed differences in proliferation, timing, and dopaminergic neuron development.

20.9 Somatosensory Nervous System in Mice at Single-Cell Resolution

The somatosensory nervous system responds to mechanical, thermal, and nociceptive stimuli. The ability to perceive and discriminate these sensations is due to the existence of specialized dorsal root ganglion (DRG). The system comprises diverse neuronal subsets with distinct conduction properties and peripheral and central innervations patterns. They include small-diameter unmyelinated C-fibers, thinly myelinated Aδ-fibers, and large-diameter thickly myelinated Aα-/β-fibers [49]. DRGs were dissected from the mouse lumbar to classify the neuronal types and to reveal the complexity of this primary sensory system. Analyzing the transcriptome of 799 single cells, four clusters could be distinguished and identified as known marker genes: first, the NF cluster expressing the neurofilament heavy chain (*Nefh*) and parvalbumin (*Pvalb*); second, the PEP cluster with active substance P (*Tac1*), TRKA (*Ntrk1*), and calcitonin gene-related peptide (*Calca*), associated with peptidergic nociceptors; the third subtype, the NP cluster, expressing *Mrgprd* and *P2rx3*, associated with nonpeptidergic nociceptors; and, lastly, the TH clusters showing expression of tyrosine hydroxylase (*Th*), associated with unmyelinated neurons. Within the main populations, a total of 11 neuronal classes were identified, NF1 to NF5, NP1 to NP3, PEP1, PEP2, and TH. This dissection illustrates the diversity of sensory neuron types and their cellular complexity [50].

20.9.1 Olfactory and Retina Neurogenesis in Mice at Single-Cell Resolution

Odor perception is the detection of odorants by olfactory receptors (ORs), located on olfactory sensory neurons (OSNs) in the epithelium of the nose. These receptors are seven-transmembrane domain G protein-coupled receptor, encoded by a large multi-gene family. In mice, odor detection is mediated by 1000 odorant receptors genes (*Olfrs*) and 350 pseudogenes [51]. Hanchate et al. evaluated how developing OSNs select *Olfr* for expression using single cell RNA-seq. Eighty-five single-cell transcriptomes were analyzed using an unsupervised algorithm that determines cellular state of differentiation in "pseudotime" which models the dynamics in gene expression during development [52]. The predicted trajectory reflected developmental progression from progenitors, precursors, and immature OSNs to mature OSNs. The results were confirmed using stage-specific markers (progenitor, *Ascl*; precursor, *Neurog1*; immature OSN, *Gap43*; mature OSN, *Omp*). Subsequently, four olfactory sensory transduction molecules downstream of odorant receptors,

Gnal, *Adcy3*, *Cnga2*, and *Cnga4*, supported the conclusions. Expression of *Olfr* appeared at a late precursor to early immature OSN stage, presenting low levels of multiple *Olfrs*. During subsequent development, the expression of single highly expressed *Olfr* overtakes the expression of the other family members. Further, coexpressed *Olfrs* overlap in zones of the nasal epithelium, suggesting regional biases. Hence, mature neurons express single *Olfrs* at elevated levels leading to two hypotheses to explain the phenomenon: First is the "winner-takes-all," when an *Olfr* becomes dominant and overtakes the other expressed *Olfrs*. A second model based on the selection of one single *Olfr* independently of *Olfrs* initially expressed [53].

The retina represents another excellent system to study neuronal diversity. The retina contains five neuronal classes defined by morphological, physiological, and molecular features that include retinal ganglion, bipolar, horizontal, photoreceptor, and amacrine. Droplet-based single-cell transcriptome sequencing was applied on 44,808 cells from the retina of a 14-day-old mouse to create a molecular atlas of retinal cells. Thirty-nine transcriptional distinct retinal cell populations were identified by unsupervised computational analysis. These matched known types and identified additional subpopulations corresponding to astrocytes (associated with the retinal ganglion cell axons exiting the retina), resident microglia, endothelial cells (intra-retinal vasculature), pericytes, and fibroblast. A further focus lied on the 21-amacrine subtypes as it represents the most morphologically diverse neuronal class since most lack a clear molecular marker profile. Single-cell analysis classified subpopulations in inhibitory (using GABA or glycine as neurotransmitter), excitatory (release of glutamate), and undefined cell types that do not express GABAergic, glycinergic, nor glutamatergic markers [7].

20.9.2 Oligodendrocyte Heterogeneity in the Mouse Central Nervous System at Single-Cell Resolution

Oligodendrocytes have been considered as a functionally homogenous population in the central nervous system (CNS). Using single-cell RNA sequencing on ten regions of the anterior-posterior and dorsal-ventral axis of the mouse juvenile and adult CNS, 13 distinct cell populations were identified. Here, clustering, differential expression, and pseudotime analysis led to the identification of the transcriptional continuum between oligodendrocyte populations. These were oligodendrocyte precursor cells (OPCs), differentiation-committed oligodendrocytes (COPs), newly formed oligodendrocytes (NFOL1 and NFOL2), myelin-forming oligodendrocytes (MFOL1 and MFOL2), mature oligodendrocytes (MOL1 to MOL6), and vascular and leptomeningeal cells (VLMCs). OPCs coexpressed *Pdgfra* and *Cspg4*, while COPs lacked the expression of the genes, but expressed *Neu4*, *Sox6*, *Bmp4*, and *Gpr17*. NFOL1 and NFOL2 expressed genes involved in early stages of differentiation. MFOL1 and MFOL2 showed activity of genes involved in the myelin formation. MOL1 to MOL6 expressed late oligodendrocytes differentiation genes (*Klk6* and *Apod*) and genes present in myelinating cells (*Trf* and *Pmp22*). VLMCs were identified as a second of *Pdgfra* population with low levels of *Cspg4* and high levels of laminins and collagens, concluding a transcriptional continuum between different oligodendrocyte populations across multiple regions of the CNS. From a translational perspective, the identification of these cell types could provide a new vision into the etiology of myelin disorders [54].

20.10 Neurodegenerative Diseases at Single-Cell Resolution

Neurodegenerative disease is the progressive loss of function and structure of neurons, including their cell death. Diseases such as Parkinson's, Alzheimer's, Huntington's, and amyotrophic lateral sclerosis result from neurodegenerative processes. At a single-cell level, Poulin et al. determined subclasses of DA, focusing on abnormalities in Parkinson's disease. Specifically, DA^{1A} neurons, located in the ventral tier of the substantia nigra, are most vulnerable to this disease. To determine the molecular peculiarities, the study evaluated the susceptibility of DA^{1A} neurons to the neurotoxin 1-methyl-4-phenyl-1,2,3,6-tetrahydropyridine (MPTP), commonly used to mimic pathological features of parkinsonism in mice [47]. The analysis of single cells revealed a decrease in *Aldh1a1*+ expression, previously reported to be downregulated in the disease [55], providing insights into disease-related cellular degeneration [48]. Cell replacement therapy is a promising avenue toward treatment of Parkinson's disease [56], and the use of transplantation of human fetal midbrain tissue, containing dopaminergic neurons, could be used as a therapeutic approach. In this regard, human pluripotent stem cell (hPSC)-derived dopaminergic neurons have been proven to recover behavior in animal models with Parkinson's disease [57]. La Manno et al. performed single-cell RNA-seq on human embryonic stem cell (hESCs) and human-induced pluripotent stem cells (hiPSCs) to evaluate their molecular composition. This approach could be used to assess the quality of stem cells for cell replacement therapy [48]. In Alzheimer's disease (AD), one of the most common forms of dementia, aneuploidy (aberrant chromosome copy numbers) has thought to be involved in the pathogenesis of this disease. To address this, single-cell whole-genome sequencing was performed on frontal cortex neurons from healthy individuals and from AD patients, concluding that aneuploidy does not play a role in the pathogenesis of this disease [58].

20.11 Neuro-oncology at Single-Cell Resolution

Cancer is a heterogeneous disease with molecular characteristics that depend on the tissue of origin. Tumors evolve from a single cell, due to the accumulation of genetic and epigenetic alterations. During the progression of a tumor, additional variations appear giving rise to different cell subpopulations and the related tumor subclonal structure. Tumor heterogeneity has many implications for clinical management since different tumor clones play different roles in disease initiation, progression, metastasis, and drug resistance [59]. Current strategies analyzing bulk tumor samples to determine tumor composition lack resolution and are insufficient to recapitulate the clonal structures of the tumors. Single-cell genomics strategies present a suitable solution to define subclonal tumor structures to unprecedented resolution. Single-cell DNA-seq was performed on glioblastoma, a common primary brain tumor with a high degree of cellular heterogeneity [60], to depict clonal diversity. The study revealed convergent evolution of EGFR mutations in different subclones from the same primary tumors [61], further supporting the complex structure tumors and related difficulties in therapeutic intervention. Single-cell RNA-seq on glioblastoma showed that cancer cells display a large range of intermediate phenotypes that do not fall into distinct classes of epithelial and/or mesenchymal cell types [62]. To characterize the cellular diversity within IDH-mutant tumors, single-cell RNA-seq was performed on oligodendroglioma and astrocytoma [63]. Both entities shared a developmental hierarchy, with most cells differentiating along two glia lineages. Tumor cells derived from a set of proliferating cancer stem cells, supporting the cancer stem model, an important finding for the management of the disease.

20.12 Conclusions and Future Directions

The cell type complexity of the brain is widely unknown and intensively debated [64]. To understand the cellular heterogeneity, cell types have to be profoundly characterized and phenotyped for functional interpretation. Single-cell techniques present a key technology to illuminate the biological complexity in a normal context and during diseases, and fundamental questions about cell identities are now being answered for the first time. The international community has gathered together to create the Human Cell Atlas project, creating a comprehensive reference map of human cells as a basis for understanding human biology and its perturbation leading to diseases.

References

1. Hawrylycz MJ, Lein ES, Guillozet-Bongaarts AL, Shen EH, Ng L, Miller JA, et al. An anatomically comprehensive atlas of the adult human brain transcriptome. Nature. 2012; 489:391–9.
2. Holmberg J, Perlmann T. Maintaining differentiated cellular identity. Nat Rev Genet. 2012;13:429–39.
3. Lein ES, Hawrylycz MJ, Ao N, Ayres M, Bensinger A, Bernard A, et al. Genome-wide atlas of gene expression in the adult mouse brain. Nature. 2007;445:168–76.
4. Gong S, Zheng C, Doughty ML, Losos K, Didkovsky N, Schambra UB, et al. A gene expression atlas of the central nervous system based on bacterial artificial chromosomes. Nature. 2003;425:917–25.
5. Frumkin D, Wasserstrom A, Itzkovitz S, Harmelin A, Rechavi G, Shapiro E. Amplification of multiple genomic loci from single cells isolated by laser micro-dissection of tissues. BMC Biotechnol. 2008;8:17.
6. Reece A, Xia B, Jiang Z, Noren B, McBride R, Oakey J. Microfluidic techniques for high throughput single cell analysis. Curr Opin Biotechnol. 2016;40:90–6.

7. Macosko EZ, Basu A, Satija R, Nemesh J, Shekhar K, Goldman M, et al. Highly parallel genome-wide expression profiling of individual cells using nanoliter droplets. Cell. 2015;161:1202–14.
8. Wang Y, Waters J, Leung ML, Unruh A, Roh W, Shi X, et al. Clonal evolution in breast cancer revealed by single nucleus genome sequencing. Nature. 2014;512:155–60.
9. Dean FB, Nelson JR, Giesler TL, Lasken RS. Rapid amplification of plasmid and phage DNA using Phi 29 DNA polymerase and multiply-primed rolling circle amplification. Genome Res. 2001;11:1095–9.
10. Spits C, Le Caignec C, De Rycke M, Van Haute L, Van Steirteghem A, Liebaers I, et al. Whole-genome multiple displacement amplification from single cells. Nat Protoc. 2006;1:1965–70.
11. Garmendia C, Bernad A, Esteban JA, Blanco L, Salas M. The bacteriophage phi 29 DNA polymerase, a proofreading enzyme. J Biol Chem. 1992;267:2594–9.
12. Zhang L, Cui X, Schmitt K, Hubert R, Navidi W, Arnheim N. Whole genome amplification from a single cell: implications for genetic analysis. Proc Natl Acad Sci U S A. 1992;89:5847–51.
13. Cheung VG, Nelson SF. Whole genome amplification using a degenerate oligonucleotide primer allows hundreds of genotypes to be performed on less than one nanogram of genomic DNA. Proc Natl Acad Sci U S A. 1996;93:14676–9.
14. Troutt AB, McHeyzer-Williams MG, Pulendran B, Nossal GJ. Ligation-anchored PCR: a simple amplification technique with single-sided specificity. Proc Natl Acad Sci U S A. 1992;89:9823–5.
15. Zong C, Lu S, Chapman AR, Xie XS. Genome-wide detection of single-nucleotide and copy-number variations of a single human cell. Science. 2012;338:1622–6.
16. de Bourcy CFA, De Vlaminck I, Kanbar JN, Wang J, Gawad C, Quake SR. A quantitative comparison of single-cell whole genome amplification methods. PLoS One. Public Library of Science; 2014;9:e105585.
17. Huang L, Ma F, Chapman A, Lu S. Xie XS. Methodology and Applications: Single-Cell Whole-Genome Amplification and Sequencing; 2015.
18. Tang F, Barbacioru C, Wang Y, Nordman E, Lee C, Xu N, et al. mRNA-Seq whole-transcriptome analysis of a single cell. Nat Methods. 2009;6:377–82.
19. Ramsköld D, Luo S, Wang Y-C, Li R, Deng Q, Faridani OR, et al. Full-length mRNA-Seq from single-cell levels of RNA and individual circulating tumor cells. Nat Biotechnol. 2012;30:777–82.
20. Picelli S, Björklund ÅK, Faridani OR, Sagasser S, Winberg G, Sandberg R. smart-seq2 for sensitive full-length transcriptome profiling in single cells. Nat Methods. 2013 Nov;10: 1096–8.
21. Islam S, Kjällquist U, Moliner A, Zajac P, Fan J-B, Lönnerberg P, et al. Highly multiplexed and strand-specific single-cell RNA 5′ end sequencing. Nat Protoc. 2012;7:813–28.
22. Hashimshony T, Wagner F, Sher N, Yanai I. CEL-Seq: Single-Cell RNA-Seq by Multiplexed Linear Amplification. Cell Rep. 2012;2:666–73.
23. Hashimshony T, Senderovich N, Avital G, Klochendler A, de Leeuw Y, Anavy L, et al. CEL-Seq2: sensitive highly-multiplexed single-cell RNA-Seq. Genome Biol. [Internet]. 2016 [cited 2016 Nov 14];17:77. Available from: http://genomebiology.biomedcentral.com/articles/10.1186/s13059-016-0938-8
24. Jaitin DA, Kenigsberg E, Keren-Shaul H, Elefant N, Paul F, Zaretsky I, et al. Massively Parallel Single-Cell RNA-Seq for Marker-Free Decomposition of Tissues into Cell Types. Science. 2014;343:776–9.
25. Sasagawa Y, Nikaido I, Hayashi T, Danno H, Uno KD, Imai T, et al. Quartz-Seq: a highly reproducible and sensitive single-cell RNA sequencing method, reveals non-genetic gene-expression heterogeneity. Genome Biol. 2013;14:R31.
26. Islam S, Zeisel A, Joost S, La Manno G, Zajac P, Kasper M, et al. Quantitative single-cell RNA-seq with unique molecular identifiers. Nat Meth.; 2014;11:163–6.
27. Baker SC, Bauer SR, Beyer RP, Brenton JD, Bromley B, Burrill J, et al. The External RNA Controls Consortium: a progress report. Nat Methods. 2005;2:731–4.

28. Guo H, Zhu P, Wu X, Li X, Wen L, Tang F. Single-cell methylome landscapes of mouse embryonic stem cells and early embryos analyzed using reduced representation bisulfite sequencing. Genome Res. 2013;23:2126–35.
29. Smallwood SA, Lee HJ, Angermueller C, Krueger F, Saadeh H, Peat J, et al. Single-cell genome-wide bisulfite sequencing for assessing epigenetic heterogeneity. Nat Methods. 2014;11:817–20.
30. Farlik M, Sheffield NC, Nuzzo A, Datlinger P, Schnegger A, Klughammer J, et al. Single-Cell DNA Methylome Sequencing and Bioinformatic Inference of Epigenomic Cell-State Dynamics. Cell Rep. 2015;10:1386–97.
31. Rotem A, Ram O, Shoresh N, Sperling RA, Goren A, Weitz DA, et al. Single-cell ChIP-seq reveals cell subpopulations defined by chromatin state. Nat Biotechnol. 2015;33:1165–72.
32. Buenrostro JD, Wu B, Litzenburger UM, Ruff D, Gonzales ML, Snyder MP, et al. Single-cell chromatin accessibility reveals principles of regulatory variation. Nature. 2015;523:486–90.
33. Cusanovich DA, Daza R, Adey A, Pliner HA, Christiansen L, Gunderson KL, et al. Multiplex single-cell profiling of chromatin accessibility by combinatorial cellular indexing. Science. 2015;348:910–4.
34. Nagano T, Lubling Y, Yaffe E, Wingett SW, Dean W, Tanay A, et al. Single-cell Hi-C for genome-wide detection of chromatin interactions that occur simultaneously in a single cell. Nat Protoc. 2015;10:1986–2003.
35. Nagano T, Lubling Y, Stevens TJ, Schoenfelder S, Yaffe E, Dean W, et al. Single-cell Hi-C reveals cell-to-cell variability in chromosome structure. Nature. 2013;502:59–64.
36. Macaulay IC, Haerty W, Kumar P, Li YI, Hu TX, Teng MJ, et al. G&T-seq: parallel sequencing of single-cell genomes and transcriptomes. Nat Methods. 2015;12:519–22.
37. Angermueller C, Clark SJ, Lee HJ, Macaulay IC, Teng MJ, Hu TX, et al. Parallel single-cell sequencing links transcriptional and epigenetic heterogeneity. Nat Methods. 2016;13:229–32.
38. Treutlein B, Brownfield DG, Wu AR, Neff NF, Mantalas GL, Espinoza FH, et al. Reconstructing lineage hierarchies of the distal lung epithelium using single-cell RNA-seq. Nature. 2014; 509:371–5.
39. Muraro MJ, Dharmadhikari G, Grün D, Groen N, Dielen T, Jansen E, et al. A Single-Cell Transcriptome Atlas of the Human Pancreas. Cell Syst. 2016;3:385–394.e3.
40. Zeisel A, Muñoz-Manchado AB, Codeluppi S, Lönnerberg P, La Manno G, Juréus A, et al. Brain structure. Cell types in the mouse cortex and hippocampus revealed by single-cell RNA-seq. Science. 2015;347:1138–42.
41. Tasic B, Menon V, Nguyen TN, Kim TK, Jarsky T, Yao Z, et al. Adult mouse cortical cell taxonomy revealed by single cell transcriptomics. Nat Neurosci. 2016;19:335–46.
42. Harris JA, Hirokawa KE, Sorensen SA, Gu H, Mills M, Ng LL, et al. Anatomical characterization of Cre driver mice for neural circuit mapping and manipulation. Front Neural Circuits. 2014;8
43. Camp JG, Badsha F, Florio M, Kanton S, Gerber T, Wilsch-Bräuninger M, et al. Human cerebral organoids recapitulate gene expression programs of fetal neocortex development. Proc Natl Acad Sci. 2015;201520760
44. Goddard CA, Butts DA, Shatz CJ. Regulation of CNS synapses by neuronal MHC class I. Proc Natl Acad Sci. 2007;104:6828–33.
45. Pollen AA, Nowakowski TJ, Shuga J, Wang X, Leyrat AA, Lui JH, et al. Low-coverage single-cell mRNA sequencing reveals cellular heterogeneity and activated signaling pathways in developing cerebral cortex. Nat Biotechnol. 2014;32:1053–8.
46. Darmanis S, Sloan SA, Zhang Y, Enge M, Caneda C, Shuer LM, et al. A survey of human brain transcriptome diversity at the single cell level. Proc Natl Acad Sci U S A. 2015;112:7285–90.
47. Poulin J-F, Zou J, Drouin-Ouellet J, Kim K-YA, Cicchetti F, Awatramani RB. Defining Midbrain Dopaminergic Neuron Diversity by Single-Cell Gene Expression Profiling. Cell Rep. 2014;9:930–43.
48. La Manno G, Gyllborg D, Codeluppi S, Nishimura K, Salto C, Zeisel A, et al. Molecular Diversity of Midbrain Development in Mouse, Human, and Stem Cells. Cell 2016;167:566–580.e19.

49. Abraira VE, Ginty DD. The Sensory Neurons of Touch. Neuron. 2013;79:618–39.
50. Usoskin D, Furlan A, Islam S, Abdo H, Lönnerberg P, Lou D, et al. Unbiased classification of sensory neuron types by large-scale single-cell RNA sequencing. Nat Publ Gr. 2014;18
51. Niimura Y, Nei M. Comparative evolutionary analysis of olfactory receptor gene clusters between humans and mice. Gene. 2005;346:13–21.
52. Trapnell C, Cacchiarelli D, Grimsby J, Pokharel P, Li S, Morse M, et al. The dynamics and regulators of cell fate decisions are revealed by pseudotemporal ordering of single cells. Nat Biotechnol. 2014;32:381–6.
53. Hanchate NK, Kondoh K, Lu Z, Kuang D, Ye X, Qiu X, et al. Single-cell transcriptomics reveals receptor transformations during olfactory neurogenesis. Science. 2015;350:1251–5.
54. Marques S, Zeisel A, Codeluppi S, van Bruggen D, Mendanha Falcao A, Xiao L, et al. Oligodendrocyte heterogeneity in the mouse juvenile and adult central nervous system. Science. 2016;352:1326–9.
55. Grünblatt E, Riederer P. Aldehyde dehydrogenase (ALDH) in Alzheimer's and Parkinson's disease. J Neural Transm. 2016;123:83–90.
56. Barrow TR. Cell replacement therapy in Parkinson's disease. Biosci. Horizons. Oxford University Press. 2015;8:hzv002-hzv002.
57. Kirkeby A, Grealish S, Wolf DA, Nelander J, Wood J, Lundblad M, et al. Generation of Regionally Specified Neural Progenitors and Functional Neurons from Human Embryonic Stem Cells under Defined Conditions. Cell Rep. 2012;1:703–14.
58. van den Bos H, Spierings DCJ, Taudt A, Bakker B, Porubský D, Falconer E, et al. Single-cell whole genome sequencing reveals no evidence for common aneuploidy in normal and Alzheimer's disease neurons. Genome Biol. 2016;17:116.
59. Alizadeh AA, Aranda V, Bardelli A, Blanpain C, Bock C, Borowski C, et al. Toward understanding and exploiting tumor heterogeneity. Nat Med. 2015;21:846–53.
60. Furnari FB, Fenton T, Bachoo RM, Mukasa A, Stommel JM, Stegh A, et al. Malignant astrocytic glioma: genetics, biology, and paths to treatment. Genes Dev. 2007;21:2683–710.
61. Francis JM, Zhang C-Z, Maire CL, Jung J, Manzo VE, Adalsteinsson VA, et al. EGFR Variant Heterogeneity in Glioblastoma Resolved through Single-Nucleus Sequencing. Cancer Discov. 2014;4:956–71.
62. Patel AP, Tirosh I, Trombetta JJ, Shalek AK, Gillespie SM, Wakimoto H, et al. Single-cell RNA-seq highlights intratumoral heterogeneity in primary glioblastoma. Science. 2014;344:1396–401.
63. Tirosh I, Venteicher AS, Hebert C, Escalante LE, Patel AP, Yizhak K, et al. Single-cell RNA-seq supports a developmental hierarchy in human oligodendroglioma. Nature. 2016;539:309–13.
64. Ascoli GA, Alonso-Nanclares L, Anderson SA, Barrionuevo G, Benavides-Piccione R, Burkhalter A, et al. Petilla terminology: nomenclature of features of GABAergic interneurons of the cerebral cortex. Nat Rev Neurosci. 2008;9:557–68.

Epigenome Editing in the Brain

21

Pavel Bashtrykov and Albert Jeltsch

Abstract

Epigenome editing aims for an introduction or removal of chromatin marks at a defined genomic region using artificial EpiEffectors resulting in a modulation of the activity of the targeted functional DNA elements. Rationally designed EpiEffectors consist of a targeting DNA-binding module (such as a zinc finger protein, TAL effector, or CRISPR/Cas complex) and usually, but not exclusively, a catalytic domain of a chromatin-modifying enzyme. Epigenome editing opens a completely new strategy for basic research of the central nervous system and causal treatment of psychiatric and neurological diseases, because rewriting of epigenetic information can lead to the direct and durable control of the expression of disease-associated genes. Here, we review current advances in the design of locus- and allele-specific DNA-binding modules, approaches for spatial, and temporal control of EpiEffectors and discuss some examples of existing and propose new potential therapeutic strategies based on epigenome editing for treatment of neurodegenerative and psychiatric diseases. These include the targeted silencing of disease-associated genes or activation of neuroprotective genes which may be applied in Alzheimer's and Parkinson's diseases or the control of addiction and depression. Moreover, we discuss allele-specific epigenome editing as novel therapeutic approach for imprinting disorders, Huntington's disease and Rett syndrome.

Keywords

Chromatin modification • DNA methylation • Epigenome editing • CRISPR/Cas • Genome targeting • Alzheimer's disease • Parkinson's disease • Huntington's disease • Rett syndrome • Imprinting disorder

P. Bashtrykov (✉) • A. Jeltsch (✉)
Institute of Biochemistry, Faculty of Chemistry, University of Stuttgart, Pfaffenwaldring 55, 70569 Stuttgart, Germany
e-mail: pavel.bashtrykov@ibc.uni-stuttgart.de; albert.jeltsch@ibc.uni-stuttgart.de

R. Delgado-Morales (ed.), *Neuroepigenomics in Aging and Disease*, Advances in Experimental Medicine and Biology, DOI 10.1007/978-3-319-53889-1_21

21.1 The Concept of Epigenome Editing

Epigenetic signals, such as DNA methylation and hydroxymethylation, posttranslational modifications of histones, and noncoding RNAs, control cell type-specific genome usage which results in the formation of the more than hundred distinct cell phenotypes with unique gene expression profiles found in the human body [1–3]. However, epigenetic mechanisms not only control the development of cells and stable preservation of differentiated phenotypes over cell divisions but also reshape the transcriptome of cells in response to external signals originating in the human body or from the environment. The most exciting example of such genome-environment interaction is found in the epigenetic mechanisms controlling the central nervous system (CNS). Intensive research over the past decade revealed that fundamental processes such as synaptic plasticity, memory formation, acquired behavior, and others are regulated by epigenetic signals [4, 5]. More to that, many psychiatric and neurological diseases are caused by an impairment of epigenetic mechanisms in neurons, for example, drug addiction, schizophrenia, epilepsy, Alzheimer's and Parkinson's diseases, and depression [6, 7].

The conventional definition of epigenetic signals is based on their heritable but reversible nature as they do not include DNA mutations [8]. Since mature neurons do not undergo cell divisions, the concept of heritability of epigenetic signals is not applicable per se. However, research over the last decade has shown that the same molecular processes, which are involved in epigenetic chromatin regulation in dividing cells and developing organisms, have central functions in cells of the CNS as well, where they are involved in stable but still reversible gene regulation. This insight has led to the generation of a new term called neuroepigenetics, which expands the classical field of epigenetics and focuses on the chromatin regulation processes in neurons and their roles in many neuron-specific functions [9].

Intensive research on the epigenetic machinery over the past two decades has led to the discovery of several dozens of chromatin modifications and (partial) elucidation of their functions, which are introduced and removed by a multitude of chromatin-modifying enzymes [1]. Chromatin marks have characteristic spatial genomic distributions, and the localization at specific genomic loci dictates their regulatory effects [10]. Decoding of three key players—a chromatin mark with its function, a writing/erasing enzyme, and the locus-specific localization of a chromatin mark—gave a birth to the emerging field of epigenome editing [11–13]. The key component of this technique is the so-called EpiEffector, a designed fusion protein assembled from two functional domains, a catalytic domain of a chromatin-modifying enzyme and a designed DNA recognition domain [13] (Fig. 21.1a). The catalytic domain introduces or erases a chromatin mark according to its intrinsic substrate specificity, whereas the DNA recognition domain controls the locus-specific delivery of the catalytic domain due to its rationally designed recognition of a unique DNA target sequence. The existing portfolio of DNA-binding systems and chromatin modifiers together with the proceeding research in this field constantly expands the toolkit of EpiEffectors for basic research and potential clinical applications.

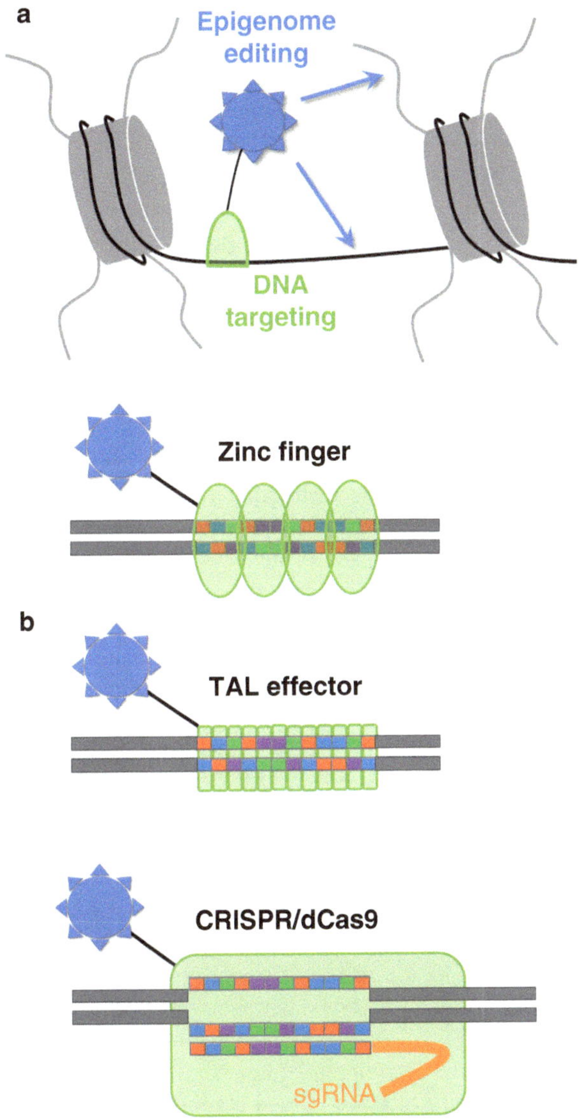

Fig. 21.1 (**a**) The concept of targeted epigenome editing. Epigenome editing is based on fusion proteins comprising a designed DNA recognition domain which targets an attached enzymatic domain to a defined genomic target sites. (**b**) Schematic drawing of the alternative DNA targeting domains currently in use. In zinc finger arrays, each zinc finger module (*green circle*) recognizes mainly three base pairs. In TAL effectors, each repeat (*green rectangle*) recognizes one base pair. In CRISPR/dCas9 (*green shade*), one strand of the target sites is recognized by Watson/Crick base pairing with a bound sgRNA (*orange*). Redrawn with modifications from [13]. Used with permission from Elsevier

21.2 DNA-Targeting Modules

The development of designed DNA-targeting modules (Fig. 21.1b) was initiated by the structural analysis of the zinc finger proteins (ZFPs) bound to specific DNA sequences in 1991, which led to an understanding of the principal rules governing the recognition of DNA sequences by ZFPs [14]. The most well-studied type of ZFPs is the Cys2-His2 zinc finger domain [15] consisting of 30 amino acids which bind three base pairs of DNA (and one in the adjacent triplet). The great features of the ZFPs, which led to their rapid adoption as tools in biotechnology, are the possibility to design finger modules with predetermined DNA-binding specificity [16, 17] and their modular structure giving an option to arrange fingers in arrays able to recognize 6, 9, 12, and more DNA base pairs. By this, individual finger motifs with known and validated sequence specificity can be combined for targeting of desired genomic loci [18]. Initially, engineered zinc finger arrays were fused to DNA cleavage domains generating zinc finger nucleases, which allow targeting a DNA double-strand break to a desired genomic locus afterward triggering gene deactivation through DNA repair [19]. An increased specificity in genome targeting was implemented in zinc finger nucleases made of modified FokI catalytic domains, which work only as a heterodimers and require the binding of two zinc finger nuclease fusion proteins in a close proximity, leading to a significantly reduction in off-target cleavage [20]. However, ZFPs were also fused to epigenetic enzymes generating EpiEffectors used for epigenetic reprogramming, for example, to introduce DNA or histone methylation [11, 13].

The next type of DNA-binding modules discovered were transcription activator-like effector (TALE) proteins. Like ZFPs, TALEs have a modular structure, where each module is built of 33–35 amino acids which recognize one base pair; they are organized in arrays of repeats recognizing a continuous stretch of base pairs [21, 22]. Binding to a single base pair makes TALEs more customizable than ZFPs and led to the generation of TALE libraries covering the entire human genome [23]. However, both ZFPs and TALEs are based on specific protein-DNA interaction for target site recognition. Hence changing the target site requires elaborate protein engineering and verification of the obtained proteins, where results turned out not always to be predictable.

The real revolution in the development of customizable DNA-targeting modules happened with the discovery of RNA-guided clustered regularly interspaced short palindromic repeats (CRISPR)/CRISPR-associated (Cas) system [24]. They originate from an adaptive immune system in bacteria and archaea which recognizes foreign DNA and cuts it via an intrinsic nuclease activity. Modification and adaptation of the CRISPR/Cas system for biotechnological applications resulted in two component systems containing a Cas9 nuclease and a single guide RNA (sgRNA), working as a complex which recognizes a 20 base pair DNA sequence complementary to the first 20 nucleotides of the sgRNA. An additional requirement for DNA binding is the presence of a short motif called the protospacer adjacent motif (PAM) next to the 20 base pairs in the DNA [25] which is bound by the Cas protein itself [26]. Redesigning of this system to a new target locus is very straightforward, since the sequence-specific interaction of the DNA-Cas9 complex outside of the PAM sequence is mediated by simple Watson-Crick base pairing between one strand of the targeted DNA and the sgRNA. Thus, it requires only the introduction of a new sgRNA into the complex to change the target specificity [27, 28].

Further rational design gave a birth to a catalytically inactive "dead" Cas9 variant, dCas9, which does not cleave DNA and can be used as DNA-targeting module. The fusion of dCas9 with epigenetic modifiers led to a vast field of applications in the control of gene expression and epigenome editing [29, 30]. Intensive search of Cas proteins discovered new orthologues of different size and different PAM sites, which significantly broadens the repertoire of targeted sequences [31–33]. Optimization of Cas9 protein by mutagenesis gave rise to variants with strongly reduced or without off-target binding [33, 34]. All these advances resulted in the recent application of dCas9-based systems in several epigenetics studies [35–38].

21.3 Chromatin-Modifying Modules

In order to use the different DNA-targeting devices in epigenome editing, they are genetically fused to chromatin-modifying modules including DNA and histone methyltransferases and demethylases, histone acetylases, and deacetylases to form fully functional EpiEffectors [13]. As described above, the main role of the epigenetic system is to orchestrate the genome-wide gene expression in accordance with the cell type-specific developmental program or in response to environmental signals, and this is often achieved by means of posttranslational modifications of histone proteins and DNA. Thus, the goal of epigenome editing is to modulate gene expression by changing chromatin marks at the targeted genomic loci using EpiEffectors [11, 13] (Fig. 21.2). Also, in a broader sense, other proteins which recruit chromatin-modifying enzymes working in complexes can be used instead of directly chromatin-modifying modules. Two directions of gene expression changes, activation of gene expression and gene silencing, can be achieved by the deposition or removal of activating or inactivating chromatin signals (Fig. 21.2), and both approaches have been successfully demonstrated in multiple studies [11, 13]. For example, DNA methylation located in promoters is a silencing chromatin mark, and targeted methylation or demethylation with EpiEffectors containing de novo DNA methyltransferase 3a [39–41] or Tet dioxygenase enzymes initiating DNA demethylation [42, 43] demonstrated predictable inactivation and activation of the genes correspondingly. Similarly, targeted deposition of H3K27 acetylation at promoters and enhancers by the p300 histone acetyltransferase fused to dCas9, ZFPs, or TALEs led to an activation of the corresponding genes [36], and methylation of H3K9 resulted in repression of gene expression [44, 45].

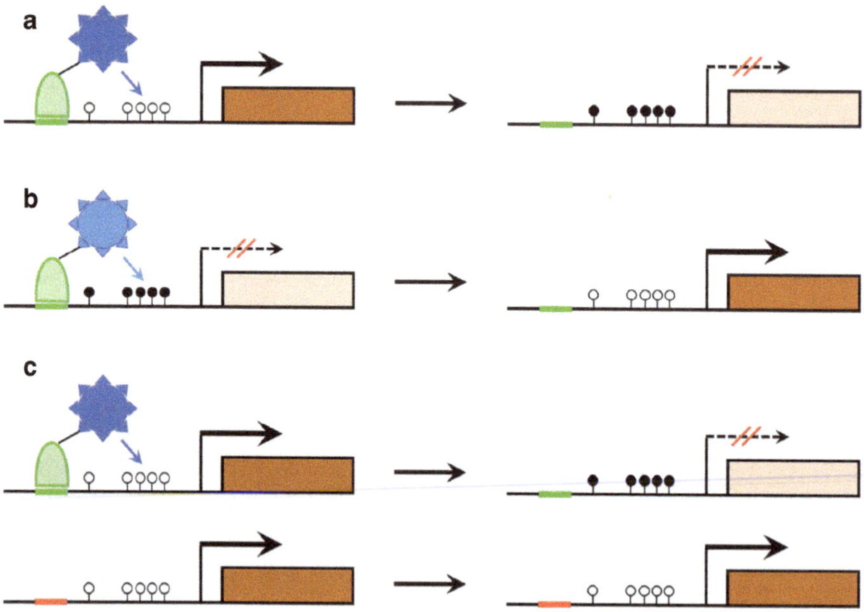

Fig. 21.2 Principles of targeted epigenome editing. (**a**) Silencing of gene expression by targeted DNA methylation. The DNA-binding domain is shown in green, the DNA methyltransferase domain in *dark blue*. The target sequence is indicated by a *green line*. Unmethylated CpG sites are indicated by open lollipops, methylated CpG sites by filled lollipops. (**b**) Activation of gene expression by targeted DNA demethylation. Elements are colored as described in panel a, except that the DNA demethyltransferase domain is *light blue*. (**c**) Allele-specific gene repression by allele-specific targeting of DNA methylation. Both alleles of the target gene are shown, one of them contains a target sequence (indicated by a *green line*), in the second allele the target sequence is disrupted by a SNP (indicated by a *red line*)

21.4 Epigenome Editing as Therapeutic Approach for Diseases of the CNS

Theoretically, many disorders of the CNS are connected to dysregulation of epigenetic mechanisms such that they may be considered as potential targets for an epigenome editing therapy. Here, we discuss some examples, which represent different types of diseases, and describe various mechanistic approaches for their potential treatment based on epigenome editing.

21.4.1 Alzheimer's Disease

Alzheimer's disease (AD) is characterized by a progressive dementia accompanied by extracellular accumulation of senile plaques made of amyloid β protein and hyperphosphorylation of tau proteins forming neurofibrillary tangles [46]. The downstream effect is a decrease of histone acetylation in hippocampal neurons, and it is known that histone acetylation is required for memory consolidation. Animal models for AD demonstrated that histone deacetylase (HDAC) inhibitors can increase the levels of histone 4 acetylation and rescue the associated memory formation [47, 48]. Since systemic administration of HDAC inhibitors has pleiotropic effects and can cause undesired outcomes, it cannot be a method of choice, but the positive results in animal models give a promise for the development of EpiEffectors for more specific deposition of histone acetylation. In addition, 2–5% of all patients have familial AD caused by mutations in the amyloid protein. In these cases, silencing of the mutant gene by targeted DNA methylation and/or H3K9 methylation is an epigenetic approach for a potential causative therapy.

21.4.2 Parkinson's Disease

The main pathophysiological pathway leading to the development of Parkinson's disease (PD) is a death of dopaminergic neurons. It has been demonstrated that an activation of the expression of glial cell line-derived neurotrophic factor (GDNF) by ZFP fused transcription factors provided a neuroprotective effect in a rat model of PD [86]. Similarly, expression of neurturin (NRTR), another member of the GDNF family, led to an increase in activity of dopaminergic neurons in monkeys, and NRTR gene transfer to putamen and substantia nigra is in the clinical trials for patients with PD [49]. Alternatively, activation of the abovementioned endogenous genes to exploit their neuroprotective or neuroregenerative potentials can be potentially achieved by designed EpiEffectors via an introduction of activating and/or removal of silencing chromatin marks at their promoters and enhancers.

21.4.3 Addiction and Depression

The nucleus accumbens (NAc) functions as a reward region in the brain and is involved in formation of drug addiction and depression [50, 51]. Data obtained in animal models and from patients showed that pathophysiological pathways leading to these disorders include changes in the activity of transcription factors and histone-modifying enzymes [52]. For example, the transcription factor deltaFosB, a key regulator of drug addiction and stress responses, is overexpressed in the NAc in cocaine addiction and downregulated by chronic stress and depression. Administration of EpiEffectors containing designed ZFPs or TALEs fused with p65 domain of the transcription factor NFkB initiated deposition of histones acetylation and led to an expression of the deltaFosB gene and potentiation of the effects of cocaine. In contrast, downregulation of deltaFosB expression by EpiEffectors depositing H3K9me2 using the catalytic domain of the G9a histone lysine methyltransferase inhibited cocaine-induced expression of deltaFosB and caused a predepressive state [53]. Thus, fine-tuning of the deltaFosB expression via epigenome editing in NAc is a very promising therapeutic approach for depression and addiction.

21.5 Allele-Specific Epigenome Editing as Therapeutic Approach for Diseases of the CNS

Next, we discuss a distinct group of disorders with different underlying pathological mechanisms, which all potentially could be treated with a specialized epigenetic therapy utilizing an approach called allele-specific epigenome editing which aims to edit epigenetic marks in only one allele of a target locus (Fig. 21.2c). This method requires the development of specially designed DNA-binding modules which utilize the presence of a single nucleotide polymorphism (SNP) in the targeted genomic locus for allelic discrimination [54]. EpiEffectors with DNA-binding modules generated to bind only one allele with a certain SNP will deposit epigenetic marks only to this targeted allele. The great flexibility of the dCas9-gRNA-based DNA-binding modules makes this nontrivial task feasible as already demonstrated in genome editing studies [55, 56]. Hence, allele-specific epigenome editing is a potential approach for treatment of neurodevelopmental disorders by specifically silencing or activating only one allele.

21.5.1 Imprinting Disorders

More than 100 genes in mammals are imprinted meaning that they are expressed based on a parent-of-origin principle, either only from the paternal or only from the maternal allele. Imprinting is regulated by the presence of an allele-specific DNA methylation at so-called imprinting centers (ICs) [57, 58]. In the absence of differential methylation of ICs, both alleles are either methylated or unmethylated leading to an aberrant expression of the genes controlled by the IC. Some genes will be overexpressed due to activity of both alleles; others will show complete loss of expression, because both alleles are inactivated. It is worth mentioning here that in some cases these imprinting disorders are only caused by this type of epimutations with no underlying genetic abnormalities. For example, in 3–5% of all patients with Angelman syndrome (AS) an epimutation was found, the loss of methylation in an IC on maternal chromosome 15, which results in silencing of both alleles of the *ube3a* gene leading to AS phenotype [59]. This pure epigenetic disorder is a tempting target for rationally designed EpiEffectors, which could deliver DNA methylation specifically to the maternal IC to restore the normal imprinting pattern. Also other imprinting disorders, like Beckwith-Wiedemann syndrome and Prader-Willi syndrome, [59] are potential targets for allele-specific epigenome editing.

21.5.2 Huntington's Disease

The neurodegenerative disorder Huntington's disease is caused by a mutation in the huntingtin gene, more precisely by the amplification of a CAG trinucleotide. Expression of the mutant misfolded protein leads to formation of cellular aggregates which affect numerous cellular processes such as vesicle trafficking and regulation of gene expression by transcription factors and chromatin remodelers [60]. Silencing of the mutant allele by targeted DNA methylation or H3K9 methylation is a very promising approach, and its effectiveness has already been demonstrated in a mouse model, where allele-specific silencing by RNA interference was applied [61]. However, the clinical perspectives of such epigenome editing therapy are yet to be elucidated. Other triplet expansion neurological diseases such as Friedreich's ataxia, fragile X syndrome, and myotonic dystrophy [62] are also potential targets for a similar approach.

21.5.3 Rett Syndrome

Another example of a neurodevelopmental disease is the Rett syndrome. It is an X-linked disease caused by mutations in a transcription regulator MECP2 [63, 64]. Studies in mouse models showed that many symptoms of the disease can be rescued by an expression of normal MECP2 in neurons [65, 66]. Since one of the MECP2 alleles in patients is normal, targeted silencing of the mutant allele together with activation of the normal allele using allele-specific EpiEffectors could be a potential therapeutic approach.

21.6 Spatial and Temporal Control of EpiEffectors

The large toolbox of DNA-binding and chromatin-modifying modules allows to assemble a remarkable variety of EpiEffectors. However, in order to use them for therapeutic or scientific purposes, they have to be delivered to the target cells for epigenome editing. Efficient gene transfer of EpiEffectors into neurons requires several technological advancements including targeted delivery to certain brain regions, to particular cell types or for systemic brain modification. Moreover, the durability of the introduced changes and the ability to trigger the modifications are of relevance. These aspects will be discussed in the next sections.

21.6.1 Targeted Delivery

Among the different vectors, viral-based systems (*Lentivirus*, adenovirus, and adeno-associated virus (AAV)) have been shown to be most efficient for gene transfer into neurons [67, 68]. However, treatment of CNS diseases is very challenging from a pharmacological point of view, since an intravenous administration of drugs, be it epigenome editing vectors or conventional chemicals, is hindered by the blood-brain barrier. Thus, more invasive but at the same time more efficient intracranial injections are in use (Fig. 21.3a). This approach is suitable for disorders like AD where local modifications of brain regions are required, since only cells in a close proximity to the injection site are targeted. In order to expand the targeted region, attempts with neurotropic virus-based delivery, such as modified herpes simplex virus 1, have been made to use the natural ability of the virus to spread in the nervous system [69]. Intensive research and significant achievements with recombinant AAVs over the last years made them the most used vectors for CNS targeting and animal model studies using AAV vectors demonstrated gene delivery to the CNS without direct manipulations in the brain or spinal cord [70] (Fig. 21.3b). For example, intramuscular-injected AAVs were transported to motor neurons in the spinal cord which are damaged in amyotrophic lateral sclerosis and represent

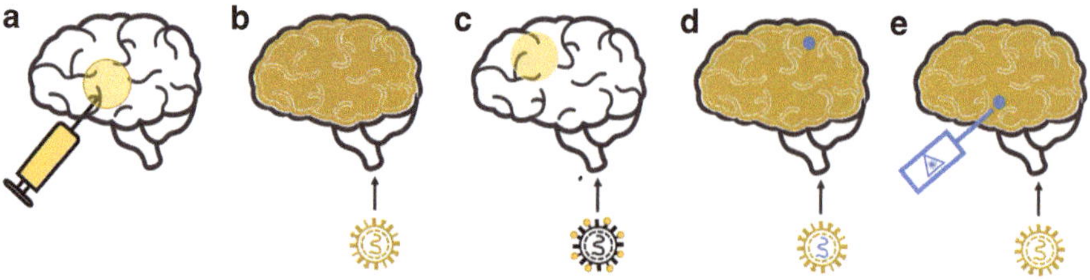

Fig. 21.3 Concepts of delivery of EpiEffectors. (**a**) Local injection in brain regions. (**b**) Systemic intravenous application. (**c**) Application of virus with cell type-specific transduction. (**d**) Application of virus with cell type-specific expression of EpiEffectors. (**e**) Local activation of light-regulated EpiEffectors with a laser

the main target for treatment [71]. Even more importantly, several AAV serotypes (e.g., AAV9 and AAVrh10) can transduce brain neurons after intravenous administration, since they can cross the blood-brain barrier [72, 73]. However, one problem of this approach is the limited transport capacity of AAV which is not sufficient for EpiEffectors based on current CRISPR/dCas9 technology.

21.6.2 Cell Line-Specific Expression of EpiEffectors

Brain tissues consist of multiple cell types, but in many applications, only certain cell types require epigenome editing. Then, a cell type-specific expression of EpiEffectors is needed, which can be achieved by using cell type-specific viral transductions (Fig. 21.3c) or cell type-specific promoters for expression of the EpiEffectors (Fig. 21.3d). For example, intracerebroventricular injection of an AAV9 vector carrying the eGFP gene under the control of a human synapsin 1 gene promoter in mice resulted in an expression of the transgene in neurons of the substantia nigra pars reticulata, motor cortex, hippocampus, cerebellum, cervical spinal cord, and ventromedial striatum [74]. Similarly, usage of promoters of the platelet-derived growth factor B-chain (PDGF-beta), calcium/calmodulin-dependent protein kinase II, tubulin alpha I, or neuron-specific enolase genes resulted in an expression of transgenes in neurons [75, 76]. Additionally, one may achieve a cell type-specific gene transfer by developing recombinant viruses targeting only certain types of neurons, for example, recombinant herpes simplex 1 viruses containing chimeric glycoprotein C surface proteins fused either with glial cell line-derived neurotrophic factor or brain-derived neurotrophic factor specifically transduce nigrostriatal neurons, which contain receptors for these proteins. Such systems can be applied for EpiEffectors delivery in PD [77].

21.6.3 Maintenance of Introduced Epigenetic Signals

Numerous studies have demonstrated the possibility to introduce or remove epigenetic signals in human chromatin. In the majority of these studies, the presence of the desired modifications at the targeted loci was verified several days after delivery of the EpiEffectors. Since the goal of epigenome editing (at least for disorders which are discussed above) is a permanent change of the local chromatin modifications, it needs a stable introduction of epigenetic marks. However, long-term persistence of newly deposited modifications has not been well studied so far. The durability of repressive DNA methylation marks was analyzed in three studies, which published contradicting results. DNA methylation was delivered by Dnmt3a methyltransferase fused to ZFP and disappeared several days after adenoviral gene transfer [44], but remained stable in a study using retroviral vectors under conditions of very low expression of the EpiEffector [40]. Stable introduction of chromatin marks was also observed in another study using stably integrated EpiEffectors with regulated promoter even after repression of the EpiEffector [78]. One has to mention that different delivery systems and target loci were used in these experiments, which may explain the differences in the results. Similar studies were done for the maintenance of H3K9 methylation and showed that the newly introduced modifications are lost within several cell divisions [44, 79, 80]. Interestingly, the work of Bintu and colleagues showed that the fraction of cells obtaining a new stable phenotype after epigenetic editing correlated with the duration of treatment, and it varied with the modifications [81]. However, all these experiments were conducted using non-neuronal cells such that the dynamics and durability of introduced chromatin marks in neurons still need to be studied. One potential advantage of epigenome editing in neurons is that neuronal cells do not divide such that the delivered chromatin marks cannot be passively lost through cell divisions. Hence there is no demand for a maintenance machinery, but active removal of the introduced modification at the target locus due to competing stimuli can be a critical issue.

21.6.4 Control of EpiEffectors by Optogenetics and Other Stimuli

Besides a rough spatial control of editing systems, for example, by injections of vectors in brain regions, some neuroepigenetic studies might need a more precise spatial targeting at microscopic levels or a temporal control of the editing activity (Fig. 21.3e). This could be achieved by using an inducible editing system based on the activation of EpiEffector proteins by external physical or chemical stimuli. One of the most promising tools are LITEs, light-inducible transcriptional effectors, which are modified TALE systems enhanced with a two-hybrid light-sensitive system from *Arabidopsis thaliana*. Cryptochrome 2 protein (fused to a TALE) interaction with CIBI (fused to an effector domain) protein requires an activation by light [82, 83]. Illumination of tissues with blue light leads to an assembly of functional EpiEffectors, which can be used to trigger epigenome editing as well as to restrict its effect to certain brain regions. In addition, there are two other two-hybrid systems, activation of which can be started by an endogenous stimulus [84] or by external chemical ligand [85]. Probably these systems can also be used to fine-tune the intensity of epigenome editing, if this is required.

21.7 Concluding Remarks

The discovery of numerous epigenetic chromatin modifications and chromatin-modifying enzymes has opened a new dimension in genome regulatory networks. The understanding of epigenetic chromatin regulation in the brain, neuroepigenetics, is still in its infancy facing great discoveries but also many challenges. However, it is already clear that epigenetic mechanisms are involved in the development of a diversity of neurological diseases. Hence, epigenome editing, the targeted rewriting of epigenetic information, offers an option to treat such diseases. Of course, it will not become a panacea and a lot challenges are to be solved, but it offers an unreplaceable toolkit for study and potential treatment of neurological and psychiatric disorders, which provides several advantages in comparison to the current pharmacological treatments. First of all, site-specific editing potentially allows an unprecedented specificity of the applied therapy. Secondly, the potential durability of introduced chromatin changes gives a chance to develop stable therapeutic effects without the need of a constant medication. Thirdly, in some cases epigenetic treatment may allow a causative therapy by activating pathologically repressed genes or silencing a dominant mutant gene. Fourthly, the new approach of allele-specific epigenome editing provides an even more specific tool for genome regulation that is of particular interest for neurological disorders. However, there are several scientific and technological challenges still on the way to a routine clinical application of epigenome editing in CNS disorders, which will require a systematic and interdisciplinary research effort to overcome them. Molecular neuroscience has to identify the genes, brain regions, and cell types, which are the best targets for up- and downregulation in different diseases. Neuroepigenetics has to understand the natural regulation of these genes and find the best approaches for stable and durable alteration of their expression in neurons. Synthetic biology has to improve the design, specificity, regulation, and activity of EpiEffectors, and biotechnology has to develop improved systems for the targeted delivery of EpiEffectors to defined brain regions and cell types. Finally, basic research results have to be translated into animal models and clinical applications. Hence, a lot of important and challenging pioneering basic and translational research work will be needed, until the fascinating perspectives of epigenome editing in the brain can come true.

Acknowledgments Work in the authors' laboratory has been supported by the BW Foundation (BWST_NCRNA_007) and the BMBF (01GM1513E).

References

1. Allis C, Jenuwein T, Reinberg D, Lachlan M. Epigenetics. 2 ed. Cold Spring Harbor Laboratory Press; 2015.
2. Bonasio R, Tu S, Reinberg D. Molecular signals of epigenetic states. Science. 2010;330(6004): 612–6. doi:10.1126/science.1191078.
3. Margueron R, Reinberg D. Chromatin structure and the inheritance of epigenetic information. Nat Rev Genet. 2010;11(4):285–96. doi:10.1038/nrg2752.
4. Guzman-Karlsson MC, Meadows JP, Gavin CF, Hablitz JJ, Sweatt JD. Transcriptional and epigenetic regulation of Hebbian and non-Hebbian plasticity. Neuropharmacology. 2014;80:3–17. doi:10.1016/j.neuropharm.2014.01.001.

5. Sweatt DJ. The emerging field of Neuroepigenetics. Neuron. 2013;80(3):624–32. doi:10.1016/j.neuron.2013.10.023.
6. Rudenko A, Tsai L-HH. Epigenetic modifications in the nervous system and their impact upon cognitive impairments. Neuropharmacology. 2014;80:70–82. doi:10.1016/j.neuropharm.2014.01.043.
7. Szyf M. Epigenetics, a key for unlocking complex CNS disorders? Therapeutic implications. Eur Neuropsychopharmacol. 2015;25(5) doi:10.1016/j.euroneuro.2014.01.009.
8. Felsenfeld G. A brief history of epigenetics. Cold Spring Harbor perspectives in biology. 2014;6(1). doi:10.1101/cshperspect.a018200.
9. Day JJ, Sweatt JD. DNA methylation and memory formation. Nat Neurosci. 2010;13(11):1319–23. doi:10.1038/nn.2666.
10. Romanoski CE, Glass CK, Stunnenberg HG, Wilson L, Almouzni G. Epigenomics: roadmap for regulation. Nature. 2015;518(7539):314–6. doi:10.1038/518314a.
11. de Groote ML, Verschure PJ, Rots MG. Epigenetic editing: targeted rewriting of epigenetic marks to modulate expression of selected target genes. Nucleic Acids Res. 2012;40(21):10596–613. doi:10.1093/nar/gks863.
12. Jurkowski TP, Ravichandran M, Stepper P. Synthetic epigenetics—towards intelligent control of epigenetic states and cell identity. Clin Epigenetics. 2015;7(1):1. doi:10.1186/s13148-015-0044-x.
13. Kungulovski G, Jeltsch A. Epigenome editing: state of the art, concepts, and perspectives. Trends Genet. 2016;32(2):101–13. doi:10.1016/j.tig.2015.12.001.
14. Pavletich NP, Pabo CO. Zinc finger-DNA recognition: crystal structure of a Zif268-DNA complex at 2.1 a. Science. 1991;252(5007):809–17.
15. Beerli RR, Barbas CF. Engineering polydactyl zinc-finger transcription factors. Nat Biotechnol. 2002;20(2):135–41. doi:10.1038/nbt0202-135.
16. Jamieson AC, Miller JC, Pabo CO. Drug discovery with engineered zinc-finger proteins. Nat Rev Drug Discov. 2003;2(5):361–8. doi:10.1038/nrd1087.
17. Pabo CO, Peisach E, Grant RA. Design and selection of novel Cys2His2 zinc finger proteins. Annu Rev Biochem. 2001;70:313–40. doi:10.1146/annurev.biochem.70.1.313.
18. Sander JD, Dahlborg EJ, Goodwin MJ, Cade L, Zhang F, Cifuentes D, et al. Selection-free zinc-finger-nuclease engineering by context-dependent assembly (CoDA). Nat Methods. 2011;8(1):67–9. doi:10.1038/nmeth.1542.
19. Handel EM, Cathomen T. Zinc-finger nuclease based genome surgery: it's all about specificity. Curr Gene Ther. 2011;11(1):28–37.
20. Szczepek M, Brondani V, Büchel J, Serrano L, Segal DJ, Cathomen T. Structure-based redesign of the dimerization interface reduces the toxicity of zinc-finger nucleases. Nat Biotechnol. 2007;25(7):786–93. doi:10.1038/nbt1317.
21. Boch J, Scholze H, Schornack S, Landgraf A, Hahn S, Kay S, et al. Breaking the code of DNA binding specificity of TAL-type III effectors. Science. 2009;326(5959):1509–12. doi:10.1126/science.1178811.
22. Jankele R, Svoboda P. TAL effectors: tools for DNA targeting. Brief Funct Genomics. 2014;13(5):409–19. doi:10.1093/bfgp/elu013.
23. Kim Y, Kweon J, Kim A, Chon J, Yoo J, Kim H, et al. A library of TAL effector nucleases spanning the human genome. Nat Biotechnol. 2013;31(3):251–8. doi:10.1038/nbt.2517.
24. Jinek M, Chylinski K, Fonfara I, Hauer M. A programmable dual-RNA–guided DNA endonuclease in adaptive bacterial immunity. Science. 2012; doi:10.1126/science.1225829.
25. Mojica FJ, Díez-Villaseñor C, García-Martínez J, Almendros C. Short motif sequences determine the targets of the prokaryotic CRISPR defence system. Microbiology. 2009;155(Pt 3):733–40. doi:10.1099/mic.0.023960-0.
26. Anders C, Niewoehner O, Duerst A, Jinek M. Structural basis of PAM-dependent target DNA recognition by the Cas9 endonuclease. Nature. 2014;513(7519):569–73. doi:10.1038/nature13579.
27. Mohr SE, Hu Y, Ewen-Campen B, Housden BE, Viswanatha R, Perrimon N. CRISPR guide RNA design for research applications. FEBS J. 2016; doi:10.1111/febs.13777.

28. Ran FA, Hsu PD, Wright J, Agarwala V, Scott DA, Zhang F. Genome engineering using the CRISPR-Cas9 system. Nat Protoc. 2013;8(11):2281–308. doi:10.1038/nprot.2013.143.
29. Qi LS, Larson MH, Gilbert LA, Doudna JA, Weissman JS, Arkin AP, et al. Repurposing CRISPR as an RNA-guided platform for sequence-specific control of gene expression. Cell. 2013;152(5):1173–83. doi:10.1016/j.cell.2013.02.022.
30. Vora S, Tuttle M, Cheng J, Church G. Next stop for the CRISPR revolution: RNA guided epigenetic regulators. FEBS J. 2016; doi:10.1111/febs.13768.
31. Kleinstiver BP, Prew MS, Tsai SQ, Topkar VV, Nguyen NT, Zheng Z, et al. Engineered CRISPR-Cas9 nucleases with altered PAM specificities. Nature. 2015;523(7561):481–5. doi:10.1038/nature14592.
32. Shmakov S, Abudayyeh OO, Makarova KS, Wolf YI, Gootenberg JS, Semenova E, et al. Discovery and functional characterization of diverse class 2 CRISPR-Cas systems. Mol Cell. 2015;60(3):385–97. doi:10.1016/j.molcel.2015.10.008.
33. Zetsche B, Gootenberg JS, Abudayyeh OO, Slaymaker IM, Makarova KS, Essletzbichler P, et al. Cpf1 is a single RNA-guided endonuclease of a class 2 CRISPR-Cas system. Cell. 2015;163(3):759–71. doi:10.1016/j.cell.2015.09.038.
34. Kleinstiver BP, Pattanayak V, Prew MS, Tsai SQ, Nguyen NT, Zheng Z, et al. High-fidelity CRISPR–Cas9 nucleases with no detectable genome-wide off-target effects. Nature. 2016;529(7587):490–5. doi:10.1038/nature16526.
35. Gimenez CA, Ielpi M, Mutto A, Grosembacher L, Argibay P, Pereyra-Bonnet F. CRISPR-on system for the activation of the endogenous human INS gene. Gene Ther. 2016;23(6):543–7. doi:10.1038/gt.2016.28.
36. Hilton IB, D'Ippolito AM, Vockley CM, Thakore PI, Crawford GE, Reddy TE, et al. Epigenome editing by a CRISPR-Cas9-based acetyltransferase activates genes from promoters and enhancers. Nat Biotechnol. 2015;33(5):510–7. doi:10.1038/nbt.3199.
37. McDonald JI, Celik H, Rois LE, Fishberger G, Fowler T, Rees R, et al. Reprogrammable CRISPR/Cas9-based system for inducing site-specific DNA methylation. Biol Open. 2016;5(6) doi:10.1242/bio.019067.
38. Vojta A, Dobrinic P, Tadic V, Bockor L, Korac P, Julg B, et al. Repurposing the CRISPR-Cas9 system for targeted DNA methylation. Nucleic Acids Res. 2016; doi:10.1093/nar/gkw159.
39. Nunna S, Reinhardt R, Ragozin S, Jeltsch A. Targeted methylation of the epithelial cell adhesion molecule (EpCAM) promoter to silence its expression in ovarian cancer cells. PLoS One. 2014;9(1):e87703. doi:10.1371/journal.pone.0087703.
40. Rivenbark AG, Stolzenburg S, Beltran AS, Yuan X, Rots MG, Strahl BD, et al. Epigenetic reprogramming of cancer cells via targeted DNA methylation. Epigenetics. 2012;7(4):350–60. doi:10.4161/epi.19507.
41. Siddique AN, Nunna S, Rajavelu A, Zhang Y, Jurkowska RZ, Reinhardt R, et al. Targeted methylation and gene silencing of VEGF-A in human cells by using a designed Dnmt3a-Dnmt3L single-chain fusion protein with increased DNA methylation activity. J Mol Biol. 2013;425(3):479–91. doi:10.1016/j.jmb.2012.11.038.
42. Chen H, Kazemier HG, Groote ML, Ruiters MHJ, Xu G-L, Rots MG. Induced DNA demethylation by targeting ten-eleven translocation 2 to the human ICAM-1 promoter. Nucleic Acids Res. 2014;42(3):1563–74. doi:10.1093/nar/gkt1019.
43. Maeder ML, Angstman JF, Richardson ME. Targeted DNA demethylation and activation of endogenous genes using programmable TALE-TET1 fusion proteins. Nat Biotechnol. 2013;31(12):1137–42.
44. Kungulovski G, Nunna S, Thomas M, Zanger UM, Reinhardt R, Jeltsch A. Targeted epigenome editing of an endogenous locus with chromatin modifiers is not stably maintained. Epigenetics Chromatin. 2015;8:12. doi:10.1186/s13072-015-0002-z.
45. Snowden AW, Gregory PD, Case CC, Pabo CO. Gene-specific targeting of H3K9 methylation is sufficient for initiating repression in vivo. Curr Biol. 2002;12(24):2159–66.
46. Hyman BT, Phelps CH, Beach TG, Bigio EH, Cairns NJ, Carrillo MC, et al. National Institute on Aging–Alzheimer's Association guidelines for the neuropathologic assessment of Alzheimer's disease. Alzheimers Dement. 2012;8(1):1–13. doi:10.1016/j.jalz.2011.10.007.

47. Fischer A. Targeting histone-modifications in Alzheimer's disease. What is the evidence that this is a promising therapeutic avenue? Neuropharmacology. 2014;80:95–102. doi:10.1016/j.neuropharm.2014.01.038.
48. Francis YI, Fà M, Ashraf H, Zhang H, Staniszewski A, Latchman DS, et al. Dysregulation of histone acetylation in the APP/PS1 mouse model of Alzheimer's disease. J Alzheimers Dis. 2009;18(1):131–9. doi:10.3233/jad-2009-1134.
49. Warren Olanow C, Bartus RT, Baumann TL, Factor S, Boulis N, Stacy M, et al. Gene delivery of neurturin to putamen and substantia nigra in Parkinson disease: a double-blind, randomized, controlled trial. Ann Neurol. 2015;78(2):248–57. doi:10.1002/ana.24436.
50. Kelz MB, Chen J, Carlezon WA, Whisler K, Gilden L, Beckmann AM, et al. Expression of the transcription factor deltaFosB in the brain controls sensitivity to cocaine. Nature. 1999;401(6750):272–6. doi:10.1038/45790.
51. Robison AJ, Vialou V, Mazei-Robison M, Feng J, Kourrich S, Collins M, et al. Behavioral and structural responses to chronic cocaine require a feedforward loop involving ΔFosB and calcium/calmodulin-dependent protein kinase II in the nucleus accumbens shell. J Neurosci. 2013;33(10):4295–307. doi:10.1523/jneurosci.5192-12.2013.
52. Renthal W, Nestler EJ. Epigenetic mechanisms in drug addiction. Trends Mol Med. 2008;14(8):341–50. doi:10.1016/j.molmed.2008.06.004.
53. Heller EA, Cates HM, Peña CJ, Sun H, Shao N, Feng J, et al. Locus-specific epigenetic remodeling controls addiction- and depression-related behaviors. Nat Neurosci. 2014;17(12):1720–7. doi:10.1038/nn.3871.
54. Bashtrykov P, Kungulovski G, Jeltsch A. Correction of aberrant imprinting by allele-specific epigenome editing. Clin Pharmacol Ther. 2016;99(5):482–4. doi:10.1002/cpt.295.
55. Smith C, Abalde-Atristain L, He C, Brodsky BR, Braunstein EM, Chaudhari P, et al. Efficient and allele-specific genome editing of disease loci in human iPSCs. Mol Ther. 2015;23(3):570–7. doi:10.1038/mt.2014.226.
56. Yoshimi K, Kaneko T, Voigt B, Mashimo T. Allele-specific genome editing and correction of disease-associated phenotypes in rats using the CRISPR–Cas platform. Nat Commun. 2014;5:4240. doi:10.1038/ncomms5240.
57. Ferguson-Smith AC, Surani MA. Imprinting and the epigenetic asymmetry between parental genomes. Science. 2001;293(5532):1086–9. doi:10.1126/science.1064020.
58. Jurkowska RZ, Jeltsch A. Genomic imprinting—the struggle of the genders at the molecular level. Angew Chem Int Ed. 2013;52(51):13524–36. doi:10.1002/anie.201307005.
59. Horsthemke B, Wagstaff J. Mechanisms of imprinting of the Prader–Willi/Angelman region. Am J Med Genet A. 2008;146A(16):2041–52. doi:10.1002/ajmg.a.32364.
60. Valor LM, Guiretti D. What's wrong with epigenetics in Huntington's disease? Neuropharmacology. 2014;80:103–14. doi:10.1016/j.neuropharm.2013.10.025.
61. Yu D, Pendergraff H, Liu J, Kordasiewicz HB, Cleveland DW, Swayze EE, et al. Single-stranded RNAs use RNAi to potently and allele-selectively inhibit mutant huntingtin expression. Cell. 2012;150(5):895–908. doi:10.1016/j.cell.2012.08.002.
62. Nageshwaran S, Festenstein R. Epigenetics and triplet-repeat neurological diseases. Front Neurol. 2015;6:262. doi:10.3389/fneur.2015.00262.
63. Amir RE, den Veyver IB, Wan M, Tran CQ, Francke U, Zoghbi HY. Rett syndrome is caused by mutations in X-linked MECP2, encoding methyl-CpG-binding protein 2. Nat Genet. 1999;23(2):185–8. doi:10.1038/13810.
64. Chunshu Y, Endoh K, Soutome M, Kawamura R, Kubota T. A patient with classic Rett syndrome with a novel mutation in MECP2 exon 1. Clin Genet. 2006;70(6):530–1. doi:10.1111/j.1399-0004.2006.00712.x.
65. Guy J, Gan J, Selfridge J, Cobb S, Bird A. Reversal of neurological defects in a mouse model of Rett syndrome. Science. 2007;315(5815):1143–7.
66. Ure K, Lu H, Wang W, Ito-Ishida A, Wu Z, He L-J, et al. Restoration of Mecp2 expression in GABAergic neurons is sufficient to rescue multiple disease features in a mouse model of Rett syndrome. elife. 2016;5 doi:10.7554/elife.14198.
67. Kantor B, McCown T, Leone P, Gray SJ. Chapter two clinical applications involving CNS gene transfer. Adv Genet. 2014;87:71–124. doi:10.1016/b978-0-12-800149-3.00002-0.

68. Lentz TB, Gray SJ, Samulski JR. Viral vectors for gene delivery to the central nervous system. Neurobiol Dis. 2012;48(2):179–88. doi:10.1016/j.nbd.2011.09.014.
69. Berges BK, Wolfe JH, Fraser NW. Transduction of brain by herpes simplex virus vectors. Mol Ther. 2007;15(1):20–9. doi:10.1038/sj.mt.6300018.
70. Murlidharan G, Samulski RJ, Asokan A. Biology of adeno-associated viral vectors in the central nervous system. Front Mol Neurosci. 2014;7:76. doi:10.3389/fnmol.2014.00076.
71. Kaspar BK, Lladó J, Sherkat N, Rothstein JD, Gage FH. Retrograde viral delivery of IGF-1 prolongs survival in a mouse ALS model. Science. 2003;301(5634):839–42. doi:10.1126/science.1086137.
72. Bourdenx M, Dutheil N, Bezard E, Dehay B. Systemic gene delivery to the central nervous system using adeno-associated virus. Front Mol Neurosci. 2014;7:50. doi:10.3389/fnmol.2014.00050.
73. Foust KD, Nurre E, Montgomery CL, Hernandez A, Chan CM, Kaspar BK. Intravascular AAV9 preferentially targets neonatal neurons and adult astrocytes. Nat Biotechnol. 2008;27(1):59–65. doi:10.1038/nbt.1515.
74. McLean JR, Smith GA, Rocha EM, Hayes MA, Beagan JA, Hallett PJ, et al. Widespread neuron-specific transgene expression in brain and spinal cord following synapsin promoter-driven AAV9 neonatal intracerebroventricular injection. Neurosci Lett. 2014;576:73–8. doi:10.1016/j.neulet.2014.05.044.
75. Hioki H, Kameda H, Nakamura H, Okunomiya T, Ohira K, Nakamura K, et al. Efficient gene transduction of neurons by lentivirus with enhanced neuron-specific promoters. Gene Ther. 2007;14(11):872–82. doi:10.1038/sj.gt.3302924.
76. Liu BH, Wang X, Ma YX, Wang S. CMV enhancer/human PDGF-β promoter for neuron-specific transgene expression. Gene Ther. 2004;11(1):52–60. doi:10.1038/sj.gt.3302126.
77. Cao H, Zhang G-R, Wang X, Kong L, Geller AI. Enhanced nigrostriatal neuron-specific, long-term expression by using neural-specific promoters in combination with targeted gene transfer by modified helper virus-free HSV-1 vector particles. BMC Neurosci 2008;9(1):1–14. doi:10.1186/1471-2202-9-37.
78. Stolzenburg S, Beltran AS, Swift-Scanlan T, Rivenbark AG, Rashwan R, Blancafort P. Stable oncogenic silencing in vivo by programmable and targeted de novo DNA methylation in breast cancer. Oncogene. 2015;34(43):5427–35. doi:10.1038/onc.2014.470.
79. Audergon PN, Catania S, Kagansky A, Tong P, Shukla M, Pidoux AL et al. Epigenetics. Restricted epigenetic inheritance of H3K9 methylation. Science. 2015;348(6230):132–5. doi:10.1126/science.1260638.
80. Ragunathan K, Jih G, Moazed D. Epigenetics. Epigenetic inheritance uncoupled from sequence-specific recruitment. Science. 2015;348(6230):1258699. doi:10.1126/science.1258699.
81. Bintu L, Yong J, Antebi YE, McCue K, Kazuki Y, Uno N et al. Dynamics of epigenetic regulation at the single-cell level. Science. 2016;351(6274):720–4. doi:10.1126/science.aab2956.
82. Konermann S, Brigham MD, Trevino A, Hsu PD, Heidenreich M, Cong L, et al. Optical control of mammalian endogenous transcription and epigenetic states. Nature. 2013;500(7463):472–6. doi:10.1038/nature12466.
83. Liu H, Yu X, Li K, Klejnot J, Yang H, Lisiero D, et al. Photoexcited CRY2 interacts with CIB1 to regulate transcription and floral initiation in Arabidopsis. Science. 2008;322(5907):1535–9. doi:10.1126/science.1163927.
84. Li Y, Moore R, Guinn M, Bleris L. Transcription activator-like effector hybrids for conditional control and rewiring of chromosomal transgene expression. Sci Rep. 2012;2:897. doi:10.1038/srep00897.
85. Mercer AC, Gaj T, Sirk SJ, Lamb BM, Iii CF. Regulation of endogenous human gene expression by ligand-inducible TALE transcription factors. ACS Synthetic Biol. 2014;3(10):723–30. doi:10.1021/sb400114p.
86. Laganiere J, Kells AP, Lai JT, Guschin D, Paschon DE, Meng X, Fong LK, Yu Q, Rebar EJ, Gregory PD, Bankiewicz KS, Forsayeth J, Zhang HS. An engineered zinc finger protein activator of the endogenous glial cell line-derived neurotrophic factor gene provides functional neuroprotection in a rat model of Parkinson's disease. J Neurosci. 2010;30(49):16469–74. doi:10.1523/JNEUROSCI.2440-10.2010.

Techniques for Single-Molecule mRNA Imaging in Living Cells

22

Kevin Czaplinski

Abstract

Typical measurement of macromolecules in a biological sample typically averages the result over all the cells or molecules within the sample, and while these types of measurements provide very useful information, they completely miss heterogeneity among the components within the sample that could be a very important aspect of the sample's function. These techniques are also limited in their ability to examine intracellular spatial orientation of molecular activity, which is often a critical component to the regulation of biological processes, particularly in cells with unique spatial relationships, such as neurons. This makes a strong case for single-cell and single-molecule analysis that allows similar novel insight into complex molecular machinery that would not be possible when pooling heterogeneous molecular states. mRNA has proven to be quite tractable to molecular analysis in single cells. Almost two decades of single-molecule studies of mRNA processing both in situ and in live cells have been facilitated by microscopy of mRNA. This has been made possible by multiplexing fluorophores in situ hybridization probes or fluorescent RNA-tag-binding protein probes. The purpose of this chapter is to describe the approaches that have made single-molecule mRNA imaging accessible, as well as to give an overview of the state of the art for techniques that are available to track mRNA in real time in living cells, highlighting the application to neuroscience.

K. Czaplinski
Center for Nervous System Disorders, Centers for Molecular Medicine, Stony Brook University, Stony Brook, NY 11794, USA

Department of Anesthesiology, Stony Brook University, Stony Brook, NY 11794, USA
e-mail: Kevin.Czaplinski@stonybrook.edu

R. Delgado-Morales (ed.), *Neuroepigenomics in Aging and Disease*, Advances in Experimental Medicine and Biology, DOI 10.1007/978-3-319-53889-1_22

Keywords
Real-time RNA imaging • RNA trafficking • RNA localization • MS2 mRNA imaging reporter • Live cell mRNA imaging • Single-molecule mRNA imaging • mRNA imaging in the nervous system

22.1 mRNA Trafficking to Axons and Dendrites Facilitates Localized Translation

The interest in studies of mRNA trafficking within neurons developed from demonstrations that mRNA translation localized to axons and dendrites is necessary for axon guidance and synaptic plasticity. There are a number of informative reviews that cover these connections [1–9]. mRNA trafficking within the cytoplasm has long been recognized as a key component of localized translation, since only the mRNAs that are trafficked to axons or dendrites can be locally translated [10–12]. Studies to identify mRNAs that are found in axons and dendrites have shown that these represent only a subset of all of the mRNAs available in the cell [13–18]. This fact implies that what reaches these compartments has been actively transported there, but the transport mechanisms remain obscure. Since active transport isn't a general property of all mRNA, then it should be due to some special features of those that are transported. The ability to traffic an mRNA can be facilitated by *cis*-acting sequence elements specific to each mRNA, called localization elements (LEs) or zipcodes (Fig. 22.1) [19–29]. By and large these elements are poorly understood and very diverse in their sequence and may also be similarly diverse in their mechanism of action [21, 30–33]. In support of this view, a combination of RNA base sequence recognizing RNA-binding proteins and RNA structural element-recognizing proteins (collectively called RNA-binding proteins, RBPs) has been associated with LE activity [10]. Yet, even for an mRNA with axonal and dendritic trafficking capacity, significant amounts of the mRNA remain in the cell soma so trafficking to axons and dendrites is not exclusive for these domains, and mechanisms of trafficking may not universally engage all substrate mRNAs at all times. As a result of this, technology that allows the isolation of that trafficking subpopulation of mRNA within a sample is critical to understand the mechanisms acting upon that subset [34, 35].

The technology that has had perhaps the most impact on our understanding of how mRNAs traffics is real-time microscopy of individual mRNAs in cells [35–37]. Through these studies, it is clear that the trafficking of axonal and dendritic mRNAs is bidirectional, anterograde, and retrograde and is almost certainly microtubule associated through molecular motors [10, 38]. The fact that LEs account for the trafficking suggests that the factors recognizing LEs link an mRNA to the motors (Fig. 22.1). Despite the strong reasoning behind this, the identity of the motors involved as well as the cargo adapters between the mRNA and the motors all remain largely unidentified. The maturation of single-molecule in vivo mRNA imaging technology enables the development assays that should be able to address these significant gaps in our knowledge in the future,

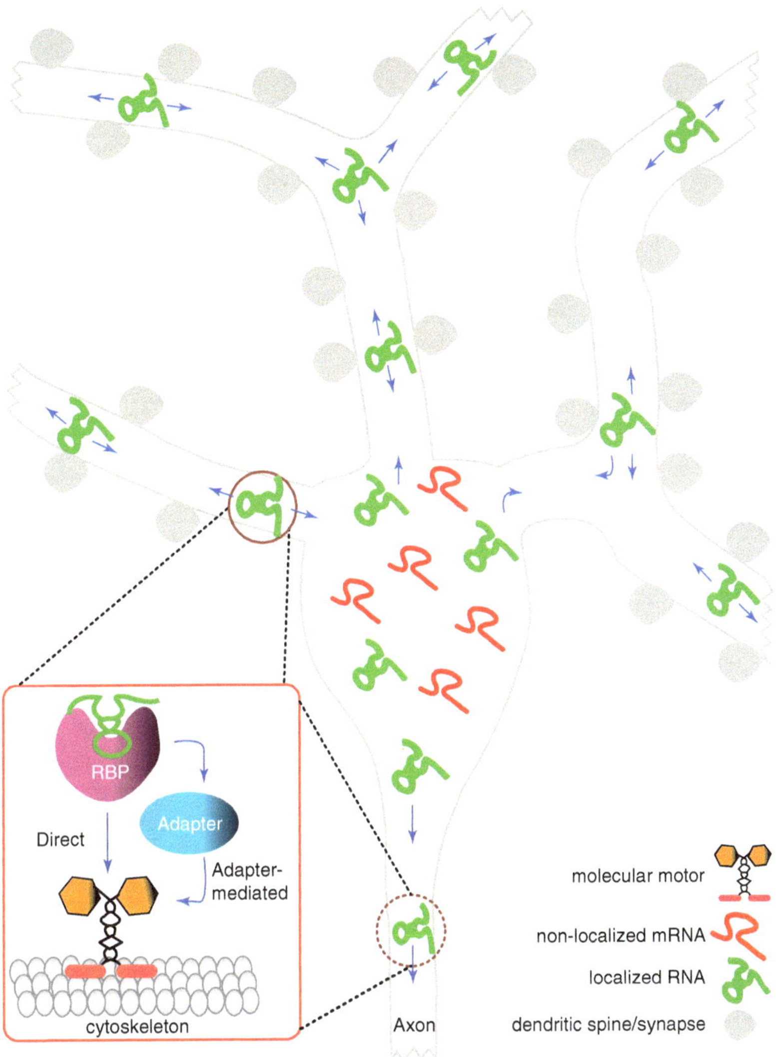

Fig. 22.1 mRNA trafficking in neurons. In this cartoon of a typical pyramidal neuron, nonlocalized mRNAs (in *red*) fail to escape from the cell soma, while subsets of mRNA are actively trafficked due to specific sequences or structures within them called localization elements (in *green*). In the inset panel, mRNAs that are actively trafficked are transported cargo of cytoskeleton-associated motor proteins. Although the nature of these complexes is not clear, RNA-binding proteins may directly interact with cargo domains of the motor protein or indirectly interact through putative adapter proteins

and the focus of this article is to illustrate the techniques that have proven successful for this purpose [19, 35]. While my emphasis is on cytoplasmic mRNA localization, it is worth to note that these approaches have been very successful at labeling nascent mRNA during transcription, allowing researchers to watch the production of mRNA as it is occurring [39].

22.2 From Single-Cell to Single-Molecule Studies of mRNA

Measuring mRNA abundance is a routine laboratory technique, performed on RNA chemically extracted from a collection of cells within a tissue or culture dish. But these types of samples are non-homogenous, even within a culture of a cell line, and the resulting ensemble measurement averages out all cells in the sample, ignoring diversity within the sample [34, 40–42]. Studying single cells has revealed surprising variability among what would be considered an otherwise homogenous collection of cells [41, 43, 44]. Single-cell studies provide critical understanding of how the variability in cell state contributes to a biological pathway. Just as single-cell studies are at the forefront of being able to describe cell states within a population, single-molecule studies hold similar promise of being able to describe multistep cellular processes with unprecedented detail [35, 39].

Single-molecule studies allow direct observations of molecular activity to be made in the absence of potential interference inherent in ensemble measurement [45]. Single-molecule studies have enhanced our understanding of many of the processes underlying gene expression [35, 39]. There are common strategies in the experiments that have allowed isolation of individual processes of gene expression, and these are necessary to overcome the challenges inherent in single-molecule observation within cells. These types of experiments have made a very strong impact on studies of regulating gene expression at the posttranscriptional level.

RNA, as the intermediate between DNA and protein, has long been exploited as a readout of gene expression, but not as often identified as the regulator of gene expression itself. Many steps of gene expression occur posttranscriptionally; mRNA splicing, polyadenylation, nuclear export, translation, mRNA trafficking, and turnover are all of these have been shown to regulate gene expression of different mRNAs. And if these processes don't give you enough to think about, processes that control the chemical nature of RNA itself are becoming increasingly recognized to strongly effect these steps of gene expression, exemplified in the study of RNA chemical modification, editing, and structure formation [46–50]. Our understanding of mRNA regulation at the posttranscriptional level is far from complete and changing rapidly.

22.3 Detecting Single Molecules of mRNA: From In Situ *Hybridization* to Live Cell Microscopy

Detecting specific RNA molecules within a sample is rather straightforward and can be performed by hybridizing antisense probes, with a Northern blot or in situ hybridization (ISH) being well known examples of this methodology. However this methodology is robust enough to work in situ down to the single-molecule level to determine the spatial distribution of mRNA within a cell, as well as to quantify mRNA levels within individual cells by counting the number of molecules observed (Fig. 22.2) [52, 55–58]. Using probes directed against introns very precisely localizes nascent mRNA at the transcription site with quite low background, enabling direct visualization of sites of active transcription. These studies have led to some of

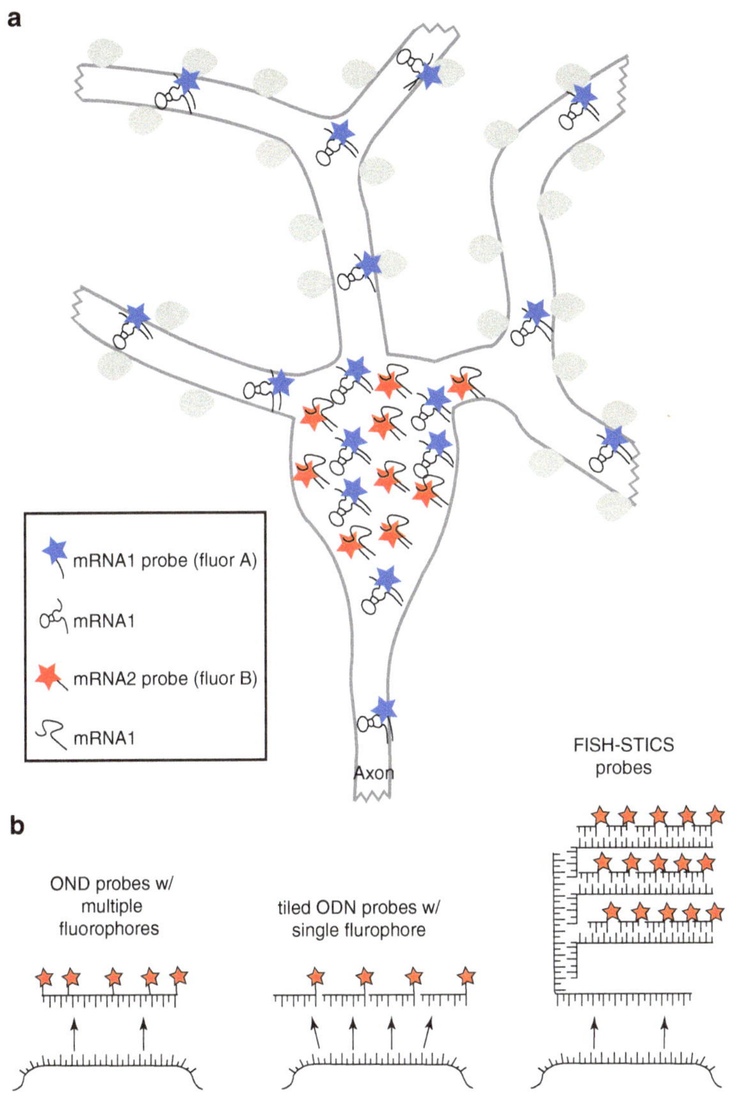

Fig. 22.2 Single-molecule RNA fluorescence in situ hybridization (smRNA-FISH). (**a**) Fluorescent oligodeoxyribonucleotide (ODN) probes can be hybridized to fixed cells on a slide or coverslip to quantify their spatial orientation within a cell. An artist rendering of an smRNA-FISH image from a pyramidal neuron is depicted here. (**b**) Single-molecule sensitivity within fixed cells is greatly enhanced through binding many fluorophores to the mRNA. Three arrangements of smRNA-FISH ODN probes are diagrammed here. A 50 nt ODN containing multiple fluorophores is sensitive enough with careful imaging, but three to five of these against the same target mRNA are used in practice [51]. Multiple singly labeled 20 nt ODN against the target mRNA (up to 20 is typical) has proven to be a very successful approach [52, 53]. Fluorescence in situ hybridization with sequential tethered and intertwined ODN complexes (FISH-STICs) uses successive rounds of hybridization with three ODN, to bind many fluorophores to a 50 nt mRNA sequence [54]. Widefield fluorescence microscopy can then be applied to quantify mRNA abundance and spatial localization of mRNAs within the sample. Multiple mRNAs can be analyzed at the same time using nonoverlapping fluorophores, and the specificity of nucleic acid hybridization allows multiple transcripts to be co-hybridized together

the most significant insight into the basic aspects of gene expression. However, what the elegant specificity of hybridization hasn't been able to achieve effectively is labeling of mRNA within live cells. For this purpose, inventive genetically encoded mRNA-tagging systems that exploit fluorescent proteins have found great success.

The most widely used RNA-tagging system derives from the bacteriophage MS2. The MS2 coat protein (MCP) nucleates capsid formation on the MCP-binding site (MBS) of the bacteriophage genome (Fig. 22.3a) [35, 59–61]. This binding site is a specific RNA stem-loop structure and can also referred to as the MS2 stem loop (MSL, Fig. 22.3a). When the MBS is placed into an mRNA that is expressed ectopically in cells, it provides a platform for the MCP to bind. MCP can be genetically fused to a fluorescent protein, such as green fluorescent protein (GFP), and the resulting MCP-GFP is able to label the MBS-tagged mRNA to be captured by fluorescence microscopy (Fig. 22.3b). In practice single mRNA-bound GFP molecules cannot be readily detected in live cells over a background of unbound molecules, so the label on the mRNA is amplified by insertion of multiple tandem MBS sequences, as many as 24 (Fig. 22.3b). These makes the mRNA much brighter than the free background of MCP-GFP so that mRNA can be seen as diffraction limited spots. Adding a nuclear localization signal (NLS) to the MCP further decreases the cytoplasmic background when cytoplasmic imaging of mRNA is needed. These particles have sufficient brightness to be tracked in real time in living cells (Fig. 22.3c). This system has been successfully applied to all of the most widely used model organisms for molecular biology.

The basic premise and components of the MS2 system have been in place since it was first described; however, there have been improvements made to the various components of this tagging system that have helped to overcome some of the practical limitations of using earlier version [62]. The MCP binds to RNA as a dimer, and fortunately the carboxy terminus of one monomer is juxtaposed to the amino terminus of the other monomer when bound to RNA, making it possible to express the MCP as a tandem dimer in a single polypeptide (Fig. 22.3b). This had the result of increasing the affinity of the MCP for the MBS making the labeling of tagged mRNAs much more robust [63]. The repetitive nature of adding 24 MBS sites makes their application more challenging; large numbers of short nucleotide repeats tend to be lost upon plasmid amplification in *E. coli* or when placed into retroviral vectors owing to the fact that reverse transcription is error-prone and causes deletion of nucleic acid repeats. Modifications to reduce the repetitive nature of the nucleic acid sequence elements encoding the system have made the current state of the art for this labeling strategy more user-friendly than ever lowering the obstacles for their routine use in any lab [62].

The MCP-MBS system is the most widespread example for real-time imaging of mRNA, but it isn't the only one since other means have been developed. A second bacteriophage coat protein related to MS2, that of PP7, has been applied in the same way as MCP. In this arrangement, a tandem dimer of PP7 coat protein (PCP) fused to a fluorescent protein (GFP) can bind a distinct stem loop in the PP7 genome, called the PCP-binding site (PBS or PSL) [60, 61]. The application of PCP-PBS is identical to that of MCP-MBS; however, the stem loops are distinct and the MCP and PCP do not cross-recognize stem loops. This is therefore very useful for

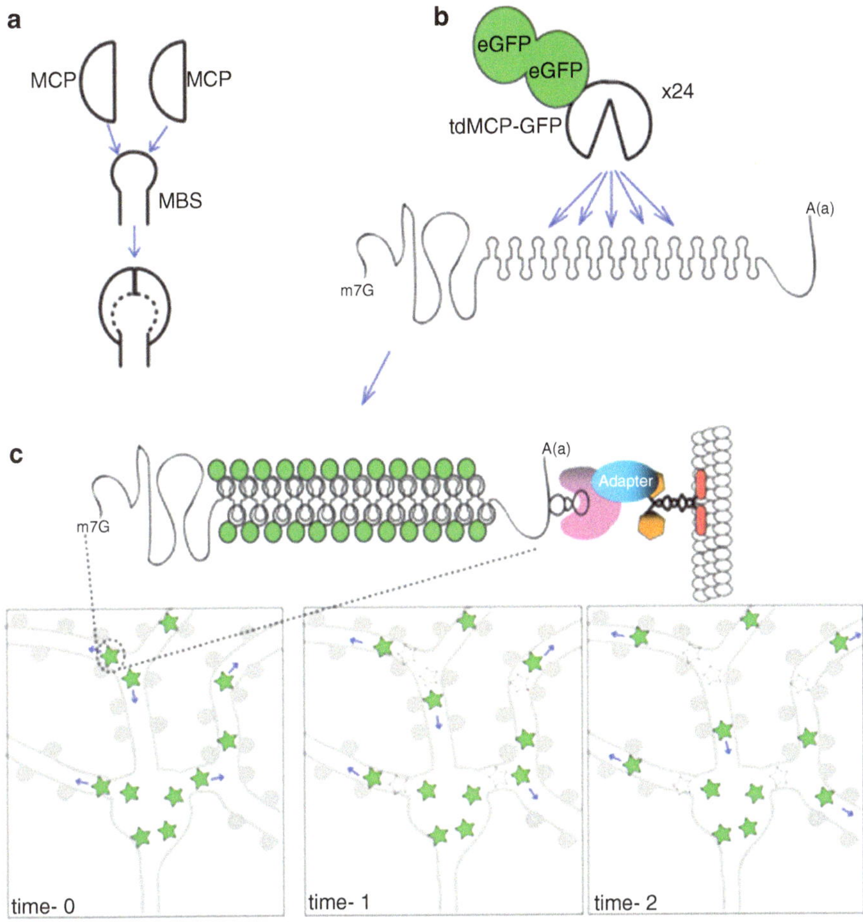

Fig. 22.3 The MS2 mRNA imaging reporter system. (**a**) The MCP protein binds to the MBS stem loop as a dimer, and fusing the MBS to a reporter mRNA allows co-expressed MCP to specifically bind these mRNAs in living cells. (**b**) Single polypeptide MCP dimers (tandem dimer MCP, tdMCP) fused to GFP are the state of the art for imaging reporter mRNAs engineered to encode 24 copies of the MBS stem loop. Multiple copies of GFP can be employed to compensate for the decrease in GFP signal that is lost through creating the tdMCP (two are shown here). (**c**) These three panels are an artist's rendering of time-lapse mRNA imaging. When the MS2 reporter system is expressed in cells, the individual mRNAs can be seen as diffraction limited spots within cells (*green stars*). The mRNAs are bright enough to allow rapid time-lapse imaging that can capture the real-time movements of mRNA throughout the cell cytoplasm. In the pyramidal neurons depicted here, only a subpopulation of the mRNA is trafficking through a motor-dependent mechanism (see Fig. 22.1); these mRNAs are depicted with arrows indicating the direction of motion, both anterograde and retrograde. The origin position of the moving mRNAs is indicated in the second and third frames by an open star

dual-color labeling RNA by placing both MBS and PBS sites and co-expressing MCP and PCP fused to different colored fluorescent proteins [64, 65].

A third bacteriophage RNA-binding protein, protein N (pN) of λ phage, has also successfully been adapted for imaging mRNA in live cells [66, 67]. pN binds to the small BOXB stem-loop RNA from the N utilization (Nut) site of the λ phage genome, using an arginine-rich motif peptide, the BOXB-binding peptide (BBP). This BBP–BOXB binding pair has been exploited in the same way as MCP and PCP. The BBP's RNA-binding properties differ from MCP and PCP so the peptide doesn't need to be dimerized before fusing it to a fluorescent protein, but multiple BOXB sites are nonetheless required for efficient labeling of mRNA. The smaller size of the BBP and smaller stem-loop give this system some advantage over the MCP system in the size of the components added for tagging; however, it doesn't have the proven track record that the MCP system has. This could also be employed in a multicolor system for labeling multiple mRNAs that have either the MBS or PBS sites in combination with multiple BOXB RNA sites.

There are other examples of RNA tagging that have also been employed, and they are designed mostly in the same way that the MS2 system works, although none have the track record of success that the MS2 system has [35]. To use these systems, you have to introduce the separate elements of the tagging system into live cells. This is typically with transfection or viral vectors in mammalian cells or genetic insertion in tractable model organisms. With all these systems, the ability to visualize an mRNA trafficking in the cytoplasm is now clearly established; however, the fact that a tagged reporter mRNA must be used will always leave open questions as to the applicability of the results to the bona fide mRNA substrates they represent. There aren't ways to address these concerns, since there is no way to image endogenous mRNA behavior at the moment, but all the available evidence suggests that there aren't serious adverse effects of adding the MS2 tag other than a decrease in diffusion rate that should accompany the corresponding increase in size.

Fig. 22.4 Recent advances in live cell mRNA imaging. Imaging endogenous mRNAs in real time is very desirable to avoid characterizing atypical trafficking that might be associated with using a reporter. Four approaches for imaging endogenous mRNA are diagrammed in panels (**a–d**). (**a**) Molecular beacons are auto-quenched fluorescent ODN hairpin probes. Hybridization of the beacon to the target mRNA is stronger than the quenching stem loop, so recognition of the target unmasks the fluorophore. (**b**) Multiply labeled tetravalent RNA imaging probes (MTRIPs) are fluorescent ODN (similar to Fig. 22.2b) that are oligomerized through binding to streptavidin. These complexes can be introduced directly into permeabilized cells where they can bind a target mRNA. (**c**) PUF-FBF RNA-binding domains can be engineered to recognize specific sequences of an mRNA, and several of these could be co-introduced in the cell to label this mRNA endogenously (three such proteins are depicted). (**d**) Adaptation of CRISPR/Cas9 for imaging endogenous cellular mRNA has been achieved by adding sequences complementary to a cellular mRNA (antisense, AS) to a co-expressed CRISPR gRNA (Cas9-AS-gRNA) and a co-introduced AS-ODN that contains sequence necessary for binding to the Cas9 DNA-binding domain (Cas9-AS-ODN). The Cas9 in nuclease inactive (*nuc-*) to maintain the stable Cas9-mRNA complex is fused to a GFP protein to facilitate fluorescence. (**e**) Imaging mRNAs are no longer the only means to assess the possibility of localized translation. The SunTag system consists of 24 copies of a GCN4 peptide (scFv-antigen) at the amino terminus of an open reading frame (ORF) combined with a co-expressed scFv-GFP fusion against it (only eight are shown). The presence of the pre-folded fluorophore brightly labels nascent polypeptides and more brightly labels polysomes owing to the multiple nascent chains that are present. Recent studies have added RNA-tagging systems in cis to mRNAs encoding SunTagged mRNAs in order to study the spatial and temporal relationship of mRNAs to their translation state

22.4 From Tagged Reporters Toward Imaging Endogenous RNAs

There is great interest in approaches that attempt to label unmodified endogenous mRNA, and there have been several systems developed with that goal in mind. Molecular beacon technology has the potential to hybridize a quenched fluorophore-containing probe to a cellular RNA that will fluoresce only upon RNA binding (Fig. 22.4a) [68, 69]. Similar in concept to molecular beacons, a

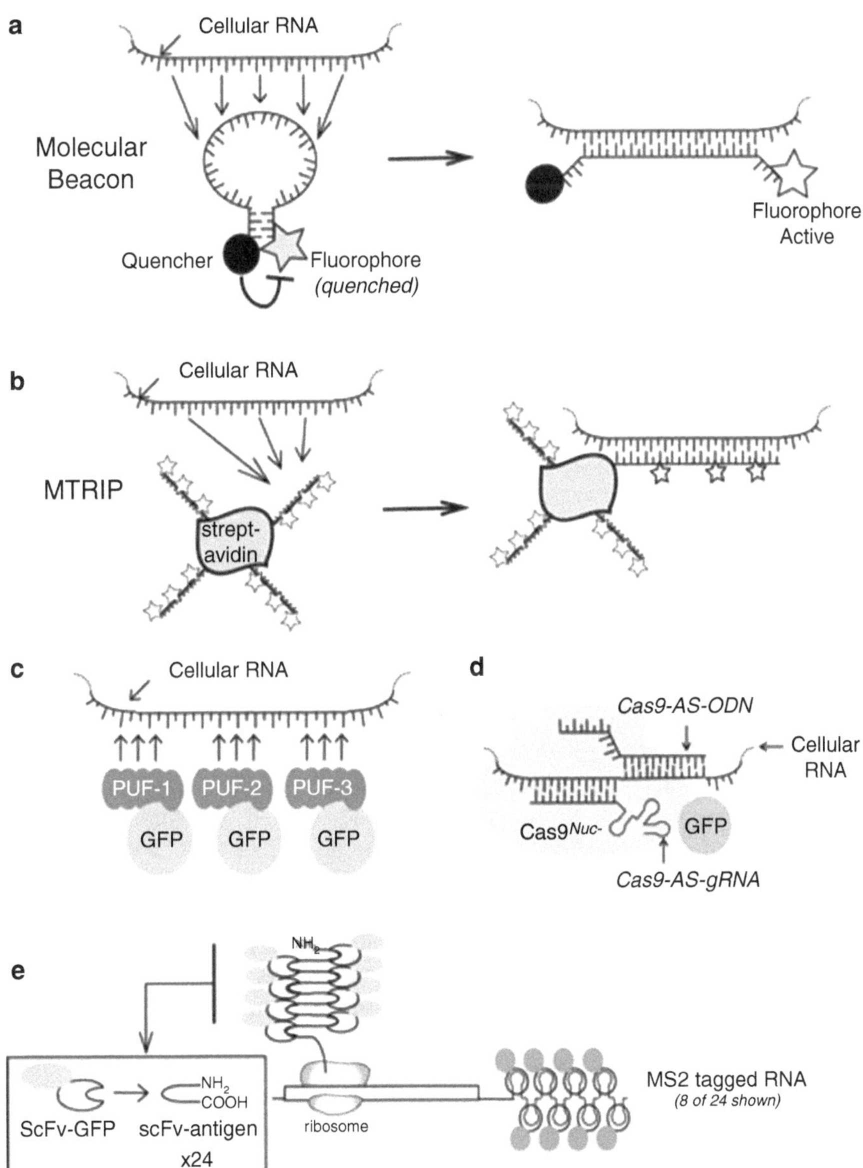

small molecule-binding RNA aptamer that can activate fluorescence of the small molecule only when hybridized to a gene specific target has been developed [70]. Multiply labeled tetravalent RNA imaging probes (MTRIP) are streptavidin tetramers that are bound to four fluorescently labeled oligodeoxyribonucleotide (ODN) probes that can be directly introduced into cells permeabilized with streptolysin O. MTRIPS can then bind target cellular mRNAs, with multiple MTRIP probes against a single target boosting signal to noise ratios in cells (Fig. 22.4b) [71, 72]. Pumilio and FBF family RNA-binding domains (PUF) can be engineered to recognize specific sequences, and combinations of these co-expressed in cells can be used to label mRNAs as well (Fig. 22.4c) [73–75]. Most of these other approaches don't offer the brightness of signal that the MS2 system does at the moment, and this makes microscopy of these probes very challenging. But these are all very sound conceptual approaches and there is data to support their function. I believe that eventually there will be a simple-to-use endogenous mRNA-labeling approach that will emerge that is effective in live cells.

The application of the CRISPR/Cas9 genome editing system has recently been shown to be capable of acting as a platform to localize endogenous nucleic acid sequences. The Cas9 protein from *Streptococcus pyogenes* is an RNA-programmed DNAse that cleaves DNA at specific target locations defined by complementarity to a Cas9-bound RNA molecule (a CRISPR RNA) [76, 77]. By manipulating the sequences in the CRISPR RNA that targets Cas9, called the guide RNA (gRNA), the Cas9/gRNA complex can be targeted to endogenous genomic sequences of any organism they are both introduced into. gRNA base pairing to target genomic DNA causes Cas9 to cut genomic DNA at these target sites, and this ability is being exploited to empower site-specific genome editing in many model organisms [76, 77].

The ability of Cas9 to very specifically target nucleic acid sequences is being exploited for the purpose of specifically visualizing them in live cells. By using multiple gRNAs that target clustered sites on genomic DNA, a nuclease-deficient Cas9-GFP can be used to visualize where these DNA sites are located in live cells [78]. A modification of this approach has been applied to target Cas9-GFP to mRNAs. A combination of a gRNA and targeting DNA oligonucleotide that are both complementary to adjacent regions of an endogenous target RNA have been used in conjunction with a nuclease-deficient Cas9-GFP fusion to successfully localize endogenous RNAs in cells (Fig. 22.4d) [79]. The sensitivity of the system was limited because of the presence a single GFP on the Cas9 protein, but this was sufficient to observe sequence-specific targeting to multi-mRNA containing structures such as p-bodies or stress granules. Other recent work has shown that Cas9 gRNA can be modified to include large RNA sequences, so whether multiple MBS or PBS containing gRNAs can be applied to this mRNA imaging application of Cas9 to improve single-molecule resolution remains to be demonstrated [80].

22.5 Beyond Imaging mRNA Lies Imaging Translation Sites

These advances in mRNA reporter imaging have recently taken a significant step further. The trafficking of mRNA within the cytoplasm serves to provide a template for localized synthesis of the encoded RNA; therefore, the presence of the mRNA is typically considered a surrogate for the site of synthesis of the encoded protein [12]. By applying the same principle of adding multiple tandem tags to visualize single molecules in living cells, a method to label nascent polypeptides while translation was occurring has been presented [81].

Using fluorescent proteins for imaging of molecular behavior has been extremely powerful. In principle using these tags to visualize the process of translating the GFP fusion protein itself was desirable; however, in living cells this cannot occur since it takes several minutes for an actively fluorescent GFP protein structure to develop. This is longer than the time it takes to synthesize the protein, so at the time fluorescence appears, the protein is no longer ribosome associated. Fast folding versions of fluorescent proteins have been applied to help achieve this goal, but these experiments are technically challenging and not widely applicable [82]. To avoid the problem of delays due to folding of the protein being synthesized, delivery of a pre-folded fluorophore to a nascent chain proved necessary to visualize translation.

Chemical dye-binding peptide tags fused to the amino terminus of a reporter mRNA have been used to target fluorophores to nascent polypeptides during translation, but lack sensitivity for single-molecule analysis [83]. Single-chain fragment variable (scFv) antibodies are single polypeptide proteins that can be expressed in the cytoplasm and retain their capacity to interact specifically with their antigen, even when fused to a protein tag, such as GFP [84]. When the affinity of interaction is high enough, the scFv-GFP fusion protein can bind and localize the antigen-containing protein in cells. Tanenbaum et al. showed a scFv-GFP fusion protein that recognizes a peptide from the GCN4 protein that was effective at localizing proteins tagged with this peptide sequence [81]. Tanenbaum et al. applied 24 copies of a peptide from GCN4 (called a SunTag) to the amino terminus of several different proteins (Fig. 22.4e). These fusions were co-expressed in cells with an anti-GCN4 scFv fused to GFP, and the sites of translation of these reporter mRNAs were clearly visible through the recognition of the multiplexed epitope on the nascent polypeptide. The fact that multiple rounds of translation can initiate on a single mRNA to create polysomes led to further signal amplification, making the visualization even more striking. The clarity of the translation sites labeled by this technology is high enough to directly measure the properties of translation of several different individual mRNAs that have never been previously possible to visualize. The brightness of the SunTag was also sufficient to visualize specific sites within the genome of live cells by adding it to catalytically inactive Cas9 and co-expressing site-specific gRNAs as described above.

Very recently, several different groups have exploited the capacity to visualize nascent translation with the SunTag approach, adding the MBS or PBS tags to mRNAs encoding different SunTagged polypeptides, one of these studies even extended the tag from 24 copies to 56 (Fig. 22.4e) [85–88]. In doing so these studies could co-label mRNA together with nascent protein and follow the spatial distribution of the mRNA at the same time as the sites of translation. In a variation on the SunTag, one study demonstrated that a repeated Flag epitope tag could replace the GCN4 peptide and be labeled in cells using exogenously introduced F_{ab} that was covalently labeled with a fluorophore [85]. This provides alternative to co-expression of the GCN4 scFv-GFP and opens the possibility for multi-translation site imaging in the same cells.

In two of these studies, the relationship between translation and mRNA trafficking was analyzed directly in neurons, and it was shown that the two processes were not coordinates; translation and trafficking could occur both simultaneously and independent of each other [87, 89]. Translation sites were either static or moving in a directed manner, showing the same types of movement that mRNAs do. Two-color labeling of the translation site and mRNA showed that translation sites moved together with its mRNA at speeds consistent with those reported for mRNA. Translation was not constant, and the translation seemed to occur in pulses. Adding the 3′ UTR of the Arc mRNA, a normally dendrite-targeted mRNA had a significant effect on the number of translation sites found much farther from the cell soma, indicating that the 3'UTR indeed accounted for the trafficking of the mRNA to this compartment and that this trafficking facilitated translation in this cellular compartment [89]. In the context of neuron function, the translating nascent proteins would be free to interact with any cellular structures they encountered, such as dendritic spines, and future work will undoubtedly be able to take advantage of the single-molecule SunTag approach to study the fate of single proteins after being synthesized within the dendrites or axons.

Conclusion

The ability to visualize an mRNA in real time is now a reality, and with that has come an opportunity to define the spatial regulation of a cellular process that is typically refractory to direct observation due to technical limitations. These advances come largely from the general approach of oligomerization of fluorescent probes on target molecules to facilitate live cell single-molecule imaging over long periods of time. The application of these multivalent artificial intracellular reporters correspondingly necessitates appropriate caution about the interpretations of the resulting data; however, the lack of available techniques to visualize untagged molecules makes this trade-off acceptable in my view, in light of what can be learned from these studies. There will hopefully also be advances in imaging of endogenous mRNAs so that the reliance on artificial reporters will decrease in the not-too-distant future.

Acknowledgments K. C. is supported by a National Science Foundation CAREER award (IOS-1254146) and is grateful for the support received from the Stony Brook University School of Medicine and Departments of Anesthesiology and Pharmacological Sciences.

References

1. Donnelly CJ, Fainzilber M, Twiss JL. Subcellular communication through RNA transport and localized protein synthesis. Traffic. 2010;11(12):1498–505. doi:10.1111/j.1600-0854.2010.01118.x.
2. Holt CE, Schuman EM. The central dogma decentralized: new perspectives on RNA function and local translation in neurons. Neuron. 2013;80(3):648–57. doi:10.1016/j.neuron.2013.10.036.
3. Hornberg H, Holt C. RNA-binding proteins and translational regulation in axons and growth cones. Front Neurosci. 2013;7:81. doi:10.3389/fnins.2013.00081.
4. Jung H, O'Hare CM, Holt CE. Translational regulation in growth cones. Curr Opin Genet Dev 2011;21(4):458–464. doi:S0959-437X(11)00072-4 [pii]. 10.1016/j.gde.2011.04.004.
5. Klann E, Richter JD. Translational control of synaptic plasticity and learning and memory. In: Sonenberg N, Matthews M, Hershey J, editors. Translational control: Cold Spring Harbor Press; 2006.
6. Martin KC, Zukin RS. RNA trafficking and local protein synthesis in dendrites: an overview. J Neurosci. 2006;26(27):7131–7134. doi:26/27/7131 [pii]. 10.1523/JNEUROSCI.1801-06.2006.
7. Richter JD, Klann E. Making synaptic plasticity and memory last: mechanisms of translational regulation. Genes Dev. 2009;23(1):1–11. doi:23/1/1 [pii]. 10.1101/gad.1735809.
8. Sutton MA, Schuman EM. Dendritic protein synthesis, synaptic plasticity, and memory. Cell. 2006;127(1):49–58. doi:10.1016/j.cell.2006.09.014.
9. Yoon YJ, Wu B, Buxbaum AR, Das S, Tsai A, English BP, et al. Glutamate-induced RNA localization and translation in neurons. Proc Natl Acad Sci U S A. 2016; doi:10.1073/pnas.1614267113.
10. Czaplinski K. Understanding mRNA trafficking: are we there yet? Semin Cell Dev Biol. 2014;32:63–70. doi:10.1016/j.semcdb.2014.04.025.
11. Doyle M, Kiebler MA. Mechanisms of dendritic mRNA transport and its role in synaptic tagging. EMBO J 2011;30(17):3540–3552. doi:emboj2011278 [pii]. 10.1038/emboj.2011.278.
12. Sinnamon JR, Czaplinski K. mRNA trafficking and local translation: the Yin and Yang of regulating mRNA localization in neurons. Acta Biochim Biophys Sin Shanghai 2011;43(9):663–670. doi:gmr058 [pii]. 10.1093/abbs/gmr058.
13. Cajigas IJ, Tushev G, Will TJ, Tom Dieck S, Fuerst N, Schuman EM. The local transcriptome in the synaptic neuropil revealed by deep sequencing and high-resolution imaging. Neuron. 2012;74(3):453–66. doi:10.1016/j.neuron.2012.02.036.
14. Gumy LF, Yeo GS, Tung YC, Zivraj KH, Willis D, Coppola G et al. Transcriptome analysis of embryonic and adult sensory axons reveals changes in mRNA repertoire localization. RNA 2010;17(1):85–98. doi:rna.2386111 [pii]. 10.1261/rna.2386111.
15. Misra M, Edmund H, Ennis D, Schlueter MA, Marot JE, Tambasco J et al. A genome-wide screen for dendritically localized RNAs identifies genes required for dendrite morphogenesis. G3 (Bethesda). 2016;6(8):2397–405. doi:10.1534/g3.116.030353.
16. Poon MM, Choi SH, Jamieson CA, Geschwind DH, Martin KC. Identification of process-localized mRNAs from cultured rodent hippocampal neurons. J Neurosci. 2006;26(51):13390–9. doi:26/51/13390 [pii]. 10.1523/JNEUROSCI.3432-06.2006.
17. Taylor AM, Berchtold NC, Perreau VM, Tu CH, Li Jeon N, Cotman CW. Axonal mRNA in uninjured and regenerating cortical mammalian axons. J Neurosci. 2009;29(15):4697–707. doi:29/15/4697 [pii]. 10.1523/JNEUROSCI.6130-08.2009.
18. Zhong J, Zhang T, Bloch LM. Dendritic mRNAs encode diversified functionalities in hippocampal pyramidal neurons. BMC Neurosci. 2006;7:17. doi:10.1186/1471-2202-7-17.
19. Buxbaum AR, Haimovich G, Singer RH. In the right place at the right time: visualizing and understanding mRNA localization. Nat Rev Mol Cell Biol. 2015a;16(2):95–109. doi:10.1038/nrm3918.
20. Hengst U, Deglincerti A, Kim HJ, Jeon NL, Jaffrey SR. Axonal elongation triggered by stimulus-induced local translation of a polarity complex protein. Nat Cell Biol 2009;11(8):1024–1030. doi:ncb1916 [pii]. 10.1038/ncb1916.
21. Jambhekar A, Derisi JL. Cis-acting determinants of asymmetric, cytoplasmic RNA transport. RNA. 2007;13(5):625–642. doi:13/5/625 [pii]. 10.1261/rna.262607.

22. Kobayashi H, Yamamoto S, Maruo T, Murakami F. Identification of a cis-acting element required for dendritic targeting of activity-regulated cytoskeleton-associated protein mRNA. Eur J Neurosci 2005;22(12):2977–2984. doi:EJN4508 [pii]. 10.1111/j.1460-9568.2005.04508.x.
23. Martin KC, Ephrussi A. mRNA localization: gene expression in the spatial dimension. Cell 2009;136(4):719–730. doi:S0092-8674(09)00126-3 [pii]. 10.1016/j.cell.2009.01.044.
24. Meer EJ, Wang DO, Kim S, Barr I, Guo F, Martin KC. Identification of a cis-acting element that localizes mRNA to synapses. Proc Natl Acad Sci U S A. 2012;109(12):4639–44. doi:10.1073/pnas.1116269109.
25. Muslimov IA, Nimmrich V, Hernandez AI, Tcherepanov A, Sacktor TC, Tiedge H. Dendritic transport and localization of protein kinase Mzeta mRNA: implications for molecular memory consolidation. J Biol Chem 2004;279(50):52613–52622. doi:10.1074/jbc.M409240200. M409240200 [pii].
26. Patel VL, Mitra S, Harris R, Buxbaum AR, Lionnet T, Brenowitz M et al. Spatial arrangement of an RNA zipcode identifies mRNAs under post-transcriptional control. Genes Dev. 2012;26(1):43–53. doi:26/1/43 [pii]. 10.1101/gad.177428.111.
27. Rehbein M, Wege K, Buck F, Schweizer M, Richter D, Kindler S. Molecular characterization of MARTA1, a protein interacting with the dendritic targeting element of MAP2 mRNAs. J Neurochem. 2002;82(5):1039–46.
28. Tubing F, Vendra G, Mikl M, Macchi P, Thomas S, Kiebler MA. Dendritically localized transcripts are sorted into distinct ribonucleoprotein particles that display fast directional motility along dendrites of hippocampal neurons. J Neurosci. 2010;30(11):4160–70. doi:30/11/4160 [pii]. 10.1523/JNEUROSCI.3537-09.2010.
29. Vuppalanchi D, Coleman J, Yoo S, Merianda TT, Yadhati AG, Hossain J et al. Conserved 3'-untranslated region sequences direct subcellular localization of chaperone protein mRNAs in neurons. J Biol Chem 2010;285(23):18025–18038. doi:M109.061333 [pii]. 10.1074/jbc.M109.061333.
30. Donnelly CJ, Willis DE, Xu M, Tep C, Jiang C, Yoo S et al. Limited availability of ZBP1 restricts axonal mRNA localization and nerve regeneration capacity. EMBO J 2011. doi:emboj2011347 [pii]. 10.1038/emboj.2011.347.
31. Mikl M, Vendra G, Doyle M, Kiebler MA. RNA localization in neurite morphogenesis and synaptic regulation: current evidence and novel approaches. J Comp Physiol A Neuroethol Sens Neural Behav Physiol. 2010;196(5):321–34. doi:10.1007/s00359-010-0520-x.
32. Muslimov IA, Patel MV, Rose A, Tiedge H. Spatial code recognition in neuronal RNA targeting: role of RNA-hnRNP A2 interactions. J Cell Biol 2011;194(3):441–457. doi:jcb.201010027 [pii]. 10.1083/jcb.201010027.
33. Shan J, Munro TP, Barbarese E, Carson JH, Smith R. A molecular mechanism for mRNA trafficking in neuronal dendrites. J Neurosci. 2003;23(26):8859–66.
34. Grunwald D, Singer RH, Czaplinski K. Cell biology of mRNA decay. Methods Enzymol 2008;448:553–577. doi:S0076-6879(08)02627-X [pii]. 10.1016/S0076-6879(08)02627-X.
35. Lampasona AA, Czaplinski K. RNA voyeurism: a coming of age story. Methods. 2016;98:10–7. doi:10.1016/j.ymeth.2015.11.024.
36. Eliscovich C, Buxbaum AR, Katz ZB, Singer RH. mRNA on the move: the road to its biological destiny. J Biol Chem. 2013;288(28):20361–8. doi:10.1074/jbc.R113.452094.
37. Park HY, Buxbaum AR, Singer RH. Single mRNA tracking in live cells. Methods Enzymol 2010;472:387–406. doi:S0076-6879(10)72003-6 [pii]. 10.1016/S0076-6879(10)72003-6.
38. Buxbaum AR, Yoon YJ, Singer RH, Park HY. Single-molecule insights into mRNA dynamics in neurons. Trends Cell Biol. 2015b;25(8):468–75. doi:10.1016/j.tcb.2015.05.005.
39. Chen H, Larson DR. What have single-molecule studies taught us about gene expression? Genes Dev. 2016;30(16):1796–810. doi:10.1101/gad.281725.116.
40. Mellis IA, Raj A. Half dozen of one, six billion of the other: what can small- and large-scale molecular systems biology learn from one another? Genome Res. 2015;25(10):1466–72. doi:10.1101/gr.190579.115.

41. Moignard V, Gottgens B. Dissecting stem cell differentiation using single cell expression profiling. Curr Opin Cell Biol. 2016;43:78–86. doi:10.1016/j.ceb.2016.08.005.
42. Symmons O, Raj A. What's luck got to do with it: single cells, multiple fates, and biological nondeterminism. Mol Cell. 2016;62(5):788–802. doi:10.1016/j.molcel.2016.05.023.
43. Levsky JM, Singer RH. Gene expression and the myth of the average cell. Trends Cell Biol 2003;13(1):4–6. doi:S0962892402000028 [pii].
44. Raj A, van Oudenaarden A. Nature, nurture, or chance: stochastic gene expression and its consequences. Cell. 2008;135(2):216–26. doi:10.1016/j.cell.2008.09.050.
45. Yu J. Single-molecule studies in live cells. Annu Rev Phys Chem. 2016;67:565–85. doi:10.1146/annurev-physchem-040215-112451.
46. Licht K, Jantsch MF. Rapid and dynamic transcriptome regulation by RNA editing and RNA modifications. J Cell Biol. 2016;213(1):15–22. doi:10.1083/jcb.201511041.
47. Satterlee JS, Basanta-Sanchez M, Blanco S, Li JB, Meyer K, Pollock J, et al. Novel RNA modifications in the nervous system: form and function. J Neurosci. 2014;34(46):15170–7. doi:10.1523/JNEUROSCI.3236-14.2014.
48. Silverman IM, Berkowitz ND, Gosai SJ, Gregory BD. Genome-wide approaches for RNA structure probing. Adv Exp Med Biol. 2016;907:29–59. doi:10.1007/978-3-319-29073-7_2.
49. Weidmann CA, Mustoe AM, Weeks KM. Direct duplex detection: an emerging tool in the RNA structure analysis toolbox. Trends Biochem Sci. 2016;41(9):734–6. doi:10.1016/j.tibs.2016.07.001.
50. Yue Y, Liu J, He C. RNA N6-methyladenosine methylation in post-transcriptional gene expression regulation. Genes Dev. 2015;29(13):1343–55. doi:10.1101/gad.262766.115.
51. Femino AM, Fogarty K, Lifshitz LM, Carrington W, Singer RH. Visualization of single molecules of mRNA in situ. Methods Enzymol. 2003;361:245–304.
52. Raj A, van den Bogaard P, Rifkin SA, van Oudenaarden A, Tyagi S. Imaging individual mRNA molecules using multiple singly labeled probes. Nat Methods 2008;5(10):877–879. doi:nmeth.1253 [pii]. 10.1038/nmeth.1253.
53. Shaffer SM, Wu MT, Levesque MJ, Raj A. turbo FISH: a method for rapid single molecule RNA FISH. PLoS One 2013;8(9):e75120. doi:10.1371/journal.pone.0075120.
54. Sinnamon JR, Czaplinski K. RNA detection in situ with FISH-STICs. RNA. 2014;20(2):260–6. doi:10.1261/rna.041905.113.
55. Batish M, van den Bogaard P, Kramer FR, Tyagi S. Neuronal mRNAs travel singly into dendrites. Proc Natl Acad Sci U S A. 2012;109(12):4645–50. doi:10.1073/pnas.1111226109.
56. Buxbaum AR, Wu B, Singer RH. Single beta-actin mRNA detection in neurons reveals a mechanism for regulating its translatability. Science. 2014;343(6169):419–22. doi:10.1126/science.1242939.
57. Femino AM, Fay FS, Fogarty K, Singer RH. Visualization of single RNA transcripts in situ. Science. 1998;280(5363):585–90.
58. Trcek T, Larson DR, Moldon A, Query CC, Singer RH. Single-molecule mRNA decay measurements reveal promoter- regulated mRNA stability in yeast. Cell. 2011;147(7):1484–97. doi:10.1016/j.cell.2011.11.051.
59. Bertrand E, Chartrand P, Schaefer M, Shenoy SM, Singer RH, Long RM. Localization of ASH1 mRNA particles in living yeast. Mol Cell. 1998;2(4):437–45.
60. Chao J, Czaplinski, K, Singer, RH. Using the bacteriophage MS2 coat protein-RNA binding interaction to visualise RNA in living cells. In: LW M, editor. Probes and tags to study Biomolecular function. Weinheim: Wiley-VCH; 2008a.
61. Chao JA, Patskovsky Y, Almo SC, Singer RH. Structural basis for the coevolution of a viral RNA-protein complex. Nat Struct Mol Biol. 2008b;15(1):103–5. doi:10.1038/nsmb1327.
62. Wu B, Miskolci V, Sato H, Tutucci E, Kenworthy CA, Donnelly SK, et al. Synonymous modification results in high-fidelity gene expression of repetitive protein and nucleotide sequences. Genes Dev. 2015;29(8):876–86. doi:10.1101/gad.259358.115.

63. Wu B, Chao JA, Singer RH. Fluorescence fluctuation spectroscopy enables quantitative imaging of single mRNAs in living cells. Biophys J 2012;102(12):2936–2944. doi:S0006-3495(12)00570-X [pii]. 10.1016/j.bpj.2012.05.017.
64. Halstead JM, Lionnet T, Wilbertz JH, Wippich F, Ephrussi A, Singer RH, et al. Translation. An RNA biosensor for imaging the first round of translation from single cells to living animals. Science. 2015;347(6228):1367–671. doi:10.1126/science.aaa3380.
65. Hocine S, Raymond P, Zenklusen D, Chao JA, Singer RH. Single-molecule analysis of gene expression using two-color RNA labeling in live yeast. Nat Methods. 2013;10(2):119–21. doi:10.1038/nmeth.2305.
66. Daigle N, Ellenberg J. LambdaN-GFP: an RNA reporter system for live-cell imaging. Nat Methods. 2007;4(8):633–6. doi:10.1038/nmeth1065.
67. Konig J, Baumann S, Koepke J, Pohlmann T, Zarnack K, Feldbrugge M. The fungal RNA-binding protein Rrm4 mediates long-distance transport of ubi1 and rho3 mRNAs. EMBO J. 2009;28(13):1855–66. doi:10.1038/emboj.2009.145.
68. Bratu DP, Catrina IE, Marras SA. Tiny molecular beacons for in vivo mRNA detection. Methods Mol Biol. 2011;714:141–57. doi:10.1007/978-1-61779-005-8_9.
69. Catrina IE, Marras SA, Bratu DP. Tiny molecular beacons: LNA/2'-O-methyl RNA chimeric probes for imaging dynamic mRNA processes in living cells. ACS Chem Biol. 2012;7(9):1586–95. doi:10.1021/cb300178a.
70. Sato S, Watanabe M, Katsuda Y, Murata A, Wang DO, Uesugi M. Live-cell imaging of endogenous mRNAs with a small molecule. Angew Chem Int Ed Engl. 2015;54(6):1855–8. doi:10.1002/anie.201410339.
71. Alonas E, Lifland AW, Gudheti M, Vanover D, Jung J, Zurla C, et al. Combining single RNA sensitive probes with subdiffraction-limited and live-cell imaging enables the characterization of virus dynamics in cells. ACS Nano. 2014;8(1):302–15. doi:10.1021/nn405998v.
72. Santangelo PJ, Alonas E, Jung J, Lifland AW, Zurla C. Probes for intracellular RNA imaging in live cells. Methods Enzymol. 2012;505:383–99. doi:10.1016/B978-0-12-388448-0.00028-0.
73. Edwards TA. Bespoke RNA recognition by Pumilios. Biochem Soc Trans 2015;43(5):801–806. doi:10.1042/BST20150072.
74. Yamada T, Yoshimura H, Inaguma A, Ozawa T. Visualization of nonengineered single mRNAs in living cells using genetically encoded fluorescent probes. Anal Chem. 2011;83(14):5708–14. doi:10.1021/ac2009405.
75. Yoshimura H, Inaguma A, Yamada T, Ozawa T. Fluorescent probes for imaging endogenous beta-actin mRNA in living cells using fluorescent protein-tagged pumilio. ACS Chem Biol. 2012;7(6):999–1005. doi:10.1021/cb200474a.
76. Hsu PD, Lander ES, Zhang F. Development and applications of CRISPR-Cas9 for genome engineering. Cell. 2014;157(6):1262–78. doi:10.1016/j.cell.2014.05.010.
77. Sternberg SH, Doudna JA. Expanding the Biologist's toolkit with CRISPR-Cas9. Mol Cell. 2015;58(4):568–74. doi:10.1016/j.molcel.2015.02.032.
78. Chen B, Gilbert LA, Cimini BA, Schnitzbauer J, Zhang W, Li GW, et al. Dynamic imaging of genomic loci in living human cells by an optimized CRISPR/Cas system. Cell. 2013;155(7):1479–91. doi:10.1016/j.cell.2013.12.001.
79. Nelles DA, Fang MY, O'Connell MR, Xu JL, Markmiller SJ, Doudna JA, et al. Programmable RNA tracking in live cells with CRISPR/Cas9. Cell. 2016;165(2):488–96. doi:10.1016/j.cell.2016.02.054.
80. Shechner DM, Hacisuleyman E, Younger ST, Rinn JL. Multiplexable, locus-specific targeting of long RNAs with CRISPR-display. Nat Methods. 2015;12(7):664–70. doi:10.1038/nmeth.3433.
81. Tanenbaum ME, Gilbert LA, Qi LS, Weissman JS, Vale RD. A protein-tagging system for signal amplification in gene expression and fluorescence imaging. Cell. 2014;159(3):635–46. doi:10.1016/j.cell.2014.09.039.
82. Tatavarty V, Ifrim MF, Levin M, Korza G, Barbarese E, Yu J, et al. Single-molecule imaging of translational output from individual RNA granules in neurons. Mol Biol Cell. 2012;23(5):918–29. doi:10.1091/mbc.E11-07-0622.

83. Rodriguez AJ, Shenoy SM, Singer RH, Condeelis J. Visualization of mRNA translation in living cells. J Cell Biol 2006;175(1):67–76. doi:jcb.200512137 [pii]. 10.1083/jcb.200512137.
84. Colby DW, Garg P, Holden T, Chao G, Webster JM, Messer A, et al. Development of a human light chain variable domain (V(L)) intracellular antibody specific for the amino terminus of huntingtin via yeast surface display. J Mol Biol. 2004;342(3):901–12. doi:10.1016/j.jmb.2004.07.054.
85. Morisaki T, Lyon K, DeLuca KF, DeLuca JG, English BP, Zhang Z, et al. Real-time quantification of single RNA translation dynamics in living cells. Science. 2016;352(6292):1425–9. doi:10.1126/science.aaf0899.
86. Pichon X, Bastide A, Safieddine A, Chouaib R, Samacoits A, Basyuk E, et al. Visualization of single endogenous polysomes reveals the dynamics of translation in live human cells. J Cell Biol. 2016;214(6):769–81. doi:10.1083/jcb.201605024.
87. Wu B, Eliscovich C, Yoon YJ, Singer RH. Translation dynamics of single mRNAs in live cells and neurons. Science. 2016;352(6292):1430–5. doi:10.1126/science.aaf1084.
88. Yan X, Hoek TA, Vale RD, Tanenbaum ME. Dynamics of translation of single mRNA molecules in vivo. Cell. 2016;165(4):976–89. doi:10.1016/j.cell.2016.04.034.
89. Wang C, Han B, Zhou R, Zhuang X. Real-time imaging of translation on single mRNA transcripts in live cells. Cell. 2016;165(4):990–1001. doi:10.1016/j.cell.2016.04.040.

Stem Cell Technology for (Epi)genetic Brain Disorders

23

Renzo J.M. Riemens, Edilene S. Soares, Manel Esteller, and Raul Delgado-Morales

R.J.M. Riemens, M.Sc
Cancer Epigenetics and Biology Program (PEBC), Bellvitge Biomedical Research Institute (IDIBELL), Duran i Reynals Hospital, 3rd floor, Gran Via de L'Hospitalet 199-203, L'Hospitalet de Llobregat, Barcelona, Catalonia 08908, Spain

Department of Psychiatry & Neuropsychology, School for Mental Health and Neuroscience (MHeNs), Maastricht University, Maastricht, The Netherlands

Institute of Human Genetics, Julius Maximilians University, Biozentrum, Am Hubland, Wurzburg 97074, Germany
e-mail: renzo.riemens@maastrichtuniversity.nl

E.S. Soares
Cancer Epigenetics and Biology Program (PEBC), Bellvitge Biomedical Research Institute (IDIBELL), Duran i Reynals Hospital, 3rd floor, Gran Via de L'Hospitalet 199-203, L'Hospitalet de Llobregat, Barcelona, Catalonia 08908, Spain
e-mail: esiqueira@idibell.cat

M. Esteller, M.D., Ph.D.
Cancer Epigenetics and Biology Program (PEBC), Bellvitge Biomedical Research Institute (IDIBELL), Duran i Reynals Hospital, 3rd floor, Gran Via de L'Hospitalet 199-203, L'Hospitalet de Llobregat, Barcelona, Catalonia 08908, Spain

Department of Physiological Sciences II, School of Medicine, University of Barcelona, Carrer de Casanova, 143, Barcelona, Catalonia, Spain

Institució Catalana de Recerca i Estudis Avançats (ICREA), Passcig Lluís Companys 23, Barcelona, Catalonia 08010, Spain
e-mail: mesteller@idibell.cat

R. Delgado-Morales, Ph.D. (✉)
Cancer Epigenetics and Biology Program (PEBC), Bellvitge Biomedical Research Institute (IDIBELL), Duran i Reynals Hospital, 3rd floor, Gran Via de L'Hospitalet 199-203, L'Hospitalet de Llobregat, Barcelona, Catalonia 08908, Spain

Department of Psychiatry & Neuropsychology, School for Mental Health and Neuroscience (MHeNs), Maastricht University, Maastricht, The Netherlands
e-mail: rdelgado@idibell.cat

R. Delgado-Morales (ed.), *Neuroepigenomics in Aging and Disease*, Advances in Experimental Medicine and Biology, DOI 10.1007/978-3-319-53889-1_23

Abstract
Despite the enormous efforts of the scientific community over the years, effective therapeutics for many (epi)genetic brain disorders remain unidentified. The common and persistent failures to translate preclinical findings into clinical success are partially attributed to the limited efficiency of current disease models. Although animal and cellular models have substantially improved our knowledge of the pathological processes involved in these disorders, human brain research has generally been hampered by a lack of satisfactory humanized model systems. This, together with our incomplete knowledge of the multifactorial causes in the majority of these disorders, as well as a thorough understanding of associated (epi)genetic alterations, has been impeding progress in gaining more mechanistic insights from translational studies. Over the last years, however, stem cell technology has been offering an alternative approach to study and treat human brain disorders. Owing to this technology, we are now able to obtain a theoretically inexhaustible source of human neural cells and precursors in vitro that offer a platform for disease modeling and the establishment of therapeutic interventions. In addition to the potential to increase our general understanding of how (epi)genetic alterations contribute to the pathology of brain disorders, stem cells and derivatives allow for high-throughput drugs and toxicity testing, and provide a cell source for transplant therapies in regenerative medicine. In the current chapter, we will demonstrate the validity of human stem cell-based models and address the utility of other stem cell-based applications for several human brain disorders with multifactorial and (epi)genetic bases, including Parkinson's disease (PD), Alzheimer's disease (AD), fragile X syndrome (FXS), Angelman syndrome (AS), Prader-Willi syndrome (PWS), and Rett syndrome (RTT).

Keywords
Brain disorders • Stem cells • Disease modeling • Regenerative medicine • Drug screening • Epigenetics • iPSCs

23.1 Introduction

For decades, the scientific community has been intensively trying to translate their preclinical findings to discover and develop potent therapeutical interventions for pandemic diseases. Although (epi)genetic brain disorders, such as Alzheimer's disease (AD), rank among the most devastating diseases and account for immeasurable socioeconomic burdens, therapeutical approaches that can prevent, stop, or even reverse them remain unidentified. This lack of effective therapeutics can be attributed to multiple factors, including the inaccessibility of human brain tissue samples, the scarcity of proper human longitudinal studies, the limitations of animal and cellular models, as well as the direct contribution of absence in understanding all brain functions together with the complex cellular heterogeneity of the brain [1, 2].

Although we have acquired an enormous body of knowledge over the years, there is an urgent and unmet demand to develop alternative model systems in order to better understand the underlying biological nature of the brain, as well as to develop new effective therapeutics to be able to reduce the suffering and costs that come along with the occurrence of these disorders.

A major challenge in modeling and treating brain disorders has been the inaccessible nature of specific human neural cell types affected by the disease. Stem cell technology has contributed to overcome these challenges, and with the recent discoveries of induced pluripotency, the field has been growing at a rapid pace [3, 4]. In fact, stem cell technology has since then been offering a promising avenue to fill the gap between animal and human research. The field combines the efforts of cell biologists, (epi)genetic researchers, and clinicians to understand human biological systems and to develop effective treatment strategies for human disease. With the use of embryonic stem cells (ESCs), neural stem cells (NSCs), and induced pluripotent stem cells (iPSCs), the field of stem cell technology is dedicated to develop adequate disease model systems, preclinical platforms for (high-throughput) drugs and toxicity screenings, and strategies for transplant therapies (Fig. 23.1) [5–7]. Theoretically, ESCs and iPSCs allow to obtain nearly every cell type of the human body in vitro and provide an inexhaustible cell source due to their pluripotent differentiation potential and capacity of self-renewal [7]. As a consequence, these stem cells can be cultured and differentiated into NSCs and functional neural cells by using directed differentiation techniques [8, 9].

Although there is still a lot to improve in terms of their efficiency and safety, stem cell-based models harbor high translational potential and are currently very appealing to study [10]. In fact, these model systems have proved instrumental to model in vitro molecular alterations associated with genetic mutations in disease-causing genes and allow mechanistic cellular studies of multifactorial (epi)genetic brain disorders [5, 11]. While it is well established that (epi)genetic alterations contribute to the pathophysiology of human disease, detailed epigenomic characterization of stem cell-based models and the role of epigenetic changes in the pathophysiology of these diseases remain underexplored and are currently just starting to become increasingly considered [2, 12]. The complex and interconnected network of epigenetic alterations, including DNA (hydroxy)methylation, histone modifications, and noncoding RNAs, has previously only been studied in animal models and in human postmortem brain samples. Nevertheless, with the rise and recent advances in stem cell technology, it should be accordingly expectable that stem cell-based neural models may represent valid tools to explore epigenetic changes involved in many brain disorders. The use of these models will undoubtedly contribute to a better understanding of human brain epigenetics and general physiology in the future.

In the current chapter, we will introduce the use of stem cell technology for human brain disorders with multifactorial and (epi)genetic bases, including Parkinson's disease (PD), AD, fragile X syndrome (FXS), Angelman syndrome (AS), Prader-Willi syndrome (PWS), and Rett syndrome (RTT). We will address the recent advances of stem cell technology either with regard to disease modeling,

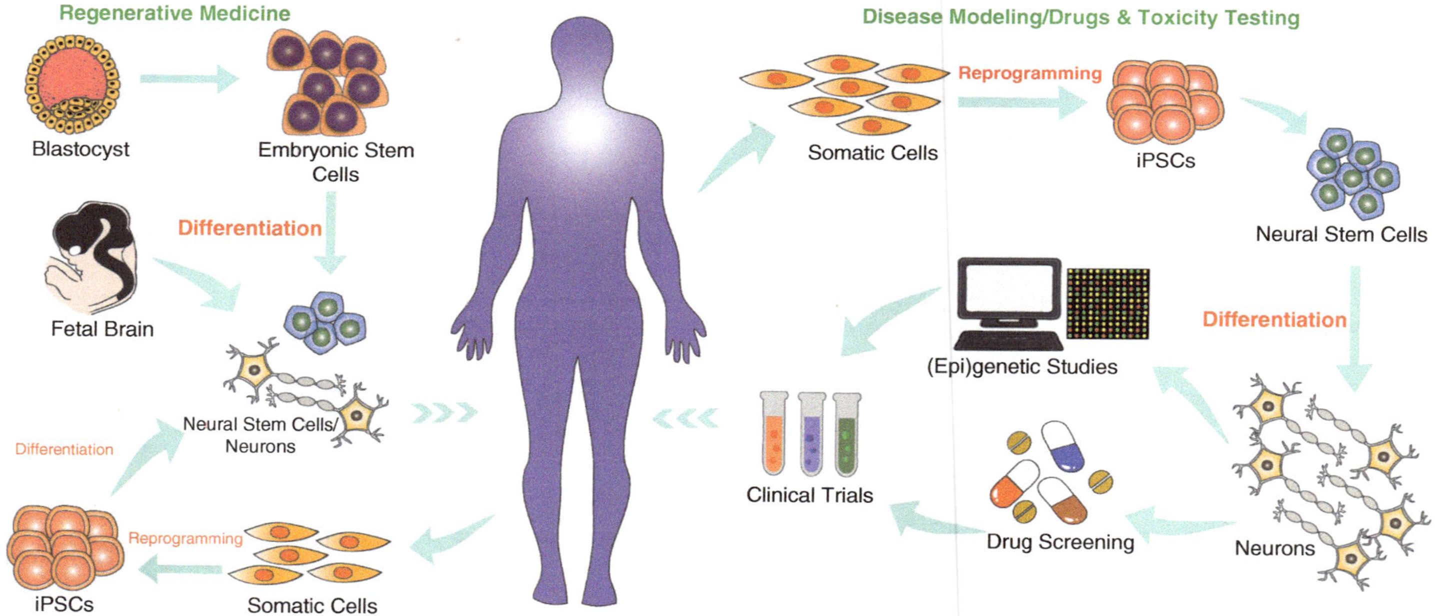

Fig. 23.1 Illustration of the different stem cell-based applications currently available for human brain disorders, including disease modeling, drugs and toxicity screening, and transplant therapies in regenerative medicine. The possible origins of the stem cells are depicted and approaches to move from bench-to-bedside specified

drugs, and toxicity testing or direct clinical applications, with the aim to provide the notion of their utility in fundamental research, as well as in the field of biomedicine.

23.2 Parkinson's Disease

PD is a neurodegenerative disorder affecting over ten million people worldwide [13]. It is estimated that the majority of PD cases (80–90%) are idiopathic with a multifactorial origin, while the minority of the cases (10–20%) are familial and linked to monogenic mutations in PD-related genes, including *PARK2*, *PARK7*, *PINK1*, *ATP13A2*, *SCNA*, *LRRK2*, *TAU*, *NURR1*, and *GBA* [6, 14]. Furthermore, there is a growing body of evidence supporting the role of epigenetics in the development and progression of PD [15, 16]. Differences in disease onset are observed, but PD pathology is common and characterized by progressive loss of dopaminergic neurons in the substantia nigra pars compacta of the mesencephalon [13, 17]. The most crucial pathological hallmarks seen in these neurons are abnormal aggregates of SNCA protein that form Lewy bodies and Lewy neurites [18, 19]. The exact molecular mechanisms underlying the initiation and progression of neurodegeneration remain elusive, but loss of dopaminergic neurons causes deficits in dopaminergic neurotransmission. These neurons are essential in the regulation of the motor functions, and their loss results in typical motor symptoms such as rest tremor, rigidity, bradykinesia, and gait abnormalities [18].

Previous therapeutic strategies have heavily relied on dopamine-enhancing drugs such as levodopa, dopamine receptor agonists, and monoamine oxidase inhibitors to compensate for the loss of dopaminergic neurotransmission [20, 21]. Alternatively, deep brain stimulation of the subthalamic nucleus represents an effective therapy in PD [22]. Even though these therapeutics are able to alleviate the symptoms, they are not able to compensate the cellular loss typical for PD. The availability of stem cell technology, however, has offered an alternative approach to treat PD and overcome this issue. Numerous studies have addressed the therapeutic potential of compensating the dopaminergic cell loss by replacing them by external cell sources, including NSCs and NSC-derived dopaminergic neurons. NSCs can be isolated from fetal brain tissue or specific regions in the adult brain, they can be obtained from differentiated ESCs and iPSCs after neural induction in vitro, or they can be derived from direct reprogramming of somatic cells [6, 23].

Although NSCs, ESCs, and iPSCs have all been studied for their therapeutic potential in PD, mesenchymal stem cells and olfactory ensheathing cells have also been considered [6]. Furthermore, the use of NSCs, ESCs, and iPSCs for disease modeling or drugs and toxicity testing in PD is another rapid-moving field of research [19]. While this area of research is also very interesting, here we will mainly focus on the direct clinical application and major problems that have been encountered when using these stem cells for the treatment of PD.

Studies using rodent and nonhuman primate models of PD have demonstrated that NSCs transplanted into animal brains differentiate into dopaminergic neurons

[24, 25]. Endogenous NSCs retain their regional specificity, and therefore, fetal grafts derived from a dopaminergic- enriched region, such as the mesencephalon, are appropriate cell sources for direct transplantation [21]. When grafts of fetal brain tissue are transplanted to the midbrain of induced parkinsonism rats or non-human primates, they can improve many of the typical motor deficits seen in these animal models [24, 26–28]. Additionally, differentiation of NSCs into dopaminergic neurons can also be enhanced in vitro prior to transplantation [29, 30]. Studies that differentiated fetal grafts into enriched populations of dopaminergic neurons have also found significant cellular and motor behavior recovery in PD animal models [31, 32]. Graft-induced amelioration of motor deficits is dependent on the ability of the grafted NSCs and neurons to restore dopaminergic neurotransmission in the affected area surrounding the transplant. The mechanisms that are thought to underlie these effects can be classified into two categories: direct repair by dopaminergic neuron replacement and indirect repair trough stem cell-derived neurotropic factors [21].

Clinical trials using fetal brain mesencephalic tissue grafts were initiated in the 1990s [33–36]. Improvements have been documented in PD patients in terms of behavior, histology, and survival of the transplanted cells, and in several cases, they were even able to eliminate their dopamine-enhancing medication [37]. Interestingly, studies have demonstrated that transplanted grafts survived and remain functional up to 14 years posttransplantation, although evidence that PD pathology may propagate from host to grafts is emerging [38, 39]. On the other hand, several mild to severe side effects were also observed. One of the most troubling side effects was the occurrence of graft-induced dyskinesia [40, 41]. Furthermore, since the midbrain tissue used to treat PD patients is derived from a genetically distinct individual, i.e., allogeneic, the transplanted grafts cause immunogenic responses that need to be repressed continuously to prevent graft rejection [42]. In addition to these side effects, the use of fetal tissue grafts for PD treatment is also challenging on a large scale, given the limited accessibility and ethical concerns behind the use of primary brain tissue from aborted fetuses [43].

ESCs derived from blastocyst embryos, on the other hand, have an intrinsic capability for infinite self-renewal and are able to differentiate into nearly any cell of the human body, including NSCs or dopaminergic neurons [44]. This theoretical inexhaustible source of cells indicated ESCs as having great promise for cell transplantation therapy in PD. Likewise to fetal grafts, the functional characteristics of transplanted ESC-derived NSCs and dopaminergic neurons have been addressed in vivo by engrafting these cells into animal models of PD [45–49]. These experiments demonstrated that ESC-derived NSCs or dopaminergic neurons were able to integrate into the host brains and to restore dopaminergic neurotransmission, as well as to improve locomotive deficits seen in these PD models. This provided preclinical evidence of the potential of ESC-derived NSCs and dopaminergic neurons for the treatment of PD [50]. Although ESCs seemed to be a very promising cell source, the efficiency of neural cell conversion from ESCs is still limited and often results in incompletely differentiated heterogeneous cellular populations containing different neural cells [51–53]. Therefore, a critical issue that must be resolved and that might even increase the functional outcome after transplantation is the improvement of directed differentiation protocols, as well as cell sorting techniques.

The potential clinical relevance of ESCs found in animal models has opened the possibility for transplantation therapy in human PD patients, but clinical trials using ESCs have not been initiated for treating PD until very recently. Main issues that prevented these cells to move from bench-to-bedside were related to their possible phenotypic instability due to incomplete directed differentiation processes. Moreover, there is a chance of transplanting residual undifferentiated ESCs that can lead to tumor formation in vivo [54]. Furthermore, and also similar to human fetal tissue grafts, ESCs are allogeneic and harbor the problem of immune rejection [55]. Despite the methodological and potential ethical caveats, the Australian Therapeutic Goods Administration has approved a Phase I clinical trial for PD using parthenogenetic ESCs (pESCs) [56, 57]. Parthenogenetic embryos are formed by chemically activating the unfertilized human ovum, which allows induction of pESC cultures that are unable to produce a viable offspring [57]. Based on two preclinical safety studies with rats and nonhuman primates [58], the idea of the International Stem Cell Corporation is to derive NSCs from these pESCs to treat PD patients [59].

With the elegant discovery of somatic cell reprogramming and induced pluripotency by Takahashi and Yamanaka in 2006 [4], another opportunity for the treatment of PD became available. iPSCs are similar if not virtually identical to ESCs in terms of their self-renewal, differentiation potential (pluripotency), morphology, surface marker expression, and in vivo teratoma formation capacity [4, 7]. In relation to transplant therapies for PD, the major advantage of iPSCs above all the aforementioned stem cell types is that iPSCs can be generated from somatic cells of the PD patient to be treated, which allows autologous transplantation [60]. As a consequence, these cells contain the genetic background of the donor, which is speculated to minimize the risk of immune rejection [61]. Moreover, the ethical issues with regard to destruction of fetuses and embryos are circumvented with the use of iPSCs. These unique characteristics support the therapeutic potential of human iPSCs for personalized cell replacement therapy of PD.

To date, several studies have differentiated iPSCs to NSCs or dopaminergic neurons and examined their clinical potential in PD animal models. Grafted iPSC-derived NSCs survived, differentiated in vivo into dopaminergic neurons, matured, and integrated into the recipients' brains [62]. In addition, transplanted human and non-primate iPSC-derived NSCs or dopaminergic neurons were found to have significant therapeutic effects in rat and non-primate PD models by alleviating PD phenotypes [62–66]. Interestingly, in one of these studies, the differentiated neural populations derived from the iPSCs were characterized and sorted based on cellular markers prior to transplantation [63]. Sorting the iPSC-derived neural populations eliminated the undifferentiated tumorigenic cells and significantly increased the number of dopaminergic neurons in the cell grafts compared to unsorted cell populations [63]. These findings demonstrate that sorted and enriched dopaminergic neuronal populations are viable, safe, and functional in vivo, as well as improve the functional impairments posttransplantation [46, 63].

Based on the animal studies, much efforts have been made to bring these potential therapeutic cells to GMP (Good Manufacturing Practice) standards so they can be translated to the clinic for treatment of PD. However, clinical trials using human iPSCs have not been reported yet. In spite of the initial positive results of iPSC-derived NSCs and dopaminergic neurons in animal models of PD, there are several hurdles that have to be elucidated to realize their full potential in regenerative medicine, including the discrepancies around the iPSCs' epigenetic memory [67], differentiation bias [68], mitochondrial dynamics [69, 70], and the appropriate choice of reprogramming technology. In fact, current somatic cell reprogramming techniques have heavily relied on genomic integrating techniques containing factors such as *c-MYC* and *KLF4* [3]. These techniques can affect the genome in a yet unspecified way and might alter the neurobiology of the derived cells, including their differentiation potential, as well as their survival and integration into the recipients' brains.

Another concern related to use of patient-derived iPSCs is that these cells potentially harbor susceptibility traits to PD phenotypes because of mutations or epigenetic markers that could be present in these patients' cells. For instance, establishing iPSC lines derived from patients that harbor PD-related genetic risk loci might make these cells more susceptible to develop PD phenotypic characteristics post-transplantation. To overcome this issue, iPSCs have been generated where the underlying mutations in the disease-causing genes were modified using genomic editing techniques [71–73]. Finally, whereas it has been generally assumed that autologous iPSCs should be immune-tolerated for the patient from whom the somatic cells are derived, several studies have reported immune rejection responses [61, 74].

Stem cell studies and in particular the use of iPSCs also allow for PD disease modeling in vitro. These stem cell-based studies offer a unique opportunity to unravel the (epi)genetic and environmental contributions of the disorder in patient-specific dopaminergic neurons [19]. Dopaminergic neurons from genetic PD cases have already been used to recapitulate disease phenotypes, such as impaired dopamine metabolism, SCNA accumulation, mitochondrial dysfunction, and oxidative stress vulnerability [75–82]. In addition, the study of epigenetic mechanisms using patient iPSC-derived PD models is expected to have a huge impact in understanding the pathophysiology of PD and to assist the development of therapeutic interventions [2]. Moreover, patient iPSC-derived dopaminergic cells could represent useful models to potentially recapitulate the environmental exposome through the patients' epigenome [2]. Accordingly, iPSC-based models are expected to be helpful for investigating epigenetic changes of disorders where the environment is supposed to play a more prominent role, such as idiopathic form of PD [2]. Although the complexity of these multifactorial disorders is expected to be high, especially when taking into account possible interactions between (epi)genetic factors that could modify pathological phenotypes, iPSC-derived models provide new opportunities to investigate epigenomic alterations associated with the disorder. A pioneer study by Fernández-Santiago et al. [83] recently provided first evidence that epigenetic deregulation is associated with monogenic and idiopathic PD in an iPSC-based model system. Interestingly, their findings suggest the presence of molecular deficits in PD somatic cells that manifest only upon differentiation into dopaminergic neurons [83]. Future comparable studies will have important implications for disease modeling, as well as transplant therapies using patient-derived iPSCs.

23.3 Alzheimer's Disease

AD is the most common neurodegenerative disorder and leading cause of age-related dementia [84]. In 2010 approximately 36 million people were diagnosed with AD worldwide, and the incidence is expected to double every 20 years to an estimated 115 million cases in 2050 [85]. The early onset autosomal-dominant form of AD, termed familial AD (FAD), generally occurs between 30 and 60 years of age and is estimated to represent less than 5% of all AD cases [86]. The average age of occurrence for the more common multifactorial late onset form, sporadic AD (SAD), is 65 years with an increasing likelihood of developing the disorder each subsequent year [86]. Both forms are characterized by progressive memory disorientation and cognitive disturbances, but remain clinically and neuropathologically heterogeneous [87]. Main hallmarks that are seen in AD brains include aggregation of amyloid-β (Aβ) peptides into extracellular senile plaques and accumulation of intracellular hyperphosphorylated tau protein into neurofibrillary tangles (NFTs) [88]. Furthermore, neuroinflammation, oxidative stress, and endoplasmatic reticulum (ER) stress have also been implicated in the disease [89]. Although many hypotheses have been proposed to explain the pathogenesis of AD, the interrelationships and causality of these hallmarks remain to be elucidated.

To date, AD studies have mainly relied on the use of transgenic mice models, the use of non-neural human cell cultures, and human postmortem tissue analyses [85, 90]. Although the significant impact of transgenic mice models on progress in understanding various aspect of the disorder is undeniable, they only reproduce specific AD hallmarks and do not reflect clinical phenotypes completely. Moreover, the use of non-neural cell lines omits unique neural features and therefore may fail to capture essential biological processes. The limited accessibility to postmortem tissues and an inadequate amount of cell subtype-specific samples add up to this and together hinder the study of its biological basis. Even in the case that samples are available, the use of postmortem tissue does simply not allow to differentiate between molecular hallmarks that are involved in the causes or consequences of the disease. For these reasons, there has been an ongoing demand for innovative and predictive model systems that closely resemble unique human neural features and which allow to study cause-effect relationships in a controlled setting.

The recent advances in stem cell technology make the availability of iPSCs for AD studies very relevant in this context. iPSCs derived from either FAD or SAD patients´ somatic cells contain a patient-specific pathogenic background, which offers a promising avenue for AD modeling [89]. In fact, the use of disease-relevant neural cells, by differentiating iPSCs along the neural lineage, offers an alternative approach to study the underlying neuropathological mechanisms in vitro in a humanized, personalized, and cell subtype-specific manner. iPSC derived neuronal populations generated from AD patients with known pathogenic backgrounds can be studied, (epi)genetically probed, and treated with drug libraries to investigate their effects on molecular and cellular responses. For these reasons, there has been a growing body of research over the past years to adopt rapidly improving iPSC-derived model systems of AD for fundamental research applications, as well as for the assessment of drugs prior to the initiation of clinical trials [91, 92]. Due to this increasing interest in patient-derived iPSCs for AD research, we will here only focus on the recent progresses of iPSC studies and demonstrate their utility for disease modeling and drug discovery.

Modeling AD using patient-derived iPSCs was initiated from FAD cases with known mutations in disease-causing genes, including amyloid precursor protein (*APP*), presenilin 1 (*PSEN1*), and presenilin 2 (*PSEN2*) [93]. The main goal of these preliminary studies using FAD patient-derived iPSCs has been the validation of their potential for AD modeling, in which they have been seeking to find AD-associated cellular phenotypes. It has previously been shown that mutations in *APP*, *PSEN1*, or *PSEN2* may cause abnormal cleavage of APP, which results in increased levels of total Aβ or increased ratios of neurotoxic Aβ42 to Aβ40 (Aβ42/40) peptides [94].

Several studies have focused on increased copy numbers of the *APP* gene. The first described was a trisomy 21 Down syndrome model using both patient ESCs and iPSC-derived neurons [95]. Down syndrome individuals have an increased risk of developing AD, which has been attributed at least in part to having three copies of *APP* [96]. In the iPSC-derived cortical neurons, increased Aβ production and Aβ42/40 ratios were observed [95]. Furthermore, phosphorylated tau and total tau levels were seen to be upregulated and mislocalized to the neuronal dendrites [95]. In a separate study, iPSC neurons from FAD patients with a duplication of APP were analyzed [97]. Compared to non-demented control individuals, these FAD neurons exhibited significantly higher levels of Aβ40, phosphorylated tau, and glycogen synthase kinase-3β (GSK-3β) activity – a physiological kinase of tau [97]. Moreover, the neurons also accumulated large RAB5-positive endosomes, which has been seen in autopsies from SAD and some forms of FAD [97, 98].

Missense mutations in *APP* that are known to cause AD have also been studied in patient-derived iPSC models. A study by Kondo et al. [99] included three iPSC lines from FAD patients harboring the *APP-E693Δ* (Osaka) mutation. The iPSC-derived cortical neurons exhibited accumulated intracellular Aβ oligomers, leading to endoplasmic reticulum and oxidative stress [99]. Interestingly, two *APP-V717 L* mutant lines in the same study were tested and produced large quantities of extracellular Aβ42 that lacked intracellular accumulation, as well as the accompanying stress response hallmarks [99]. In iPSC-derived forebrain glutamatergic neurons from AD patients harboring the *APP-V717I* (London) mutation, significant higher levels of Aβ42 were also found [100]. The fold increase in Aβ42/40 ratio reported was highly similar to that observed in plasma from human subjects with the same mutation [100, 101]. Taken together, the pathological phenotypes found in these AD-iPSC studies mimic those that have been previously defined in mice and/or cellular models carrying the same mutations [99, 100], demonstrating the validity of iPSC-based model system.

In addition to *APP* mutations, studies have also focused on patient-derived iPSCs that contain mutations in *PSEN1* or *PSEN2*. Yagi et al. [84] demonstrated that iPSC-derived neurons harboring the *PSEN1-A246E* and *PSEN2-N141I* mutations have increased levels of Aβ42. Both mutations have also been reported to induce elevated levels of Aβ42 in human plasma, as well as in animal and other cellular models [84, 101–103]. Sproul et al. [104] analyzed iPSC lines with the *PSEN1-A246E* and *PSEN1-M146L* mutations, also demonstrating that mutant neural precursor cells (NPCs) presented increased Aβ42/40 ratios. Molecular profiling in this latter study identified 14 genes differently regulated in mutant *PSEN1* NPCs, of which five targets were previously shown to be differentially expressed in late and intermediate AD patients [104]. While the gene expression changes identified in this study are intriguing, they also emphasize the power of AD-iPSC studies to replicate additional phenotypic features. However, future mechanistic studies in both human cells and animal models are required to determine whether they indeed play a role in AD.

APP mutants, but not *PSEN1* and *PSEN2* mutants' iPSC lines, demonstrated elevated total levels of tau and tau phosphorylation in a recent study by Moore et al. [105]. They compared different iPSC-derived cortical neurons from AD patients harboring *APP* mutations (*APP-V717I* and duplication of *APP*) or *PSEN1* mutations (*PSEN1-Intron/Δ4*, *PSEN1-Y115C*, *PSEN1-M146I*). While these differences might be related to unknown effects of culture conditions or timing, another testable possibility is that *APP* and *PSEN1* or *PSEN2* mutations differ in their pathological phenotypes.

Another interesting approach for studying genetic-based disorders like FAD is with the use of isogenic lines created by genome editing of ESCs or iPSCs [73, 92]. Isogenic cell lines can be generated by inducing or correcting AD mutations in the wild-type (WT) cell line or patient-derived iPSCs, respectively [106]. The use of isogenic lines with the same genetic background reduces the intrinsic variability that comes from comparing cells from different individuals and allows to determine how a single targeted mutation affects molecular and cellular mechanisms [106]. The first study to do this for FAD mutations was conducted by Woodruff et al. [107]. They were able to generate an allelic series of heterozygous and homozygous *PSEN1-ΔE9* knock in iPSC lines, as well as heterozygous null mutants with WT and *PSEN1-ΔE9* alleles. The authors demonstrated that *PSEN1-ΔE9* mutant neurons increased the Aβ42/40 ratio in a gene dosage-dependent manner by significantly decreasing the amount of Aβ40, while moderately increasing the amount of Aβ42 [107]. The results found in this study emphasize the use of isogenic cell lines as potential promising tool for modeling AD.

iPSC-derived neurons from SAD patients have also been used to study AD, by comparing the phenotypic characteristics of these cells with iPSC-derived FAD neurons. In some of the previously mentioned reports, i.e., Israel et al. [97] and Kondo et al. [99], iPSC lines from two random SAD patients were analyzed in parallel. Interestingly, in each case only one of the two iPSC-derived neuronal populations demonstrated phenotypes consistent with the FAD lines. This heterogeneity corresponds with the complex origin of SAD and its influence on disease causation and progression. SAD is thought to be multifactorial, defined by a lack of autosomal-dominant inheritance and arises due to a complex interplay of (epi)genetic and environmental risk factors [108–110]. It is estimated that at least 60–80% of SAD may have a genetic underpinning [109, 111, 112]. Recent developments in genomic technologies, on the other hand, have allowed for high-throughput interrogation of the epigenome, and epigenome-wide association studies (EWAS) have identified unique epigenetic signatures that play a role in AD [15, 16, 113].

To date, only two studies have been published that directly focused on patient-derived iPSC lines that harbor known SAD genetic risk factors [114, 115]. For many years, *APOE3/E4* was the only known robust genetic risk factor, but as a result of several genome-wide association studies (GWAS) collaborations, increasing sample sizes and meta-analysis, at least 20 other risk, as well as protective, loci have been reported [108, 112, 116, 117]. Duan et al. [114] analyzed iPSC lines derived from three *APOE3/E4* SAD patients next to two *PSEN1* FAD mutant lines (*PSEN1-A246E* and *PSEN1-M146 L*). The basal forebrain cholinergic neurons derived from AD-*APOE3/E4* patients' iPSCs showed typical AD hallmarks, including increased Aβ42/40 ratios [114]. A second study by Young et al. [115] focused on *SORL1*, which encodes an endocytic trafficking factor whose levels modulate the processing of APP to Aβ and other proteolytic products implicated in SAD [118]. Loss of *SORL1* expression has been documented in SAD cases [119], and the *SORL1* locus has been associated with SAD in both candidate gene and GWAS analyses [112]. By studying patient iPSC-derived neural cells, this latter study confirmed the importance of the SORL1/APP pathway in SAD, and their findings corroborate most previous studies in cell and animal models [115].

In another study by Hossini at el. [120], SAD iPSC-derived neurons were analyzed to assess the reflection of disease phenotype in gene expression and to examine the expression of typical AD proteins. The differentiated neuronal cells seemed to reflect the SAD phenotype by the expression of phosphorylated tau proteins and the upregulation of GSK-3β [120]. Further analysis of the neuronal cells also revealed significant changes in the expression of other genes associated with AD, including subunits of the proteasome complex [120]. Moreover, a disease-specific protein association network that models AD pathology on the transcriptome level could be generated from the AD-iPSCs [120]. Taken together, these studies have demonstrated that SAD patients' iPSC-derived neuronal cells are able to recapitulate neuropathological processes of the disease.

Unfortunately, the contribution of epigenetic signatures, as well as environmental factors, has not been addressed yet in iPSC neurons derived from SAD, as well as FAD patients. Nonetheless, iPSC-derived neurons offer a platform to examine the casual relationships between environmental insults and the generation of molecular, cellular, and epigenomic responses in AD-relevant neuronal populations. In theory, one could test how the derived neuronal populations aggravate AD characteristic phenotypes when exposed to environmental risk factors or pathological hallmarks, such as stress hormones [121]. Furthermore, the use of iPSCs also provides the opportunity to examine the contribution of AD-associated epigenetic signatures.

Based on recent advances in the field of epigenetic editing [122, 123], these AD-associated epigenetic signatures can be modified at any given locus in order to normalize the cellular phenotypes in patient iPSC-derived neurons, as well as to induce AD characteristics in control iPSC lines.

In addition to evaluating the potential of iPSC-derived neurons to model typical AD phenotypes, substantial work has been done in order to assess the possible medical relevance of AD iPSC models in terms of drug discovery and selection of appropriate therapeutics. Many of the aforementioned studies have tried to normalize the AD-associated phenotypes by previous studied therapeutics for AD. In the iPSC model of Down syndrome, for example, Shi et al. [95] speculated whether Aβ40 and Aβ42 peptide generation could be reduced by pharmacological inhibition of the γ-secretase complex. Compounds that inhibit γ-secretase, as well as β-secretase, are potential therapeutics for AD, and inhibition of these protease complexes has been shown to reduce Aβ level in mice models [124, 125]. When a γ-secretase inhibitor was administered for 4 consecutive days to the Down syndrome iPSC-derived neurons, Aβ40 and Aβ42 peptide production was reduced by almost half, whereas longer-term treatment (21 days) reduced secretion of both Aβ peptides below detectable levels [95]. Also Yagi et al. [84] found a dose-dependent reduction in Aβ42 and Aβ40 in iPSC-derived neurons form AD patients treated with γ-secretase inhibitors and modulators. A γ-secretase inhibitor was also sufficient to block Aβ production in both control and *PSEN1* mutant NPCs [104], as well as in *PSEN1* mutant neuronal cells [107]. Surprisingly, γ-secretase inhibitor treatment paradoxically increased Aβ40 secretion in the *APOE3/E4* SAD lines [114]. The reason for this latter finding is not clear yet and will need to be addressed properly in future studies.

In other studies, FAD and SAD patient iPSC-derived neurons were treated with γ-secretase inhibitors or β-secretase inhibitors, and Aβ, GSK-3β, phosphorylated tau, and total tau levels were assessed. It was shown that β-secretase inhibitors, but not γ-secretase inhibitors, could significantly reduce the levels of phosphorylated tau and GSK-3β, while γ-secretase inhibitors only reduced the level of Aβ40 [97]. These findings suggested that APP proteolytic processing had a direct relationship with GSK-3β activation and tau phosphorylation in these neuronal models [97]. In line with these findings, manipulation of APP metabolism by β-secretase and γ-secretase inhibition/modulation also affected tau protein levels in the study by Moore et al. [105]. Furthermore, inhibition of γ-secretase significantly reduced the production of extracellular Aβ38, Aβ40, and Aβ42 in neurons of all genotypes [105]. In a separate study, also significant reductions of phosphorylated tau and tau expression were found in neuronal cells differentiated from a SAD patient after treatment with a γ-secretase inhibitor [120] .

Kondo et al. [99] evaluated β-secretase inhibitors and three additional drugs that have been reported to improve ER stress or to inhibit reactive oxygen species (ROS) generation, including docosahexaenoic acid (DHA). Intracellular accumulation of Aβ oligomers disappeared and ROS formation decreased after treatment with β-secretase inhibitors in both FAD and SAD iPSC-derived neurons [99]. DHA treatment, on the other hand, decreased the generation of ROS in AD neural cells harboring the *APP-E693Δ* mutation, whereas the amount of Aβ oligomers in cell lysates was not altered [99]. The clinical effectiveness of DHA treatment is still controversial and, interestingly, only one of two sporadic AD neurons accumulated intracellular Aβ oligomers and showed cellular phenotypes that could respond to DHA, while the other did not [99]. This result may explain why DHA treatment is only effective for some subpopulations of SAD patients, although disease stage and timing of treatment could be other critical factors to explain this phenomenon. These patient-specific iPSCs might, therefore, provide a chance to reevaluate the effect of a drug that failed in AD clinical trials, depending on the subpopulation of patients. Finally, immunotherapy is one of the alternative strategies being studied for the treatment of AD [126]. In the study of Muratore et al. [100], Aβ-specific antibodies were able to reverse the phenotype of increased total tau in AD iPSC-derived neurons harboring the *APP-V717I* mutation.

To conclude, current studies represent critical first steps in assessing the potential of using iPSCs in AD research. Patient-specific iPSCs-derived neural cells have demonstrated validity on modeling AD pathological molecular alterations, such as increased Aβ42/40 ratios and tau hyperphosphorylation. Moreover, these studies have addressed the benefits of these iPSC systems for testing therapeutic intervention strategies and drug libraries. Furthermore, iPSC technology might also be a valuable tool in exploring the complex heterogeneous nature in the etiology of SAD through interrogation of functional effects of (epi)genetic variants linked to risk and protective factors. With the availability of epigenetic editing systems, we might be able to decipher how epigenetic alterations participate in AD, which may also offer opportunities for future epigenetic-based pharmacological interventions. Nevertheless, there is enormous promise in the utility of iPSC technology to predict how individual epigenetic and cellular phenotypic variation contributes to the etiology and pathophysiology of AD, as well as to pharmacological responses at clinically relevant levels.

23.4 Fragile X Syndrome

FXS is considered an autism spectrum disorder and the most frequent form of inherited intellectual disability with a penetrance of 1 in 2500 males and 1 in 4000 females [127]. Patients suffer from multifactorial symptoms such as learning deficits, low IQ, autism-like behavior, obesity, hypotonia in childhood, and seizures in adult life [128]. The disease is caused by the loss of fragile X mental retardation protein (FMRP), which is a cytoplasmic RNA-binding protein involved in transport and translocation of mRNA and proteins from the nucleus to neuronal dendrites [129, 130]. FMRP plays an important role in regulating synaptic development, plasticity [131], and vesicular dynamics [132]. FMRP absence is usually caused by an aberrant epigenetic silencing of *fragile X mental retardation 1* (*FMR1*) gene due to promoter inhibition, as a consequence of CGG repeat expansion in proximity to the 5-UTR region [133, 134]. The increase in length of CGG repeats marks the region for hypermethylation. Non-affected individuals typically have 6–50 CGG repeats in the CpG island close to *FMR1* locus. For reasons still unknown, some subjects are susceptible to an expansion of those repeats. Between 50 and 200 CGG repeats are found in this region in a pre-mutation state, whereas most FXS patients carry >200 repeats and characterize the full mutation state, likely leading to CpG methylation, chromatin condensation, and *FMR1* transcriptional silencing [135].

Mice models mimicking the disease phenotype were developed using *FMR1* gene knockout (KO), mirroring several pathological features of FMRP impairment as *FMR1* mRNA targets and FMRP functions [136–138]. Although very interesting, mice lack the epigenetic silencing of *FMR1* in large CGG repeats, therefore, limiting the use of mice models for fully understanding the molecular mechanism of the disease. Furthermore, the mechanism of epigenetic alterations on *FMR1* locus and the consequent loss of FMRP during development are still uncertain [139]. The use of human stem cells has collaborated to a better comprehension of the disease mechanisms [140], and for the purpose of this section, we will, therefore, explore the recent discoveries using iPSC-based models for FXS.

ESCs with >200 CGG repeats present unmethylated CpGs and normal levels of *FMR1* gene in early stages of development and progressively become methylated during differentiation [141]. Patient-derived iPSCs, on the other hand, maintain their methylation status during the reprogramming process, which demonstrates that in iPSCs the methylation status is not reversible [140, 142]. The epigenetic silencing of *FMR1* is believed to occur in a differentiation-dependent manner, although the maintenance of methylation during reprogramming of iPSC-derived neurons from FXS patients supports their use for modeling the disorder [143]. Nonetheless, iPSCs generated from patients' fibroblasts usually show that the repeat expansion is not equally present in all the cells, thus modeling the mosaicism observed in patients [144]. This mosaicism contributes greatly to disease pathogenesis and individual phenotypic variability [127, 135, 144]. Moreover, by using iPSCs, scientists were able to illustrate the phenomenon of CGG repeat expansion. After reprogramming, fibroblasts from mutated individuals (>200 repeats) generated cells with full-length expansions and fully methylated status. However, fibroblasts with a pre-mutation genotype (between 50 and 200 repeats) generated cells with either a normal genotype (<50 repeats) or full-mutated clones [141, 145]. These data show that pre-mutation expansion length is genetically instable and that the pre-mutated state is critical for disease pathogenesis [146].

Apart from that, a growing body of evidence suggested that the epigenetic state of *FMR1* locus, rather than the extension of the CGG repeats, is critical for transcriptional silencing [135, 144, 147]. In agreement, de Esch et al. [148] studied iPSC derived from unmethylated full-mutated individuals (presenting >200 repeats but no methylation on *FMR1* locus). Interestingly, after fibroblast reprogramming, the obtained iPSC-derived neurons presented a full methylation status on *FMR1* region. This data demonstrated that methylation is a standard mechanism in CGG repeats expansions and that individuals harboring repeats without methylation had a normal phenotype, likely attributed to the unmethylated status of *FMR1*.

Likewise, Park et al. [149] also illustrated the importance of (CGG)n repeats in methylation and silencing of *FMR1*. In their study, genetic editing with CRISPR/

Cas9 technology was used to eliminate CGG repeats upstream of the promoter in FXS iPSC-derived neurons. CRISPR/Cas9 technology works as a sequence-specific nuclease inducing a double cleavage in the DNA in targeting regions [150]. After deleting a large portion of CGG repeats, the authors described that the chromatin opened and demethylation occurred, leading to transcriptional activation of *FMR1*. These findings do not only suggest that transcriptional silencing of *FMR1* can be reverted, but also that methylation status in this region is constantly monitored.

Finally, FXS patient-derived iPSCs are also starting to be used for high-throughput drug testing and harbor high potential for drug discovery research. For instance, iPSCs have been treated with drugs that could increase *FMR1* levels [151], but unfortunately no clinical relevant results were obtained yet. The possibility of reverting FXS phenotype after reestablishment of *FMR1* levels is also still uncertain and might partly explain these findings. To conclude, although all the pathophysiological alterations in FXS remain to be elucidated, stem cell technology is unraveling the role of epigenetic alterations in the disorder, such as the methylation dynamics of the *FMR1* locus. Research along this line will greatly facilitate the development of therapeutic interventions and will allow future studies using epigenetic editing techniques.

23.5 Angelman Syndrome and Prader-Willi Syndrome

AS was first described in 1965 by the pediatrician Harry Angelman who noticed autistic-like features in three different individuals [152]. AS patients present neurological problems, abnormally motor condition, severe mental retardation, epileptic seizures, and episodes of inappropriate laughter [153]. The prevalence is estimated between 1/10,000 and 1/20,000 [154]. PWS is characterized by hypotonia, hypogonadism, intellectual disability, a tendency to develop compulsive and obsessive behavior, and hyperphagia-causing obesity [155]. The incidence, on the other hand, is estimated between 1/15,000 and 1/30,000 [155]. Both disorders are caused by imprinting alterations on chromosome 15q11–13 region [156] although AS is caused by maternal inherited alterations, while PWS is caused by paternal ones.

During development, specific genes on a variety of chromosomes are subjected to silencing by a process known as genomic imprinting, which is dependent on the location of the gene on the chromosome, as well as its parent origin [157, 158]. The reason for silencing one of the parental genes is not fully understood, and estimations point to only 1% of human genes to have inherited repression markers [159]. Silencing is mediated by a set of epigenetic alterations such as methylation of promoters and histone modifications [160, 161]. An important region of chromosome 15 (15q11–13) that is critical for many cellular processes is also subjected to genomic imprinting. This region compromises a variety of genes that are exclusively expressed in a parental fashion: *MKRN3*, *MAGEL2*, *NDN*, *C15orf2* and *PWRN1 SNURF-snrpn* for paternal origin, and *UBE3A* and *ATP10A* for maternal origin [162, 163]. Because many of these genes in this region are imprinted, the loss of function of one copy leads to vigorous alterations [164].

AS is caused by the reduction or loss of the maternal allele coding for *UBE3A* gene, while PWS results from partial deletion of the 15q11.2-q13 region of the paternal allele and affects seven genes [160]. However, not all AS or PWS diagnosed patients present the same classical alterations on 15q11.2-q13 region [153]. Both disorders and other related pathologies are also known to be caused by other genetic alterations [160], including large and small deletions and duplications of 15q11.2-q13, mutations in imprinting centers, and uniparental disomy, among others [165]. As for many epigenetic disorders, the variety of genomic alterations affects the clinical severity of the disease and the possibility of modeling it for studying.

Mice models targeting chromosome 15 have been used to study AS dynamics. The most commonly used model is the deletion of exon 2 from *Ubea3*, since it is the only model that induces inherited maternal loss of *Ubea3* [166, 167]. Using *Ubea3*-null mice, researchers were able to dissect the protein function. In the brain, UBEA3 contributes to synapses formation and neuronal circuitry [168, 169] probably by downregulating the expression of other proteins [167]. As a consequence of UBEA3 loss, synapses development is impaired [169]. PWS has also been mimicked using mice models since human pattern alleles found in 15q11.2-q13 also occur in a well-conserved region in chromosome 7 in mice. Although very similar, there are differences regarding centromere distance and the absence of *C15ORF2* gene and two noncoding snoRNAs in mice [170]. PWS mice targeting pattern genes in this homologous region conveniently recapitulate some PWS symptoms but still remain an incomplete model. The complexity of the human chromosomal region with genes being paternally or maternally expressed, together with the variety of genotypes/phenotypes that emerge from all the possible alterations occurring in this chromosome range, makes it difficult to model, study, and develop therapeutic approaches for both disorders [171].

To date, only a few groups have explored the potential of iPSCs in modeling pathological mechanisms in AS or PWS and addressed their potential to find therapeutic interventions for both disorders. Similar to FXS, differentiated iPSCs from AS and PWS recapitulate imprinting and methylation patterns unlike ESC, making it very suitable to study disease progression and pathological mechanisms [172]. Recently, the development of AS- and PWS iPSC-derived neurons showed that imprinting process occurs during neuronal differentiation and that this model can successfully recapitulate some of the disorder's mechanism [173]. Although this latter study did not find neuronal differences between normal individuals and AS and PWS iPSC-derived neurons, large analyses of human iPSCs carrying chromosome 15 alterations have shown common pathological pathways for AS (deletion of 15q11–13) and other disorders harboring duplications of 15q11–13 [174].

Interestingly, most human tissues express both paternal and maternal alleles of *UBE3A*, whereas in neurons the paternal *UBE3A* is usually epigenetically silenced by the long noncoding RNA (ncRNA) UBEA3-ATS [175, 176]. Studies using iPSCs from PWS patients with microdeletion in a region critical for paternal imprinting showed that the activation of this ncRNA alone can alter imprinting patterns in *UBE3A* paternal allele, suggesting that in neurons UBEA3-ATS is sufficient for

UBE3A paternal silencing [177, 178]. A similar model was used in an attempt to clarify the underlying mechanisms that trigger UBEA3-ATS expression in neurons [179], but the exact molecular mechanisms remain unclear. In parallel, therapeutic interventions for targeting UBEA3-ATS have been assessed in mice, with the aim of reestablishing normal levels of the protein by activating the paternal allele [180]. Furthermore, the administration of selected drugs have also been used to increase paternal *Ubea3* expression in a mice model of AS [181]. Both strategies successfully reduced UBEA3-ATS levels and consequently recovered *Ubea3* pattern allele expression, leading to amelioration of the cognitive deficits observed in mice [180]. Although not addressed yet for AS or PWS, human stem cells could offer a valuable platform to study similar therapeutic strategies for a variety of candidate genes that are inherently repressed in the same region. Especially the use of patient-derived iPSCs for these studies could potentially help to better understand the mechanisms behind specific gene silencing, both in imprinting, as well as in disease situations, and will likely contribute to develop therapeutic interventions in AS, PWS and other related disorders.

23.6 Rett Syndrome

RTT is a neurodevelopment disorder affecting mostly girls, with a prevalence of 1 in 10,000 [182]. Patients first have an apparent normal intrauterus development, and symptoms only start to appear between the sixth and eightieth months of postnatal development. Typical symptoms include motor and language impairment, as well as cognitive regression. Although the symptomatology stabilizes approximately by 5 years of age, the life span of girls is severely reduced [183]. In 1999, the syndrome was correlated to de novo mutations in the methyl-CpG-binding gene (*MECP2*) [184], which is an important regulator of the epigenetic state highly expressed in developing neurons [185, 186]. MECP2 is involved in regulating chromatin structure and acts as a transcriptional repressor, as well as activator [187]. Although RTT is mainly associated to *MECP2*, several other mutations in different genes have been described in RTT-like phenotypes, such as *CDKL5*, *TC4*, *JMJD1C*, and *FGX1*, among others [188, 189].

MECP2 levels are precisely controlled during development, and studies have shown that both overexpression [190, 191] and downregulation [192] affect brain normal development. The affinity of MECP2 to target regions in the DNA is highly dependent on its sequence [193]. As a consequence, even small alterations in the protein sequence can affect the interaction of MECP2 domains with the chromatin [194–196], thereby altering the epigenetic state of the genome. To date, over 800 mutations have been mapped for the *MECP2* locus in RTT patients [197, 198]. Several alterations, including missense and nonsense mutations, deletions, and duplications, can dramatically affect MECP2 function and contribute to a large range of clinical variability [203]. Furthermore, female embryos undergo X chromosome inactivation during development to compensate for the presence of two parental copies. Since *MECP2* is located on the X chromosome (Xq28), RTT patients, therefore, can show mosaicism, which also contributes to the clinical observation of different phenotypes within the disorder [199, 200].

Mice models of RTT are widely used for modeling the disorder [201–204] and recapitulate several pathological hallmarks of the disease, such as abnormalities in dendritic morphology and neuronal connectivity [205, 206]. Interestingly, mice models have also shown that distinct neuronal populations are differently altered due to *MECP2* loss [207–209]. Furthermore, those models have demonstrated that RTT phenotypes can be rescued by different treatment strategies [210–212]. One of the disadvantages of these RTT mice models, however, is that they are usually generated by *MECP2* knockouts, while RTT patients likely present missense mutations in *MECP2*. Another caveat in these models is that the defects only appear very late during development in contrast to the early onset seen in humans [213]. Moreover, there are many biological differences in neurodevelopment and brain functions among mice and humans, which makes the use of humanized models more appealing for understanding RTT [11, 214].

The use of stem cells and specifically iPSC has shed a new light in the comprehension of RTT mechanisms [215]. The first in vitro neuronal model derived from RTT iPSCs with *MECP2* mutations demonstrated most of the cellular abnormalities found in human and mice brains, such as lack of complex synapses (e.g., defects in synaptic outgrowth) and reduced number of dendrites [216]. Also, this study showed abnormalities in glutamatergic synapses and impaired electrophysiological properties [216]. Neurons obtained from iPSC from RTT patients also demonstrated defects in synaptic transmission [217] and connectivity [218], a phenomenon that is likely related to a reduced number of dendrites and which could be ameliorated through the administration of choline [217]. It has also been shown that iPSCs from RTT patients have maturation defects [219] and harbor a possible early deviation into the astrocytic lineage during differentiation [220], compromising neuronal commitment and proliferation. Furthermore, it is important to point out that cells derived from patients likely recapitulate the random X chromosome inactivation during development and differentiation [218, 221], providing construct validity to the model.

The possible contribution of glial cells in RTT pathology has also been addressed by studying iPSC-derived astrocytes [222, 223]. Mutated astrocytes presented impairment in vesicular transport [223] and interacted with normal neurons affecting neuronal morphology and maturation [222]. By using a 3D culture model of iPSCs from RTT patients, Zhang et al. [224] assessed the consequences of *MECP2* impairment in neuronal migration and in the interaction of neurons and astrocytes [224]. In this 3D RTT model, normal neurons and astrocytes were combined with mutated cells in a set of experiments for cellular migration. In the control group, normal astrocytes and neurons migrated toward each other, establishing cohesion. Interestingly, when normal astrocytes and RTT-derived neurons were combined, the migration impairment was similar to the impairment noticed in RTT-derived astrocytes and normal neurons. This shows that *MECP2* can be important to cellular migration during developing of the brain and that the involvement of glia cells in the process shouldn't be overlooked. One interesting conclusion from the abovementioned studies, however, is that the implication of *MECP2* in migration process contradicts the general assumption that the lack of complex connections on *MECP2*-impaired brain occurs due to poor synapses development, rather than migration problems.

Besides *MECP2* mutations as a cause of RTT, several other mutations have been identified in RTT-like phenotypes [186, 188, 189]. For example, *CDKL5*-mutated iPSC were recently studied [225, 226], not only in the context of RTT but also for modeling other *CDKL5*-related disorders, such as epileptic encephalopathy or West syndrome. Cellular abnormalities caused by disruption of *CDKL5* are similar to *MECP2* mutations, but underlying mechanisms are different and CDLK5 protein appears to have a direct role in synaptic control [225, 226].

Most studies modeling *MECP2* loss are not only interested in the basic pathological pathways of RTT but also in the function of *MECP2* itself as a key component of epigenetic regulatory mechanisms in the brain [227], affecting gene expression [228, 229] and chromatin structure [230]. The analysis of in vitro phenotypes of RTT-derived iPSCs will collaborate to understanding the function of each MECP2 protein domains in RTT and potentially enable the development of mutation specific therapy [182]. By using iPSC-based approaches or allogenic stem cell [231] populations in combination with CRISPR/Cas9-induced mutations, we can nowadays assess specific consequences of selected mutations in neuronal function. In this perspective, iPSC-derived neurons from RTT patients will greatly contribute to unveil molecular targets and cellular process within neuronal development.

23.7 Discussion and Future Perspectives

In the current chapter, we have addressed the growing body of scientific interest in stem cells and their utility in terms of their potential to increase our general knowledge of several (epi)genetic brain disorders. A universal challenge in the field of translational neuroscience has been the development of animal and cellular models that effectively recapitulate the biology of the human brain. Stem cells, including ESCs, NSCs, and iPSCs, have entered the field as a potential breakthrough to overcome this hurdle and presented an alternative model for human brain studies. The field has been developing at a rapid pace, and methods to obtain, maintain, and differentiate these cells are in continuous adaptation and optimization. Stem cell technology and directed neural differentiation techniques allow to examine the broad repertoire of neural cells found in the human brain, with the goal to elucidate developmental, cellular, and molecular features that were previously inaccessible in animal models or clinical studies. Aside from contributing to the understanding of underlying neurobiology and consequences of personal molecular variations on healthy brain functioning and disease, stem cell-based studies have the potential to enhance the development of new and effective therapeutic interventions. The use of stem cells and derivatives for high-throughput screening of compound libraries and toxicological analysis before the initiation of clinical trials becomes increasingly favored for the aim of drug discovery. Moreover, fully functional neural cells or neural precursors that are differentiated from human stem cells have direct therapeutic potential in the field of regenerative medicine and might be employed as a cell replacement therapy for multiple neurodegenerative disorders such as Parkinson's disease. In summary, stem cell technology has been offering unprecedented possibilities to investigate unique human biological features in a cell subtype-specific, as well as personalized matter, and is expected to greatly contribute to the development of therapeutic interventions [232].

Especially relevant in the light of current chapter, these stem cell-based studies offer a unique opportunity to further reveal the (epi)genetic and environmental contributions of brain disorders. In fact, stem cell-based studies of (epi)genetic influences in combination with the environmental exposome are expected to have a huge impact in enhancing our understanding of brain disorders and will also assist the development of (epi)genetic-based therapeutic intervention. Moreover, with the current advances in genetic [233] and epigenetic [234] editing techniques, human stem cells and derivatives can be (epi)genetically altered in a controlled setting to induce, aggravate, or recover cellular disease phenotypes in vitro. This directly allows to study the role of individual (epi)genetic signatures on disease causation and progression. In fact, a unique opportunity of human stem cell-based systems is the ability to probe how (epi)genomic architectures predispose individuals and how they influence the behavior of various, participating cell types. A consequent translational contribution, therefore, is the possibility of early detection of brain disorders that will improve timeliness and efficacy of diagnostic and therapeutic interventions. Furthermore, there is enormous promise in the utility of human neural culture systems to predict how individual (epi)genetic and cellular phenotypic variations contribute to the response of pharmacological interventions at clinically relevant levels. Finally, the (epi)genetic technologies available may also allow the deciphering of how (epi)genomic architectures found in individual humans act together to generate susceptibility and variation in response to the environmental insults that may also contribute to, or pharmacologically modify, disease phenotypes in patients. Taken together, these study strategies provide an elegant and dynamic tool for modeling, following, and understanding various essential pathological mechanisms of multiple brain disorders.

Although considerable progress has been made in validating these stem cell-based applications, further research is necessary to realize their full potential [235]. To date, there remains widespread lab-to-lab variability in culture methods and a lack in differentiation techniques to obtain homogeneous neural populations, as well as a lack of data regarding the desired cell types for modeling or treating certain brain disorders. Cellular heterogeneity, in combination with unidentified effects of reprogramming processes, can act as a potential confounder in epigenetic research and may lead to an under- or overestimation of the observed epigenetic differences. Furthermore, the simplicity of an in vitro model system is both an advantage and a significant disadvantage when culture purity in this case becomes a liability. Such models lack the complex mixture of neural cells found in the human brain, the complex extracellular matrices, and their integrated 3D organizations, which might also cause confounding effects on experimental outcomes. Although the latter might be circumvented by combined cultures, 3D culture systems, or the use of organoids

[236], the future development of stem cell-based approaches will require extensive analysis and standardization, and any translation from bench to bedside must be undertaken gradually, with great caution and based on profound experimental data [1]. Altogether, stem cell technology is hypothesized to revolutionize the field of neuroscience the next 10 years, contributing substantially to improve our knowledge of brain functions and epigenetic dysregulation in disease and to identify new druggable targets.

Acknowledgments Renzo J. M. Riemens is supported by Maastricht University (Maastricht, The Netherlands) and Julius Maximilians University (Würzburg, Germany). Edilene Siqueira Soares is supported by the National Council for Scientific and Technological Development (CNPQ; grant n. 202074/2015-3), Ministry of Science, Brazil. The work of laboratory is supported by, among other institutions, the EU Joint Programme – Neurodegenerative Disease Research (JPND), Cellex Foundation, the Health and Science Departments of the Catalan Government (Generalitat de Catalunya), the E-Rare (ERA-Net for research programs on rare diseases), and EuroRETT (a European network on Rett syndrome, funded by the European Commission under its 6th Framework Program since 2006). Dr. Manel Esteller is an ICREA Research Professor.

References

1. Horvath P, Aulner N, Bickle M, Davies AM, Del Nery E, Ebner D, et al. Screening out irrelevant cell-based models of disease. Nat Rev Drug Discov. 2016;15:751–69.
2. Fernández-Santiago R, Ezquerra M. Epigenetic research of neurodegenerative disorders using patient iPSC-based models. Stem Cells Int. 2016;2016:1–16.
3. Takahashi K, Yamanaka S. A decade of transcription factor-mediated reprogramming to pluripotency. Nat Rev Mol Cell Biol. 2016;17:183–93.
4. Takahashi K, Yamanaka S. Induction of pluripotent stem cells from mouse embryonic and adult fibroblast cultures by defined factors. Cell. 2006;126:663–76.
5. Rubin LL. Stem cells and drug discovery: the beginning of a new era? Cell. 2008; 132:549–52.
6. Han F, Baremberg D, Gao J, Duan J, Lu X, Zhang N, et al. Development of stem cell-based therapy for Parkinson's disease. Transl Neurodegener. 2015;4:16.
7. Yap MS, Nathan KR, Yeo Y, Lim LW, Poh CL, Richards M, et al. Neural differentiation of human pluripotent stem cells for nontherapeutic applications: toxicology, pharmacology, and in vitro disease modeling. Stem Cells Int. 2015;2015:105172.
8. Ma L, Liu Y, Zhang SC. Directed differentiation of dopamine neurons from human pluripotent stem cells. Methods Mol Biol. 2011;767:411–8.
9. Santos DP, Kiskinis E. Generation of spinal motor neurons from human pluripotent stem cells. Synap Dev Methods Protoc. 2017;53–66
10. Payne NL, Sylvain A, O'Brien C, Herszfeld D, Sun G, Bernard CCA. Application of human induced pluripotent stem cells for modeling and treating neurodegenerative diseases. New Biotechnol. 2015;32:212–28.
11. Wen Z, Christian KM, Song H, Ming G. Modeling psychiatric disorders with patient-derived iPSCs. Curr Opin Neurobiol. 2016;36:118–27.
12. Portela A, Esteller M. Epigenetic modifications and human disease. Nat Biotechnol. 2010;28:1057–68.
13. Badger JL, Cordero-Llana O, Hartfield EM, Wade-Martins R. Parkinson's disease in a dish - Using stem cells as a molecular tool. Neuropharmacology. 2014;76(Pt A):88–96.
14. Pu J, Jiang H, Zhang B, Feng J. Redefining Parkinson's disease research using induced pluripotent stem cells. Curr Neurol Neurosci Rep. 2012;12:392–8.

15. Sanchez-Mut JV, Heyn H, Vidal E, Moran S, Sayols S, Delgado-Morales R, et al. Human DNA methylomes of neurodegenerative diseases show common epigenomic patterns. Transl Psychiatry. 2016;6:e718.
16. Lardenoije R, Iatrou A, Kenis G, Kompotis K, Steinbusch HW, Mastroeni D, et al. The epigenetics of aging and neurodegeneration. Prog Neurobiol. 2015;131:21–64.
17. Damier P, Hirsch EC, Agid Y, Graybiel AM. The substantia nigra of the human brain. II. Patterns of loss of dopamine-containing neurons in Parkinson's disease. Brain. 1999;8: 1437–48.
18. Beitz JM. Parkinson's disease: a review. Front Biosci (Schol Ed). 2014;6:65–74.
19. Byers B, Lee HL, Reijo PR. Modeling Parkinson's disease using induced pluripotent stem cells. Curr Neurol Neurosci Rep. 2012;12:237–42.
20. Nishimura K, Takahashi J. Therapeutic application of stem cell technology toward the treatment of Parkinson's disease. Biol Pharm Bull. 2013;36:171–5.
21. Fu MH, Li CL, Lin HL, Chen PC, Calkins MJ, Chang YF, et al. Stem cell transplantation therapy in Parkinson's disease. Springerplus. 2015;4:597.
22. Kocabicak E, Tan SK, Temel Y. Deep brain stimulation of the subthalamic nucleus in Parkinson's disease: why so successful? Surg Neurol Int. 2012;3:S312–4.
23. Kornblum HI. Introduction to neural stem cells. Stroke. 2007;38:810–6.
24. Nishino H, Hida H, Takei N, Kumazaki M, Nakajima K, Baba H. Mesencephalic neural stem (progenitor) cells develop to dopaminergic neurons more strongly in dopamine-depleted striatum than in intact striatum. Exp Neurol. 2000;164:209–14.
25. Redmond DE, Bjugstad KB, Teng YD, Ourednik V, Ourednik J, Wakeman DR, et al. Behavioral improvement in a primate Parkinson's model is associated with multiple homeostatic effects of human neural stem cells. Proc Natl Acad Sci U S A. 2007; 104:12175–80.
26. Lee CS, Cenci MA, Schulzer M, Björklund A. Embryonic ventral mesencephalic grafts improve levodopa-induced dyskinesia in a rat model of Parkinson's disease. Brain. 2000;123:1365–79.
27. Kondoh T, Pundt LL, Low WC. Development of human fetal ventral mesencephalic grafts in rats with 6-OHDA lesions of the nigrostriatal pathway. Neurosci Res. 1995;21:223–33.
28. Redmond Jr DE, Vinuela A, Kordower JH, Isacson O. Influence of cell preparation and target location on the behavioral recovery after striatal transplantation of fetal dopaminergic neurons in a primate model of Parkinson's disease. Neurobiol Dis. 2008;29:103–16.
29. Yang H, Wang J, Wang F, Liu X, Chen H, Duan W, et al. Dopaminergic neuronal differentiation from the forebrain-derived human neural stem cells induced in cultures by using a combination of BMP-7 and Pramipexole with growth factors. Front Neural Circuits. 2016;10:1172.
30. Park CH, Kang JS, Shin YH, Chang MY, Chung S, Koh HC, et al. Acquisition of in vitro and in vivo functionality of Nurr1-induced dopamine neurons. FASEB J. 2006;20:2553–5.
31. Parish CL, Castelo-Branco G, Rawal N, Tonnesen J, Sorensen AT, Salto C, et al. Wnt5a-treated midbrain neural stem cells improve dopamine cell replacement therapy in parkinsonian mice. J Clin Invest. 2008;118:149–60.
32. Studer L, Tabar V, McKay RD. Transplantation of expanded mesencephalic precursors leads to recovery in parkinsonian rats. Nat Neurosci. 1998;1:290–5.
33. Hagell P, Schrag A, Piccini P, Jahanshahi M, Brown R, Rehncrona S, et al. Sequential bilateral transplantation in Parkinson's disease. Brain. 1999;122:1121–32.
34. Levivier M, Dethy S, Rodesch F, Peschanski M, Vandesteene A, David P, et al. Intracerebral transplantation of fetal ventral mesencephalon for patients with advanced Parkinson's disease. Stereotact Funct Neurosurg. 1998;69:99–111.
35. Freed CR, Breeze RE, Rosenberg NL, Schneck SA, Kriek E, Qi J, et al. Survival of implanted fetal dopamine cells and neurologic improvement 12 to 46 months after transplantation for Parkinson's disease. N Engl J Med. 1992;327:1549–55.
36. Freeman TB, Olanow CW, Hauser RA, Nauert GM, Smith DA, Borlongan CV, et al. Bilateral fetal nigral transplantation into the postcommissural putamen in Parkinson's disease. Ann Neurol. 1995;38:379–88.

37. Lindvall O, Sawle G, Widner H, Rothwell JC, Björklund A, Brooks D, et al. Evidence for long-term survival and function of dopaminergic grafts in progressive Parkinson's disease. Ann Neurol. 1994;35:172–80.
38. Mendez I, Viñuela A, Astradsson A, Mukhida K, Hallett P, Robertson H, et al. Dopamine neurons implanted into people with Parkinson's disease survive without pathology for 14 years. Nat Med. 2008;14:507–9.
39. Chu Y, Kordower JH. Lewy body pathology in fetal grafts. Ann N Y Acad Sci. 2010;1184:55–67.
40. Hagell P, Piccini P, Björklund A, Brundin P, Rehncrona S, Widner H, et al. Dyskinesias following neural transplantation in Parkinson's disease. Nat Neurosci. 2002;5:627–8.
41. Olanow CW, Goetz CG, Kordower JH, Stoessl AJ, Sossi V, Brin MF, et al. A double-blind controlled trial of bilateral fetal nigral transplantation in Parkinson's disease. Ann Neurol. 2003;54:403–14.
42. Nauta AJ, Westerhuis G, Kruisselbrink AB, Lurvink EGA, Willemze R, Fibbe WE. Donor-derived mesenchymal stem cells are immunogenic in an allogeneic host and stimulate donor graft rejection in a nonmyeloablative setting. Blood. 2006;108:2114–20.
43. Turner DA, Kearney W. Scientific and ethical concerns in neural fetal tissue transplantation. Neurosurgery. 1993;33:1031–7.
44. Kirkeby A, Grealish S, Wolf DA, Nelander J, Wood J, Lundblad M, et al. Generation of regionally specified neural progenitors and functional neurons from human embryonic stem cells under defined conditions. Cell Rep. 2012;1:703–14.
45. Muramatsu S, Okuno T, Suzuki Y, Nakayama T, Kakiuchi T, Takino N, et al. Multitracer assessment of dopamine function after transplantation of embryonic stem cell-derived neural stem cells in a primate model of Parkinson's disease. Synapse. 2009;63:541–8.
46. Kriks S, Shim JW, Piao J, Ganat YM, Wakeman DR, Xie Z, et al. Dopamine neurons derived from human ES cells efficiently engraft in animal models of Parkinson's disease. Nature. 2011;480:547–51.
47. Takagi Y, Takahashi J, Saiki H, Morizane A, Hayashi T, Kishi Y, et al. Dopaminergic neurons generated from monkey embryonic stem cells function in a Parkinson primate model. J Clin Invest. 2005;115:102–9.
48. Kikuchi T, Morizane A, Doi D, Onoe H, Hayashi T, Kawasaki T, et al. Survival of human induced pluripotent stem cell-derived midbrain dopaminergic neurons in the brain of a primate model of Parkinson's disease. J Parkinsons Dis. 2011;1:395–412.
49. Falkner S, Grade S, Dimou L, Conzelmann KK, Bonhoeffer T, Götz M, et al. Transplanted embryonic neurons integrate into adult neocortical circuits. Nature. 2016;539:248–53.
50. Grealish S, Diguet E, Kirkeby A, Mattsson B, Heuer A, Bramoulle Y, et al. Human ESC-derived dopamine neurons show similar preclinical efficacy and potency to fetal neurons when grafted in a rat model of Parkinson's disease. Cell Stem Cell. 2014;15:653–65.
51. Arenas E, Denham M, Villaescusa JC. How to make a midbrain dopaminergic neuron. Development. 2015;142:1918–36.
52. Brederlau A, Correia AS, Anisimov SV, Elmi M, Paul G, Roybon L, et al. Transplantation of human embryonic stem cell-derived cells to a rat model of Parkinson's disease: effect of in vitro differentiation on graft survival and teratoma formation. Stem Cells. 2006;24:1433–40.
53. Ben-Hur T, Idelson M, Khaner H, Pera M, Reinhartz E, Itzik A, et al. Transplantation of human embryonic stem cell–derived neural progenitors improves behavioral deficit in parkinsonian rats. Stem Cells. 2004;22:1246–55.
54. Blum B, Benvenisty N. The Tumorigenicity of human embryonic stem cell. Adv Cancer Res. 2008;100:133–58.
55. Grinnemo KH, Kumagai-Braesch M, Mânsson-Broberg A, Skottman H, Hao X, Siddiqui A, et al. Human embryonic stem cells are immunogenic in allogeneic and xenogeneic settings. Reprod Biomed Online. 2006;13:712–24.

56. Trounson A, McDonald C. Stem cell therapies in clinical trials: progress and challenges. Cell Stem Cell. 2015;17:11–22.
57. Trounson A, DeWitt ND. Pluripotent stem cells progressing to the clinic. Nat Rev Mol Cell Biol. 2016;17:194–200.
58. Gonzalez R, Garitaonandia I, Crain A, Poustovoitov M, Abramihina T, Noskov A, et al. Proof of concept studies exploring the safety and functional activity of human parthenogenetic-derived neural stem cells for the treatment of Parkinson's disease. Cell Transplant. 2015;24:681–90.
59. Fikes BJ. Parkinson's stem cell therapy OK'd for testing. 2015. http://www.sandiegouniontribune.com/business/biotech/sdut-international-stem-cell-parkinsons-australia-2015dec14-story.html. Accessed 20 Nov 2016.
60. Hallett PJ, Deleidi M, Astradsson A, Smith GA, Cooper O, Osborn TM, et al. Successful function of autologous iPSC-derived dopamine neurons following transplantation in a non-human primate model of Parkinson's disease. Cell Stem Cell. 2015;16:269–74.
61. Zhao T, Zhang ZN, Rong Z, Xu Y. Immunogenicity of induced pluripotent stem cells. Nature. 2011;474:212–5.
62. Han F, Wang W, Chen B, Chen C, Li S, Lu X, et al. Human induced pluripotent stem cell-derived neurons improve motor asymmetry in a 6-hydroxydopamine-induced rat model of Parkinson's disease. Cytotherapy. 2015;17:665–79.
63. Sundberg M, Bogetofte H, Lawson T, Jansson J, Smith G, Astradsson A, et al. Improved cell therapy protocols for Parkinson's disease based on differentiation efficiency and safety of hESC-, hiPSC-, and non-human primate iPSC-derived dopaminergic neurons. Stem Cells. 2013;31:1548–62.
64. Rhee YH, Ko JY, Chang MY, Yi SH, Kim D, Kim CH, et al. Protein-based human iPS cells efficiently generate functional dopamine neurons and can treat a rat model of Parkinson disease. J Clin Invest. 2011;121:2326–35.
65. Hargus G, Cooper O, Deleidi M, Levy A, Lee K, Marlow E, et al. Differentiated Parkinson patient-derived induced pluripotent stem cells grow in the adult rodent brain and reduce motor asymmetry in parkinsonian rats. Proc Natl Acad Sci U S A. 2010;107:15921–6.
66. Emborg ME, Liu Y, Xi J, Zhang X, Yin Y, Lu J, et al. Induced pluripotent stem cell-derived neural cells survive and mature in the nonhuman primate brain. Cell Rep. 2013;3:646–50.
67. Kim K, Doi A, Wen B, Ng K, Zhao R, Cahan P, et al. Epigenetic memory in induced pluripotent stem cells. Nature. 2010;467:285–90.
68. Kim DS, Ross PJ, Zaslavsky K, Ellis J. Optimizing neuronal differentiation from induced pluripotent stem cells to model ASD. Front Cell Neurosci. 2014;8:109.
69. Parker GC, Acsadi G, Brenner CA. Mitochondria: determinants of stem cell fate? Stem Cells Dev. 2009;18:803–6.
70. Khacho M, Clark A, Svoboda DS, Azzi J, MacLaurin JG, Meghaizel C, et al. Mitochondrial dynamics impacts stem cell identity and fate decisions by regulating a nuclear transcriptional program. Cell Stem Cell. 2016;19:232–47.
71. Soldner F, Laganiere J, Cheng AW, Hockemeyer D, Gao Q, Alagappan R, et al. Generation of isogenic pluripotent stem cells differing exclusively at two early onset Parkinson point mutations. Cell. 2011;146:318–31.
72. Ross CA, Akimov SS. Human-induced pluripotent stem cells: potential for neurodegenerative diseases. Hum Mol Genet. 2014;23:R17–26.
73. Maeder ML, Gersbach CA. Genome-editing technologies for gene and cell therapy. Mol Ther. 2016;24:430–46.
74. Wang L, Cao J, Wang Y, Lan T, Liu L, Wang W, et al. Immunogenicity and functional evaluation of iPSC-derived organs for transplantation. Cell Discov. 2015;1:15015.
75. Cooper O, Seo H, Andrabi S, Guardia-Laguarta C, Graziotto J, Sundberg M, et al. Pharmacological rescue of mitochondrial deficits in iPSC-derived neural cells from patients with familial Parkinson's disease. Sci Transl Med. 2012;4:141ra90–0.

76. Nguyen HN, Byers B, Cord B, Shcheglovitov A, Byrne J, Gujar P, et al. LRRK2 mutant iPSC-derived DA neurons demonstrate increased susceptibility to oxidative stress. Cell Stem Cell. 2011;8:267–80.
77. Jiang H, Ren Y, Yuen EY, Zhong P, Ghaedi M, Hu Z, et al. Parkin controls dopamine utilization in human midbrain dopaminergic neurons derived from induced pluripotent stem cells. Nat Commun. 2012;3:668.
78. Mazzulli JR, Xu YH, Sun Y, Knight AL, McLean PJ, Caldwell GA, et al. Gaucher disease glucocerebrosidase and α-synuclein form a bidirectional pathogenic loop in synucleinopathies. Cell. 2011;146:37–52.
79. Devine MJ, Ryten M, Vodicka P, Thomson AJ, Burdon T, Houlden H, et al. Parkinson's disease induced pluripotent stem cells with triplication of the α-synuclein locus. Nat Commun. 2011;2:440.
80. Reinhardt P, Schmid B, Burbulla LF, Schöndorf DC, Wagner L, Glatza M, et al. Genetic correction of a LRRK2 mutation in human iPSCs links parkinsonian neurodegeneration to ERK-dependent changes in gene expression. Cell Stem Cell. 2013;12:354–67.
81. Woodard CM, Campos BA, Kuo SH, Nirenberg MJ, Nestor MW, Zimmer M, et al. iPSC-derived dopamine neurons reveal differences between monozygotic twins discordant for Parkinson's disease. Cell Rep. 2014;9:1173–82.
82. Sánchez-Danés A, Richaud-Patin Y, Carballo-Carbajal I, Jiménez-Delgado S, Caig C, Mora S, et al. Disease-specific phenotypes in dopamine neurons from human iPS-based models of genetic and sporadic Parkinson's disease. EMBO Mol Med. 2012;4:380–95.
83. Fernández-Santiago R, Carballo-Carbajal I, Castellano G, Torrent R, Richaud Y, Sánchez-Danés A, et al. Aberrant epigenome in iPSC-derived dopaminergic neurons from Parkinson's disease patients. EMBO Mol Med. 2015;7:1529–46.
84. Yagi T, Ito D, Okada Y, Akamatsu W, Nihei Y, Yoshizaki T, et al. Modeling familial Alzheimer's disease with induced pluripotent stem cells. Hum Mol Genet. 2011;20:4530–9.
85. Wojda U, Kuznicki J. Alzheimer's disease modeling: ups, downs, and perspectives for human induced pluripotent stem cells. J Alzheimers Dis. 2013;34:563–88.
86. Bekris LM, Yu CE, Bird TD, Tsuang DW. Genetics of Alzheimer disease. J Geriatr Psychiatry Neurol. 2010;23:213–27.
87. Lam B, Masellis M, Freedman M, Stuss DT, Black SE. Clinical, imaging, and pathological heterogeneity of the Alzheimer's disease syndrome. Alzheimers Res Ther. 2013;5:1.
88. Serrano-Pozo A, Frosch MP, Masliah E, Hyman BT. Neuropathological alterations in Alzheimer disease. Cold Spring Harb Perspect Med. 2011;1:a006189.
89. Yang J, Li S, He XB, Cheng C, Le W. Induced pluripotent stem cells in Alzheimer's disease: applications for disease modeling and cell-replacement therapy. Mol Neurodegener. 2016;11:39.
90. Sullivan SE, Young-Pearse TL. Induced pluripotent stem cells as a discovery tool for Alzheimers disease. Brain Res. 2015; doi:10.2016/j.brainres.2015.10.005.
91. Inoue H, Nagata N, Kurokawa H, Yamanaka S. iPS cells: a game changer for future medicine. EMBO J. 2014;33:409–17.
92. Sproul AA. Being human: the role of pluripotent stem cells in regenerative medicine and humanizing Alzheimer's disease models. Mol Asp Med. 2015;43–44:54–65.
93. Mohamet L, Miazga NJ, Ward CM. Familial Alzheimer's disease modelling using induced pluripotent stem cell technology. World J Stem Cells. 2014;6:239–47.
94. Weggen S, Beher D. Molecular consequences of amyloid precursor protein and presenilin mutations causing autosomal-dominant Alzheimer's disease. Alzheimers Res Ther. 2012;4:9.
95. Shi Y, Kirwan P, Smith J, MacLean G, Orkin SH, Livesey FJ. A human stem cell model of early Alzheimer's disease pathology in Down syndrome. Sci Transl Med. 2012;4:124ra9.
96. Prasher VP. Down syndrome and Alzheimer's disease: biological correlates: Radcliffe Publishing; 2006.
97. Israel MA, Yuan SH, Bardy C, Reyna SM, Mu Y, Herrera C, et al. Probing sporadic and familial Alzheimer's disease using induced pluripotent stem cells. Nature. 2012;482: 216–20.

98. Cataldo A, Rebeck GW, Ghetri B, Hulette C, Lippa C, Van Broeckhoven C, et al. Endocytic disturbances distinguish among subtypes of Alzheimer's disease and related disorders. Ann Neurol. 2001;50:661–5.
99. Kondo T, Asai M, Tsukita K, Kutoku Y, Ohsawa Y, Sunada Y, et al. Modeling Alzheimer's disease with iPSCs reveals stress phenotypes associated with intracellular Abeta and differential drug responsiveness. Cell Stem Cell. 2013;12:487–96.
100. Muratore CR, Rice HC, Srikanth P, Callahan DG, Shin T, Benjamin LN, et al. The familial Alzheimer's disease APPV717I mutation alters APP processing and tau expression in iPSC-derived neurons. Hum Mol Genet. 2014;23:3523–36.
101. Scheuner D, Eckman C, Jensen M, Song X, Citron M, Suzuki N, et al. Secreted amyloid beta-protein similar to that in the senile plaques of Alzheimer's disease is increased in vivo by the presenilin 1 and 2 and APP mutations linked to familial Alzheimer's disease. Nat Med. 1996;2:864–70.
102. Borchelt DR, Thinakaran G, Eckman CB, Lee MK, Davenport F, Ratovitsky T, et al. Familial Alzheimer's disease–linked presenilin 1 variants elevate Aβ1–42/1–40 ratio in vitro and in vivo. Neuron. 1996;17:1005–13.
103. Tomita T, Maruyama K, Saido TC, Kume H, Shinozaki K, Tokuhiro S, et al. The presenilin 2 mutation (N141I) linked to familial Alzheimer disease (Volga German families) increases the secretion of amyloid β protein ending at the 42nd (or 43rd) residue. Proc Natl Acad Sci U S A. 1997;94:2025–30.
104. Sproul AA, Jacob S, Pre D, Kim SH, Nestor MW, Navarro-Sobrino M, et al. Characterization and molecular profiling of PSEN1 familial Alzheimer's disease iPSC-derived neural progenitors. PLoS One. 2014;9:e84547.
105. Moore S, Evans LD, Andersson T, Portelius E, Smith J, Dias TB, et al. APP metabolism regulates tau proteostasis in human cerebral cortex neurons. Cell Rep. 2015;11:689–96.
106. Young JE, Goldstein LSB. Alzheimer's disease in a dish: promises and challenges of human stem cell models. Hum Mol Genet. 2012;21:R82–9.
107. Woodruff G, Young JE, Martinez FJ, Buen F, Gore A, Kinaga J, et al. The presenilin-1 ΔE9 mutation results in reduced γ-secretase activity, but not total loss of PS1 function, in isogenic human stem cells. Cell Rep. 2013;5:974–85.
108. Tanzi RE. The genetics of Alzheimer disease. Cold Spring Harb Perspect Med. 2012;2:a006296.
109. Gatz M, Reynolds CA, Fratiglioni L, Johansson B, Mortimer JA, Berg S, et al. Role of genes and environments for explaining Alzheimer disease. Arch Gen Psychiatry. 2006;63:168–74.
110. Delgado-Morales R, Esteller M. Opening up the DNA methylome of dementia. Mol Psychiatry. 2017; doi:10.1038/mp.2016.242.
111. Lord J, Lu AJ, Cruchaga C. Identification of rare variants in Alzheimer's disease. Front Genet. 2014;5:369.
112. Van Cauwenberghe C, Van Broeckhoven C, Sleegers K. The genetic landscape of Alzheimer disease: clinical implications and perspectives. Genet Med. 2016;18:421–30.
113. Iatrou A, Kenis G, Rutten BPF, Lunnon K, van den Hove DLA. Epigenetic dysregulation of brainstem nuclei in the pathogenesis of Alzheimer's disease: looking in the correct place at the right time? Cell Mol Life Sci. 2016;1–15
114. Duan L, Bhattacharyya BJ, Belmadani A, Pan L, Miller RJ, Kessler JA. Stem cell derived basal forebrain cholinergic neurons from Alzheimer's disease patients are more susceptible to cell death. Mol Neurodegener. 2014;9:3.
115. Young JE, Boulanger-Weill J, Williams DA, Woodruff G, Buen F, Revilla AC, et al. Elucidating molecular phenotypes caused by the SORL1 Alzheimer's disease genetic risk factor using human induced pluripotent stem cells. Cell Stem Cell. 2015;16:373–85.
116. Alagiakrishnan K, Gill SS, Fagarasanu A. Genetics and epigenetics of Alzheimer's disease. Postgrad Med J. 2012;88:522–9.
117. Medway C, Morgan K. The genetics of Alzheimer's disease; putting flesh on the bones. Neuropathol Appl Neurobiol. 2014;40:97–105.

118. Andersen OM, Reiche J, Schmidt V, Gotthardt M, Spoelgen R, Behlke J, et al. Neuronal sorting protein-related receptor sorLA/LR11 regulates processing of the amyloid precursor protein. Proc Natl Acad Sci U S A. 2005;102:13461–6.
119. Dodson SE, Gearing M, Lippa CF, Montine TJ, Levey AI, Lah JJ. LR11/SorLA expression is reduced in sporadic Alzheimer disease but not in familial Alzheimer disease. J Neuropathol Exp Neurol. 2006;65:866–72.
120. Hossini AM, Megges M, Prigione A, Lichtner B, Toliat MR, Wruck W, et al. Induced pluripotent stem cell-derived neuronal cells from a sporadic Alzheimer's disease donor as a model for investigating AD-associated gene regulatory networks. BMC Genomics. 2015;16:84.
121. Xu W, Tan L, Wang HF, Jiang T, Tan MS, Tan L, et al. Meta-analysis of modifiable risk factors for Alzheimer's disease. J Neurol Neurosurg Psychiatry. 2015;86:1299–306.
122. Kungulovski G, Jeltsch A. Epigenome editing: state of the art, concepts, and perspectives. Trends Genet. 2016;32:101–13.
123. Thakore PI, Black JB, Hilton IB, Gersbach CA. Editing the epigenome: technologies for programmable transcription and epigenetic modulation. Nat Methods. 2016;13:127–37.
124. Barten DM, Meredith Jr JE, Zaczek R, Houston JG, Albright CF. Gamma-secretase inhibitors for Alzheimer's disease: balancing efficacy and toxicity. Drugs R D. 2006;7:87–97.
125. Ghosh AK, Osswald HL. BACE1 (beta-secretase) inhibitors for the treatment of Alzheimer's disease. Chem Soc Rev. 2014;43:6765–813.
126. Yiannopoulou KG, Papageorgiou SG. Current and future treatments for Alzheimer's disease. Ther Adv Neurol Disord. 2013;6:19–33.
127. Lozano R, Rosero CA, Hagerman RJ. Fragile X spectrum disorders. Intractable Rare Dis Res. 2014;3:134–46.
128. Saldarriaga W, Tassone F, González-Teshima LY, Forero-Forero JV, Ayala-Zapata S, Hagerman R. Fragile X syndrome. Colomb Med. 2014;45:190–8.
129. Eberhart DE, Malter HE, Feng Y, Warren ST. The fragile X mental retardation protein is a ribonucleoprotein containing both nuclear localization and nuclear export signals. Hum Mol Genet. 1996;5:1083–91.
130. Ashley CT, Wilkinson KD, Reines D, Warren ST. FMR1 protein: conserved RNP family domains and selective RNA binding. Science. 1993;262:563–6.
131. Antar LN, Afroz R, Dictenberg JB, Carroll RC, Bassell GJ. Metabotropic glutamate receptor activation regulates fragile X mental retardation protein and FMR1 mRNA localization differentially in dendrites and at synapses. J Neurosci. 2004;24:2648–55.
132. Broek JAC. Lin Z, de Gruiter HM, van 't Spijker H, Haasdijk ED, cox D, et al. Synaptic vesicle dynamic changes in a model of fragile X Mol Autism. 2016;7:17.
133. Kremer EJ, Pritchard M, Lynch M, Yu S, Holman K, Baker E, et al. Mapping of DNA instability at the fragile-X to a Trinucleotide repeat sequence P(Ccg)N. Science. 1991;252: 1711–4.
134. Sutcliffe JS, Nelson DL, Zhang F, Pieretti M, Caskey CT, Saxe D, et al. DNA methylation represses FMR-1 transcription in fragile X syndrome. Hum Mol Genet. 1992;1:397–400.
135. Mailick MR, Hong J, Rathouz P, Baker MW, Greenberg JS, Smith L, et al. Low-normal FMR1 CGG repeat length: phenotypic associations. Front Genet. 2014;5:309.
136. Miyashiro KY, Beckel-Mitchener A, Purk TP, Becker KG, Barret T, Liu L, et al. RNA cargoes associating with in cellular functioning in Fmrl Fmr1 reveal deficits null mice. Neuron. 2003;37:417–31.
137. Primerano B, Tassone F, Hagerman RJ, Hagerman P, Amaldi F, Bagni C. Reduced FMR1 mRNA translation efficiency in fragile X patients with premutations. RNA. 2002;8:1482–8.
138. Suhl JA, Warren ST. Single-nucleotide mutations in FMR1 reveal novel functions and regulatory mechanisms of the fragile X syndrome protein FMRP. J Exp Neurosci. 2015;9:35–41.
139. Brouwer JR, Mientjes EJ, Bakker CE, Nieuwenhuizen IM, Severijnen LA, Van der Linde HC, et al. Elevated Fmr1 mRNA levels and reduced protein expression in a mouse model with an unmethylated fragile X full mutation. Exp Cell Res. 2007;313:244–53.

140. Mor-Shaked H, Eiges R. Modeling fragile X syndrome using human pluripotent stem cells. Genes (Basel). 2016;7:77.
141. Urbach A, Bar-Nur O, Daley GQ, Benvenisty N. Differential modeling of fragile X syndrome by human embryonic stem cells and induced pluripotent stem cells. Cell Stem Cell. 2010;6:407–11.
142. Khalfallah O, Jarjat M, Davidovic L, Nottet N, Cestèle S, Mantegazza M, et al. Depletion of the fragile X mental retardation protein in embryonic stem cells alters the kinetics of neurogenesis. Stem Cells. 2016; doi:10.1002/stem.2505.
143. Ben-Reuven L, Reiner O. Modeling the autistic cell: iPSCs recapitulate developmental principles of syndromic and nonsyndromic ASD. Develop Growth Differ. 2016;58:481–91.
144. Sheridan SD, Theriault KM, Reis SA, Zhou F, Madison JM, Daheron L, et al. Epigenetic characterization of the FMR1 gene and aberrant neurodevelopment in human induced pluripotent stem cell models of fragile X syndrome. PLoS One. 2011;6:e26203.
145. Doers ME, Musser MT, Nichol R, Berndt ER, Baker M, Gomez TM, et al. iPSC-derived forebrain neurons from FXS individuals show defects in initial Neurite outgrowth. Stem Cells Dev. 2014;23:1777–87.
146. Usdin K, Kumari D. Repeat-mediated epigenetic dysregulation of the FMR1 gene in the fragile X-related disorders. Front Genet. 2015;6:192.
147. Liu J, Kościelska KA, Cao Z, Hulsizer S, Grace N, Mitchell G, et al. Signaling defects in iPSC-derived fragile X premutation neurons. Hum Mol Genet. 2012;21:3795–805.
148. de Esch CEF, Ghazvini M, Loos F, Schelling-Kazaryan N, Widagdo W, Munshi ST, et al. Epigenetic characterization of the FMR1 promoter in induced pluripotent stem cells from human fibroblasts carrying an unmethylated full mutation. Stem Cell Rep. 2014; 3:548–55.
149. Park CY, Halevy T, Lee DR, Sung JJ, Lee JS, Yanuka O, et al. Reversion of FMR1 methylation and silencing by editing the triplet repeats in fragile X iPSC-derived neurons. Cell Rep. 2015;13:234–41.
150. Ran FA, Hsu PD, Wright J, Agarwala V, Scott DA, Zhang F. Genome engineering using the CRISPR-Cas9 system. Nat Protoc. 2013;8:2281–308.
151. Kumari D, Swaroop M, Southall N, Huang W, Zheng W, Usdin K. High-throughput screening to identify compounds that increase fragile X mental retardation protein expression in neural stem cells differentiated from fragile X syndrome patient-derived induced pluripotent stem cells. Stem Cells Transl Med. 2015;4:800–8.
152. Angelman H. "Puppet" children a report on three cases. Dev Med Child Neurol. 1965;7:681–8.
153. Van Buggenhout G, Fryns JP. Angelman syndrome (AS, MIM 105830). Eur J Hum Genet. 2009;17:1367–73.
154. Trillingsgaard A, ØStergaard JR. Autism in Angelman syndrome: an exploration of comorbidity. Autism. 2004;8:163–74.
155. Cassidy SB, Driscoll DJ. Prader-Willi syndrome. Eur J Hum Genet. 2009;17:3–13.
156. Kalsner L, Chamberlain SJ. Prader-Willi, Angelman, and 15q11-q13 duplication syndromes. Pediatr Clin N Am. 2015;62:587–606.
157. Koerner MV, Barlow DP. Genomic imprinting - an epigenetic gene-regulatory model. Curr Opin Genet Dev. 2010;20:164–70.
158. Sanchez-Delgado M, Court F, Vidal E, Medrano J, Monteagudo-Sánchez A, Martin-Trujillo A, et al. Human oocyte-derived methylation differences persist in the placenta revealing widespread transient imprinting. PLoS Genet. 2016;12:e1006427.
159. Wilkinson LS, Davies W, Isles AR. Genomic imprinting effects on brain development and function. Nat Rev Neurosci. 2007;8:832–43.
160. Horsthemke B, Wagstaff J. Mechanisms of imprinting of the Prader-Willi/Angelman region. Am J Med Genet. 2008;146A:2041–52.
161. Constância M, Pickard B, Kelsey G, Reik W. Imprinting mechanisms. Genome Res. 1998;8:881–900.

162. Butler MG. Prader-Willi syndrome: obesity due to genomic imprinting. Curr Genomics. 2011;12(3):204–15.
163. Cavaillé J, Buiting K, Kiefmann M, Lalande M, Brannan CI, Horsthemke B, et al. Identification of brain-specific and imprinted small nucleolar RNA genes exhibiting an unusual genomic organization. Proc Natl Acad Sci U S A. 2000;97:14311–6.
164. Sanchez-Delgado M, Riccio A, Eggermann T, Maher ER, Lapunzina P, Mackay D, et al. Causes and consequences of multi-locus imprinting disturbances in humans. Trends Genet. 2016;32:444–55.
165. LaSalle JM, Reiter LT, Chamberlain SJ. Epigenetic regulation of UBE3A and roles in human neurodevelopmental disorders. Epigenomics. 2015;7:1213–28.
166. Hulbert SW, Jiang YH. Monogenic mouse models of autism spectrum disorders: common mechanisms and missing links. Neuroscience. 2016;321:3–23.
167. Jiang YH, Armstrong D, Albrecht U, Atkins CM, Noebels JL, Eichele G, et al. Mutation of the Angelman ubiquitin ligase in mice causes increased cytoplasmic p53 and deficits of contextual learning and long-term potentiation. Neuron. 1998;21:799–811.
168. Scheiffele P, Beg AA. Neuroscience: Angelman syndrome connections. Nature. 2010;468:907–8.
169. Greer PL, Hanayama R, Bloodgood BL, Mardinly AR, Lipton DM, Flavell SW, et al. The Angelman syndrome protein Ube3A regulates synapse development by Ubiquitinating arc. Cell. 2010;140:704–16.
170. Bervini S, Herzog H. Mouse models of Prader–Willi syndrome: a systematic review. Front Neuroendocrinol. 2013;34:107–19.
171. Tan WH, Bird LM, Thibert RL, Williams CA. If not Angelman, what is it? A review of Angelman-like syndromes. Am J Med Genet A. 2014;164:975–92.
172. Yang J, Cai J, Zhang Y, Wang X, Li W, Xu J, et al. Induced pluripotent stem cells can be used to model the genomic imprinting disorder Prader-Willi syndrome. J Biol Chem. 2010;285:40303–11.
173. Chamberlain SJ, Chen PF, Ng KY, Bourgois-Rocha F, Lemtiri-Chlieh F, Levine ES, et al. Induced pluripotent stem cell models of the genomic imprinting disorders Angelman and Prader – Willi syndromes. Proc Natl Acad Sci U S A. 2010;107:17668–73.
174. Germain ND, Chen PF, Plocik AM, Glatt-Deeley H, Brown J, Fink JJ, et al. Gene expression analysis of human induced pluripotent stem cell-derived neurons carrying copy number variants of chromosome 15q11-q13.1. Mol Autism. 2014;5:44.
175. Rougeulle C, Cardoso C, Fontés M, Colleaux L, Lalande M. An imprinted antisense RNA overlaps UBE3A and a second maternally expressed transcript. Nat Genet. 1998;19:15–6.
176. Vu TH, Hoffman AR. Imprinting of the Angelman syndrome gene, UBE3A, is restricted to brain. Nat Genet. 1997;17:12–3.
177. Martins-Taylor K, Hsiao JS, Chen PF, Glatt-Deeley H, De Smith AJ, Blakemore AIF, et al. Imprinted expression of UBE3A in non-neuronal cells from a Prader-willi syndrome patient with an atypical deletion. Hum Mol Genet. 2014;23:2364–73.
178. Chamberlain SJ, Germain ND, Chen PF, Hsiao JS, Glatt-Deeley H. Modeling genomic imprinting disorders using induced pluripotent stem cells. Methods Mol Biol. 2016;1353:45–64.
179. Chen PF, Hsiao JS, Sirois CL, Chamberlain SJ. RBFOX1 and RBFOX2 are dispensable in iPSCs and iPSC-derived neurons and do not contribute to neural-specific paternal UBE3A silencing. Sci Rep. 2016;6:25368.
180. Meng L, Ward AJ, Chun S, Bennett CF, Beaudet AL, Rigo F. Towards a therapy for Angelman syndrome by targeting a long non-coding RNA. Nature. 2015;518:409–12.
181. Huang HS, Allen JA, Mabb AM, King IF, Miriyala J, Taylor-Blake B, et al. Topoisomerase inhibitors unsilence the dormant allele of Ube3a in neurons. Nature. 2011;481:185–9.
182. Katz DM, Bird A, Coenraads M, Gray SJ, Menon DU, Philpot BD, et al. Rett syndrome: crossing the threshold to clinical translation. Trends Neurosci. 2016;39:100–13.
183. Chahrour M, Zoghbi HY. The story of Rett syndrome: from clinic to neurobiology. Neuron. 2007;56:422–37.

184. Amir RE, Van den Veyver IB, Wan M, Tran CQ, Francke U, Zoghbi HY. Rett syndrome is caused by mutations in X-linked MECP2, encoding methyl-CpG-binding protein 2. Nat Genet. 1999;23:185–8.
185. Dunn KW, Kamocka MM, McDonald JH. A practical guide to evaluating colocalization in biological microscopy. AJP Cell Physiol. 2011;300:C723–42.
186. Dragich JM, Kim YH, Arnold AP, Schanen NC. Differential distribution of the MeCP2 splice variants in the postnatal mouse brain. J Comp Neurol. 2007;501:526–42.
187. Della Ragione F, Vacca M, Fioriniello S, Pepe G, D'Esposito M. MECP2, a multi-talented modulator of chromatin architecture. Brief Funct Genomics. 2016;15:420–31.
188. Sáez MA, Fernández-Rodríguez J, Moutinho C, Sanchez-Mut JV, Gomez A, Vidal E, et al. Mutations in JMJD1C are involved in Rett syndrome and intellectual disability. Genet Med. 2016;18:378–85.
189. Lucariello M, Vidal E, Vidal S, Saez M, Roa L, Huertas D, et al. Whole exome sequencing of Rett syndrome-like patients reveals the mutational diversity of the clinical phenotype. Hum Genet. 2016;135:1343–54.
190. Petazzi P, Akizu N, García A, Estarás C. Martínez de Paz a, Rodríguez-Paredes M, et al. an increase in MECP2 dosage impairs neural tube formation. Neurobiol Dis. 2014;67:49–56.
191. Meins M, Lehmann J, Gerresheim F, Herchenbach J, Hagedorn M, Hameister K, et al. Submicroscopic duplication in Xq28 causes increased expression of the MECP2 gene in a boy with severe mental retardation and features of Rett syndrome. J Med Genet. 2005;42:e12–2.
192. Gemelli T, Berton O, Nelson ED, Perrotti LI, Jaenisch R, Monteggia LM. Postnatal loss of methyl-CpG binding protein 2 in the forebrain is sufficient to mediate behavioral aspects of Rett syndrome in mice. Biol Psychiatry. 2006;59:468–76.
193. Ausió J. Martinez de Paz a, Esteller M. MeCP2: the long trip from a chromatin protein to neurological disorders. Trends Mol Med. 2014;20:487–98.
194. Kucukkal TG, Yang Y, Uvarov O, Cao W, Alexov E. Impact of Rett syndrome mutations on MeCP2 MBD stability. Biochemistry. 2015;54:6357–68.
195. Brown K, Selfridge J, Lagger S, Connelly J, De Sousa D, Kerr A, et al. The molecular basis of variable phenotypic severity among common missense mutations causing Rett syndrome. Hum Mol Genet. 2016;25:558–70.
196. Stuss DP, Cheema M, Ng MK, Martinez de Paz A, Williamson B, Missiaen K, et al. Impaired in vivo binding of MeCP2 to chromatin in the absence of its DNA methyl-binding domain. Nucleic Acids Res. 2013;41:4888–900.
197. Christodoulou J, Grimm A, Maher T, Bennetts B. RettBASE: the IRSA MECP2 variation database-a new mutation database in evolution. Hum Mutat. 2003;21:466–72.
198. Gold WA, Christodoulou J. The utility of next-generation sequencing in Gene discovery for mutation-negative patients with Rett syndrome. Front Cell Neurosci. 2015;9:266.
199. Naidu S, Bibat G, Kratz L, Kelley RI, Pevsner J, Hoffman E, et al. Clinical variability in Rett syndrome. J Child Neurol. 2003;18:662–8.
200. Bao X, Jiang S, Song F, Pan H, Meirong Li WX-R. X chromosome inactivation in Rett syndrome and its correlations with MeCP2 mutations and phenotype. J Child Neurol. 2008;23:22–5.
201. Huang TW, Kochukov MY, Ward CS, Merritt J, Thomas K, Nguyen T, et al. Progressive changes in a distributed neural circuit underlie breathing abnormalities in mice lacking MeCP2. J Neurosci. 2016;36:5572–86.
202. Shahbazian M, Young J, Yuva-Paylor L, Spencer C, Antalffy B, Noebels J, et al. Mice with truncated MeCP2 recapitulate many Rett syndrome features and display hyperacetylation of histone H3. Neuron. 2002;35:243–54.
203. Stearns NA, Schaevitz LR, Bowling H, Nag N, Berger UV, Berger-Sweeney J. Behavioral and anatomical abnormalities in Mecp2 mutant mice: a model for Rett syndrome. Neuroscience. 2007;146:907–21.
204. Pelka GJ, Watson CM, Radziewic T, Hayward M, Lahooti H, Christodoulou J, et al. Mecp2 deficiency is associated with learning and cognitive deficits and altered gene activity in the hippocampal region of mice. Brain. 2006;129:887–98.

205. Calfa G, Percy AK, Pozzo-Miller L. Experimental models of Rett syndrome based on Mecp2 dysfunction. Exp Biol Med. 2011;236:3–19.
206. Rietveld L, Stuss DP, McPhee D, Delaney KR. Genotype-specific effects of Mecp2 loss-of-function on morphology of layer V pyramidal neurons in heterozygous female Rett syndrome model mice. Front Cell Neurosci. 2015;9:145.
207. Medrihan L, Tantalaki E, Aramuni G, Sargsyan V, Dudanova I, Missler M, et al. Early defects of GABAergic synapses in the brain stem of a MeCP2 mouse model of Rett syndrome. J Neurophysiol. 2008;99:112–21.
208. Chao HT, Zoghbi HY, Rosenmund C. MeCP2 controls excitatory synaptic strength by regulating glutamatergic synapse number. Neuron. 2007;56:58–65.
209. Viemari JC, Roux JC, Tryba AK, Saywell V, Burnet H, Peña F, et al. Mecp2 deficiency disrupts norepinephrine and respiratory systems in mice. J Neurosci. 2005;25:11521–30.
210. Szczesna K, de la Caridad O, Petazzi P, Soler M, Roa L, Saez MA, et al. Improvement of the Rett syndrome phenotype in a MeCP2 mouse model upon treatment with levodopa and a dopa-decarboxylase inhibitor. Neuropsychopharmacology. 2014;39:2846–56.
211. Ward CS, Arvide EM, Huang TW, Yoo J, Noebels JL, Neul JL. MeCP2 is critical within HoxB1 derived tissues of mice for normal lifespan. J Neurosci. 2011;31:10359–70.
212. Kim S, Broströmer E, Xing D, Jin J, Chong S, Ge H, et al. Probing allostery through DNA. Science. 2013;339:816–9.
213. Lyst MJ, Bird A. Rett syndrome: a complex disorder with simple roots. Nat Rev Genet. 2015;16:261–75.
214. Beltrão-Braga PCB, Muotri AR. Modeling autism spectrum disorders with human neurons. Brain Res. 2016;1–8
215. Dajani R, Koo SE, Sullivan GJ, Park IH. Investigation of Rett syndrome using pluripotent stem cells. J Cell Biochem. 2013;114:2446–53.
216. Marchetto MCN, Carromeu C, Acab A, Yu D, Yeo GW, Mu Y, et al. A model for neural development and treatment of rett syndrome using human induced pluripotent stem cells. Cell. 2010;143:527–39.
217. Chin EWM, Marcy G, Yoon SI, Ma D, Rosales FJ, Augustine GJ, et al. Choline ameliorates disease phenotypes in human iPSC models of Rett syndrome. NeuroMolecular Med. 2016;18:364–77.
218. Djuric U, Cheung AYL, Zhang W, Mok RS, Lai W, Piekna A, et al. MECP2e1 isoform mutation affects the form and function of neurons derived from Rett syndrome patient iPS cells. Neurobiol Dis. 2015;76:37–45.
219. Kim KY, Hysolli E, Park IH. Neuronal maturation defect in induced pluripotent stem cells from patients with Rett syndrome. Proc Natl Acad Sci U S A. 2011;108:14169–74.
220. Andoh-Noda T, Akamatsu W, Miyake K, Matsumoto T, Yamaguchi R, Sanosaka T, et al. Differentiation of multipotent neural stem cells derived from Rett syndrome patients is biased toward the astrocytic lineage. Mol Brain. 2015;8:31.
221. Cheung AYL, Horvath LM, Carrel L, Ellis J. X-chromosome inactivation in Rett syndrome human induced pluripotent stem cells. Front Psychiatry. 2012;3:24.
222. Williams EC, Zhong X, Mohamed A, Li R, Liu Y, Dong Q, et al. Mutant astrocytes differentiated from Rett syndrome patients-specific iPSCs have adverse effects on wild-type neurons. Hum Mol Genet. 2014;23:2968–80.
223. Delépine C, Meziane H, Nectoux J, Opitz M, Smith AB, Ballatore C, et al. Altered microtubule dynamics and vesicular transport in mouse and human MeCP2-deficient astrocytes. Hum Mol Genet. 2016;25:146–57.
224. Zhang ZN, Freitas BC, Qian H, Lux J, Acab A, Trujillo CA, et al. Layered hydrogels accelerate iPSC-derived neuronal maturation and reveal migration defects caused by MeCP2 dysfunction. Proc Natl Acad Sci U S A. 2016;113:3185–90.
225. Amenduni M, De Filippis R, Cheung AYL, Disciglio V, Epistolato MC, Ariani F, et al. iPS cells to model CDKL5-related disorders. Eur J Hum Genet. 2011;19:1246–55.

226. Ricciardi S, Ungaro F, Hambrock M, Rademacher N, Stefanelli G, Brambilla D, et al. CDKL5 ensures excitatory synapse stability by reinforcing NGL-1-PSD95 interaction in the postsynaptic compartment and is impaired in patient iPSC-derived neurons. Nat Cell Biol. 2012;14:911–23.
227. Bienvenu T, Chelly J. Molecular genetics of Rett syndrome: when DNA methylation goes unrecognized. Nat Rev Genet. 2006;7:415–26.
228. Aldinger KA, Plummer JT, Levitt P. Comparative DNA methylation among females with neurodevelopmental disorders and seizures identifies TAC1 as a MeCP2 target gene. J Neurodev Disord. 2013;5:15.
229. Li Y, Wang H, Muffat J, Cheng AW, Orlando DA, Lovén J, et al. Global transcriptional and translational repression in human-embryonic-stem-cell-derived Rett syndrome neurons. Cell Stem Cell. 2013;13:446–58.
230. Akbarian S, Huang HS. Epigenetic regulation in human brain-focus on histone lysine methylation. Biol Psychiatry. 2009;65:198–203.
231. Ananiev G, Williams EC, Li H, Chang Q. Isogenic pairs of wild type and mutant induced pluripotent stem cell (iPSC) lines from Rett syndrome patients as in vitro disease model. PLoS One. 2011;6:e25255.
232. Kimbrel EA, Lanza R. Current status of pluripotent stem cells: moving the first therapies to the clinic. Nat Rev Drug Discov. 2015;14:681–92.
233. Hendriks WT, Warren CR, Cowan CA. Genome editing in human pluripotent stem cells: approaches, pitfalls, and solutions. Cell Stem Cell. 2016;18:53–65.
234. Liu XS, Wu H, Ji X, Stelzer Y, Wu X, Czauderna S, et al. Editing DNA methylation in the mammalian genome. Cell. 2016;167:233–47.e17.
235. Tapia N, Schöler HR. Molecular obstacles to clinical translation of iPSCs. Cell Stem Cell. 2016;19:298–309.
236. Passier R, Orlova V, Mummery C. Complex tissue and disease modeling using hiPSCs. Cell Stem Cell. 2016;18:309–21.

Technologies for Deciphering Epigenomic DNA Patterns

24

Sebastian Moran

Abstract

DNA methylation, consisting on the covalent addition of a methyl group in cytosines, plays a vital role for the development and correct functioning of cells. It constitutes a mechanism by which cell genome is regulated, allowing from a common genome of an individual to obtain all the different cell types that constitute the individual. Nowadays, we understand how the epigenetic machinery works; however, this critical mechanism might promote the appearance of certain diseases if dysregulated, thus the importance of studying the epigenetic patterns on both normal and disease tissues. During the last decades, huge advances on techniques to measure the level of DNA methylation have occurred; we have passed from measuring it with more rudimentary and expensive techniques to nowadays the ability to measure DNA methylation at a single-base resolution in an affordable manner. In this chapter we will cover all the main technologies available, with a special emphasis on the microarray technology, as it supposes a perfect choice taking into account its price as well as the amount of cytosines interrogated, the compatibility with formalin-fixed paraffin-embedded samples, and its standardized procedure.

Keywords

Methylation • Epigenetics • Microarrays • 5-Methylcytosine • 5-Hydroxymethylcytosine • Formalin fixed paraffin embedded • FFPE

S. Moran, M.Sc.
Cancer Epigenetics and Biology Program (PEBC), Bellvitge Biomedical Research Institute (IDIBELL), Duran i Reynalds Hospital, 3rd floor, Gran Via de L'Hospitalet 199-203, Barcelona, Catalonia 08908, Spain
e-mail: smoran@idibell.cat

R. Delgado-Morales (ed.), *Neuroepigenomics in Aging and Disease*, Advances in Experimental Medicine and Biology, DOI 10.1007/978-3-319-53889-1_24

24.1 Introduction

All those phenomena which have a direct impact on the final outcome of a given locus without modifying the underlying DNA sequence are known as epigenetics. Therefore, the epigenome acts as a bridge between the genotype and the phenotype [1]. Having this definition in mind, then the epigenome, or epigenetic information, is the one responsible for differing from cell type to cell type, regulating in each cell type the expression in a different number of ways (modulating the organization of the nuclear genomic material, blocking/allowing the access of gene transcription factors to the DNA, etc.). Consequently, the epigenome represents a second-level regulation layer on top of the genomic sequence information, and it is a key mechanism by which in early embryogenesis, as cells differentiate, the epigenome of the different cell types differentiates too, maintaining cell-type-specific gene expression patterns and enabling the creation of various cell types from the same genomic information [2].

Different epigenetic phenomena have been described (histone modification, microRNAs, DNA methylation, etc.); however, the most studied epigenetic mechanism is the DNA methylation. DNA methylation consists of the covalent addition of a methyl group ($-CH_3$) to the fifth carbon of the pyrimidine ring of cytosine nucleotides. Although DNA methylation can occur in all cytosines of the genome, in mammals, this modification mainly takes place when a cytosine is contiguous to a guanine nucleotide, in what is called a CpG dinucleotide (~28 million CpGs on the human genome). Interestingly, CpG distribution is not random over the genome, but rather, CpGs are enriched in gene promoter regions known as CpG islands, defined as a region of at least 200 bp with a GC content greater than 50% and with an observed-to-expected CpG ratio that exceeds 60% [3]. It is well known that genes being actively transcribed normally lack DNA methylation in their CpG-rich promoter regions (CpG islands). Counter-wise, gene bodies of actively transcribed genes usually show dense DNA methylation, which might promote the transcription elongation rather than repress transcription initiation, as DNA methylation does in promoter regions [4–7]. Then, it is not unusual that DNA methylation is involved in many cellular processes such as silencing of repetitive and centromeric sequences, X-chromosome inactivation in female mammals, mammalian imprinting, chromosome stability, and transcriptional repression, among others [1].

The physical incorporation of the methyl group into the cytosines is performed by proteins of the methyltransferase family. Multiple proteins comprise this family, being DNMT3A and DNMT3B the ones in charge of the de novo DNA methylation [8, 9], while DNA methylation maintenance during replication is performed by DNMT1 [4]. Different studies have shown that methyltransferases are essential for normal mammalian development. For example, deletion of DNMT1 or DNMT3B genes leads to embryonic lethality, while homozygous knockout of the DNMT3a gene in the mouse results in mice surviving only up to the fourth week after birth [9, 10]. Moreover, defects on a member of the DNA methylation machinery, DNMT3B, cause immunodeficiency–centromeric instability–facial anomalies (ICF) syndrome [9, 11]. Besides, aberrant methylation profiles may lead to multiple diseases: the Prader–Willi/Angelman syndrome (PWS/AS) [12] or the Beckwith–Wiedemann syndrome [2], where aberrant methylation of imprinted genes occurs; cancer, where multiple aberrations in DNA methylation profiles have been observed [13–17]; neurodegenerative diseases [18]; and cerebrovascular events [19, 20], among others. However, DNA methylation aberrancies not only may lead to a disease, but it may also be present in as many aspects as those cellular processes where DNA methylation is involved, such as drug response [21], aging [22, 23], and much more. These examples clearly demonstrate that precisely regulation of the DNA methylation and its machinery is crucial for the proper functioning of the cells.

Although methylcytosine (5mC), considered as the fifth DNA base, is the more abundant nucleotide modification, other nucleotide modifications are being discovered. For example, 5-hydroxymethylcytosine (5hmC), which is the hydroxylated form of 5mC, has been reported to be present in bacteriophage DNA as a defense mechanism for restriction enzymes of the infected hosts [24, 25]. In humans, hydroxylation of the 5mC is performed by the TET1 protein [26] to create the 5-hydroxymethylcytosine, and it has been reported to play a major role in neural development, where 5hmC is highly enriched [27]. Besides cytosine modifications, another nucleotide modification which is being actively investigated is the N6-methyladenine, a modification deeply studied in prokaryotes where it has a defensive role, but without a clearly known function in eukaryotes [28, 29].

24.2 DNA Methylation Techniques

The huge improvements made in understanding how the epigenetic machinery works, interacts, and acts, the implications it has for the proper functioning of cells, and how epigenetic dysregulation may lead to developmental diseases, cancer, or other malignancies could have not been possible without the incredible advances made in the field of strategies to detect DNA methylation levels. Numerous methods have been developed to interrogate DNA methylation, making impossible to cover them all, not just because of the number, but because it is not the purpose of this chapter. However, and for broadly understanding them, we could organize them mostly if they are sodium bisulfite based, restriction enzyme based, affinity based, or based on physical properties. Each method has its own advantages and disadvantages, and the election of the method to use would come from different variables such as cost, time, if we require a quantitative or qualitative method, the amount of DNA available, global- or locus-specific targets, genome wide or locus specific, etc.

Among the methods based on physical properties, high-performance liquid chromatography (HPLC) and matrix-assisted laser desorption–ionization time of flight (MALDI–TOF) mass spectrometry have been successfully used. These methods are highly quantitative and reproducible techniques. However, on the counterpart, they require large amounts of high-quality DNA and are technically challenging, thus not making them suitable for high-throughput approaches. Other methods to interrogate DNA methylation exploit the use of high-affinity specific antibodies. These methods, although not providing a single-base resolution, can identify regions where DNA methylation is enriched, for example, methylated DNA immunoprecipitation (MeDIP) where an antibody against 5-methylcytosine precipitates highly methylated DNA fragments to latter sequence and identifies those regions or the use of enzyme-linked immunosorbent assay (ELISA).

The third approach to the examination of DNA methylation is the one using restriction enzymes to achieve this purpose. Within this category, and just as an example, restriction landmark genomic scanning (RLGS) is one of those techniques that uses different restriction enzymes, some of them having a recognition sequence affected by methylation status of the cytosines, in conjunction with 2D electrophoresis; it is capable of rapid and simultaneous interrogation of thousands of restriction enzyme sites for its methylation status. But there are also many methods that could be classified as restriction based while at the same time also are classifiable as sodium bisulfite based, which is the fourth approach to the interrogation of DNA methylation. Among these methods which benefit from both approaches, reduced representation bisulfite sequencing (RRBS) is probably the most common and known one. With this approach, methyl-sensitive restriction enzymes are used to create DNA fragments highly enriched in DNA methylation, and then those fragments are subject to sodium bisulfite conversion and to latter sequence and obtain single-base methylation resolution of the fragments.

Finally, and probably the group including the methods nowadays more used, are the strategies based on a chemical reaction, the sodium bisulfite modification [30]. This reaction serves as the foundation for the more recent methylation techniques. In the presence of sodium bisulfite, cytosines are subject to a reversible sulfonation reaction. This reaction is shifted toward the formation of cytosine sulfonate under acidic environment (ph 5.0). Moreover, cytosine sulfonate in the presence of a water molecule quickly suffers a hydrolytic deamination, obtaining as a consequence a uracil sulfonate intermediate. As this last deamination reaction is irreversible, and by maintaining the pH of the reaction under acidic conditions, we can guarantee that all cytosines will be transformed into uracil sulfonate (Fig. 24.1a).

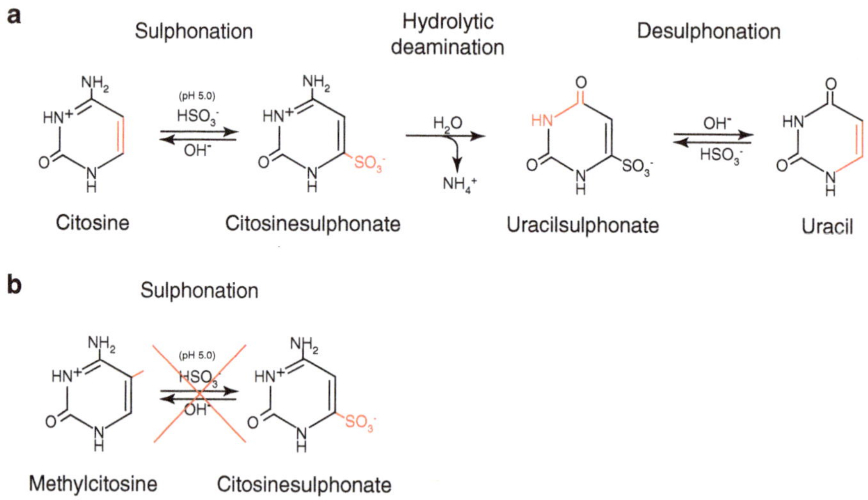

Fig. 24.1 Sodium bisulfite conversion reaction by which unmethylated cytosines are transformed into uracil (**a**), while methylcytosines (**b**) are protected from such transformation

Finally, after eliminating all the residual bisulfite reactive from the solution and modifying the conditions of the reaction toward alkaline conditions, the uracil sulfonate is transformed into uracil, in a desulfonation reaction. Conversely, the methylated group of methylcytosines prevents the methylcytosine to be subject to the sulfonation reaction (Fig. 24.1b) and hence not allowing the conversion from methylcytosine into uracil by the aforementioned reaction. Summarizing this reaction, unmethylated cytosines will be transformed into uracil, while methylated cytosines will remain unchanged. With such reaction, what initially was just a methylation status (methylated cytosine/unmethylated cytosine) has now been transformed into a single nucleotide variation (C/U), which after an amplification reaction will become a C/T variation. Now that we have obtained this nucleotide modification, we just have to interrogate which base is present in an original cytosine position after being subject to bisulfite conversion reaction, to know its methylation status.

Multiple techniques exploit this bisulfite conversion reaction. Methylation-specific PCR (MSP) [31] is a rapid and highly sensitive method for interrogating DNA methylation. It uses two sets of primers to amplify either the methylated or unmethylated allele after bisulfite treatment of the DNA. The advantage of this method is the low input (both in quantity and quality) of the DNA, as well as its high sensitivity. However, in counterpart is a nonquantitative method. Another commonly used method, which was the gold standard in pre-next-generation sequencing (pre-NGS) era, but is still being used nowadays as confirmatory technique, is the allele-specific bisulfite sequencing [32], by which after bisulfite conversion, the DNA is amplified by PCR to latter transfect the fragments of DNA with the use of a plasmid infection into competent cells. Then, transformed cells are cultured in agar plates with antibiotic selection. Plasmids of individual growing colonies (representing a single DNA molecule of the PCR product) are then isolated and sequenced. Although this technique per se is not quantitative, sequencing multiple clones, we can obtain in a semiquantitative manner the methylation level of a locus. The disadvantages of this method are that it is not a real quantitative method and becomes very labor and cost intensive for medium or large set of samples. Pyrosequencing has also been used to analyze bisulfite-treated DNA (bs-DNA). After PCR amplification over bs-DNA, the resulting fragments of DNA are sequenced by the synthesis of the complementary strand performed by a polymerase, which has been coupled with a sulfurylase, luciferase, and apyrase enzymes. As a result, when adding a dideoxynucleotide to the reaction, if the polymerase incorporates it to the elongating chain, as a subproduct, a pyrophosphate group is released. This pyrophosphate group acts as a substrate for the sulfurylase, to transform it into ATP, which will be used by the luciferase to emit light. Finally, the apyrase degrades all dideoxynucleotides which have not been incorporated. In this way, and selecting the order of dispensation of the different dideoxynucleotides, we can reconstitute the sequence by observing in which dideoxynucleotide additions a light signal is created.

Today's gold standard method for bisulfite-converted DNA is the whole-genome bisulfite sequencing (WGBS). After bisulfite conversion of the DNA, the bs-DNA is subject to library preparation, and the entire genome is sequenced, obtaining a single-base resolution measurement of all CpG sites in the human genome (~28 million CpGs). The advantages of this method are evident since we obtain all the DNA methylation status of the genome. However, due to the cost of the method, as well as the computational requirements needed in order to analyze the obtained data, this approach, although every day being more adopted, has not extensively been used in large cohorts.

24.3 DNA Methylation Microarrays

Large cohorts of samples have been widely studied DNA methylation microarrays. Although different manufacturers of methylation microarrays exist, Illumina's Infinium microarrays have become the dominant solution in the scientific arena. The chemistry behind Infinium methylation microarrays is the same one Illumina has extensively been using for single nucleotide polymorphism (SNP) microarrays, with the only addition of previously treating the DNA with sodium bisulfite. After bisulfite conversion of the DNA, the bs-DNA is amplified in a whole-genome amplification approach using a proprietary chemistry that uses random hexamer primers and a Φ29 DNA polymerase at a constant temperature. Compared to conventional PCR amplification, this approach renders largest-sized products with a lower error frequency and less bias. Amplified material is then subject to a proprietary enzymatic fragmentation process which avoids over fragmentation of the DNA. As a result, DNA fragments range between 300 and 500 bp. After exchanging the solvent in which the fragments of DNA are suspended, DNA fragments are denatured and hybridized to the microarray, which contains millions of 3-micron silica beads self-assembled in microwells of the microarray slide. Each silica bead is coated with an oligonucleotide 50 bases long, which has been specifically designed to have complementarity to the region adjacent to a certain CpG site. Once we have allowed the hybridization of the single-stranded DNA molecules with its corresponding 50 bases oligonucleotide, unhybridized molecules, are washed away and the two-step process by which C/T alleles are interrogated begins. The first step involves the use of a DNA polymerase and the four DNA nucleotides (A, C, T, G). These nucleotides carry a blocking agent that only allows the elongation of a single base into the 50 bases long oligonucleotides, using the bs-DNA as a template, in what is called the single nucleotide extension reaction. Interestingly, the blocking agent preventing the incorporation of a second base is different between the four nucleotides, being a biotin-blocking molecule for cytosine and guanine nucleotides, while adenines and thymines carry a 2,4-dinitrophenol blocking molecule. The second step toward CpG interrogation entails the use of specific antibodies fluorescently labeled against biotin (green) and 2,4-dinitrophenol (red) molecules. The fluorescent signal of these antibodies is then amplified by using rounds of secondary/primary antibodies stains, allowing the multiplication of the fluorescent signal, in order to make it detectable by the scanner's cameras. Green and red fluorescent intensities ratios per bead are then computed as a direct measurement of the methylation level of that CpG site being interrogated by each one of the million beads present on the microarray. The advantages of this platform are the single nucleotide resolution of the platform, its cost, which is one or two orders of magnitude cheaper than NGS, while at the same time being able to interrogate thousands of CpG sites simultaneously (27 k: ≈27,000 CpG sites [33]; 450 K: 485,577 CpG sites [34, 35]; and EPIC (850 K): 853,307 CpG sites [36]), for an adaptable cohort size (from tens to hundreds of samples). However, its disadvantage over NGS approach is that arrays do not cover all the 28 million CpG sites of the human genome.

Moreover, another crucial advantage of the methylation microarrays is that they are compatible with samples preserved as formalin fixed paraffin embedded (FFPE). DNA obtained from FFPE samples, because of the fixation process, tends to be highly fragmented, hindering downstream applications. Standard Infinium microarray's procedure on FFPE samples will fail to obtain accurate methylation measurements, as a consequence of the inefficacy of the whole-genome amplification reaction of the procedure [37]. This restricts the potentially analyzable samples to those whose DNA has been extracted from fresh-frozen (FF) tissues, blood samples, or in vitro cultured cells. Conversely, samples with previous associated clinical and follow-up data are routinely formalin fixed and paraffin embedded (FFPE) for histopathological diagnosis, which is known to affect DNA integrity and to prevent these valuable samples being subjected to methods, such as Infinium, that are sensitive to DNA fragmentation. This restricts the potentially analyzable samples to those whose DNA has been extracted from fresh-frozen (FF) tissues, blood samples, or in vitro cultured cells. To overcome these limitations, a combination of DNA repair and ligation reactions (FFPE restoration procedure) can be performed to the bisulfite-converted DNA. The use of such procedure in the Infinium HumanMethylation27 DNA methylation microarray [37, 38], as well as in the Infinium HumanMethylation450 [39–42] and the newer Infinium MethylationEpic DNA methylation microarrays, has been reported [36]. Essentially, most of these studies have compared the methylation profiles obtained from fresh-frozen samples, with the ones obtained from their match FFPE samples of the same donor, obtaining highly correlated measurements. Thus, FFPE samples render data of the same quality as the one obtained from fresh-frozen samples, bypassing the handicap caused by the formalin/paraffin fixation and preservation.

Interestingly, DNA methylation microarrays have been adapted to interrogate besides 5- methylcytosine (5mC), the levels of 5-hydroxymethylcitosine (5hmC) in the genome, with the same single-base resolution provided by the Infinium platform. To achieve such purpose, a modification of the protocol is required. After performing the bisulfite conversion of the DNA and prior to processing it throughout the Infinium procedure, bs-DNA is subjected to an oxidation reaction. As a result, hydroxymethylated cytosines will lose their protection against its conversion when sodium bisulfite is present, resulting in both unmethylated cytosines and hydroxymethylated ones being converted into uracils when bisulfite conversion is carried out over oxidized genomic DNA. On the contrary, methylated cytosines still are protected against conversion during bisulfite reaction and will remain unchanged. Then, for a given CpG, if we subtract the methylation value obtained in the oxidative procedure (oxidative bisulfite conversion) to the one obtained with the regular procedure (regular bisulfite conversion), the result is the level of 5-hydroxymethylcitosine [43–45]. Interestingly, it has been observed an enrichment of 5-hydroxymethylcitosine in certain brain structures [46], demonstrating its role in neural development.

DNA methylation patterns have been widely explored in neurological disorders, and DNA methylation microarrays have been chosen in many of those multiple approaches. Numerous studies have centered their aim in finding the presence of epigenetic alteration in Alzheimer's disease [47–53], schizophrenia [54–57], autism [58], childhood psychiatric disorders [59], or depression [60]; nevertheless, previous chapters of this book have extensively and deeply covered those aspects.

24.4 Summary

DNA methylation plays a critical role in cell development and homeostasis. Its interrogation in many clinical settings has allowed the identification of multiple causal alterations of different diseases, for example, cancer and neurodegenerative or aging diseases. Nonetheless, it has also allowed the discovery of biomarkers for predicting disease progression and response to treatments and for disease diagnosis. Such advances have been only possible by the incredible advance contribution of the analytical methods for measuring DNA methylation. Nowadays, whole-genome bisulfite sequencing is considered as the gold standard method for measuring the levels of 5-methylcytosine in the genome; however, its cost and its computational requirements, both in terms of equipment and know-how, are the main limitations toward massive adoption on larger cohort studies. DNA methylation microarrays are a suitable option for such studies, with whom an elevated amount of CpG sites (Infinium MethylationEPIC (850 K): 853,307 CpG sites) are interrogated at a cost which is two orders of magnitude cheaper than NGS. Moreover, its standardized procedure and its compatibility with samples preserved as FFPE have allowed during the last years the exponential growth in their use, not only on the cancer research but also in the neurological research.

References

1. Goldberg AD, Allis CD, Bernstein E. Epigenetics: a landscape takes shape. Cell. 2007;128:635–8. doi:10.1016/j.cell.2007.02.006.
2. Gosden RG, Feinberg AP. Genetics and epigenetics--nature's pen-and-pencil set. N Engl J Med. 2007;356:731–3. doi:10.1056/NEJMe068284.
3. Gardiner-Garden M, Frommer M. CpG Islands in vertebrate genomes. J Mol Biol. 1987;196:261–82. doi:10.1016/0022-2836(87)90689-9.
4. Baylin SB. A decade of exploring the cancer epigenome - biological and translational implications. Nat Rev Cancer. 2012;11:726–34. doi:10.1038/nrc3130.A.

5. Meissner A, Mikkelsen TS, Gu H, Wernig M, Hanna J, Sivachenko A, et al. Genome-scale DNA methylation maps of pluripotent and differentiated cells. Nature. 2008;454:766–70. doi:10.1038/nature07107.
6. Suzuki MM, Bird A. DNA methylation landscapes: provocative insights from epigenomics. Nat Rev Genet. 2008;9:465–76. doi:10.1038/nrg2341.
7. Lister R, Pelizzola M, Dowen RH, Hawkins RD, Hon G, Tonti-Filippini J, et al. Human DNA methylomes at base resolution show widespread epigenomic differences. Nature. 2009;462:315–22. doi:10.1038/nature08514.
8. Okano M, Xie S, Li E. Cloning and characterization of a family of novel mammalian DNA (cytosine-5) methyltransferases. Nat Genet. 1998;19:219–20. doi:10.1038/890.
9. Okano M, Bell DW, Haber DA, Li E. DNA methyltransferases Dnmt3a and Dnmt3b are essential for de novo methylation and mammalian development. Cell. 1999;99:247–57.
10. Li E, Bestor TH, Jaenisch R. Targeted mutation of the DNA methyltransferase gene results in embryonic lethality. Cell. 1992;69:915–26.
11. Xu GL, Bestor TH, Bourc'his D, Hsieh CL, Tommerup N, Bugge M, et al. Chromosome instability and immunodeficiency syndrome caused by mutations in a DNA methyltransferase gene. Nature. 1999;402:187–91. doi:10.1038/46052.
12. Angulo MA, Butler MG, Cataletto ME. Prader-Willi syndrome: a review of clinical, genetic, and endocrine findings. J Endocrinol Investig. 2015;38:1249–63. doi:10.1007/s40618-015-0312-9.
13. Heyn H, Vidal E, Ferreira HJ, Vizoso M, Sayols S, Gomez A, et al. Epigenomic analysis detects aberrant super-enhancer DNA methylation in human cancer. Genome Biol. 2016;17:11. doi:10.1186/s13059-016-0879-2.
14. Stefansson OA, Moran S, Gomez A, Sayols S, Arribas-Jorba C, Sandoval J, et al. A DNA methylation-based definition of biologically distinct breast cancer subtypes. Mol Oncol. 2015;9:555–68. doi:10.1016/j.molonc.2014.10.012.
15. Øster B, Linnet L, Christensen LL, Thorsen K, Ongen H, Dermitzakis ET, et al. Non-CpG island promoter hypomethylation and miR-149 regulate the expression of SRPX2 in colorectal cancer. Int J Cancer. 2013;132:2303–15. doi:10.1002/ijc.27921.
16. Heyn H, Carmona Javier F, Gomez A, Ferreira HJ, Bell JT, Sayols S, et al. DNA methylation profiling in breast cancer discordant identical twins identifies DOK7 as novel epigenetic biomarker. Carcinogenesis. 2013;34:102–8. doi:10.1093/carcin/bgs321.
17. Sandoval J, Mendez-Gonzalez J, Nadal E, Chen G, Carmona FJ, Sayols S, et al. A prognostic DNA methylation signature for stage I non-small-cell lung cancer. J Clin Oncol. 2013;31:4140–7.
18. Sanchez-Mut JV, Heyn H, Vidal E, Moran S, Sayols S, Delgado-Morales R, et al. Human DNA methylomes of neurodegenerative diseases show common epigenomic patterns. Transl Psychiatry. 2016;6:e718. doi:10.1038/tp.2015.214.
19. Zaina S, Gonçalves I, Carmona FJ, Gomez A, Heyn H, Mollet IG, et al. DNA methylation dynamics in human carotid plaques after cerebrovascular events. Arterioscler Thromb Vasc Biol. 2015;35:1835–42. doi:10.1161/ATVBAHA.115.305630.
20. Zaina S, Heyn H, Carmona FJ, Varol N, Sayols S, Condom E, et al. DNA methylation map of human atherosclerosis. Circ Cardiovasc Genet. 2014;7:692–700. doi:10.1161/CIRCGENETICS.113.000441.
21. Vizoso M, Ferreira HJ, Lopez-Serra P, Carmona FJ, Martínez-Cardús A, Girotti MR, et al. Epigenetic activation of a cryptic TBC1D16 transcript enhances melanoma progression by targeting EGFR. Nat Med. 2015;21:741–50. doi:10.1038/nm.3863.
22. Heyn H, Li N, Ferreira HHJ, Moran S, Pisano DG, Gomez A, et al. Distinct DNA methylomes of newborns and centenarians. Proc Natl Acad Sci U S A. 2012;109:10522–7. doi:10.1073/pnas.1120658109/-/DCSupplemental.www.pnas.org/cgi/doi/10.1073/pnas.1120658109.
23. Heyn H, Moran S, Esteller M. Aberrant DNA methylation profiles in the premature aging disorders Hutchinson-Gilford Progeria and Werner syndrome. Epigenetics. 2013;8:28–33.
24. Wyatt GR, Cohen SS. A new Pyrimidine Base from bacteriophage nucleic acids. Nature. 1952;170:1072–3. doi:10.1038/1701072a0.

25. Vrielink A, Rüger W, Driessen HP, Freemont PS. Crystal structure of the DNA modifying enzyme beta-glucosyltransferase in the presence and absence of the substrate uridine diphosphoglucose. EMBO J. 1994;13:3413–22.
26. Lorsbach RB, Moore J, Mathew S, Raimondi SC, Mukatira ST, Downing JR. TET1, a member of a novel protein family, is fused to MLL in acute myeloid leukemia containing the t(10;11)(q22;q23). Leukemia. 2003;17:637–41. doi:10.1038/sj.leu.2402834.
27. Guibert S, Weber M. Functions of DNA methylation and hydroxymethylation in mammalian development. Curr Top Dev Biol. 2013;104:47–83. doi:10.1016/B978-0-12-416027-9.00002-4.
28. Luo G-Z, Blanco MA, Greer EL, He C, Shi Y. DNA N(6)-methyladenine: a new epigenetic mark in eukaryotes? Nat Rev Mol Cell Biol. 2015;16:705–10. doi:10.1038/nrm4076.
29. Heyn H, Esteller M. An adenine code for DNA: a second life for N6-methyladenine. Cell. 2015;161:710–3. doi:10.1016/j.cell.2015.04.021.
30. Clark SJ, Harrison J, Paul CL, Frommer M. High sensitivity mapping of methylated cytosines. Nucleic Acids Res. 1994;22:2990–7.
31. Herman JG, Graff JR, Myohanen S, Nelkin BD, Baylin SB. Methylation-specific PCR: a novel PCR assay for methylation status of CpG islands. Proc Natl Acad Sci. 1996;93:9821–6. doi:10.1073/pnas.93.18.9821.
32. Wong H-L, Byun H-M, Kwan JM, Campan M, Ingles SA, Laird PW, et al. Rapid and quantitative method of allele-specific DNA methylation analysis. BioTechniques. 2006;41:734–9.
33. Bibikova M, Le J, Barnes B, Saedinia-melnyk S, Zhou L, Shen R, et al. Genome-wide DNA methylation profiling using Infinium® assay. Epigenomics. 2009;1:177–200. doi:10.2217/epi.09.14.
34. Sandoval J, Heyn HA, Moran S, Serra-Musach J, Pujana MA, Bibikova M, et al. Validation of a DNA methylation microarray for 450,000 CpG sites in the human genome. Epigenetics. 2011;6:692–702. doi:10.4161/epi.6.6.16196.
35. Dedeurwaerder S, Defrance M, Calonne E, Denis H, Sotiriou C, Fuks F. Evaluation of the Infinium methylation 450 K technology. Epigenomics. 2011;3:771–84. doi:10.2217/epi.11.105.
36. Moran S, Arribas C, Esteller M. Validation of a DNA methylation microarray for 850,000 CpG sites of the human genome enriched in enhancer sequences. Epigenomics. 2015;6:epi.15.114. doi: 10.2217/epi.15.114
37. Thirlwell C, Eymard M, Feber A, Teschendorff A, Pearce K, Lechner M, et al. Genome-wide DNA methylation analysis of archival formalin-fixed paraffin-embedded tissue using the Illumina Infinium HumanMethylation27 BeadChip. Methods. 2010;52:248–54. doi:10.1016/j.ymeth.2010.04.012.
38. Jasmine F, Rahaman R, Roy S, Raza M, Paul R, Rakibuz-Zaman M, et al. Interpretation of genome-wide infinium methylation data from ligated DNA in formalin-fixed, paraffin-embedded paired tumor and normal tissue. BMC Res Notes. 2012;5:117. doi:10.1186/1756-0500-5-117.
39. Moran S, Vizoso M, Martinez-Cardús A, Gomez A, Matías-Guiu X, Chiavenna SM, et al. Validation of DNA methylation profiling in formalin-fixed paraffin-embedded samples using the Infinium HumanMethylation450 microarray. Epigenetics. 2014;9:829–33. doi:10.4161/epi.28790.
40. Dumenil TD, Wockner LF, Bettington M, McKeone DM, Klein K, Bowdler LM, et al. Genome-wide DNA methylation analysis of formalin-fixed paraffin embedded colorectal cancer tissue. Genes Chromosomes Cancer. 2014;53:537–48. doi:10.1002/gcc.22164.
41. Siegel EM, Berglund AE, Riggs BM, Eschrich SA, Putney RM, Ajidahun AO, et al. Expanding epigenomics to archived FFPE tissues: an evaluation of DNA repair methodologies. Cancer Epidemiol Biomark Prev. 2014;23:2622–31. doi:10.1158/1055-9965.EPI-14-0464.

42. de Ruijter TC, de Hoon JPJ, Slaats J, de Vries B, Janssen MJFW, van Wezel T, et al. Formalin-fixed, paraffin-embedded (FFPE) tissue epigenomics using Infinium HumanMethylation450 BeadChip assays. Lab Investig. 2015;95:833–42. doi:10.1038/labinvest.2015.53.
43. Stewart SK, Morris TJ, Guilhamon P, Bulstrode H, Bachman M, Balasubramanian S, et al. OxBS-450 K: a method for analysing hydroxymethylation using 450 K BeadChips. Methods. 2015;72:9–15.
44. Field SF, Beraldi D, Bachman M, Stewart SK, Beck S, Balasubramanian S. Accurate measurement of 5-methylcytosine and 5-hydroxymethylcytosine in human cerebellum DNA by oxidative bisulfite on an array (OxBS-Array). PLoS One. 2015;10:e0118202. doi:10.1371/journal.pone.0118202.
45. Houseman EA, Johnson KC, Christensen BC. OxyBS: estimation of 5-methylcytosine and 5-hydroxymethylcytosine from tandem-treated oxidative bisulfite and bisulfite DNA. Bioinformatics. 2016;32:2505–7. doi:10.1093/bioinformatics/btw158.
46. Lunnon K, Hannon E, Smith RG, Dempster E, Wong C, Burrage J, et al. Variation in 5-hydroxymethylcytosine across human cortex and cerebellum. Genome Biol. 2016;17:27. doi:10.1186/s13059-016-0871-x.
47. Bakulski KM, Dolinoy DC, Sartor MA, Paulson HL, Konen JR, Lieberman AP, et al. Genome-wide DNA methylation differences between late-onset Alzheimer's disease and cognitively normal controls in human frontal cortex. J Alzheimers Dis. 2012;29:571–88. doi:10.3233/JAD-2012-111223.
48. Sanchez-Mut JV, Aso E, Panayotis N, Lott I, Dierssen M, Rabano A, et al. DNA methylation map of mouse and human brain identifies target genes in Alzheimer's disease. Brain. 2013;136:3018–27. doi:10.1093/brain/awt237.
49. Sanchez-Mut JV, Aso E, Heyn H, Matsuda T, Bock C, Ferrer I, et al. Promoter hypermethylation of the phosphatase DUSP22 mediates PKA-dependent TAU phosphorylation and CREB activation in Alzheimer's disease. Hippocampus. 2014;24:363–8. doi:10.1002/hipo.22245.
50. Sanchez-Mut JV, Gräff J. Epigenetic alterations in Alzheimer's disease. Front Behav Neurosci. 2015;9:347. doi:10.3389/fnbeh.2015.00347.
51. Bennett DA, Yu L, Yang J, Srivastava GP, Aubin C, De Jager PL. Epigenomics of Alzheimer's disease. Transl Res. 2015;165:200–20. doi:10.1016/j.trsl.2014.05.006.
52. Smith AR, Smith RG, Condliffe D, Hannon E, Schalkwyk L, Mill J, et al. Increased DNA methylation near TREM2 is consistently seen in the superior temporal gyrus in Alzheimer's disease brain. Neurobiol Aging. 2016;47:35–40. doi:10.1016/j.neurobiolaging.2016.07.008.
53. Watson CT, Roussos P, Garg P, Ho DJ, Azam N, Katsel PL, et al. Genome-wide DNA methylation profiling in the superior temporal gyrus reveals epigenetic signatures associated with Alzheimer's disease. Genome Med. 2016;8:5. doi:10.1186/s13073-015-0258-8.
54. Wockner LF, Noble EP, Lawford BR, Young RM, Morris CP, Whitehall VLJ, et al. Genome-wide DNA methylation analysis of human brain tissue from schizophrenia patients. Transl Psychiatry. 2014;4:e339. doi:10.1038/tp.2013.111.
55. Wockner LF, Morris CP, Noble EP, Lawford BR, Whitehall VLJ, Young RM, et al. Brain-specific epigenetic markers of schizophrenia. Transl Psychiatry. 2015;5:e680. doi:10.1038/tp.2015.177.
56. Montano C, Taub MA, Jaffe A, Briem E, Feinberg JI, Trygvadottir R, et al. Association of DNA methylation differences with schizophrenia in an epigenome-wide association study. JAMA Psychiat. 2016;73:506–14. doi:10.1001/jamapsychiatry.2016.0144.
57. Alelú-Paz R, Carmona FJ, Sanchez-Mut JV, Cariaga-Martínez A, González-Corpas A, Ashour N, et al. Epigenetics in schizophrenia: a pilot study of global DNA methylation in different brain regions associated with higher cognitive functions. Front Psychol. 2016;7:1496. doi:10.3389/fpsyg.2016.01496.
58. Ladd-Acosta C, Hansen KD, Briem E, Fallin MD, Kaufmann WE, Feinberg AP. Common DNA methylation alterations in multiple brain regions in autism. Mol Psychiatry. 2014;19:862–71. doi:10.1038/mp.2013.114.

59. Fisher HL, Murphy TM, Arseneault L, Caspi A, Moffitt TE, Viana J, et al. Methylomic analysis of monozygotic twins discordant for childhood psychotic symptoms. Epigenetics. 2015;10:1014–23. doi:10.1080/15592294.2015.1099797.
60. Córdova-Palomera A, Fatjó-Vilas M, Gastó C, Navarro V, Krebs M-O, Fañanás L. Genome-wide methylation study on depression: differential methylation and variable methylation in monozygotic twins. Transl Psychiatry. 2015;5:e557. doi:10.1038/tp.2015.49.

Bioinformatics Tools for Genome-Wide Epigenetic Research

25

Vladimir Espinosa Angarica and Antonio del Sol

Abstract

Epigenetics play a central role in the regulation of many important cellular processes, and dysregulations at the epigenetic level could be the source of serious pathologies, such as neurological disorders affecting brain development, neurodegeneration, and intellectual disability. Despite significant technological advances for epigenetic profiling, there is still a need for a systematic understanding of how epigenetics shapes cellular circuitry, and disease pathogenesis. The development of accurate computational approaches for analyzing complex epigenetic profiles is essential for disentangling the mechanisms underlying cellular development, and the intricate interaction networks determining and sensing chromatin modifications and DNA methylation to control gene expression. In this chapter, we review the recent advances in the field of "computational epigenetics," including computational methods for processing different types of epigenetic data, prediction of chromatin states, and study of protein dynamics. We also discuss how "computational epigenetics" has complemented the fast growth in the generation of epigenetic data for uncovering the main differences and similarities at the epigenetic level between individuals and the mechanisms underlying disease onset and progression.

Keywords

Computational epigenetics • Histone code • Epigenetics regulation • Transcriptional regulation • Next generation sequencing analysis • Single cell neuroepigenetics

V.E. Angarica, Ph.D. (✉) • A. del Sol, Ph.D.
Computational Biology Group, Luxembourg Centre for Systems Biomedicine, University of Luxembourg, 6 Avenue du Swing, 4366 Belvaux, Luxembourg
e-mail: antonio.delsol@uni.lu

R. Delgado-Morales (ed.), *Neuroepigenomics in Aging and Disease*, Advances in Experimental Medicine and Biology, DOI 10.1007/978-3-319-53889-1_25

25.1 Chromatin Structure, Combinatorial Complexity of Histone Modifications, and Mechanisms of Epigenetic Regulation

Epigenetic phenomena constitute a very important regulatory checkpoint in many key cellular processes such as DNA maintenance and repair [1, 2], epigenetic inheritance [3, 4], and gene expression [5, 6]. While the genome underlying structure – i.e., DNA sequence – is highly stable, epigenetic signatures are dynamic [7–9], with different epigenetic phenomena having different degrees of stability and variability, causing most of the phenotypic differences across cells in multicellular organisms. Fluctuations in DNA condensation, and the establishment of heterochromatic or euchromatic regions, are determined by covalent modifications of chromatin, including DNA methylation of CpG islands [10–12], and a wide range of histone modifications [9, 13, 14], which form complex combinatorial networks of histone marks, that constitute the "histone code" [15]. Moreover, DNA methylation and histone modification pathways are significantly interconnected [16–18], and the cross talk between DNA and histone epigenetic modifications significantly increases the combinatorial complexity of the mechanisms of epigenetic regulation. Although not yet fully understood, there are two characterized mechanisms by which epigenetic modifications exert their function [9]: the first is the disruption of contacts between nucleosomes in order to "unravel" chromatin, and the second is the recruitment of nonhistone proteins [9]. A wide family of epigenetic signaling proteins – i.e., readers, writers, and erasers – [19–22] recognize the complex code of epigenetic modifications, controlling the condensation levels of genomic regions, and the susceptibility of these regions to be transcribed [5, 6], to be subject of DNA repair [1, 2] or be involved in other cellular processes. The central role of epigenetics in the regulation of a broad range of key cellular processes explains their implication in multiple common and serious human pathologies [23–25], such as developmental diseases [26–28], cancer [29–32], and neurological disorders [33–37]. Despite technological advances for the study of mechanisms of epigenetic regulation, we still lack a systematic understanding of how the epigenomic landscape contributes to cellular circuitry, lineage specification, and the onset and progression of human disease [38]. Due to the significant complexity of the mechanisms of epigenetic regulation, computational and bioinformatics approaches have been essential for disentangling these mechanisms at the genome-wide level and for answering important questions such as how the epigenetic level senses environmental cues during lineage specification and development and which are the interactions among different chromatin modifications to control transcription.

In this chapter, we review the state of the art of computational approaches and bioinformatics tools for genome-wide epigenetic research. We cover the field of "computational epigenetics" and discuss recent advances in computational methods for processing and quality control of different types of epigenetic data, the prediction of chromatin states and the study of the dynamics of chromatin, and the analysis of 3D structure of chromatin. We also address the status of different collaborative projects and databases comprising a wealth of genome-wide epigenetic data. We discuss how the fast growth in the generation of epigenetic data, boosted by the development of high-throughput sequencing (HTS) experimental technologies and inter-institutional public/private collaborative projects, has been complemented and prompted by the development of computational methods for analyzing and rationalizing huge quantities of epigenetic data. The steady decrease in the cost of technologies for generating epigenetic data has also opened the possibility of performing epigenetic surveys in human populations. In this regard, we also examine the recent development in the computational approaches used to perform these studies for uncovering the main differences and similarities at the epigenetic level between individuals and their implication in cellular differentiation, gene regulation, and disease.

25.2 Whole Genome Annotation of Histone Modifications: Computational Tools for Data Quality Control and Mapping of Epigenetic Data

The characteristics and specificities of the wide range of computational methods commonly used for the analysis of epigenetic data depend significantly on the particularities of the experimental techniques used to perform epigenomic profiling. The techniques available for profiling histone modifications (and the other epigenetic phenomena described in the next sections of this chapter) are described in detail in a previous chapter of this book, but it would be important to summarize their commonalities and differences to discuss the different computational approaches used to analyze the epigenetic data generated in each case. The most commonly used experimental approaches to profile histone posttranslational modifications are ChIP-on-chip [39–41], ChIP-seq [42–44], and mass spectrometry [45–48]. In ChIP-on-chip, histone modification-specific antibodies, bound to chromatin regions bearing the corresponding modification, are cross-linked to DNA by treatment with formaldehyde. Next, chromatin is collected and fragmented by sonication or using nucleases, and the fragments bearing the histone modification are enriched by using an antibody matrix specific to the histone modification-specific antibody – i.e., immunoprecipitation. The DNA in the enriched fragments is released reverting cross-linking by increasing temperature, and purified DNA fragments are amplified and labeled with fluorescent dyes for further quantitation. Finally, purified DNA is hybridized to a tilling microarray, which allows the identification of regions over-represented in the immunoprecipitated DNA relative to control DNA – i.e., regarded as epigenetically modified. ChIP-seq shares the initial steps of the ChIP-on-chip technique, but unlike the former, it relies on HTS DNA sequencing rather than on microarrays for identifying the sequences enriched in histone marks. Unlike immunoprecipitation techniques, proteomic profiling using mass spectrometry (MS) allows the detailed characterization of histone tail posttranslational modifications. This technique relies on the chromatographic separation of histones from cell lysates, followed by enzymatic digestion of individual histones for the accurate assignment and quantification of the amino acids bearing different kinds of post-translational modifications [9, 13, 14], following top-down, bottom-up, or middle-down approaches [47, 49].

Immunoprecipitation techniques are by far the most commonly used, thanks to their high-throughput capabilities and the developments in the production of highly specific histone modification-specific antibodies. The main bioinformatics problem for the analysis of ChIP-on-chip data is establishing a ranking of genomic regions overrepresented on the arrays from raw probe intensities. In this regard, many different approaches have been specifically developed for performing peak calling from ChIP-on-chip experiments. In general, these methods have a set of common steps, encompassing the normalization of the intensities of hybridized fragments, assessment of the statistical significance of the intensities of each peak with respect to the whole array, and finally merging overlapping overrepresented regions [39–41, 50]. The list of peak-calling packages for processing ChIP-on-chip data is fairly ample and diverse, including Tilescope [51], an automated data processing toolkit for analyzing high-density tiling microarray data that integrates data normalization, combination of replicate experiments, tile scoring, and feature identification in an easy-to-use online suite. Tilemap [52] is a stand-alone package that provides a flexible way to study tiling array hybridizations under multiple experimental conditions in Affymetrix ChIP-on-chips. Ringo [53] is an R package devised for NimbleGen microarrays, which facilitates the construction of automated programmed workflows and enables the scalability and reproducibility of the analyses in comparison to other ChIP-on-chip peak callers. The abovementioned list of bioinformatics tools for processing ChIP-on-chip microarray data is by no means exhaustive, and there is a wide spectrum of other approaches, including ACME [54], HGMM [55], ChIPOTle [56], HMMTiling [57], and MAT [58], among others. Notwithstanding the diversity of tools for processing ChIP-on-chip data, the bioinformatics analysis of tiling microarrays shares the same drawbacks of the algorithms for analyzing DNA arrays, as they fail to accurately estimate histone modifications spanning extended genomic regions and underestimate weak binding events [50].

The key bioinformatics challenge in the analysis of ChIP-seq data is the fast and accurate mapping of thousands to millions of short reads, corresponding to the regions bearing a specific histone modification, to the reference genome. Many sequence aligners for solving the problems of mapping short sequence reads have been developed, such as Bowtie [59], BWA [60], SOAP [61], and BLAT [62], among a wide list of others (for a detailed review on short-read alignment methods see [63]). Other methods with alignment strategies optimized for reads obtained with specific sequencing platforms have been developed, including commercial suites such as ELAND that form part of the SOLEXA pipeline (http://www.solexa.com/), and the Broad Institute sequencing platform [64] (http://genomics.broadinstitute.org/). While mapping short reads to a reference genome, special care should be taken to the quality control of sequencing data. For instance, random fragmentation of ChIP-seq samples treated with sonication renders an array of overlapping reads corresponding to the same genomic regions, and these duplicated reads should be removed, using for example SAMtools [65]. This requirement for quality control is not necessary, however, while analyzing ChIP-seq data generated from samples treated with nucleases, because the likelihood of the generation of overlapping reads is rather low. The assessment of "uniquely mapped" and "unique reads" is also a very important step in the quality control of ChIP-seq data. The former correspond to reads that aligned to specific regions, excluding repetitive genomic *loci* and non-repetitive regions with highly similar sequences, while the latter correspond to de-duplication PCR reads. In this regard, depending on the specificities of the ChIP-seq dataset, removal of duplicated reads to reduce amplification artifacts could result in an underestimation of real binding events. On the other hand, not removing duplicate reads could cause the inclusion of a significant amount of false positives, which could have strong implications in the downstream analysis of ChIP-seq data. Therefore, alignment of short sequence reads to the reference genome, and quality control of sequencing data, still remains a bioinformatics challenge. The analysis of the signal-to-noise ratio of sequencing signals also constitutes an important step on ChIP-seq quality control. The estimation of the "fraction of reads in peaks" (FRiP) – i.e., number of reads per region – and cross-correlation profiles (CCPs), i.e., read clustering prior to peak calling [66], are very useful for assessing the signal-to-noise ratio. Based on these metrics, different approaches for estimating the signal-to-noise ratio of ChIP-seq sequencing data have been developed [67].

The procedures for performing peak calling from ChIP-seq samples are different from those commonly used for ChIP-on-chip experiments. There exists a myriad of different peak callers based on different statistical criteria, which cannot be covered here in detail (for a detailed review, please see [68]). The general procedure followed by all of these algorithms includes the identification of enriched sequence read density for different chromosome *loci*, relative to a background sequence read distribution. The first step common to all ChIP-seq peak callers is the generation of a signal profile by integrating reads mapped to specific genomic regions. Different tools rely on sliding-window approaches for smoothing the discrete distribution of read counts into a continuous signal profile distribution. Tools such as CisGenome [69] follow this rationale, estimating the number of reads above a predefined peak cutoff, and others like SISSRs [70], Peakzilla [71], and SPP [72] also take into account the correspondence of read counts in positive and negative strands to improve peak resolution. Other tools use more sophisticated approaches for integrating the signals in sequence windows. For example, MACS [73] uses the local Poisson model to identify local biases in genomic positions, F-Seq [74] and QuEST [75] rely on kernel density estimations, and PICS [76] uses a Bayesian hierarchical *t*-mixture model for smoothing count reads in the genomic signal profile. The HOMER program suite [77] has been also widely used for peak calling and is specially useful for analyzing broad peak corresponding to histone modifications – e.g., H3K9me3 – spanning large chromosome regions. Other tools such as JAMM [78] and PePr [79] integrate information from biological replicates to determine enrichment site widths in neighboring narrow peaks, whereas GLITR [80] and PeakSeq [81] use tag extension –i.e., extension of ChIP-seq tags along their strand direction – to identify genomic regions enriched in sequence reads. The selection of the background distribution used in the comparison with the sample analyzed is also an essential step in peak calling. Although there is no consensus on which is the best background distribution, different datasets have been used as control sample, such as ChIP-seq data for histone H3, or from experiments using a control antibody for nonbinding proteins, such as immunoglobulins [66, 82]. The following steps during peak calling include the selection of the statistical criteria for identifying enriched peaks, which usually correspond to a specific cutoff for the enrichment of peaks relative to the background, or estimating metrics with more statistical support, such as the false discovery rate (FDR). Once enriched peaks are identified for a selected number of genes, or genome wide, most peak calling algorithms allow ranking and selection of the more significant peaks by estimating their corresponding p-values and q-values. Despite the great variety of peak calling toolkits for analyzing ChIP-seq data, the comparison of the performance of different approaches shows that different programs produce very different peaks in terms of peak size, number, and position relative to genes [83, 84] when presented to the same input dataset. Thus, as different tools usually generate significantly different epigenomic profiles, peak calling of ChIP-seq data remains a difficult task, and the selection of the best performing methods usually depends on the species, sample conditions, and target proteins [43].

The bioinformatics analysis of histone posttranslational modification profiles obtained with MS is significantly dependent on the specific MS approach used – e.g., top-down, bottom-up, or middle-down approaches [47, 49]. The preprocessing of MS data for removing false fragment ion assignments can be performed with different programs, such as Thrash [85], MS-Deconv [86], or YADA [87]. These approaches can also be used to deconvolute ion signals with multiple charges into mono-charged ion mass values from bottom-up MS profiles, but are unable to produce good results for other approaches generating longer peptides [88]. Unlike immunoprecipitation techniques, in which PTM-specific antibodies are used to profile one histone modification per experiment, the analysis of cell lysates with MS has the added difficulty of having to deal with the genome-wide profile of all the histone modifications. Due to the huge combinatorial complexity of this problem, current approaches concentrate on the most common histone PTM [47], which might overlook unknown, but functionally relevant modifications. Top-down and middle-down proteomics strategies require specialized search algorithms and annotation tools, due to the great complexity of the MS spectra generated for intact or large polypeptides [89]. Methods such as ProSight PTM [90], MX-Align+ [91], ROCCIT [92], and MLIP [93] are tools specifically suited for performing database sequence searches from neutral mass lists of precursor and fragment ions obtained with top-down approaches. Different implementations of the THRASH [85] algorithm have been adapted for top-down histone modification profiling [94, 95], as well as MS-Deconv tool [86], developed specifically to analyze MS spectra from complete proteins. These methods offer a number of different functionalities for guiding the search for specific modifications that allows a significant reduction of the search space, which can increase the significance of assigned peaks. Other tools allow tackling the complex problem of identifying different histone PTM fragments with fairly similar ion masses [93, 96]. The software VEMS is included in this category [97], which can discriminate acetyl and trimethyl lysine histone modifications. In summary, mass spectrometry constitutes a very powerful approach for the genome-wide profiling of histone modifications, but there is still a need for the development of more accurate bioinformatics approaches to allow a more comprehensive and thorough study of MS histone modification spectra.

25.3 Bioinformatics Approaches for Analyzing Genome-Wide Methylation Profiling

DNA methylation, which is the only epigenetic phenomena involving the direct modification of genome underlying structure, can be profiled experimentally with bisulfite sequencing [98, 99], bisulfite microarrays [100, 101], and enrichment methods, such as MeDIP-seq and MethylCap-seq [102–104]. Different computational approaches have been developed for processing genome-wide profiling data obtained with each of the abovementioned techniques. In the case of bisulfite sequencing data, methylated cytosines are protected from chemical modification – i.e., sulfonation – induced by treatment with bisulfite, while unmethylated cytosines are sulfonated and appear as thymines after sequencing. Following, the reads obtained at the sequencing stage are mapped back to the reference genome, and the ratios of Cs and Ts are measured, representing the methylation levels of genomic regions. In principle, aligners such as those currently used for mapping ChIP-seq reads (please see in the previous section in this chapter) can be used for processing bisulfite sequencing reads, but in this case it is necessary to account for the underrepresentation of unmethylated Cs. Moreover, different approaches specifically suited for analyzing this data have been developed, comprising RRBSMAP [105], RMAP [106], GSNAP [107], and Segemehl [108], among others, which have been coined as wildcard aligners. These tools offer multiple functionalities for wildcarding Cs in the sequencing reads during the alignment and also adjusting the matrices used for scoring tag alignment for accommodating base mismatches. Furthermore, wildcard aligners allow the efficient and fast alignment to large genomic regions, although they tend to overestimate highly methylated regions. A second group of tools (MethylCoder [109], BRAT [110], and Bismark [111]) follow a more straightforward strategy, leveraging from well-established short-read alignment tools, and use a three-letter alphabet – i.e., considering T, G, and A – in the alignment. Three-letter alignment approaches are not very efficient for scanning large genomic regions, as a significant proportion of regions are filtered out of the alignment due to lack of sequence complementarity, caused by an increased alignment ambiguity. Once bisulfite sequence reads are aligned to the reference genome, the methylation levels of specific genomic regions can be estimated by using variant caller algorithms, which allow the quantitation of the frequency of Cs and Ts. For instance, Bis-SNP [112] relies on a Bayesian inference approach to evaluate strand-specific base calls and base call quality scores, and experiment-specific bisulfite conversion efficiency to derive fairly accurate DNA methylation estimates. Faster variant callers have been developed, including MethylExtract [113] that implements a modified version of the VarScan algorithm [114], and BS-SNPer [115] based on a "dynamic matrix algorithm" and Bayesian modeling, which are able to process large quantities of genomic sequences.

The most widely used bisulfite microarrays are Illumina® Infinium Methylation Assay [100], which allows single-CpG-site resolution quantitative measurement of genome-wide methylation profiles. In this assay, cytosine methylation at CpG islands is detected by multiplexed genotyping of bisulfite-converted genomic DNA, upon treatment with bisulfite (this technique also relies on bisulfite selective DNA modification of unmethylated regions, as described above). The assay uses two site-specific probes, one for methylated and another for the unmethylated *loci*. The Infinium MethylationEPIC BeadChip Kit enables quantitative genome-wide profiling of almost 900,000 methylation sites at the single-nucleotide resolution, encompassing expert-selected coverage of up to 99% of RefSeq genes, 95% of CpG islands, and ENCODE enhancer regions. In addition to the great potential of this technology, it has been the focus of intense research for the development of proprietaries and open-source bioinformatics tools for processing Illumina Methylation Arrays. The GenomeStudio software developed by the chip supplier enables differential methylation analysis for small-scale studies, also including advanced tools for visualization of large amounts of data, plotting, and statistical analysis. The R/Bioconductor BeadArray toolkit [116] is also available for performing large-scale stand-alone analysis requiring more intense calculations or parallel computing infrastructures. Infinium® arrays include multiple probes for performing sample-dependent and sample-independent data quality control, which is the input of packages like IMA [117] and LumiWCluster [118]. These tools use different approaches for removing noisy probes from the chip data, which are straightforwardly filtered out based on the median detection p-value cutoff in the case of IMA, while LumiWCluster relies on a more sophisticated weighted likelihood model based on clustering methylation data. Background correction should also be performed for removing nonspecific signals and differences between replicates. This step can be performed with the GenomeStudio Infinium integrated package, but also with many other toolkits, such as lumi [119], limma [120], and BeadArray [116]. After the initial quality control, microarray data need to be normalized to remove random noise, technical artifacts, and measurement variation inherent to microarrays. Normalization should be performed between different replicate array measurements, i.e., between array, and internally for each array, i.e., within array. This can be accomplished with HumMethQCReport [121] and lumi [119], which use spline and weighted scatter smoothing for normalizing methylation data, but there are also many other alternative approaches based on different statistical approaches [122]. Special interest should also be put on scaling the signal obtained for the two different probes used in this technique – i.e., probes for methylated and unmethylated *loci* – that produce rather different signal distributions, due to the bias towards CpG islands in the genome [100]. Peak rescaling is usually performed with methods such as SWAN [123] that implements a sub-quantile within-array normalization (SQN) procedure, similar to the rationale followed in another study implementing a pipeline for processing Illumina® Infinium Methylation BeadChip [124]. Other approaches use variations of this procedure, such as the mixture quantile normalization method to rescale the distributions of the methylation and unmethylation probes into distributions that can be compared statistically [125, 126]. Batch effects, which are also common on DNA methylation arrays, can be corrected with toolkits like CpGassoc [127], MethLAB [128], and ISVA [129] R/Bioconductor packages.

Enrichment techniques, such as MeDIP-seq and MethylCap-seq [102–104], are based on the use of proteins that specifically bind to methylated DNA regions – e.g., 5-methylcytosine-specific antibodies [104, 130] (methylated DNA immunoprecipitation (MeDIP)) or methyl-binding domain proteins [131, 132] (MethylCap) – to enrich hypermethylated fragments that are subject to HTP or microarray sequencing. The bioinformatics processing of methylation data generated with these approaches can be performed with the same methods describe above for processing sequencing or microarray platforms. Moreover, there are some methods exclusively tailored for enrichment data, like MEDIPS [133], an R/Bioconductor suite that enables processing multiple replicates and performing a great variety of statistical analyses. Another toolkit, coined as Batman [102], which stands for "Bayesian tool for methylation analysis" relies on the knowledge that almost all DNA methylation in mammals occurs at CpG dinucleotides and uses a standard Bayesian inference approach to estimate the posterior distribution of the methylation state parameters from data to generate quantitative methylation profiles. A very interesting study built on a thorough comparison of more than 20 different software tools has resulted in the development of RnBeads [134], an integrative suite that supports all genome-scale and genome-wide DNA methylation assays, implemented to facilitate standalone running of complex pipelines in high-performance computing infrastructures. With this toolkit, it is possible to perform all the steps of DNA methylation data analysis, ranging from data visualization, quality control, handling batch effects, correction for tissue heterogeneity, and differential DNA methylation analysis.

25.4 Computational Analysis of Chromatin Accessibility Data

The chromatin accessibility of genomic regions can be profiled with methodologies such as DNase-seq [135], FAIRE-seq [136], and ATAC-seq [137], which rely on different experimental principles and produce rather different data outputs. DNase-seq and ATAC-seq are based on the use of endonucleases – i.e., DNase I and engineered Tn5 transposase, respectively – to fragment DNA, while FAIRE-seq is a physical fragmentation method, in which DNA is treated with formaldehyde to cross-link chromatin. The differences between DNA fragmentation procedures used in each technique – i.e., DNase I and engineered Tn5 transposase have a tendency to cleave some DNA sequences more efficiently than others, and sonication could produce under and over sonicated chromatin depending on the sonication parameters used – cause that each technique generates rather different accessibility profiles [138]. In accordance, these differences should be taken into consideration while performing the downstream bioinformatics processing of sequencing data. Chromatin accessibility peaks are generally different from peak signals generated with histone modification ChIP-seq experiments, which are in general broad sequence read peaks. Hence, peak callers designed for ChIP-seq need some fine-tuning for processing chromatin accessibility data [138, 139]. Furthermore,

ChIP-seq data usually shows a higher signal-to-noise ratio compared to DNase-seq, making ChIP-seq peaks easier to detect [140]. Different peak callers have been developed to process accessibility data, including F-Seq [74] toolkit, which can be used for ChIP-seq and FAIRE-seq data [141], and ZINBA [142], which relies on a mixture regression approach for probabilistically identifying real and artifact peaks and can also handle ChIP-seq and FAIRE-seq data. Moreover, the Hotspot program [143] has been developed as part of the ENCODE project specifically for analyzing DNase-seq data, and follows a similar rationale to ChIP-seq sliding-window peak callers described above, using a probabilistic model to classify peaks by assessing the differences between the sample and a background distribution. MACS [73], which is commonly used for ChIP-seq data, and ChIPOTle [56], suited for processing ChIP-on-chip data as described above, have also been used for DNase-seq [144] and FAIRE-seq [136], respectively. In general, most of these tools have also been applied for ATAC-seq data analysis, but there are some other tools specifically implemented for this novel technique, such as I-ATAC (https://www.jax.org/research-and-faculty/tools/i-atac). This tool integrates multiple methods for quality check, preprocessing, and running sequential, multiple-parallel, and customized data analysis pipelines into a cross platform and open-source desktop application. Interestingly, the selection of the peak caller of use could play a key role in peak assignment output, as a comparison of the most common tools for processing accessibility data has shown that there is little overlap among called peaks obtained for the same chromatin accessibility dataset [140].

25.5 Epigenomic Databases and Epigenome Mapping Initiatives

The great developments of high-throughput sequencing technologies have allowed the steady generation of great quantities of epigenomic data in different cell types/lines and multiple organisms. This has been boosted by many large-scale epigenome mapping projects, such as the ENCODE project [145], the NIH Roadmap Epigenomics [146], the International Human Epigenome Consortium (http://ihec-epigenomes.org/), and the HEROIC European project (http://cordis.europa.eu/project/rcn/78439_en.html), among others. Other resources, such as the MethBase database (http://smithlabresearch.org/software/methbase/) [147], encompassing hundreds of methylomes from different organisms allow comparing the methylation profiles of genomics regions in different animal and plant genomes. There exist other more specialized epigenomic projects and databases encompassing information of the brain. These neuroepigenomic resources include MethylomeDB database (http://www.neuroepigenomics.org/methylomedb) [148] that includes genome-wide DNA methylation profiles of human and mouse brain and is integrated with a genome browser which allows surfing through the genome and analyzes the methylation of specific *loci*, searches for specific methylation profiles, and compares methylation patterns between individual samples. The Brain Cloud (http://braincloud.jhmi.edu/) [149] compiles methylation data from human postmortem dorsolateral prefrontal cortices from normal subjects across the life span, also integrating single-nucleotide polymorphism data. The great amount of data generated in these projects has prompted the development of a great variety of computational tools for the analysis of epigenetic data, some of which have been described in detail in previous sections of this chapter. Moreover, the wealth of data in these databases has enabled groundbreaking studies, such as one recent report [38] encompassing a thorough integrative study of different epigenetic phenomena – e.g., chromatin accessibility, DNA methylation, chromatin marks, gene expression – in different reference epigenomes. In this study, the authors profile cells from different tissues and organs in more than 100 adult and fetal epigenomes and were able to identify epigenetic differences arising during lineage specification and cellular differentiation, which are the modules of regulatory regions with coordinated activity across cell types, and the role of regulatory regions in human disease associated with common traits and disorders [38]. This study shows that genomic regions vary greatly in their association with active marks, with approximately 5% of each epigenome marked by enhancer or promoter signatures, showing increased association with expressed genes and increased evolutionary conservation, while two-thirds of each reference epigenome are quiescent and enriched in gene-poor stably repressed regions [38]. Furthermore, the authors find that genetic variants associated with complex traits are highly enriched in epigenomic annotations of trait-relevant tissues, and genome-wide association enrichments are significantly strongest for enhancer-associated marks, consistent with their high tissue-specific nature [38]. However, promoter-associated and transcription-associated marks were also enriched, implicating several gene-regulatory levels as underlying genetic variants associated with complex traits [38].

25.6 Epigenetic Differential Analysis and Integration of Epigenomic and Gene Expression Data

Despite the great wealth of epigenomic data, we still lack a systematic understanding of how the epigenomic landscape regulates gene expression and which are the epigenetic signatures that control the most important regulatory circuitry in the transcriptional level. Differential analysis of ChIP-seq genome-wide profiles obtained for different cellular phenotypes is a rather challenging problem, due to the significant heterogeneity in peak calling between different measurements and the lack of overlap between peak assignments obtained with different peak callers [140]. The diffReps program [150] has been designed to detect differential sites from ChIP-seq data, with or without biological replicates, and implements a sliding-window approach to estimate the statistical significance of differential peaks based on a binomial distribution model across samples. The differential histone modification profiles generated with diffReps can be used to try to superimpose the epigenetic differential profile with gene expression data. The GeneOverlap R/Bioconductor tool implements different statistical models for estimating the significance of the overlap of histone modification and gene expression profiles. However, the great complexity of the histone code, and the cross talk established between different histone marks to cooperatively regulate gene expression, makes it difficult to capture the regulatory epigenetic mechanisms just by superimposing histone modification and gene expression data. More complex computational models for predicting gene expression from complex histone modification profiles have been proposed [151, 152]. In order to reproduce the quantitative relationship between gene expression levels and histone modifications, these approaches combine information from many different data tracks of repressive and activating chromatin modifications, which are processed with machine learning approaches and were able to explain a fairly high proportion of the gene expression profiles in different organisms [151, 152]. In more complex expression datasets, such as brain tissues, similar approaches for combining histone modification data [153] have not been able to obtain a good correlation with the observed gene expression profiles, which could be related to the great complexity of gene regulation in these heterogeneous tissues, and the regulatory role of other histone marks not included in the study.

The prediction of epigenetic states has also been the focus of intense research. Several computational approaches have been devised for predicting promoter regions (extensively reviewed in [154]), prediction of CpG islands [155, 156], DNA methylation [157, 158], and nucleosome positioning [159, 160]. However, with the advent of next-generation sequencing (NGS), which is used in combination with techniques for profiling chromatin accessibility, histone modifications, and DNA methylation that have allowed the generation of huge quantities of genome-wide epigenetic data, the prediction of epigenetic states has lost relevance. Nevertheless, a different group of approaches has been developed for leveraging from genome annotation data at the epigenetic level for predicting the chromatin states – e.g., poised or strong enhancers, active promoters, and heterochromatin, among others – from histone modification data [161, 162]. ChromHMM [161] relies on a multivariate hidden Markov model that represents the observed combination of chromatin marks as the product of independent Bernoulli random variables for segmenting the genome into regions with different chromatin states. Segway [162] can also input histone modification data, but also DNA methylation and chromatin accessibility data, and implements a Dynamic Bayesian Network model for hierarchical genome segmentation. Interestingly, ChromHMM and Segway can be used to process fairly complex datasets of experimental data and perform chromatin state assignments, which have provided key insights in transversal epigenomic studies in different cell types, tissues, or human populations [38, 163, 164].

25.7 Systems Biology Approaches and Reconstruction of Multilevel Regulatory Networks

The availability of highly detailed annotation of human and mouse genomes [38, 145, 146] has paved the way for performing studies for integrating multilevel biological data, encompassing epigenetics, DNA sequence variation, gene expression, and clinical data. The regulatory events triggering phenotypic transitions such as cellular differentiation, and the dysfunctions associated to disease onset and progression are usually mediated by multiple genes, which establish complex interaction networks. Thus, in order to gain understanding of the regulatory mechanisms at the epigenetic and transcriptional levels involved in the regulation of these cellular phenotypes, it is necessary to derive more comprehensive systems-level computational models. For such large-scale molecular datasets, several network approaches have been developed to identify and dissect the underlying "interactomes" for discovering key mechanisms and causal regulators in normal or pathological biological systems [165]. Gene regulatory Boolean network models have been very useful for conducting systems-level modeling of complex high-throughput biological data enabling the construction of complex interaction networks for studying disease mechanisms [166]. Disease network models have been essential for predicting disease-related genes based on the analysis of different topological characteristics, such as node connectivity [167, 168], gene-gene interaction tendency in specific tissues [169], or network neighbors of disease-related genes [170, 171]. A different group of approaches tries to model cellular phenotypes as attractors in the gene expression landscape, and phenotypic transitions are modeled by identifying nodes destabilizing these attractors [172–174], and disease perturbations, such as chemical compounds or mutations, can cause a switch from a healthy to a disease attractor state [175–177]. Co-expression-based network inference approaches [178, 179] have also been used to build regulatory network models from HTS data. Weighted gene co-expression models (WGCNA) [180] – i.e., there exists a widely used and very efficient R/Bioconductor package to build WGCNA network models [181] – which allow embodying important information of the underlying relationships and interactions among genes have been widely used to identify disease-causing genes in multigene human pathologies, such as autism [182–184] and Alzheimer's disease [185, 186]. These WGCNA formalisms allow the generation of fairly complex network representations – e.g., eigengene networks [187, 188], in which the nodes are composite network modules. WGCNA models have enabled the identification of an age-related co-methylation module present in multiple human tissues, including the blood and brain from the analysis of up to 2442 Illumina DNA methylation arrays [189]. Similarly, these approaches have been used to identify common methylation patterns correlated with age in identical twins [190], the identification of the upstream epigenetic control and the downstream cellular physiology associated with alcohol dependence and neuroadaptive changes in alcoholic brain [191] and the prediction of the co-methylation modules associated with the Huntington's disease pathogenesis [192]. The developments of the abovementioned integrative and other multiscale network modeling approaches for trying to integrate complex and multidimensional biological data to infer regulatory relationships linking different regulatory levels – e.g., DNA sequence variations, epigenetic, transcriptional, and metabolic – will be key for gaining a deeper understanding of disease onset and progression, or other important biological processes, such as development.

25.8 The Advent of the Single-Cell Era in Neuroepigenetics: Challenges for Analyzing Single-Cell Epigenomic Data

The great technological advances in the methodologies for generating high-quality genome-wide epigenomic data have caused a revolution in the study of the epigenetic mechanisms regulating gene expression, stem cell differentiation, disease onset and progression, and other key biological phenomena. These developments have also contributed to the emergence of the field of "neuroepigenetics," aimed at studying the epigenetic regulatory mechanisms in cells from the central nervous system. It has been shown that in neurons, which live throughout most of the life span of an animal, epigenetic mechanisms play a key role in the regulation of the complex metabolic and gene expression these cells must go through upon synaptic input or interactions with other nervous system cells [193, 194]. One of the main problems for studying cells from the mammalian nervous systems is trying to disentangle the great cellular heterogeneity of bran tissues [195–197]. In this regard, most of the neuroepigenomic studies conducted so far have been performed with the traditional techniques for profiling chromatin accessibility, histone modifications, and DNA methylation described in this and other chapters of this book. These approaches require as input samples containing hundreds of thousands or millions of cells, encompassing highly heterogeneous cell populations. In recent years, different experimental techniques have been developed for studying heterogeneous cell populations. Gene expression single-cell transcriptional profiling techniques first developed 20 years ago [198] have become a very popular technique conventionally used in most laboratories, thanks to great technological developments in cell capture and next-generation sequencing approaches. The application of single-cell gene transcriptomics techniques has been central in the study of gene expression and functional diversity in somatosensory neurons from the dorsal root ganglia [199, 200], in different cortical regions [197, 201, 202], and developing retina [203].

Different single-cell epigenomic approaches have been recently developed for high-throughput genome-wide mapping of DNA methylation, histone modifications, and chromatin accessibility. The single-cell reduced-representation bisulfite sequencing (scRRBS) technique [204] is highly sensitive and can detect the methylation status of up to 1.5 million CpG sites within the genome of an individual cell. This technique is very efficient for profiling promoter regions, though it has poor coverage in enhancer regions. Bisulfite single-cell sequencing approaches enable genome-wide profiling of single cells or very small cell populations, although with a rather low sequencing coverage [205, 206]. Histone modification single-cell profiling can be measured with different barcoding approaches, taking advantage of techniques for indexing regions bearing the posttranslational modification in individual cells with specific sequence tags, and then performing ChIP-seq measurement after pooling cells from different wells – i.e., the heterogeneous population – which reduces the problem associated to input sample requirement of ChIP-seq [207, 208]. A different technique has been developed (the nano-ChIP-seq protocol) [209], which combines a high-sensitivity small-scale ChIP assay tailored for HTS libraries from scarce amounts of ChIP DNA. Recently, the single-tube

DNA amplification method (LinDA) has been conceived, enabling ChIP-seq measurements of picogram DNA amounts obtained from a few thousand cells [210]. Chromatin accessibility single-cell profiling can be performed with a modification of the ATAC-seq approach, based on combinatorial indexing for barcoding populations of nuclei in different wells, and then performing chromatin accessibility after pooling [211]. There exists another methodology available for single-cell chromatin accessibility profiling, based on a programmable microfluidics platform for capturing and analyzing cells in specific microfluidic chambers [212]. These methodologies are still under development for improving single-cell isolation [203, 213] and single-molecule sequencing techniques [214, 215], to try to increase the reliability of the measurements and sequencing coverage. The application of these approaches to study central nervous system samples will be essential for obtaining a clearer picture of the epigenetic regulatory mechanisms in neurons from different brain regions and how the heterogeneity at the epigenetic level defines different circuitries at the transcriptional regulatory level in central nervous system cells. However, the computational analysis of single-cell epigenomic data poses many computational challenges that will be the focus of intense research in the next years to match the great developments of experimental techniques. Currently, the computational tools and approaches used for processing single-cell epigenomic data are essentially those developed for bulk measurements, which have been thoroughly discussed in this chapter. Nevertheless, it is crucial to develop computational methods that are tailored specifically for processing single-cell data for tackling the problems associated with normalization and cell-type identification and for dissecting variability levels across cells [216]. It is expected that such methods will be developed in the next few years, leading to new discoveries in areas ranging from the physiology of tissues to systems biology [216].

References

1. Gehring M, Reik W, Henikoff S. DNA demethylation by DNA repair. Trends Genet. 2009;25(2):82–90.
2. Loizou JI, Murr R, Finkbeiner MG, Sawan C, Wang ZQ, Herceg Z. Epigenetic information in chromatin: the code of entry for DNA repair. Cell Cycle. 2006;5(7):696–701.
3. Probst AV, Dunleavy E, Almouzni G. Epigenetic inheritance during the cell cycle. Nat Rev Mol Cell Biol. 2009;10(3):192–206.
4. Richards EJ. Inherited epigenetic variation--revisiting soft inheritance. Nat Rev Genet. 2006;7(5):395–401.
5. Grewal SI, Moazed D. Heterochromatin and epigenetic control of gene expression. Science. 2003;301(5634):798–802.
6. Jaenisch R, Bird A. Epigenetic regulation of gene expression: how the genome integrates intrinsic and environmental signals. Nat Genet. 2003;33(Suppl):245–54.
7. Heard E, Martienssen RA. Transgenerational epigenetic inheritance: myths and mechanisms. Cell. 2014;157(1):95–109.
8. Smith ZD, Chan MM, Mikkelsen TS, Gu H, Gnirke A, Regev A, et al. A unique regulatory phase of DNA methylation in the early mammalian embryo. Nature. 2012;484(7394):339–44.
9. Kouzarides T. Chromatin modifications and their function. Cell. 2007;128(4):693–705.
10. Bird A. DNA methylation patterns and epigenetic memory. Genes Dev. 2002;16(1):6–21.

11. Jones PA. Functions of DNA methylation: islands, start sites, gene bodies and beyond. Nat Rev Genet. 2012;13(7):484–92.
12. Smith ZD, Meissner A. DNA methylation: roles in mammalian development. Nat Rev Genet. 2013;14(3):204–20.
13. Strahl BD, Allis CD. The language of covalent histone modifications. Nature. 2000;403(6765): 41–5.
14. Tessarz P, Kouzarides T. Histone core modifications regulating nucleosome structure and dynamics. Nat Rev Mol Cell Biol. 2014;15(11):703–8.
15. Rando OJ. Combinatorial complexity in chromatin structure and function: revisiting the histone code. Curr Opin Genet Dev. 2012;22(2):148–55.
16. Du J, Johnson LM, Jacobsen SE, Patel DJ. DNA methylation pathways and their crosstalk with histone methylation. Nat Rev Mol Cell Biol. 2015;16(9):519–32.
17. Guo X, Wang L, Li J, Ding Z, Xiao J, Yin X, et al. Structural insight into autoinhibition and histone H3-induced activation of DNMT3A. Nature. 2015;517(7536):640–4.
18. Ooi SK, Qiu C, Bernstein E, Li K, Jia D, Yang Z, et al. DNMT3L connects unmethylated lysine 4 of histone H3 to de novo methylation of DNA. Nature. 2007;448(7154):714–7.
19. Arrowsmith CH, Bountra C, Fish PV, Lee K, Schapira M. Epigenetic protein families: a new frontier for drug discovery. Nat Rev Drug Discov. 2012;11(5):384–400.
20. Musselman CA, Lalonde ME, Cote J, Kutateladze TG. Perceiving the epigenetic landscape through histone readers. Nat Struct Mol Biol. 2012;19(12):1218–27.
21. Torres IO, Fujimori DG. Functional coupling between writers, erasers and readers of histone and DNA methylation. Curr Opin Struct Biol. 2015;35:68–75.
22. Bannister AJ, Kouzarides T. Regulation of chromatin by histone modifications. Cell Res. 2011;21(3):381–95.
23. Handel AE, Ebers GC, Ramagopalan SV. Epigenetics: molecular mechanisms and implications for disease. Trends Mol Med. 2010;16(1):7–16.
24. Ordovas JM, Smith CE. Epigenetics and cardiovascular disease. Nat Rev Cardiol. 2010;7(9): 510–9.
25. Portela A, Esteller M. Epigenetic modifications and human disease. Nat Biotechnol. 2010; 28(10):1057–68.
26. Feinberg AP. Phenotypic plasticity and the epigenetics of human disease. Nature. 2007; 447(7143):433–40.
27. Haas J, Frese KS, Park YJ, Keller A, Vogel B, Lindroth AM, et al. Alterations in cardiac DNA methylation in human dilated cardiomyopathy. EMBO Mol Med. 2013;5(3):413–29.
28. Pujadas E, Feinberg AP. Regulated noise in the epigenetic landscape of development and disease. Cell. 2012;148(6):1123–31.
29. Feinberg AP, Ohlsson R, Henikoff S. The epigenetic progenitor origin of human cancer. Nat Rev Genet. 2006;7(1):21–33.
30. Ahuja N, Easwaran H, Baylin SB. Harnessing the potential of epigenetic therapy to target solid tumors. J Clin Invest. 2014;124(1):56–63.
31. Baylin SB, Jones PA. A decade of exploring the cancer epigenome - biological and translational implications. Nat Rev Cancer. 2011;11(10):726–34.
32. Jones PA, Baylin SB. The epigenomics of cancer. Cell. 2007;128(4):683–92.
33. Jakovcevski M, Akbarian S. Epigenetic mechanisms in neurological disease. Nat Med. 2012;18(8):1194–204.
34. Day JJ, Sweatt JD. Epigenetic mechanisms in cognition. Neuron. 2011;70(5):813–29.
35. Dulac C. Brain function and chromatin plasticity. Nature. 2010;465(7299):728–35.
36. Jensen LR, Amende M, Gurok U, Moser B, Gimmel V, Tzschach A, et al. Mutations in the JARID1C gene, which is involved in transcriptional regulation and chromatin remodeling, cause X-linked mental retardation. Am J Hum Genet. 2005;76(2):227–36.
37. Miller CA, Gavin CF, White JA, Parrish RR, Honasoge A, Yancey CR, et al. Cortical DNA methylation maintains remote memory. Nat Neurosci. 2010;13(6):664.
38. Roadmap Epigenomics C, Kundaje A, Meuleman W, Ernst J, Bilenky M, Yen A, et al. Integrative analysis of 111 reference human epigenomes. Nature. 2015;518(7539):317–30.

39. Barski A, Cuddapah S, Cui K, Roh TY, Schones DE, Wang Z, et al. High-resolution profiling of histone methylations in the human genome. Cell. 2007;129(4):823–37.
40. Buck MJ, Lieb JD. ChIP-chip: considerations for the design, analysis, and application of genome-wide chromatin immunoprecipitation experiments. Genomics. 2004;83(3):349–60.
41. Huebert DJ, Kamal M, O'Donovan A, Bernstein BE. Genome-wide analysis of histone modifications by ChIP-on-chip. Methods. 2006;40(4):365–9.
42. Furey TS. ChIP-seq and beyond: new and improved methodologies to detect and characterize protein-DNA interactions. Nat Rev Genet. 2012;13(12):840–52.
43. Nakato R, Shirahige K. Recent advances in ChIP-seq analysis: from quality management to whole-genome annotation. Briefings Bioinform. 2016.
44. Park PJ. ChIP-seq: advantages and challenges of a maturing technology. Nat Rev Genet. 2009;10(10):669–80.
45. Bartke T, Borgel J, DiMaggio PA. Proteomics in epigenetics: new perspectives for cancer research. Brief Funct Genomics. 2013;12(3):205–18.
46. Garcia BA, Shabanowitz J, Hunt DF. Characterization of histones and their post-translational modifications by mass spectrometry. Curr Opin Chem Biol. 2007;11(1):66–73.
47. Moradian A, Kalli A, Sweredoski MJ, Hess S. The top-down, middle-down, and bottom-up mass spectrometry approaches for characterization of histone variants and their post-translational modifications. Proteomics. 2014;14(4–5):489–97.
48. Tian Z, Tolic N, Zhao R, Moore RJ, Hengel SM, Robinson EW, et al. Enhanced top-down characterization of histone post-translational modifications. Genome Biol. 2012;13(10):R86.
49. Olsen JV, Mann M. Status of large-scale analysis of post-translational modifications by mass spectrometry. Mol Cell Proteomics. 2013;12(12):3444–52.
50. Bock C, Lengauer T. Computational epigenetics. Bioinformatics. 2008;24(1):1–10.
51. Zhang ZD, Rozowsky J, Lam HY, Du J, Snyder M, Gerstein M. Tilescope: online analysis pipeline for high-density tiling microarray data. Genome Biol. 2007;8(5):R81.
52. Ji H, Wong WH. TileMap: create chromosomal map of tiling array hybridizations. Bioinformatics. 2005;21(18):3629–36.
53. Toedling J, Skylar O, Krueger T, Fischer JJ, Sperling S, Huber W. Ringo--an R/Bioconductor package for analyzing ChIP-chip readouts. BMC Bioinform. 2007;8:221.
54. Scacheri PC, Crawford GE, Davis S. Statistics for ChIP-chip and DNase hypersensitivity experiments on NimbleGen arrays. Methods Enzymol. 2006;411:270–82.
55. Keles S. Mixture modeling for genome-wide localization of transcription factors. Biometrics. 2007;63(1):10–21.
56. Buck MJ, Nobel AB, Lieb JD. ChIPOTle: a user-friendly tool for the analysis of ChIP-chip data. Genome Biol. 2005;6(11):R97.
57. Li W, Meyer CA, Liu XS. A hidden Markov model for analyzing ChIP-chip experiments on genome tiling arrays and its application to p53 binding sequences. Bioinformatics. 2005;21(Suppl 1):i274–82.
58. Johnson WE, Li W, Meyer CA, Gottardo R, Carroll JS, Brown M, et al. Model-based analysis of tiling-arrays for ChIP-chip. Proc Natl Acad Sci U S A. 2006;103(33):12457–62.
59. Langmead B, Trapnell C, Pop M, Salzberg SL. Ultrafast and memory-efficient alignment of short DNA sequences to the human genome. Genome Biol. 2009;10(3):R25.
60. Li H, Durbin R. Fast and accurate short read alignment with Burrows-Wheeler transform. Bioinformatics. 2009;25(14):1754–60.
61. Li R, Yu C, Li Y, Lam TW, Yiu SM, Kristiansen K, et al. SOAP2: an improved ultrafast tool for short read alignment. Bioinformatics. 2009;25(15):1966–7.
62. Kent WJ. BLAT--the BLAST-like alignment tool. Genome Res. 2002;12(4):656–64.
63. Li H, Homer N. A survey of sequence alignment algorithms for next-generation sequencing. Brief Bioinform. 2010;11(5):473–83.
64. Mikkelsen TS, Ku M, Jaffe DB, Issac B, Lieberman E, Giannoukos G, et al. Genome-wide maps of chromatin state in pluripotent and lineage-committed cells. Nature. 2007;448(7153):553–60.
65. Li H, Handsaker B, Wysoker A, Fennell T, Ruan J, Homer N, et al. The Sequence Alignment/Map format and SAMtools. Bioinformatics. 2009;25(16):2078–9.

66. Landt SG, Marinov GK, Kundaje A, Kheradpour P, Pauli F, Batzoglou S, et al. ChIP-seq guidelines and practices of the ENCODE and modENCODE consortia. Genome Res. 2012;22(9):1813–31.
67. Hansen P, Hecht J, Ibrahim DM, Krannich A, Truss M, Robinson PN. Saturation analysis of ChIP-seq data for reproducible identification of binding peaks. Genome Res. 2015;25(9):1391–400.
68. Pepke S, Wold B, Mortazavi A. Computation for ChIP-seq and RNA-seq studies. Nat Methods. 2009;6(11 Suppl):S22–32.
69. Ji H, Jiang H, Ma W, Johnson DS, Myers RM, Wong WH. An integrated software system for analyzing ChIP-chip and ChIP-seq data. Nat Biotechnol. 2008;26(11):1293–300.
70. Jothi R, Cuddapah S, Barski A, Cui K, Zhao K. Genome-wide identification of in vivo protein-DNA binding sites from ChIP-Seq data. Nucleic Acids Res. 2008;36(16):5221–31.
71. Bardet AF, Steinmann J, Bafna S, Knoblich JA, Zeitlinger J, Stark A. Identification of transcription factor binding sites from ChIP-seq data at high resolution. Bioinformatics. 2013;29(21):2705–13.
72. Kharchenko PV, Tolstorukov MY, Park PJ. Design and analysis of ChIP-seq experiments for DNA-binding proteins. Nat Biotechnol. 2008;26(12):1351–9.
73. Zhang Y, Liu T, Meyer CA, Eeckhoute J, Johnson DS, Bernstein BE, et al. Model-based analysis of ChIP-Seq (MACS). Genome Biol. 2008;9(9):R137.
74. Boyle AP, Guinney J, Crawford GE, Furey TS. F-Seq: a feature density estimator for high-throughput sequence tags. Bioinformatics. 2008;24(21):2537–8.
75. Valouev A, Johnson DS, Sundquist A, Medina C, Anton E, Batzoglou S, et al. Genome-wide analysis of transcription factor binding sites based on ChIP-Seq data. Nat Methods. 2008;5(9):829–34.
76. Zhang X, Robertson G, Krzywinski M, Ning K, Droit A, Jones S, et al. PICS: probabilistic inference for ChIP-seq. Biometrics. 2011;67(1):151–63.
77. Heinz S, Benner C, Spann N, Bertolino E, Lin YC, Laslo P, et al. Simple combinations of lineage-determining transcription factors prime cis-regulatory elements required for macrophage and B cell identities. Mol Cell. 2010;38(4):576–89.
78. Ibrahim MM, Lacadie SA, Ohler U. JAMM: a peak finder for joint analysis of NGS replicates. Bioinformatics. 2015;31(1):48–55.
79. Zhang Y, Lin YH, Johnson TD, Rozek LS, Sartor MA. PePr: a peak-calling prioritization pipeline to identify consistent or differential peaks from replicated ChIP-Seq data. Bioinformatics. 2014;30(18):2568–75.
80. Tuteja G, White P, Schug J, Kaestner KH. Extracting transcription factor targets from ChIP-Seq data. Nucleic Acids Res. 2009;37(17):e113.
81. Rozowsky J, Euskirchen G, Auerbach RK, Zhang ZD, Gibson T, Bjornson R, et al. PeakSeq enables systematic scoring of ChIP-seq experiments relative to controls. Nat Biotechnol. 2009;27(1):66–75.
82. Flensburg C, Kinkel SA, Keniry A, Blewitt ME, Oshlack A. A comparison of control samples for ChIP-seq of histone modifications. Frontiers in genetics. 2014;5:329.
83. Laajala TD, Raghav S, Tuomela S, Lahesmaa R, Aittokallio T, Elo LL. A practical comparison of methods for detecting transcription factor binding sites in ChIP-seq experiments. BMC Genomics. 2009;10:618.
84. Malone BM, Tan F, Bridges SM, Peng Z. Comparison of four ChIP-Seq analytical algorithms using rice endosperm H3K27 trimethylation profiling data. PLoS One. 2011;6(9):e25260.
85. Horn DM, Zubarev RA, McLafferty FW. Automated reduction and interpretation of high resolution electrospray mass spectra of large molecules. J Am Soc Mass Spectrom. 2000;11(4):320–32.
86. Liu X, Inbar Y, Dorrestein PC, Wynne C, Edwards N, Souda P, et al. Deconvolution and database search of complex tandem mass spectra of intact proteins: a combinatorial approach. Mol Cell Proteomics. 2010;9(12):2772–82.
87. Carvalho PC, Xu T, Han X, Cociorva D, Barbosa VC, Yates 3rd. JR. YADA: a tool for taking the most out of high-resolution spectra. Bioinformatics. 2009;25(20):2734–6.

88. Huttenhain R, Hess S. A combined top-down and bottom-up MS approach for the characterization of hemoglobin variants in Rhesus monkeys. Proteomics. 2010;10(20): 3657–68.
89. Sidoli S, Cheng L, Jensen ON. Proteomics in chromatin biology and epigenetics: Elucidation of post-translational modifications of histone proteins by mass spectrometry. J Proteomics. 2012;75(12):3419–33.
90. Zamdborg L, LeDuc RD, Glowacz KJ, Kim YB, Viswanathan V, Spaulding IT, et al. ProSight PTM 2.0: improved protein identification and characterization for top down mass spectrometry. Nucleic acids research. 2007;35(Web Server issue):W701–6.
91. Liu X, Sirotkin Y, Shen Y, Anderson G, Tsai YS, Ting YS, et al. Protein identification using top-down. Mol Cell Proteom. 2012;11(6):M111 008524.
92. Kalli A, Sweredoski MJ, Hess S. Data-dependent middle-down nano-liquid chromatography-electron capture dissociation-tandem mass spectrometry: an application for the analysis of unfractionated histones. Anal Chem. 2013;85(7):3501–7.
93. DiMaggio Jr PA, Young NL, Baliban RC, Garcia BA, Floudas CA. A mixed integer linear optimization framework for the identification and quantification of targeted post-translational modifications of highly modified proteins using multiplexed electron transfer dissociation tandem mass spectrometry. Mol Cell Proteomics. 2009;8(11):2527–43.
94. Pesavento JJ, Mizzen CA, Kelleher NL. Quantitative analysis of modified proteins and their positional isomers by tandem mass spectrometry: human histone H4. Anal Chem. 2006; 78(13):4271–80.
95. Siuti N, Roth MJ, Mizzen CA, Kelleher NL, Pesavento JJ. Gene-specific characterization of human histone H2B by electron capture dissociation. J Proteome Res. 2006;5(2):233–9.
96. Guan S, Burlingame AL. Data processing algorithms for analysis of high resolution MSMS spectra of peptides with complex patterns of posttranslational modifications. Mol Cell Proteomics. 2010;9(5):804–10.
97. Matthiesen R, Trelle MB, Hojrup P, Bunkenborg J, Jensen ON. VEMS 3.0: algorithms and computational tools for tandem mass spectrometry based identification of post-translational modifications in proteins. J Proteome Res. 2005;4(6):2338–47.
98. Booth MJ, Branco MR, Ficz G, Oxley D, Krueger F, Reik W, et al. Quantitative sequencing of 5-methylcytosine and 5-hydroxymethylcytosine at single-base resolution. Science. 2012;336(6083):934–7.
99. Deng J, Shoemaker R, Xie B, Gore A, LeProust EM, Antosiewicz-Bourget J, et al. Targeted bisulfite sequencing reveals changes in DNA methylation associated with nuclear reprogramming. Nat Biotechnol. 2009;27(4):353–60.
100. Bibikova M, Barnes B, Tsan C, Ho V, Klotzle B, Le JM, et al. High density DNA methylation array with single CpG site resolution. Genomics. 2011;98(4):288–95.
101. Schumacher A, Kapranov P, Kaminsky Z, Flanagan J, Assadzadeh A, Yau P, et al. Microarray-based DNA methylation profiling: technology and applications. Nucleic Acids Res. 2006;34(2):528–42.
102. Down TA, Rakyan VK, Turner DJ, Flicek P, Li H, Kulesha E, et al. A Bayesian deconvolution strategy for immunoprecipitation-based DNA methylome analysis. Nat Biotechnol. 2008;26(7):779–85.
103. Serre D, Lee BH, Ting AH. MBD-isolated Genome Sequencing provides a high-throughput and comprehensive survey of DNA methylation in the human genome. Nucleic Acids Res. 2010;38(2):391–9.
104. Weber M, Davies JJ, Wittig D, Oakeley EJ, Haase M, Lam WL, et al. Chromosome-wide and promoter-specific analyses identify sites of differential DNA methylation in normal and transformed human cells. Nat Genet. 2005;37(8):853–62.
105. Xi Y, Bock C, Muller F, Sun D, Meissner A, Li W. RRBSMAP: a fast, accurate and user-friendly alignment tool for reduced representation bisulfite sequencing. Bioinformatics. 2012;28(3):430–2.
106. Smith AD, Chung WY, Hodges E, Kendall J, Hannon G, Hicks J, et al. Updates to the RMAP short-read mapping software. Bioinformatics. 2009;25(21):2841–2.

107. Wu TD, Nacu S. Fast and SNP-tolerant detection of complex variants and splicing in short reads. Bioinformatics. 2010;26(7):873–81.
108. Otto C, Stadler PF, Hoffmann S. Fast and sensitive mapping of bisulfite-treated sequencing data. Bioinformatics. 2012;28(13):1698–704.
109. Pedersen B, Hsieh TF, Ibarra C, Fischer RL. MethylCoder: software pipeline for bisulfite-treated sequences. Bioinformatics. 2011;27(17):2435–6.
110. Harris EY, Ponts N, Le Roch KG, Lonardi S. BRAT-BW: efficient and accurate mapping of bisulfite-treated reads. Bioinformatics. 2012;28(13):1795–6.
111. Krueger F, Andrews SR. Bismark: a flexible aligner and methylation caller for Bisulfite-Seq applications. Bioinformatics. 2011;27(11):1571–2.
112. Liu Y, Siegmund KD, Laird PW, Berman BP. Bis-SNP: combined DNA methylation and SNP calling for Bisulfite-seq data. Genome Biol. 2012;13(7):R61.
113. Barturen G, Rueda A, Oliver JL, Hackenberg M. MethylExtract: High-Quality methylation maps and SNV calling from whole genome bisulfite sequencing data. F1000Research. 2013;2:217.
114. Koboldt DC, Chen K, Wylie T, Larson DE, McLellan MD, Mardis ER, et al. VarScan: variant detection in massively parallel sequencing of individual and pooled samples. Bioinformatics. 2009;25(17):2283–5.
115. Gao S, Zou D, Mao L, Liu H, Song P, Chen Y, et al. BS-SNPer: SNP calling in bisulfite-seq data. Bioinformatics. 2015;31(24):4006–8.
116. Dunning MJ, Smith ML, Ritchie ME, Tavare S. beadarray: R classes and methods for Illumina bead-based data. Bioinformatics. 2007;23(16):2183–4.
117. Wang D, Yan L, Hu Q, Sucheston LE, Higgins MJ, Ambrosone CB, et al. IMA: an R package for high-throughput analysis of Illumina's 450 K Infinium methylation data. Bioinformatics. 2012;28(5):729–30.
118. Kuan PF, Wang S, Zhou X, Chu H. A statistical framework for Illumina DNA methylation arrays. Bioinformatics. 2010;26(22):2849–55.
119. Du P, Kibbe WA, Lin SM. lumi: a pipeline for processing Illumina microarray. Bioinformatics. 2008;24(13):1547–8.
120. Wettenhall JM, Smyth GK. limmaGUI: a graphical user interface for linear modeling of microarray data. Bioinformatics. 2004;20(18):3705–6.
121. Mancuso FM, Montfort M, Carreras A, Alibes A, Roma G. HumMeth27QCReport: an R package for quality control and primary analysis of Illumina Infinium methylation data. BMC Res Notes. 2011;4:546.
122. Marabita F, Almgren M, Lindholm ME, Ruhrmann S, Fagerstrom-Billai F, Jagodic M, et al. An evaluation of analysis pipelines for DNA methylation profiling using the Illumina HumanMethylation450 BeadChip platform. Epigenetics. 2013;8(3):333–46.
123. Maksimovic J, Gordon L, Oshlack A. SWAN: Subset-quantile within array normalization for illumina infinium HumanMethylation450 BeadChips. Genome Biol. 2012;13(6):R44.
124. Touleimat N, Tost J. Complete pipeline for Infinium((R)) Human Methylation 450 K BeadChip data processing using subset quantile normalization for accurate DNA methylation estimation. Epigenomics. 2012;4(3):325–41.
125. Pidsley R, CC YW, Volta M, Lunnon K, Mill J, Schalkwyk LC. A data-driven approach to preprocessing Illumina 450 K methylation array data. BMC Genomics 2013;14:293.
126. Teschendorff AE, Marabita F, Lechner M, Bartlett T, Tegner J, Gomez-Cabrero D, et al. A beta-mixture quantile normalization method for correcting probe design bias in Illumina Infinium 450 k DNA methylation data. Bioinformatics. 2013;29(2):189–96.
127. Barfield RT, Kilaru V, Smith AK, Conneely KN. CpGassoc: an R function for analysis of DNA methylation microarray data. Bioinformatics. 2012;28(9):1280–1.
128. Kilaru V, Barfield RT, Schroeder JW, Smith AK, Conneely KN. MethLAB: a graphical user interface package for the analysis of array-based DNA methylation data. Epigenetics. 2012;7(3):225–9.
129. Teschendorff AE, Zhuang J, Widschwendter M. Independent surrogate variable analysis to deconvolve confounding factors in large-scale microarray profiling studies. Bioinformatics. 2011;27(11):1496–505.

130. Zhang X, Yazaki J, Sundaresan A, Cokus S, Chan SW, Chen H, et al. Genome-wide high-resolution mapping and functional analysis of DNA methylation in arabidopsis. Cell. 2006;126(6):1189–201.
131. Cross SH, Charlton JA, Nan X, Bird AP. Purification of CpG islands using a methylated DNA binding column. Nat Genet. 1994;6(3):236–44.
132. Kangaspeska S, Stride B, Metivier R, Polycarpou-Schwarz M, Ibberson D, Carmouche RP, et al. Transient cyclical methylation of promoter DNA. Nature. 2008;452(7183):112–5.
133. Lienhard M, Grimm C, Morkel M, Herwig R, Chavez L. MEDIPS: genome-wide differential coverage analysis of sequencing data derived from DNA enrichment experiments. Bioinformatics. 2014;30(2):284–6.
134. Assenov Y, Muller F, Lutsik P, Walter J, Lengauer T, Bock C. Comprehensive analysis of DNA methylation data with RnBeads. Nat Methods. 2014;11(11):1138–40.
135. Song L, Crawford GE. DNase-seq: a high-resolution technique for mapping active gene regulatory elements across the genome from mammalian cells. Cold Spring Harbor protocols. 2010;2010(2):pdb prot5384.
136. Giresi PG, Kim J, McDaniell RM, Iyer VR, Lieb JD. FAIRE (Formaldehyde-Assisted Isolation of Regulatory Elements) isolates active regulatory elements from human chromatin. Genome Res. 2007;17(6):877–85.
137. Buenrostro JD, Giresi PG, Zaba LC, Chang HY, Greenleaf WJ. Transposition of native chromatin for fast and sensitive epigenomic profiling of open chromatin, DNA-binding proteins and nucleosome position. Nat Methods. 2013;10(12):1213–8.
138. Meyer CA, Liu XS. Identifying and mitigating bias in next-generation sequencing methods for chromatin biology. Nat Rev Genet. 2014;15(11):709–21.
139. Madrigal P, Krajewski P. Current bioinformatic approaches to identify DNase I hypersensitive sites and genomic footprints from DNase-seq data. Front Genet. 2012;3:230.
140. Koohy H, Down TA, Spivakov M, Hubbard T. A comparison of peak callers used for DNase-Seq data. PLoS One. 2014;9(5):e96303.
141. Gaulton KJ, Nammo T, Pasquali L, Simon JM, Giresi PG, Fogarty MP, et al. A map of open chromatin in human pancreatic islets. Nat Genet. 2010;42(3):255–9.
142. Rashid NU, Giresi PG, Ibrahim JG, Sun W, Lieb JD. ZINBA integrates local covariates with DNA-seq data to identify broad and narrow regions of enrichment, even within amplified genomic regions. Genome Biol. 2011;12(7):R67.
143. Thurman RE, Rynes E, Humbert R, Vierstra J, Maurano MT, Haugen E, et al. The accessible chromatin landscape of the human genome. Nature. 2012;489(7414):75–82.
144. Wang YM, Zhou P, Wang LY, Li ZH, Zhang YN, Zhang YX. Correlation between DNase I hypersensitive site distribution and gene expression in HeLa S3 cells. PLoS One. 2012;7(8):e42414.
145. Consortium EP. An integrated encyclopedia of DNA elements in the human genome. Nature. 2012;489(7414):57–74.
146. Bernstein BE, Stamatoyannopoulos JA, Costello JF, Ren B, Milosavljevic A, Meissner A, et al. The NIH Roadmap Epigenomics Mapping Consortium. Nat Biotechnol. 2010;28(10):1045–8.
147. Song Q, Decato B, Hong EE, Zhou M, Fang F, Qu J, et al. A reference methylome database and analysis pipeline to facilitate integrative and comparative epigenomics. PLoS One. 2013;8(12):e81148.
148. Xin Y, Chanrion B, O'Donnell AH, Milekic M, Costa R, Ge Y, et al. MethylomeDB: a database of DNA methylation profiles of the brain. Nucleic Acids Res. 2012;40(Database issue):D1245–9.
149. Colantuoni C, Lipska BK, Ye T, Hyde TM, Tao R, Leek JT, et al. Temporal dynamics and genetic control of transcription in the human prefrontal cortex. Nature. 2011;478(7370):519–23.
150. Shen L, Shao NY, Liu X, Maze I, Feng J, Nestler EJ. diffReps: detecting differential chromatin modification sites from ChIP-seq data with biological replicates. PloS one. 2013;8(6):e65598.
151. Cheng C, Yan KK, Yip KY, Rozowsky J, Alexander R, Shou C, et al. A statistical framework for modeling gene expression using chromatin features and application to modENCODE datasets. Genome Biol. 2011;12(2):R15.

152. Dong X, Greven MC, Kundaje A, Djebali S, Brown JB, Cheng C, et al. Modeling gene expression using chromatin features in various cellular contexts. Genome Biol. 2012; 13(9):R53.
153. Feng J, Wilkinson M, Liu X, Purushothaman I, Ferguson D, Vialou V, et al. Chronic cocaine-regulated epigenomic changes in mouse nucleus accumbens. Genome Biol. 2014;15(4):R65.
154. Bajic VB, Tan SL, Suzuki Y, Sugano S. Promoter prediction analysis on the whole human genome. Nat Biotechnol. 2004;22(11):1467–73.
155. Bock C, Walter J, Paulsen M, Lengauer T. CpG island mapping by epigenome prediction. PLoS Comput Biol. 2007;3(6):e110.
156. Feltus FA, Lee EK, Costello JF, Plass C, Vertino PM. Predicting aberrant CpG island methylation. Proc Natl Acad Sci U S A. 2003;100(21):12253–8.
157. Das R, Dimitrova N, Xuan Z, Rollins RA, Haghighi F, Edwards JR, et al. Computational prediction of methylation status in human genomic sequences. Proc Natl Acad Sci U S A. 2006;103(28):10713–6.
158. Fang F, Fan S, Zhang X, Zhang MQ. Predicting methylation status of CpG islands in the human brain. Bioinformatics. 2006;22(18):2204–9.
159. Peckham HE, Thurman RE, Fu Y, Stamatoyannopoulos JA, Noble WS, Struhl K, et al. Nucleosome positioning signals in genomic DNA. Genome Res. 2007;17(8):1170–7.
160. Segal E, Fondufe-Mittendorf Y, Chen L, Thastrom A, Field Y, Moore IK, et al. A genomic code for nucleosome positioning. Nature. 2006;442(7104):772–8.
161. Ernst J, Kellis M. ChromHMM: automating chromatin-state discovery and characterization. Nat Methods. 2012;9(3):215–6.
162. Hoffman MM, Buske OJ, Wang J, Weng Z, Bilmes JA, Noble WS. Unsupervised pattern discovery in human chromatin structure through genomic segmentation. Nat Methods. 2012;9(5):473–6.
163. Andersson R, Gebhard C, Miguel-Escalada I, Hoof I, Bornholdt J, Boyd M, et al. An atlas of active enhancers across human cell types and tissues. Nature. 2014;507(7493):455–61.
164. Kircher M, Witten DM, Jain P, O'Roak BJ, Cooper GM, Shendure J. A general framework for estimating the relative pathogenicity of human genetic variants. Nat Genet. 2014; 46(3):310–5.
165. Maze I, Shen L, Zhang B, Garcia BA, Shao N, Mitchell A, et al. Analytical tools and current challenges in the modern era of neuroepigenomics. Nat Neurosci. 2014;17(11):1476–90.
166. Schadt EE, Friend SH, Shaywitz DA. A network view of disease and compound screening. Nat Rev Drug Discov. 2009;8(4):286–95.
167. Jonsson PF, Bates PA. Global topological features of cancer proteins in the human interactome. Bioinformatics. 2006;22(18):2291–7.
168. Wachi S, Yoneda K, Wu R. Interactome-transcriptome analysis reveals the high centrality of genes differentially expressed in lung cancer tissues. Bioinformatics. 2005;21(23): 4205–8.
169. Goh KI, Cusick ME, Valle D, Childs B, Vidal M, Barabasi AL. The human disease network. Proc Natl Acad Sci U S A. 2007;104(21):8685–90.
170. Barabasi AL, Gulbahce N, Loscalzo J. Network medicine: a network-based approach to human disease. Nat Rev Genet. 2011;12(1):56–68.
171. Lage K, Karlberg EO, Storling ZM, Olason PI, Pedersen AG, Rigina O, et al. A human phenome-interactome network of protein complexes implicated in genetic disorders. Nat Biotechnol. 2007;25(3):309–16.
172. Crespo I, Del Sol A. A general strategy for cellular reprogramming: the importance of transcription factor cross-repression. Stem Cells. 2013;31(10):2127–35.
173. Crespo I, Perumal TM, Jurkowski W, del Sol A. Detecting cellular reprogramming determinants by differential stability analysis of gene regulatory networks. BMC Syst Biol. 2013;7:140.
174. Del Sol A, Buckley NJ. Concise review: a population shift view of cellular reprogramming. Stem Cells. 2014;32(6):1367–72.
175. del Sol A, Balling R, Hood L, Galas D. Diseases as network perturbations. Curr Opin Biotechnol. 2010;21(4):566–71.

176. Huang S, Ernberg I, Kauffman S. Cancer attractors: a systems view of tumors from a gene network dynamics and developmental perspective. Semin Cell Dev Biol. 2009;20(7):869–76.
177. Zickenrott S, Angarica VE, Upadhyaya BB, del Sol A. Prediction of disease-gene-drug relationships following a differential network analysis. Cell Death Dis. 2016;7:e2040.
178. Marbach D, Costello JC, Kuffner R, Vega NM, Prill RJ, Camacho DM, et al. Wisdom of crowds for robust gene network inference. Nat Methods. 2012;9(8):796–804.
179. Marbach D, Prill RJ, Schaffter T, Mattiussi C, Floreano D, Stolovitzky G. Revealing strengths and weaknesses of methods for gene network inference. Proc Natl Acad Sci U S A. 2010;107(14):6286–91.
180. Horvath S. Weighted Network Analysis: Applications in Genomics and Systems Biology: Springer; 2011.
181. Langfelder P, Horvath S. WGCNA: an R package for weighted correlation network analysis. BMC bioinformatics. 2008;9:559.
182. Parikshak NN, Luo R, Zhang A, Won H, Lowe JK, Chandran V, et al. Integrative functional genomic analyses implicate specific molecular pathways and circuits in autism. Cell. 2013;155(5):1008–21.
183. Luo R, Sanders SJ, Tian Y, Voineagu I, Huang N, Chu SH, et al. Genome-wide transcriptome profiling reveals the functional impact of rare de novo and recurrent CNVs in autism spectrum disorders. Am J Hum Genet. 2012;91(1):38–55.
184. Voineagu I, Wang X, Johnston P, Lowe JK, Tian Y, Horvath S, et al. Transcriptomic analysis of autistic brain reveals convergent molecular pathology. Nature. 2011;474(7351):380–4.
185. Miller JA, Horvath S, Geschwind DH. Divergence of human and mouse brain transcriptome highlights Alzheimer disease pathways. Proc Natl Acad Sci U S A. 2010;107(28):12698–703.
186. Miller JA, Woltjer RL, Goodenbour JM, Horvath S, Geschwind DH. Genes and pathways underlying regional and cell type changes in Alzheimer's disease. Genome Med. 2013;5(5):48.
187. Langfelder P, Horvath S. Eigengene networks for studying the relationships between co-expression modules. BMC Syst Biol. 2007;1:54.
188. Oldham MC, Konopka G, Iwamoto K, Langfelder P, Kato T, Horvath S, et al. Functional organization of the transcriptome in human brain. Nat Neurosci. 2008;11(11):1271–82.
189. Horvath S, Zhang Y, Langfelder P, Kahn RS, Boks MP, van Eijk K, et al. Aging effects on DNA methylation modules in human brain and blood tissue. Genome Biol. 2012;13(10):R97.
190. Bocklandt S, Lin W, Sehl ME, Sanchez FJ, Sinsheimer JS, Horvath S, et al. Epigenetic predictor of age. PLoS One. 2011;6(6):e14821.
191. Ponomarev I, Wang S, Zhang L, Harris RA, Mayfield RD. Gene coexpression networks in human brain identify epigenetic modifications in alcohol dependence. J Neurosci. 2012;32(5):1884–97.
192. Horvath S, Langfelder P, Kwak S, Aaronson J, Rosinski J, Vogt TF, et al. Huntington's disease accelerates epigenetic aging of human brain and disrupts DNA methylation levels. Aging. 2016;8(7):1485–512.
193. Gage FH, Temple S. Neural stem cells: generating and regenerating the brain. Neuron. 2013;80(3):588–601.
194. Satterlee JS, Beckel-Mitchener A, Little R, Procaccini D, Rutter JL, Lossie AC. Neuroepigenomics: Resources, Obstacles, and Opportunities. Neuroepigenetics. 2015;1:2–13.
195. Pollen AA, Nowakowski TJ, Shuga J, Wang X, Leyrat AA, Lui JH, et al. Low-coverage single-cell mRNA sequencing reveals cellular heterogeneity and activated signaling pathways in developing cerebral cortex. Nat Biotechnol. 2014;32(10):1053–8.
196. Poulin JF, Tasic B, Hjerling-Leffler J, Trimarchi JM, Awatramani R. Disentangling neural cell diversity using single-cell transcriptomics. Nat Neurosci. 2016;19(9):1131–41.
197. Tasic B, Menon V, Nguyen TN, Kim TK, Jarsky T, Yao Z, et al. Adult mouse cortical cell taxonomy revealed by single cell transcriptomics. Nat Neurosci. 2016;19(2):335–46.
198. Brady G, Barbara M, Iscove NN. Representative in vitro cDNA amplification from individual hemopoietic cells and colonies. Methods Mol Cell Biol. 1990;2(1):17–25.

199. Li CL, Li KC, Wu D, Chen Y, Luo H, Zhao JR, et al. Somatosensory neuron types identified by high-coverage single-cell RNA-sequencing and functional heterogeneity. Cell Res. 2016;26(8):967.
200. Usoskin D, Furlan A, Islam S, Abdo H, Lonnerberg P, Lou D, et al. Unbiased classification of sensory neuron types by large-scale single-cell RNA sequencing. Nat Neurosci. 2015;18(1):145–53.
201. Fuzik J, Zeisel A, Mate Z, Calvigioni D, Yanagawa Y, Szabo G, et al. Integration of electrophysiological recordings with single-cell RNA-seq data identifies neuronal subtypes. Nat Biotechnol. 2016;34(2):175–83.
202. Zeisel A, Munoz-Manchado AB, Codeluppi S, Lonnerberg P, La Manno G, Jureus A, et al. Brain structure. Cell types in the mouse cortex and hippocampus revealed by single-cell RNA-seq. Science. 2015;347(6226):1138–42.
203. Macosko EZ, Basu A, Satija R, Nemesh J, Shekhar K, Goldman M, et al. Highly Parallel Genome-wide Expression Profiling of Individual Cells Using Nanoliter Droplets. Cell. 2015;161(5):1202–14.
204. Guo H, Zhu P, Wu X, Li X, Wen L, Tang F. Single-cell methylome landscapes of mouse embryonic stem cells and early embryos analyzed using reduced representation bisulfite sequencing. Genome Res. 2013;23(12):2126–35.
205. Farlik M, Sheffield NC, Nuzzo A, Datlinger P, Schonegger A, Klughammer J, et al. Single-cell DNA methylome sequencing and bioinformatic inference of epigenomic cell-state dynamics. Cell Rep. 2015;10(8):1386–97.
206. Smallwood SA, Lee HJ, Angermueller C, Krueger F, Saadeh H, Peat J, et al. Single-cell genome-wide bisulfite sequencing for assessing epigenetic heterogeneity. Nat Methods. 2014;11(8):817–20.
207. Lara-Astiaso D, Weiner A, Lorenzo-Vivas E, Zaretsky I, Jaitin DA, David E, et al. Immunogenetics. Chromatin state dynamics during blood formation. Science. 2014; 345(6199):943–9.
208. Rotem A, Ram O, Shoresh N, Sperling RA, Goren A, Weitz DA, et al. Single-cell ChIP-seq reveals cell subpopulations defined by chromatin state. Nat Biotechnol. 2015;33(11):1165–72.
209. Adli M, Bernstein BE. Whole-genome chromatin profiling from limited numbers of cells using nano-ChIP-seq. Nat Protoc. 2011;6(10):1656–68.
210. Shankaranarayanan P, Mendoza-Parra MA, Walia M, Wang L, Li N, Trindade LM, et al. Single-tube linear DNA amplification (LinDA) for robust ChIP-seq. Nat Methods. 2011;8(7):565–7.
211. Cusanovich DA, Daza R, Adey A, Pliner HA, Christiansen L, Gunderson KL, et al. Multiplex single cell profiling of chromatin accessibility by combinatorial cellular indexing. Science. 2015;348(6237):910–4.
212. Buenrostro JD, Wu B, Litzenburger UM, Ruff D, Gonzales ML, Snyder MP, et al. Single-cell chromatin accessibility reveals principles of regulatory variation. Nature. 2015;523(7561): 486–90.
213. Klein AM, Mazutis L, Akartuna I, Tallapragada N, Veres A, Li V, et al. Droplet barcoding for single-cell transcriptomics applied to embryonic stem cells. Cell. 2015;161(5):1187–201.
214. Eid J, Fehr A, Gray J, Luong K, Lyle J, Otto G, et al. Real-time DNA sequencing from single polymerase molecules. Science. 2009;323(5910):133–8.
215. Flusberg BA, Webster DR, Lee JH, Travers KJ, Olivares EC, Clark TA, et al. Direct detection of DNA methylation during single-molecule, real-time sequencing. Nat Methods. 2010;7(6): 461–5.
216. Stegle O, Teichmann SA, Marioni JC. Computational and analytical challenges in single-cell transcriptomics. Nat Rev Genet. 2015;16(3):133–45.

Index

R. Delgado-Morales (ed.), *Neuroepigenomics in Aging and Disease*, Advances in Experimental Medicine and Biology, DOI 10.1007/978-3-319-53889-1

I

K

L

T

Kolibri 360 GmbH
Ferdinand-Jühlke-Straße 7
99095 Erfurt, Deutschland
produktsicherheit@kolibri360.de